Regional Climate Studies

Series Editors: H.-J. Bolle, M. Menenti, I. Rasool

Hans-Jürgen Bolle, Matthias Eckardt,
Dirk Koslowsky, Fabio Maselli, Joaquín Meliá
Miralles, Massimo Menenti, Folke-Sören Olesen,
Ljiljana Petkov, Ichtiaque Rasool,
Adriaan Van de Griend (Eds.)

Mediterranean Land-surface Processes Assessed From Space

With 442 Figures, 320 in colour

Springer

Editors

Hans-Jürgen Bolle, Dr., Prof. a. D.
Stücklenstrasse 18 c
81247 München, Germany

Matthias Eckardt, Dipl. Met.
Freie Universität Berlin, Institut für Meteorologie
Carl-Heinrich-Becker-Weg 6-10
12165 Berlin, Germany

Dirk Koslowsky, Dr.
Freie Universität Berlin, Institut für Meteorologie
Carl-Heinrich-Becker-Weg 6-10
12165 Berlin, Germany

Fabio Maselli, Dr.
CNR - Istituto di Biometeorologia (IBIMET)
Piazzale delle Cascine 18
50144 Firenze, Italy

Joaquín Meliá Miralles, Dr., Prof.
Universitat de Valencia, Facultat de Física
Dpt. de Termodinàmica, Remote Sensing Unit
Calle Dr. Moliner, 50, Burjassot
46100 Valencia, Spain

Massimo Menenti, Dr., Prof.
CNR - Istituto Per I Sistemi Agricoli E Forestali
Del Mediterraneo - ISAFoM,
P.O. Box 101
80040 S. Sebastiano al Vesuvio (NA), Italy

Folke-Sören Olesen, Dipl. Met.
Forschungszentrum Karlsruhe
Institut für Meteorologie und Klimaforschung
Postfach 3640
76133 Karlsruhe, Germany

Ljiljana Petkov, Dr.
Adriaan van Ostadelaan, 5
2343 EL Oegstgeest, The Netherlands

Ichtiaque Rasool, Dr.
60 Quai Louis Blèriot
75016 Paris, France

Adriaan A. Van de Griend, Dr., Prof. Em.
Guido Gezellelaan 38
3705 AT Zeist, The Netherlands

ISBN 10 3-540-40151-2 Springer Berlin Heidelberg New York
ISBN 13 978-3-540-40151-3 Springer Berlin Heidelberg New York

Library of Congress Control Number: 2006927434

Springer is part of Springer Science+Business Media
springeronline.com

Cover design: E. Kirchner, Heidelberg
Production: A. Oelschläger
Typesetting: Camera-ready by the Editors

Printed on acid-free paper 30/2132/AO 54321

Contributing Authors

Heiner Billing

Freie Universität Berlin, Institut für Meteorologie
Carl-Heinrich-Becker-Weg 6-8
12165 Berlin, Germany

Anatoly Gitelson

Ben-Gurion University of the Negev
Remote Sensing Laboratory
Sede-Boker Campus, 84990 Israel

Frank Göttsche

Hermann-von-Helmholtz-Platz 1
76344 Eggenstein-Leopoldshafen, Germany

Anne Jochum-Osann

Alphaclim
Zapateros, 15-7
02005 Albacete, Spain

Ernesto López-Baeza

Remote Sensing Unit, Dptm. de Termodinàmica
Facultat de Física, Universitat de Valencia
Calle Dr. Moliner, 50, Burjassot
46100 Valencia, Spain

Francesco Meneguzzo

CNR - Istituto di Biometeorologia (IBIMET)
via Giovanni Caproni 8, 50145 Firenze, Italy

Jose Moreno

Department of Thermodynamics, Faculty of Physics
University of Valencia
46100 Burjassot, Valencia, Spain

Françoise Nerry

Laboratoire des Sciences de l'Image
et de la Télédétection, ENSPS/LSIIT
University L. Pasteur, Strasbourg 1
Boulevard Sebastien Brandt
67400 Illkirch, France

Paolo Rossini

Studio PAN Ricerche
Viale Regina Margherita 270
00198 Roma, Italia

Frank Veroustraete

Flemish Institute for Technological Research - VITO
Centre for Remote Sensing and Atmospheric Processes TAP
Boeretang 200, 2400 Mol, Belgium

Roland Vogt

University of Basel, Department of Geography
Institut für Meteorologie, Klimatologie.
und Fernerkundung., MCR Lab
Spalenring 145
4055 Basel, Schweiz

Peter J. Van Oevelen

European Space Agency
European Space Research & Technology Centre
ESA - ESTEC, Mission Experts Division
Land Surfaces Unit (EOP - SML)
P.O. Box 299, 2200 AG Noordwijk zh
The Netherlands

Preface

Observations from space have been an important component in a number of research projects of the European Commission aiming, in the context of "Global Change", at the assessment of land-surface processes and their changes in the Mediterranean area. With the new generation of satellites, carrying improved instrumentation, these observations will gain in importance in the future. Changes in the Mediterranean environment are linked to the global climate system which is characterized by a strong inherent interannual variability but may, in addition, undergo trends that develop slowly in time. To assess, to which degree Mediterranean land-surface processes such as aridification, desertification, soil quality and changes of water resources are affected by the development of the global climate system, it is necessary to extend such studies over long time periods which would allow to average over the "noise" in the signals caused by its natural variability. Presently time series of thirty years are found adequate to distinguish shorter term fluctuations from long term trends and to draw reliable conclusions from those data.

For two reasons it seems now timely to summarize recent experience in dealing with satellite data when studying changes at the land surfaces. Firstly, to document the results obtained so far and secondly to pave the ground for a smooth transition from old sensor systems to the advanced ones which are already available or will soon become operational. From the new sensor systems, reliable long time series will become available only thirty years from now. In combination with existing data sets this goal can be accomplished in fifteen years from now. To create a coherent data set of the required length it is mandatory to fit the new measurements with their different instrumental parameters to the present data series.

The information content of measurements made from space can only fully be understood and applied if the physical and - in the case of vegetation - also the biological limitations are kept in mind. This knowledge sometimes gets lost as the applications diverge from the objective of original data. It was therefore found formative to combine in one volume background information of both the measuring systems and the objects of investigation with the methodology that leads to applications. Newcomers and students in this field may also be interested in how research can be organized to validate and support the inferred information by corroborative measurements made at the surface. Experiences gained during field experiments therefore are described to some detail and useful supplemental information is given in appendices.

Amsterdam Berlin München Firenze Karlsruhe Napoli Oegstgeest Paris Valencia

August 2006

The Editors

Acknowledgements

Most of the results reported here originate from a series of EC projects that started in 1987 with *Remote Sensing of Energy and Mass Exchanges at the Surface of the Earth* (Twinning project no. ST 2 J 0222) and continued with *The ECHIVAL Field Experiment in a Desertification-threatened Area - EFEDA* (EPOC-CT 90-0030), *Remote Sensing and Radiometric Properties of the Surface: Assessment of Desertification from Space - EFEDA Phase II* (EV5V-CT93-0284), *An Integrated Monitoring and Modelling Study of Desertification and Climatic Change Impacts in the Messara Valley of Crete* (EV5V-CT94-0466), *Remote Sensing of Mediterranean Desertification and Environmental Changes - RESMEDES* (ENV4-CT95-0094), and finally *Synthesis of Change Detection Parameters Into A Land-surface Change Indicator for Long Term Desertification Studies in the Mediterranean Area - RESYSMED* (ENV4-CT97-0683). EFEDA was launched by the EC under the Directorate of Scientific Research and Development, Director R.Fantechi, as European contribution to the International Satellite Land-surface Climatology Project (ISLSCP) which was established 1983 and developed its momentum at the International ISLSCP Conference held in Rome, December 1985. The activities were continued under the Director for Research, C. Patermann, and the Head of the Unit for Global Change, DG Research, A. Ghazi, about whose early death in 2005 we deeply mourn for. The projects were supervised by the Officers P. Balabanis and D. Peter

The research community participating in the European activities is deeply indebted to the EC for sponsoring this research into which in addition substantial national funds were invested from the participating countries. More than 35 research groups, including one of the U.S.A., participated in this research of which about one half used satellite data. The book is a recognition of the dedicated work of the many scientists, technicians and administrators that led these EC projects to success. The list of authors includes those scientists, who, in addition to the editors, wrote substantial parts of the book. They were supported as documented in the official project reports by the work of many colleagues to whom editors and authors express their sincere thanks.

These are, in addition to the editors and authors, the initial EFEDA participants: J.-C. André, J. L. Arrué, H. K. Barth, G. Bergkamp, P. Bessemoulin, A. Brasa, J. Bromley, H. A. R. de Bruin, J. Cruces, G. Dugdale, E. T. Engman, D. L. Evans, F. Fiedler, G. Folloni, J.-P. Goutorbe, R. Harding, A. C. Imeson, A.M. Jochum, P. Kabat, T. Kratzsch, J. P. Lagouarde, I. Langer, R. Llamas, L. S. Muniosguren, J. Noilhan, H. R. Oliver, R. Roth, S. S. Saatchi, J. Sanchez Diaz, F. Martin de Santa Olalla, W. J. Shuttleworth, H. Sogaard, H. Stricker, J. Thornes, E. Todini, I.

Vardavas, M. Vauclin, and D. Wickland. Special thanks go to Jutta Wedel for administering the sequence of projects over many years.

In addition we would like to thank the following colleagues who have contributed to reports from which the authors have drawn information, and for assisting editors and authors in many other ways: W. G. M. Bastiaanssen, M. Berger, F. H. Berger, M. Bindi, G. Bitelli, K. Blümel, L. Bottai, J. Cermak, M. Chiesi, R. A. M. De Jeu, J. J. De Vries, P. Dubois, M. Frezzotti, B. T. Gouweleeuw, R. Haverkamp, G. Markl, C. Martellacci, M. Owe, G. Pandiscia, E. Parlow, H. Pelgrum, S. W. M. Peters, D. Praetorius, G. Rana, A. Raschi, F. Rembold, G. Roerink, H. Schmidt, E. Seyhan, B. J. M. Van den Hurk, J. Van Zyl, A. Verhoef and many others who collaborated over the years with editors and authors. We would like to include in our thanks the technical staff and students who build the basis for the extensive field experiments and the data evaluation.

Our gratitude goes to the local administrators who helped to establish the necessary logistic and especially to the land owners who generously allowed their land to be used for the installation the measuring devices and the execution of field measurements.

We acknowledge the support provided by national research establishments and founding organizations which built the basis to receive EC funds and to carry out the large experiment in Castilla-La Mancha and the following evaluationod the data. Spanish researchers are especially thankful to the Regional Ministry for Environment (Directorate General for Environmental Quality) for the maintenance of the Valencia Anchor Station station which is the result of an infrastructure project co-financed by the Regional Government of the Valencia Region and the University of Valencia. The land was indefinitely lent by Bodegas Iranzo and acknowledge also the support provided by the Town Halls of Utiel and of Caudete de las Fuentes. Also thankfully acknowledged is the confidence and scientific support from the GIST (GERB International Science Team) and the SMOS Science Team who count on the Valencia Anchor Station as a reference station for their validation activities, the generous PAPS acquisitions over the validation site by the CERES Science Team and the current collaboration of the Spanish Institute for Meteorology, Institute for Space Studies of Catalonia, Polytechnic University of Catalonia and Field Radiometry Group of the University of Valencia.

Contents

Symbols

Symbol	Definition	Dimension, Value
↓	downwelling (incident) radiation flux	-
↑	upwelling (reflected) radiation flux	-
*	" net" in connection with fluxes	-
α	penetration depth (of heat wave in soils)	m
α	line halfwidth	cm^{-1}
α	absorptance (Φ_a/Φ_0)	-
α	polarizability	$A\ s\ m^2\ V^{-1}$
β	Bowen ratio (Φ_{SH}/Φ_{LH})	-
γ	aerosol type factor	-
γ	psychometric constant	0.67 $hPa\ K^{-1}$ at STP
$\gamma\ (\zeta)$	instrumental response factor for direct solar radiation	-
γ_m	mean atmospheric temperature gradient	$K\ m^{-1}$
δ_λ, $\delta(\lambda)$	(vertical)optical depth at wavelength λ of scattering (δ_{scat}) and Rayleigh (δ_R) atmospheres	-
ΔT	temperature amplitude	K
ε	specific kinetic energy dissipation	$m^2 s^{-3}$
ε	permittivity, dielectric constant	$A\ s\ V^{-1}\ m^{-1}$
ε	emissivity ($M/M_{Black\ Body}$); $\varepsilon_{\mu w}$ microwave emissivity, ε_{IR} thermal infrared emissivity	-
ε_0	permittivity of vacuum	$8.8542 \cdot 10^{-12}$ $F\ m^{-1} = A\ s\ V^{-1}\ m^{-1}$
ζ	$\equiv \zeta_{sun}$, solar zenith angle	degree
η	characteristic length of the inner scale (turbulence theory)	mm
ϑ'	zenith angle of the reflected radiation	degree

ϑ	zenith angle of the received radiation	degree
ϑ_{obs}	observation nadir angle	degree
θ	temperature in degree Celsius	°C
θ	specific volumetric soil water content	$m^3\ m^{-3}$
$\varkappa$	instrumental response factor for diffuse radiation	-
$\varkappa$	attenuation index, extinction coefficient	-
λ	wavelength	nm, µm
λ'	local longitude	degree
λ_{ref}	longitude from where the reference time is counted	degree
μ_0, (μ)	cosine of solar(observation) angle	-
μ	magnetic permeability	$V\ A^{-1}\ s\ m^{-1}$
μ_0	permeability of vacuum	$1.25664 \cdot 10^{-6}$ $H\ m^{-1} = V\ s\ A^{-1}\ m^{-1}$
ν	kinematic viscosity	$m^2\ s^{-1}$
$\prod \varepsilon_i^{ai}$	temperature Independent Spectral Index (TISI)	-
ρ $[\rho(\lambda), \rho_\lambda]$	general symbol for reflectance [spectral reflectance] and albedo (Φ_r/Φ_0) [spectral albedo] [1])	-
$\rho(\Omega, \Omega')$	BDRF (Bi-Directional Reflection Factor)	-
$\rho(d\Omega, d\Omega')$	BRDF (Bi-directional Reflectance Distribution Function)	sr^{-1}
$\rho(z)$	air density at height z above mean sea level	$kg\ m^{-3}$
ρ_{air}, ρ	dry air density (0.06% CO_2) $[1.2932/(1+0.00367\ \theta)](p/p_0)$	1226 $kg\ m^{-3}$ at 15°C, 1013 hPa
ρ_p	planetary albedo	-
ρ_{sa}	spherical albedo of the atmosphere	-
σ	specific conductivity	$A\ V^{-1}\ m^{-1}$
$\sigma_R(\vartheta)$	Rayleigh scattering coefficient	$1.06 \cdot 10^{-8}\ \lambda^{-4.09}\ m^{-1}$
τ	transmittance (Φ_t/Φ_0)	-
$\tau_\lambda(m)$	transmittance at wavelength λ and airmass m	-
τ_{atm}	transmittance of the atmosphere	-
φ_{sun}, φ_{obs}	azimuth angle of sun and observation	degree
φ', (φ)	azimuth angle of observed (incident) radiation	degree

Φ	flux	W m^{-2}
Φ_H	buancy flux (flux of turbulent energy)	W m^{-2}
Φ_{LH}, LH	latent heat flux, evaporation	W m^{-2}
Φ_R, R	radiation flux	W m^{-2}
Φ_{SH}, SH	sensible heat flux	W m^{-2}
Φ_{soil}, G	soil (ground) heat flux	W m^{-2}
Ψ	matric potential or suction (soil water pressure/specific weight of water)	Pa kg^{-1} m^2 s^2 = m
ω	$= 15[t_{ref} + (\lambda' - \lambda_{ref})/15 + Z - 12]$, hour angle	hours
ω	circular frequency ($2\pi/N$)	s^{-1}
$\tilde{\omega}$	single scattering albedo	-
Ω	solid angle (into which radiation is reflected)	sr
Ω'	solid angle from which a target receives radiation	sr
a	absolute humidity	kg m^{-3}
a	$= 4\pi\varkappa/\lambda$ absorption coefficient	m^{-1}
A	area (m^{-2})	m^{-2}
A	area	m^2
$A0$	offset coefficient or spectral radiance at QCAL = 0	W m^{-2} sr^{-1}
$A0_\lambda$	post-calibration offset coefficient	mW cm^{-2} sr^{-1} µm^{-1}
$A1$	gain coefficient in units of	W m^{-2} sr^{-1}
$A1_\lambda$	post-calibration gain coefficient	mW cm^{-2} sr^{-1} µm^{-1} DN^{-1}
$\boldsymbol{B}$	magnetic flux density vector	T = kg A^{-1} s^{-2}
$B(T)$	$= L_{BB}(T)$ black body radiance	W m^{-2} sr^{-1}
c	speed of light in vacuum	$2.997925 \cdot 10^8$ ms^{-1}
c	specific heat capacity of soils	J kg^{-1} K^{-1}
c_p	specific heat at constant pressure of air	1005 J kg^{-1} K^{-1}
c_1	$= 2\pi hc^2$, first Planck radiation constant	$3.7418 \cdot 10^{-16}$ Wm^2
c_2	$= hc/k$, second Planck radiation constant	$1.4388 \cdot 10^{-2}$ mK,
C_n^2	structure parameter (or constant) of the refractive index of air	m$^{-2/3}$
C_u^2	structure parameter for velocity	(m/s)^2 m$^{-2/3}$

C_L, C_M, C_H	fractional cloud cover for low, middle and high clouds	-
C_s	volumetric heat capacity	$J\ m^{-3}\ K^{-1}$
d	instantaneous Sun-Earth distance	km
d	path length	m
$\bar{d}$	1 Astronomical Unit (A. U., mean Sun-Earth distance)	$1.496 \cdot 10^8$ km
$d\varepsilon$	vegetation structure parameter	-
D	structural function	$m^2\ s^{-2}$
$\boldsymbol{D}$	displacement vector	$C\ m^{-2} = A\ s\ m^{-2}$
D	$= \lambda\ C_s^{-1}$, thermal diffusivity	$m^2\ s^{-1}$
e	water vapour pressure	hPa
e_0	saturation water vapour pressure	hPa
$\boldsymbol{E}$	electric field strength vector	$V\ m^{-1}$
E	irradiance (at a target)	Wm^{-2}
f	Cabannes factor	1.054
f	areal fraction covered by vegetation	-
$F_i(\lambda)$	filter function	-
g	acceleration due to gravity	$9.8062\ m\ s^{-2}$ at 45 ° latitude
g_i	weight	-
h	empirical roughness parameter	m
h	height	m
h	Planck constant	$6.6262 \cdot 10^{-34}$ J s,
H	index, horizontal polarization	-
$\boldsymbol{H}$	magnetic field strength vector	$A\ m^{-1}$
H	$= RT_0/Mg$, scale height	8435 m
H_d	extraterrestrial solar radiant exposure (irradiance at a horizontal plane integrated over specified exposure time)	$J\ m^{-2}\ day^{-1}$
i	(index) isotropically reflecting (Lambertian) surface	-
i	index denotes spectral channel	-
I	$= (\rho c \lambda \omega)^{1/2}$, thermal inertia	$W\ m^{-2}\ K^{-1}$

k	time constant of clear sky surface temperature post maximum exponential decay	hours
k	$= 2\pi/\lambda$ wavenumber	m^{-1}, nm^{-1}, μm^{-1}
k	von Kármán constant	0.4
k	Boltzmann constant	$1.3807 \cdot 10^{-23}$ JK^{-1}
K	force constant	$N\ m^{-1}$
$K(\theta)$	hydraulic (or capillary) conductivity	$m\ s^{-1}$
L	latent heat of water vaporization (A.5)	≈2.465 MJ kg^{-1} at 15°C
L	radiance	$Wm^{-2}sr^{-1}$
L_λ	spectral radiance	$mW\ cm^{-2}\ sr^{-1}\ \mu m^{-1}$
$LMAX_\lambda$	maximum spectral radiance	$mW\ cm^{-2}\ ster^{-1}\ \mu$ at QCAL = 255 DN
$LMIN_\lambda$	minimum spectral radiance	$mW\ cm^{-2}\ ster^{-1}\ \mu m$ at QCAL= 0 DN
L_{MO}	$= -\ c_p\ T_v\ \rho\ u_*^3 /\ g\ k\ \Phi_H$, Monin-Obukhov length	m
m	mass	kg
m	relative airmass	-
M	radiation flux density across a unit area	Wm^{-2}
M_a	molar mass of air up to about 90 km	0.028964 kg $mole^{-1}$
n	refractive index	-
n	index "natural" reflector	-
N	molecules per unit volume	m^{-3}
N	number of molecules per unit volume (Loschmidt number)	$2.6867 \cdot 10^{25}$ m^{-3}
n_0	refractive index of air at 700 nm and sea level	$2.76 \cdot 10^{-4}$
p	pressure of air	hPa
$\mathbf{p}, \mathbf{P}$	electric dipole momentum	A s m
P_v	fractional vegetation cover	-
p_0	standard air pressure	1013.25 hPa
Ph	energy consumed for photosynthesis	$W\ m^{-2}$
$p_R\ (\vartheta)$	Rayleigh scattering phase function	-
q	specific humidity	$g\ kg^{-1}$

QCAL	quantised and calibrated scaled radiance	DN
$q_l^*(T)$	the temperature dependent value of the specific humidity at saturation that is assumed to be reached within the leaves	g kg^{-1}
q_s	the specific humidity at the bare soil surface or if this is dry	g kg^{-1}
$q_s\,(SST)$	specific humidity at saturation defined by the sea surface temperatur	g kg^{-1}
r	position vector	m
r	resistance to energy transfer r_a aerodynamic resistance, r_s resistance of unsaturated soil, r_{st} stomatal resistance	s m^{-1}
R	gas constant	8.314 J $mole^{-1}$ K^{-1}
R	mean earth radius	6356766 m
R_0	range (ΔR_0 pixel size)	m
$S\,(T)$	line strength (integrated absorption)	cm^{-1} sec^{-1}
S_0	exo-atmospheric solar flux density ("solar constant")	1368±1 W m^{-2}
$S_\lambda\,(\varphi,\delta,t)$	solar spectral flux density at the top of the atmosphere at latitude φ, declination δ and time t through a horizontal surface	W m^{-2} µm
T	temperature	K
T	$= \delta_{scat}/\delta_R$, turbidity factor (Linke)	-
t_{ref}	reference time, time in hours measured in zonal mean time	hours
$t(T_m)$	time of clear sky surface temperature maximum	hours
T_0	reference temperature, e.g. 288.15 K (15 °C)	K
T_{BB}	black body emission temperature	K
T_{eff}	effective radiative surface temperature	K
T_v	$= T(1+0.61q)$, virtual temperature	K
u_{10}	wind speed at 10m height	m s^{-1}
u_*	friction velocity	m s^{-1}
V	index, vertical polarization	-
w	precipitable amount of water vapour	cm
W	band-width	µm

W	equivalent width	cm^{-1}
z_0	roughness length	cm
z_{ref}	reference height	m
Z	equation of time	hours

[1]) Use of the reflectance symbol ρ:

Symbol	Definition	Dim.
ρ (dΩ, dΩ')	bi-directional [spectral] reflectance distribution function (BRDF), ratio of the directional radiance (W m^{-2} sr^{-1}) reflected from a target within an infinitesimally narrow solid angle to the incident flux density (irradiance, W m^{-1}) generated by a point source	sr^{-1}
ρ (Ω, Ω') [ρ_λ (Ω, Ω')]	bi-directional [spectral] reflectance factor (BDRF) defined as ratio of the radiance reflected from a target to that one reflected from a white Lambertian reference panel under identical illumination and observation conditions. The directional radiance originating from a white Lambertian surface equals 1/π of its irradiance. The BRDF therefore can be expressed as (radiance target)/(π·radiance reference) or π·BRDF = BDRF	-
ρ (dΩ, 2π)	ratio of directional radiance reflected from a target to the irradiance from a hemispherical source(directional-hemispherical reflectance)	sr^{-1}
ρ (2π, 2π)	albedo (= bi-hemispherical reflectance factor): ratio of hemispherically reflected to incident radiation flux (sometimes the overbar $\bar{\rho}$ is used to indicate broad band albedo)	-

Chapter 1 Introduction

1.1 Space View and Ground Observations

The approach followed in this publication is based upon available long term data series of NOAA-AVHRR and Nimbus-SMMR and occasional Landsat-TM, SPOT, Meteosat, and ERS1/2 scenes. The spectrum ranges from the visible to microwaves. This broad approach was found to be advantageous for the following reasons: (i) information inferred from medium resolution satellite data can be validated by stepwise scaling up from point measurements made at the ground first to high resolution satellite data and then, by aggregation of pixels, to measurements made by NOAA-AVHRR and Meteosat, (ii) the different observation times and different spatial resolutions of satellite systems supplement each other, and (iii) information gained from sensors with different spectral characteristic mutually support each other. For the entire lifetime of the new European Envisat mission, for example, measurements from a variety of instruments are simultaneously available for the first time. This, nowadays greatly enhances the synergy effect of the measurements.

Most of the presented data result from research projects of the European Commission, DG Research, starting 1991 with the ECHIVAL[1] Field Experiment in Desertification-threatened Areas (EFEDA)[2]. These projects were initiated to study the causes of land degradation and desertification, their relationship to climate change and man's activities, and to develop indices to quantify these changes. One aim was to explore the role which observations from space can play to analyse the processes that occur at the land surfaces and to overview the whole Mediterranean basin. Because of the complex topographical structure of the Mediterranean landscape it seems impossible to obtain such an overview for a longer time period exclusively by measurements at the surface. Long term observations are necessary to assess trends superimposed by large annual fluctuations as is the case in the Mediterranean area. Some climate state variables indicate a quasi-periodicity of about 23 years. Consequently the aim must be to extend the use of measurements from space to such time scales. Because different satellites with varying instruments must be used to cover such a long period, great care has to be taken to construct homogeneous data series. Only then the analysis of remote sensing data gains weight

[1]European International Project on Climatic and Hydrological Interactions between Vegetation, Atmosphere, and Land-surfaces

[2]A short description of these activities can be found in Appendix 1.

in this research. The outcome for the Mediterranean area may also be useful for investigations and applications in other parts of the world.

The remote sensing component of the above mentioned EC research projects did not aim at an assessment of land-surface changes by means of repetitive classification of land-use units. They were rather meant to relate measurements from satellites to physical and biological quantities that drive land surface - atmosphere interactions and change due to man's activities and climate variability. One important question is whether these processes may be subject to irreversible trends due to global change.

Changes at the land surfaces are driven by the annual sequence of weather situations, extreme weather events, long term global climate change, and the activities of man in response to ecological and economical forcing. The various processes involved are briefly reviewed in the following sections of this introduction.

1.2 Mediterranean Climatic Environment

Mediterranean climate occurs in a number of western continental coasts between 30° and 45° north and south (Strahler 1975). Research in the about 4 Mkm^2 large European, African and Levantine land masses around the Mediterranean Sea therefore finds its congruity in other regions of the world, such as at the fringe of the subtropics in the south-west of the United States, Mexico, Chile, Australia, and South Africa. Notwithstanding this climatic correspondence, the situation of the European-African-Levantine Mediterranean Basin differs in some respect from that of the other Mediterranean regions. Its highly structured landmasses border a nearly closed large inland sea of about 2.5 Mkm^2 stretching over 42° longitude or as much as 3.02 Mkm^2 if, in addition, one counts the Marmora and the Black Seas as part of the Mediterranean area. Because of this longitudinally elongated water mass enclosed by land, the Mediterranean climate stretches further eastward to south of the Caspian Sea. This would enlarge the extend of the Mediterranean area to about 10 Mkm^2, which is unique in the world.

The Mediterranean climate and its variability was recently described by Lionello et al. (eds.) (2005). Here only a few processes are recalled in connection to remote sensing opportunities. The topographically complex Mediterranean Basin, positioned between the subtropical Hadley circulation system[3] and the westerlies, is characterized by strong climatic gradients and several specific phenomena (see Fig A.2.1 in Appendix 2). The latter are caused by the seasonal variability of the latitudinal position of the "polar front", land-sea circulation systems, travelling cyclones, low pressure systems caused by convection from hot surfaces, and across and around mountain air-flows. Mineral dust from the neighboured Saharan desert as well as from Mediterranean areas with bare soils and air pollution generated locally or imported from central Europe are blown over the area. The Mediterranean basin therefore is ideal for research into the interaction of processes between land,

[3]Meteorological terms used in this book are explained in Appendix 2

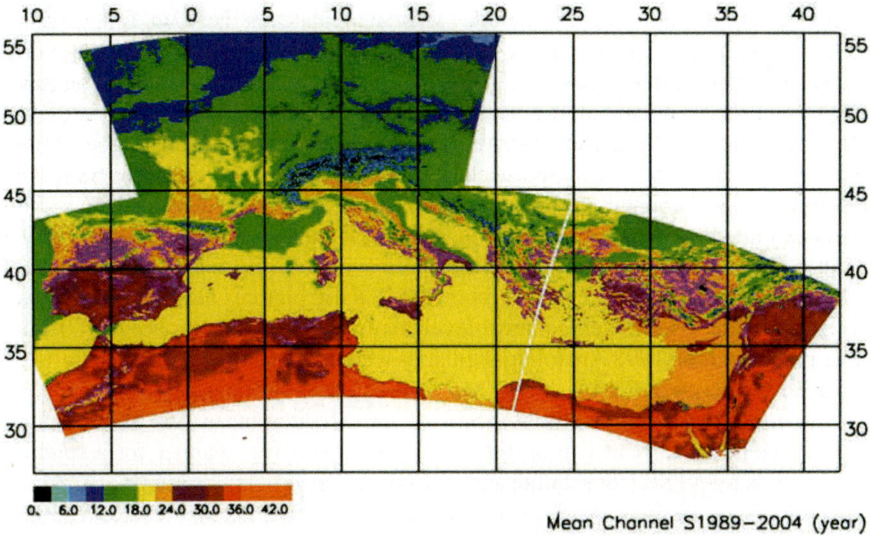

Fig. 1.1. Long term 1989 - 2004 annual mean of the radiative surface temperature for cloudless days and at the time of the NOAA satellites overpasses (the scale is in °C)

atmosphere and sea under variable climatic conditions and at different scales.

The south - north climate gradient shows up in measurements of the radiance in the thermal infrared spectral bands converted to equivalent temperatures at the top of the atmosphere (TOA) and then corrected with the split-window technique for atmospheric effects. The annual mean temperature averaged over the years 1989 to 2004 as obtained by the AVHRR instrument for cloudless days at the time of the NOAA satellite overpass (e. g. nominal 15:50 UT for NOAA-14 which was the same for NOAA-11 at launch but by March 1995 NOAA-11 had drifted to 17:33 UT) is shown in Fig. 1.1. Seasonally averaged TOA temperatures for cloudless conditions during the years 1989 - 1998 are presented in section 6.9. The colours indicate temperatures in steps of five degrees ranging from black (0 °C) to red (≥ 42 °C).

Higher spatial resolution is obtained with Landsat-TM images as shown for south-western Tuscany in Fig. 1.2. Here, the relationship between surface temperature and land cover (but also altitude) becomes evident. Areas with a high vegetation index are cooler than harvested fields or quarries with normalized difference vegetation indices (section 4.6) of typically < 0.2 that indicate bare soils. Most of the high vegetation index sites are at hills which in addition are affected at their windward side by the sea breeze. Solar radiation at inclined hilly terrain, various types of land-use, and the sea breeze cause a diversity of microclimates.

The average land-surface maximum temperature gradient across the basin is of the order of 20 K. As Fig. 1.2 shows, temperature differences of this magnitude also occur at single days in heterogeneous terrain. During winter, the southern European countries are close to sea surface temperature near noon and, therefore, one can hardly detect the coastlines in thermal infrared satellite images (see Fig. 6.9.1). In spring, lowlands and bare to sparsely vegetated plateaux heat up first. During summer, the heat is nearly equally distributed across the basin though the North African and Levantine areas are on average about 15 K warmer than the most southern European countries. In autumn, the south of Spain, the Anatolian highlands, the east of Greece, the chain of central Mediterranean Islands and Puglia remain warm longer than the rest of southern Europe which tends towards the SST. The contrast between mountainous areas and plains is considerable. This leads to locally complex valley-mountain circulation systems or katabatic winds as known for the north-eastern Adriatic coast, where cold air descends from the mountains to sea level ("Bora"). Though regional contrasts show up in these pictures nearly the same way every year, the average temperature level may change from year to year. As an example, in autumn 1998 Anatolian highlands were remarkably warmer than in 1997 and 1999 (see section 6.9.1).

The sea surface temperatures (SST) of the Mediterranean Sea show spatial differences up to 15 K as can be seen in a more distinctive manner for a summer month taken by ATSR (for satellite and instruments specifications see Chapter 2 and Appendix 4) on ERS-2 (Fig. 1.3). There are different reasons that lead to a patchy distribution of the SST: Differential solar heating, upwelling of cooler deep water due to internal circulations, water exchange with the Atlantic Ocean, run-off of cooler river water into the sea, and the intrusion of water from the Black Sea through the Sea of Marmara. A cold water surge into the Aegean in some years occurs in the early spring when the Black Sea has the lowest temperatures in the region due to the inflow of water from northern rivers. On an annual basis, the general features of the SST reappear year by year. During summer, the central and eastern parts of the sea are warmer than the western part and the Aegean.

During winter, vigorous synoptic scale weather systems imbedded in the westerlies are the overriding weather phenomena. High mountain barriers such as the Atlas, the Pyrenees, the Alps, and the Balkan mountains modify or even generate these weather systems which develop in the middle troposphere and gain their momentum by internal energy transfer processes. Well known is the Genova cyclone generated by the interaction of the westerly airstream with the bow of the Alps. Behind higher mountains often chinook-like pattern ("Föhn") develop. Smaller topographic obstacles are less important for these synoptic scale processes.

The picture changes completely during summer when the westerlies pass further north and only seldom, in "blocking" situations, affect the Mediterranean area. At this time of the year, mesoscale and regional topographic effects gain in importance and interact with the now much smoother large scale pressure distribution in the Mediterranean area. It is mainly the land-sea circulation that becomes responsible for the exchange of dry and humid air between land and sea. Already during spring, large thermal contrasts build up during daytime between sea and land causing warm air to rise in coastal zones. This generates low surface pressure entraining cooler, wetter, and heavier air from the sea which warms up rapidly when arriving over land.

Uprising may develop into vigorous thunderstorms if the moisture is available to generate deep convective systems that are fed by the latent heat. The air over the sea which in summer has surface temperatures of 22 - 27°C, locally even higher, can take up large amounts of water vapour but it often needs additional lifts by near

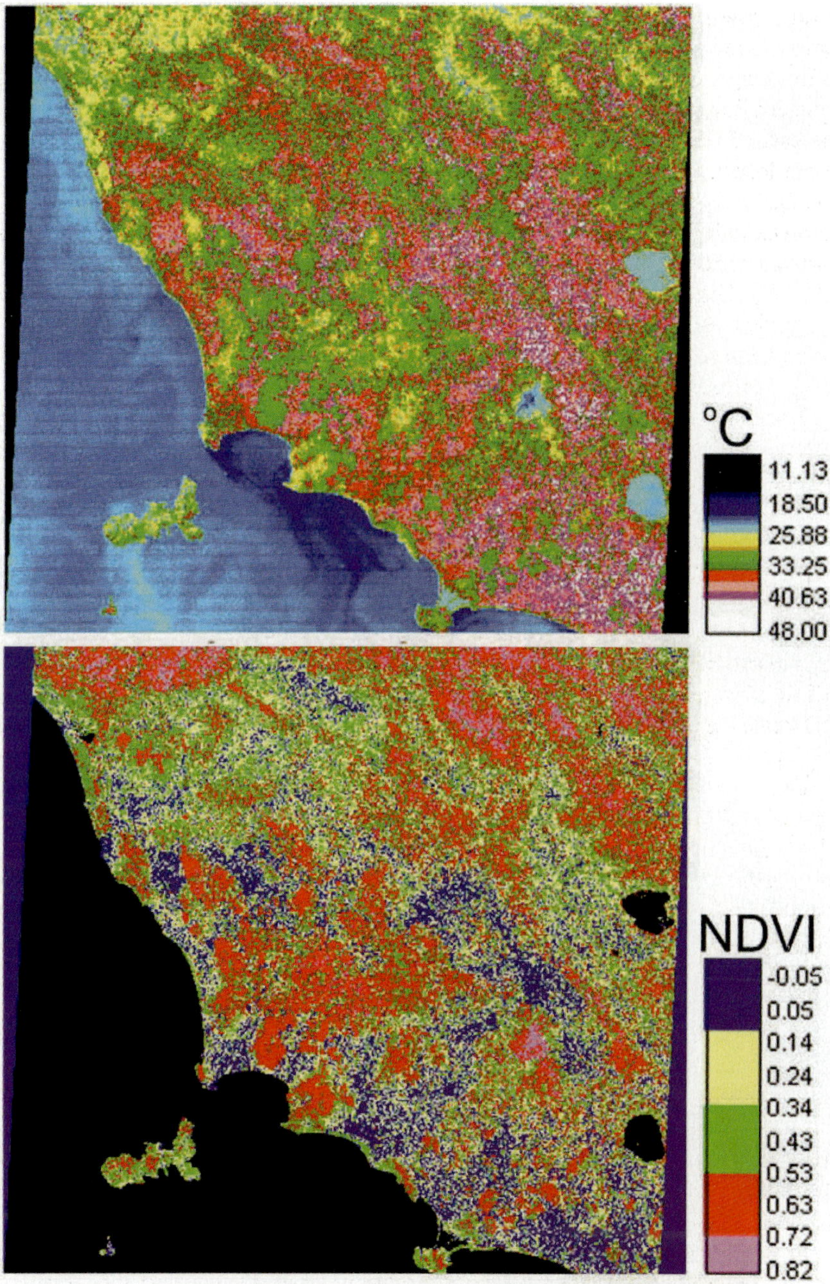

Fig. 1.2. Top: Landsat-TM channel 6 thermal image of south-western Tuscany, 10 August 1998. Colour code on the right in degree Celsius. Bottom: Normalized Difference Vegetation Index (NDVI - normalized near infrared to red signal difference) of the same scene with scale on the right

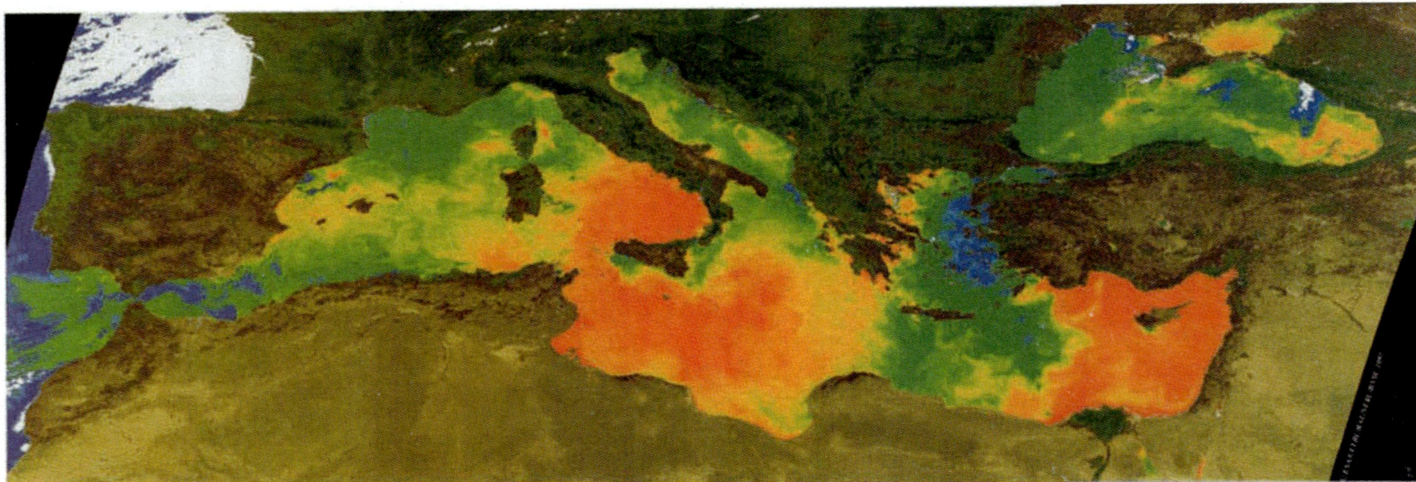

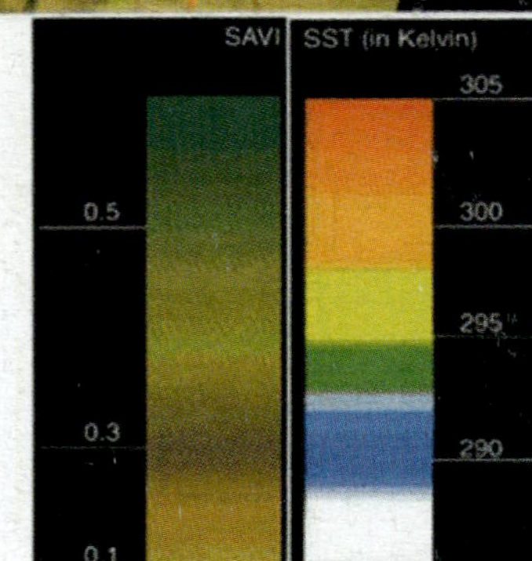

Fig. 1.3. Mediterranean sea surface temperature and vegetation cover over land. The image was derived from 400 scenes of the Along Track Scanning Radiometer (ATSR) on board of the ERS-2 satellite recorded between 17 July and 8 August 1996. The vegetation index used over land is the Soil Adjusted Vegetation Index (SAVI). Courtesy ESA

coastal mountains to move this air up to the condensation level. Pfister (1999) therefore made such local effects responsible for severe floods in the Mediterranean area. These occur often in autumn when the air is cooling more rapidly than the ocean. Millán et al. (1995) and Millán (2000) reported cases of torrential rainfall near Valencia that occurred in autumn when a high pressure cell was over France and a low pressure cell over north-western Africa. The air took up its moisture from very warm water near Tunisia. This was led quickly towards the Spanish coast and released its water when it was lifted upward by the mountains near the coast.

Buzzi et al. (1994, 1998) and Pfister (1999) could attribute such flood events *inter alia* to processes which occurred several days earlier over the North Atlantic Ocean. They found that the atmospheric flow over the North Atlantic and Mediterranean area intensifies as the south to north temperature gradient increases in autumn which would transport considerable additional amounts of moisture into these regions. Pinto et al. (2001) investigated 30 cases of intense rainfall and showed that tropical systems and tropical-extratropical interactions indeed can play an important role in these processes. They detected three mechanisms that may influence the development of extreme Mediterranean precipitation events:

A. Tropical systems over the eastern-central North Atlantic curve directly towards the Mediterranean, undergo a transition into an extra-tropical cyclone and unload their moisture in the western Mediterranean area.
B. Tropical systems over the western and central North Atlantic become extra-tropical cyclones and advect moisture from the subtropics to the extra-tropics. Part of this moisture is then transported by the converted former tropical or other systems along the southern rim of the upper tropospheric main flow towards southern Europe.
C. The tropical system over the western North Atlantic curves east and connects with an approaching upper-tropospheric mid-latitude trough system. The eastern trough of this Rossby wave over the Iberian peninsula induces a south-westerly flow of this moist air over the western Mediterranean area and directs this flow against the south side of the Alps.

The authors investigated in detail the heavy precipitation event of 13-16 October 2000 when the Po level reached record heights. A tropical storm ("Leslie") became an extra-tropical cyclone near Newfoundland. It followed and joined the westerlies and, positioned in a strong baroclinic zone, it crossed the North Atlantic towards the British Isles. Over Spain the general air flow formed a trough which directed a secondary system that inherited part of Leslie's moisture towards the Strait of Gibraltar and from there to the Alps. The updraft due to the mountain barrier then caused the heavy rainfall.

In summer, the westerlies are positioned more northward. Then the these processes do not play a role. Fig. 1.4 gives an impression of phenomena that can be observed in early summer. Over the north-eastern Iberian Peninsula and southern France the development of a large cloud field can be observed that is related to a low pressure system embedded in the north-easterly flow of the westerlies. This is a situation similar to what has been described for the autumn rainfall period but the north-eastward flow occurs more westerly towards France. Over Italy and further east the land-sea circulation leads to cumulus convection over mountain chains such

Fig. 1.4. Three Meteosat scenes of 18.06.96, 12:00, 14:30, and 17 hours. North of Spain passes the warm front of a low pressure system that is dissolved over the Iberian peninsula into a number of convective cells of which some are developing into large deep convective systems topped by cirrus and probably with strong showers underneath. Over Italy and the Balkan the sea breeze generates a line of cumulus first near the coast which is moving into the interior. Some of these convective cells develop into cumulonimbus with tops that reach high into the troposphere which can be seen in the southward extensions of these cloud systems over Italy and the north-easterly stretched clouds over Greece in the afternoon indicating the local wind directions in the free troposphere

as the Apennines. The humid and still relatively cool air that is imported from the sea needs an updraft either by mountains or by mixing with dry convection over warm land surfaces to reach the condensation level. Here precipitation may start if enough water vapour is present. In early autumn the situation changes insofar as the sea remains relatively warm in comparison to the atmosphere and the land-surfaces. At this time of the year, convection starts already over sea and strong latent heat fluxes lead to thunderstorms which then enter the land. Superimposed to these local effects the large scale processes described above determine the weather in the Mediterranean basin.

There exists a marked difference between topographically highly structured areas like the central Mediterranean and the Aegean archipelago and the large, nearly quadratic, landmass of the Iberian Peninsula. While the circulation systems over Italy and Greece are strongly modified by coastlines, mountains, and hills, a large low pressure system often develops during summer near the surface over the central plateau of Spain (Gaertner et al. 1993, Portela and Castro 1996) centred over the southern central part of the Extremadura.

An important question remains to what degree the large scale processes described above are affected by global climate change and how a changed large scale circulation would transform into the regional scale, where topography becomes crucial. In other words, whether or not it has to be accounted for a change of land-surface - atmosphere interactions in the Mediterranean as a consequence of global warming (such as, e.g., a change of the occurrence of extreme events) and responses of the land surfaces to such a change. This prompts the question what feed-backs to climate can be expected from changing land surfaces. Some of these questions are dealt with in Bolle (ed., 2003). Here, in relation to the use of satellite data for land-surface process studies, the focus lies on the interactions between the atmosphere and the Earth's surface. The processes which link these two systems to each other are the net radiation flux at the surface, precipitation, evaporation, sensible heat flux, momentum transfer, and the exchange of gaseous and solid substances such as carbon dioxide as well as other trace gases and aerosols.

Transport, condensation, and precipitation processes in the free troposphere determine the regional "input" from the climate system to the surface. Transports in synoptic scale systems, to a large degree, are independent of processes at the underlying land-surfaces. Condensation processes depend on the presence of aerosols arising from the surface or produced locally from the gas phase. In the Mediterranean area three types of aerosols are of specific importance: Industrial aerosols, sea-spray aerosols, and sand particles. They act in different ways. Salt particles risen from the sea favour the condensation as do many of the industrial aerosols which, in addition, are absorbers of solar radiation. Desert sand particles are not hygroscopic and due to absorption of solar radiation they may warm the dusty atmosphere and thus prevent condensation. The presence of the sand particles is an outstanding feature of the Mediterranean atmosphere. During *scirocco* events, warm dry air from Africa loaded with Saharan dust takes up water vapour while travelling over the Mediterranean sea and small amounts of rain, enriched with sand particles, fall over the southern European areas. These particles have a climatic effect since they scatter back solar radiation to space and thus diminish the radiative flux that reaches the surface. This counteracts the greenhouse effect since it causes cooling at the surface. But this effect may be compensated by their impact on cloudiness. More serious in this respect are sulphate particles.

Precipitation is a function of the available water vapour and growth rate of the condensation droplets. The water vapour is either imported from the Atlantic Ocean (affecting mainly the Iberian Peninsula), evaporated from the Mediterranean Sea (affecting mainly the central and eastern parts of the Mediterranean area) or recycled by evaporation processes over land (affecting inland areas).

According to Trenberth (1997) the Mediterranean area is one region of the globe where the moisture recycling is geographically most variable. He considers the annual turnover of water at a scale of 1,000 km. Over North Africa only 2% of the local precipitation comes from regions within 1,000 km distance, the rest is imported from more distant areas. For the Iberian Peninsula the number is about 18% while in central and eastern Mediterranean regions the values are more around 26% with "hot spots" of 34%. This result strongly underscores the effect of regional topographically induced weather systems for precipitation in the European part of the Mediterranean area, as well as of soil moisture and evaporation for regional precipitation.

On the other hand, the Mediterranean occasionally exports large amounts of water vapour that can cause heavy rains and floods in central Europe. During the flood event 11 - 14 August 2002 in the upper Elbe region trajectories of water vapour bearing air masses have been traced back to the western Mediterranean. The air of many trajectories was curving from the Ligurian Sea (south of France) eastwards over the Mediterranean and then northward via the Balkans and the Black Sea to eastern Europe. Additional water vapour may be picked up over the Black Sea where unusually high SST's have been observed before the air entered Saxony and the surrounding areas from the east (James et al. 2002).

1.3 Processes at Surfaces

1.3.1 Deforestation and Land-use Changes

Major land surface changes in the Mediterranean started ages ago (see e.g. Bolle 2003) and continue to present. Man-made changes *inter alia* are deforestation to a degree of exposing soils to water and wind erosion, large scale irrigation, urbanization, migration of agriculture from terraced hilly terrain to flat coastal plains, abandonment of degraded land, levelling of terraces and structures caused by long time erosion processes to facilitate the application of large machines in agriculture. Many of the human activities concentrate in coastal areas, for example: agriculture because of the flat areas at the mouth of rivers and, as in Italy, the Maremma, which allow more machine work; industry, because of the easier transportation of goods; tourism and secondary houses close to the seaside; urbanization because of better climatic conditions than in polluted cities. This makes the coastal areas more vulnerable to land degradation.

The destruction of the primarily woods, mostly oaks, leads to macchia (matorral, Fig. 1.5), a mixture of partly thorny shrubs which can, if left alone, regrow to the high macchia with small trees. If climatic and soil conditions are less favourable or if fires occur, it will remain a low macchia (montebajo, xerovumi) under which soil erosion may start. Further destruction leads to the sparsely vegetated garrigue (phrygana, tomillares, batha) with dwarf bushes, herbs, some forms of heath and patches of bare soil in between. This can lead to wasteland with very little potential for regeneration.

Portugal, as an example, decided in the 1930's to become autarkical. Forests were cleared e.g. in the Baixo Alentejo to grow cereals. Due to decreasing market prizes and increasing fuel and fertilizer cost, cultivation is decreasing and strong winter rains erode the area which carries only a thin soil layer.

The plains of Castilla - La Mancha (see also Appendix 1) have been nearly completely cleared of trees to give room for agriculture - mainly vine (Figs. 1.6 and 5.60), cereals and olive trees. Under the pressure of heavy water problems part of the land is either abandoned since dry agriculture leads only to marginal cereal yields or heavily irrigated which can cause a continuous fall of the ground water level by one metre per year or more in some cases. This makes dry agriculture even more difficult since the unsaturated zone on top of the groundwater becomes thicker. If the ground water level gets lower than river beds then the rivers may not feed the surrounding area with water anymore but drain water from these soils. Flat lakes, lagunas (Fig. 1.7), are drying out in dry years leaving behind dissolved salts (Figs. 1.8 and 1.9). Thus a concentration of productivity takes place in areas with irrigation potential while other areas become abandoned land where matorral takes over.

In Andalusia, land use changes are closely related to economy. Population goes down with the decline of mining and since 1960 agriculture on marginal land was largely abandoned or changed into sugar beet, sunflower and wheat. Economic pressures also reduced the exploitation of timber and cork. Abandoned land is partly afforested or used for animal reserves or grazing (García Novo and Jiménez Cañete, 1955). In Spain the land use changed from 1750 to 1980 in the following way: Forest from 57% to 32%, dry agriculture from 40% to 47%, and irrigated agriculture from 4% to 21%.

In different parts of Italy the so-called *calanchi* (Fig. 1.10) are present: steep erosion features such as the well known "*balze*" of Volterra (see: Bazzoffi et al.1997). *Calanchi* are small hydrographic basins with a well developed drainage system formed by surface processes such as rillwash, gullying, and frequent shallow land-sliding on very steep slopes. *Biancane* are small clay domes up to 20 m in height and grouped in clusters. Their southern side generally is deprived of vegetation and dominated by incised rills and micropipes. They are surrounded by miniature pediments. Balze are free vertical faces or crags which are continuously subject to falls causing their progressive retreat.

One of the areas formed by these erosional processes is the Orcia valley south of Siena which is a "graben" that was filled with Pliocene marine sediments (8 - 2.5 million of years before present). In 1953 a general reclamation plan was developed which involved an area of 353 km^2 of which 80% were hills and 20% mountains. Calanchi, biancane and mass movements occurred in an area of 50 km^2 . Degradation processes were specifically active in the upper Orcia valley covering a total area of 280 km^2. During the period 1965 - 1985 the activity of the reclamation departments

Fig. 1.5. View from the hilltop seen in Fig. 1.15 towards the small village of Rada de Haro showing a matorral landscape with marginal dry agriculture. In the foreground the car track can be seen which is discussed in the text

Fig. 1.6. Grapevine monoculture east of Tomelloso, Castilla - La Mancha, Spain, during the field experiment "EFEDA", 1991. The picture shows a dense network of masts equipped with instruments for temperature, humidity, wind and flux measurements

Fig. 1.7. Laguna at wet season near Pedro Muñoz (Castilla la Mancha, Spain)

Fig. 1.8. Dried out laguna near Pedro Muñoz, Castilla-La Mancha, Spain, summer 1994

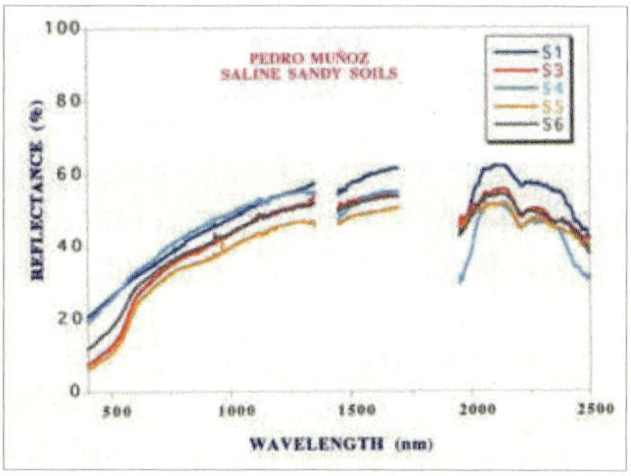

Fig. 1.9. Reflection spectra from different samples of Pedro Muñoz saline sandy soils

Fig. 1.10. Erosion near Abbazia di Monte Oliveto Maggiore, Tuscany, Italy

Fig. 1.11. Harvested wheat fields near Orciano, Tuscany, Italy

Fig. 1.12. Rock removal in progress northeast of Matera, Basilicata, Italy. Large rocks have been crushed. The remnants are piled up to walls.

Fig. 1.13. A field obtained after removing the larger stones. The soil still covered with debris of rocks is ready for agricultural use. The location is close to that one of Fig. 1.12

became less active. Old and new landowners, partly immigrants from Sardinia, erased most of the isolated gullies and biancana fields by levelling the area with bulldozers. In the well known Le Crete south of Siena the now soft slopes are used mainly for durum wheat cultivation. During winter and spring these are treeless green soft hills which, like in other locations (Fig. 1.11), turn into desert-like areas in summer after harvest. They can be identified in Fig. 1.2 by low vegetation indices and high temperatures (compare Fig. 6.5.1).

With the improvement of technology similar changes take place nowadays in rocky areas. To gain arable land, heavy machinery is used to crush rocks and to smooth the surface (Figs. 1.12 and 1.13).

The desire to intensify agriculture also led to a completely new category of land cover during the last decades: the increasing area covered by plastic greenhouses. They range from flat plastic covers to sophisticated constructions as can be seen in Fig. 1.14 which is from the western part of the Messara valley in Crete, Greece. Near Almeria in southern Spain a whole Peninsula (Campo die Dalias) is covered by such greenhouses which makes the peninsula disappear in vegetation index images (Fig. 6.5.2). Only recently two Landsat images from 1997 and 2004, which document the dramatic change which took place here, were published as an example for overexploitation of nature by UNEP in the atlas “One Planet, many Humans”. The coverage of such a large area with plastic and the change of albedo is one side. More risky in view of the droughts which frequently strike the area is the water consume needed to grow the legumes here. This water has to be transported to the south from far away in the north. In cases of strong droughts like that one in 1995 therefore the impact of water shortness immediately spreads over wide areas and can lead to disastrous situations. For the year 2005 a similar situation as 1995 was experienced in Portugal and Spain. Here the year started with an extremely dry spring with wild fires and food shortage for animals already during this season.

Fig. 1.14. Solid greenhouse constructions which constitute a new type of land cover can be found in different parts of the Mediterranean area. Here in the Messara Valley, Crete

After this excursion to the impact of modern technology on land surfaces one example of the fragility of matorral landscapes will be given here. At the beginning of this section a description of this land-surface type was given and illustrated by Fig. 1.5 which shows the landscape near Rada de Haro in Castilla - La Mancha, Spain, one of the experimental sites of EFEDA. The hill from which this picture was taken is shown in Fig. 1.15 from three nearly identical positions in different years. The car track visible in all three pictures was laid during the field experiment in 1991 (top). The picture in the centre of Fig. 1.15 was taken in late summer of 1997 and the one at the bottom during a rainy period at the end of October 2003. The still visible car track demonstrates the very low restoration potential of the matorral after compaction. The trail to the left is part of the "andalusia", an ancient and still used sheep trail (prior to 1997 also used by a vehicle). It can also be seen that during the rainy period (2003) there is some low green vegetation along the "andalusia" which is fertilized by the sheep excrements. Bushes and trees at the scene did hardly grow during these 12 years. The soil seen in the foreground of the upper part of the figure is classified as "Lithic Leptosol" (for a description of the soil types see Appendix 3).

1.3.2 Water Related Problems

Collin (1995) presented numbers for the water resources of the countries Spain, France, Italy and Greece. In these four countries respectively 39%, 2.7%, 7.9% and 6.3% of the total resources are consumed. In Spain 3.3 Mha (12% of the national farmland) are irrigated, the total water resources are $31 * 10^9$ m^3/a, the groundwater holds $9.1 * 10^9$ m^3/a, and $12.2 * 10^9$ m^3/a are consumed which is more than the groundwater reservoir holds. 6000 m^3 water are spent per ha and per year on irrigated surfaces. The South-East of Spain is an area in which agriculture is not possible without intensive irrigation. For this purpose water is exported from the *Embalse de Alarcón* through the about 250 km long *Canal de Trasvase Tajo Segura* (Fig. 1.16) which adds to the water shortage in central Spain. The Embalse de Alarcón in turn is connected by the Trasvase Tajo Segura with *Rio Záncara* and the *Embalse de Entrepeñas* (Mar de Castilla) further north.

Similar problems of sinking ground water levels are reported from other areas such as the Messara valley (Fig. 1.17) of Crete (Vardavas 2000, Croke et al. 2000). Körner et al. (2005) report a long term increase of dryness at the island of Samos evidenced by a successive drying up of springs. Some of the main agricultural areas which need irrigation are situated close to the sea where the extensive irrigation lowers the ground water level and allows sea water to enter the aquifers. Salinization takes place that reduces the quality of the soils like at the Spanish lagunas (Fig. 1.8).

In the Argolid area of Greece (north of Mykenes) the irrigated area increased from 4.000 ha between 1939 and 1945 to 20.000 ha in 1990. The water consumed for irrigation rose from about 45 to 145 Mm^3/a and is presently used mainly for citrus. Around 1960 water demand and recharge were still balanced. By the nineties the groundwater level had dropped in the average by about 100 m, during the last years of this period by about 9 m per year in worst cases. In peripheral areas boreholes drilled in the nineties to over 400 m below ground level failed to locate water. Salt

Fig. 1.15. Degraded matorral landscape near Rada de Haro, Castilla-La Mancha, Spain. Top 1991, centre 1997, and bottom 2003. Up hill a car track can be seen which was laid in June 1991 for transportation and maintenance of instruments installed at the top of the hill. This car track survived for more than 12 years as can be seen in the lower picture

Fig. 1.16. Two views at the "Canal de Trasvase Tajo Segura" near the village of Barrax, Castilla - La Mancha, Spain

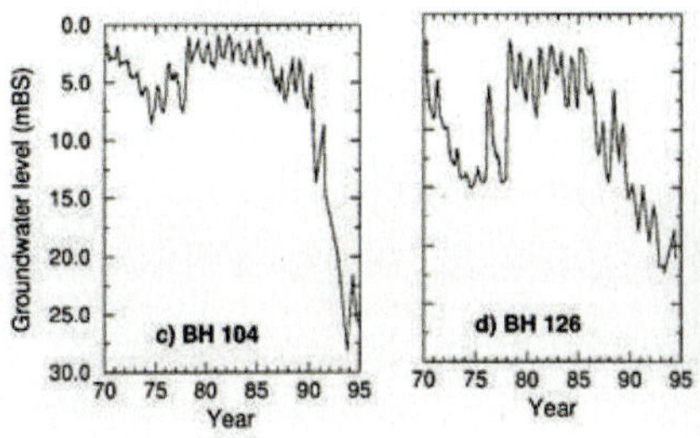

Fig. 1.17. Two examples for the decline of the groundwater level between 1970 and 1995 at two boreholes along the Geropotamos river in the Messara valley, Crete, Greece after Croke et al. (2000). Extensive Olive grove irrigation started 1985

Fig. 1.18. One of the *Lagunas de Ruidera*, Castilla - La Mancha, at its very low level 1994 during a severe drought period

Fig. 1.19. The laguna of Fig. 1.18 nearly refilled in 1997 (left) and in October 2003 (right)

Fig. 1.20. Construction of a water pool near Rada de Haro, Castilla - La Mancha, Spain. The large pool built by walls of red soil is covered by a plastic foil. The pool has to be filled by pumping.

Fig. 1.21. Water reservoir near the monastery Arkádi (Moni Arkadiou)

water intrudes from the sea into some of the boreholes (Alexandris et al. 1998 and Lemon et al. 1995).

The need for a seasonally balanced supply with water prompted the governments of most Mediterranean countries to develop water plans which include construction of dams, water reservoirs, and aqueducts. Two of the largest reservoirs presently are filled, the Atatürk dam in Turkey and the Alqueva dam in the Alentejo region in Portugal (area 250 km^2, volume up to 4.15 km^3). In the meantime, older reservoirs are already filling up with sediments. It has been calculated (Wagner 2000) that even the many medium sized reservoirs in Morocco will not satisfy the requirements of a growing population within some years. After a few dry years in the mid nineties some Spanish reservoirs were nearly at the end of their capacity (Fig. 1.18 and 1.19).

In addition to large dam constructions and natural lakes, agricultural communities and individual farmers built smaller reservoirs ranging from little water ponds to football field sized basins. Two examples are shown here. Fig. 1.20 shows a reservoir near Rada de Haro built by heaping up huge earth walls of local red soil. In 1991 when the picture was taken, the reservoir was still under construction. Half of it is covered with white plastic material. In later Landsat-TM and SPOT scenes the pool can be localized as a small black spot. Fig. 1.21 shows a modern reservoir built in a valley close to the monastery of Arkádi on Crete.

1.3.3 Fire, Grazing, and Land Degradation

Fires affect large areas in Spain, France, Italy and Greece. Two percent of the surface of Sardinia burned down alone in the 1980s. The island of Samos experienced a severe fire event in 2001. It depends on the intensity of the fire how long lasting the impacts are. If the fire is quickly rushing through very dry thin undergrowth, then trees are less affected and the area recovers within even one year. If the undergrowth is dense, flames reach high and trees are burning down even into their root systems. This damages the area sustainably. Nevertheless olive trees may regrow within 5 to 10 years but often as wild species. Also seeds laying in the ground may be burned depending on the species. It has been observed that after the complete burning of a forest of *pinus pinea* the species *pinus halepensis* reappeared which obviously was dormant for many decades. With success *Pinus nigra* is planted for reforestation in some areas after wildfires.

In the DeMon study site of Xilokastron at the Peloponnese, Greece, land condition changes have been investigated with LANDSAT images of 1985 to 1990 and ground checks in 1994 (Hill, 1996, Lacaze et al. 1996). Careful corrections and adjustments of the two scenes revealed a reduction of photosynthetic active vegetation and degradation of the soil on steep hill-slopes of neogene marls resulting in a complete loss of vegetation while, in other areas, the vegetation quickly recovered. Similar observations have been made e.g. in the area between Brádano and Basento NW of Metaponto, Basilicata, Italy. These changes are attributed to

Fig. 1.22. Eroded hills at the Basento, Basilicata, Italy

Fig. 1.23. Erosion in Morocco north of route N13 near Tiflèt.

Fig. 1.24. Desert and marginal agricultural land use in the area between Abanilla and La Alguena north of Fortuna and further north of Murcia at about 500 m altitude with view at the Sierra de la Pila. Orchards are irrigated from artificial water ponds as can be seen at the right of the centre of the picture (see also Fig. 1.25)

burning with successive water erosion (Fig. 1.22). Such degradation/erosion features can be found in many Mediterranean regions. An example of Morocco is shown in Fig. 1.23. A scenery most spectacular for southern Europe is found in the vicinity of Murcia, Spain (Fig. 1.24).

Grazing is developing differently throughout the Mediterranean countries. In Sardinia it seems to have stabilized while in Greece it still is increasing. Specifically the damage caused by fires sustains if afterwards the new young shots are eaten by goats or sheep.

The combined impact on the shaping of the Mediterranean landscapes of forest fires, goat and sheep herds, the cutting of trees for the production of timber, the removal of firewood, and the erosion caused by torrential rainfalls as well as wind is very large and leads together with the expansion of agriculture to the sequence of land degradation described in section 1.3.1.

1.3.4 Drought, Floods, Frost, and Desertification

A change of global climate not necessarily affects regional annual mean values of meteorological quantities. It may also manifest itself in a change of the frequency, phase, and intensity of the extreme events. These always happen from time to time but their intensity and frequency seem to increase during the last decades. Even if they are of short duration, they can affect lastingly land-surface processes and economy at a local and regional scale. Apparently these extreme meteorological events are regionally as well as temporarily irregularly distributed. It is difficult to detect pattern and to identify the "driving force behind the scene". A few recent events may be recalled. During the winter of 1996/97 a series of cold days destroyed many Eucalyptus trees in central Italy. These trees play a role in local water budgets and, because of their height, as well for the structure of the atmospheric boundary layer. Until they are regrown the wind effect on the surface will locally be larger and the water budget will be changed. During the same period the Iberian peninsula suffered from heavy rains. Around the Guadiana basin in Portugal, where in the thirties the forest have been cleared in favour of crops, large amounts of soils were washed away irreversibly. Repetition of such extreme short duration events at a high rate will lead to desertification if the soil layer already is thin. The year 1999 started dry in Portugal as well as in the Middle East where it continued to early summer (Annual Bulletin on the Climate in WMO Region VI 1999). March was extremely rainy in Portugal as well as northern Spain (Coruña: 246 mm). Sicily had the first snow since in February 1981. The April to July temperatures of the Spanish mainland reached their highest value since 1961 and in Portugal, maximum temperatures reached new records. August was hot and dry in southern Europe. On the tenth of August the temperatures climbed up to 45 °C in Catania Signorella, which is a new record. In September the weather was very unsettled. Portugal had severe damage due to flooding because the rain was much above normal. In Valladolid it rained 106.8 mm which is the highest monthly total in the last 130 years. Segovia had the highest number of rainy days during the last century. Three severe tornados damaged Alicante, Murcia, and Granada. Up to 230 mm rain per day fell in north-eastern Italy which has a statistical return period of more than 50 years. Extreme weather conditions continued in October. The central Mediterranean experienced temperature anomalies up to 2°C above normal (e.g. 37.4 °C in Palermo, 38.6 °C in Catania). In the south-east of Spain the monthly precipitation totals reached 400 % of normal and torrential rainfalls occurred in the mountain watersheds of the Venetian plains and along the northern Apennines (23 and 26 October). In November temperatures in Israel were 5 °C above normal for eight consecutive days. Heavy storms led to floods and landslides in Greece with heavy losses in infrastructure and crops, four people died. An exceptionally rainy period occurred in south-western France with rainfall up to 620.2 mm in 48 hours, causing estimated losses of more than 500 million €, and 196 mm in 24 hours in Cagliary. On 18/19 November rainstorms with strong winds and 100 mm precipitation/10 hours went over Tuscany, inundated industrial areas and settlements, uprooted trees and blocked roads. Also Turkey was affected by storms. In Campania, east of Naples, 324 mm of rain were recorded within 2 days in December . In the south-east it was again dry. Cyprus experienced the fifth consecutive year below normal precipitation. Israel reported one of the hottest summers of the last century (1998 was the record in Bet Dagan with even one degree more) and, as in 1958/9, one of the driest September to December periods. In the south-east of Spain and the Balearic Islands the year 1999 ended with 50% less rain than normal. After a dry period in places such as the east coast of Spain, the south coast of France and northern and middle Italy, a rainy period was experienced in late autumn and winter 2000/2001 with disastrous torrential rain amounts causing flash floods and land-slides which killed people, destroyed houses, eroded fertile soil, and obstructed farmers to bring out seed at the right time. Venice had "aqua alta" the first time for several weeks continuously. In November 2001 Algiers suffered from extreme rainfall and a severe storm hit the Balearic Islands. In early 2002 snow fell over Crete and a short frost period in March 2003 nearly completely prevented the production of olives in some parts of Tuscany, Italy, which already in the eighties had been affected during an even stronger cold event and needed many years to regrow. Then in 2003 the Mediterranean area as well as large parts of northern Europe suffered from the extreme dry and hot summer episode.

Fig. 1.25. Marginal agriculture with small water reservoirs. Same area as Fig. 1.24

Prolonged or more frequent drought periods can cause extreme water shortages. With the need of enforced irrigation and other water consumption the ground water level is continuously reduced and the ground water volumes become marginal. Soil is degraded by the salty minerals which are pumped with the irrigation onto the fields.

Marginal land has to be abandoned. If in addition fires occur and if after a long dry period torrential rainfalls carry away soils, the degradation process may lead to desertification which means, in its ultimate stage, irreversible degradation. Such areas sometimes are found not far from densely populated regions along rivers like the Ebro or the Segura. One of these areas, shown in Fig. 1.19, exists about 26 km north of Murcia between the villages Abanilla and La Alguena, north of Fortuna. Water reservois filled by winter rain support marginal agriculture (Fig. 1.25).

There exist many definitions of the term "desertification" (see e.g. Mainguet, 1991, United Nations 1992). The official definition introduced by the United Nations *Convention on Desertification* is the following: *"'Desertification' means land degradation in arid, semi-arid and dry sub-humid areas from various factors, including climatic variations and human activities"*. The *Convention* refers to the following particular conditions of the northern Mediterranean area:

(a) Semi-arid climatic conditions affecting large areas, seasonal droughts, very high rainfall variability and sudden and high-intensity rainfall;

(b) poor and highly erodable soils, prone to develop surface crusts;

(c) uneven relief with steep slopes and very diversified landscapes;

(d) extensive forest coverage losses due to frequent wildfires;

(e) crisis conditions in traditional agriculture with associated land abandonment and deterioration of soil and water conservation structures;

(f) unsustainable exploitation of water resources leading to serious environmental damage, including chemical pollution. Salinization and exhaustion of aquifers; and

(g) concentration of economic activity in coastal areas as a result of urban growth, industrial activities, tourism and irrigated agriculture.

Desertification can also be considered as a *process* causing a *gradual loss of the capacity of land to support the sustainable development of existing ecosystems or, more specifically, the growth of vegetation that is essential for human and animal life*. This in fact is a definition of desertification by its impact, not of the process itself which for the time being is ill-defined because no precise and integrative formulation exists.

Climatically induced desertification according to this definition would occur, if

(i) due to water and wind erosion the soil gets too thin to carry deeper rooting vegetation,

(ii) during a sequence of vegetation periods there is not enough soil moisture near the surface to let vegetation and seeds survive,

(iii) the root zone becomes disconnected from water in the conducting deeper soil layers, the aquifers, and the surface soil layers do not receive enough rain at the right time of the year to support vegetation growth,

(iv) due to hot and dry weather conditions the probability of fires is enhanced, or

(v) the lack of rain enforces an overexploitation of water resources.

In view of temporary water deficits it is of utmost importance that a hydrological equilibrium is restored at a time scale shorter than the response time of the vegetation-soil system to water stress. For Mediterranean macchia-type ecosystems this response time seems to lie in the range of three to five years as could be observed from a dry period of three years duration in Tuscany, Italy. Towards the end of this dry spell, trees lost their leaves already in early summer; the soil was exhausted of water, wells and rivers fell dry, but nature recovered after first heavy autumn rains.

Of overriding importance for the desertification process therefore is the duration of dry periods. During droughts, vegetation periods are shortened, leaf-area-indices are reduced, grasses, herbs and dry farming crops are stressed and/or thinned out; only deep rooting plants will be less affected. Wildfires will add to the stress. The energy budget of the surface changes because the surface albedo and temperature change. Dry vegetation as well as bare soil generally have a higher albedo and lower emissivity than biologically active vegetation. The radiative energy absorbed at the surface is reduced and converted mainly into sensible heat. The reduction of evaporable soil moisture leads to a larger diurnal temperature amplitude. This increases the instability of the lower atmosphere and dry convection further reduces the probability for condensation processes and rainfall. The surface roughness is also reduced which increases the surface wind stress and favours wind erosion (Sud et al. 1988). The "Charney mechanism" (Charney 1975, Charney et al. 1975, Ellsaesser et al. 1976, Rasool 1984) becomes active which maintains the drought (increased albedo → reduced rainfall → less plant growth → further increased albedo → enhanced desertification). With respect to temperature the feedback is negative (increased albedo → reduced absorption of solar energy → lower temperature than a dry surface of the same albedo would have → reduced evaporation → moderating desertification). The worst situation would occur if a dry spell that lasts long enough to substantially reduce the leaf area is followed by extreme precipitation events at a date which allows the soil to dry again prior to the start of the next vegetation period. Then wind erosion would be replaced by water erosion and soil would be lost at an increased rate.

Sud and Smith (1985) and Sud et al. (1988) investigated the influence of changing surface roughness on the atmospheric circulation and precipitation. In a study with a global general circulation model they reduced the roughness length z_0 (the effective height of the obstacles at the surface) over the continents from 45 cm to 0.02 cm. The reduction of z_0 resulted in an increase of the wind speed in the atmospheric boundary layer by a factor of two, and at the same time a two-fold decrease in the magnitude of the surface stress. This caused a corresponding large change in the horizontal convergence of the water vapour transport in the atmospheric boundary layer and in the distribution of rainfall. Due to a decrease of pressure over northern Africa this led to a strong increase of westerly winds over the Sahel and to an increase of the easterly wind component over the Mediterranean in the lowest model layer.

Ellsaesser et al. (1976) summarized the causes put forward from various authors for the response of the land-surface system to human impact:

- precipitation can be reduced through stabilization of clouds due to overseeding by condensation nuclei lofted from the surface during the denudation process (Squires, 1958, Schaefer, 1968);
- increased airborne dust risen from the denuded surface emits additional infrared radiation to space, thus cooling the atmosphere, which induces additional subsidence and suppresses the precipitation mechanism (Bryson and Baerreis, 1967);
- increased albedo reduces the surface temperature and causes a "thermal

depression" with the same effect on convective precipitation (Otterman, 1974);

- increased albedo reduces the net radiation at the surface and contributes to a net radiative loss in the free atmosphere with the result that horizontal temperature gradients induce a frictionally controlled circulation with subsidence tending to restore the thermal equilibrium with adjacent areas, and the subsidence depresses convective processes (Charney, 1975);
- the decreased availability of biogenic ice nuclei in denuded areas reduces convective precipitation (Schnell, 1975).

The authors added their own model results, which confirmed the Charney feedback mechanism with respect to precipitation.

Otterman (1989) pointed out, that vegetation at the fringe of deserts reduces the radiation reaching the soil surface, thus reducing the soil heat flux and increasing the absorption of solar radiation of the soil-vegetation system because its albedo is lower than that of the bare soil. The result is an increase of sensible heat flux into the atmosphere, which increases convection and by this the probability of precipitation, if moisture is advected. Removal of this vegetation will lead to increased aridity.

1.3.5 Coupling Between Surface and Atmosphere: The Role of the Atmospheric Boundary Layer

Land-surface processes are linked to the global climate system through the water and nutrients cycles as well as the exchange of energy and momentum with the atmosphere. It is one of the major challenges for remote sensing to assess these fluxes from space respectively to supply data and information that support the computation of these fluxes. To understand this challenge one has to recall how atmosphere and surface interact. The partition of fluxes at the surface not only is a matter of the surface properties which are observable from space. The sub-surface hydrological regime, as well as the relatively thin Atmospheric Boundary Layer (ABL) on top of the surface, strongly influence this process. At its bottom, the ABL is intimately connected to the surface by the vegetation that protrudes into this layer and the roughness of land surface itself. The top of the ABL is the altitude at which the direct influence of the surface on the dynamics of the atmosphere ceases. The exchange processes between the ABL and the free troposphere above are of completely different nature than those at the surface and will not be considered here (see e.g. Kossmann et al. 1999).

Surface and atmosphere are intimately coupled by the net radiation flux at the surface which during daytime is directed towards the surface (the downward direction is denoted as negative) and upward (positive) during night. The spatial deposition of this energy ultimately is responsible for the land-surface processes. It causes differential heating of the surface depending on albedo. The radiative energy deposited at the surface evaporates water and heats the surface. This drives the motion of the air that in turn transfers momentum to the surface due to friction and carries latent and sensible heat away. Part of the radiative energy is responsible for the photosynthesis which in turn causes changes of the albedo due to the growth of vegetation.

The partitioning of the radiative energy into heat fluxes depends essentially on the availability of moisture that can be evaporated, and on the structure of the atmospheric boundary layer. Directly at the sea surface the air can be considered as saturated with moisture. Locally the evaporation from the Mediterranean Sea is controlled by the humidity gradient over the sea, established by the moisture content of advected air and measured at a reference height, q_{ref}, and the aerodynamic resistance:

$$\Phi_{LH} = L\rho \frac{q_s(SST) - q_{ref}}{r_{as}}. \tag{1.1}$$

L (J kg^{-1}) is the latent heat of vaporization of water (see Appendix 5), ρ (kg m^{-3}) the air density, $q_s(SST)$ is the specific humidity at saturation defined by the sea surface temperature, and r_{as} (s m^{-1}) is the aerodynamic resistance over the sea. According to Francey and Garratt (1978) and Bill *et al.* (1980), r_{as} for latent heat is $(0.0018u_{10})^{-1}$ respectively $(0.0012u_{10})^{-1}$ if the reference height is 10m and u_{10} is the wind speed at this height. The sea surface temperature (SST) is a function of the flux of net radiation but depends in addition on currents in the sea and the influx of fresh water from the surrounding landmasses and thus from the large scale water import as well. As can be seen from Eq. 1.1 the water vapour flux in principle can also be directed towards the sea surface if warm humid air of average humidity is floating over relatively cold water. If, for example, the SST is 25°C and the air temperature 40°C then for a relative humidity of more than 42% the numerator in Eq. 1.1 becomes negative. Such cases have been computed with models for the eastern near coastal part of the Mediterranean e.g. for August 1987 and August 1988 (Castellari et al., 1998).

Over vegetated land surfaces the situation is more complex. The moisture can come from the uppermost soil layers or may be pumped to the surface from deeper layers by plants in the course of the assimilation process. In addition, the case has to be considered that the water is evaporated by the diurnal heat wave at a "waterfront" deeper in the soil (Menenti 1984) and that the water vapour has to penetrate the unsaturated zone. A number of conceptional models have also been developed for the processes which are governed by biological mechanisms in the plants ranging from the water suction by the roots to the evaporation through stomata. Furthermore, surface wetness of the leaves and the structure of the atmosphere surrounding the plants have to be considered in the description of the heat and moisture exchange between surface and atmosphere. The basis for the mathematical representation of most of these processes can be found in Brutsaert (1982/84).

Based on these conceptional models and mathematical parametrizations of the individual processes, a number of so-called Soil-Vegetation-Atmosphere-Transfer (SVAT) models have been developed, *inter alia* by Dickinson (1984), Sellers et al. (1986), Dickinson et al. (1986), Van de Griend and Van Boxel (1989), Braud et al. (1993), and Blümel (1992, 1998). These models can be used as stand-alone versions in connection with experiments or with the investigation of the sensibility of the fluxes with respect to changing environmental conditions and may serve as modules in climate and weather forecast models. The performance of SVAT models applied in the EFEDA project have been tested and compared with measurements made in the field (Linder et al. 1995, Calvet et al. 1998).

It is not possible to infer the many parameters needed to run the more complex models from remote sensing measurements. Only few of the quantities can be obtained from measurements made from space. The estimate of area averaged fluxes using measurements made from satellites without intensive accompanying ground

Box 1. Results of a sensitivity study for the latent heat flux over dense and sparse vegetation after Blümel (2004). The number of X stands for the importance of the parameter (the magnitude of the slope of the flux derivative), the + and – signs indicate positive and negative correlation of the flux derivative with the parameter change. In the "large deviation" column only those parameters are listed which in addition to those marked in the first column have to be taken into account respectively which change importance or direction

Parameter	Dense vegetation		Sparse veget., dry soil	
	Near Reference	Large Deviation	Near reference	Large Deviation
Leaf Area Index [*LAI*]	XX+		XX+	
Stem Area Index [*SAI*]	X–		X–	
Minimum Stomata Resistance [$r_{s,\, min}$]	XXX–		XX–	
Near Infrared Canopy Albedo [ρ_{NIR}]	XX–		X–	
Maximal Possible Transpiration [γ_{r00}]		X+ [1])	XX+	
Shelter factor [*shelt*]	XX+		XX+	
Emissivity of the vegetation [ε_f]	XX–	X–	XX–	X–
Fractional vegetation coverage [σ_f]	XX+	X– (!)	XXXX+	
Leaf thickness [d_f]		+		
Soil moisture profile [$s(z)$]		XXXX+ [1])	XXXX+	
Soil temperature profule [$T_g(z)$]	X+			
Ratio of root length densities in the upper and lower root layer [*ROOTF*]				X–
Roughness length of the vegetation [$z_{0,f}$]		X+		X–
Soil texture [*TEX*]		XXXX– [1])		XXX–

[1]) important for negative deviations only

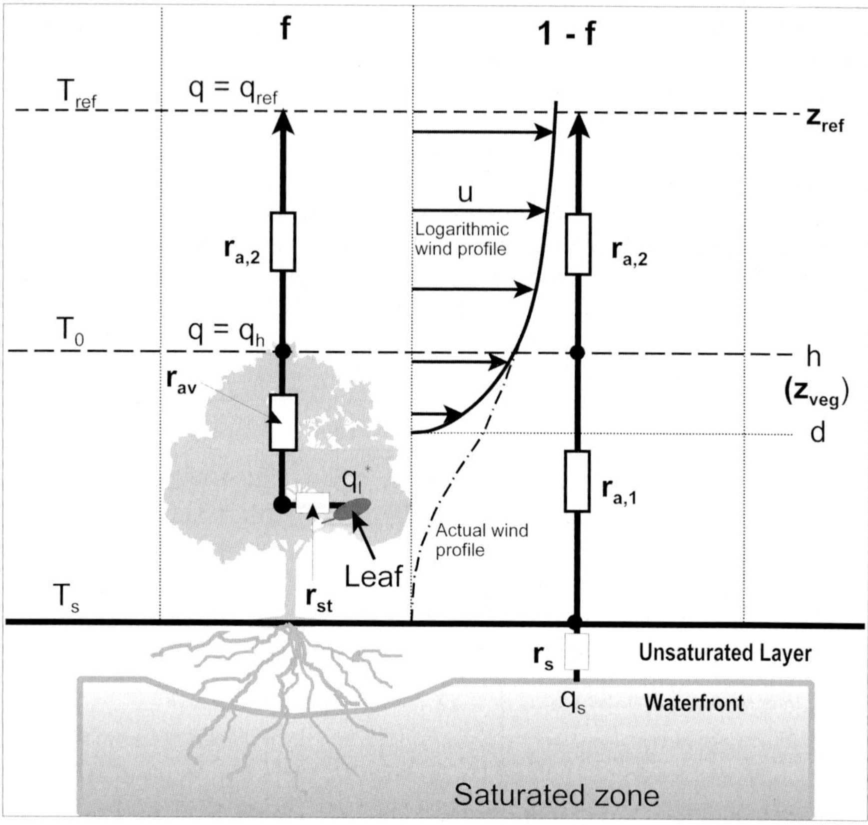

Fig. 1.26. Idealized water vapor pathes from vegetated and bare soils to the atmosphere. Symbols are explained in the text. The "logarithmic wind profile" exists only above the layer where the air is disturbed by roughness elements and for neutral atmospheric boundary layers

measurements therefore is mostly based upon models that can be operated with a small number of bulk parameters. In fact, Blümel showed (see box 1.1) that to estimate evaporation only few parameters need to be known accurately, the others can be taken from available empirical data sets or even treated as constants without affecting the result too much.

In a very simplified manner, the latent heat flux from a partly vegetated surface may be represented by a resistance formulation (compare Fig. 1.26):

$$\Phi_{LH} = L\rho_{air}\left[f\frac{q_l^*(T) - q_{ref}}{r_{st} + r_{av} + r_{a,2}} + (1-f)\frac{q_s - q_a}{r_a} \right] \tag{1.2}$$

Here ρ_{air} is the air density, f the areal fraction covered by vegetation, L the latent heat of evaporation, $q_l^*(T)$ the temperature dependent value of the specific humidity at saturation that is assumed to be reached within the leaves, q_{ref} the specific humidity at a reference height, z_{ref}, in the atmospheric boundary layer, q_s the specific humidity at the bare soil surface or, if this is dry, at the "waterfront" in the soil (Menenti 1984), r_{st} the resistance for the water vapour flux through the leaf stomata, r_{av} the resistance of the air within the canopy between the evaporating leaf and the top of the canopy, $r_{a,2}$ the resistance between the top of the canopy, h, and the reference height, z_{ref}, and r_a, the aerodynamic resistance between the bare soil and the reference height consisting of two parts, $r_{a,2}$ as before, and $r_{a,1}$ that reaches from the bare soil to the canopy height $h = z_{veg}$. In case the upppermost soil layer is dry, the resistance r_s of the unsaturated soil layer must be added. The term "resistance" used here can be considered as a short-hand paraphrase for the dependance of the fluxes on other parameters than the water vapour gradient that is explicitly written down in Eq. (1.2), namely the stability of the atmospheric boundary layer, wind velocity, the surface roughness (affecting the aerodynamic resistance), and the state of the vegetation (affecting the stomata and canopy resistances). These parameters determine the intensity of the latent heat flux.

Eq. (1.2) is an approximation that may be applicable only if the areas of vegetated and bare soils are large enough to establish their own boundary layers. Otherwise the fluxes will mix, internal boundary layers may develop and the total flux will become a non-linear combination of the two pure flux approximations suggested in Fig. 1.26 and Eq. (1.2). Three dimensional models or bulk formulas based upon observations would then have to be used to describe the fluxes. The basic question remains the determination of the magnitude of the resistances and their variability with climate and changing soil properties. This can not be computed from first principles and is currently determined for the various ecosystems in field experiments.

For the sensible heat fluxes, similar relationships as Eq. 1.1 and Eq. 1.2 exist. The sum of both heat fluxes into the atmosphere must follow the rule for energy conservation:

$$\Phi_R^* + \Phi_G^* + \Phi_{LH} + \Phi_{SH} + \Phi_{Ph} = 0 \tag{1.3}$$

where Φ stands for flux, R for radiation, G for heat flux into the "ground" which consist of the soil and the organic material of the vegetation, LH is the latent heat, SH the sensible heat, and Ph the energy consumed for photosynthesis. The * stands for "net", net radiation and net ground heat flux. The net surface heat flux is consumed exclusively for warming the soil and the vegetation. It equals the total heat flux at the surface of the soil and the vegetation minus that part which is used for the evaporation inside the soil-vegetation system, e.g. the upper 10 - 30 cm of the soil. The evaporated water escapes as water vapour and is measured as latent heat flux or evaporative flux at a reference level above soil and canopy. The energy balance equation in principle is difficult to handle since in general the radiation energy is not deposited at a flat surface but in a volume. It is very useful in cases where there is no vegetation ($\Phi_{Ph} = 0$) and if the fraction of the soil heat flux that is used for evaporation can be neglected ($\Phi_G^* = \Phi_G$). Then it can be applied in the form

$$\Phi_R + \Phi_G + \Phi_{LH}(1+\beta) = \Phi_R + \Phi_G + \Phi_{SH}\left(1+\frac{1}{\beta}\right) = 0. \tag{1.4}$$

Here β is the *Bowen ratio* Φ_{SH}/Φ_{LH}, a quantity that can be measured relatively easy. Under the condition that the turbulent transfer of moisture can be described by the same "universal functions" (see Brutsaert, 1982/84, pp. 65/66) as the turbulent transfer of heat, β can be written $\beta = \rho_a c_p \Delta T/\rho_a L \Delta q$ where $q = \rho_v/\rho_a$ is the specific humidity. The density of the air can be written $\rho_v = p/RT$, the density of the water vapour $\rho_v = e/RT$. With this notation follows $\beta = (c_p p/0.622 L)(\Delta T/\Delta e)$. Thus all what is needed to determine β are two psychrometers mounted in two heights and a barometer.

In chapter 5 it will be shown how remote sensing techniques make use of these basic relationships to estimate the heat fluxes into the atmosphere. Fed with experimental data from field experiments, SVAT models can then be used to validate these estimates for specific sites.

1.4 Role and Capabilities of Measurements Made From Space

1.4.1 Research Programmes

One of the most challenging questions of our age is whether direct and indirect impacts of man´s activities may have irreversible consequences for living conditions on Earth. Direct impacts are changes made at the earth surface by changing land use and water ways, exploiting natural resources, and releasing trace substances, pollutants, into the atmosphere and the sea. The indirect impacts result from the emission of the so-called "greenhouse gases" and aerosols which cause climate change as predicted by the IPCC (International Panel of Climate Change, see e.g. Houghton et al. 1992, 1994, 2001) and confirmed by world wide temperature measurements. The global dimensions of the interactions between the indirect and the direct impacts on vegetation, soils, hydrology, agriculture, bio-diversity and social life are investigated by international research programmes like the International Geosphere-Biosphere Programme (IGBP), the International Human Dimension Programme (IHDP), and the programme of biodiversity changes, DIVERSITAS, which together with the World Climate Research Programme (WCRP) are grouped together as "Global Change" research programmes. The UNEP Convention to Combat Desertification (UNCCD) fosters the translation of the scientific results into concerted national actions against further degradation of the land surfaces (United Nations 1992).

The Mediterranean is a specifically sensible area with respect to man induced changes, climate change, and climate variability. This includes the impact of extreme events such as droughts as a precursor of desertification, floods that cause erosion, storms, fires, extreme temperatures and also short term land-surface modifications made by man. The European Commission and national institutions initiated in recent years a number of research projects to study the state and the changes of Mediterranean landscapes. These studies necessarily were limited to very small areas of specific interest where the infrastructure exists to conduct research into the interaction between atmosphere and land surfaces, the soils and the vegetation as well as sub-surface hydro-geology. Instrumental and human resources restrict the extent of these investigations to areas of 1 ha to 10^4 km^2.

It is the ultimate goal to describe and predict such changes by models. These models would have to translate forcing that occurs at the global scale into impacts at the regional scale. It is evident that these models need thorough testing and validation since the scaling reaches over five to ten orders of spatial magnitude.

Research projects, therefore, have two goals. Firstly, a diagnostic one, to analyse the present state of the land-surface and, retrospectively, the development that led to this state. Secondly, to improve the tools by which future developments can be predicted. This second goal is strongly connected to the development and validation of models which, with appropriate boundary conditions, can be run into the future. To meet this goal the processes that led to observed changes have to be studied in much detail and cast into mathematical formulations which can be implemented into existing model structures.

Fiedler et al. (2000) defined seven principal categories of experiments:

- phenological experiments serving a diagnosis of a state,
- long time observations of individual state variables to study the presence of trends (changes of the state with time),
- measurements to prove the reliability of a method,
- process studies,
- experiments to obtain data for the development of a parameterization,
- determination of mass budgets, and
- verification of a model.

With respect to the model improvement, which here is under discussion, the following types of experiments have to be considered:

(i) *Budget experiments* aiming at an accurate determination of e.g. the integral water *input and output of a watershed* to estimate an instantaneous area-averaged hydrological budget and its development enforced by the general atmospheric circulation system over a certain experimental period. This type of experiments may be compared with model simulations and serve the validation of the applied hydrological model modules. If the area is connected to surrounding areas by exchange processes the experiments have to be expanded into:

(ii) *"Cage" experiments* to investigate the *interaction of vertical and advective fluxes*, e.g. the interdependence of land-surface processes and large scale atmospheric circulations or the water budget of soils in a limited area that is connected to other areas by sub-surface aquifers. This type of experiments is specifically useful for studies of the representation of dynamical processes in models such as long distance atmospheric transports or sub-surface water movements. It includes "hand-over"-processes (Kossmann et al. 1999) of substances such as water and water vapour from one layer or one column to the next one. This experiment type serves *inter alia* the improvement of the coupling between global and mesoscale models and needs higher order experiments in an 'enclosed' region with many instrumented sites.

(iii) *Flux experiments* to determine mesoscale integrals of *vertical energy and mass fluxes* for comparison with and validation of the output of models over complex terrain, ultimately serving the improvement of process descriptions in global and mesoscale models.

All three types of experiments are components of the Global Energy and Water Cycle Experiment (GEWEX) which is a branch of the WCRP.

1.4.2 Expected Information

At this point arises the question, to what degree remote sensing methods - especially measurements made from space - can support and supplement such "Global Change" studies. To use satellite data in this context has another dimension than the "conventional" use of these remote sensing data. For land-surface studies satellite images generally are used to map land-use types and to monitor land-use changes. Typical applications are to register topographic changes such as the expansion of urban areas, land sealing, and deforestation by counting the changes of the number of pixels in individual classes. Other applications are directed towards agricultural problems such as crop inventories, vegetation density, changes of cultures, and yield estimates. The state of forests can be monitored and the damage caused by wild-fires is investigated. Some research activities go into the direction to obtain very detailed information of soil and vegetation quality by analysing pixel by pixel spectral signatures. This is achieved by classification of land-surface units according to their spectral characteristics or their geometrical pattern. The classes are related to the real land-use types by comparing them with the characteristics of known units.

For "Global Change" studies the goal is to analyse, in addition to such "visible" land-use and land-cover changes, land-surface *processes* which ultimately provide means to estimate area averaged energy and mass fluxes between the surface and the atmosphere over long time periods. The aspect of area averaging (e.g. over the grid area of a model) over inhomogeneous landscapes immediately evokes a scaling problem because not all quantities can be averaged linearly. Specific questions are how measurements made in space

- contribute to a better understanding of the spatial dimensions of land-surface processes including desertification,
- help to expand the results obtained by intensive field experiments in certain small areas to the whole Mediterranean Basin,
- detect synergies in space and time of the variability of these processes,
- assess long term changes at the land surfaces by appropriate indicators or by inferring physical properties and their changes in time from the primary signals measured in space,
- determine area averaged quantities which can be used as input parameters for climate and other models and serve for validation of these models.

Several projects were initiated in the past to develop the methods by which these questions can be answered. The goal of this publication is to summarize at the advent of a new generation of satellites the progress made in reaching these objectives and to provide guidelines on how to use observations made in space in support of the study of land-surface processes and land-surface climatology. To meet this goal it was found to be necessary to inspect critically the whole chain of operations necessary to convert signals measured in space into meaningful information, which can be used by modellers or be applied to diagnose state and changes of the soil-vegetation system.

1.4.3 Research Strategy

The underlying strategy of satellite data evaluation followed here is to infer information at three levels from the signals received from space. The first level requires a rather straightforward but nevertheless already elaborate processing of the raw data, leading to a calibrated and normalized extraterrestrial or top of the atmosphere (TOA) data set of a few quantities such as spectral reflectances or radiative temperatures. These data already reflect some aspects of the state of the surface-atmosphere system. Under the assumption that over a certain time period changes of the cloudless atmosphere average out, they can already be used to characterize and monitor changes of the surface. Indices can be computed which minimize the influence of the atmosphere on the signal. A subsequent step within this first level is to "clean" these data from the influence of the atmosphere. If that is done, one obtaines a set of spectral reflectances, equivalent radiative temperatures, or indices that are directly valid for the surface but otherwise identical with those obtained for the top of the atmosphere. These data sets may be called *"primary satellite data products"*.

To study quantitatively processes at the land surfaces and the causes of changes it is necessary in most cases to *combine* the information inferred from different spectral channels and even from different sensor systems. In addition *a priori* information about the surfaces may have to be used such as digital topographical maps. After the application of complex evaluation schemes, called algorithms, one arrives at quantitative information such as of the fluxes between the surface and the atmosphere. This level of information could be called secondary or, depending on the complexity, *"higher level products"*.

At a third level, all obtained physical-biological data may then be transformed into tools and recommendations for users such as policy makers coping in a practical way with the identified problems. This third level would be a problem-related *information* to be disseminated to the public domain.

Fig. 1.27 summarises the different steps and tasks necessary to arrive - starting from the signals measured by satellite instruments - finally at "Global Change" related research issues concerning the trends of land-surface cover and use, the impact of climate changes on these trends and the feedback of changes at the land surfaces to the climate system. Blue colour symbolizes the flow of satellite data from the raw signals through the different processing stages to information that is needed to tackle problems such as shown by the brown box in the upper right corner and to conduct research into the "Global Change" themes, as indicated in the lower right corner. The way towards this goal generally goes through three stages (red boxes). Some individual processes can be studied by inspecting carefully *selected* multi-temporal satellite data to which *elaborate algorithms* may have to be applied. Evaluation algorithms can be tested by comparison with results of field experiments

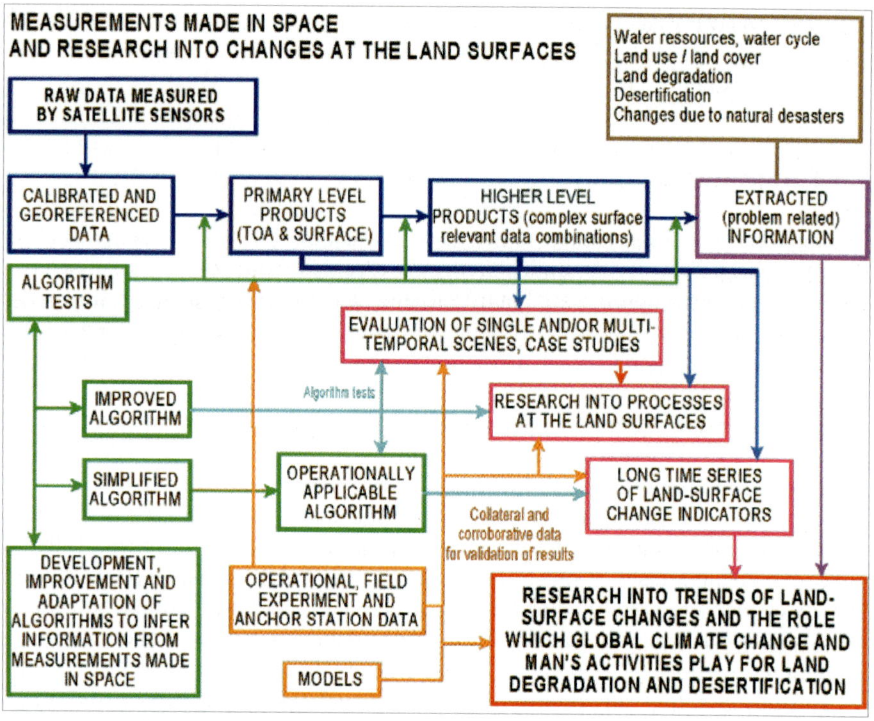

Fig. 1.27. Steps that lead from raw signals measured in space to information about land-surface processes. For details see the text

(light brown). First, mostly regional, information on processes is gained by this procedure. In parallel runs the compilation of *long term data series* for the whole Mediterranean basin. These are needed to identify trends and to separate them from the interannual "noise" of the climate system. This task needs *simplified algorithms* which can be applied to large amounts of data operationally without losing much of the accuracy achieved in case studies. The steps needed for the development of adequate *algorithms* are plotted in green at the left side of the diagram. The algorithms have to be tested and successively be improved which needs interaction with the data production at various stages and the inclusion of models at different hierarchical levels.

The validation of the products obtained by applying these algorithms is an important task that needs collateral *in situ* data which only partly can be gained from operational networks. It therefore needs *"Anchor Stations"* in representative and critical areas where measurements of quantities are performed which can directly be compared with the products obtained from the satellite data analysis. Among these are fluxes at the surface, vegetation parameters, and optical depth of the atmosphere. Model generated atmospheric fields may be used for the correction of the data for atmospheric effects.

1.4.4 Observation of Changes in Heterogeneous Landscapes: Spatial and Temporal Scales

In the preceding sections it was referred to scales in space and time. Unfortunately there exist no unique system of scales. Every discipline uses its own specifications (see section 6.3). The scales which are relevant for the observation of Mediterranean features observed from space range from local to sub-continental and from hourly to decadal. The *local scale* in this context is represented by a uniform area up to 10^4 m^2 size (1 ha) which can be recognized in high resolution satellite images by about $4*4$ pixels (image elements). The repetition rate for observing such a site, under cloudless conditions, is of the order of months. The *regional* or *meso-scale* has an extent of the order of 10^4 km^2 and may be represented by a small watershed. The *synoptic scale* is the scale of travelling eddies which cover an area of the order of 10^6 km^2. These larger scales are observed half hourly from geostationary satellites (spatial resolution 4 - 8 km) and daily from polar orbiting satellites (spatial resolution 0.5 - 1 km). The order of the *sub-continental scale* is 10^7 km^2.

The basis for determining cumulative fluxes, such as of evaporation, are daily sums. Satellites - with the exception of geostationary meteorological satellites - provide only data at certain fixed hours, within a repetition period slightly oscillating. In some cases observation times are shifting due to slowly drifting overpass times during the years. Daily integrals of land-surface variables therefore have to be inferred from relatively sparse satellite information if high spatial resolution is required. These have to be combined with frequent lower resolution data from geostationary satellites, with the results obtained from soil-vegetation-atmosphere-transfer (SVAT) models or with cycles measured during short term experiments.

To compute energy transfers, *slowly* variable quantities such as albedo as well as *quickly* changing variables like cloudiness have to be accounted for. For "slowly" changing quantities such as reflectance in semi-arid and forested areas, one observation within a month may serve the purpose, but such quantities may change after rainfall very rapidly. The variable cloudiness can be observed at half hourly intervals by geostationary satellites at medium spatial resolution of the order of a few km.

One of the problems remote sensing is confronted with, therefore, relates to the different time and spatial scales at which changes and observations occur. The high spatial resolution instruments of Earth observation satellites provide information at low repetition rates. Only in the case of SPOT with its pointable telescope the observation cycle of a specific scene may be enhanced to daily observations at the expense of other areas. Clouds reduce the useful number of these observations. Repetitive observations therefore rely more on the NOAA-AVHRR satellites that guarantee daily observations, which in most cases allow to compose weekly cloudless scenes. The problem with these data is that they are taken by a scanning radiometer under variable illumination and observation angles. Accordingly, they have to be normalized to produce a consistent and intercomparable data sets. Interpretation problems further arise in the Mediterranean area because of the mixture of surface types that may be present in pixel sizes of about one square kilometer. The study of land-surface processes from space therefore requires a twofold approach: the use of medium resolution satellite data for frequent overviews and high resolution satellite data to zoom from time to time into critical and vulnerable areas in order to validate, more accurately interpret, and, if necessary, re-calibrate the medium resolution data.

This brings the spatial aggregation problem into focus which is dealt with in greater detail in section 6.5. Area averages at the scale of the grid widths of models are necessary as input for model simulations. Quantities like radiances can easily be averaged. Whether this also counts for derived quantities depends on the linearity of the relationship (Raffy, 1992). In the case of the vegetation index and the temperature these relationships are not linear. Even in the case of reflectances ("albedo") atmospheric corrections could introduce a non-linear component.

Homogeneous areas of 1 ha size may be found in heterogeneous terrains, but a 1 km square often contains already a variety of land cover types. Because of non-linear relationships the averaging of quantities inferred from high resolution data may therefore give a different result than the inference of the same information from medium resolution data. As an example, an agricultural area may contain small sized fields with different crops (see Figs. 6.5.7 - 6.5.13). The spectral albedo, temperature and a vegetation index would normally change from one field to the other and each field would have a specific annual cycle. If one looks at all fields simultaneously the area may look rather uniform over the vegetation period. But the integrative view may provide an incorrect evaporation estimate. A different effect may be obtained if one looks at an evergreen forest in a hilly terrain. According to its orientation each patch gets different insolation which changes over the day but the spatial average may remain fairly constant.

If the range of the values within a scene is not too large, a linear approximation may be sufficient in view of other errors. As an example, with AVHRR data of the island of Fehmarn (crops and pasture; February 1992 - August 1992) the area-averaged NDVI computed pixelwise from reflectances remains in an interval of 0 - 5% below the NDVI computed after the reflectances first are averaged (Koslowsky 1996). A similar result was obtained for the Castilla - La Mancha area with TM data (Langer 1997). In this case the difference was season-dependent. Best agreement was obtained at the time of the minimum NDVI in early June. Maximum deviation of 3.4% in the same direction as in the Fehmarn case was found in April, values up to 4% in the opposite direction at the end of June and in August.

1.4.5 Land-surface Change Indicators Observable from Space

Spectral Characteristics of Vegetation and Soils

Generally land surfaces are characterized by measurements in the visible and near infrared part of the spectrum. Due to their chlorophyll, green plants absorb solar radiation in the visible and at the transition from the visible to the near infrared (NIR) part of the spectrum (Fig. 1.28) and then again beyond 2.5 μm at the same order of magnitude as between 600 and 700 nm. At about 700 nm a steep increase of reflectance occurs. Chlorophyll has two components, a and b with slightly different absorption bands. All green plants contain both types of chlorophyll and the a/b ratio generally has been estimated to be 3:1 but also ratios between 1.2 and 4 have been found. The specific absorption coefficients of the two species are shown in green and blue colour. In addition the sum of both spectra, mixed at the ratio 3:1, is shown in red colour. The maxima of the absorption coefficient of the mixture are determined by chlorophyll a and lie at 430 nm and 659 nm. For comparison, the absorption coefficient of carotene is shown. When chlorophyll decays in autumn the absorption of carotene becomes dominant and the leaves turn into golden brown to red colour due to the fact that carotene absorbs most of the green light.

In the reflection spectra of leaves at 553 nm, between the chlorophyll absorption bands, a small peak appears (see Fig. 1.29). In the near infrared, up to 1.2 μm, the reflectance of vegetation is generally high and decreases beyond this wavelength towards 2.5 μm. The more biologically active vegetation is, the more pronounced are the water absorption bands at 980 nm and 1.18 μm. These bands are situated in the long-wave wing of the atmospheric water vapour bands at 947 nm (ρστ-band), and 1.137 μm (Φ-band) and are therefore difficult to detect by remote sensing methods To study the water content of the leaves the measurements need very careful corrections for atmospheric water vapour.

There is another water band partly hidden by the Ψ-band of water vapour at 1.38 μm. This band extends to about 1.6 μm and may be used as an indicator for water at the surface of soils. If vegetation is stressed by water deficiency or is ripening as in the case of

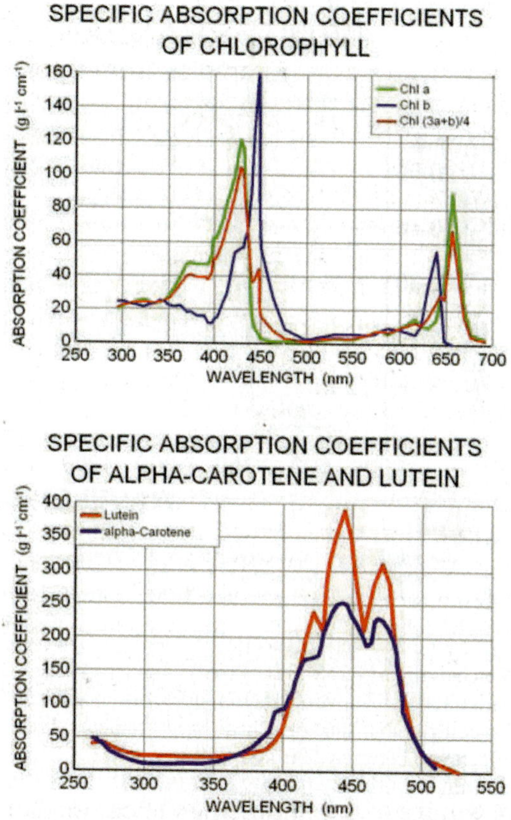

Fig. 1.28. Specific absorption coefficients of a solution of chlorophyll a and b (top) as well as of carotene and lutein (bottom) after data given by Gates (1980). In addition the spectrum of a 3:1 mixture of chlorophyll a and b is given in the upper graph (red curve). The actual absorption spectrum of a leaf depends on the variable mixture of the pigments to which others not shown here have to be added

crops in summer, the reflectance in the short-wave range is increasing, especially the deep depression at 690 nm is rising, the water bands vanish and the reflectance increases at longer wavelengths in the near infrared like for the matorral vegetation shown in Fig. 1.29 as brown line. Finally, if the vegetation is thinned out, the characteristics of the underlying soils become more and more prominent. The reflectances of soils spread over a wide range (Fig. 1.30) and their spectral dependence is generally smoother than those of vegetation. In the visible, soil reflectances generally increase with wavelength. This increase often continues throughout the near infrared part of the spectrum up to about 1.6 μm and then the

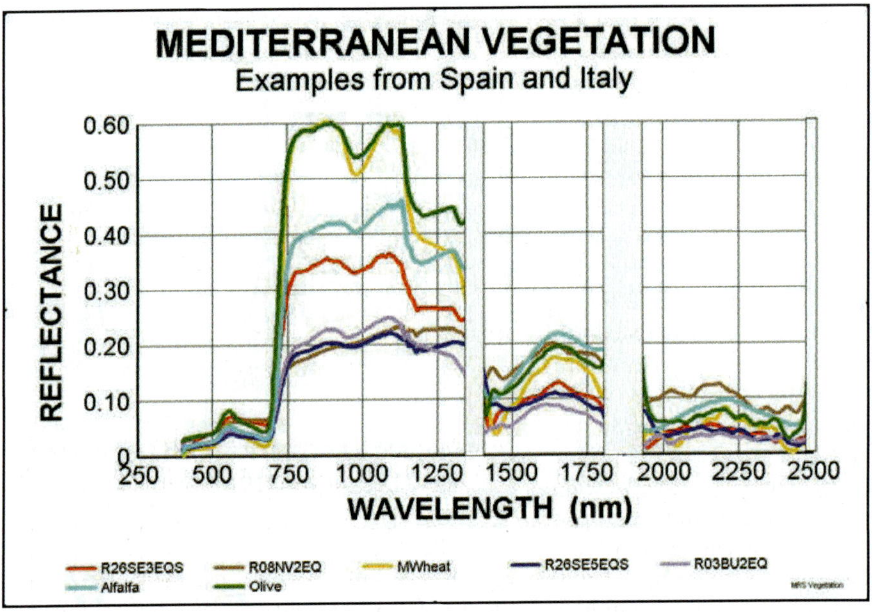

Fig. 1.29. Reflection spectra of Mediterranean vegetation. Rada de Haro, Castilla La Mancha, Spain: R26SE3EQS (red) - holm oak tree, young shoots, 12:05 UT; R26SE5EQS (dark blue) - small wood of holm oaks, distance ca. 500 m, both spectra taken on 26.06.1991 (for comparison, spectral albedo measurements directly on top of holm oak resulted in 5% reflectance between 500 and 700 nm and 24% around 1000 nm; R08NV2EQ (brown) - matorral vegetation, 08.06.91, 12:32 UT; R03BU2EQ (magenta) - small bush in sun, 03.06.91, 08:54 UT; Alfalfa (bright blue) - irrigated alfalfa, 21.06.91, 11:00UT. Matera, Basilicata, Italy: MWheat (yellow) - wheat on the road to Laterza, 29.04.95. Cafaggio, Tuscany, Italy: Olive (green) - mean of two olive tree spectra, 07. and 18.8.98, 12:00 MEST. The spectra are similar in their general features but there are remarkable differences: The red and brown curves have a less pronounced dip at 690 nm which indicates a lower chlorophyll absorption than is e.g. measured for the wheat. The olive has a generally higher reflectance in the visible part of the spectrum than the other plants, which is a result of its one sided silvery leaves. The plants with high biological activity show very pronounced liquid water bands which are very weak for the typical matorral vegetation. Wavelength ranges covered by grey rectangles are those of water vapour bands that block the signal from the surface to the spectrometer under open air conditions

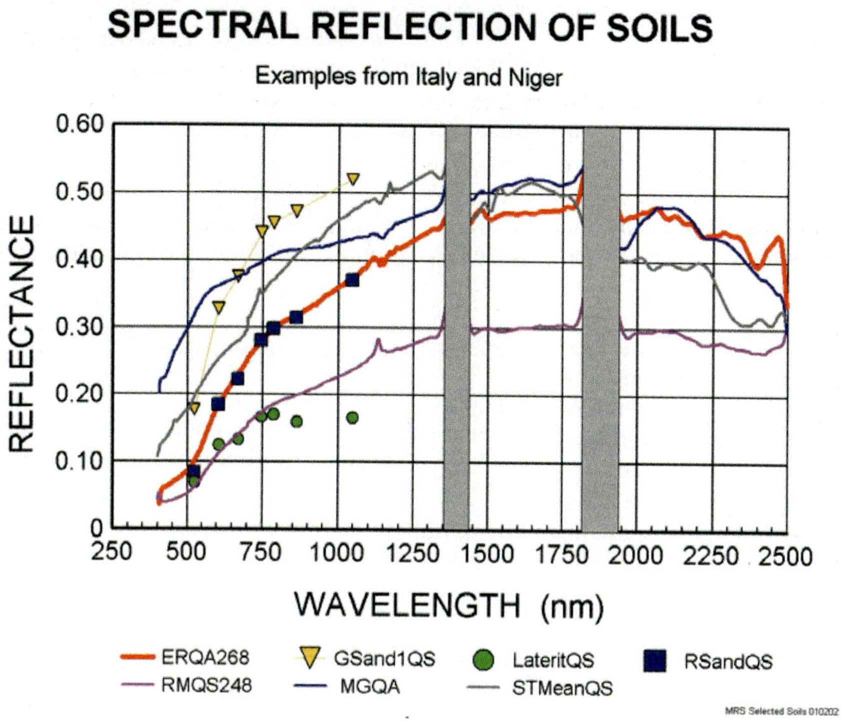

Fig. 1.30. Reflection spectra of soils demonstrating the wide range of brightness. The data points up to 1060 nm denote reflection measurements in Niger, Africa, on yellow desert sand (yellow triangles), red desert sand (blue squares) and dark brown lateritic soils (green circles). The grey line (STMeanQS) stands for the extremely sparsly vegetated soil near Matera, Italy, shown in Fig. 1.31, which is partly covered by whitish calcic stones. Its sparse vegetation can be identified by the small dip at 700 nm, compare Fig. 1.29. The red line ERQA268 is a measurement at a plane surface of a larger bloc of red sandy soil from the maremma in Tuscany, Italy. The pink line RMQS248 stems from the same material but crumbled into small grains which demonstrates what happens if the surface gets rougher. The blue line is a spectrum of a grey sludge soil from Orciano, Tuscany. The characteristics of these two soils are for grey and red soil respectively: Total sand 33/77%, sludge 65/16%, clay 2/7%, total lime 18.1%/traces, active lime 6.75/-%, pH in water 7.5/7.1, conductivity 2110/235 μS/cm, structure traces/17

reflectance slightly decreases. The reflectance of soils depends on the soil moisture but under arid conditions the soil surfaces are quickly drying at their surfaces after a rain.

In addition to these shortwave features land surfaces can also be discerned by their thermal properties measured in the longwave infrared spectrum and by their scattering potential in the microwave region. In the thermal infrared (8-13 μm) vegetated and bare land surfaces can be discerned by their emissivities and their diurnal temperature wave. Surfaces covered by biologically active vegetation generally have emissivities close to one and lower daytime temperatures than bare dry soils because of the cooling effect of the evaporation.

Fig. 1.31. Bright calcic stones covering the soil at the experimental site near Matera, Basilicata, Italy (compare Fig. 1.31). In between grows sparse, mostly dry, vegetation

Responses of Remote Sensing Signals to Changes of Land-surface Properties

Changes of land surfaces which manifest themselves in geometrical patterns like clearings of forests, urbanization, and wildfires or in changes of pixel cluster contents like changes of crops can be measured with the accuracy of the size of a pixel. No unique measure exists for the degradation of a landscape as the term itself is not precisely defined. Denudation of the surface seems to be a good and easily measurable quantity but it also takes place if uncultivated land with dense vegetation is converted into agriculturally used land, e.g. with drip-irrigated fruit trees and gaps between them through which bare soil is viewed. This can not necessarily be classified as degradation.

Even more difficult to quantify is aridification which clearly is a precursor of desertification. It shows up in various ways in remote sensing signals. If vegetation thins out, changes are observed in the backscattered microwave signal, in the reflectance of the shorter wavelengths, and in the thermal infrared due to an increase of temperature. But in the Mediterranean the land is arid during summer and wet in winter. Whether aridification takes place or not can only be determined over a sequence of years. To enhance reliability of conclusions drawn from remote sensing data, not only one but different indicators should be consulted.

Ultimately a measure is desired to quantify the observed changes of the characteristics of the land surfaces as well as their coupling with the climate system. Though satellite data probably never can be the only source of information for land degradation, they can contribute to its assessment in conjunction with models and ground measurements performed at the already mentioned *"anchor stations"*. Such a "measure" will not be a straight forward scale of land degradation attached to one data product, but will consist of a number of components which fall into two categories: *Indices* - not necessarily absolute measures - which allow to quantify at a relative scale the state of the land surface and its development in time, *physical* as well as *biological quantities* which relate to the interactive processes between the surface and the atmosphere, and can be used for diagnostic purposes and as input or validation data for models.

Useful as indices are *primary data products* like spectral reflectance, equivalent radiative temperature, functional vegetation types[4], and different kinds of vegetation indices. The second level of information, which is built on top of the primary products, contains the physical and biological state quantities such as broadband albedo which is responsible for the absorbed solar energy, thermodynamic surface temperature which determines the sensible heat flux from the surface into the atmosphere, moisture in the uppermost soil layer, and biological quantities like vegetation cover density, leaf area index and biomass production. Taking into account some *a priori* knowledge about the observed landscapes, collateral measurements made at the surface, and correlations between the measured primary quantities, it becomes possible to estimate the energy budget at the surface which includes the net radiation flux, the soil heat flux, the sensible and the latent heat fluxes into the atmosphere.

Consequently the basic approach is to combine a wide range of available satellite and collateral data. The joint application of the individual remote sensing quantities is then referred to as Land-surface Change Indicator (LCI). The candidates for this indicator are listed in Table 1.1.

For a number of reasons the assessment of quasi irreversible desertification requires long term and continuous data sets. We are not talking here about transient desertification like harvested fields, wildfires in areas where the vegetation recovers after a few years or of forests which suffer from drought events lasting one or two years but about a continuous process which finally will lead to the loss of fertile land. These trends have to be filtered out from the large variability observed in the Mediterranean region. Since observed climate variability in the Mediterranean seems to contain a period of 20 - 25 years it has to be aimed at data sets of more then this length in order not to be mislead by natural fluctuations.

In the shortwave and infrared spectral bands land surfaces can only be observed if there are no clouds. To compose decadal averages daily data are needed to filter out the clouds (Koslowsky, 1998). Time series of satellite data processed according to the latest quality standards are not yet available for longer than 15 years (see Appendix 8). Even these data have not yet been analysed for the whole basin with respect to trends.

Less affected by clouds are microwave signals which, therefore, are an interesting tool for land-surface observations. Table 1.1 contains microwave change indicators. Their interpretation can sometimes be rather complex because they depend strongly

---

4 The meaning of the term "functional" is that ecosystems populated with different species have the same or very similar functions in the climate system with respect to energy budget and surface - atmosphere transfers. They are not described by their individual plants but, e.g., by SVAT model variables.

Table 1.1. Relationship between the tendency of radiative quantities measured from space or indicators derived from such measurements and the deviation of surface characteristics from a hypothetical "mean" situation towards dryer and degraded conditions respectively towards an improvement of the state of vegetation which in most cases is related to wetter conditions. The quantities used in this table are discussed in detail in the following chapters. The NDVI is explained below in connection with Fig. 1.26, the Microwave Polarization Difference Index (MPDI) is the normalized difference of the vertically and horizontally polarized brightness temperatures measured at a given frequency (see Ch. 4).

Change Indicator Component	Symbols	Degrading Land/ Dry Situation	Developing Vegetation
Directional nadir-normalized spectral shortwave reflectance	$\rho_{\lambda,\ VIS}$	↑	↓
Directional nadir-normalized near infrared reflectance	$\rho_{\lambda,\ NIR}$	↓	↑
Directional nadir-normalized middle infrared reflectance	$\rho_{\lambda,\ MIR}$	↑	↓
Broadband albedo	ρ	Tendency towards ρ_{soil}, region dependent	Tends to ρ_{veg}, species dependent
μwave backscatter signal: SAR / WSC	$\rho_{\mu w}$ and phase	to be coupled to other components before the direction can be determined	
Vegetation Index: NIR/VIS	NDVI	↓	↑
Vegetation Index: μwave (37 resp.6.6 GHz)	MPDI	↑	↓
Clear sky daytime surface temperature	T or T_{eff}	↑	↓
Clear sky diurnal surface temperature amplitude	ΔT	↑	↓
Time of clear sky surface temperature maximum	$t(T_m)$	delayed	
Time constant of clear sky surface temperature post maximum exponential decay	k	↑	↓

Table 1.1 continued

Thermal infrared emissivity	ε_{IR}	↓	↑
Temperature Independent Spectral Index (TISI)	$\prod \varepsilon_i^{a_i}$	↑ varies analog to NDVI at high levels	Strong variability from high to very low values
μwave emissivity 6.6 GHz 37 GHz	$\varepsilon_{\mu w}$	direction depends on relative importance of moisture change/biomass change	
Angular dependence of microwave emissivity	$\varepsilon_{\mu w}$	$\varepsilon_{\mu w,\ horiz.} < \varepsilon_{\mu w,\ vert.}$ $\varepsilon_{\mu w,\ horiz.}$ ↓ with increasing scan angle	$\varepsilon_{\mu w,\ horiz.} = \varepsilon_{\mu w, vert.}$ no scan angle dependence
37 GHz polarization difference	P	↑	↓
Surface net radiation flux	Φ_R, R_n	depends on ρ and T	
Soil heat flux	Φ_{soil}, G	↓	↑
Sensible heat flux	Φ_{SH}, SH	↑	↓
Latent heat flux, evaporation	Φ_{LH}, λE	↓	↑
Land use/Land cover	Classes defined by combination of spectral radiances and seasonal variability	max $L_{\lambda,\ LW\text{-}IR}$ high $\rho_{\lambda,\ SW}$ high $\rho_{\lambda,\ NIR}$ low $\rho_{\lambda,\ MIR}$ high small seasonal variability	max $L_{\lambda,\ LW\text{-}IR}$ high $\rho_{\lambda,\ SW}$ low $\rho_{\lambda,\ NIR}$ high $\rho_{\lambda,\ MIR}$ low large seasonal variability except for evergreen forests

on the water content. As an example, a dry soil has a high emissivity and thus a high brightness temperature. But, a dense vegetation also has a high emissivity and high brightness temperature. A desert can therefore not easily be distinguished from a tropical forest in the microwave region. The indication of a change from arid to less arid is therefore not possible in this way without additional information such as combination with other indicators.

With increasing aridification and land-surface degradation the radiative

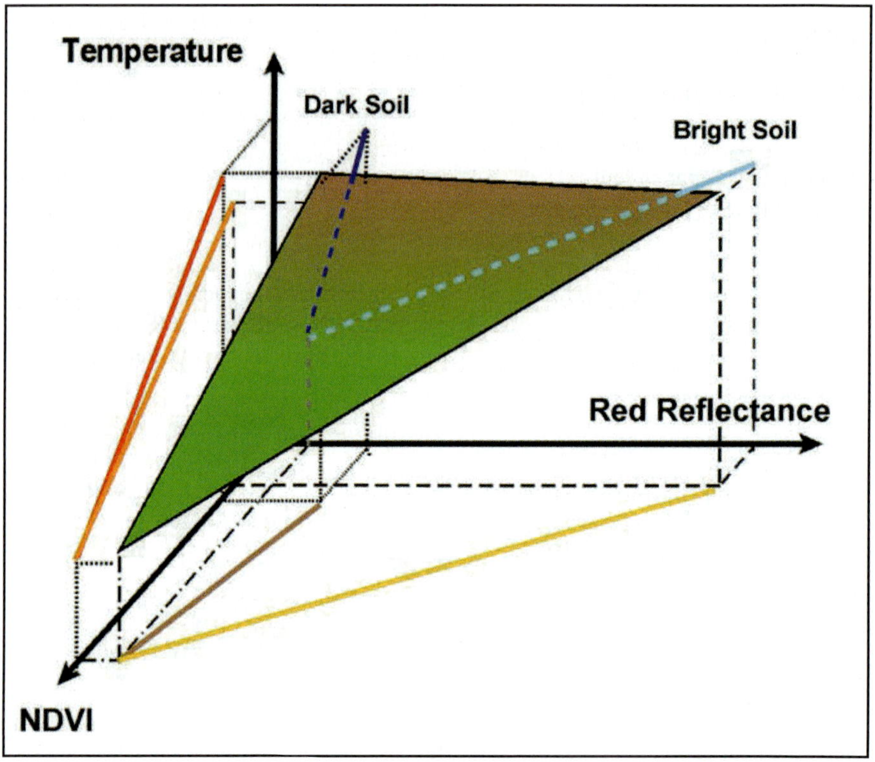

Fig. 1.32. Characterization of land-surface types in the three-dimensional space of Landsat-TM channels 3 (red reflectance), 3 and 4 combined to the NDVI, and 6 (temperature). The coloured lines indicate the projections of the triangle into the three two-dimensional plains

characteristics of biologically productive vegetation give way to those of bare soil: The "greenness" as measured by "vegetation indices" is reduced, the amplitude of the diurnal surface temperature increases but emissivity decreases, spectral reflectance is approaching the value of the underlying soil and becomes less variable in the course of the year but strongly if it rains, and the scattering, emission and polarization of microwaves reflect more and more the properties of the ground. It is therefore feasible to follow a land-surface degradation process by observing shifts of these quantities.

The individual change indicators can be looked at as components of an n-dimensional vector that describes the state of the land surface in time. Fig. 1.32 demonstrates this for the three-dimensional space spanned up by three channels, one in the red spectral range, one in the near infrared, and one in the thermal infrared band. These will be either AVHRR channels 1 (red), 2 (near infrared), and 4 and 5 (both thermal infrared) or Landsat-TM channels 3 (red), 4 (near infrared), and 6 (thermal infrared). In Fig. 1.32, instead of the near infrared channel its combination with the red channel into the "Normalized Difference Vegetation Index" (NDVI) is

used as y-coordinate. The NDVI is defined as the ratio of the difference divided by the sum of the red and near infrared reflectances ρ: $[\rho(IR) - \rho(RED)]/[\rho(IR) + \rho(RED)]$, see chapter 2. The vector changes its position and length as the surface experiences modifications. The situation is stable if this vector circles, year by year, around an "attractor" that is fixed in this space. If the centre or the shape of this movement varies, it is an indication that the surface undergoes changes. If the position of the “attractor” changes during the years into one direction, one can speak of a trend which the land surface encounters. If all components of the vector move into the direction of a more arid and less vegetated surface one can speak of desertification. The directions the individual components would take are indicated in Table 1.1 by arrows. The direction of changes can be determined if the extreme positions are known, such as complete vegetation cover, respectively bare soil. Whether this change is significant or not depends on the size
of the difference vector in relation to the variance inherent in the data. If $\mathbf{r}(t)$ denotes the vector at the time t then the difference vector between the times t_1 and t_2 is $\Delta\mathbf{r}(t_1, t_2) = \mathbf{r}(t_2) - \mathbf{r}(t_1)$.

If the spectral characteristics of surfaces are described by the reflectances as measured, e.g. in the six Landsat-TM channels, a vector of the type

$$\mathbf{r}(Landsat - TM) = \{\rho(CH1), \rho(CH2), \rho(CH3), \rho(CH4), \rho(CH5), \rho(CH7)\} \quad (1.5)$$

may be defined.

With image processing systems it is easy to determine the statistical characteristics of a given area, e.g. by applying the "maximum likelihood" method. The mean values of the reflectances define the average state of the surface, the variances define the volume in which the vector oscillates as a result of the natural variations from pixel to pixel.

For the area of Rada de Haro, Castilla - La Mancha, Spain, where an EFEDA experiment was performed in 1991, the vector at the beginning of summer had the value:

$$\mathbf{r}(27\,May) = \{4.6, 9.3, 12.4, 31.0, 26.6, 22.9\} \quad (1.6)$$

Towards the end of summer it was:

$$\mathbf{r}(31\,August) = \{7.1, 12.5, 17{,}4, 28.6, 30.8, 26.0\} \quad (1.7)$$

The difference vector is:

$$\Delta\mathbf{r} = \{2.5, 3.2, 5.0, -2.4, 4.2, 4.0\} \quad (1.8)$$

and the averaged variances are $\{1.1, 2.0, 3.1, 4.0, 3.8, 4.3\}$. This demonstrates that the seasonal change can be measured and that the change signal is significant in the spectral channels 1, 2, 3, and 5. The change points into the direction of less biologically active vegetation since the difference is positive in the channels 1, 2, 3, 5, and 7 and negative in channel 4. For comparison, the vectors of holm oak trees and the underlying soil are respectively 27{0.092, 0.19, 0.15, 0.82, 0.45, 0.26} and 84{0.12, 0.21, 0.34, 0.43, 0.59, 0.53}. The directions of the unity vectors {} and the lengths of the vectors are quite different. From space such changes can only be

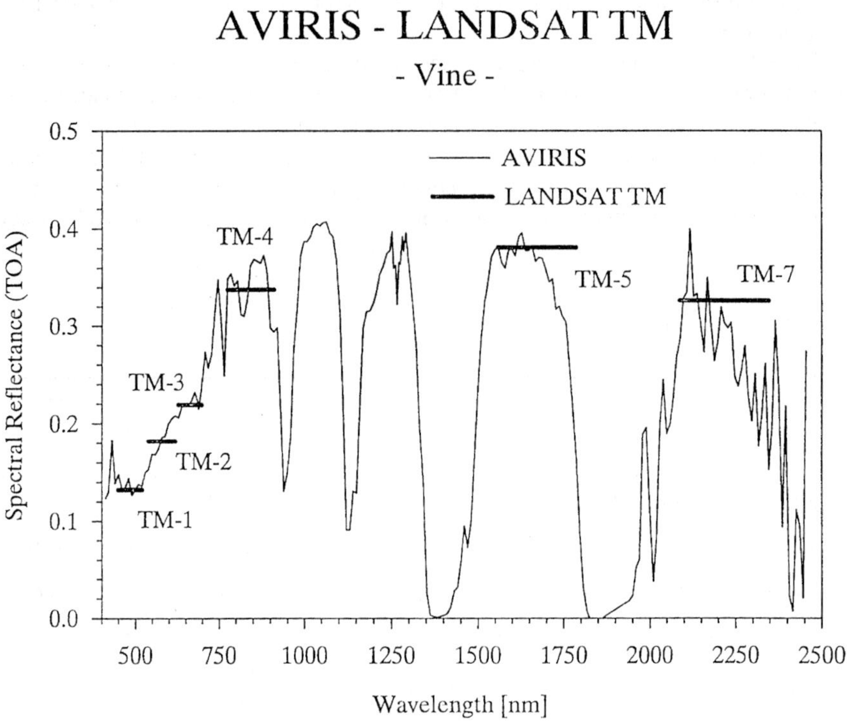

Fig. 1.33. AVIRIS spectrum over grapewine with indication of the Landsat-TM spectral bands

detected by multichannel instruments which are present in the Landsat and SPOT satellites and will in a much more refined version be flown on future space platforms. A precursor of these instruments is the airborne AVIRIS that provides spectra as shown in Fig. 1.33. Spectral reflectances inferred from high spatial resolution satellite measurements, such as available from Landsat-TM and SPOT, must be validated by comparison with albedo and spectral reflectance measurements performed at the surface.

In common medium resolution instruments flown in meteorological satellites so far only two channels had been available. These can be combined in different ways to a measure of "greenness" of the surface, which can be related to the leaf area index or biomass (see, e.g. Goel and Qin, 1994) and used as a measure for the biologically active vegetation. The simplest way is to compute the ratio of the signals in the near infrared (ρ_2) to those in the red (ρ_1): $(\rho_2)/(\rho_1)$, or by normalizing the difference between these two channels which leads to the before mentioned NDVI. In operational applications the chosen signal is that of the radiance at the satellite but also the atmospherically corrected reflectances at the surface can be used. In addition to these simple vegetation index formulations first introduced by Rouse et al. (1974), various modifications have been proposed to correct for the signal of the underlying soil (see chapter 4.5).

The NDVI was and still is widely used because it compensates - in an easy way but not completely - for atmospheric effects. In desertification studies it can be used to indicate vulnerable areas and to assess in a qualitative way changes of land-surface characteristics. A quantitative determination of the vegetation is more complicated. The reason is that over sparse vegetation the signal of one channel is composed of three components: vegetation, underlying soil, and shadowed surface.

Table 1.2. Residual NDVI values of soils

Type of soil		ρ 650 nm	ρ 860 nm	Virtual NDVI value
Bright yellow desert sand (Niger)		0.369	0.478	0.129
Grey sludge soil (Orciano, Tuscany, Fig. 1.24)		0.374	0.418	0.056
Soil with bright white stones (Matera, Fig. 1.24 and 1.25)		0.272	0.41	0.202
Soil with small stones (Matera, Italy)		0.275	0.363	0.138
Plane surface of red soil (Maremma, Tuscany, Fig. 1.24)		0.224	0.32	0.177
Red soil rough surface (Maremma, Tuscany, as well as red desert sand, Niger, Fig. 1.24)		0.132	0.201	0.207
Lateritic soil (Niger, Fig. 1.24)		0.128	0.162	0.117
Dry soil of arable land (hilly site, Cafaggio, Tuscany)		0.148	0.227	0.21
Wet soil of arable land (hilly site, Cafaggio, Tuscany)		0.101	0.142	0.167
	"characteristic soil"	0.166	0.249	0.2
	dry soil	0.137	0.222	0.279
	wet soil (field capacity)	0.083	0.125	0.247
	dry cultivation soil	0.252	0.397	0.223
	degraded matorral	0.202	0.311	0.213
	pasture	0.119	0.226	0.310
Castilla - La Mancha soils (laboratory measurements)	2.1% organic material	0.257	0.355	0.160
	3.6% organic material	0.210	0.321	0.210
	8.8% organic material	0.105	0.198	0.307
	17.2% clay	0.317	0.425	0.146
	33.7% clay	0.258	0.355	0.157
	57.0% clay	0.100	0.148	0.200
	9.11% calcium carbonate	0.111	0.186	0.254
	12.8% calcium carbonate	0.153	0.228	0.198
	18.5% calcium carbonate	0.313	0.412	0.136

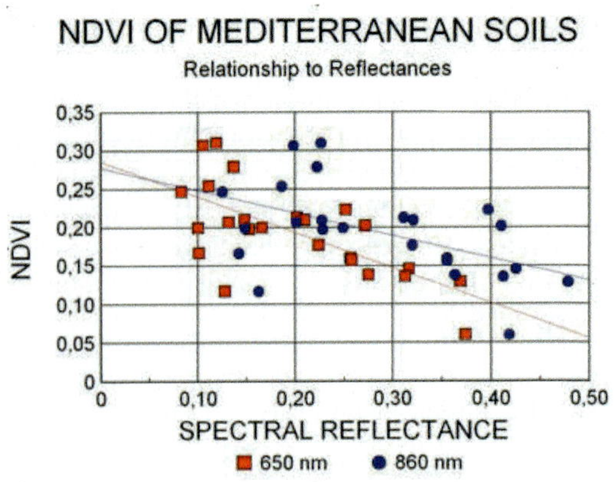

Fig. 1.34. Relationship between spectral soil reflectance at the wavelengths 650 nm (red) and 860 nm (blue) and NDVI

Also for most soils the vegetation index has positive values. Only if the reflectances of the individual components are known, it is possible to isolate the vegetation component from the mixed signal and to determine the fractional coverage of the vegetation and its changes in time. Another possible use of the NDVI is to determine variations of the length of the vegetation period which allows to draw conclusions with respect to the level of aridity.

In using the NDVI one has to be careful at the lower end of its range because this is marked by the residual value of the vegetation index determined by the soil rather than the zero value. Different soils have different virtual "NDVIs" as shown in Table 1.2 from which must be concluded that NDVI values below 0.2 indicate bare soils in most cases. If more organic matter is contained, also NDVI values above 0.2 may result from bare soils. These are often dark soils which can be recognized by low reflectance especially in the near infrared. The graphical representation of these values (Fig. 1.34) shows that the correlation between reflectance and the NDVI of soils is very poor. Therefore, without corroborative information about the soil type, it is difficult to apply a "soil correction" to the NDVI as the density of the vegetation gets low.

1.5 About this Book

This book summarizes the experience gained in dealing with the above described problems and presents the results obtained in field experiments, primarily sponsored by the European Commission, which have been carried out during the nineties at several places of the Mediterranean basin and of consecutive satellite data analysis up to the first years of the 21st century. It may as well be read as a tutorial introduction which familiarize newcomers to the field with the various problems involved in the quantitative use of satellite data and the physical limits of the methods. Consequently its chapters follow the steps which lead from raw signals to information. The results of the evaluation of satellite data are documented by a rich amount of maps and to bring abstract diagrams of ground based measurements into relation with visible features of the environment it was attempted to illustrate them, where possible, with photographs or of the targets.

Chapter two explains the steps that lead from the reception of signals to the compilation of calibrated and geographically registered products and long term data series. This is a more technical part dealing with the experiences gained during the work with satellite data of different provenances. Even after more than 20 years of practice most of the satellite data are not delivered to users in a format that allows direct application. More details about the satellites and their instrumentation used to observe the land surfaces are given in Appendix 4. The following nomenclature is applied for the satellites and their sensors: Names of satellites which are composed of the initials of descriptive programs or institutions are printed capital (like SPOT), given names with capital first letters only (like Landsat).

Chapter three recalls some facts of the physics of remote sensing in order to sharpen the awareness for the meaningfulness of information which can be inferred. Experts may tend to skip this chapter but non-experts may wish to improve their understanding of the interactions between electromagnetic radiation and land-surface features.

Chapter four introduces the quantities which are defined as "primary" according to the definition given in section 1.4.5. In the course of the projects which led to this publication great care has been taken that the derived quantities do not start an uncontrolled life of their own but are tied to direct measurements at the ground, a process which is called "validation". In this respect the projects followed the guidelines given by the International Satellite Land-surface Climatology Project (ISLSCP) (see e.g. Rasool and Bolle, 1984; Becker, Bolle, and Rowntree, 1988). Firstly the satellite data products stemming from different systems are compared and secondly, during larger or smaller field experiments, measurements have been made in representative areas which could be compared with information inferred from measurements made from space. This led to an improvement of the calibration of satellite instruments as well as of the algorithms used to infer the primary level data.

Chapter five then addresses the problem to derive higher order products leading to flux estimates, by combining different spectral information and *a priori* supplemental information. The validation of these quantities and of the algorithms used to infer them from processed satellite data is detailed in this chapter.

Chapter six deals with the third level of satellite data interpretation, i.e. the transformation of remote sensing data into information in various fields of application. In the first sections the relationship between the remote sensing radiative quantities and those used in the description of land-surface processes by ground observations are discussed. Then, examples are given how the satellite data are used to assess changes at the surface. In many cases the observations made from space have to be combined with ground observations, *a priori* knowledge, or models. There are three major strings of applications: The use of satellite data for diagnostic purposes, for the spatial inter- and extrapolation of ground measurements, and in conjunction with models.

Remote sensing data can either be used to support the adaptation of models for specific areas by providing input data like albedo and vegetation types with equal functions in the land surface - atmosphere exchange processes or to validate the output of the models with moisture of the top soil, temperature, energy budget, and vegetation data, averaged over the grid size of the respective model. As mentioned earlier, changes at land surfaces can be assessed by studying changes of the population of land-use classes or of land-use patterns which is not the central topic of this book. Here, the range of change indicators is considerably extended to quantities which provide information about changes of physical and biological parameters and ultimately of processes. In addition, changes manifest themselves in the phase shifts and the length of a phenomenon such as the vegetation period.

An ultimate goal of the application of satellite data is to determine trends, changes that continue into the same direction over longer time spans and linked to the "Global Change" issue. One result of research addressing this problem is that the inter-annual variability of land surface characteristics in the Mediterranean area is so high that trends can only be distilled out of long observation series (20 years or longer). Extended time series of the same data used in conjunction with models are also required for diagnostic purposes. Thus it is now feasible to identify the impact of inter-annual climate variability on the land surfaces and their vegetation by a number of indicators such as vegetation index, maximum land-surface temperatures, spectral reflectances, surface emissivity, radiation, heat and water vapour fluxes, information which can be put together to compute water budgets of watersheds. These data now become available for the whole Mediterranean area at time scales of more than ten years. This allows to study spatial as well as temporal synergies across the basin and fosters the understanding of the climate system, specifically its regional variability. To study the relationship between forcing factors and responses of the surface it seems not to be sufficient to describe land surfaces in terms of classical land use types like forest, cereal, orchards, settlements and so on. More efficient for a systems approach would be a description by functional types which cause e.g. the same amount of evaporation during the seasons, having the same course of albedo change during the year or causing the same temperature anomalies.

Practical, supplemental, and background information is given in appendices. *Appendix 1* is a short description of the structure of the EFEDA pilot project which can be seen as a model for the organization of large scale land-surface process and satellite data validation studies. *Appendix 2* explains some of the meteorological terms used in this book for readers who are not familiar with this nomenclature. In *Appendix 3* the same is done for the soils which are mentioned in the context of reflection measurements. *Appendix 4* supplements the information about space systems of Chapter 2 and describes the instruments used for land-surface research. In *Appendix 5* some formula are compiled which are used in the context of the evaluation of satellite data and gived an example for the UTM system. *Appendix 6* is devoted to spectral measurements. Five classes of spectra are considered: Spectra of the atmospheric radiation, spectral atmospheric transmissivity, laboratory data of long-wave infrared reflexion measurements, reflexion spectra of natural surfaces (most of them documented by photographs), and discrete spectral reflexion and albedo measurements made over various surfaces. *Appendix 7* gives some information about the relatively new scintillometer technique for heat flux measurements. Finally, in *Appendix 8*, long-time data series of AVHRR measurements up to the year 2005 are discussed with respect to the reflectances in the two AVHRR channels 1 and 2, the NDVI, the broadband albedo derived from the primary data set, and of radiative temperature. These data set provide a baseline for the end of the last and the beginning of the new century.

Chapter 2 Processing and Archiving of Satellite and Ancillary Data

2.1 Introduction

Studies of land-surface processes related to research into "Global Change" require satellite data that are available for long time periods and to which users have relatively easy access. Three categories of satellite data can be distinguished:

- Data that can be received by small and not very expensive receiving stations, such as the freely available measurements made from NOAA satellites, or require slightly larger systems and an agreement with the meteorological services as in the case for Meteosat data.
- Data from research satellites such as the former NASA satellite system Nimbus or the ESA-ERS satellites and new systems, which can be obtained from principal investigators and/or specific archives.
- Data that are only commercially available like those of SPOT and Landsat.

In the first part of this chapter a short overview is given of the remote sensing systems used for land-surface studies in the past and of those from which data are presently becoming available. More detailed characteristics of the sensors, most commonly used for research into land-surface processes, can be found together with the sources for these data in Appendix 3.

The question may be asked, why data from *different* satellites are necessary for research into land-surface processes. The reason is that in the past no one single satellite could deliver the suite of mutually supplementing information necessary to conduct these studies. Only Envisat presently provides, for its lifetime, a wide range of spectral information simultaneously. In the future, smaller satellites are planned that may again impede the fusion of data from different instruments of which each supplements the other.

The criteria for selecting data from specific satellites are wavelength range, spatial resolution, temporal and spatial coverage, and the cost factor. Some of the instruments carried by different satellites are optimized for the shorter wavelengths of the solar spectrum, others have improved potential in the thermal infrared or in the microwave region. Large antennas needed for the reception of microwaves, which are essential for soil moisture estimates, often do not allow other heavy instruments beside them. The temporal coverage plays, as will be demonstrated, an essential role: geostationary satellites provide half-hourly overviews of a large area at the expense of spatial resolution while high spatial resolution instruments, on polar orbiting satellites, have a much less temporal coverage. The need for many images to overcome the cloud problem can become very expensive even if one single scene may be rather inexpensive as it is the case for Meteosat scenes.

Ideally the technical conditions of Earth observation from space should match the spatial and temporal scales of land-surface processes under investigation. Even if the whole hierarchy of satellite systems is considered, this still is difficult to achieve. Land-surface processes are governed by the net radiation flux at the surface which is modulated by clouds, starting at a very small time scale. Within minutes the net radiation flux may change by orders of magnitude at spatial scales determined by the size of the clouds. The highest frequency at which observations are possible from geostationary satellites is of the order of 15 - 30 minutes at a resolution of a few kilometres. Slower land-surface processes such as the change of reflectance, emissivity, or vegetation have time scales of the order of ten days if abrupt effects due to rainfall are not taken into account. Daily observations are necessary to follow these low-frequency changes because cloudless pixels will in many regions only be obtained every few days. Furthermore, averaging over a few days can also reduce aspect angle effects if the target is imaged day by day under different angles which is the case for the operational NOAA satellites. Daily observations with full global coverage are only possible at a spatial resolution of a few hundred metres to one kilometre. If one goes to smaller spatial scales like 20 - 30 m, the observation becomes immediately less frequent, although the SPOT operational system has the capability to programme the view in a way that one target may be observed under different nadir angles at consecutive days but at the expense of observing other targets.

To obtain cumulative values for land-surface processes, such as areal evaporation, certain parameters can be deduced from multispectral measurements made once a day but their diurnal development can only be estimated with the aid of the much more frequent observations from a geostationary orbit. To match these two data sets either the cloudiness has to be scaled down to the pixel size of the higher resolution data set or one has to reduce spatial resolution and aggregate the higher spatial resolution data set into the size of the geostationary one. For cloudless scenes the general course of the diurnal variability of specific land-surface processes often is known from land-surface experiments. In such cases it may be possible to determine its actual amplitude by only one observation from space per day.

Microwave remote sensing is less affected by clouds. Passive microwave sensors presently have a spatial resolution of the order of 50 km (at some wavelength down to 16 km) and a temporal resolution of twice per day. Active microwave sensors have a much higher spatial resolution but the swath is narrow so that a full coverage can only be achieved after a number of days.

An example may illustrate the situation in the Mediterranean. If one assumes that the aim is to investigate the interannual variability and, ultimately, trends of the state of the land surface in a topographically structured region at high spatial resolution. It may be desirable for the investigation of erosion processes at slopes, regrowth of burned areas, or agricultural development. In such case, one has to rely on data from Landsat, SPOT or other instruments with similar high spatial resolution. To keep the amount of scenes to be analysed in reasonable limits - which is a matter of resources - one could select images e.g. of the same month for a sequence of years. The interannual variability is superimposed by the seasonal variability, the phase of which may change from year to year. It therefore is desirable to test whether the data of one month in a sequence of years mirror interannual variability or whether they randomly reflect short term weather conditions. In addition, it therefore has to be investigated how the seasons develop and how representative the situation in a selected month is for the interannual variability. In the simplest approach one would select the month with lowest cloudiness, e.g. August. For a test area in central Italy one could obtain cloudless Landsat-TM scenes for the following dates (day of August/year): 4/1990, 7/91, 25/92, 12/93, 15/94, 2/95, 20/96, and 23/97. This is as close as one can get to a specific date of the year. To control, how much the situation at these dates is affected by the seasonal development, the daily data series of spatially lower resolution NOAA-AVHRR can be inspected. As will be shown in Chapter 4, from these data e.g. the phase of the vegetation period can be deduced. How these lower resolution data transform into the high resolution scale can be studied by comparison with high resolution data of different months of one year. For the same test area as above, a maximum of cloudless scenes was available in 1990: 30 April, 1 June, 4 August, and 8 September. This demonstrates the rarity of easily processable data. Only with much more effort one could build up a composite of cloudless pixels from partially cloudy scenes.

Another problem is to assimilate *in situ* measured surface variables with measurements made from space. This problem comes up, e.g., if surface fluxes have to be calculated that should be representative for a heterogeneous area. *In situ* measurements are representative for a small area only comparable to one pixel. Due to the complex nature of the biophysical processes which occur at the land-atmosphere interface, linear spatial averaging of input parameters to flux algorithms may lead to erroneous results (Michaud and Shuttleworth 1998). From such examples it follows that "scaling" between various remote sensing sources, and with respect to available *in situ* measurements made at certain sites at the surface, becomes a central task.

In the future the situation may improve with the advent of new satellite systems carrying more problem focussed integrated payloads as well as cheaper data handling and archive systems. The principal problem, nevertheless, will remain that the observation of processes which develop at high frequencies need observations from geostationary orbits, while for low frequency processes daily or weekly observations may be sufficient. Furthermore, improving only one of the observational scales is not sufficient. If, e.g., the spectral bandwidth is narrowed as desired for a number of applications, then the full impact of this improvement is only obtained, if the spatial resolution is adequately high. If the field of view is too large, a pixel will contain a variety of targets the individual signatures of which can hardly be entangled even if the spectral resolution is high.

Following the description of the space systems some principal aspects of data treatment are discussed in this chapter and examples are given for the present status of data processing. The corrections necessary to make remote sensing data a powerful tool for research into land-surface processes are summarized in Table 2.1.

These correction methods have been refined considerably during recent years. The NOAA-AVHRR short-wave channels are calibrated with degradation functions derived at the Free University of Berlin in cooperation with NOAA. The sensor

Table 2.1. Corrections to be applied to measurements made from space for land-surface change studies

Type of Correction	Tasks to be Performed
Rectification of sensor degradation	Continuous updating of calibration coefficients according to trend analysis and control over known test areas such as desert and sea, where reflectivity is more or less constant
Geometrical rectification	Re-sampling of the signals of scanning instruments into a grid with constant pixel size. To compare images obtained from two different satellites they have, in addition, to be aligned so that they can be superimposed
Geographical registration	The re-sampled image has to be geographically located using orbital parameters and/or landmarks
Cloud-screening	To obtain an undisturbed image of the surface, cloudy pixels, as well as pixels with cloud shadows have to be removed and eventually be replaced by cloudless pixels by means of appropriate interpolation
Normalization for illumination and observation geometry	To normalize the signal with respect to the insolation which changes with solar zenith and azimuth angles and for the variable direction of observation, due to which the satellite sensor receives radiation from differently illuminated facets, the bidirectional reflection function of the landscape must be known to normalize the signal to a specific geometry
Radiometric correction for atmospheric effects	To relate observations made under different atmospheric conditions with each other, the signals have to be corrected for aerosol scattering and absorption, the absorption by trace gases (ozone, water vapour, carbon dioxide, etc.) and, in the thermal infrared, in addition for the emission of trace gases
Temperature correction	The measured radiometric temperature must be corrected for the emissivity of the surface to obtain the thermodynamic temperature. To obtain a basin wide map of surface temperatures for a specific local time, temperatures measured at different swath angles have to be corrected for the local time respectively local solar azimuth angle. In the case of Meteosat one may select data of a specific solar azimuth angle from a sequence of slots .

degradation is monitored by observing bright desert targets in the Great Erg Occidental that can be assumed to be stable in their reflectance properties. Recent comparison to other studies proved them to be most reliable not only for use in the Mediterranean but also for global application (Koslowsky 1999). Landsat data are calibrated by standard procedures. The Nimbus-SMMR and ERS-1/2 data referred to in later chapters are delivered calibrated by the Vrije Universiteit Amsterdam

(Faculty of Earth and Life Sciences) and the Agricultural University of Wageningen.

Because cloud screening algorithms cited in literature did not work accurately enough in the Mediterranean area, a new algorithms was developed that uses the history of the surface reflectance, as well as temperature, to decide whether a cloud is present or not. Various methods are in use to compose cloud free scenes from consecutive data. The most advanced algorithm is the "dynamic" method which uses previous information on surface reflectance and temperature to identify clouds.

2.2 The Remote Sensing Data Base

2.2.1 Satellite Instruments

The NOAA Observing System

NOAA Satellites. The NOAA satellite system consists of a Geostationary Orbiting Satellite (GEOS) component and a Polar Orbiting Satellite System (POES). The GEOS satellites, of which the first one was launched 1975, plays no role for the observation of the Mediterranean area. The POES component includes two systems, the NOAA satellites and the Defense Meteorological Satellite Program (DMSP). The NOAA satellite system is based on the first meteorological satellite, TIROS, which was equipped with television cameras and took the first image on 1 April 1960. An advanced version of the NOAA satellite system was put into orbit with NOAA-4 (TIROS-N) on 13 October 1978 and was again improved with NOAA-7. Successors of this satellite type were numbered NOAA-8 until NOAA-14 (NOAA-13 failed in orbit shortly after launch), whereas the so-called NOAA KLM-series followed in May 1998 with NOAA-15 as the first one of a new generation. NOAA-M was launched on 24 June 2002. The next launch is expected for March 2005 (NOAA-N).

During the time period which is considered here only the satellites NOAA 11, 12, 14 and possibly 15 are of interest. But as the data set is still updated, also NOAA 16 and NOAA 17 have to be considered. NOAA satellites are continuously sending data to numerous receiving stations throughout the world. No restrictions for receiving data exist, but most of pre-processing work has to be done by the user. The instruments on board provide images of the Earth in the visible, near and far infrared (AVHRR) and vertical temperature and water vapour profiles (TOVS).

AVHRR. The AVHRR instrument is one of the most powerful instruments for investigating long-term trends on global as well as regional scales so far. Since its first operation in 1978 the characteristics of the instruments remained nearly unchanged and it is planned to have AVHRR instruments in orbit at least until 2011 and it will be flown on METOP, the European operational meteorological satellite system, with minor changes. Basic information and all relevant calibration data are given by Schwalb (1978), Kidwell (1995) and Goodrum et al.(2000).

The AVHRR is a radiation-detection imager. It uses 5 (in older versions 4) detectors, that collect within the same field of view, radiation of different wavelength bands (see Appendix 3). This permits multispectral analysis. One channel monitors reflected solar energy in the visible band, a second one in the near-infrared part of the electromagnetic spectrum. Various combinations of the data of these two channels provide information on vegetation ("Vegetation Indices") and/or allows snow and ice detection. The other three channels operate in the infrared band. Especially the bands 4 and 5 in the 10–12 micron region are useful for deriving sea surface temperatures through split window techniques. For land surfaces the derivation of surface temperatures is as well possible but more complicated. Channel 3 in the medium infrared receives emitted thermal infrared radiation which is mixed with reflected solar radiation during daytime.

Until NOAA-14, the AVHRR was flown as 4-channel (morning passes) or 5-channel instrument (afternoon passes). In the case of "morning satellites" one of the thermal infrared channels was omitted. Starting with NOAA-15, the first satellite of the KLM-series, an additional channel is available, which will be switched on and off alternating with channel 3. Both channels are now named channel 3a and channel 3b. Users are aware of the fact that AVHRR channel 3 data on each of the NOAA spacecraft have been very noisy due to a spacecraft problem and can be useless, especially when the satellite is in daylight.

The only sources for calibration are internal calibration targets for the IR channels. During each scan the instrument views cold space (zero radiance) and a black target in its housing (~290 K). Four Platinum Resistance Thermometers are embedded in this "black body" to monitor its temperature. These temperature measurements are included in the data stream. Therefore the instruments response curve can easily be ascertained from the received data by interpolation between the deep space and black body signal. Starting with NOAA-15 also nonlinear radiance corrections for AVHRR thermal channels 4 and 5 are incorporated in the new Level 1b data stream. Direct readout users find correction formulae in the NOAA-K, -L, and -M Users Guide (Goodrum et al. 2000). Unfortunately there exists no in-flight calibration source for the visible and near infrared channels of the AVHRR (see Chapter 5).

TOVS. Under the name of TOVS, three instruments are combined: HIRS (High Resolution Infrared Sounder), SSU (Stratospheric Sounding Unit), and MSU (Microwave Sounding Unit). They provide information on mainly vertical temperature distribution. TOVS-derived profiles are widely used in numerical modelling and are useful in data-sparse regions such as oceans.

The main instrument is the HIRS, a 20 channel scanning radiometric sounder. Nineteen bands are in the IR and one at the high frequency end of the visible. There have been only minor changes between HIRS/2 (until NOAA-14) and HIRS/3 (NOAA-15 ff). The instantaneous FOV is between 1.25 and 1.4 degrees, corresponding to a horizontal resolution at nadir between 18 and 20 km.

The SSU provides measurements in three channels using a pressure-modulated gas (CO_2) cell to accomplish selective bandpass filtration of sampled radiances from which temperatures in the stratosphere are derived. Synchronously with the scan of the HIRS, once every 25 seconds, an in-flight calibration is performed, where the instruments first look into space and then at an internal black body target.

The MSU detects and measures microwave energy from the troposphere up to 20 km height in four frequency channels. The advantage of MSU is that the measurements are not seriously affected by clouds. The antenna support structure has a microwave black body calibration load where the antenna dwells periodically for 1.9 seconds during each scan (25.6 seconds). The calibration loads consist of parallel rows of knife edge material produced by casting an iron-filled epoxy onto an aluminium backing plate. These essentially isothermal targets, with embedded temperature sensors, provide one of two calibration points, the second is the cold deep space view.

The SSU and MSU on NOAA satellites, until NOAA-14, were replaced by the advanced AMSU-A and, beginning with NOAA-15, by AMSU-B .

ATOVS. The Advanced TIROS Operational Vertical Sounder (ATOVS) combines data from HIRS/3 and AMSU. The AMSU-A is designed to measure scene radiation in the microwave spectrum between 23 and 89 GHz (15 channels) to calculate global temperature profiles from the Earth`s surface to the upper stratosphere (48 km), but also includes three channels in the atmospheric "window" for deriving precipitation, surface emissivity, liquid water content, snow cover and sea ice cover. The AMSU-B is designed to allow the calculation of vertical water vapour profiles up to about 12 km. Three channels are in the water vapour absorption bands around 183 GHz, whereas two channels are window channels, but influenced by cloud water. Each part has one channel at 89 GHz, but with different resolution.

Besides the complete change of instruments (replacement of SSU and MSU) the horizontal resolution of the AMSU instruments is now greatly improved to a nadir resolution of 48 km (AMSU-A) and 16 km (AMSU-B).

The calibration procedures for the AMSU-A and AMSU-B are the same except a few minor differences due to different antennas. Radiances for both AMSU-A and AMSU-B Earth views are derived from the measured counts and the calibration coefficients inferred from the internal black body and space view data. Both instruments were tested and calibrated prior to launch. Each instrument has four options for the viewing direction of the space view, which can be selected by ground command. Detailed information on calibration of the AMSU is given by Mo (1995).

Meteosat

Meteosat First Generation. The European Meteosat system consists of a series of operational satellites in a geostationary orbit. The first satellite of this series was launched 1977. Since June 1998 Meteosat-7 is in the nominal position at 0° longitude providing operational service. Meteosat-6 still is in a back-up status and Meteosat-5 at a position over the Indian Ocean where it had been positioned in support of the INDOEX experiment.

The Meteosat spin scan radiometer MVIRI (Meteosat Visible and InfraRed Imager) has been unchanged and operates in three spectral bands. To scan the full Earth disk it needs 25 minutes, which allows a 30-min repetition rate. Scanning from east to west is achieved through the spin of the satellite. With every rotation of the spacecraft the telescope scans a new line on the earth approximately 5 km north of the previous scan line. For in-flight monitoring of the performance of the infrared and water vapour detectors two black bodies are located on opposite sides of the main optical path. One is kept at ambient temperature of about 290 K as a cold reference and the other heated to about 50 degrees higher as a warm reference. The temperatures of the black bodies are telemetered to the Primary Ground Station together with the corresponding blackbody imaging values. This allows to detect calibration trends due to sensor degradation. The view of cold deep space obtained during normal imaging is a supplement to the onboard calibration. Since for this purpose the whole optical system is used, the cold space view helps to calibrate the entire radiometer not only the detector systems.

The routine calibration of the thermal infrared and water vapour channels is achieved through statistical comparison of spacecraft data with conventional measurements at the ground. The calibration process assumes a linear relationship between the counts and the relevant physical unit radiance. The actual retrieval of calibration coefficients in the count radiance domain, that is the digital signal steps, is achieved by a linear regression technique using as input the conventional measurements, spacecraft observations, and the offset of the radiometer when looking to cold deep space. These coefficients are used internally within the MPEF (Meteosat Products Extraction Facility) and are also transmitted to the end-users.

Meteosat Second Generation (MSG). Meteosat Second Generation satellites follow the original system with advanced sensors on board. The first MSG was launched on 28 August 2002. It went into its operational phase on 29 January 2004 and was renamed Meteosat 8. A second satellite, MSG-2, is scheduled for 18 months later. From the two operational spacecraft one will act as operational and the other one as back-up. About 7 years after the launch of MSG-1, it is planned to launch MSG-3.

The imaging mission, that will be the main mission of MSG, will be performed by the SEVIRI (Spinning Enhanced Visible and Infrared Imager) instrument that has 12 spectral channels located in different window and absorption bands in the solar and terrestrial spectrum. Three Visible and Near Infrared (VNIR) channels were defined following the heritage of NOAA-AVHRR and one high spatial resolution visible (HIV) channel that is heritage of the existing Meteosat generation. The infrared channels are either similar to AVHRR or will be in the water vapour and ozone absorption bands as on HIRS and in the CO_2 absorption band as on the GOES VAS sounder.

The absolute radiometric accuracy of the VNIR channels is based on a vicarious calibration process by observation of landmarks with well-known radiative properties at different periods of the year. The signal offset, including the instrument background and the detection dark signal, is measured at each line scan during deep space viewing and is subsequently removed from the detector output signal.

Beside the SEVIRI is another instrument, that is assigned as a scientific payload in contrast to the operational meteorological payload. This is the Geostationary Earth Radiation Budget (GERB) instrument, a broad-band, three mirror-telescope housed in the Instrument Optical Unit. It views the Earth with a black-and-white detector array, providing measurements of the Earth`s radiation output in a long-wave and a short-wave band.

Nimbus-7

Although satellites of NASA's Nimbus series are not any longer operational, there have been some instruments on board which were precursors of later versions used in the NOAA weather satellite program. The first generation of passive microwave sensors, ESMR, SCAMS and SMMR, proved to be suited especially for the detection of sea ice and hydrological quantities. The data of the SMMR (Scanning Multichannel Microwave Radiometer) have been used in research of soil moisture determination.

The SMMR is a ten-channel passive microwave instrument delivering orthogonally polarized (i.e. H and V) antenna temperature data at five microwave wavelengths, 0.81, 1.36, 1.66, 2.8 and 4.54 cm (6.63, 10.69, 18.0, 21.0 and 37.0 GHz).

It consists of six separate Dicke-type radiometers. Those operating at the four longest wavelengths measure alternate polarizations during successive scans of the antenna. Those, at the shortest wavelength, operate continuously for each polarization. The antenna is a 42° offset parabolic reflector. The SMMR operated on board of the Nimbus 7 satellite from October 1978 through August 1987 which presents one of the longest continuous periods of observation by satellite-borne passive microwave radiometers.

For SMMR, first, a pre-launch calibration was carried out in a thermal vacuum chamber at JPL. The response of each SMMR channel was measured for a variety of temperatures ranging from 100K to 350K. Secondly, a semiempirical correction was made for the polarization mixing. A third correction, described in detail in Gloersen et al. (1992), is an adjustment to the pre-launch calibration. Normalized counts were calculated from the antenna counts, the warm reference counts and the space-horn reference counts. From normalized counts brightness temperatures are calculated. The data are then further adjusted for two additional corrections, one for a long-term instrument drift and one for instrument errors dependent on the ecliptic angle.

Landsat

The first Landsat satellite (Landsat 1) was launched in 1972 with the objective to test technology for remote sensing of earth resources from space. The series of satellites was designed to provide repetitive global coverage of the Earth's continents. With Landsat 4 a second generation came into orbit, followed by Landsat 5. Since Landsat 6 failed to get in orbit, the next generation started with Landsat 7 in spring 1999.

The operation of the Landsat system was transferred from NASA to NOAA in 1983 and commercialized in 1985. Throughout all changes, the USGS Earth Resources Observation Systems EROS Data Center (EDC) retained primary responsibility of the governmental archive of Landsat data.

All six operating Landsats have been in sun-synchronous orbits with equatorial crossing times ranging from 8:30 a.m. for Landsat 1 to approximately 9:45 a.m. for Landsat 7. The Landsat system provides global coverage between 81 degrees north latitude and 81 degrees south latitude. They are imaging the same 185 km ground swath every 16 days.

The first Landsat satellites carried the Multispectral Scanner (MSS), Landsat 4 and 5 were equipped with two sensors, the MSS and the Thematic Mapper (TM). Landsat 7 carries an advanced TM sensor, called Enhanced Thematic Mapper Plus (ETMP) with an additional spectral band (0.50-0.90 μm) and a higher spatial resolution of 15 m.

Landsat 7 uses the Landsat 4 & 5 land remote sensing image Worldwide Reference System (WRS) with consecutive 233 paths and 248 rows. Path and row number therefore will be the scene characterizing information also for future .

TM data are received directly by a network of 16 worldwide ground stations. Prior transmissions to TDRS (Tracking Data and Relay Satellite) ended due to K-band transmission failure on both Landsat 4 and 5.

At the equator, the ground track separation is 172 km, with a 7.6% overlap. This overlap gradually increases as the satellites approach the poles, reaching 54 % at 60° latitude.

As a first processing step at the EROS Data Center TM data, systematically corrected, are generated as follows:

- Correcting and validating the mirror scan and payload correction data.
- Providing for image framing by generating a series of scene centre parameters.
- Synchronizing telemetry data with video data.
- Estimating linear motion deviation of scan mirror/scan line corrections.
- Generating benchmark correction matrices for specified map projections.
- Producing along- and across-scan high-frequency line matrices.

Data resampling begins after the completion of satellite orbit model generation/refinement and radiometric calibration. Data resampling consists of three distinct components: coarse resampling, resampling table densification and two-dimensional convolution resampling.

Radiometric calibration is the process of converting raw digital numbers (DNs) transmitted by the sensor into physical units. The radiometric calibration of Landsat TM data is performed in two steps:

- Absolute Calibration: recovery of radiance (as observed by the sensor) from the raw digital number recorded. This involves modelling of the characteristics of the optics and electronics of the sensor.
- Relative Calibration: removal of residual errors in the calibrated data to improve the qualitative appearance of the imagery.

Geometric correction removes geometrical distortions in the image based on the knowledge of the satellite and sensor and remaps the image to a regular grid in a standard map projection. This is accomplished by constructing a relationship between pixel coordinates in the images and geographic coordinates on the surface of the earth.

Finally these two processing formats are offered to the public:

- Map registered data: These data are radiometrically and geometrically corrected. The image data are rotated and aligned to a user-defined map projection, using ground control points for improving the satellite model: These data are also called "Precision Corrected".
- Terrain corrected data: As above but additionally using a digital terrain model,

which should improve the geodetic inaccuracy caused by the parallax error due to local terrain elevation.

Information on all processing steps can be found in US Geological Survey homepage under http://edcwww.cr.usgs.gov/glis/hyper/guide/landsat_tm.

At ESA's Earthnet facility in Fucino, Italy, most standard tapes are calibrated based on pre-flight data and not with the aid of the internal calibration source or ground measurements during the flight. The method in detail is described by Epema (1990) for both EOSAT and Earthnet. At present there is a lookup-table applied to the image digital count. The gains and biases for each channel are given in the product and they are applied to get a linear correction of the data (private communication P. Goryl from ESRIN).

SPOT

The SPOT satellites (Systeme Probatoire de l´Observation de la Terre) have been developed in France and commercialized like Landsat TM products. The first satellite was launched in 1986. SPOT 1, 2, and 3 belong to the first generation. With SPOT 4 the first satellite of the second generation was launched in 1998. The instrumentation of the SPOT satellites offers some new capabilities due to the swivel mounted radiometers. This allows to obtain stereoscopic pairs of images and to preselect targets which are not at the nadir field of view. Furthermore two identical instruments are on board.

Each of the HRV/HRVIR instruments covers a swath of 60 km at the nadir or 2 x 60 km when the two identical instruments are in a simultaneous vertical viewing configuration. Each instrument offers oblique viewing capability with a viewing angle adjustable through ±27° with respect to nadir. This improves the revisit capability. At a latitude of 45° an area may be imaged once every 2.5 days providing stereo pairs of the same area. Areas as far as 400 km off the sub-satellite track can be observed. Each instrument has two telemetry channels, one containing the panchromatic data, the second one the multispectral data.

Beginning with SPOT-4 in addition an imaging radiometer called VEGETATION is flown with characteristics similar to the AVHRR.

Pre-processing (in the SPOT terminology "processing" means data manipulation undertaken by end-users) includes geometric and radiometric corrections, resulting in five standard pre-processing levels available to the user:

- Level 1A - Radiometric corrections. Information from the detectors are processed to normalize each spectral band. Coefficients for calibration between bands and absolute values are specified.
- Level 1B - Radiometric and geometric corrections. Corrected for systematic errors with regard to radiometry and geometry, the Earth's rotation and curvature and angle of registration.
- Level 2A - Ortho-corrected SPOTView product using satellite ephemeris data as ground control. This results in a geometrically precisely corrected product.
- Level 2B - Ortho-corrected SPOTView product using customer-supplied ground control.

- Level 3A - SPOTView coverage for the U.S. using US Geological Survey 7.5min quadrangle maps for ground control along with USGS DEM 3-arc-second data for ortho-correction.

VEGETATION products fall into two main categories:

VGT-P products are for scientific applications requiring highly accurate physical measurements. In these images the pixel count is the apparent surface reflectance as seen at top of atmosphere (TOA). User can specify standard map projections. Auxiliary data delivered with the products will allow to process the original reflectance values with their own algorithms.

VGT-S products are geocoded and corrected for the effects of the earth`s atmosphere on daily or ten-day basis. VGT-S provide either NDVI or data from all four spectral bands plus the NDVI and auxiliary data on acquisition parameters.

DMSP

The Defense Meteorological Satellite Program (DMSP) is operated by US Department of Defense and the first satellite of this program was launched in 1987. In addition to equipment for measurements in the visible and infrared part of the spectrum the DMSP carries a new microwave instrument, the Special Sensor Microwave Imager (SSM/I). In contrast to temperature sounders the microwave imager contains six channels in the window region (19, 37, 85 GHz) with dual polarization and a seventh channel centred at the 22.23 GHz water vapour line with vertical polarization. In addition to SSM/I, a temperature sounder SSM/T was first launched in 1989 containing seven channels in the 50-60 GHz oxygen band followed in 1991 by the latest development of this sensor: SSM/T2. It operates with window channels at 91 and 150 GHz and three channels around the strong water vapour line at 183.31 GHz. The three instruments are flying together aboard the F-14 satellite. But all have different scan geometry, so that it is very difficult to combine the information gained in the various spectral channels. This will be alleviated by a new sensor being developed called SSM/IS, however, the AMSU-A and AMSU-B developed by NOAA also provide coincident observations over the full spectral region of 19 - 183 GHz.

SSM/I products, which will become available from AMSU data as well after completion of the algorithm development, are rain rate, snow water content, cloud liquid water, sea ice concentration, ocean wind speed, land-surface temperature, land-surface moisture.

ERS-1 and ERS-2

The first European Remote Sensing Satellite ERS-1 was launched as a forerunner of a new generation of satellites for environmental monitoring in July 1991. The ERS-mission uses advanced microwave techniques to acquire measurements and images regardless of cloud and sunlight conditions. The second and last satellite of this series, ERS-2, followed 4 years later on 21 April 1995 with a nearly identical instrumentation. Only the Global Ozone Monitoring Experiment (GOME) had been added.

The Active Microwave Instrument (AMI) of the ERS satellites has several modes, those which will be of interest in land-surface processes research are the ERS SAR and ERS Windscatterometer. The ERS SAR is a single frequency and polarization radar operating at C-Band (5.3 GHz) and VV-polarization. Products from SAR are available as Single Look Complex Image (SLC) or as Precision Image (PRI). SLC is a single-look, complex, slant-range, digital image generated from raw SAR image mode data with up-to-date auxiliary parameters. A minimum number of corrections and interpolations are performed on the data in order to allow end-users maximum freedom to derive higher-level products. Complex output data are being retained to avoid loss of information. PRI is a multi-look, ground range, digital image generated from raw SAR image mode data using up-to-date (at time of processing) auxiliary parameters and are corrected for antenna elevation gain and range spreading loss. ERS SAR.PRI has been specified for users wishing to perform application-oriented analysis. Engineering corrections and relative calibration are applied to compensate for well-understood sources of system variability. Absolute calibration parameters, if available, will be provided in the product annotations, but this depends on external calibration activities.

The Windscatterometer (WSC) measures the echo power of a signal transmitted by the satellite and returned from the ocean surface. The echo power is affected by the surface wind conditions. Three antennas are used to obtain information about the wind at the ocean surface. Each antenna points in a different direction and hence for a particular point on the ocean surface three echo power values, obtained from three different angles, are used to calculate the surface wind vectors. Measurements are also made over land and here these three antennae can be used to infer surface quantities. Scatterometer applications have e.g. been reported in ESA (1998). Soil moisture and vegetation assessment using both ERS1/2 SAR and Windscatterometer (WSC) has been performed by the Department of Water Resources of the Wageningen Agricultural University (van Oevelen 1998).

In addition to the microwave payload both ERS-1 and ERS-2 have been equipped with an Along Track Scanning Radiometer (ATSR) that allows very accurate sea surface temperature measurements. Measurements are made in four spectral channels, 1.6 μm, 3.7 μm, 11 μm and 12 μm. The ATSR-2 of ERS-2 is an enhanced version with three additional visible channels for vegetation monitoring (555 nm, 670 nm, and 870 nm wavelength). Due to a unique conical scanning technique, the instruments view a surface target at nadir as well as from other directions. As the two views of the same scene are taken through different atmospheric path lengths it becomes possible to calculate a correction for the atmospheric absorption in the thermal infrared channels without the need for ancillary data.

TRMM

The Tropical Rainfall Measuring Mission (TRMM) satellite was launched in fall 1997 to measure tropical rainfall by microwave instruments. The range of its orbit is

limited to ± 35° latitude and thus covers only the southern part of the Mediterranean basin. It carries three rain measuring experiments:
- the Precipitation Radar (PR) operating at 13.8 GHz;
- the TRMM Microwave Imager (TMI), a nine channel, five-frequency mircowave radiometer operating with dual polarization at 10.65, 19.35, 37, and 85.5 GHz and with vertical polarization at 21 GHz;
- the Visible/Infrared Scanner (VIRS) with five channels at 0.63, 1.6, 3.75, 10.8, and 12 μm.

The data products can be obtained from the TRMM Science Data and Information System (TSDIS) at GSFC http://tsdis02.nascom.nasa.gov). Information about the TRMM mission can be obtained under http://trmm.gsfc.nasa.gov and the data can be accessed under http://lake.nascom.nasa.gov/data/dataset/TRMM/.

Terra

Terra is the name for NASA's first Earth Observing System (EOS) AM-1 satellite that was launched on 18 December1999. The satellite is equipped with five state-of-the-art instruments for monitoring global environmental changes over time:
- ASTER (Advanced Spaceborne Thermal Emission and Reflection Radiometer) consists of three distinct telescope subsystems. It is a high spatial (15 to 90 m), spectral and radiometric resolution instruments with 14 spectral bands separated by bandpass filters and dichroics. Each subsystem operates in a different spectral region, has its own telescope and is built by a different Japanese company. The ASTER instrument operates only for a limited time during the day and night portions of an orbit (full configuration ~8 min). Calibration for the visible and the near infrared bands is performed by onboard halogen or calibration lamps and in the thermal infrared by an onboard blackbody.
- CERES (Clouds and the Earth`s Radiant Energy System) consists of two broadband scanning radiometers with very high radiometric measurement precision and accuracy. Two identical instruments are operating in a cross-track scan mode and in a biaxial scan mode. The cross-track instrument continues the ERBE as well as the TRMM mission. Each CERES instrument has three channels in the shortwave, thermal infrared and for total radiation. Onboard calibration includes a tungsten lamp system and a pair of blackbody sources. Cold space and internal calibration looks are performed during each scan.
- MISR (Multiangle Imaging Spectro-Radiometer) is a new type of instrument and will view the Earth with cameras pointed at nine different angles. Special attention has been paid to providing highly accurate absolute and relative radiometric calibrations using onboard deployable solar diffuser plates and several types of photodiodes.
- MODIS (Moderate Resolution Imaging Spectroradiometer) is a scanning imaging radiometer consisting of a cross-track scan mirror, collecting optics and a set of linear arrays with spectral interference filters. Observations are made on global scale every 1-2 days. Spatial resolution ranges from 250 m to 1 km. The broad spectral coverage of the instrument is divided in 36 bands of various bandwidths.

The MODIS instrument has comprehensive calibration subsystems. The calibration hardware includes a solar diffuser, a solar diffuser stability monitor, a spectroradiometric calibrations assembly, a plate-type black body, and a space viewport.

- MOPITT (Measurements Of Pollution In The Troposphere) is a scanning radiometer employing gas correlation spectroscopy to measure infrared radiance in three absorption bands of carbon monoxide and methane. Calibration by onboard black body and space views occurs during each normal scan.

Envisat

The Envisat mission continues and improves upon measurements that began with ERS-1 and ERS-2 and at the same time extends the range of the observed geophysical quantities by including the atmosphere through three instruments, MIPAS, GOMOS, and SCIAMACHY, and the marine biology with MERIS. Envisat is the largest free-flying and probably most complex satellite ever built in Europe. The payload consists of the following instruments:

- ASAR (Advanced Synthetic Aperture Radar) ensures continuity with the image mode (SAR) and the wave mode of the ERS1/2 AMI. It has enhanced capability in terms of coverage, range of incidence angles, polarisation, and modes of operation.
- MERIS (Medium Resolution Imaging Spectrometer) measures solar reflected radiation in the visible and near infrared range during daytime. The 1150 km wide swath is divided into five segments covered by five individual cameras. Each camera images a stripe of the Earth`s surface onto an imaging optical grating spectrometer. The entrance slit is imaged through the spectrometer onto a two-dimensional CCD array. The calibration is performed at the orbital south pole, where the calibration diffuser is illuminated by the sun. MERIS will produce an extremely large amount of data.
- AATSR (Advanced Along Track Scanning Radiometer) continues data sets gained by the ERS 1 and ERS 2 ATSR's of precise sea surface temperature (SST) and land-surface characteristics.
- MIPAS (Michelson Interferometer for Passive Atmospheric Sounding) is a Fourier transform spectrometer for the measurement of high-resolution gaseous emission spectra at the earth`s limb.
- GOMOS (Global Ozone Monitoring by Occultation of Stars) enables simultaneous monitoring of ozone and other trace gases as well as the temperature distributions in the stratosphere in the 20 -100 km altitude range.
- RA-2 (Radar Altimeter) is derived from the ERS 1 and 2 Radar Altimeters to determine ocean topography as well as wave height and wind speed and mapping sea ice and land surfaces.
- MWR (Microwave Radiometer) enables the measurement of integrated water vapor column and cloud liquid water content needed e.g. as correction terms for the radar altimeter signal. Furthermore surface emissivity and soil moisture over land can be determined.

- SCIAMACHY (Scanning Imaging Absorption Spectrometer for Atmospheric Cartography) performs global measurements of trace gases in the troposphere and stratosphere.
- LRR (Laser Retro Reflector) is a passive device that reflects ground-based laser pulses.

Aqua

Aqua is as Terra part of NASA's Earth Observing System (EOS). It was launched on 4 May 2002. Its primary goal is to gather information about water in the Earth system. One of the six sensors aboard Aqua is the Advanved Microwave Scanning Radiometer (AMSR-E). It is a passive microwave radiometer sensing microwave radiation at 12 channels and 6 frequencies ranging from 6.9 to 89 GHz, both in horizontal and vertical polarization. From an altitude of 705 km, it has a swath of 1445 km and a spatial sampling interval of 10 km. AMSR-E is part of the Earth Observing System (EOS) program and is built by Mitsubishi Electric Company with NASDA funding.

The broad frequency range allows the monitoring of various water related processes that exert a strong influence on climate and weather. The variables to be measured include precipitation, oceanic water vapor, cloud water, near-surface wind speed, sea surface temperature, snow cover, sea ice parameters and soil moisture. The C-band (6.925 GHz) radiometer is capable of retrieving surface moisture under conditions of low vegetation biomass, i.e. with an anticipated precision of 0.06 g/cm3 for vegetation biomass less than 1.5 kg/m^2 (Njoku et al. 2003).

The special Sensor Microwave Imager (SSM/I) flying on Department of Defense weather satellites is the heritage instrument of AMSR-E. AMSR-E will therefore extend and enhance the SSM/I data record, that began in 1987, and the earlier Nimbus 7 Scanning Multichannel Microwave Radiometer (SMMR) data record that began in 1978, allowing studies of decadal-scale climate change.

2.2.2 Aircraft Instruments

The Use of Aircraft for Land-surface Process Studies

Aircraft remote sensing can be applied only for short-term studies or repetitive surveys of a given land area. It seems therefore that the operation of aircraft for the purpose of land-surface process studies is not always very efficient. Nevertheless airborne instruments play an important role in this context. Only by aircraft one can investigate the structure of the atmospheric boundary layer which is important for the coupling of the surface with the atmosphere. As an example the distribution of water vapour or dust released from the surface can only accurately be traced by aircraft. Airborne remote sensing instruments also expand the scale of spatial resolution to very small units. Thus if a zoom into an area becomes necessary, aircraft may solve the problem. Aircraft also can be used to follow certain processes in the course of a day which is impossible from satellites because of overpass of the area only once during daytime (with the exception of courser resolution geostationary equipment). Last, but not least, airborne instruments are applied to re-calibrate satellite sensors from time to time by performing sub-satellite flights. Disadvantages are that a precise superposition of satellite and aircraft remote sensing data sometimes is difficult to achieve. The operation of aircraft has to be planned many months in advance and is expensive which results in a low repetition rate of remote sensing measurements from aircraft. Some of the instruments operated from aircraft are described below.

AVIRIS

The Airborne Visible-Infrared Imaging Spectrometer (AVIRIS) is operated by the NASA/Jet Propulsion Laboratory (JPL). It became operational in 1989 with continuing improvements in system performance and reliability. The instrument has a spectral coverage from 0.4 - 2.5 μm. The average bandwidth of the 224 spectral bands is 10 nm. Flight lines up to 100 km length are typically flown on the NASA high altitude aircraft U-2 or ER-2. These aircraft, as platforms for AVIRIS, have the advantage that they are generally very stable in flight, have a relatively low velocity-to-height (V/H) ratio and acquisition sites can be accessed as far as 2100 km from the aircraft departure site. More details and description of the AVIRIS are given by Vane et al. (1994). Measurements with the AVIRIS onboard the airplane ER-2 were taken over the Castilla-La Mancha region in June and July 1991 from an altitude of 20 km.

The pre-processing of these hyperspectral data is somewhat complicated. Although the AVIRIS Data Facility can provide calibrated and pre-processed data, most of the required pre-processing steps are to be done by the end-user because of the different requirements of each application, including geometrical registration in order to properly account for solar illumination and viewing geometry in spectral reflectance modelling. The radiometric "system" calibration is sufficient for most applications. Because of the high stability of the ER-2 platform and the roll-angle compensation of the AVIRIS instrument, geometric registration of AVIRIS data is relatively easy as compared to other airborne systems.

A problem encountered in the retrieval of parameters from AVIRIS data is the presence of spatial interference pattern due to electronic noise. This does not turn out in original images but becomes apparent in the retrieval of some parameters like atmospheric water vapour. An algorithm for eliminating this noise has been improved by Moreno (Moreno and Green 1996), it works over the power spectrum in the Fourier transform of the image. The noise is characterized by systematic spikes. Each one is modelled as a double Gaussian, and the centre position and width of the Gaussians are empirically determined from the display of the power spectrum by assuming an exponential relationship between the distances of the spike centres to the origin of spatial frequencies and the widths and amplitudes of the spikes.

The most critical problem in the pre-processing of AVIRIS data is the instability in spectral channel positions. An algorithm was developed which is based on the derivation of water vapour maps on a channel-by-channel basis for all the channels included in the spectral range of 0.85 - 1.10 μm. The image-based auto-calibrations are preferable to provide temporal series of consistent data.

TMS

The Daedalus AADS-1268 multispectral scanner is a NASA-developed Thematic Mapper Simulator (TMS), flown on the NASA ER-2 aircraft. This sensor simulates the performance of the Thematic Mapper multispectral scanners orbiting on Landsat 4 and Landsat 5, with a slightly higher spatial resolution. The TMS sensor collects the data in 12 channels, including Landsat TM bands 1-7 (band 6 is collected in both low-gain and high-gain modes) and four additional bands between the Landsat TM bands.

Spectral calibration of the sensor is performed prior to each major scientific campaign (typically twice per year). This calibration is performed at the NASA Ames Research Center sensor calibration laboratory, using a scanning monochromator and a Fourier transform infrared interferometer, which measure the spectral response function (SRF) of each band. The sensor radiometry for the visible/shortwave infrared (Vis/SWIR) bands is measured by viewing a 30-inch integrating sphere, calibrated against a National Institute of Standards and Technology standard lamp. The resulting digital counts from the TMS sensor are convolved with the sphere radiance data and the SRF to produce a radiance-per-count coefficient for each band. The thermal band is calibrated by viewing a laboratory extended-area black body source of known temperature.

Inflight calibration of the thermal band consists of two onboard black body sources of known temperature that are viewed once per scan. The black body temperatures also are recorded for each scan, together with their corresponding digital counts from the sensor, providing a two-point calibration. There is no onboard calibrator for the Vis/SWIR bands, although the black bodies are used as a "zero radiance" reference to track changes in offset, which are then applied together with the laboratory-measured radiance coefficients, to obtain at-sensor radiance.

Data from TMS were used during EFEDA 1991 for testing new algorithms, for computation of sensible and latent heat fluxes, and for scale integration purposes.

AIRSAR

The NASA/JPL AIRSAR is a side-looking imaging radar system which utilizes the SAR (Synthetic Aperture Radar) principle to obtain images of high resolution. The radar backscatter of the imaged surface is represented at different frequencies and polarizations. The coordinates of the scene are usually given as slant or ground range and azimuth. Slant range is the radial distance of the image object from the radar, while ground range is the distance of the object from a point directly below the aircraft. Near range in the images is closest to the aircraft and far range is furthest away. The azimuth dimension is parallel to the line described by the aircraft trajectory. The spatial resolution of the AIRSAR images is 12 x 12 metres.

The information contained in each pixel represents the radar backscatter for all possible combinations of horizontal and vertical transmit and receive polarizations. The data is compressed in Stokes matrix format with 10 bytes per pixel.

The radar backscatter measurement is a strong function of incidence angle, which is defined as

$$\theta_i = \cos^{-1}\left(\frac{h}{R_0 + i\Delta R_0}\right) \tag{2.1}$$

where h is the altitude of aircraft (typically 8 km), R_o is near range in the image (typically 8-10 km), ΔR_o the size of the (slant) range pixels in metres and i is the line number of the object in the image. This equation is strictly valid only for a flat earth. The radar backscatter has a tendency to decrease as the incidence angle increases. Most of the provided data are calibrated, so that radar backscatter measurements are in normalized radar cross section format (m^2/m^2).

DIAL

An airborne water vapour DIAL (DIfferential Absorption Lidar) has been developed by the German Aerospace Research Establishment (DLR). The instrument was first used in an operational mode during EFEDA 1991, measuring aerosol and water vapour distributions in the atmospheric boundary layer. The DIAL technique is based on a two-wavelength lidar measurement using narrowband laser transmitters. During the "on-line" measurement, an extinction enhancement occurs when the laser beam penetrates the atmosphere because of atmospheric water vapour absorption. A second "off-line" lidar sounding in the wing of a water vapour line with small or negligible water vapour absorption is needed as a reference measurement. This second measurement relates to aerosol backscattering and extinction as known from conventional lidar techniques.

The laser transmitter of the DIAL system consists of a narrowband tunable dye laser pumped by a frequency-doubled Nd:YAG laser. The bandwidth of the narrowband peak and its wavelength stability is approximately in the order of 0.001 nm. Thus systematic errors in the determination of the absorption cross sections can be neglected. However the narrowband central peak of the dye laser is accompanied by additional unwanted broadband background emission. This leads to errors in an accurate DIAL calculation when the laser line profile is only approximately known. To overcome these difficulties the necessary absorption cross sections as function of altitude are determined by fitting the DIAL calculations to a nearby radiosonde measurement. Additional errors may be caused by aircraft cabin pressure and temperature variations which affect the wavelength stability. The on-line wavelength is therefore calibrated to the maximum absorption of the selected water vapour line.

The backscattered photons are collected by a telescope and mirror and detected by a photomultiplier after having passed a narrowband interference filter. The photocurrent is digitized with 12 bit resolution at 10 MHz sampling rate which corresponds to 15 m vertical resolution. The horizontal resolution is also 15 m for typical shot frequencies and aircraft speed. A detailed description of the instrument and signal improvement is given by Ehret et al. (1993).

2.3 Reception, Acquisition and Availability of Satellite Data

2.3.1 Reception of AVHRR (HRPT) Data

The AVHRR data stream is part of the High Resolution Picture Transmission (HRPT) system of the TIROS-N series satellites (see section 2.2.1). The selection of special orbital parameters allows one to operate these satellites in nearly sun-synchronous orbits in order to achieve daily global coverage at nearly identical local observation times. The instrument produces, at the full resolution, Local Area Coverage (LAC) mode about $4 * 10^9$ bit per orbit. This type of satellite was not designed to store all data. Only up to 11.5 minutes of LAC data or 115 minutes of Global Area Coverage (GAC) data, with reduced resolution, can be stored on one of the five onboard tape recorders and transmitted to the Command and Data Acquisition (CDA) stations. Four out of every five samples along the scan line are used to compute an average value and the data from only every third scan line are processed, yielding 1.1 km by 4 km resolution at nadir. One of the receiving stations is located in Wallops Island, Virginia and one in Fairbanks, Alaska, at a high latitude to facilitate data download during 11 of the 14 passes per day. Additional receiving stations are necessary to gather the data for full global coverage. Thus data of more than 20 receiving stations around the world are used to process the global 1-km data set (Eidenshink and Faundeen 1994).

Real-time data acquisition of non-encrypted data is possible without any restrictions at all receiving stations that match the technical specifications (Schwalb 1978). A great number of receiving stations for local applications exists around the globe. Depending on the gain of the antenna and receiving conditions, they can gather data from satellite 'rise' to satellite 'set' for a maximum of 16 minutes (ca. 1/6 of a pass). Combining all possible passes of a daytime or nighttime observation period an area of about 1/18 of the earth's surface can be covered. Two examples for this type of receiving stations are given here, an operationally working station and a research facility.

For meteorological applications an operational type of a receiving station is operating since 1965 at the Free University of Berlin where since 1982 HRPT data are received and processed. Since August 1988, when high density data cartridges became available, archiving of the full HRPT data stream and processing of all AVHRR channels at full resolution became possible. Since then, daily seven full passes of AVHRR data of the afternoon satellites NOAA-11, NOAA-14 and NOAA-16 as well as six passes of the morning satellites NOAA-10, NOAA-12 and NOAA-15 are received and processed.

A 3.3 m diameter parabola antenna mounted on the top of a water reservoir tower 80 m above ground (Fig. 2.1) receives data from horizon to horizon and allows to gather the maximum amount of data possible for a local station. Using the TBUS

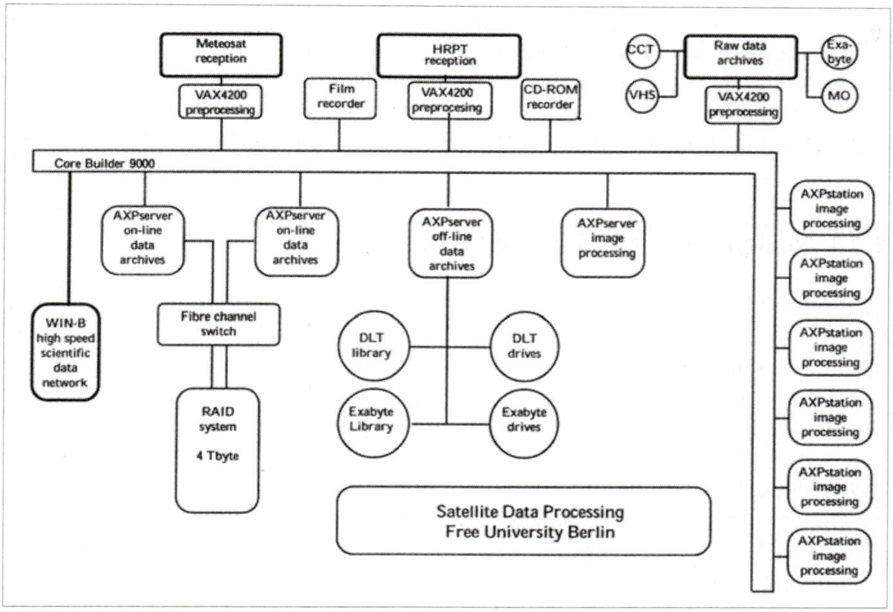

Fig. 2.2. Block diagram of the satellite data receiving and processing facility at the Free University of Berlin

bulletins provided by NOAA via the Global Telecommunication System (GTS) the antenna tracking and the pre-processing chain up to geographically remapped AVHRR products are performed fully automatically. HRPT raw data as well as geographically remapped single passes and multi-pass composites are archived on magnetic tape cartridges. Fig.2.2 is a schematic diagram of the whole data receiving and processing facility.

Smaller NOAA-AVHRR receiving stations nowadays are very common at research facilities and universities. A low-cost version of a NOAA/Meteosat receiving station was designed and installed at the University Louis Pasteur-Strasbourg which is used to receive AVHRR raw

Fig. 2.1. NOAA-AVHRR receiving antenna on top of the water tower in Berlin-Zehlendorf

data for the research objectives of the G.R.T.R. (Groupment de Recherche en Télédétection et Radiometrie, Université Louis Pasteur, Strasbourg). The NOAA receiving system is composed of a 1.2 m diameter antenna controlled by a personal computer and a UNIX workstation performing the pre-processing including geocoding and geographic remapping (http://www-grt.u-strasbg.fr).

2.3.2 Reception of Meteosat Data

To receive full resolution digital Meteosat data, a "Primary Data User Station" (PDUS) is necessary. As an example again the station built in 1989 in Berlin is described briefly. A 3 m parabolic antenna fixed to the nominal position of the Meteosat is used for data reception. The pre-amplified signal, similar to the HRPT system, is sent to a pre-processing unit consisting of amplifier, demodulator, bit synchronizer, and frame synchronizer. The resulting reconstructed digital data stream is linked to a DEC VAX front-end computer system (Fig. 2.2). A decryption unit had to be included in 1994 when data encryption was added to the Meteosat data stream. The necessary full access key is granted by the Deutscher Wetterdienst (DWD) to the institute for research application.

The PDUS data rates are by a factor of eight lower than those of the HRPT data transmission but due to a half-hourly scan repetition rate the daily archived data of the northern hemisphere amount to about 500 Mbyte due to the nearly continuous data stream for 24 hours.

2.3.3 Acquisition of Landsat TM Data

For Europe and the Mediterranean area the best coverage is archived by the ESA Fucino Receiving Station in central Italy. Part of the area is additionally covered by the Kiruna Receiving Station (Sweden) and Maspalomas (Canary Islands, Spain) and Neustrelitz (Germany). Table 2.2 gives the acquisition periods for LANDSAT 4, 5 and 7.

Table 2.2. Acquisition of Landsat data by ESA receiving stations

Ground Station	Start and End of Receiving Period		
	Landsat-4	Landsat-5	Landsat-7
Fucino/Matera, Italy	8/82 - 7/87	since 3/84	since 7/1999
Kiruna, Sweden	8/82 - 9/93	since 3/84	
Maspalomas, Canary Islands, Spain	- -	7/84 - 12/90	
Neustrelitz, Germany			since 7/1999

2.3.4 Acquisition of SPOT Data

SPOT HRG (Haute Résolution Géométrique) and VEGETATION data are commercially available since 1986 through SPOT IMAGE. More than 20 HRG receiving stations guarantee a world wide coverage.

SPOT satellites transmit image data to ground stations in two ways, depending on whether the satellite is within the range of a receiving station or not. As the satellite proceeds along its orbit, imagery acquisition is possible in four operating modes:

- The satellite is within the range of a Direct Receiving Station (DRS): Imagery is down-linked in real-time. SPOT's worldwide ground station network comprises 16 DRS.
- The satellite is not within the range of a SPOT DRS: Programmed acquisitions are executed. The images are stored on onboard recorders.
- The satellite is within the range of a main receiving station (Kiruna or Toulouse): Data can be down-linked in real-time or data stored before are played back from onboard recorders.
- The satellite is not in the range of one of these ground stations: The satellite data acquisition system remains in standby, ready to acquire imagery in accordance with uplink commands.

For the Vegetation instrument onboard SPOT-4 the main receiving station is Kiruna (Sweden), which receives and records image data acquired in global observation mode (GO) several times a day. After having passed the Vegetation image quality center (QIV) and the Vegetation image processing center (CTIV) products are distributed via Spot Image's four distributors (including subsidiaries and its worldwide commercial network): SPOT IMAGE CLEO (Cluster for Earth Observation - Horizon Centre, 1930 Zaventem, Belgium), SSC (Swedish Space Corporation), Telespazio and JPL (Pasadena).

2.3.5 Acquisition of Nimbus SMMR and NOAA-AVHRR GAC Data

The Nimbus-7 Scanning Multichannel Microwave Radiometer (SMMR) level 1b Pathfinder data set was created at the Jet Propulsion Laboratory (JPL) from level 1a (definition of processing levels see section 2.2) data set provided by the Goddard Space Flight Center (GFSC). The SMMR operated continuously from 25 October 1978 through 16 November 1987 at which date an alternate-day operation started due to power management problems. The SMMR Pathfinder Brightness Temperatures data set contains global brightness temperatures in swath format (level 1b) from 25 October 1978 to 20 August 1987. The instrument obtained near-global coverage at five frequencies (6.6, 10.7, 18, 21, and 37 GHz) in both horizontal and vertical polarizations, at a constant incidence angle of 50.3 degrees, every six days.

As the active life-span of the SMMR ended before the beginning of the EFEDA period, all analysis referred to later in this volume are based on retrospective data. SMMR observes a particular area at mid-latitudes approximately four times during a one week period, i.e. twice during the nighttime (near 00 UTC) and twice during the day (near 12 UTC).

NOAA-AVHRR data used in conjunction with the long-term microwave study, based on Nimbus/SMMR, are from a time filtered Global Area Coverage data set (GAC), provided by NASA as 10-day and monthly values. This data set has been empirically corrected for viewing angle, instrument degradation etc. (Los 1998).

2.3.6 Acquisition of ERS 1/2 Data

The European Remote Sensing Satellites ERS-1 operated regularly from July, 1991 to March, 2000, while ERS-2 is operating since May, 1995. The satellites are tracked by the Mission Management Control Centre (MMCC) of the European Space Operations Centre (ESOC) with its network of ESA tracking stations (ESTRACK network) distributed around the globe. The International Ground Station (IGS) network enables acquisition of full resolution data of almost all of the land surface, and much of the oceans. The user interface and exploitation of the payload data has been implemented within the Earthnet ERS-1 Central Facility (EECF) at ESRIN in Frascati. Data can be ordered via Eurimage (http://www.eurimage.com).

2.4 Calibration of Satellite Data

2.4.1 Calibration of the AVHRR Short Wave Channels

Detection and Causes of Signal Degradation

At the time of the design of the AVHRR instrument the necessity of in-flight calibration for a quantitative analysis of reflectance data was not considered (Schwalb 1978). The two shortwave channels were included in order to get images of the cloud distribution on a global scale for use in weather analysis and prediction. The increasing interest in global coverage of albedo and vegetation information linked the interest to AVHRR data as the only available data source, prior to the lauch of Terra on 18 December 1999, with daily global coverage and acceptable spatial resolution (Gutman and Ignatov 1995). This type of instrument, first flown on the NOAA-6 satellite in 1978, will remain, with minor changes, an operational data source for some time. This challenges to establish one of the most valuable long-term sets of remote sensing data for global change research. Therefore it is worthwhile to continue the efforts to improve these time series by different approaches in order to reduce the error bars for derived products and strengthen the representability for the analysis of global change. Great efforts were undertaken to prepare the coarse resolution pathfinder data set (James and Kalluri 1994). Work is now in progress to establish a similar data set retaining the full AVHRR resolution (1 km data) (Eidenshink and Faundeen 1994). Unfortunately there are different artificial long-term trends affecting the time series of reflectance measurements and products derived from these data, like vegetation indices, and conceal the real changes of surface properties due to the danger of misinterpretations.

The sensors of the AVHRR are calibrated prior to the launch of the satellite (Abel

1990). These coefficients are found to be inapplicable to the instrument after launch because they lead to unrealistic reflectances of known targets (Rao and Chen 1994). The decrease of the sensitivity of the instrument is ascribed to contamination caused by the engine exhaust of the last stage of the launch vehicle. This effect cannot be measured directly due to missing on-board calibration tools (Rao and Chen 1994). There is still no knowledge about the possible impacts on the filter functions of the instrument (Vermote and Kaufman 1995). Targets with known reflectances have to be used to get reliable post-launch calibration coefficients.

Many authors dealt with the combined calibration and validation task. Pioneering work has been done by Holben et al. (1990), Staylor (1990), Rao and Chen (1987, 1993, 1994), Kaufman and Holben (1993), Vermote and Kaufman (1995). Extraordinary efforts were made by NOAA and NASA to get absolute calibration in the case of NOAA-9 and inter-calibration coefficients for the other AVHRR instruments. Ground observations as well as measurements by an absolutely calibrated AVHRR from high flying aircraft were used for time congruent data intercomparison. This resulted in a reasonable absolute calibration of archive data, which can be used as anchor values for inter-satellite calibrations (Smith et al. 1988, Abel et al. 1993, Whitlock et al. 1990). Efforts were also made to predict changes in sensor sensitivity in order to reduce errors for information extraction in real-time (Kaufman and Holben 1993, Rao and Chen 1994). The large differences in the obtained results highlight the complexity of this task (Vermote and Kaufman 1995).

It will be shown later in this chapter that a precise inter-calibration of the instruments flown on different satellites can be achieved after half a year of operation. This is important for Near Real-Time (NRT) applications. In the case of NOAA-14 AVHRR the results and experiences gathered by the interpretation of NOAA-11 data could be used efficiently because the geometric observation conditions of NOAA-11 in 1989 and NOAA-14 in 1995 nearly coincide at the same day of the year. For NOAA-16 AVHRR reliable in-flight calibration coefficients could be derived after about half a year of operation.

First inter-comparisons between the AVHRR sensor and the on-board calibrated MODIS and ATSR sensors show major discrepancies, mainly caused by the different spectral characteristics (Heidinger et al. 2002, Rao et al. 2003).

Effective Signal Degradation of the NOAA-11 AVHRR Shortwave Channels

It is easy to find a low calibration target. The space view count is included to the Tiros Information Processor (TIP) data and found to be very stable for all instruments operated so far. Water surfaces with very low reflectances especially in the near infrared channel can be used as reference. To find a reliable high calibration target is more difficult. Cloud tops or the ice shields of Greenland or the Antarctica or large glaciers in lower latitudes show high reflectances but are affected by an anisotropic reflectance characteristic like other surfaces and by changing albedo of aging snow cover. Furthermore, a disadvantage is that an identification of clouds above the ice shields is nearly impossible. This already lead NOAA to White Sands as calibration standard. Consequently a bright sand desert area of the Western Great Erg was chosen as target for calibration of the Mediterranean data set. The daily data were screened for clouds (Koslowsky 1996). The reliable calibration coefficients for the beginning of the mission of NOAA-11 anchored to the absolute in-flight calibration of NOAA-9 (Rao and Chen 1994) were used and kept constant for the full period. The time series of daily reflectances (TOA) and NDVI data are shown in Fig. 2.3a. The considerable scattering of the data with a period of about nine days is caused by the reflection anisotropy of the land surface. The periodic change reflects the length of the geometric repetition cycle of the satellite that views the site each day at a different angle and illumination geometry. The sequence is repeated after less than nine days. Channel 1 and channel 2 series show a nearly coincident slope and a nearly constant NDVI over the full life time data set.

Without the knowledge of the aerosol distribution a reliable atmospheric correction is impossible. Only at the neutral point (reflectances of about 15%) the diminution of the Earth surface signal by aerosols is compensated by the backscatter of solar radiation from the aerosols. Volcanic aerosol increase the reflectances below the neutral point and decreasing it above this point (Rao and Chen 1993). Evaluating mid-latitude data there is no certain indication for the influence of the Pinatubo event in 1991 in the records of land surface reflectances.

The magnitude of the observed signal degradation shows no evident change. The change in local observation time of about 40 minutes per year on the other hand causes increasing mean zenith distances of the sun. This results in an increasing attenuation of the incoming radiation (Fig. 2.3b). The long-term trend found in Fig. 2.3a is totally removed for the channel 2 and even over-compensated for the channel 1 data. As sensor aging results in signal degradation, there has to exist another long-term trend, which causes an increase of the reflectances of nearly the same magnitude.

The reflectances in Fig. 2.3b even after atmospheric corrections show periodic changes with high values in winter time and low values at large solar elevation angles in the summer. The corrections of the changing distance of the earth to the sun is already included and can be ruled out as an explanation. From local measurements it is known that the albedo of most land surfaces is increasing with increasing solar zenith angle. No measurements have been made in the Great Erg but similar conditions have locally been found in the Sahel (Niger) where the reflection measurements of Figs. 2.4a and b have been made that show the dependence of the directional reflection of solar zenith angle as well as of the difference between the solar azimuth angle and the azimuth to which the radiometer is directed. Fig. 2.4a features individual measurements made under various solar zenith angles (ζ) and azimuth differences (Az) between the directions towards sun and towards radiometer as seen from the target which is yellow Sahel sand. The radiometer points nearly towards south and under a nadir angle of 30° at the target. During the afternoon, the difference between the azimuth angles of the sun and the direction of observation changes between 2° and 67°and the solar zenith angle between 33° and 64°. The reflectance increases as the solar zenith angle increases. An observed deviation from this observation on 4.11.85 at a solar zenith angle of 38° and 55° probably is due to variable thin cloudiness. At the time of the measurement at 39° solar zenith angle a

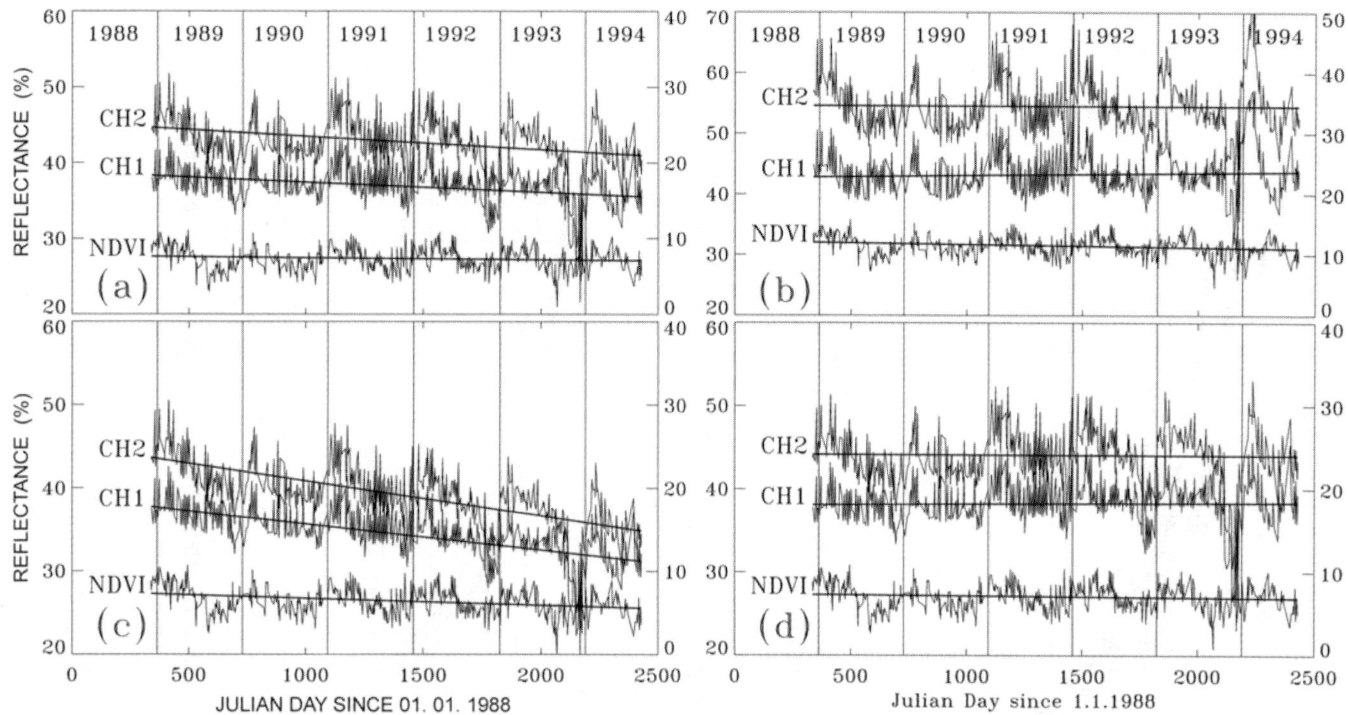

Fig. 2.3. Reflectances and NDVI for a desert test site. Contributions to signal degradation: (a) constant initial calibration values after Rao and Chen (1995), (b) corrected for atmospheric effects, (c) normalized to nadir view, (d) result of effective signal degradation (detail see text)

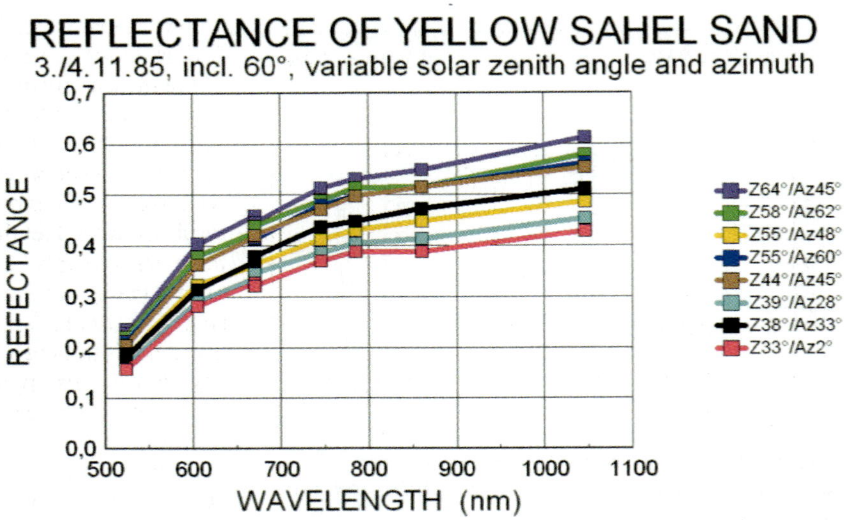

Fig. 2.4a. Reflectance of yellow Sahel sand. Z denotes the solar zenith angle, Az the difference between solar and observation azimuth angle. The target is observed under an inclination angle of 60° towards south

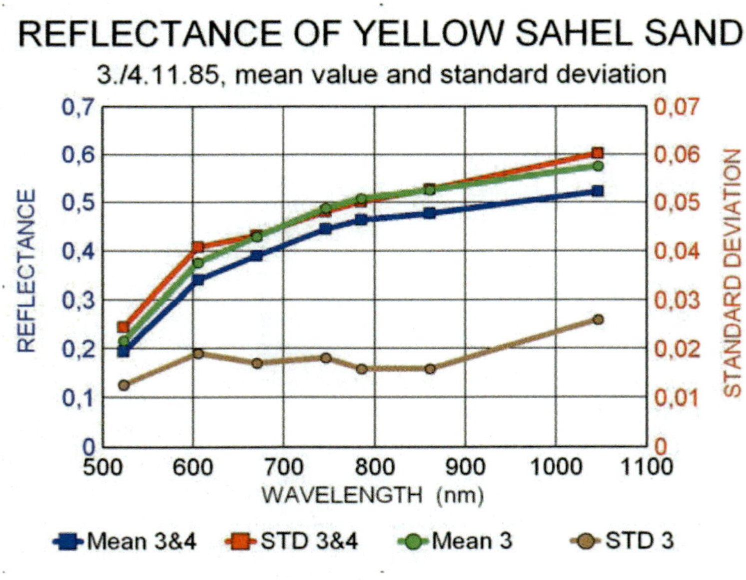

Fig. 2.4b. Average of the reflection measurements over yellow Sahel sand (blue) and standard deviation of the measurements (red) for all data shown in Fig. 2.4a and mean value (green) as well as STD (brown) of the data of 3.11.85 (Z = 64°, 55°, 44°) alone. The reflectances on 3. 11. are higher than those of the whole assembly shown in Fig. 2.4.a and their STD is much smaller

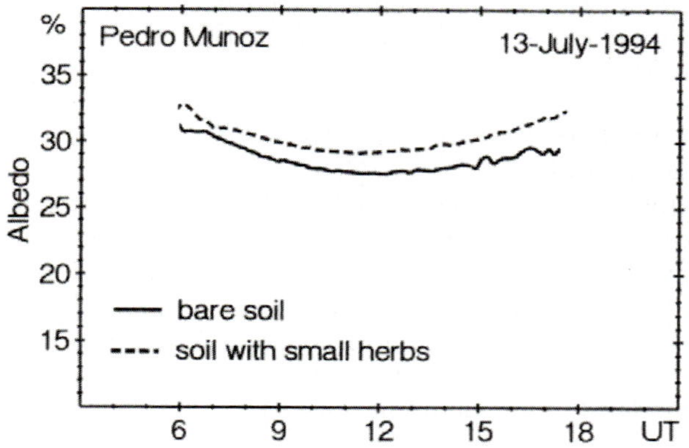

Fig. 2.4c. Albedo measurements over bare soil and soil with small herbs near Pedro Munoz, Spain, under very dry conditions (compare Figs. 1.8 and 1.9)

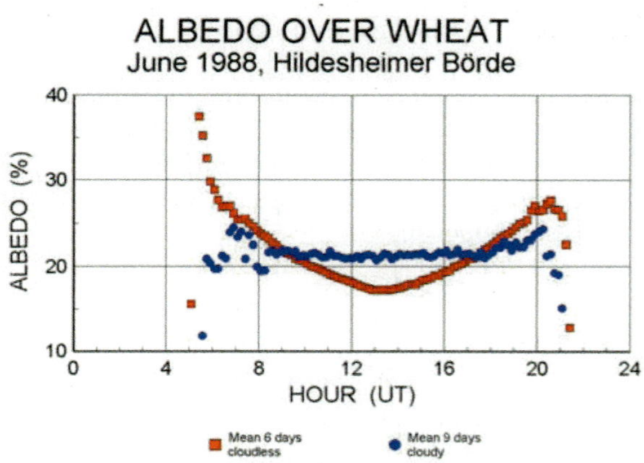

Fig. 2.4d. Albedo of a wheat field under clear sky conditions (red) and cloudy skies (blue)

halo was observed. Fig. 2.4b shows the average of the measurements of Fig. 2.4a and the standard deviation of the measurements. The wide spread of the data result from the fact that the measurements have been made at two slightly different spots and thin cloudes at one day. Further examples are given in Appendix 6, section 6.

The increase of reflectance with increasing solar zenith angle can also be recognized in albedo measurements (Fig. 2.4c) of a bare soil near Pedro Muñoz (Castilla - La Mancha, Spain). At that time the soil was very dry, similar to a desert scenario, though the surface was crusted. For comparison also albedo values for a wheat field are added in Fig. 2.4d. It can be seen that the dependence of the albedo of the solar zenith angle is smaller for the flat desert-like surface than for the wheat. Furthermore under a cloudy sky, when the dependence on the solar zenith angle vanishes, the albedo around mid-day is larger than under the cloudless sky.

In conclusion, the sun height variations together with the anisotropic behaviour of the desert surface cause annual variations in the time series of the reflectance. The magnitude of this effect was estimated from the mean differences between summer and winter reflectances. The variation of the solar zenith angle between summer and winter of about 30° is as high as the mean increase during the six year's period caused by the delay of the local observation time from about 13:00 to 16:30 hours local time. If effects caused by changes of the relative azimuth angle are neglected and therefore the mean annual variations should lead to a comparable increase of the surface reflectances as the increasing solar zenith distance during the full operation period. In Fig. 2.3c this influence is removed. Since the atmospheric influence was already removed previously the slope of the regression line in Fig. 2.3c reflects the real sensor degradation of the AVHRR instrument of NOAA-11. Possible effects of long-term trends in the relative azimuth of illumination and viewing direction are not removed. The knowledge of the full bidirectional reflectance distribution function (BRDF) of the targets would be required to separate these effects more precisely.

The contributions to the three components are given in Table 2.3 together with the final values reported by Rao and Chen (1994) for NOAA-11. The combined influence observed by the AVHRR at the top of the atmosphere over unchanging land surfaces is addressed as *effective signal degradation* in order to emphasize that the *sensor degradation* is only one contributing part to the problem. It is important to mention, that data sets which are calibrated using such effective degradation values include the mean influence of increasing atmospheric attenuation with the increasing solar zenith angle. That has to be taken into account, when detailed atmospheric corrections are applied. Time series of the reflectance with the correction for the effective signal degradation applied are shown in Fig. 2.3d.

Neither in the reflectance nor in the NDVI time series long-term trends for the desert calibration target are found anymore .

The agreement of the effective degradation coefficients with those reported by Rao and Chen is very good for channel 1 but shows a slight discrepancy for channel 2. Using the higher values of Rao and Chen would result in an increasing NDVI for bare surfaces and therefore would show a greening of the desert (for a more detailed discussion of long term data series see Appendix 8).

Table 2.3. Sensor degradation of channels 1 and 2 of NOAA-11-AVHRR converted to linear degradation rates in percent per year

Total and contributing effects	Channel 1	Channel 2
Rao and Chen (1994)	1.25	2.13
Contribution of the atmosphere	1.93	1.57
Contribution of the BRDF	-2.13	-2.59
Contribution of physical sensor degradation	1.73	2.51
Resulting effective sensor degradation (independently determined)	1.48	1.55

Intercalibration of NOAA-14 and NOAA-11 Instruments

Figs. 2.3a-d show considerable periodic variations of the reflectances throughout the year. Similar variations can be found in the time series of reflectance of the Libyan desert (Rao and Chen 1994). They show no symmetry and cannot be explained exclusively by sun height variations. The annual variations of the distance to the sun may cause a thermal effect to the sensor response. As well annual variations in the aerosol and water vapour content will not be symmetrical to the sun's illumination conditions. But there is no doubt that the main contribution has to be addressed to the bidirectional reflectance distribution function (BRDF) of the test areas. The high reflectances in late winter go along with observation conditions near to the principal plane, while the very low values in the early winter of 1993 have to be ascribed to shadows caused by extremely low sun elevation and surface roughness. Besides the periodic variations in viewing conditions with a repetition cycle of about nine days the viewing conditions of the radiometer show no trends. This offers the possibility to sample, parts of the BRDF for all viewing angles ranging from +60° to −60° with a time resolution up to nine days (depending on cloud cover) for which one can assume that the land surfaces do not change. By interpolation of the measured data points the data can be normalized to nadir conditions. A *nadir correction function* (NCF) is derived which eliminates the variations of viewing conditions from the time series. Due to the short sampling period they account for real variations of the surface characteristics in the course of the year due to the vegetation cycle (Koslowsky 1996). If such NCFs are applied, most of the short periodic variations in the time series are removed. Unlike statistical smoothing this method is based on physics and in the case of non-equally distributed viewing angles no artificial bias is added.

The local observation times of the NOAA-14 satellite, launched on 30 December 1994, and NOAA-11, launched on 24 September 1988, were nearly identical and even in fall 1999 differed by only 0.5 hours. This fact was used to fit the NOAA-11 time series of reflectances with the new data of NOAA-14, starting with the same date but 6 years apart, to account for the previous mentioned residual annual variations of the reflectance of the desert target. Both time series were normalized to nadir viewing conditions and then smoothed by a 30 day median filter. Then a least square fit of the differences was constructed. From a linear regression to the fitted NOAA-14 data the initial post-launch calibration coefficients are given by the intercept, and the signal degradation rates by the slopes of the regression line.

Fig. 2.5 shows the results of this technique after different numbers of days of operation of the NOAA-14 spacecraft. The first fit was done in October 1995 after about 300 days and provided to NOAA/NESDIS as a first result. A later one, about 500 days after launch (Koslowsky 1997), resulted not only in a more precise intercalibration at that date but also in degradation rates that proved to be precise and stable and can be used for the next two years of operation as it can be seen from the results obtained 1200 days after launch.

By adding data of the years 1998 to 2000 to the linear regression fits of the differences between the NOAA-11 and the NOAA-14 time series a completely different situation was found for NOAA-14. The verified linear degradation within

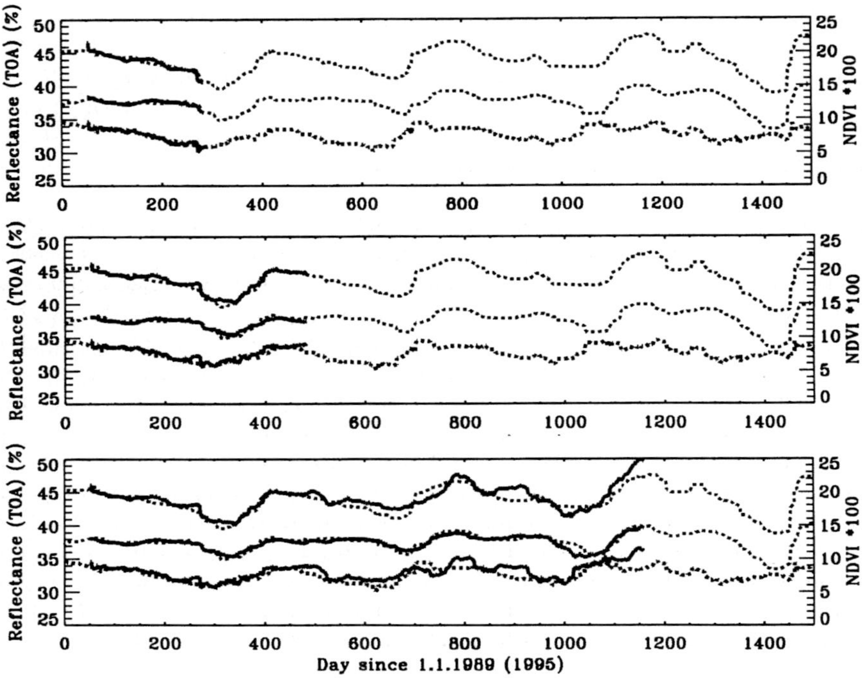

Fig. 2.5. Intercalibration by applying RMS-fit for nadir corrected time series of NOAA-11 and NOAA-14 reflectances channel 1, 2 and NDVI.

the first three years of operation cannot be expanded to the end of the mission. In 1998 the degradation decreases strongly and nearly vanishes for 1999 and 2000. This comes out more clearly when fits between the NOAA-11 and NOAA-14 time series are done for different starting points and period lengths rather than beginning with the first day of operation.

Linear regressions to the fitted NOAA-14 data were applied therefore for different period lengths and showed that the high degradation rates for the first two to three years had to be considerably reduced for the next years (Fig. 2.6). If the slopes of the first three years are used to predict degradation rates for 1998 and even 1999 the final relative error will be about 7% for channel 1 and 9% for channel 2. If reflectances of 20% for channel 1 and 30% for channel 2 are considered the absolute error will be 1.5% and 3% respectively, which is much more than can be tolerated.

In comparison to NOAA-11 the degradation rates of NOAA-14 are considerably higher and comparable to those of NOAA-7 and NOAA-9. A large difference is also found between the pre-launch calibration coefficients and the observed in-flight values, too. Thus the apparent sensor degradation in the first weeks of operation was strong and possibly caused by contamination of the radiometric optics by outgassing of the rocket engine. Therefore data of the first three weeks are skipped in the intercalibration fitting procedure.

A formula to correct the level1b archive can be found in the NOAASIS web site (http://noaasis.noaa.gov/NOAASIS/ml/aboutn14vis.html, Rao and Chen 1999)[1].

Comparison with Other Calibrations Studies

Till the end of 2001 not many studies were published concerning the NOAA-14 AVHRR. This may be the consequence of final overall agreement with the results of NOAA (Rao and Chen 1994) and the resulting conviction that calibration and inter-calibration of different AVHRR instruments are settled and the coefficients used by NOAA/NESDIS are applicable. But already in 1996 it was felt that the applied degradation rates for calibration of the level 1b data seem to be too high (Koslowsky 1997). Revised calibration coefficients are used for this product since December 1998 and formulae to correct the level 1b archive data are given (Rao 1999).

Most studies on signal degradation used desert targets for calibration. A comprehensive study was published by Gutman (1999). The approach to use global data separated in six land surface classes allowed to demonstrate the different behaviour in resulting trends of reflectance time series. The overall view which allows to quantify this problem for global impact is an outstanding result of this study. Appraising the different results, calibration coefficients and degradation rates are presented and compared to each other, including un-published studies.

These results are compared with those of this study, which seems to be the only one using daily and high resolution AVHRR data. Only one bright desert target with surface characteristics not known by local observation is used. The data revised by Rao and Chen (c), Gutman (e), Meteo France (g), and FUB (d) are compared in Fig. 2.6. The slopes of (c) and (d) show no significant differences, while the slope of (g) seems to be too low in the case of channel 2. The bias between Gutman (e) and FUB (d) for channel 2 nearly vanishes, but is somewhat more pronounced for channel 1, and takes the middle position between (c) and (e) coinciding with (g) for the beginning of 1997.

From this comparison the (d) values lying between the extremes seem to be reliable. The difference to the other results is smaller than ±4% and for reflectances of 30% the absolute error is of the order of 1%. Only the evaluation (d) accounts for the decreasing degradation rate of the last two years. The test site for (d) seems to represent the differences in land-surface classes reported by Gutman better than the test site for (c) and can be regarded as a global reference. The decreasing sensor degradation shown in the result of the FUB data (d) was verified by inter-calibration of the on-board calibrated VIRS (Visible Infrared Scanner) flown on TRMM (Tropical Rainfall Measuring Mission) and NOAA-14 channel 1 (Nguyen et al. 2001, Doelling et al. 2001) with a nearly perfect coincidence. This further confirms the reliability of the method used in Berlin. For the desert calibration target long-term trends are not anymore found in the reflectance or NDVI time series.

---

[1]Referenced internet links were checked and confirmed during the editing phase. They might have been moved, changed or even deleted in the meantime.

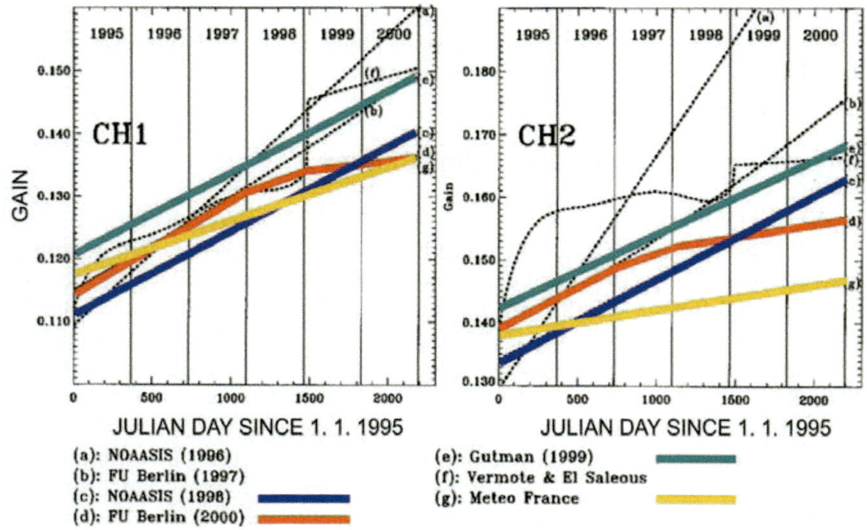

Fig. 2.6. Signal degradation of channel 1 and 2 of the AVHRR of NOAA-14 by different authors (see text)

The agreement of the effective degradation coefficients with those reported by Rao and Chen is very good for channel 1 but shows a slight discrepancy for channel 2. Using the higher values of Rao and Chen would result in an increasing NDVI for bare surfaces and therefore would show a greening of the desert.

One reason for the lower values of (c) and (d) can be explained by the use of BRDF corrections to nadir view conditions. Fig. 2.7 shows the 12-years time series of NOAA-11, NOAA-14, and NOAA-16 daily data of the calibration target used for FUB. Besides the individual observations (top) the nadir corrected time series (bottom) are included. Mean nadir correction factors (NCF) as a function of viewing angle, which is the factor to be used for individual measurements to correct to nadir viewing conditions, can be found in chapter 4. While at FUB for the normalization *empirical* NCFs sampled directly from daily data sets were used, Rao [(a) respectively (c)] uses a BRDF *model*. This may explain the differences between the two data sets.

Intercalibration Between NOAA-16 and NOAA-11 Instruments

Beginning with NOAA-15 the AVHRR/3 instrument is flown on the NOAA-K, -L, -M and -N spacecrafts. Two major changes are noticeable: The switching between channel 3a (daytime) and 3b (nighttime) and the split gain used for channel 1, 2 and 3a. The split gain option increases the digital resolution for the low albedo range from 0% to 25% by a factor of two while the resolution for the range 25% to 100%

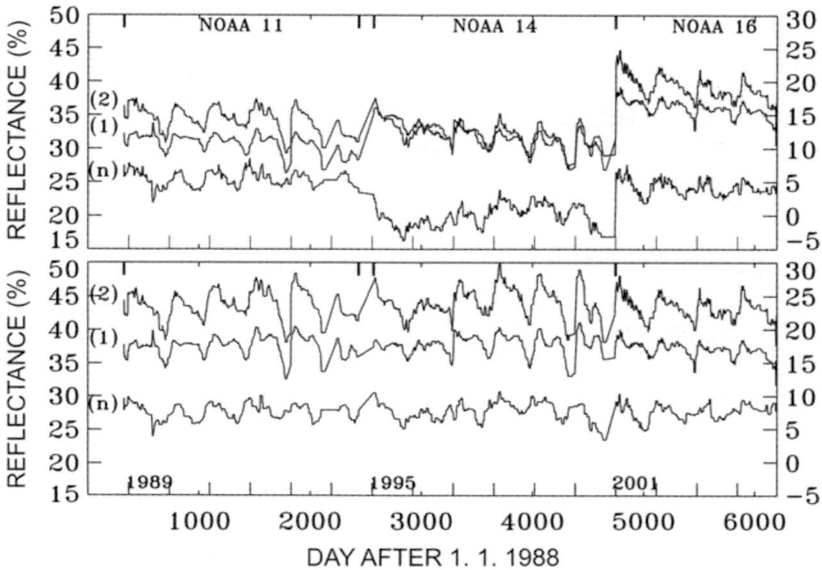

Fig. 2.7. Time series of daily reflectances derived from AVHRR channel 1 (1), channel 2 (2) and NDVI (n) of the satellites NOAA-11 to NOAA-16 of a calibration test site in the Great Erg Occidental in Algeria. Data are normalized to nadir observation condition and smoothed by a 9-day median filter. Upper panel: constant preflight gains are used showing the calibration problem. Lower panel: corrections for sensor degradation and satellite inter-calibration are applied.

is reduced. The break or switch count, sc, for both channels is close to digital count 500 (Table 2.4). The preflight gains for the low albedo range are 0.0523 and 0.0513 respectively assigning 26.0% and 25.7% to the break values. If 100% is assigned to count 1023 the gains for the high albedo range are 0.1528 and 0.1510. This equals the low albedo gains multiplied by the factors 2.9216 and 2.9435 which can be treated as fixed instrumental constant. Thus the formulae of Table 2.4 can be applied for converting counts to technical albedo. For possible sensor degradation it is only necessary to modify the low albedo gains.

At least one year of operation is required to get reliable degradation rates. But nevertheless as for NOAA-14 (Fig. 2.5) corrected inflight gains for channel 1 and channel 2 were derived by inter-comparing the BRDF corrected time series of NOAA-11 and NOAA-16 (Fig. 2.8).

The presented method for inter-calibration of the shortwave channels of the different AVHRR instruments enable a near real-time monitoring of the sensor gains. Additionally this procedure assures not only to remove artificial long term trends from the time series but also a re-adjustment to the absolute calibrations of the NOAA-9 satellite. The uncertainty by possible long-term trends due to atmospheric

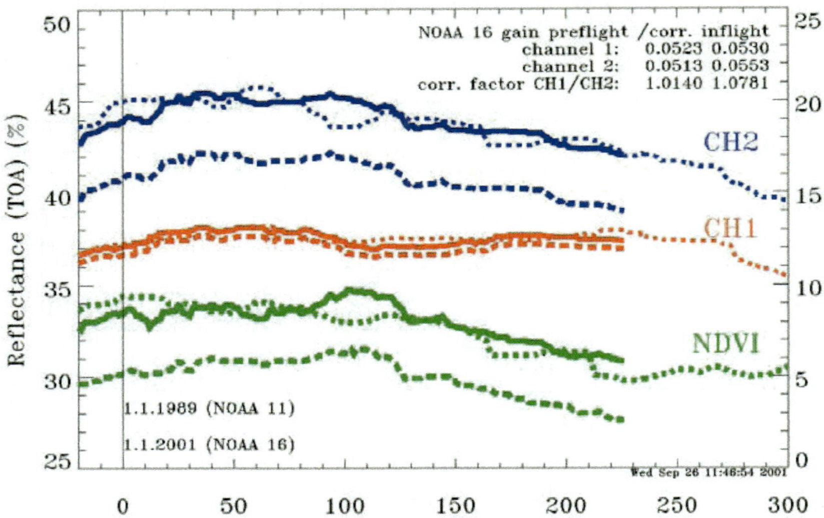

Fig. 2.8. Time series for the days 0 to 300 after launch with preflight (dashed) and corrected inflight (solid) gains for channel 1 and 2 of the NOAA-16 AVHRR compared with the NOAA-11 times series (dotted). The numbers at the abscissa are days after the first of January following the launch in September 1988 of NOAA-11 respectively in October 2000 of NOAA-16

attenuation however still exists and can only be removed by a network of anchor stations performing continuous measurements of the aerosol optical depth.

The resulting corrections to the pre-flight gains are 1.4 % for channel 1 and 7.8% for channel 2 of NOAA-16. The considerable discrepancies between the corrections generate large errors with consequences for the derived NDVI values.

The test site used for deriving corrected calibration formulae are inside the main application area of MEDOKADS (Mediterranean Extended Daily One Km AVHRR Data Set, see section 2.8.3) which covers the whole Mediterranean since 1989 (Koslowsky 1998). Thus it can be expected that possible atmospheric long-term trends are common to the whole region and removed by this procedure.

Using pre-flight calibration coefficients for the new NOAA-16 AVHRR instrument results in errors for the derived NDVI in the order of 10%. Since January 2004 linear degradation rates (Table 2.5) are used at FUB which are equivalent to 0.5 %/a for channel 1 and 2 %/a for channel 2.

Table 2.4. Conversion of preflight split gain AVHRR/3 counts (C) into technical albedo (Goodrum et al. 2000)
Gain low: $\rho = C * g_l - I_l$ for $0 < C <$ sc
Gain high: $\rho = C * g_h - I_h$ or $\rho = \rho_{sw} + g_h * (C - sc)$ for $sc < C < 1023$

Satellite	NOAA-16		NOAA-17		NOAA-18	
Channel	1	2	1	2	1	2
gain low g_l	0.052	0.051	0.056	0.054	0.054	0.053
intercept I_l	-2.016	-1.943	-2.219	-2.123	-2.213	-2.084
switch count sc	497.5	500.3	497.53	500.32	500.54	500.4
gain high/ low g_h/g_l	2.9216	2.9435	2.9315	2.9853	2.9815	3.0006
intercept I_h	-51.91	-51.77	-55.96	-56.22	-55.37	-55.31

Table 2.5. Initial in-flight gains (g_i) and degradation rates (r_d) for NOAA-11/14/16 after launch and for the following years of operation (L = low gain, H = high gain)

Spacecraft	NOAA-11 1.1.1989		NOAA-14 1.1.1995		NOAA-16 1.1.2001	
Channel	Ch. 1	Ch. 2	Ch. 1	Ch. 2	Ch. 1	Ch. 2
Gain after launch (g_i)	0.106	0.11	0.115	0.138	L 0.0533 H 0.1556	L 0.0549 H 0.1616
Year of operation	Degradation rates (r_d) per day $* 10^5$ for the year of operation and day d_a since 1.1.1989/1995/2001 for NOAA 11/14/16. The actual gain is $g_a = g_i\,(1 + r_d\, d_a)$					
1 - 2	4.3	4.55	14.8	14	1.5	4.4
3	4.3	4.55	14.8	9.4	1.5	4.4
4	4.3	4.55	7.4	4.7	1.5	4.4
>4	4.3	4.55	4	4		

2.4.2 Calibration of AVHRR Thermal Channels

Images derived from NOAA-AVHRR data produced at the University of Strasbourg are calibrated with respect to the NOAA recommendations as defined in the NOAA POD Users's Guide (Kidwell 1995). A description of the calibration procedure can be obtained under http://www-grtr.u-strasbg.fr. The procedure requires two steps: Firstly the computation of the linear calibration coefficients for each line of the raw image and secondly the conversion from radiance to brightness temperature.

In the first step it is assumed that the relation between the numerical counts and the radiance measured by the sensor is linear:

$$L_i = S_i C + I_i \; . \tag{2.2}$$

L_i is the radiance value in mW/m^2 sr cm^{-1}, C is the digital count (ranging from 0 to 1023), and S_i respectively I_i are the scaled slope and intercept values. To calculate slope and intercept, the telemetry data from the HRPT stream are used. The two points of the calibration line are given by the digital counts of the space view and of the radiance from the internal black body which is derived from its temperature (Planet ed. 1988, Goodrum ed al. 2000, and NOAA KLM User's Guide with NOAA-N, -N' Supplement, September 2000, which can be obtained from www2.ncdc.noaa.gov/docs/klm/).

The slope S_i in Eq. (2.2) is calculated according to $S_i = (L_{BB} - Ls)/(C_{BB} - Cs)$ and the intercept is defined by $I_i = Ls - [(L_{BB} - Ls)/(C_{BB} - Cs)] * Cs$. L_{BB} and Ls are the radiance values for the internal black body and space respectively, C_{BB} and Cs the corresponding counts.

Prior to NOAA-13 the conversion from radiance to brightness temperature was performed using the inverse Plank radiation equation

$$T(L) = \frac{c_2 v}{\ln\left(1 + \frac{c_1 v^3}{L}\right)} \tag{2.3}$$

where T is the absolute temperature (K) for the radiance value L, v is the central wavenumber of the channel (cm^{-1}), $c_1 = 1.1910659 \times 10^{-5}$ mW/(m^2 sr cm^{-4}) and $c_2 =$ 0.01438833 m K are constants. The central wave numbers are provided by NOAA as a function of temperature and channel for each spacecraft. Non-linearity corrections are then applied for channels 4 and 5 using tables provided by NOAA. These tables give brightness temperature correction terms as a function of channel and baseplate temperature.

The radiometric effective black body temperature T_{BB}* deviates slightly from the measured temperature T_{BB} of the black body housing and is defined as

$$T^*_{BB} = A + BT_{BB} \, . \tag{2.4}$$

The corresponding black body radiance is:

$$L_{BB} = B_v(T^*_{BB}) = \frac{c_1 v^3}{e^{\frac{c_2 v}{T^*_{BB}}} - 1} \tag{2.5}$$

From NOAA-15 on an alternate method was supplied by NOAA to handle the non-linearity which can be applied to radiances instead of brightness temperatures. The following quadratic equation is used to compute the corrected radiance, L_{cor}, from the "linear" radiance, L_{lin}:

$$L_{cor} = b_0 + b_1 L_{lin} + b_2 L_{lin}^2 \tag{2.6}$$

The values of b_0, b_1, b_2 as well as the radiance of space *Ls* used in the slope and intercept computation are given for each channel of the NOAA-15 and NOAA-16 instruments in Table 2.6. The numerical values of *A* and *B* are given in Table 2.7 together with the nominal central wavenumbers ($\nu = 1/\lambda$, measured in cm^{-1}; λ being the wavelength measured in nm or µm) for channel 3B, 4, and 5. The central wavenumbers are temperature dependent as shown in Table 2.8.

Although the data for space view and for the internal black body are given in each scan line of the record and the values from the four Platinum Resistance Thermometers (PRT) C_{BB} are given every fifth scan line, in Berlin, tables are calculated for averages of these values and their standard deviation for every minute.

Table 2.6. NOAA-15 and NOAA-16 radiance of space Ls and coefficients for nonlinear radiance correction quadratic of Eq. 2.6

Channel	NOAA-15				NOAA-16			
	Ls	b0	b1	b2 *E-7	Ls	b0	b1	b2 *E-7
4	-4.5	4.76	-0.0932	4524	-2.47	3	-0.054	25
5	-3.6	3.83	-0.0659	2811	-2.01	2.3	-0.037	15

Table 2.7. NOAA-15 and NOAA-16 AVHRR/3 thermal channel temperature conversion coefficients (http://www2.ncdc.noaa.gov/docs/klm/html/d/app-d1.htm and http://www2.ncdc.noaa.gov/docs/klm/html/d/app-d2.htm)

Spacecraft	NOAA-15			NOAA-16		
Coefficients of Eq. 2.2	ν_c	A	B	ν_c	A	B
Channel 3B	2695.9743	1.621256	0.998015	2700.1148	1.592459	0.998147
Channel 4	925.4075	0.33781	0.998719	917.2289	0.33238	0.998522
Channel 5	839.8979	0.304558	0.999024	838.1255	0.674623	0.998363

Table 2.8. Central wavelength for different temperature ranges for NOAA-14

Temperature Range	Channel 3	Channel 4	Channel 5
190 - 230	2638.652	928.2603	834.4496
230 - 270	2642.807	928.8284	834.8066
270 - 310	2645.899	929.3323	835.1647
290 - 330	2647.169	929.5878	835.374

Only data with small standard deviations are taken and averaged for the whole orbit. From these data a conversion table is calculated, again for the whole orbit. It turnes out that the conversion tables for the central day and night orbit give the best results. Taking the calibration counts from each line could introduce errors due to noise.

As an example, the conversion from PRT-counts into target black body temperature T_{BB} in degree Kelvin is given for NOAA-14 (NOAA-J/14 AVHRR FM-204) by the polynomial

$$T_{BB} = a_0 + a_1 C_{BB} + a_2 C_{BB}^2 + a_3 C_{BB}^3 + a_4 C_{BB}^4 \tag{2.7}$$

with $a_0 = 276.597$, $a_1 = 0.051275$, $a_2 = 1.363\text{E-}06$ and $a_3 = a_4 = 0$ for all four PRT. For NOAA-15 and 16 these values can be found in the NOAA-KLM-user's guide in tables D.1-8/D.2-9.

By applying the inverse Planck function, the central wave numbers ν_c give the minimum error of the four specified temperature ranges. In the case of data received in Berlin, the filter functions (also given in the Appendices of the AVHRR User's Guide) are explicitly used to convert radiance into temperature.

2.4.3 Conversion of Meteosat Signals to Absolute Values

Visible Channel

No on-board calibration is performed for the visible channel. Vicarious procedures are therefore necessary using clear sky, sea surface and bright desert targets. Large numbers of images are used to derive mean calibration coefficients. Atmospheric correction is done by the 6S code (see Chapter 3) using water vapour data provided by ECMWF, total ozone content from TOMS (Total Ozone Mapping Spectrometer) and a constant value of the aerosol optical thickness.

Radiances can be calculated by the formula

$$L_s = c_s (DC - DC_0) \tag{2.8}$$

where c_s is the calibration coefficient, DC the observed digital count, and DC_0 the space count. The equivalent spectral width W, the integral over the spectral response, and the corresponding spectral solar irradiance allow to convert radiances into reflectances. Samples of the calibration coefficients for Meteosat 5 to 7 are summarized in Table 2.9 (stated 1999). In the case of Meteosat 5 and 6 detailed information about calibration coefficients, including a linear degradation formula can be found under http://www.eumetsat.de.

Table 2.9. Calibration coefficients for METEOSAT (from www.eumetsat.de, 1999)

Period	Slots	Meteosat	#Obs.	Coeff. c_s	Std. Dev.	Space Count DC_0	W	S_0
1-12 1995	24	5	489	0.857	0.04	4.7	0.501	692.2
1-12 1996	24	5	500	0.873	0.03	4.7		
3-12 1997	24	6	423	0.862	0.04	5.7	0.501	692.2
71998	17, 24, 28	6	542	0.86	0.05	5.6		
71998	12, 14, 18, 22, 24, 28, 30, 32, 34	7	418	0.873	0.06	5	0.501	693.2

Table 2.10. Meteosat offset and gain coefficients

Offset	Gain
-0.06662	0.04197
-0.15732	0.10345
-0.11269	0.065
-0.23286	0.11705
-0.0864	0.02727
-0.05114	0.01692

Infrared Channel

Since 29 May 2000 a Black Body Calibration Method is used for Meteosat 7 and from 31 May 2001 onwards a Satellite Cross Calibration method is used for Meteosat 5. Before these dates backwards to the start of the Meteosat program, the Vicarious Calibration Method (VCM) has been used. Data of the National Center for Environmental Prediction (NCEP) are in this case used for the calibration of the infrared channel and radiosonde observations for the water vapour channel.

The calibration coefficients and space counts can be found on the above mentioned internet site. Actual calibration values are included in the header of the direct read-out PDUS data. Eq. 2.2 (respectively the equivalent of the following Eq. 2.10) can be used to convert digital counts to radiances with the offset (intercept) and gain (slope) coefficients given in Table 2.10. In the Meteosat data archive in Berlin brightness temperatures are derived from radiances using Planck's law and re-converted to 8 bit digital numbers using the formula

$$DN = 2\,(T_B - 50) \tag{2.9}$$

which covers the temperature range from 50.0 °C to -77.5 °C with a resolution of 0.5 °C. This conversion is done for the water vapour channel similarly.

Cross calibration between Meteosat 7 and 5 data is performed since 31 May 2001 when both satellites can get data from overlapping parts of the earth surface with viewing angle differences less than 5°.

2.4.4 Conversion of Landsat Data to Absolute Values

Calibration information can be found on the Landsat home page http://mtpe.gsfc.nasa.gov/landsat/default.htm. The file DOCUMENTATION.TXT includes a guide how to calibrate the data. The page L7.html gives a short overview about the satellite Landsat 7 and satellite.html information about the instruments and data. Chapter 11 of the Landsat 7 Science Data User´s Handbook provides detailed information.

The conversion of the so-called "quantized calibrated radiance" QCAL of a specified pixel in units of digital numbers (DN) to spectral radiance is called the "Dynamic Range Method" (EOSAT, Landsat Technical Notes, No 1. August 1986). The spectral radiance is calculated according to the formula

$$L_\lambda = LMIN_\lambda + \frac{LMAX_\lambda - LMIN_\lambda}{255} * QCAL = A0_\lambda + A1_\lambda * QCAL \tag{2.10}$$

The symbols used here and in the following formulae have the meaning and dimensions:

L_λ	spectral radiance for a specific pixel in units of mW cm^{-2} sr^{-1} µm^{-1}
$A0_\lambda$	post-calibration offset coefficient in units of mW cm^{-2} sr^{-1} µm^{-1}
$A1_\lambda$	post-calibration gain coefficient in units of mW cm^{-2} sr^{-1} µm^{-1} DN^{-1}
$LMIN_\lambda$	Minimum spectral radiance in units of mW cm^{-2} sr^{-1} µm at QCAL= 0 DN
$LMAX_\lambda$	Maximum spectral radiance in units of mW cm^{-2} sr^{-1} µm at QCAL = 255 DN

QCAL quantized and calibrated scaled radiance of a specified pixel in units of DN
A0 offset coeficient or spectral radiance at QCAL = 0 in units of W m^{-2} sr^{-1}
A1 gain coefficient in units of W m^{-2} sr^{-1}
W band-width in units of µm.

QCAL gives already the raw data counts corrected for sensor degradation. The calibration constants are given in the "LEADER File" at the beginning of each scene in form of an *offset A0* and a *gain A1* for converting QCAL for a specific pixel from DN to spectral radiance in each band. The offset and gain factors are related to the spectral post-calibration values by:

$$A0_\lambda = LMIN_\lambda = \frac{A0}{W} \tag{2.11}$$

$$A1_\lambda = \frac{LMAX_\lambda - LMIN_\lambda}{255} = \frac{A1}{W} \tag{2.12}$$

Depending on how W is determined, different types of calibration factors for Landsat data are given in the literature, for example by Epema (1990), Slater (1987), and Earthnet. In Table 2.11 the band-specific $LMIN_\lambda$ and $LMAX_\lambda$ parameters are listed that have been used for the ground processing of TM data at NASA Goddard Space Flight Center (EOSAT Landsat Technical Notes No. 1, August 1986). The values used by ESA Earthnet are reproduced in Table 2.12 and Table 2.13.

The thermal Channel 6 is calibrated in equivalent temperature according to

$$T = \frac{K2}{\ln\left(\frac{K1}{L_\lambda} + 1\right)} \tag{2.13}$$

T is the effective at-satellite temperature in Kelvin, K and L_λ the spectral radiance in mW cm^{-2} sr^{-1} μm^{-1}. For Landsat-5 the calibration constants have the values: K1 = 60.776 K and K2 = 1260.56 mW cm^{-2} sr^{-1} μm^{-1}.

2.4.5 Conversion of SPOT Data to Absolute Values

The SPOT "radiance counts *X*" are converted to "equivalent radiance *L*" by

$$L = \frac{X}{A} + B \tag{2.13}$$

where
A = absolute calibration gain in W^{-1} m^2 sr µm (radiance counts), and
B = absolute calibration offset in W m^{-2} sr^{-1} μm^{-1}.

These coefficients are recomputed to take into account the on board gain used for imaging the SPOT scene, as well as the possible digital dynamic stretching which has been applied to the data. *B* generally is zero. The user gets with each scene the value for the absolute calibration gain for each channel (the dimension erroneously is given as W/m^2/sr/µm).

Table 2.11. Landsat-5-TM calibration parameters according to EOSAT LANDSAT TECHNICAL NOTES No. 1, August 1986. Presented are the following quantities: TM Solar Exoatmospheric Spectral Irradiance $ESUN_\lambda$ (mW cm^{-2} µm^{-1}), Quadratic Moment Band-Width W (nm), $LMIN_\lambda$ and $LMAX_\lambda$ in units of mW cm^{-2} sr^{-1} µm^{-1}

Ch.	$ESUN_\lambda$	W	$LMIN_\lambda$	$LMAX_\lambda$
TM 1	195.7	70.1	-0.15	15.21
TM 2	182.9	88.9	-0.28	29.68
TM 3	155.7	76.6	-0.12	20.43
TM 4	104.7	134.3	-0.15	20.62
TM 5	21.93	226.5	-0.037	2.719
TM 7	7.452	268.2	-0.015	1.438

Table 2.12. Band-specific calibration quantities for Landsat 5 as in Table 2.9 but as used for ground processing of TM data at Earthnet and in addition ESUN in W m^{-2} based on quadratic moment method .

Ch.	$ESUN_\lambda$ mW cm^{-2} µm^{-1}	ESUN W m^{-2} (Quadratic moment method)	W nm (Quadratic moment band-width)	$LMIN_\lambda$ mW cm^{-2} sr^{-1} µm^{-1}	$LMAX_\lambda$ mW cm^{-2} sr^{-1} µm^{-1}
TM 1	195.7	137.2	70.1	-0.09	15.17
TM 2	182.9	162.6	88.9	-0.18	29.49
TM 3	155.7	119.3	76.6	-0.15	21.49
TM 4	104.7	140.6	134.3	-0.17	19.81
TM 5	21.93	49.67	226.5	-0.038	3.031
TM 7	7.452	19.99	268.2	-0.019	1.588

Table 2.13. Band-specific calibration quantities as in Table 2.9 but based on full-width half-maximum method after EOSAT after Epema (1990)

Ch.	ESUN Wm^{-2} full-width at half maximum method	W nm Full-Width at Half Maximum Band-Width	$LMIN_\lambda$ in units of $mW\ cm^{-2}\ sr^{-1}\ \mu m^{-1}$	$LMAX_\lambda$ in units of $mW\ cm^{-2}\ sr^{-1}\ \mu m^{-1}$
TM 1	129.2	66	-0.10	16.12
TM 2	150.0	82	-0.19	31.97
TM 3	104.3	67	-0.17	24.57
TM 4	134.0	128	-0.18	23.13
TM 5	47.6	217	-0.040	3.164
TM 7	18.8	252	-0.020	1.69

2.4.6 Calibration of Nimbus 7 SMMR Data and Orbit Stability

The procedure for converting raw SMMR antenna signals (digital counts) to microwave radiances entails taking the instrument output data stored as digital counts on tapes called TATs (Antenna Temperature Tapes) and applying a series of calibrations and corrections. First, a pre-launch calibration was carried out based on tests done in a thermal vacuum chamber at the Jet Propulsion Laboratory. During these tests the response of each of the ten SMMR channels is measured against a black body target operated at a variety of temperatures ranging from 100 K to 350 K. Second, a semiempirical correction procedure was applied and its results were stored on tape as calibrated brightness temperatures (TCTs). Furthermore the data are adjusted by two additional corrections, one for long-term instrument drift and the other for instrument errors dependent on ecliptic angle (Gloersen et al. 1992).

For long-term change analysis, knowledge of the stability of the satellite orbit is crucial. For Nimbus/SMMR a long-term drift of the orbit is difficult to determine because of the large pixel dimensions. Therefore a separate study was devoted by the Vrije Universiteit Amsterdam to this subject mainly focussing on coastal pixels with a sharp contrast between the sea surface and the land surface.

During the day the drift had a northwest direction with a horizontal shift of about 4 km and a vertical shift of about 2 km in 10 years. During the night the shift had the same magnitude but in the opposite direction (southeast). This may be due to tilt of the sub-orbital track of the satellite. Njoku and Brown (1993) already mentioned that the SMMR instrument on the Nimbus-7 spacecraft, and mission designs caused data artifacts caused by calibration drifts, instrument on/off cycling, and antenna pointing stability. With careful analysis most of these problems have been understood and quantified; however these small changes are still included in the data set. If this drift-hypothesis is true the drift will reach its maximum in the Polar regions.

The drift accelerated between February 1984 and October 1984. This acceleration could not be identified in the nighttime measurements, which show more noise/seasonal effects than the daytime measurements. This was possibly caused by the low power supply of the spacecraft during this period (pers. comm. NASA). As compared to the large footprint (~ 130 km) of the SMMR at 6.6 GHz, it is expected that the drift has negligible effect on the detection of long-term changes in surface moisture. Moreover, comparing the daytime values with the nighttime values can eliminate this error.

2.5 Georeferencing and Geographical Registration

2.5.1 Georeferencing of AVHRR Data

A first requisite for georeferencing of data acquired by radiometers onboard of satellites is knowledge of the satellite position at a specific time. The TBUS bulletin contains information related to satellite tracking and performance as well as Brower mean orbital parameters valid for the third day after the date on which the bulletin is prepared and transmitted. The bulletins are provided by NESDIS and transmitted through the National Weather Service Telecommunications Gateway (KWBC) to major meteorological centers and relay points around the world, which comprise the Global Telecommunications System (GTS).

Furthermore the scanning law of the radiometer and the attitude of the spacecraft is required to link the digital count of the radiometer, characterised by scan line number since ascending node of the satellite and sample number in the scan line, to geographic coordinates. Georeferencing and registration to geographic map projections is performed using orbital models within the accuracy of the predicted orbital parameters and the knowledge of the satellite attitude.

It is beyond the scope of this book to go into details, because there is no specific impact on the Mediterranean data sets. A good overview can be found in the NOAA-KLM User's Guide (http://www2.ncdc.noaa.gov/docs/klm) and software like the AAPP (ATOVS and AVHRR Processing Package) to process NOAA-HRPT data which can be downloaded from http://www.eumetsat.de/en/area4/aapp. Software developed in Berlin was firstly applied to scanning radiometer data (Koslowsky 1977) and has been continuously updated.

2.5.2 Map Projections and Registration Methods

To correctly present spatial data on a flat medium, such as paper or a computer screen, raw images and vector layers referring to spherical earth quantities and measurements must undergo a mathematical transformation known as geographical registration. Projecting refers to the representation of one surface onto another. Since it is impossible to transform spherical measurements into plane systems without a consequent distortion in distance, area, or angular relationships, it is necessary to know which specific property (e.g., the preservation of areal or angular relationships) is maintained with a certain projection. There are 3 different geographic projections to be considered:

Equivalents: Projections for which represented areas are proportional to those on the earth.

Equidistant: Projections which preserve the length of certain lines on the map.

Orthomorphic or Conformal: Angular relationships are correctly preserved in the near vicinity of any location which allows to measure accurately distances over an irregular terrain. This is the most important projection for Geographic Information Systems (GIS) and high resolution remote sensing.

For large scale overviews as provided by Meteosat and NOAA-AVHRR the stereographic projection is generally used for higher latitude regions, while for low latitude the Mercator projection or the latitude-longitude representation are more appropriate. High resolution data with nominal resolution of about 30 m like Landsat TM or SPOT data are often presented in the Universal Transverse Mercator Projection (UTM), because they offer a nearly distortion free presentation that fits to high resolution geographic maps. The region of validity is limited to ±3° relative to the main reference longitudes (3^0, 9^0, 15^0, ...) and is therefore is not adequate for large scale overviews.

For large scale representation at low or mid latitudes the latitude/longitude representation, often called geographic projection even though it is not a real projection, is the most appropriate one and is widely used. Each map position can be directly addressed by its geographic coordinates without any conversion formula and overcomes problems with incompatible geographic map projections, that might use different projection planes, principal longitudes or earth ellipsoid data.

As the daily AVHRR data built up the most comprehensive data set of the Mediterranean area, map presentations of all other remote sensing data have to be compatible with it. The nominal ground resolution of the AVHRR along the sub-satellite track of about 1 km suggests a map resolution of 0.01° in both directions. Data presented in such a manner are commonly called 1-km data sets.

To minimize problems which could complicate data exchange caused by the different computer platforms and software packages, the pre-processed AVHRR data were converted from internal data format to standardized 1 km data sets. A plane raster image format was chosen for each individual channel and collateral information is given in pixel congruent form.

In order to cover all areas which were selected for special evaluation, as test sites in Tuscany, LaCrau, Tunisia, Algeria, upper Rhine valley and Israel as well as the whole Mediterranean basin, which will be subject for the application of the derived evaluation algorithms, the MEDOKADS (Mediterranean Extended Daily One Km AVHRR Data Set - ref. to chapter 2.8.3) has been established starting on January, 1, 1989.

Data with considerably higher/lower resolution (i.e. Landsat, Meteosat, AMSU) should be remapped to latitude/longitude representation with raster resolutions, which should differ preferably by factors or quotients of powers of 2 compared to the resolution of these 1-km data sets. To avoid resolution losses or uncertainties by superimposing data with different resolution pixel elements are defined by the coordinates of the upper left corner and its nominal resolution.

With raster images, a common form of georeferencing is to indicate the reference system (e.g., latitude/longitude), the reference units (e.g., degrees), and the coordinate positions of the left, right, top and bottom edges of the image. For small scale application or if distortions between raw image and referencing system is not too large, raw images are often registered to a chosen grid referencing system by polynomial equations to establish what is called a "rubber sheet transformation". The actual process is one in which a new grid is constructed and a set of polynomial equations is developed to describe the spatial mapping of data from the old grid into the new one. The new grid is then filled with data values by resampling the old grid and estimating, if necessary, the new value. Resampling methods include a nearest neighbour option in which the new grid value is the same as that in the nearest cell in the old grid, and an interpolation option in which the new value is a distance weighted average of the nearest neighbours in the old grid.

The process requires information for a well distributed set of points (Ground Control Points), about their position in the old grid system and the position they should take in the new one. The number of GCP needed depends on the order of polynomial fit desired; linear (first order), quadratic (second order), or cubic (third order). In general it should be chosen to the lowest order of polynomial that provides a reasonable solution on the basis of an adequate number of control points. Linear warping requires a minimum of 3, quadratic requires 6 and cubic requires 10. In practice however, a reasonable fit should require at least twice the minimum required number.

Interpolation is generally assessed through a bilinear algorithm (distance weighted average of the nearest 4 pixels) or a cubic convolution algorithm (distance weighted average of the nearest 16 pixels). As a general rule, it can be assumed that if the input image contains quantitative values, the nearest neighbour method should be used. For qualitative data, however, it is possible to use either one. Interpolation produces smoother results, but radiometric output values that are not any more identical with original ones. The nearest neighbour method retains the original radiometric values within the new spatial pattern, which is the best choice if different spectral channels have to be combined.

For large scale registration of AVHRR data a method only based on GCPs is not applicable. An orbital model is required to refer the raw data coordinates, given by line numbers starting with the ascending node and sample numbers in the scan line, to the cartesian coordinates of the map projection. Displacements of GCP in those pre-navigated images can be used to adjust orbital and attitude parameters in order to get fine-navigated products. This may be necessary if data of different dates have to be combined, e.g. time composites to remove cloudy pixels, or compare with data from other sources.

2.5.3 Universal-Transversal-Mercatorprojection (UTM)

High spatial resolution satellite images of limited areal extend are often presented in the Universal-Transversal-Mercatorprojection (UTM). This projection allows to image small portions of the Earth surface with a minimum of distortion. The Earth surface is projected at a cylinder, the axis of which lies in the equatorial plain and which cuts through the surface of the Earth at two lines ± 180 km distant from a selected central meridian. These two lines are projected in their true length. For the central meridian between these two lines the scaling error is only 0.9996 and going out for another 180 km at both sides the scaling error becomes 1.0015. To keep the error within these boundaries, one has to limit the zone defined by the edges of the usable cylinder section to 6° longitude. Consequently the Earth must be subdivided into 60 zones, each 6° wide. They are counted from the 180° meridian onwards. The first zone is the one between 180° and 174° west, the last one that of 174° east to 180°. The area from 80° S to 84°N is subdivided into 20 bands of 8° width denominated from 80° south to 84° north by the letters C to X. The letters I and O are left out and the letter X spans 12 degree of latitude from 72 to 84° N.

To each of the 1200 boxes defined by a zone number from 1 - 60 and a band letter from C to X (without I and O to avoid confusion with one and zero) a grid with 100 km x 100 km large squares is superimposed. The origin of the 100-km-grid is aligned to the centre meridian of each zone and to the equator. Because the zones are limited by meridians and not by grid lines, there remain at the borders of each zone residual rectangular areas which are also assigned letters. Within each of the 100 km square boxes the position of a point is defined by 4 to 10 digits starting from west to east and south to north. Four digits allow to position any point with an accuracy of 1 km and with 10 digits it is positioned to 1 m.

Because of the alignment of the 100 km squares the coordinates of any point can be expressed in distances from the centre meridian, called *eastings*, and the equator, called *northings*. To avoid negative values for the eastings each central meridian gets the value 500 000 m. An easting of zero never occurs since a 6° wide zone is never more than 674 000 metres wide. Northings are measured from the equator in the northern hemisphere and for the southern hemisphere the value of 10 000 000 is assigned to the equator to avoid negative numbers.

The starting point of an UTM zone depends on the ellipsoid used to describe the shape of the Earth. Mostly used is now the World Geodetic System 1984 (WGS 84) on which GPS is based. Conversion applets between latitude/longitude and UTM coordinates, which are available on the internet, offer the use of different reference ellipsoids (e.g. www.cellspark.com/UTM.html, or www.dmap.co.uk/ll2tm.htm). The difference between WGS84 and other systems is that WGS84 is based on satellite navigation while former systems are fixed to terrestrial marks such as stable parts of terrestrial plates. Examples for the differences which occur between systems are given together with an explanation of the Military Grid Reference System (MGRS) and a numerical example in Appendix 5.

2.6 Cloud Detection and Elimination

2.6.1 Definition of the Cloudless Atmosphere and "Cloud Screening"

Remote sensing of land-surface properties in the visible and infrared spectral range requires a free view of the landscape. This, in general, is only the case if the interfering atmosphere is "clear" which means that it only contains the permanent gases and a "standard" amount of aerosols for which correction tools are available. The transition from a clear to a cloudy (or dusty) atmosphere sometimes can not be defined very well. Most clouds of course have sharp boundaries by which they can be identified if they are filling a number of pixels. But the transition from a clear to a cloudy atmosphere often goes through a "pre-condensation" state during which water droplets start to form and build up thin clouds which are barely visible by the eye. In the Mediterranean area apart from an occasional interference of volcanic aerosol it often happens that sand is blown out of the Sahara forming dust clouds which have a similar, but spectrally different, effect as compared to thin water clouds. They are more diffuse than water clouds and their recognition is difficult. A general guideline for the determination of the threshold between a clear and a cloudy atmosphere may be that the radiance measured by an Earth pointing sensor should not deviate from the measurements made under the best possible conditions for more than the standard deviation known for measurements made with this sensor. With respect to aerosol and dust, the "ideal" conditions may be defined by a meteorological range of more than, say, 50 km. On the one hand, this would point to an atmosphere with a background amount of aerosols and on the other hand to a relatively dry troposphere in which no condensation occurs. It does not exclude condensation near the tropopause where clouds may form in contrails and spread over the sky as thin cirrus. This may be a growing handicap for remote sensing from space and must be considered carefully. For "ideally clear" an additional condition may therefore be that the field of view should not cross a major air traffic zone.

In the microwave range the "clear" sky may contain clouds with small droplets. The threshold depends on the wavelength of the microwaves. The larger rain droplets the stronger the interaction with the microwave radiation. If an atmosphere is found to be "clear" in the shortwave and infrared range it will certainly be so in the microwave range.

2.6.2 Previous Cloud Detection Algorithm

Three categories of cloud screening algorithms can be distinguished: Statistical methods, which are based on histogram analysis (Phulpin et al. 1983), methods with thresholds for different spectral channels (Saunders and Kriebel 1988; Gutman et al. 1994), and methods on the basis of pattern recognition, which analyse the surrounding texture (Garand and Weinmann 1986, Ebert 1987, Berger 1992). Often only the criterion of maximum NDVI is used but this favours off-nadir viewing conditions (Gutman 1991). Methods that account for the spatial variability and variations in time require the evaluation of daily data to extract maps of reflectances with high temporal resolution (Derrien et al. 1993).

The AVHRR with 5 spectral channels offers various ways of cloud detection. Threshold or thin cirrus checks can be performed without geolocation. More advanced methods like spatial coherence need geolocation in order to sort out pixel with land, water and coast (i.e.pixels with partly water and land). All these methods can be applied to individual AVHRR passes. For further improvement the temporal information has to be taken into account. This requires accurate location and registration of the data into a common grid.

The software-package called “AVHRR Processing scheme Over cLouds, Land and Ocean” (APOLLO) was originally developed at the U.K. Meteorological Office (Saunders and Pescod 1988). The purpose of the system is to derive surface- and cloud-parameters from AVHRR-counts (Gesell 1989, Kriebel et. al. 1989). During the quasi-operational application of the cloud screening method of Saunders and Kriebel, some shortcomings were detected in the APOLLO processing chain. This was also found during the build-up of the MEDOKADS data set (see section 2.8.3) as the main source for land surface studies in the Mediterranean. Therefore a special cloud screening scheme was developed that is described below. Recent changes to the APOLLO cloud analysis could not be tested so far but are reviewed in the section “Improved APOLLO cloud analysis” for information, because in the revised form, this scheme seems to be a robust cloud analysis tool.

2.6.3 Cloud Mask Generation Used for the MEDOKADS Product: The Dynamic Threshold Algorithm

Most cloud screening algorithms, this being also true for the APOLLO algorithm (Saunders and Kriebel 1988), use fixed thresholds for reflectances and temperatures and are not applicable to regions with a great variability in surface reflectance characteristics as found in the Mediterranean. The transition zone from high NDVI values accompanied by low reflectances in the visible range to the high reflectances found in the Sahara desert proved to be a critical one. Additionally strong variations throughout the year due to the vegetation cycle and variation of solar height must be accounted for. An additional problem for AVHRR data is the strong variability of the reflectances caused by the broad range of viewing angles and the surface anisotropy in reflection (see chapter 4).

A dynamic threshold algorithm was developed for application to a broad spectrum of earth surfaces, such as desert areas with very high reflectances and areas with dense vegetation. No *a priori* information is needed to activate the unsupervised algorithm. Using composites of the maximum Normalized Difference Vegetation Index (NDVI) for a ten-day period, the corresponding reflectance and temperature maps are used as dynamic thresholds. Empirically derived nadir correction factors normalize the reflectances to nadir viewing condition, reduce the influence of surface anisotropy and tighten the algorithm considerably. Even the exclusive use of reflectance data results in reliable, surface independent, cloud detection. Additional tests using the thermal channels, help to detect semitransparent and sub-pixel clouds.

The algorithm primarily was developed for NDVI time series analysis. If the algorithm is too tight, the time resolution or the range of areas of application is too poor. If it is not sharp enough, cloud contamination distorts the data sets. Thus not all cloudy or partially cloudy pixels are masked out and tools to clean up the time series have to be applied.

All AVHRR channels are included, with varying weights, in this algorithm. The main criterion is the comparison of reflectance data in the shortwave channels with maps of cloud-free reflectance data covering the ten-day period preceding the day of application as a dynamic threshold. Those maps are first created by selecting the reflectances and temperatures matching the ten-day maximum NDVI.

For areas of 16 x 16 pixels, about 18 x 18 km^2 (depending on latitudes in a polar stereographic map projection), the actual, mean and the RMS values are derived. The RMS values are used as a bias. The bias is very low for homogeneous regions and high for regions with a large variation in surface characteristics. The bias for the radiation temperatures is enlarged considerably, to account for possible real variations of the surface temperature that might be realistic from one day to the next, such as in the case of occurrence of precipitation. The Fig. 2.9 shows the flow chart for the algorithm.

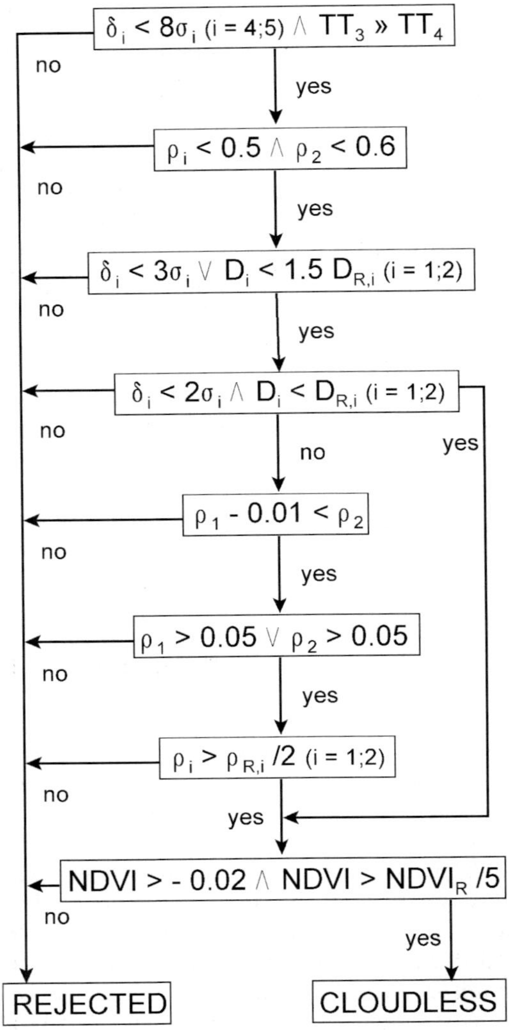

Fig. 2.9. Flow diagram for cloud mask generation

ρ_i	calibrated value channel i
$\rho_{R,i}$	reference value channel i
$\delta_i = \rho_i - \rho_{R,i}$	absolute difference
σ_i	scatter of the reference value
$D_i = (\rho_i - \rho_{Ri}) / \rho_{Ri}$	normalized difference
$\wedge$	"and"
$\vee$	"or"
$NDVI_R$	reference NDVI

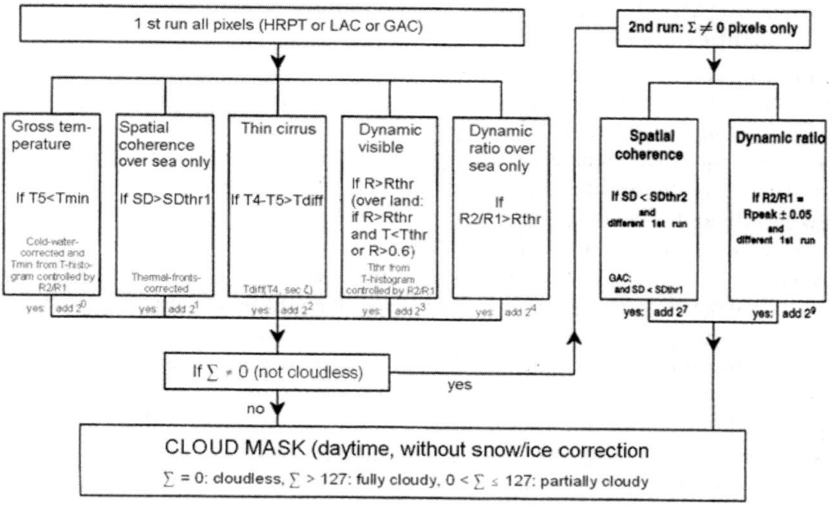

Fig. 2.10. APOLLO daytime algorithm after Kriebel et al. (1999)

2.6.4 Improved APOLLO Cloud Analysis

The APOLLO cloud screening algorithm has been revised in 1999 (Kriebel et al. 1999) with the main advantage that fixed thresholds were replaced by dynamic ones. This leads to a globally applicable procedure except in the polar regions, bright desert areas, and sun elevations between $\pm 5^0$, where errors are not acceptable.

The daytime procedure consists of five tests using AVHRR channel 1, 2, 4, and 5: dynamic visible threshold test, infrared gross temperature test, reflectance ratio test, temperature difference test, and a spatial coherence test. The dynamic thresholds for reflectances and temperatures are taken from regional histograms taking into account some differences in the strategy over land and sea. The spatial coherence test is applied only over sea. Some modifications have been introduced in the use and combination of the different tests.

Improvements have been included to the night time algorithm too. Block diagrams for the daytime algorithm is presented in Fig. 2.10. The algorithms are used to establish a cloud climatology of Europe and will be adopted to the Meteosat Second Generation (MSG) data.

2.6.5 Reconstruction of Cloudless Time Series with the Aid of Fourier Components

Time Series Analysis

Data to be analysed over long time periods should be relatively free of errors, cloudless and equidistant in time. In practice however, time series of images often contain still some cloudy pixel, even if maximum NDVI composites are taken as input. The detection and masking of clouds always results in data gaps and the area of interest may not be covered completely anymore. To recover a data series which can be used for time series analysis these gaps have to be closed and a data series with equal time steps has to be reconstructed. One technique that leads to the desired result is to search for the principal temporal components of the variability of the observed phenomena. The application of these components produces a new set of image-components that are uncorrelated with each other and explain progressively less of the variance found in the original data set. The technique is used for data compression since the first two components explain the vast majority (typically more than 95%) of the variance in the original set of bands. In these cases, the components explaining less than a certain percent of the variance can be dropped and the analysis can be concentrated on the most significant components.

Software has been developed to decompose the time signal for each individual pixel into its Fourier components, i.e. the series of harmonic sine and cosine waves from which the original signal can be reconstructed. Here the term harmonic is used to indicate that a whole number of sine and cosine waves fits into the total period, which also implies that the time series is periodic, i.e. that it repeats itself after a complete cycle, over all samples.

This assumption of periodicity is a rather fundamental choice that has been made. One could apply all kinds of other types of time series analysis to the data, but in that case the analysis would no longer be based on the assumption of *harmonic* components. The choice for a periodic model is inspired by the fact that time series of yearly and daily sets of remotely sensed images often clearly display the influence of the diurnal and annual course of the sun on images of the earth. The annual cycle is expressed in the seasonal growth of vegetation on global scales, and, for instance, a daily series of Meteosat images, taken each half hour, clearly shows the influence of the varying local solar elevation angle on optical imagery and the daily temperature course in thermal infrared images.

Of course the actual signals are not purely periodic, but yet the periodic model can be applied successfully, because it allows more insight in the dynamics of land-surface processes. It is much easier to talk about the amplitude and phase of the annual frequency, than it is to talk about the complete time series of observations. Besides, often a significant data reduction can be realized by discarding the components associated with high frequencies, which usually represent merely noise. To overcome the problem of incomplete data series the Harmonic Analysis of Numerical Time Series (HANTS) algorithm was developed at the Dutch National Aerospace Laboratory (NLR). The proposed method (Verhoef 1996) only works on time series of images and is based upon the Fourier transform.

Fourier Transform

Fourier transform generally is defined for continuous functions but can also be applied for finite length time series as so-called discrete Fourier transform. If a time series of N equidistant samples of a variable y is given, and y_i indicates the i-th sample, then the periodic model implies that the series can be described by means of a Fourier series as follows:

$$y_i = a_0 + \sum_{j=1}^{N/2} a_j \cos(\omega_j t - \varphi_j) \tag{2.15}$$

where t is the time at which the i-th sample is taken. The amplitudes a_j, the frequencies ω_j, and the phases φ_j, are numbered from zero to $N/2$ (if N is even). As the zero frequency has no phase, and the cosine becomes one, the amplitude associated with the zero frequency is called a_0 and is written in front of the summation as an additive term equal to the average signal. The unknown parameters of the periodic model are the amplitudes a_j and the phases φ_j Due to the requirement of harmonic components, the frequencies can only take on certain values, namely only those generating waves that repeat themselves after N samples. This is the case for zero frequency, the so-called base frequency, and for all integer multiples of the base frequency. In order to determine the base frequency, it is necessary to consider the units of time that are applied. But as in the cosine term only the product $\omega_j t$ is relevant, time units are arbitrary, and t might be given by a series of numbers going from 0 to N - 1. In that case the base frequency equals

$$\omega_1 = 2\pi / N \tag{2.16}$$

and all other frequencies follow directly from multiplying this value with j.

Determination of the amplitudes and phases of a series is called Fourier analysis. The reverse process of reconstructing the series of original data is called (Fourier) synthesis. As in Fourier analysis one speaks of a transformation when going from the time domain to the frequency domain. The same data can be represented in both domains. This transformation can also be expressed mathematically by writing Eq. 2.15 in matrix-vector form. Suppose the time series is given by a data vector **y**, and the amplitudes are represented by a vector **a**, then it can be written

$$\mathbf{y} = \mathbf{F}\mathbf{a} \tag{2.17}$$

where **F** is a matrix of which the elements depend only on the value of N. If the elements of **y**, **F**, and **a** can be considered as complex numbers, it becomes easy to find the solution. in this case multiplying **F** by itself gives N times the identity matrix, so that

$$\mathbf{F}\mathbf{y} = N\mathbf{a} \tag{2.18}$$

from which **a** can be obtained. The matrix multiplication on the left is now called the Fourier transform of **y**. The Fast Fourier Transform (FFT) is a very efficient algorithm to execute this matrix multiplication. This algorithm is based on the finding that a FFT of length N can be decomposed in FFTs of lengths equal to the factors making up N. For instance, an FFT of length 1024 can be decomposed into 10 FFTs of length 2. This saves a good deal of processing time. Most FFT computer codes are based on the assumption of N being a power of 2, but Singleton (1969) published a Fortran routine for a so-called mixed radix FFT, which allows other radix numbers than 2. Using Singleton's approach, at NLR a program was implemented für the radix numbers 2, 3, 4, and 5. By means of these radix numbers an FFT of length 360 is decomposed as 2 x 4 x 3^2 x 5. Radix number 4 is not strictly necessary, but it is included because a single transform step with radix 4 is faster to execute than two with radix 2. Therefore, radix 4 is always favoured, and 2 is only used when the exponent of 2 is odd, in which case still one step with radix 2 is necessary. Actually in the implemented version the number $N/2$ is factored, because real data of length N can be transformed with a complex FFT of length $N/2$.

The output of the NLR version of the FFT algorithm consists of $N/2 + 1$ amplitudes (including the zero frequency) and $N/2 - 1$ phases (the zero and the maximum frequency have zero phase).

A few problems are associated with application of the straightforward FFT algorithm:
- The restriction of N being factorable in the radix numbers 2, 3, 4, and 5;
- problems with missing data, especially when in certain parts of the area image data are missing;
- cloud cover for single samples or periods in the time series.

These problems can be avoided by replacing the FFT with classical Fourier analysis (by which the restrictions to N are removed) and by attaching weights to the different observations. In that case the Fourier analysis can be formulated as a curve fitting problem. In doing so, it is also possible to reject certain observations by attaching a weight of zero to them, and the parameters describing the curve are found from a weighted least squares fit. Suppose the observations y_i, with i= 1,. . ., N, are expressed in $M \leq N$ basis functions of time $f_j(t)$, with j = 1,. .., M. Then it can be written a matrix-vector equation:

$$\begin{bmatrix} y_1 \\ \vdots \\ y_{n.} \end{bmatrix} = \begin{bmatrix} f_1(t_1) & \dots & f_M(t_1) \\ \vdots & & \vdots \\ f_1(t_N) & \dots & f_M(t_N) \end{bmatrix} \begin{bmatrix} a_1 \\ \vdots \\ a_M \end{bmatrix} \tag{2.19}$$

or

$$\mathbf{y} = \mathbf{Fa} \tag{2.20}$$

which is identical to Eq. (2.17), with the difference that matrix **F** is not necessarily square anymore. Pre-multiplication by the transposed of **F** gives

$$\mathbf{F}^T\mathbf{y} = \mathbf{F}^T\mathbf{Fa} \tag{2.21}$$

or

$$\mathbf{a} = (\mathbf{F}^T\mathbf{F})^{-1}\mathbf{F}^T\mathbf{y} \tag{2.22}$$

From this it follows that the coefficients **a** can be determined from a matrix inversion and a matrix-vector multiplication. Introducing weights associated with the observations can be accomplished by means of a diagonal weight matrix **W**. For the weighted observations it can be written

$$\mathbf{Wy} = \mathbf{WFa} \tag{2.23}$$

and

$$\mathbf{F}^T\mathbf{Wy} = \mathbf{F}^T\mathbf{WFa} \tag{2.24}$$

with the solution

$$\mathbf{a} = (\mathbf{F}^T\mathbf{WF})^{-1}\mathbf{F}^T\mathbf{Wy} \tag{2.25}$$

This is the weighted least squares solution of the curve fitting problem. In case all weights are equal to one and the basis functions chosen are the complete set of sines and cosines corresponding to the Fourier frequencies, then the columns of matrix F contain these functions for the different points in time. If the time samples are equidistant, the matrix product $\mathbf{F}^T\mathbf{F}$ is a diagonal matrix, and the solution becomes identical to the one obtained directly by means of a Fourier transform. So it can be concluded that Eq. (2.25) is a generalization of Fourier analysis, that allows non-equidistant time samples, weighted observations, and even arbitrary (e.g. non-periodic) basis functions. However, the latter option is not considered here, as an almost infinite number of basis functions might be chosen. Alternatively, the basis functions are restricted to those that are identical to or form a subset of the basis functions involved in Fourier analysis, the constant function and the sines and cosines at the base frequency plus the overtones (integer multiples of the base frequency). From Eq.(2.25) it becomes already clear that this curve fitting requires much more processing power than FFT, as it involves several matrix multiplications and a matrix inversion, whereas FFT involves only a single matrix multiplication, which besides is optimized for a minimum number of floating point operations by means of the mixed radix decomposition. However, the curve fitting using harmonic basis functions is more flexible and can give much more reliable results, as invalid data can be assigned a weight of zero, and the observed samples are no longer required to be equidistant in time.

Harmonic Analysis of Numerical Time Series (HANTS)

Based on the approach outlined above, the HANTS (Harmonie ANalysis of Time Series) algorithm has been designed with the application to time series of NDVI images in mind. Its basic idea is that the temporal behaviour of a land surface property can be described by a Fourier analysis. In that case cloud and error free images can be reconstructed. In the first instance a Fast Fourier Transform (FFT) algorithm is used to perform the Fourier analysis (Fourier 1820, Main 1990). Only the most significant frequencies are considered which can be expected to be present in the time series as, for instance, determined in preceding analyses. For each frequency the amplitude and phase (see Fig. 2.11) of the cosine function is determined during an iterative procedure.

NDVI images are usually composited by means of the so-called Maximum Value Compositing (MVC) algorithm in order to suppress cloud cover effects. Clouds always have a negative influence on the NDVI and therefore taking the maximum value of the NDVI over a limited period tends to remove most of the cloud

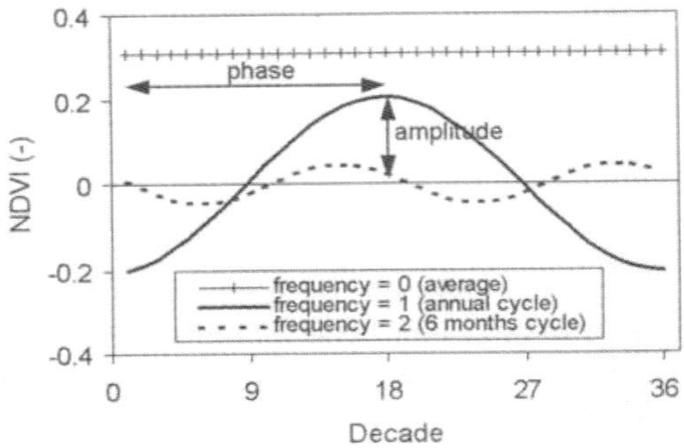

Fig. 2.11. Fourier coefficients of the reconstructed cloudless scene

contaminated observations. The HANTS algorithm also exploits this negative effect of clouds on the NDVI, but in a different way. In HANTS a least squares curve fitting procedure based on harmonic components (cosines) is applied iteratively (Verhoef et al. 1996), i.e. first a least squares curve is computed based on all data points, and next the observations are compared to the curve. Observations that are clearly below the curve are candidates for rejection due to cloud cover, and the points that have the largest negative deviation from the curve therefore are removed first. Next a new curve is computed based on the remaining points and the process is repeated. Pronounced negative outliers are removed by assigning a weight of zero to them, and a new curve is computed. This iteration eventually leads to a smooth curve that approaches the upper envelope over the data points. In this way cloudy observations have been removed and the amplitudes and phases computed are much more reliable than those based on a straightforward FFT.

In the implemented algorithm the iteration can stop in two ways. First, it is possible that the number of remaining data points approaches a certain threshold, so that the system of equations might come close to the state of under determinedness. In that case the iteration is stopped and the amplitudes and phases of the situation just after the last point was rejected is taken as output. Second, it is possible that before the above threshold is reached, the fit through the remaining points is already so good that the maximum error of fit becomes less than a preset threshold. Also in that case the iteration is terminated.

As each rejection of a data point involves a change of the weight matrix **W**, and therefore recalculation of Eq. (2.25), removing one point at a time might lead to excessive computation times. Therefore, whole groups of points are collectively removed after one iteration step, namely all points that deviate more than 50% of the maximum deviation found, as long as they deviate more than the error tolerance, and in the direction chosen. For NDVI images this is the negative direction, but also visible channel Meteosat images might be processed, and in that case clouds lead to positive outliers, so in that case the deviation is measured in positive direction. Note that obvious errors in the opposite direction are always considered being associated with valid data, so it must be taken care that such errors are absent, as they have a devastating effect on the quality of the analysis results. In order to help identify such errors, the user can specify a threshold to flag such data as invalid, but of course this only helps for very obvious errors.

The curve fitting process is controlled by 5 parameters, which have to be set at the beginning of each HANTS run. These 5 control parameters are:

- Number Of Frequencies (NOF). A curve is described by means of (2 x NOF - 1) parameters (amplitude and phase), where NOF is the number of frequencies including the zero frequency (mean), which has no phase. Here also the periodic size of the frequency is indicated.
- Hi/Lo Suppression Flag (SF). This flag indicates whether high or low values (outliers) should be rejected during curve fitting.
- Invalid Data Rejection Threshold (IDRT). In some cases one might know that digital numbers below or above a certain threshold should be considered invalid.
- Fit Error tolerance (FET). During curve fitting the root mean square error (rmse) of the remaining (i.e. not rejected) data points with respect to the current curve is determined after each iteration. Iteration stops when the rmse becomes smaller than the FET. The FET value should not be set to low, as otherwise the fit might be based on too few points, which gives unreliable results.
- Degree Of Over-Determinedness (DOD). The number of valid observations must always be larger than or equal to the number of parameters that describe the curve (2 x NOF-1). How much the number of valid observations have to be above the number of parameters is given by the DOD value.

All these parameters have an influence on the final results and this is also one of the disadvantages of HANTS as compared to FFT: there are no objective rules that might be applied to choose these parameters except the first (period) and the third (HiLo flag) one. The selection is done on the basis of experience and after running several combinations of control parameters. Discussing their effects on the result might be helpful to select suitable values for them.

The number of frequencies determines the amount of detail that can be described in a curve. A low number of frequencies leads to a very smooth curve with little details, a high number of frequencies to a less smooth curve with much more detail. When the number of frequencies specified is equal to the number of Fourier frequencies, the curve goes through all data points, and no detail is lost. In this case the result is identical to the one obtained with FFT, provided the data were sampled equidistant in time.

However, removal of cloudy data is impossible in this case. Recognition of outliers is only possible if the curve is described by means of a limited number of frequencies.

In some cases invalid data are already marked in an image by assigning an "impossible" value to them, for instance 255 in 8 bit data. Now specifying an IDRT of 254 would automatically lead to immediate rejection of these data. When the user specifies a negative value for IDRT, data less than the absolute value of this threshold are considered invalid. For instance, specifying IDRT = –5 fenders all data less than 5 invalid.

The fit error tolerance (FET) specifies which absolute deviation in the chosen direction is still acceptable. When a large FET is specified by the user, only the most obvious outliers are removed during the curve fitting procedure, and there will be a great chance that the iteration will stop soon because all remaining points meet the FET criterion. When a small FET is specified, many points will be rejected, and there will be a great chance that the iteration stops because finally too few points are left.

The degree of over determinedness (DOD) is a safety measure in order to avoid under determinedness, which occurs if the number of remaining points becomes less than the number of parameters by which the curve is described. Suppose the number of frequencies above the zero frequency is NF, then the number of parameters describing the curve by means of harmonic functions is 2NF + 1, as the zero frequency only has an amplitude, and all other frequencies have an amplitude and a phase. So the number of remaining points should always stay greater than or equal to 2NF + 1, otherwise the system is under determined and the matrix inversion in Eq. (2.25) will be impossible. How far we stay away from this condition is given by DOD. The iteration in HANTS is terminated if the number of remaining points becomes less than DOD + 2NF + 1, if it was not already stopped because the FET criterion was met.

The HANTS algorithm offers greater flexibility in the choice of frequencies and the length of the time series than does the FFT algorithm. It is quite easy to exclude certain images from the time series if so desired, since the samples are not required to be equidistant in time. Due to the removal of obvious errors, the computed amplitudes and phases at the selected frequencies are much more reliable than with the direct FFT algorithm. The price paid is a processing time of roughly ten times that of the FFT, but this is still acceptable (approx. ten hours for a series of 36 images of 1536 x 2800 pixels on a work station, if two frequencies are selected).

As an example the HANTS algorithm was applied to one year of calibrated AVHRR 10-days-max-NDVI composites of 1996 (Koslowsky 1996). The covered area is from 10° West to 20° East and from 29° to 55° North with a resolution of 0.01°. The following control parameters were used:

- NOF = 3; the selected frequencies are the zero frequency (mean NDVI) and the frequencies with time periods of 1 year and six months. So the output consisted out of 5 Fourier components (3 amplitudes and 2 phases).
- SF = Low; for NDVI images this flag is set at low, as cloudy observations lead to low NDVI values.
- IDRT = + 0.7; this means that NDVI values higher than 0.7 are rejected. In the input data a few high outliers existed, which are excluded in this way.
- FET = 0.05 NDVI units; the rmse of the remaining observations should be smaller than 0.05 NDVI units with respect to the currently fitted curve.
- DOD = 13; together with the minimum of 5 time steps this gives that each fitted curve is based on a minimum of 18 observations in time, which

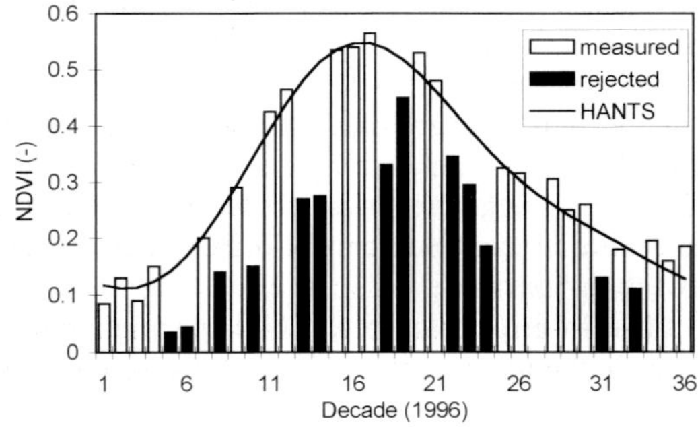

Fig. 2.12. NDVI time series of a pixel in northern France for 1996. Black bars: decadal means identified by application of HANTS to be affected by clouds. White bars: used to reconstruct the cloudless scene

is half of all data points.

Fig. 2.12 shows the original and HANTS reconstructed NDVI time series for a pixel in Northern France. Iteration stopped when 14 out of the 36 original NDVI values were rejected, i.e. classified as cloud affected data points. The remaining 22 values have a rmse smaller than 0.05 NDVI units from the resulting fit. Fig. 2.11 shows the Fourier components of the 3 different frequencies from which the cloudfree profile is reconstructed. The zero frequency is represented only by amplitude, while the other two are defined by amplitude and phase. The arrows represent the values of the annual profile (frequency = 1).

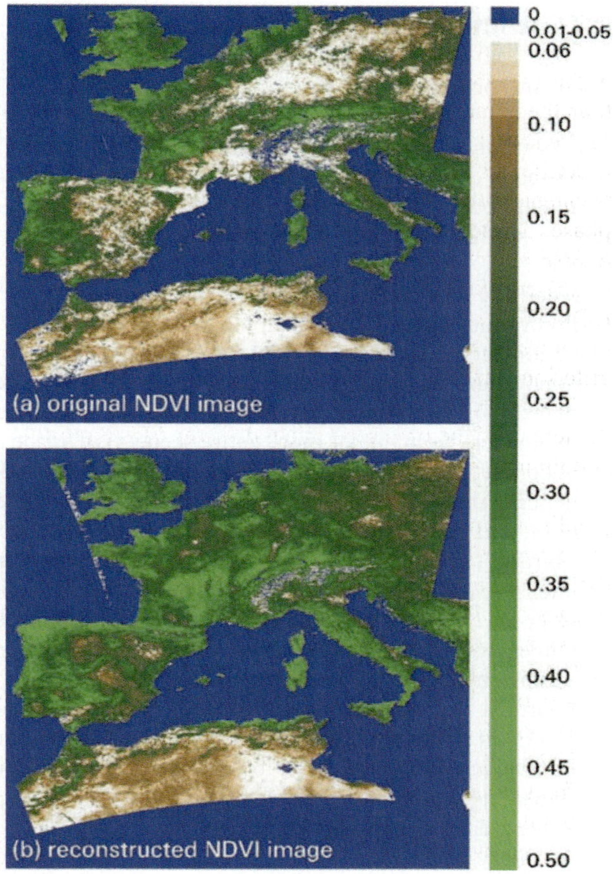

Fig. 2.13. Maps of an original NDVI composite (upper part) and the reconstructed cloudless scene after application of the HANTS algorithm (lower part). The data are of the decade 33 (9-13 November) 1996

The method is especially useful for seasons with long lasting cloudiness when the cloud screening algorithm is applicable only with limitations. As an example, in Fig. 2.13 upper part the original NDVI 10-days-composite of decade 33, 1996, is shown with residual clouds scattered all over Europe. By using the 5 Fourier components the cloudless NDVI image is reconstructed in Fig. 2.13 lower part. The clouds disappeared and the natural land cover patterns become visible again, while for the original cloudless areas in the upper part of Fig 2.13, the original NDVI values are retained.

2.7 Ancillary Data

2.7.1 Introduction

The term "ancillary data" (or "collateral data") primarily is defined as "non-space data used in conjunction with space data in a complementary way". Here, we use an extended definition in the sense of "data which do not originate from the primary remote sensing sensor (as described in Section 2.2) used to create a specific product".

Ancillary data are used for several purposes:

- Firstly they are needed as input data for algorithms to retrieve biogeophysical parameters from measurements made in space (e.g. radiometric correction for atmospheric and topographic effects), where the required input generally consists of two- or three-dimensional data fields to retrieve two-dimensional fields of parameters from one or more entire scene(s) of space data.
- Secondly, they are needed for the development and local validation of non-generic algorithms that are based on empirical relationships. Here, in general, point data are sufficient.
- Thirdly, for the purpose of development and validation of algorithms, data from intensive field work are required. These data are limited in space and time but offer a very complete data set. The area for which the data are representative ranges from plot size up to heavily instrumented "supersites" of about 10x10 km^2 extension and the duration is typically a few weeks.
- Fourthly, the synergistic use of both space and non-space (ancillary) data is crucial in obtaining important, but "difficult" secondary parameters like fluxes.
- Finally, ancillary data serve to describe the synoptic and climatological context (including climate trends and variability) and thus, to put the analyses of remote sensing data in a broader perspective.

Ancillary parameters cover a wide range of time-scales: Some, like topography and soil texture can be considered invariable (except for very long-term change analysis). Others like land-use and land-cover vary slowly (over years). Occasional rapid variations need to be accounted for in the phenological state of the vegetation and all related parameters, as well as in the soil parameters, affected by precipitation events. Finally, the atmospheric processes (described by temperature, wind, moisture, aerosol type and concentration, ozone concentration, precipitation, radiation and surface energy budget) is characterized by diurnal and seasonal cycles or even shorter time scales, if clouds play a role.

Depending on the spatial scales of direct and ancillary data involved, the aggregation problem may need to be solved. Aggregation refers to spatial averaging of surface variables in order to obtain effective surface flux values representative for a heterogeneous area. Due to the complex nature of the biophysical processes which occur at the land-atmosphere interface, linear spatial averaging of input parameters to flux algorithms may lead to erroneous results (see e.g. Michaud and Shuttleworth 1998).

2.7.2 Elevation of Land Surface

The elevation of the land surface is an important ancillary quantity. The length of the path of the radiance from the surface to satellite altitude is determined by the earth`s elevation. The slope and aspect of the surface can be calculated and is essential for the calculation of the solar irradiance of the surface. The interpretation takes elevation as well as slope and aspect values into account. The GLOBE (Global Land One-km Base Elevation) is freely available. The 30-arc-second latitude-longitude grid spacing of the data set is somewhat finer than one kilometer spacing at the ground. It is based on U.S. Defense Mapping Agency Digital Terrain Elevation Data (DTED). DTED is compiled into a 3-arc-second latitude-longitude grid, showing elevation above sea-level based on World Geodetic System (http://www.ngdc.mnoaa.gov/:seg/) and is continuously updated. Source materials for the DTED are best available data, which may include digital elevation models, stereographic imagery or topographic maps. The full resolution DTED is not freely distributed.

The U.S. Geological Survey provides the global digital elevation model GTOPO30. Elevations in GTOPO30 are regularly spaced at 30 arc seconds (approximately 1 km). The horizontal coordinate system is in decimal degrees of latitude and longitude. The vertical units represent elevation in metres above mean sea level. The data set for most of Europe was derived from a raster topographic data base with horizontal grid spacing of 3 arc seconds (approximately 90 m) produced by the National Imagery and Mapping Agency (formerly the Defense Mapping Agency). These data were validated in parts of Spain and found to be of excellent quality.

An inventory of all available topography data sets was compiled for the Centre of Earth Observation (CEO) of the Joint Research Centre of the European Commission. This inventory contains a wide range of different datasets for all parts of the world, at varying resolutions and for varying sizes and kinds of areas. It gives the name and address of the data holder and, when applicable, the cost to acquire any given data set. Information concerning the availability of elevation models and other topographic data sets can be obtained from the internet under the address http://www.eurogeographics.org/eng/04_gdd.asp.

A regional digital terrain model for Tuscany and for the two test areas Radicondoli and Campiglia respectively is used in CeSIA. A low resolution model (pixels of 250m) is taken from the archive of Nuova Telespazio S.p.A. and a high resolution model (pixels of 20m) has been purchased from Italian Geographic Military Institute. For the test site near Piano di Rosia a DEM had been used, which was created by digitizing maps. A complete digital elevation model (250m x 250m-grid) of the Basilicata Region is in use at Nuova Telespazio in connection with field work in southern Italy in 1994.

Geographic information like coastlines, rivers and location of other landmarks is essential for any geographic registration and is the best way to register time series, or data from different satellite systems. Automatic or semi-automatic satellite navigation is based on land-sea masks, that can be generated from coastline data sets. The World Data Bank (WDB) II data set provides accurate coastlines, rivers and international boundaries.

At the Free University in Berlin such a data set is used for navigation of original (raw) data as well as for controlling the results of processed data.

Further static data are land use or land cover data sets, which sometimes are derived from national inventories (e.g. forest inventories) or based on remote sensing data. So it has to be proved which data set could be of highest value for the interpretation of satellite derived quantities. Some use of digitized land use and forest maps has been undertaken at CeSIA in Florence for the region of Tuscany. The Staring Center in Wageningen, The Netherlands, used land cover and the CORINE database. The latter is developed with support from EU and is being completed for most parts of Europe.

2.7.3 State of the Atmosphere

Operational Data Sources

Atmospheric parameters can be obtained from distinctly different types of data sources, either directly from observations (unevenly distributed in space) or from gridded data sets (produced by assimilation of observational and numerical model data into a regular grid) or from climatologies.

Operational *in-situ* networks of National Weather Services include near-surface and upper air data. Near-surface data are routinely measured at synoptic and climatological stations distributed over all countries. At the principal synoptic stations a complete set of parameters is measured at the four main hours (0, 6, 12, 18 UTC) and at the intermediate hours (3, 9, 15, 21 UTC): wind direction and speed, cloud type and cover, cloud base height, visibility, air temperature, moisture, pressure, hours of sunshine. Observations at secondary synoptic stations are less complete, mainly with respect to the number of observations per day. The synoptic stations form part of the global net of the World Meteorological Organization. The net of climatological stations is much denser (e.g., about 5000 in Spain), but the observation times are different (0, 7, 13, 18 UTC) and at most of them only temperature and precipitation are measured.

Upper air data, i.e. vertical profiles of pressure, wind speed and direction, temperature and humidity, are measured routinely (in general twice daily, at 0 and 12 UTC) by means of radiosoundings. This network is rather sparse and unevenly distributed. There are, e.g., a total of eight radiosonde stations on the Iberian peninsula. The radiosonde network is much less dense in the Southern Mediterranean countries.

The term “gridded data” refers to data sets which provide values of each variable at fixed, regularly spaced locations (horizontally and vertically, if applicable) and at fixed regular time intervals. These data are produced either through direct interpolation of observational data on a two-dimensional, horizontal grid or through a process called four-dimensional data assimilation (4DDA). Methods for 4DDA all involve numerical modeling. In a first step, a numerical forecast model is used to propagate information concerning the state of the global atmosphere from data rich areas to data sparse areas. Output from this forecast is then used, together with routine observations and forcing fields, as input to the analysis, the results of which are again used as input for the next forecast.

Three different types of gridded data sets are of interest: Operational analysis and forecast model output, reanalysis model output, and special observational gridded data sets.

Operational weather forecast and analysis models in different configurations are used by several National Weather Services and the European Centre for Medium-Range Weather Forecast (ECMWF). Generally these are global at a horizontal resolution of 1° - 2°, some of them in high-resolution limited-area version (e.g. the High-Resolution Limited Area Model HIRLAM used by the Spanish Weather Service at a resolution of 0.3°- 0.5°). They provide continuously data in (near) real-time at the standard forecast intervals (0, 6, 12, 18 UTC). The numerical models used here have evolved with time, with new parameterizations and/or 4DDA schemes being incorporated whenever available. Therefore, these data are not suitable for long-term change analysis.

For the purpose of providing more consistent data sets re-analysis projects have been conducted in various international weather forecast and climate centers, basically by repeating the analysis for a number of years and using the same model for all years. Most widely used are the U.S. National Climate and Environmental Prediction (NCEP) reanalysis data (freely available, but at a spatial resolution of 2°) and the ECMWF Reanalysis Project (ERA, Gibson et al. 1997) that is of restricted access, at a spatial resolution of 1°. They provide consistent time series of 4DDA and forecast data for 15 (ERA) to 40 (NCEP) years, which are adequate for change analysis. ECMWF is preparing another reanalysis project ERA-40 which will cover 40 years.

For some variables, global gridded data sets were also derived from a combination of observational data with simple two-dimensional assimilation schemes:

- The NASA Water Vapor Project (NVAP, Randel et al. 1996) was produced by combining three independent data sources. These are, the TOVS and ATOVS onboard NOAA satellites (Wittmeyer and Vonder Haar 1994), the Special Sensor Microwave Imager (SSM/I, Greenwald et al. 1993), and radiosonde observations (Ross and Elliott 1996 and 2001). The basic data set includes daily total column water vapor and values for three atmospheric layers at a spatial resolution of 1° x 1° for 8 years.
- The Global Precipitation Climatology Project (GPCP) provides precipitation data at a spatial resolution of 2° x 2° for 5 years.
- The Global Aerosol Climatology Project (GACP) was established in 1998 as part of the NASA Radiation Sciences Program and the Global Energy and Water Cycle Experiment (GEWEX). Its main objectives are (1) to analyse satellite radiance measurements and field observations in order to infer the global distribution of aerosols, their properties, and their seasonal and interannual variations, and (2) perform advanced modelling studies of the aerosol formation, processing, and transport. The resulting data sets and analysis products are used to improve the understanding and modeling of the climate forcing due to

changing aerosols, including both the direct radiative forcing by the aerosols and the indirect radiative forcing caused by effects of changing aerosols on cloud properties.

Given the continuous availability at regular space and time intervals, these gridded data can directly be used for operational algorithm input. There is, however, great caution required because they contain the inaccuracies of the models. Thus, prior to using them in a specific area, they need to be validated for that area. Jochum et al. (1998) have validated ERA and NVAP data for the Iberian peninsula. The main result is that there are significant errors in the moisture fields generated by the ERA 4DDA scheme. This is mostly attributed to the complex topography and the vicinity of the sea. These errors obviously propagate, e.g., into the radiometrically corrected surface variables.

Several global climatologies of atmospheric parameters are available (e.g. Willmott et al. 1981, Legates 1987, Legates and Wilmott 1990), most of them only include temperature and precipitation. Recently a new climatology (Hulme et al. 1995a, b, c) has been developed for a larger European region which includes most of the Mediterranean. It has higher spatial resolution than previous ones (0.5°) and includes the following variables: mean, maximum and minimum temperatures, humidity, precipitation, wind speed, cloud cover, and sunshine hours.

Other meteorological data sets of local or regional dimensions are kept by research institutions and regional forecast services (see e.g. Enne et al. 2001). As an example, 30 years of digitized meteorological data like maximum and minimum temperature and daily rainfall are kept by La.M.M.A. (Laboratorio per la Meteorologia e la Modellistica Ambientale) for Tuscany.

Water Vapour

The importance of the accurate determination of the atmospheric water vapour content for radiometric correction is demonstrated here with an example carried out in connection with the EFEDA field experiment in La-Mancha, Spain, 1991 and 1994 by Jochum et al. (1998).

The 12th of June, 1991, was an exemplary clear day during the field experiment. In order to obtain the best possible information about the spatial distribution of the column atmospheric water vapour amount a total of 30 radiosonde stations in and around the Iberian peninsula have been used for this analysis (eight data points on the Iberian Peninsula). Various statistical techniques have been applied to derive the variability of the total water vapour amount over the peninsula. Fig. 2.14 presents the results obtained with two of these methods, the minimum curvature on top and the cubic polynomial regression at the bottom. The second one captures quite well the features seen in a high resolution model analysis and seems to be the most adequate interpolation method for this kind of data set (de Dios del Pino Corredera 1997).

Another important consideration is the temporal variability. The results shown in Fig. 2.14 reflect the situation at noon. Operational radiosonde ascents are only available at 12:00 UTC and 00:00 UTC. During some days of the EFEDA experiment a series of radiosonde ascents have been carried out at three stations

(Rada de Haro, Tomelloso, and Barrax) that allowed to analyse the variability of the vertical water vapour profile. The total precipitable column water vapour amount over the area of Tomelloso varied between 14 l m^{-2} in the morning and 17 l m^{-2} in the evening.

Additional information about the spatial variability of the water vapour in the atmospheric boundary layer was obtained from the DIAL instrument flown at the DLR research aircraft Falcon. Figs. 2.15 show results of two flight sections over irrigated and mostly dry terrain respectively. The flight on 28 June (11:55) was about two hours later than the flight on 29 June (10:05) and consequently the convection was more developed.

From climatological data it can be seen that the horizontal as well as the temporal variability is large in this bottom layer of the atmosphere. Using one mean profile for the whole Iberian Peninsula resulted in 1.3 g kg^{-1} in April and 3.4 g kg^{-1} in August 1996.

Very seldom such a data base is available for the radiometric correction of measurements made from space. In most cases model atmospheres, climatological data, or water vapour profiles inferred from vertical sounders such as TOVS, in operational meteorological satellites are used for the correction. Table 2.14 gives an overview of operational data sets that are available to perform the necessary radiative transfer calculations. The comparison of water vapor products from ERA, NVAP and operational radiosondes on climatological time scales shows overall satisfactory results.

For the NVAP data assimilation over the whole Iberian Peninsula however only very few data points were used (7 radiosoundings plus 15 TOVS retrievals). This requires specific attention with respect to topography. Yet the NVAP data set is better capable of better reproducing the horizontal variability of PWC (precipitable water content) across the peninsula. Due to the different nature of the assimilation process (4D in ERA, as opposite to 2D in NVAP) some of the variability gets lost in the ERA PWC fields. Sensitivity tests indicate that the adequate incorporation of high-resolution topography in mountainous terrain can be more important than high resolution mapping of PWC.

The results of the analysis of radiosonde as well as model data for two years are shown in Fig. 2.16. The year 1994 was the one with the lowest rainfall in the period 1991 to 1997 and the year 1996 the one with the highest rainfall during the same period. It is interesting to note that the difference in atmospheric specific humidity between the two years characterized as "dry" and "wet" are very small. In the lower part of the figure the standard profiles for the midlatitude summer and winter atmospheres are plotted as dotted lines. These standard profiles are commonly used as default values in radiative transfer codes (Tanré et al. 1992, Vermote et al 1997, McClatchey et al. 1971, Kneizys et al. 1989). It is noted that the standard profiles never come close to the monthly averages except for the month of August and the summer standard, with a low level difference of 2.5 g kg^{-1}.

Concluding from the combined analysis of radiosonde and meteorological model data, the instantaneous total column atmospheric water vapor over the Iberian Peninsula can range from 10 to 30 l m^{-2}, though in most cases the values remain below 25 l m^{-2}. Much of this variability is due to the complicated topography of the

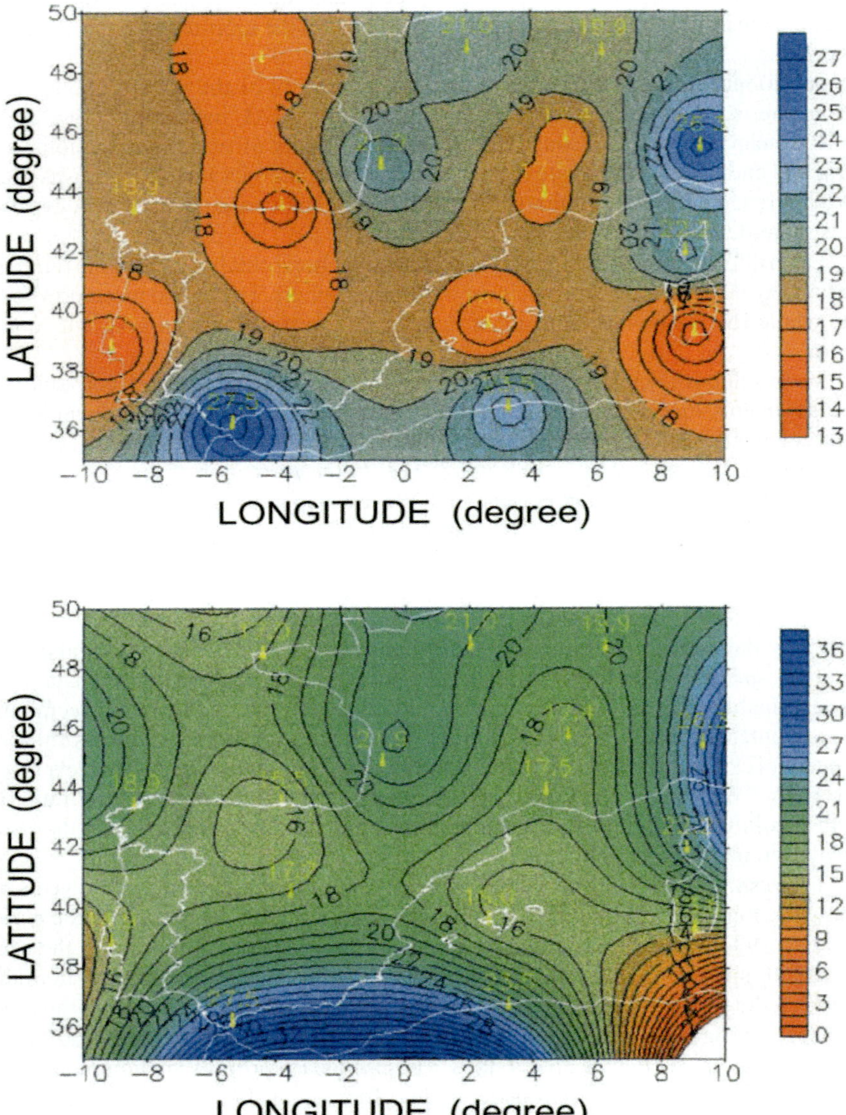

Fig. 2. 14. Spatial variability of the total column of atmospheric water vapour in litre per square meter derived from the analysis of all available radiosonde data for 12 June 1991, 12:00 UTC. The top map was obtained using the minimum curvature technique for interpolation. In the bottom map the third order polynomial regression method was applied, which captures quite well the features seen in the high resolution model analysis. The yellow numbers show the amount of water vapour derived from the local radiosondes

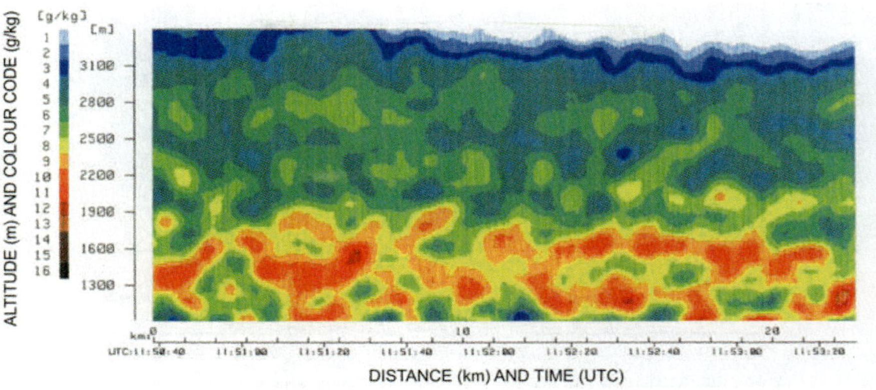

Fig. 2.15a. Water vapour mixing ratio cross section over a distance of 23 km near Barrax, Castilla - La Mancha, Spain, 28 June, 1991, measured with the DLR DIAL instrument. The water vapour distribution is very irregular resulting from the inhomogeneity of the surface which is partly irrigated

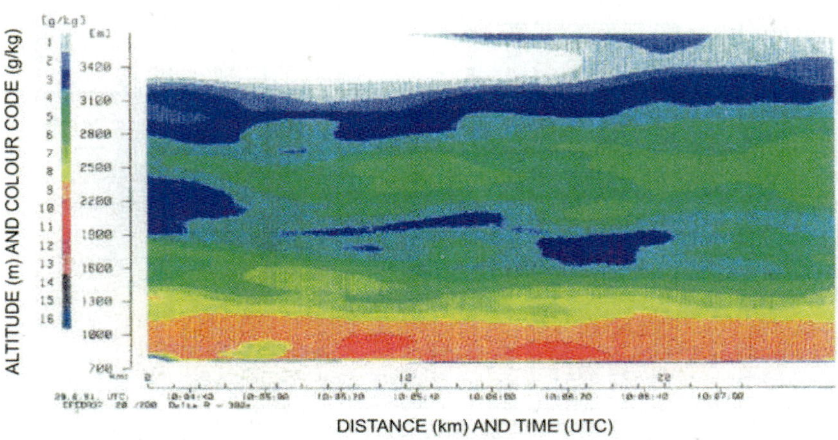

Fig. 2.15b. Water vapour mixing ratio cross section over the north-western part of the EFEDA experimental area over mostly dry terrain and matorral on 29 June 1991. The vertical scale ranges from 700 m NN (ground level) to 3000 m. The ground signal is not represented in the figure due to the DIAL evaluation method

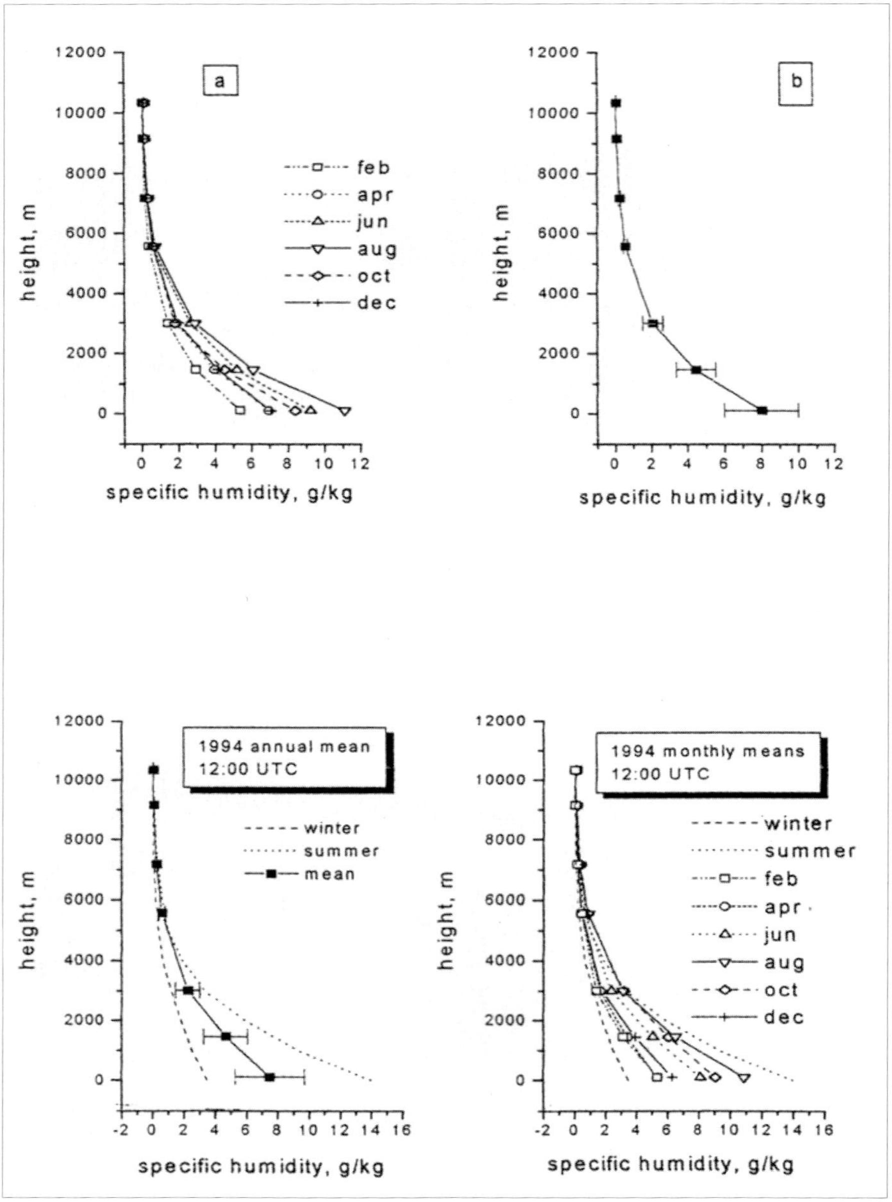

Fig. 2.16. Vertical water vapour profiles derived from radiosone ascents and averaged over the Iberian Peninsula for the years 1996 (top) and 1994 (bottom). The upper figures show selected monthly (left) and mean annual profiles for 1996, the lower figures seasonal and annual means (left) and seasonal as well as selected monthly means (right) for 1994. The dotted (dashed) lines represent the standard profiles for a midlatitude summer (winter) atmosphere as used e.g. in the 6S radiative transfer code

Table 2.14 Data sources for atmospheric water vapor and their main characteristics. NCEP: National Centers for Environmental Prediction; NCAR: National Center for Atmospheric Research; ISCCP: International Cloud Climatology Project, ECMWF: European Centre for Medium Range Weather Forecast; INM: Instituto National de Meteorología; HIRLAM: High Resolution Limited Area Model. P = profiles; PWC = precipitable water content; LW = layered water content (number of layers). For the Iberian peninsula 1° latitude (longitude) corresponds to roughly 123 km (96 km).

Data source	Grid spacing	Data points	Time period	Products
radiosonde	300-700 km irregular	grid spacing	data points	P, PWC
NCEP reanalysis	2°	28	1958(73)-present	P, PWC
ISCCP TOVS	2°	18	1978-present	PWC
ECMWF re-analysis	1.12°x1.125°	112	1981-1993	P, PWC
NVAP	1°	112	1988-1995	PWC, LW(3)
INM/HIRLAM	0.5° (0.3°)	448 (2800)	1995-present	P, PWC

Iberian Peninsula, but the particular meteorological conditions also play a major role over the synoptic scale. The effect of a variability is estimated to range between 10 and 20 l m^{-2} in atmospheric water vapour on the relevant AVHRR channels and the parameters derived from them.

In the AVHRR channel 2, the one-way atmospheric transmittance changes from 0.88 - 0.85 when the column atmospheric water vapour increases from 10 to 20 l m^{-2}. For a surface with a reflectance of 0.3 the error can amount to 6% in the retrieved surface reflectance.

In the thermal range, under vertical viewing conditions, an increase in total column water vapour from 1 to 2 g cm^{-2} produces a reduction in atmospheric transmittance of 12-14% (e.g. from 0.89 to 0.78) for channel AVHRR 4, and of 21-27% (e.g. from 0.84 to 0.66) for channel AVHRR 5. These changes are significant since they can introduce errors of several degrees in the retrieved temperature, depending on the air temperature and on the surface emissivity. The effect is increased for observations far from nadir, due to the longer atmospheric path between the surface and the sensor.

The conclusions of the study can be summarized as follows:

- The full information about the atmospheric water vapour spatial variability is achieved only by the integration of numerical meteorological model data with all available radiosonde observations using an improved statistical interpolation which also includes a detailed topographic model of the surface.

- Applying a standard (summer or winter midlatitude) atmosphere may lead to an error of up to 200% in the vertical specific humidity profile at any height level (although the absolute values of the error decrease significantly with height). In terms of the column water vapour, the actual winter mean is about 200% higher than the value obtained from the standard midlatitude winter profile. The situation is less dramatic in the other seasons, but the differences are still up to 40%. As a first step, using the seasonal or annual mean profiles instead of the standard profiles can already improve the results significantly.
- Using one radiosounding, ‘somewhere in the area’ (because this is the only one available), may introduce an error of a factor of 3 (winter) or of less than 50% (summer) in the vertical specific humidity profile. Using a radiosounding from a slightly different day in the same area may lead to an error in the same order of magnitude. Using several radiosoundings or model profiles across the area is more important than using data from the very same day and does not make any difference in winter. The observed instantaneous spatial variability in water vapour may generate artificial spatial differences across a single image which can be more significant than temporal effects in a series of multiple images.
- Using higher resolution model data improves the accuracy, but the complete set of radiosoundings available on the Iberian peninsula already captures about 70% of the moisture variability.

Ozone

The main sources of ozone information are local measurements (total column amount or detailed ozon profiles) available only at few locations and measurements from satellites (GOME on ERS-2, TOMS on Nimbus and SCIAMACHY as well as MIPAS on Envisat).

Aerosol

To compute the effect of aerosols on the remotely sensed signal, information is required on the type of particles, their optical characteristics (size distribution, refractive index, single scattering albedo) and their concentration. These properties may vary strongly with geographical location depending on the sources of the aerosols (e.g. sea spray, desert dust, rural or urban). Some information about aerosol concentrations is contained in visibility reports from airports and meteorological stations but this information is not directly usable and has to be converted. Sunphotometer measurements of direct, global and diffuse solar radiation could overcome the problem, but the instruments are still expensive, especially the hyperspectral ones, and are not available in satisfactory areal coverage. In this respect the AERONET (Aerosol Robotic Network) project of the Goddard Space Flight Center, NASA, is a breakthrough (http://aeronet.gsfc.nasa.gov). The network of sunphotometer stations is being built up successively since 1993 and maintains

Table 2.15. AERONET stations relevant for research in the Mediterraneran area in 2003 between 14° N - 47° N and 16° W - 34° E

Site	Latitude	Long.	Site	Latitude	Long.
Ahi de Cara	37° N	3° W	Ispra	46° N	9° E
Armilla	37° N	3° W	Kishiner	47° N	29° E
Avignon	44° N	4° E	Lampedusa	36° N	13° E
Azores	38° N	28° W	Lecce	40° N	18° E
Blida	37° N	3° E	Modena	44° N	10° E
Bordeux	45° N	0° W	Moldova	47° N	28° E
Bucarest	44° N	27° E	Mount Etna	38° N	15° E
Capo Verde	16° N	22° W	Nes-Ziona	31° N	34° E
Carpentras	44° N	5° E	Palencia	42° N	5° E
Dahkla	24° N	16° W	Pic du Midi	43° N	0° E
Dakar	14° N	17° W	Rome Tor Vergata	42° N	13° E
El Arenosillo	37° N	7° W	Sede Boker	31° N	35° E
Evora	39° N	8° W	Tarbes	43° N	0° E
FORTH-Crete	35°N	25° E	Toulon	43° N	6° E
IMC Oristano	40° N	8° E	Toulouse	44° N	1° E
IMS-METU-Erdemli	36° N	34° E	Venice	45° N	12° E
ISDGM CNR	45° N	12° E	Villefranche	44° N	7° E

presently world wide about 175 instruments deployed at 125 permanent sites. It uses the standard sunphotometer Cimel CE 318-1 and the polarized sunphotometer Cimel CE 318 which are re-calibrated every 2 - 3 months using the Langley plot technique at the Mauna Loa Observatory. The spectral bands used for the measurement of the atmospheric optical thickness (AOT) are 340 nm, 380 nm, 440 nm, 500 nm, 670 nm, 870 nm, and 1020 nm that are. An additional 1600 nm channel will be added. The total amount of atmospheric water vapour is inferred from the depth of an additional water vapour band relative to the signal in the surrounding window channels.

All data are transmitted hourly from the field sites by e-mail or a Vitel VX 1004 transmitter via GOES or Meteosat to the Wallops Receiving Station and from there to the Goddard facility where the aerosol optical thickness, the aerosol size distribution and the total amount of atmospheric water vapour is inferred and made available as hourly, daily and monthly data to users on the above cited website. The stations relevant for Mediterranean research are listed in Table 2.15.

In Europe, 19 lidar groups formed a network of 21 stations, the EARLINET, to establish a quantitative data base of aerosol extinction profiles and aerosol type. Eight of the stations are located in the Mediterranean area.

During the EFEDA experiment in Spain, 1991, the DIAL instrument on board of the DLR-Falcon not only determined the water vapour mixing ratio but in addition provided information about the aerosol distribution. Fig. 2.17 shows examples that demonstrate the distribution of aerosol in the atmospheric boundary layer. The first impression is the high reaching atmospheric boundary layer, sometimes up to about

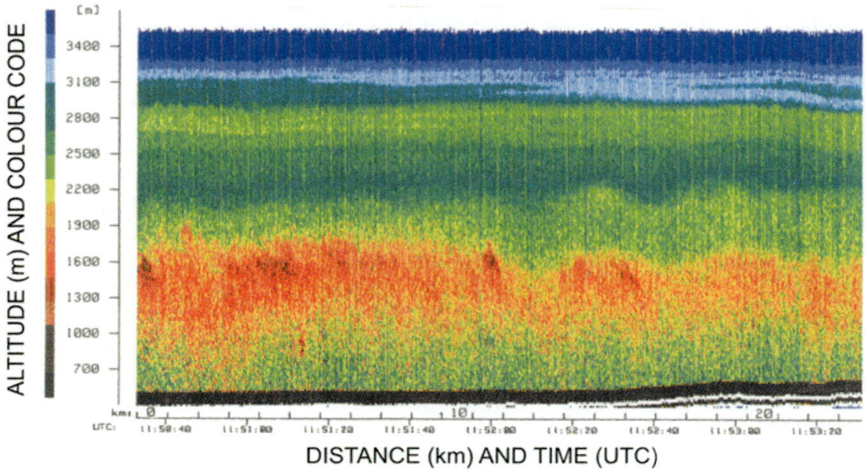

Fig. 2.17a. Aerosol backscatter cross section as a vertical cut through the lower troposphere near Barrax, Castilla-La Mancha, Spain, on 28 June 1991 at the same time as Fig. 2.15a (11:52 UTC). The flight section in this case extends over 23 km. The vertical extent in this and the following Figs. 2.17b-e range from the flight altitude around 4 km down to about 700 m NN, the ground level. The ground signal is the black line at the bottom of the picture. The colour coded intensity scale at the left is linear and relative. It ranges from blue for weak backscatter over green, yellow, and red to black for strong backscatter (as in the case of clouds)

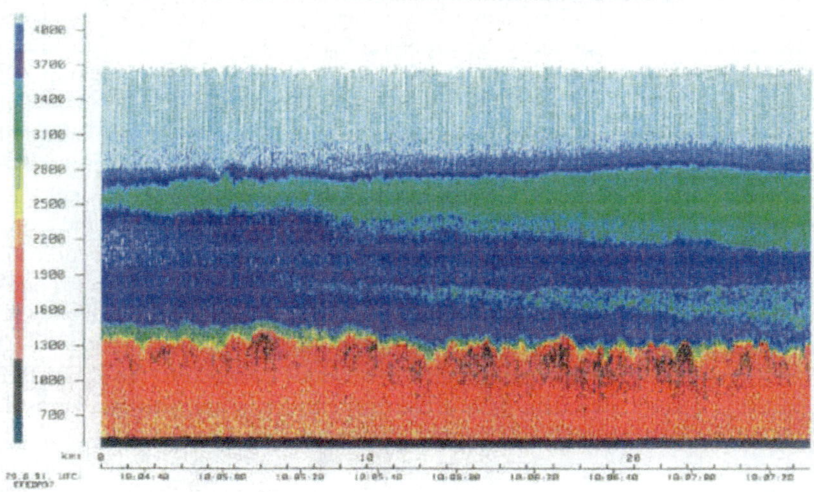

Fig. 2.17b. Aerosol backscatter distribution as a vertical cut through the lower troposphere on 29 June 1991 around 10:05 UTC in the same area as Fig. 2.15b

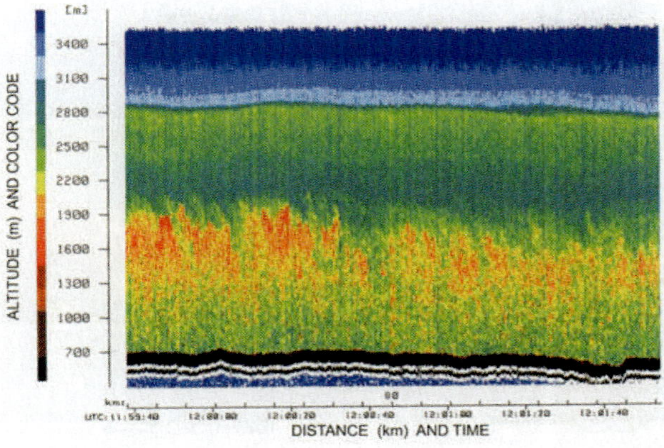

Fig. 2.17c. As Fig. 2.17a but near Tomelloso on 28 June, 1991

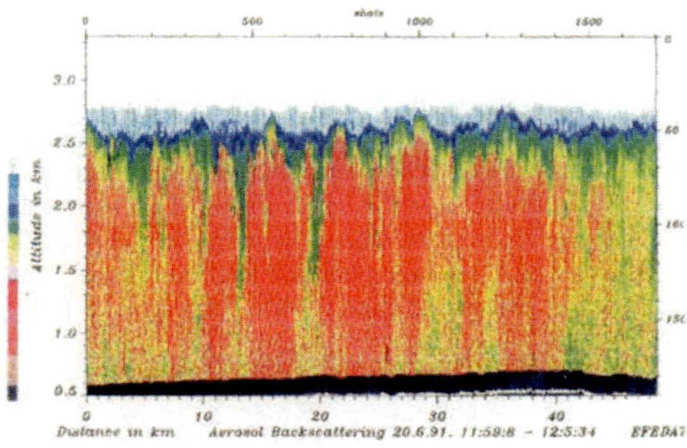

Fig. 2.17d. As Fig. 2-17a but on 20.06.91

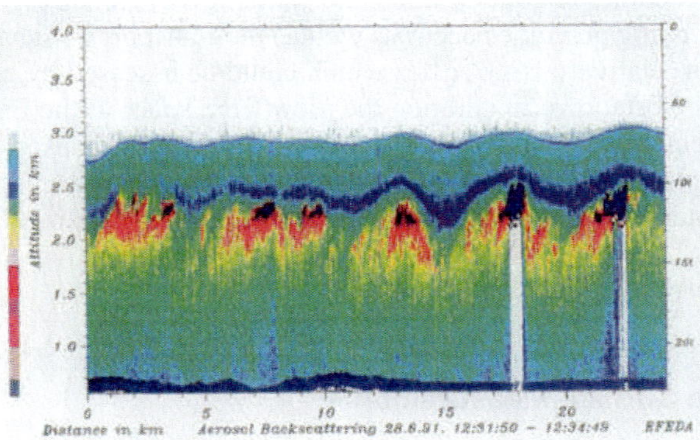

Fig. 2.17e. As Fig. 2.17a but on 28 June 1991, in the northwest of the EFEDA experimental site

Fig. 2.18. Plumes of stubble field fires in the Cornia valley, Italy, which could be traced for more than eight kilometres from the source. The upper image is from 15 September 2004. Its source area is shown enlarged at the bottom to the left. The lower image is from 20 September 2004. In the background of the upper image are dark clouds. In the original of the central picture one can recognize a second layer of pollution at the height of the tops of the chimneys which are almost invisible in the upper image. This second layer results from the industrial area of Piombino (at the right, where the otherwise clear sky becomes darker) and the power plant. The filters of this power plant are at certain times "cleaned" causing black clouds as can be seen in the lower right hand picture from August 2005

3000 m. This is due to the strong heating at the surface which favours convection and the building of a thermal low pressure system which entrains surrounding air near the surface. The pictures also show how convection develops in parcels within some of them it comes to condensation and formation of clouds (Fig. 2.17e).

These dust rising processes occur as well as the transport of Sahara dust into the region occur at relatively large scales which could be assessed by an appropriately dense network of stations. In addition the view from space at the Earth surface can locally be affected in an in-assessable way by air pollution raised from industrial areas, power plants and burning of agricultural debris after harvest which occurs mainly in September (Fig. 2.18). During morning hours, underneath the inversion layer, the fume of even relatively small fires can easily extend over many kilometres and then dissolve into wide spread turbidity.

Comparative Surface Measurements

In conjunction with remote sensing studies many measurements at or from the surface have been conducted to validate information inferred from the measurements

Table 2.16. EFEDA types of remote sensing and ancillary data. The numbers of participating groups refer to the EFEDA pilot study 1991 in Castilla-La Mancha, Spain, and follow up experiments in Spain, Italy, and Crete

Type of measurement	Data	Involved groups
Airborne *in situ* measurements	Temperature, humidity, sensible and latent heat fluxes	3 (7 aircraft involved including remote sensing studies)
Aerological ascents	Pressure, temperature, dewpoint, wind speed and direction	3
Digital terrain models	Altitude and inclination for experimental areas	1
Hydrology	Groundwater level, rainfall, run-off, soil water content, hydraulic conductivity, geo-hydrological structure (aquifers)	5
Modelling	Soil Vegetation Atmosphere Transfer models, mesoscale models, vegetation growth models	5
Radiative properties of the surface	Spectral surface reflectance, surface albedo, surface emissivity, surface temperature	5
Remote sensing from aircraft	Spectral radiation fluxes, water vapour and aerosol distribution; multispectral microwave radiometer	3
Remote sensing from satellites	Visible, infrared and microwave radiances from which surface radiative state variables, clouds, vegetation indices and fluxes were derived	6
Soils	Soil type maps around experimental sites, physical and chemical soil characteristics, soil moisture	5
Meteorological measurements at and near the surface	Temperature, humidity, pressure, wind, sensible, latent and soil heat fluxes, solar irradiance, net radiation flux, infrared radiation fluxes, atmospheric optical depth	10
Vegetation	Land cover and land use, leaf area index, sap water fluxes, vegetation structure (roughness), stomatal resistance	5

made from space or from aircraft. The so far most comprehensive investigation of this type was conducted within the framework of the ECHIVAL (European International Climatic Hydrological Interactions of Vegetation, Atmosphere and Land-surfaces) Field Experiment in Desertification-threatened Areas (EFEDA) that took place in Castilla-La Mancha, Spain, 1991, and which was followed by smaller projects of the European Commission in Spain, Italy and Greece (Crete). Most of the achievements reported here resulted from these activities. The goal of the remote sensing component of EFEDA was to contribute to the understanding of processes by applying physical and statistical methods beyond just monitoring. The large amount of ancillary data has been used to achieve this goal. The EFEDA project is briefly described in Bolle et al. 1993, Bolle 1998, Bolle (ed.) 2003, and Appendix 1.

The data collected during these activities have partly been made available on a CD-ROM. Table 2.16 provides an overview of the type of data gathered during these projects. Where necessary, it will be referred to them in the following chapters. More information especially about supplemental data sets can be found in Appendix 1.

2.8 Satellite Data Archiving

2.8.1 Development of Physical Archives Media

Recovery and archiving of remote sensing data for global change research and monitoring is a complex and expensive task. Only two aspects, as experienced at the Free University of Berlin, will be discussed here in some detail: the physical constraints of long-term archiving and the accessibility of the archived data.

In the sixties, data were recorded on Computer Compatible Tapes (CCT) at a density of 200 bpi and later on at 6250 bpi density. An industry standard (ANSI) allowed to read and interchange tapes throughout the full spectrum of computer manufacturers. Large CCT archives still exist and require readout devices which are available only at larger computer centres. During the late seventies digital cartridge tapes became available, which were mostly used for software distribution and backup purposes rather than for data archiving. In the late eighties the first high capacity cartridge tapes using conventional VHS cartridges became available with capacities of 2.5 Gbyte as a byproduct of the broad use of video recorders. This development is still continuing because of requirements in multi-media applications. Only technical storage products that can be used in home electronics as well as data processing can be provided with high performance at low costs. The VHS technique was only a short intermezzo on the storage market and was replaced by recorders using 8 mm video tapes or digital audio tapes (DAT).

More expensive devices with considerably higher storage security were introduced in recent years in response to the needs of huge data archives, such as in the field of remote sensing. Digital Linear Tapes (DLT) are currently the most often used storage medium in the high end sector and allow to store up to 80 Gbyte on one cartridge. With LTO (Linear Tape Open), AIT (Advanced Intelligent Tapes), and SDLT (Super DLT) data capacity is increased up to 200 Gbyte per cartridge. Robot systems with multiple tape drives and a great number of tape slots make data archives of hundreds of Tbytes on-line accessible.

Optical storage media became available with the development of the CD-ROM with the WORM (write once read many) technique. Discs with various diameters enabled capacities up to several Gbytes but without industry standards for data exchange purposes. A standardized exchange medium with broad use on all computer platforms is given only by the ISO 9660 norm for the common 12 cm CD-ROM with 650 to 800 Mbyte storage capacity per disc.

Magneto-optical discs using both the erasability of magnetic storage together with the data security of optical recording became available in the mid eighties, but only with limited usage. Therefore besides the more complicated construction of the required media for such devices in comparison to CD-ROM, the media prices are still high and not suited for large archives. This technique is now used in audio electronics as Mini-Disc devices.

High capacity CD-ROM, the digital versatile disc (DVD), has recently been developed for multi media application and designed for capacities up to 20 Gbyte per disc (double sided, two laser frequencies) having the same dimensions as the CD-ROM. Missing industry standards still hampered the use for digital data storage for years. DVD-recorders with 4.7 Gbyte capacity (single sided, one laser) are now on the market.

The increasing capacity of the national and international data networks is used to overcome the data storage incompatibility. The file transport protocol (FTP) available on each computer platform allows not only to access data on distributed local storage devices but also overcomes most of the problems with proprietary data and file formats of the different computer manufactures.

Following the technical development not only the data processing but also the data storage strategies changed in the course of time. High initial costs forced the build-up of computing and storage centres. The development of large scale integrated circuits enabled the design of low cost workstations and especially personal computers with high numbers of sold units and lead to decentralized computing and storage. But there is a new need for centralized data processing and storage. In contrast to the former super computers new concepts based on variable arrays of low cost computers fit the specific requirements of such centres. Sophisticated software development allows massive parallel computing previously only capable for super computers.

Random access to data with acceptable access speed is a prerequisite for complex evaluation task especially for near real-time applications. The required access speed can be obtained only if the data reside on large magnetic disk units, which even in the near past were too expensive to hold large data sets. But nowadays high capacity Redundant Arrays of Inexpensive Disks (RAID) with capacities of several Tbytes are available at moderate prices. With such large RAID systems it is now possible to provide even long-term sets of high resolution remote sensing data not only for complex time series analysis but also for concurrent access for users via high speed data links. Such a disk array with an actual capacity of about 8 Tbyte was installed recently in Berlin and is capable to store the full 16 years of daily AVHRR data including derived products.

2.8.2 Examples of Data Archives

Raw Data Archive At the Free University of Berlin

Since August 1988 the use of a digital VHS tape recorder enabled to store the full HRPT data stream (Schwalb 1978) of four daytime and 3 nighttime passes of the "afternoon" satellites NOAA-9 and 11. From the "morning" satellites at first only central passes and since October 1994 all passes, which were not in conflict with the data stream from other satellites, were archived as well. VHS tapes were used until October 1991 when Exabyte 8500 became available. The VHS data were rewritten to video-8 tapes but data from June 1990 to March 1991 were not recoverable. The raw data archives until now amount to about 30 Tbyte stored on about 5000 video-8 cartridges. A catalogue of the raw data archive is under construction and will be provided on the web site http://www.met.fu-berlin.de. All raw data archives are copied to DLT (Digital Linear Tapes) with higher data storage reliability.

Preprocessed AVHRR Data Remapped to Polar Stereographic Map Projection

Daytime and nighttime passes of NOAA-11/-14/-16 of the whole reception area covered by the station in Berlin are archived since March 1994 in full resolution and geographically remapped to polar stereographic map projection with a resolution of 1.2 km in 60°N, which keeps the full resolution of the AVHRR signal. The remapped data are stored in an internal *container* format with 6400 $*$ 6400 raster positions. This internal *container* format includes unchanged original counts from the AVHRR data stream, collateral data such as orbital parameters as well as information about the illumination and observation geometry. Calibration data are provided in a trailer section. The container files are the basis for further processing steps.

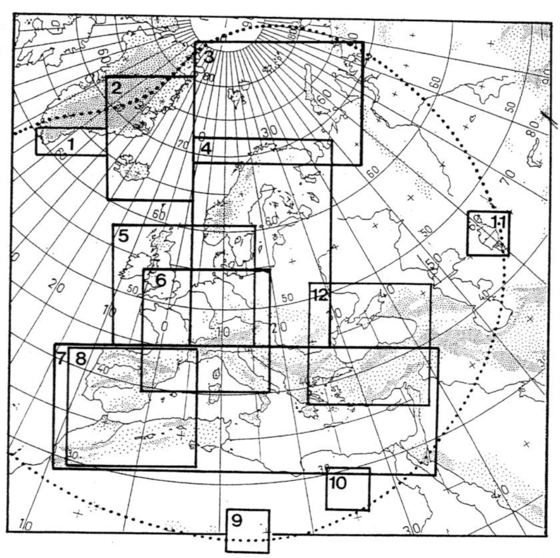

Fig. 2.19. Area for which AVHRR data are processed at the receiving station of the Free University of Berlin (full frame) and windows for specialized archives. The dotted circle represents the limit of the reception area

Each of the individual remapped passes as well as mosaics of up to four adjacent passes are archived separately. Thus the overlapping regions of subsequent passes gathered under different viewing angle can be used for studies on viewing geometry. Besides full swaths, smaller areas of interest as indicated in Fig. 2.19 are extracted and archived separately as a basis for different research projects.

The sub-areas covering the whole Mediterranean and central Europe each with 1536∗1536 raster positions are stored separately containing only data of the most appropriate pass. These areas were subjected to interactively supervised fine-navigation and are available for daytime data since January 1989.

2.8.3 The Mediterranean Extended Daily One-km AVHRR Data Set (MEDOKADS) of the Free University of Berlin

The fine-navigated subareas are used to create the MEDOKADS product, the "Mediterranean Extended Daily One-km AVHRR Data Set". This data set consists of full resolution AVHRR channel data and collateral data in „geographic" projection, i.e. latitude - longitude presentation, with a resolution of 0.01^0 in both directions.

The area of interest for the studies in the Mediterranean is defined by its corner coordinates

55^0N/10^0W	55^0N/42^0E
27^0N/10^0W	27^0N/42^0E

and amounts to 2800 lines by 5200 columns with 14,560,000 raster positions. The geographic extent is shown in Fig. 2.20.

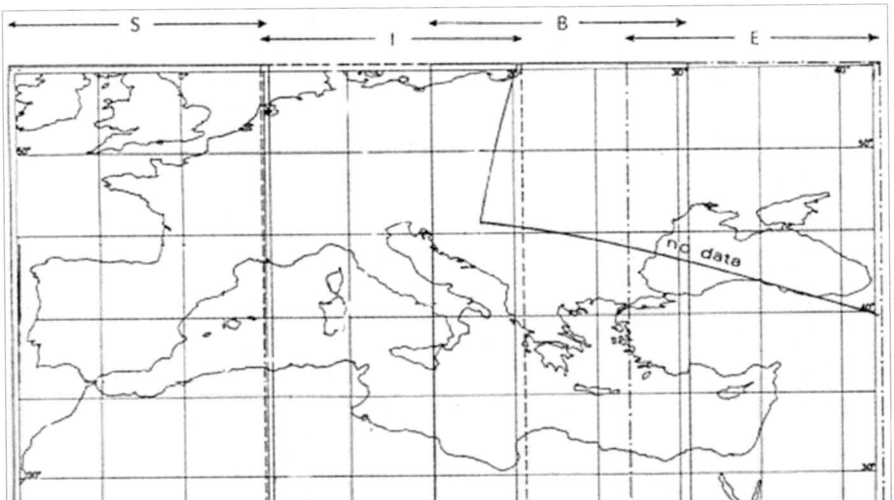

Fig. 2.20. Area covered by the MEDOKADS. **S, I, B** and **E** show the four subsets stripes for the Iberian Peninsula, Italy, Balkan, and Eastern Mediterranean. No data are processed so far for the northeast corner covering eastern Europe and the northern Black Sea

Table 2.17. Channels and pseudo-channels of the MEDOKADS data bank

No.	Content	No.	Content
1	calibrated AVHRR-CH1	9	local sun azimuth
2	calibrated AVHRR-CH2	10	local scattering angle sun to satellite
3	calibrated AVHRR-CH3	11	"origin indicator" and bitmap
4	calibrated AVHRR-CH4	12	TOA broad-band reflectance
5	calibrated AVHRR-CH5	13	NDVI
6	local satellite zenith distance	14	sea/land surface temperature (split window)
7	local sun zenith distance	15	local time since ascending node
8	local satellite azimuth	16	cloud classification

Each of the four sub-regions marked in Fig. 2.19 forms a special storage unit covering the Iberian Peninsula (S), Italy (I), the Balkans (B), and the Eastern Mediterranean area (E) respectively.

All stripes have the same size of 2800 lines and 1536 columns. Smaller windows can be selected and are completely defined by the geographic coordinates of the upper left corner and the number of lines and columns. These data sets are easy to merge if larger areas or the whole Mediterranean have to be considered. Areas for which data are not yet available in supervised form are kept blank. The coordinates of the upper left corners of the sub-areas are:

S: Iberian Peninsula	55^0 N/10^0 W
I: Italy	55^0 N/ 5^0 E
B: Balkan Peninsula	55^0 N/15^0 E
E: East Mediterranean	55^0 N/27^0 E

There is some overlapping between the stripes to cover the four main land surface sub-areas. To avoid problems of digital data representation, all values are converted to signed 16-bit integers. The formal digital resolution is 0.01 of units (i.e. % for reflectances, degree for temperatures and angles) multiplied by a factor of 100 without any offset. The calibrated data of the five AVHRR channels at top of atmosphere (TOA) form the first five files. They are completed by a number of pseudo-channels containing collateral data or derived products coincident pixel by pixel: local satellite and sun zenith distance and azimuth, local scattering angle, broadband albedo, NDVI, sea/land surface temperature, local time, and a bitmap. The upper byte of the bitmap channel represents the *origin indicator*, a 6 bit bitmap inserted into the spare bits of each HRPT superpixel at data reception time, which allows to identify the date and the orbit number of each individual pixel, even in the case of multi-day composites like the maximum NDVI. It allows to recover the original digital counts and to apply sophisticated calibration and correction algorithms later on to each individual pixel without a loss in digital resolution. The other byte contains a land/ water bit and a cloud mask and allows to switch between different evaluation algorithms. Thus the MEDOKADS builds up a 16 channel pixel-congruent data set (Table 2.17). Daily as well as ten day composite data sets according to the maximum NDVI are generated.

The west to east extent of the area is 52^0 equivalent to about 4400 km (10°W - 42°E). Thus 2 to 3 adjacent satellite passes are necessary to cover the whole region with the result that the absolute time difference can be up to 3½ hours between the western and eastern part. The total amount of data for a complete set with 16 pseudo-channels for one day of the whole basin is about 580 Mbyte. Thus up to now the MEDOKADS of 16 years from 1989 to 2004 amounts to more than 2 Tbyte.

2.8.4 Other Satellite Data Archives

AVHRR Data

Geometrically rectified and geographically registered AVHRR data can also be obtained from other sources such as the University of Strasbourg (France), Telespazio (Italy), the German Remote Sensing Data Center of the German Aerospace Center (DLR) in Oberpfafenhofen (Germany), and the station in Dundee (UK).

At the University Louis Pasteur-Strasbourg since 1995, the images from the satellites NOAA-12, NOAA-14, and since May 1998 also NOAA-15 day and night image data have been received and archived. While in the MEDOKADS data set only daytime data are processed with fine-navigation, Strasbourg put emphasis on the navigation of nighttime data as well because for the TISI algorithm (see section 5.2.1) precisely navigated pairs of day- and nighttime images are required. The "navigate" software provided by the National Center of Atmospheric Research, Boulder, Colorado, U.S.A. is used for this task (Rosborough et al. 1994).

The areas defined for research in the Mediterranean area are similar to the MEDOKADS data set as given in Table 2.18. Geographic projection was chosen in

Table 2.18. Windows of the Mediterranean region used for AVHRR data processing in Strasbourg

Area	Italy	North Africa	Spain	Med. Basin
Image Center	41^0N, 13^0E	34^0N, 10.5^0E	40^0N, 3^0W	36^0N, 15^0E
Pixel size	0.01	0.01	0.01	0.01
No. of Pixels	$1200*1000$	$900*800$	$1400*1000$	$5000*1800$

order to simplify data exchange with others groups. The archives represent now more than 8000 overpasses and is larger than 500 Gbytes. All the images are archived on vidio8-cartridges.

In summer 1996 Nuova Telespazio produced an AVHRR/NDVI data base on Tuscany conterminous area. Channels 1 and 2 have been processed radiometrically following Rao & Chen's recommendations (NOAA Technical Report NESDIS 78 1994). Images have been re-sampled with cubic convolution procedure and referenced in latitude/longitude projection with a pixel resolution of 0.01 degree at ground. A sea mask has been applied to every image. The geometric accuracy of images does not exceed the value of 3-4 pixels. Of the original 1024 digital counts values (10 bit), 256 (8bit) have been retained starting from the value of 40 (radiometer zero level) up to the value of 295. For each channel and the NDVI, negative values have been forced to 0, while positive values have been scaled between 0 and 255. For each month the Maximum Value Composite (Holben 1986) has been applied from all the available images. The original data base is composed of 2137 daily images from July 1985 till June 1994. The average number of images for one year is 237.4.

The data archive covers the period 1985 to September 1994. While from 1985 to 1990 data were acquired mainly from the DLR in Oberpfaffenhofen (Germany) and from the Dundee station in Scotland, since 1991 data were acquired directly by the Telespazio receiving station in Scanzano (Sicily).

Archives of Meteosat Data

A comprehensive archive of Meteosat data for research in the Mediterranean area exists at the Free University of Berlin since 1989 starting with hourly and, continuing later, with half-hourly data for main parts of Europe. Since 1994 all transmitted data north of the Equator are stored as raw data, and data for Europe and the Mediterranean basin are remapped to polar stereographic map projection with a nominal resolution of 4 km in 60^0 N.

Windscatterometer Data

Information obtained in the visible and thermal infrared part of the spectrum are insufficient to describe the state of the land surfaces completely. As an example, instruments such as the AVHRR on board of the NOAA series of satellites are not designed to supply data about the variability of soil moisture which is one of the most crucial information required by climate modelers. As long as such data are not available from the new generation of land applications satellites for periods which are adequately long for climate studies, sensors on board of other satellites, which are not specifically developed to study land surfaces, have been and are still used as substitute. One of these data sets is that of the Windscatterometer on board of the two ERS satellites. The ERS-1 Wind Scatterometer (WSC) instrument was designed to measure the near-surface wind field over the ocean by inferring the wind speed and direction from measurements of the surface radar backscatter. Measurements are also made, however, over land and several studies have highlighted the relationship between the low resolution scatterometer data and surface geophysical parameters such as vegetation cover, surface roughness and surface soil moisture content. While the nominal ground resolution of the data is coarse (» 45 km) the ability to provide global coverage within 4 days, regardless of local weather conditions or solar illumination, makes the WSC instrument ideally suited for meso-scale monitoring of global land-surface characteristics. The data from this instrument are available on CD-ROM.

As an example, the Wageningen Agricultural University used these data, which are processed by the Physics and Electronics Laboratory of the Netherlands Organization for Applied Scientific Research (TNO) for multi-pass interferometry. A simple, mixed target model has been developed for extracting geophysical products out of ERS-Windscatterometer data. A sub-pixel level inversion technique is possible using scatterometer data, thanks to the high radiometric accuracy and range of measurement incidence angles. The method enables observations of geophysical relevance, having different time-constants, at the resolution of a pixel.

Towards a Distributed On-line Access Archive

In order to run algorithms developed in the framework of research projects it is desirable to provide an on-line access to main parts of the whole archive. Only in this case it will be possible to extract and analyze the full range of components that will contribute to a land-surface change indicator and its impact to regional land-use planning. Parts of the MEDOKADS have therefore been copied to large tape systems at the Potsdam Institute for Climate Impact Research (PIK) and at the Forschungszentrum Karlsruhe. The full 16 years' daily data set of the MEDOKADS is in automatic direct access at the Free University of Berlin and parts of it are available on-line at the PIK whereas at Karlsruhe a considerable amount of the Meteosat data received and processed in Berlin is made available. More than 3 TBytes of data have been sent via FTP. Tools have been developed to handle such a huge data exchange task properly.

A copy of all raw data archives of NOAA-HRPT and Meteosat, received and archived in Berlin up to now, is built up at the Forschungszentrum Karlsruhe (FZK). This allows on-line access to this data set which amounts to 30 Tbyte.

Maintaining such big archives is rather expensive. Data transfer to possible new storage technology is required almost annually and therefore it seems to be impossible to grant access to these archives without sharing the expenses. As a result of this exchange experiment, a distributed data storage concept seems to be feasible if a number of institutions give storage space on disk or tape servers at reduced or no cost, within the framework of scientific collaboration. To overcome this main limitation for extensive exploitation of remote sensing data, a solution of these problems is required. Especially the use for global change research is still mainly hampered by data access and refunding problems, which are generally negligible compared to development and operating costs of satellites and sensors.

Chapter 3 Radiative Processes of the Surface - Atmosphere System

3.1 Introduction

In this chapter the theoretical background of the interaction of electromagnetic radiation with the Earth's surface and the interfering atmosphere is discussed as far as it serves the understanding of the information content of the signals which are received with remote sensing techniques and of its physical limitations. Experimental results relevant for the Mediterranean environment illustrate this short discussion. The reader who is interested in a more basic treatment of radiation transfer is referred to the standard literature like Born and Wolf (1959), Herzberg (1945 and 1950), Penner (1959) Chandrasekhar (1950), Unsöld (1955), Goody (1964), Lenoble (1985), and Ulaby, Moore, and Fung (1986). More recent literature will be quoted in the text.

The application of remote sensing techniques to land-surface processes covers a broad spectrum ranging from visible wavelength to microwaves but is limited to spectral bands in which the correction for atmospheric effect is still feasible. Because of this wide spectral range it is convenient to use different units to characterize the spectral intervals. Up to 1 µm one usually counts the wavelength, λ, in nanometre (nm) and above 1 µm in micrometre (µm) which is used throughout the infrared part of the spectrum. Spectroscopists prefer the unit of wavenumber, ν, the number of wavelength per centimetre. It is defined as $\nu = 1/\lambda$ and measured in cm^{-1}. This is more convenient for displaying line spectra. In the microwave region wavelengths range in the millimetre and centimetre domain but, as the technique in this region was adopted from radio engineering, it is common to characterize the spectrum here by its frequency $f = c/\lambda$ which, at these long wavelengths, is measured in gigahertz (GHz). c is the speed of light and 1 Hz = 1 sec^{-1}. A conversion graphic is provided in Appendix 5. Throughout the electromagnetic spectrum the same physical laws apply. However, different simplifying approximations can be introduced as the interaction shifts from processes in which bounded electrons are involved at short wavelengths to those in which vibration and rotation of molecules play the major role and, finally, to those governed by free electrons at the longest wavelength. The formulation of the interaction processes will therefore be generalized as far as possible and the spectral bands be treated in parallel. The consequence is that the nomenclature is unified and specialists in one or the other field may have to accustom to different notifications as they are used to.

The nomenclature used here to describe radiation quantities was documented by WMO (1983) and follows the SI system. The basis is the radiant energy, Q, measured in Joule = Watts times seconds (1 J = 1 Ws). The radiant power or radiant flux $\Phi = dQ/dt$ is measured in Watt (W). For the flux (area) density $d\Phi/dA$ (A = area) two symbols are in use: M for the flux density emerging from or crossing an area element dA (exitance) and E for the flux density incident onto an area element (irradiance). The radiance $L = d^2\Phi/(d\Omega \cos\theta\, dA)$ is the flux across the projection of an area A perpendicular to the direction of propagation. It is a conservative quantity in an optical system and is measured in W m^{-2} sr^{-1}. Ω is the solid angle under which the radiation source is seen and cos θ dA the effective area element as seen under the angle θ with respect to the direction of the line of sight. If the area element stands perpendicular to the line of sight, cos θ = 1. Net radiation quantities generally are designated by an asterisk (*), such as the net flux density $M^* = M_{downwelling} - M_{upwelling}$. This formulation brings up the question of the sign. Normally, in atmospheric research, the upward direction towards zenith is counted positive. This would require a negative sign for the downwelling solar radiation. More often the gain of energy at the surface is counted positive, the loss negative, so that, in the case of the net solar radiation, M^* would be positive.

Reflectance generally is denoted by the letter ρ but it has to be specified which geometry is applied in its measurement. Because of the non-isotropic nature of the reflectance of most natural surfaces, it has to be distinguished between directional and hemispherical reflectance for directionally respectively multidirectionally illumination of the target (see e.g. Bolle 1982). This will be discussed in more detail in the next section. An important quantity is the bi-hemispherical reflectance factor which is called *albedo* if one deals with a horizontal surface. If the albedo is determined in a specific spectral band one deals with the spectral albedo, $\rho(\lambda)$. If the bi-hemispherical reflectance is integrated from 0.3 to 2.5 μm one speaks about broad-band or meteorological albedo because it is measured at meteorological stations and determines most of the fraction of solar radiation that is available for energy conversion processes at the surface. 98% of the solar energy is confined in the spectral range between 0.3 μm and 4 μm (see Appendix 5). The band between 2.5 μm and 4 μm contributes only a small portion of the energy reaching the surface because of the strong absorption by atmospheric gases in this region.

The shortwave radiation incident at the surface from the upper hemisphere which is the direct and scattered solar radiation is called *global radiation*. It consists of a visible part and the near infrared (NIR). The energy emitted by the atmosphere in the thermal or longwave infrared part of the spectrum is the *atmospheric radiation*. Its downwelling component sometimes is also called atmospheric counter-radiation. Its longwave end merges into the microwave radiation.

In most cases (remote sensing with longwave microwaves is an exception) a remote sensor does not exactly measure the radiation reflected or emitted by the target but that of the surface-atmosphere system. To reconstruct the surface properties it is necessary to correct the measured signal for atmospheric effects. Thus two processes of different nature have to be regarded here: The interaction of the electromagnetic waves with the surface and with the atmosphere. These will be the topics of the following two sections.

To obtain information about the surface it is necessary to correct remote sensing data for atmospheric effects. This topic is addressed in the last section of this chapter. The desired information can regard surface "properties" or state variables, which requires different measuring sequences. It is clear that soil moisture, surface temperature and surface "greenness" (as indicated by vegetation indices) are state variables. Surface properties are conservative quantities like soil composition and structure or topography. Surface albedo and surface emissivity are the result of the interplay between surface properties and the growing and decaying vegetation as well as top soil moisture. In some cases, if the surface does not change over a certain time period like in the case of deserts or urban areas, they can be considered as conservative. Otherwise they must be regarded as state variables as well.

3.2 Interactions of Electromagnetic Waves with the Surface

3.2.1 Elementary Processes

Plain Surfaces

At short wavelengths, the part of the spectrum which can be used for remote sensing is limited to wavelengths above 330 nm because of the strong ozone absorption below this wavelength. The electromagnetic waves between this shortwave threshold and microwave wavelengths in the centimetre to decimetre range interact with matter by exciting molecules, ions, atoms or electrons to oscillations and rotations. Roughly speaking, the visible radiation excites bounded electrons, the near to medium range long wave infrared radiation molecular vibration-rotation levels, the far infrared radiation molecular rotation lines, and the radio or microwaves free electrical charges. In the energy range of these wavelengths the ionization of molecules or atoms is not possible and only few electronic bands can be excited.

To understand the interaction between an electromagnetic wave and matter, the term polarization plays a key role. It describes the displacement of electrical charges of opposite sign by the influence of the oscillations of the electric vector $\boldsymbol{E}$ of the wave. The field vectors change their direction periodically thus determining the frequency of the wave. This allows the coupling with charged particles that are bound to another or a number of other particles with opposite charge. If the charges are displaced in a way that a dipole or multipole is generated or its strength is changed, then the electromagnetic wave can interact with the particle and transfer energy to it. As an example, a nitrogen molecule N-N is completely symmetric and cannot interact with the electromagnetic wave except if it is deformed by collisions with other molecules or strong external fields. A water molecule is bent and has a permanent dipole momentum caused by the opposite charges of the hydrogen and oxygen atoms. In this case the interaction is very strong if the energy of the photon which hits the molecule corresponds to the difference of its possible energetic states.

At this interaction, a photon is absorbed and the electromagnetic wave which represents the flux of photons is reduced in amplitude. The state of the particle hit by the photon is lifted into a higher electronic, vibrational, or rotational state. If a photon of an energy which matches precisely the difference of energy levels in the matter is absorbed it would vanish leaving a molecule in an excited state while the amplitude of the wave which represents the flux of the photons is reduced. One single wave is always polarized in the optical sense of this term which means that the electric and magnetic vectors oscillate in a very regular manner either within a plane or in a circular or elliptical pattern. If in the manifold of all waves all directions of polarization are present permanently, the radiant flux may be un-polarized like in the direct solar beam or light from a light bulb. Radiation scattered or reflected once almost always is to some degree polarized. If many reflection and/or scattering processes are involved the degree of polarization again decreases. An example is the light of a blue sky which is nearly 100% polarized at 90 degree off the solar direction but if the atmosphere is turbid the polarization is less. The degree of polarization can be looked at as a signature of the scattering or reflecting matter. To take full advantage of this effect the radiation emitted by some active remote sensing systems intentionally is a polarized wave as in the case of active microwave and lidar sounding instruments.

An excited molecule will not remain for a long time in an excited state. Either it loses the absorbed energy by emitting a wave at the same wavelength or, if the absorbed energy was large, by cascading down to its ground state. Alternatively its excited oscillation respectively vibration is diminished by impacts with other molecules which transfers the energy into linear movements thus transforming it into heat, a statistical movement of molecules rather than an oscillation at a sharply defined frequency.

The discrete excitation of atoms, molecules and crystal lattices follow strict rules. According to quantum theory only specific vibrational and rotational states and transitions between them are “allowed”. Spectral lines of pure rotational states generally are observed beyond 10 μm wavelength. The excitation of vibrational levels requires more energy which is available at higher frequencies. If a vibrational level is exited at wavelength shorter than, say, 20 μm, then this transition is accompanied by a number of simultaneously excited rotational states. Thus not one single absorption line appears but a group of lines called a vibration-rotation band. Each line has a specific shape characterized by the line intensity and its halfwidth. The line intensity depends on the probability of the transition which is a function of temperature. The half width is a function of pressure and temperature. At high temperatures and/or pressures the lines and the extension of bands widen (higher levels are excited) until they merge with neighbours. The reason is that the single energy levels broaden due to increasing intensity and frequency of collisions with other molecules. By this way more photons of slightly different energy have the opportunity to excite the states. Finally a broadband feature is approached in the absorption spectrum.

A similar phenomenon is observed if the molecules are imbedded in crystal structures and loose their molecular individuality. The whole lattice can be excited to vibrations. In these cases the energy levels are not anymore sharp and broad band features evolve such as the *Reststrahlen* bands of crystals in the far infrared (see Fig. 3.1 and Appendix 6). In the most simple case of two types of ions of the masses m_1 and m_2 ordered in a linear chain, the lattice can oscillate like a linear harmonic oscillator with a frequency $\nu = 1/T$ where T is the period of the oscillation. The corresponding circular frequency in this case is

$$\omega = 2\pi\nu = \sqrt{2K\left(\frac{1}{m_1} + \frac{1}{m_2}\right)}\ . \tag{3.1}$$

K is the force constant between the ions (Möller and Rothschild 1971).

These vibrations can only be excited if the frequency of the exciting electromagnetic wave is low because the ions due to their inertia cannot follow quick oscillations, in that case they just would remain motionless. Therefore, only long infrared wavelengths are able to interact with pure crystals which could be one photon processes in ion-crystals, multi-photon processes in ion-crystals and homopolar crystals, or defect-induced one photon processes.

Electromagnetic waves are characterized by the vector of electric field strength $\boldsymbol{E}$ measured in V m^{-1} and the accompanying vector of the magnetic field $\boldsymbol{H}$ which has the dimension A m^{-1}. The interaction of electromagnetic waves with matter is governed by Maxwell´s equations (see e.g. Born and Wolf 1959):

$$\nabla \times \boldsymbol{H} = \frac{\partial \boldsymbol{D}}{\partial t} + \sigma\boldsymbol{E} = \varepsilon\frac{\partial \boldsymbol{E}}{\partial t} + \sigma\boldsymbol{E} \qquad \nabla \times \boldsymbol{E} = -\frac{\partial \boldsymbol{B}}{\partial t} = -\mu\frac{\partial \boldsymbol{H}}{\partial t} \tag{3.2}$$

$\boldsymbol{D}$ is the electrical displacement vector and $\boldsymbol{B}$ the magnetic flux density. The medium is described by its permittivity or dielectric constant, ε (A s V^{-1} m^{-1}), its magnetic permeability, μ (V A^{-1} s m^{-1}), and its specific conductivity, σ (A V^{-1} m^{-1}). ∇ is the Nabla operator.

The displacement of electrical charges in vacuum is proportional to the electric field: $\boldsymbol{D} = \varepsilon_0\boldsymbol{E}$, where $\varepsilon_0 = 8.854\ 10^{-12}$ A s V^{-1} m^{-1} is the permittivity or electric constant of the vacuum. If a dielectric material is present, an additional displacement P takes place due to charge separation within this medium, which is called *polarization*. The displacement vector caused by this interaction with matter leads to the definition of the generalized dielectric constant or permittivity $\varepsilon = \varepsilon_r\varepsilon_0$, where ε_r is a dimensionless relative number that equals 1 in the vacuum:

$$\boldsymbol{D} = \varepsilon_0\boldsymbol{E} + \boldsymbol{P} = \varepsilon_r\varepsilon_0\boldsymbol{E} \tag{3.3}$$

More precisely, $\boldsymbol{P}$ consists of two components, the oscillations of electrons and the oscillations of the molecular nucleus. If the contribution of the oscillation of the electrons is denoted by a and that one of the molecular nucleus by b, it can be written $\varepsilon_r = (1+a+b)$.

The permittivity is related to the strength of the dipole momentum which builds up in the medium under the influence of an external field. The dipole momentum is the charge multiplied by the distance of charges of opposite signs. For each molecule in an isotropic dielectric medium the electric dipole momentum $\boldsymbol{p}$ (A s m) is generated under the influence of the effective local field strength $\boldsymbol{E}'$: $\boldsymbol{p} = \alpha\boldsymbol{E}'$. If under α the mean polarizability of a volume with N molecules per unit volume is understood, then its total effective dipole momentum $\boldsymbol{P}$ is proportional to this local electric field vector $\boldsymbol{E}'$, the molecular electrical polarizability α (A s m^2 V^{-1}), and the number density N (m^{-3}):

$$\boldsymbol{P} = N\alpha\boldsymbol{E}'. \tag{3.4}$$

The molecule at which the effective field $\boldsymbol{E}'$ acts can be assumed to be positioned within an otherwise empty sphere. The surface of this sphere is electrically charged as a result of the polarization of the surrounding dielectric medium. The vector of polarization has the same direction as the external field $\boldsymbol{E}$. Of the charges at the surface of the sphere only the fraction determined by the cosine of the position of the charge with respect to the central molecule and the direction of the external field is effective. By integration of the field over the sphere it results in an additional field strength $\boldsymbol{P}/3$ in the direction of the external field. The total local field becomes

$$\boldsymbol{E}' = \boldsymbol{E} + \frac{\boldsymbol{P}}{3\varepsilon_0} = \boldsymbol{E} + \frac{Na\boldsymbol{E}'}{3\varepsilon_0} \tag{3.5}$$

By combination of the equations (3.3-3.5) with the definition of the electric displacement vector, one obtains the Lorentz (1880) - Lorenz (1881) formula:

$$\varepsilon_r = \frac{1 + \frac{2}{3}\frac{N\alpha}{\varepsilon_0}}{1 - \frac{1}{3}\frac{N\alpha}{\varepsilon_0}} \tag{3.6}$$

In the Gaussian system the factor 4π would have to be added to α.

An equation for the electric field vector $\boldsymbol{E}$ is obtained, if one eliminates $\boldsymbol{H}$ from the second equation of Eq. (3.2) by applying again the operator ∇ and replacing $\nabla\times\boldsymbol{H}$ by the first equation. If the rule $\nabla\times(\nabla\times\boldsymbol{E}) = (\nabla\boldsymbol{E})\nabla - \nabla^2\boldsymbol{E} = -\nabla^2\boldsymbol{E}$ is applied for a homogeneous medium, then

$$\nabla^2\boldsymbol{E} = \mu\sigma\frac{\partial\mathbf{E}}{\partial t} + \mu\varepsilon\frac{\partial^2\mathbf{E}}{\partial t^2}. \tag{3.7}$$

A general solution for this equation is a wave function. Considering first a standing wave, its amplitude can be represented by sin ωt. The pattern reproduces always after a time T and the sine-function after a period of 2π: $\sin \omega t = \sin(\omega t + 2m\pi) = \sin \omega(t + \{2m\pi/\omega\}) = \sin \omega(t + mT)$, m being an integer, from which follows: $\omega = 2\pi/T$. A wave progressing in the z - direction without oscillating would repeat its pattern after a wavelength λ. This can be represented by $\sin kz = \sin(kz + 2m\pi) = \sin k(z + \{2m\pi/k\}) = \sin k(z + m\lambda)$, from which follows: $k = 2\pi/\lambda$. Consequently, a wave traveling along the z-axis and oscillating, can be represented by $\sin(\omega t - kz)$. The sign in front of kz depends on the direction. For easier mathematical treatment this function can be regarded as the real part of the complex expression $e^{i(\omega t - kz)}$. The electric field vector would then become the real part of the function

$$E(z,t) = E_0 e^{i(\omega t - kz)} \tag{3.8}$$

where E_0 is the amplitude of the wave. The following equation results for k if this wave function is substituted into Eq. (3.7) :

$$k^2 = \omega^2 \mu \left(\varepsilon - i \frac{\sigma}{\omega} \right) = \omega^2 \mu \varepsilon \left(1 - i \frac{\sigma}{\omega \varepsilon} \right) = \omega^2 \mu \varepsilon_{eff} \tag{3.9}$$

It can be seen that the effective permittivity ε_{eff} is a complex number. μ equals 1 for non-magnetic materials which, in practise, is mostly the case. k is a complex function as well and may be written $k = k' + k''$. Applying the rule (see e.g. Wörle 1978)

$$\sqrt{x - iy} = \left(\sqrt{\frac{\sqrt{x^2 + y^2} + x}{2}} - i \sqrt{\frac{\sqrt{x^2 + y^2} - x}{2}} \right) \tag{3.10}$$

to (1 - iσ/ωε) in Eq. 3.9, the imaginary part of k, k'', is obtained, which provides the real part of the exponent of $E(z)$ when introduced in Eq. (3.8) :

$$k'' = \omega \sqrt{\frac{\mu \varepsilon}{2}} \sqrt{\sqrt{1 + \left(\frac{\sigma}{\omega \varepsilon} \right)^2} - 1} \tag{3.11}$$

k″ determines the size of the amplitude of the wave and is related to the absorption in the medium. Two special cases may be considered. If $\sigma/\omega\varepsilon \ll 1$, which is the case for low absorption, it follows according to $(1 + x)^{1/2} \cong 1 + (x/2)$:

$$k'' \approx \frac{\sigma}{2} \sqrt{\frac{\mu}{\varepsilon}} \tag{3.12a}$$

If $\sigma/\omega\varepsilon \gg 1$ the conductance and the absorption are high (like in metals) and it results

$$k'' \approx \sqrt{\frac{\mu \sigma \omega}{2}} \tag{3.12b}$$

The phase of the electric field is constant whenever $\omega t - kz =$ constant. Considering Eq. 3.9 the velocity of the wave can be expressed by

$$c = \frac{dz}{dt} = \frac{d}{dt} \left(\frac{const + \omega t}{k} \right) = \frac{\omega}{k} = \frac{\omega}{\omega \sqrt{\mu \varepsilon_{eff}}} = \frac{1}{\sqrt{\mu \varepsilon_{eff}}} \tag{3.13}$$

In the optical domain the index of refraction n can be measured as ratio of the sinuses of the angles of incidence and refraction at the surface between two media. This equals the ratio of the phase velocities in the two media: $n = c_1/c_2$. If medium one is the vacuum, $c_1 \rightarrow c_0 = \varepsilon_0^{-1/2}$, and if medium two is the denser one it follows $n = (\mu \varepsilon_{r,eff})^{1/2}$. For a non-magnetic medium it results $n^2 = \varepsilon_{r,eff}$. The real part of this relationship can be approximated numerically by $n^2 = \varepsilon_r = 1 + a + b$, a and b standing for the movements of electrons and heavy nuclei, respectively. In the shortwave part of the spectrum the movement of the charged heavy particles is not important for the reasons given above. Here $n = (1 + a)^{1/2}$. Since n is only measured in this part of the spectrum and ε_r in the longwave part of the spectrum it follows $\varepsilon_r = n^2 + b$ which is the reason that the numerical values of $\varepsilon^{1/2}$ and n differ from each other (e.g. NaCl: $\varepsilon^{1/2} = 2.41$, $n = 1.52$).

In analogy to the complex quantity $k = \omega \varepsilon_{eff}^{1/2} = \omega(\varepsilon_0 \varepsilon_{r,eff})^{1/2}$ for an absorbing non-magnetic medium the complex refractive index $(n' + in'') = n'(1 + i\kappa) = (\varepsilon_{r,eff})^{1/2}$ is defined which is related to k by:

$$k = k' + k'' = \frac{\omega}{c_0} n'(1 + i\kappa) = \frac{2\pi c}{\lambda c_0} n'(1 + i\kappa) = \frac{2\pi}{\lambda}(1 + i\kappa) \tag{3.14}$$

Here use have been made of the relationships $c_0 = \varepsilon_0^{-1/2}$, $c_0/c = n'$ and $\omega = 2\pi c/\lambda$. The quantity $\kappa = n''/n'$ is called *attenuation index* or extinction coefficient.

It can be shown that the energy of the wave is represented by the time average of the square of the electric vector of the wave. Inserting the last expression of Eq. 3.14 into Eq. 3.8, separating real and imaginary terms and averaging over time yields:

$$|\mathbf{E}|^2 = \mathbf{E}_0^2 \left| e^{-k''z} e^{i\omega t} \right|^2 = \mathbf{E}_0^2 e^{-\frac{4\pi\kappa}{\lambda} z} = \mathbf{E}_0^2 e^{-az} \quad . \tag{3.15}$$

$a = 4\pi\kappa/\lambda = 4\pi n\kappa/\lambda_0$ is the *absorption coefficient* which has the dimension of a reciprocal length. After the wave has advanced a distance $1/a$ in a medium, exp $(-az)$ = 1/e and the energy falls to 1/e.

The re-formulation of Maxwell's equations with n instead of $(\mu\varepsilon)^{1/2}$ leads to the following formulas of Fresnel for the reflected fraction of polarized electromagnetic waves (r_h = horizontal component, r_v = vertical component) falling on the plane surface of a non-absorbing medium with the refractive index n and the incidence angle θ:

$$r_v(\theta) = \frac{\left(n^2 \cos\theta - \sqrt{n^2 - \sin^2\theta}\right)^2}{\left(n^2 \cos\theta + \sqrt{n^2 - \sin^2\theta}\right)^2} \tag{3.16}$$

$$r_h(\theta) = \frac{\left(\cos\theta - \sqrt{n^2 - \sin^2\theta}\right)^2}{\left(\cos\theta + \sqrt{n^2 - \sin^2\theta}\right)^2} = \frac{\left(-\cos\theta + \sqrt{n^2 - \sin^2\theta}\right)^4}{(n^2 - 1)^2} \tag{3.17}$$

These formulas are only valid for specular reflection for which the angle of incidence is numerically identical with the angle of reflection. The complex refractive index sometimes is written $n = n' + in'' = n'(1 + i\kappa)$. Its imaginary part is responsible for the absorption of the wave inside the medium. By introducing this complex index of refraction it follows for vertical incidence:

$$r = \frac{(n' - 1)^2 + (n'\kappa)^2}{(n' + 1)^2 + (n'\kappa)^2} \tag{3.18}$$

If the absorption is very strong, the wave cannot enter the medium very deeply. Already at the molecules of its uppermost layers, which act as scattering centres, spherical waves develop which, by interference, form the reflected wave. The wave does not penetrate more than a fraction of the wavelength. At this short optical path not much energy can be absorbed. Most of the energy is reflected. This is the case for a number of minerals that generate "Reststrahlen" bands at longwave infrared wavelengths with large absorption indices (Fig. 3.1). If on the other hand the imaginary part of the refractive index is small most of the radiation passes through the medium except when the path in this medium is long enough to absorb the energy. Only a medium-sized imaginary part of the refractive index allows the wave to penetrate more than one wavelength, or several molecular layers, into the medium where a considerable amount of the energy can be absorbed, and therefore not much radiation is reflected.

In accordance with the Eq. (3.15) the transmission through a layer of thickness d can be written as

$$\frac{I(D)}{I_0} = \exp\left(-\frac{4\pi\kappa d}{\lambda_1}\right) = \exp\left(-\frac{4\pi n'\kappa d}{\lambda}\right) \qquad (3.19)$$

where λ = wavelength in vacuum, λ_1 = wavelength in the medium, d = path length. For water, as an example, the refractive index ranges between n' = 1.3307040 at 706.52 nm, 10°C, and 1.3425508 at 404.66 nm , 22°C. $n'\kappa = 0.716 \cdot 10^{-9}$ at 450 nm and $0.1989 \cdot 10^{-2}$ at 2.5μm. This still is small as compared to n - 1 and results in a reflectance of about 0.02 throughout the shortwave range.

Scattering and Reflection From Complex Surface Structures

The idealized conditions described in the preceding section almost never occur in nature, except, maybe, for deep and undisturbed inland lakes. Generally, the surface is neither smooth nor does it consist of uniform mineral crystals. In the case of flat water surfaces the underground plays an important role as reflector and almost all natural waters are disturbed by suspended particles which serve as scattering centres. As the water surface becomes more and more stirred up by wind, the reflected beam first opens to a cone with a pronounced specular component but the diffuse component broadens with increasing wave height. Land surfaces, in general, have a predominantly diffuse reflection indicatrix but in some cases, especially if the

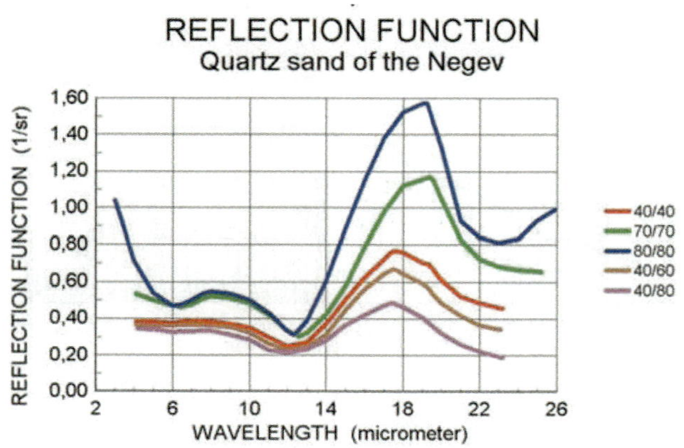

Fig. 3.1 Reflection spectra of quartz sand from the Negev. The first number in the legend is the incidence angle, the second one the reflection angle

surface is wet, the reflection also gets a specular component. In these cases it is possible, in principle, to consider each small facet of the surface as a reflecting surface and sum over all positions of these fragments. Each grain and inhomogeneity acts as a scattering centre. The fractionation of the surface finally opens small cavities in which multiple reflection takes place and which therefore act like "black bodies" in which the radiation fully can be absorbed even if the absorption for a single scattering process is small. In practice it is impossible to model these processes because neither the grain sizes nor their composition and thus their index of absorption are well known. If one would try to measure the structure of a surface in order to obtain the parameters needed to model its interaction with the radiation, one would probably destroy it at the same time, except for the microwave region where the application of macro-tools becomes possible. Even a flat stone is not uniform in its structure and if one turns it upside down one gets another result because the upper side is weathered and the lower side may be affected by soil chemical processes (Fig. 3.2).

An empirical approach, therefore, is necessary to study the behaviour of solid surfaces. Clever instruments have been designed and constructed to measure the reflection functions of surfaces from elevated platforms (Deering et al. 1992a and b). Even this is not in complete agreement with the geometrical conditions of observations made from space and few experiments are known where the same piece of surface was observed under different angles, e.g. from an aircraft flying in circles of different height around the site (Kriebel 1974).

The situation becomes even more complicated if one deals with vegetation. Since Gates and Tantraporn (1952) and Gates et al. (1965) published the first measurements of leaf reflectance and transmittance many experimental investigations have been performed on plant elements with spectrometers and on whole plants.

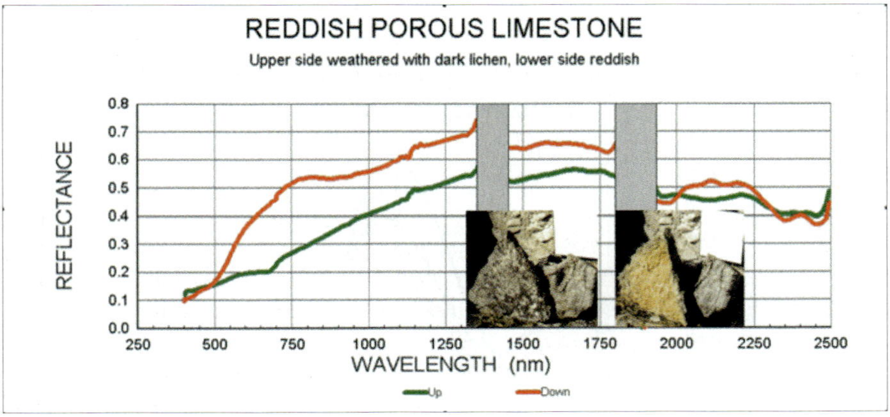

Fig. 3.2. Reflection spectra of reddish limestone. The weathered upper side is partly covered with dark lichen (left figure). The not affected downward looking side of the triangularly shaped stone is shown at the right. For comparison both figures show a stone wall in the back, a white reference in the upper right and a piece of an olive tree stem supporting it

A big step forward was the modelling of the interaction of plants with radiation (Myneni and Ross 1991). The treatment of the problem is the solution of the equation of radiative transfer for a reflecting and scattering medium and can be approached in a similar way as that of absorption and scattering by aerosols. The difference is that the aerosol particles are small and nearly spherical and the leaves are not. All mathematical tools developed for particle scattering in the atmosphere and the sea have also been applied for leaf canopies such as iterative methods in which the successive orders of scattering are computed, viz. Monte Carlo methods, discrete ordinates methods, and statistical photon transport theory. Descriptions of these methods *inter alia* are given by Lenoble (1985), Van de Hulst (1980), Myneni and Ross, (1991). In these models, generally the following components have to be considered: Diffuse single scattering of the foliage, single specular reflection from the foliage, single reflection from the underlying soil, hot spot of the foliage, multiple scattering reflectance.

The "hot spot" phenomenon describes the strong maximum of radiation reflected in the backward direction from an assemble of irregularly distributed reflecting surfaces such as leaves or needles of vegetation canopies. It is known as the "Heiligenschein" in meteorology, the "opposition effect" in astronomy, and the hot spot effect in aerial photography and remote sensing. The cause of this phenomenon is that particles of the reflecting/scattering medium cast shadows which cannot be seen if one looks along the incident radiation. As one moves with the line of sight outside the direction of the incident beam one partly looks into these shadows from which less radiation is reflected. Let us assume the following geometry: A sensor is directed at its shadow at an object consisting of small scattering centres distributed in space which are illuminated from behind the sensor. In the case of the "Heiligenschein" a solid body (mountain peak, aircraft, head) cuts out by its shadow the central point of the back-scattering feature caused by aerosols. Around the shadow the intensity is maximum as it would be in its centre if the shadow would shrink to zero.

The treatment of this effect can be based upon the determination of a Bi-Directional Gap Probability (BDGP). It depends on shape and position of the scattering objects which may be leaves or needles. If these canopy elements are oriented in a way that on the average there is a strong backward reflection, then the hot spot can be rather large (twice as large as the reflection in directions away from the backscatter direction in the visible). If one moves with the line of sight out of the direction of the shadow one would not anymore look into the specular reflectance but still through the gaps between the uppermost leaves into the darker lower layers. In the near infrared this effect is not as strong as in the visible part of the spectrum because the lower levels are more illuminated due to the in the near infrared spectral range (NIR) more transparent foliage.

A scanning instrument mounted on a satellite eventually would look at the shadow of the satellite which is negligible as compared with the instantaneous field of view of the instrument. Consequently the instrument would sense a higher backscattered signal then looking a few degrees away from this direction.

Reflectance Terminology

If measurements are carried out at a number of distinct directional incidence and observation angles, the *Bi-directional Reflectance Distribution Function* (BRDF), $\rho(\mathrm{d}\Omega, \mathrm{d}\Omega')$, is obtained. This function is defined as the ratio of the radiance (measured in W m^{-2} sr^{-1}) emerging from a target within an infinitesimal angle $\mathrm{d}\Omega = \sin\vartheta\,\mathrm{d}\vartheta\,\mathrm{d}\varphi$ (φ = azimuth angle and ϑ = zenith angle) and the irradiance or flux density (measured in W m^{-2}) which the target receives from a given direction Ω' within the solid angle $\mathrm{d}\Omega'$. The BRDF can be measured with a point source and a small detector device. This can be done in the laboratory or with some effort in the field where one uses the sun as source but has to eliminate the effect of the sky radiation.

To formulate the relationship between reflected and incident radiation the following notations are adopted:

$\rho(\Omega, \Omega')$	BDRF (Bi-Directional Reflection Factor)
$\rho(\mathrm{d}\Omega, \mathrm{d}\Omega')$	BRDF (Bi-directional Reflectance Distribution Function, sr^{-1})
φ	azimuth angle of the reflected radiation
ϑ	zenith angle of the received radiation
φ'	azimuth angle of the received radiation
ϑ'	zenith angle of the reflected radiation
ζ	$\equiv \zeta_{sun}$, solar zenith angle
A	area (m^{-2})
L	radiance (Wm^{-2}sr^{-1})
E	irradiance at a target (Wm^{-2})
M	radiation flux density across a unit area (Wm^{-2})
Φ	radiation flux (W)
Ω	solid angle in which radiation is reflected (sr)
Ω'	solid angle from which the target receives radiation (sr)
S_0	"Solar Constant", solar flux density outside the atmosphere
$\bar{d}$	1 Astronomical Unit (A. U., mean Sun-Earth distance = $1.496*10^8$ km)
d	instantaneous Sun-Earth distance
τ_{atm}	transmission of the atmosphere
$\uparrow$	upwelling (reflected) radiation
$\downarrow$	downwelling (incident) radiation
i	isotropically reflecting (Lambertian) surface
n	"natural" reflector

For reasons of comprehensibility, the spectral dependency of the radiation quantities is omitted here and in the following equations. The reader should keep in mind that, e.g., in the case of spectral radiation L should read $L(\lambda)$ or L_λ and that in the dimension μm^{-1} or nm^{-1} must be added.

The element of a horizontal plane (dA) which is illuminated out of an incremental solid angle $\mathrm{d}\Omega'$ receives the incremental (spectral) irradiance or radiation flux density

$$dE(\Omega') = \frac{d\Phi(\Omega')}{dA} = \frac{d\{A\cos\vartheta' L(\Omega')d\Omega'\}}{dA} = \cos\vartheta' L(\Omega')d\Omega' \quad (3.20)$$

or in the case of illumination by the sun: $= \cos\zeta\, L(\Omega')\, d\Omega'$.

The BRDF is defined as

$$\rho(d\Omega, d\Omega') = \frac{L(\vartheta, \varphi; d\Omega)}{L(\vartheta', \varphi'; d\Omega')\cos\vartheta' d\Omega'} \quad (3.21)$$

Its dimension is sr^{-1}.

A direct relationship between laboratory values and the angular dependence of natural surface targets of pixel size is difficult to establish because the small spot measured in the laboratory very seldom represents a natural surface. Apart from such scaling problems under natural conditions the illumination of the surface in the short wavelength range always is a combination of the direct solar radiation and the diffuse sky radiation. Only if the atmosphere is clear and the sun is high up in the sky, the diffuse sky radiation, at least in the near infrared, becomes small in comparison to the direct solar radiation and may in a first approximation be neglected. This is not the case for the blue end of the spectrum where there is always a substantial contribution due to molecular (Rayleigh) scattering. Here one faces the fundamental problem of the validation of remote sensing data by matching them with measurements made at the surface or even in the laboratory. The measurement would require to measure the irradiance at the target stemming from a point source and the reflected radiance with identical detectors (or to replace the target by a mirror of known reflectance).

Under natural conditions one deals with the illumination of the scene by the global radiation, the nearly parallel beam of the sun and the diffuse sky radiation which is distributed over the whole hemisphere. A remote sensing instrument measures the radiance reflected into a narrow solid angle. In this case a directional reflection from a hemispherically illuminated source is measured.

Typically for field (and sometimes also laboratory) measurements is the determination of the ratio of the radiance reflected by the target, a “natural” surface, and the radiance reflected by a plain reference panel of known optical properties positioned, if possible, at or near the target area. The reference panel should have a Lambertian (Lambert 1760) reflexion characteristic and should be non-absorbing. For a Lambertian reflector the radiance reflected into a specific direction is proportional to the cosine of the angle between this direction and the normal of the reflecting plane surface. The radiation emerging from the Lambertian surface can also be called isotropic. From the viewpoint of the observer, a sensor viewing the surface under a constant solid angle receives the same amount of radiation under all observation angles as long as the Lambertian reference panel is large enough to fill the field of view because then the observed area grows with the cosine of the angle between the normal of the surface and the direction of observation. Ideal non-absorbing plane Lambertian reflectors are not available but some products come close to it and corrections can be applied if the panel is calibrated by comparison with a truly isotropically emitting hollow sphere.

The radiance reflected from such an ideal isotropic reflector into each unit solid angle is π^{-1} of the incident flux density, the irradiance. From a non-absorbing

Lambertian surface, therefore, emerges the flux density

$$M_i^{\uparrow}(\Omega) = \int_{\Omega} L_i^{\uparrow}(\Omega)d\Omega = \int_{\Omega} \frac{E(\Omega')}{\pi} d\Omega \tag{3.22}$$

into a cone with the finite solid angle Ω. For a natural surface $E(\Omega')/\pi$ has to be replaced by $E(\Omega')\rho(d\Omega, d\Omega')$.

If a finite cone of incoming radiation is considered,

$$E(\Omega') = \int_{\Omega'} \cos\vartheta L(\Omega')d\Omega' \tag{3.23}$$

and the ratio of the flux density from the natural (index *n*)and the non-absorbing Lambertian (index i) surface becomes:

$$\rho(\Omega, \Omega') = \frac{M_n^{\uparrow}}{M_i^{\uparrow}} = \frac{\int_{\Omega}\int_{\Omega'} L(\Omega')\rho(d\Omega, d\Omega')\, d\Omega' d\Omega}{\frac{1}{\pi}\int_{\Omega}\int_{\Omega'} L(\Omega')\, d\Omega' d\Omega} \tag{3.24}$$

If the solid angle $\Omega' = 2\pi$ (the whole upper hemisphere) then two extreme cases can be considered, firstly the source may be the direct sun only (scattered sky radiation can be neglected) and secondly the source is the cloud covered sky (which applies for aircraft remote sensing and surface measurements exclusively). In the first case the radiance from the non-absorbing Lambertian surface equals

$$L_i^{\uparrow} = E_s / \pi = \frac{S_0(\bar{d}/d)^2 \cos\zeta\, \tau_{atm}(\zeta)}{\pi} \tag{3.25}$$

and the BRDF is determined by the position of the sun which may result in strong azimuthal differences of the reflected radiation. In the second extreme case, the cloudy sky, the illuminating radiation stemming from all over the hemisphere produces a BRDF which is less dependent on the observation angle. An example for these two cases is given in Fig. 2.4d.

If the solid angles, under which source and radiometer are seen from the target, become small, then $\rho(\Omega, \Omega') \rightarrow \rho(\Delta\Omega, \Delta\Omega')$ is called the *Bi-Directional Reflectance Factor* (BDRF). This would be the case with the sun as the only illuminating source and a radiometer with a small field of view as detector. It is also required that the product of viewed area d*A* and ΔΩ is the same for both the target and the reference measurement. If it can be neglected that the sky radiation is distributed over the whole sky, $\rho(\Delta\Omega, \Delta\Omega')$ is approximating the BRDF.

In the field in most cases the reference for reflection measurements is a white panel illuminated by the sun and the sky which is the natural condition for remote sensing. These measurements can not be regarded to yield either the BRDF or the BDRF but a *directional reflectance factor for hemispherical incidence* (of mostly unknown spatial distribution). The "reflectance" obtained this way is the ratio of the radiance emerging from the target into a specific direction divided by radiance from the reference panel. If this panel can be regarded as non-absorbing Lambertian, then the referene signal is E/π and ρ/π (sr^{-1}) is the ratio of the radiance emerging from the target and the irradiance at this surface. The reflectances reported in this book are mostly of this type. Only in the near infrared under clear sky conditions this factor approaches the BDRF. The *albedo* $\overline{\rho}$, defined as the ratio of the flux from the target into the hemisphere and the flux incident at the target, can be obtained by integrating $(\rho/\pi)d\Omega$ over the hemisphere.

3.2.2 Solar Radiation

Interaction With Soils

The reflectance of soils is determined by its chemical composition, grain size, and water content. Absorption features identified in some mineral soils have been collected by Ferrier et al. (1994), some of which are presented in Table 3.1.

Younis et al. (1997) investigated soils typical for Castilla - La Mancha, Spain, in the laboratory (Fig. 3.3): Leptosols, Calcisols, and Solonchak (saline soil). Water vapour bands and OH bonds can clearly be detected in hygroscopic soils. Carbonate traces are to be seen near 2300 nm. The short wave end of the spectra is strongly influenced by the presence of iron oxides which give the soil its red color, while the organic matter reduces the reflectance. Some typical composition data are presented in Table 3.2. which are average values. If the amount of the components varies clear correlations with the reflectance in some Landsat-TM channels exist, except for carbonates.

Table 3.1. Absorption features of soils after Ferrier et al. (1994), Hill (1991, 1996), Hill et al. (1991, 1994, 1997), Grove et al. (1992)

Wavelength (nm) (read from top to bottom)	650	662	836	930	1394	1428	1472	1745	1830	1904	1919	2143	2161	2183	2200	2216	2229	2305	2312	2340	2342	2238	2380
AlOH															×								
Alunite							×	×				×							×				
Calcite														×								×	×
Pyrophyllite					×								×					×					
Kaolinite					×								×	×	×								
Illite					×					×					×	×					×		
Vermiculite									×		×						×						
Fe-oxides	×	×	×	×																			
$CaCO_3$														×						×			

Table 3.2. Characteristics of some Mediterranean soils

Composition, electrical conductivity and characteristics of the location of acquisition	Soil Location, Type, and Characteristics				
	Castilla - La Mancha			Tuscany	
	Calcisol	Leptosol	Arenosol Solonchak	Red Soil	Grey Soil
General characteristics	high $CaCO_3$ content	high organic matter content	high salt content	high amount of sand	high amount of sludge
Organic matter	0.016	0.086	0.01		
Carbonate	0.35	0.32	0.2		
Sand	0.27	0.56	0.81	0.77	0.33
Silt	0.48	0.22	0.08	0.16	0.65
Clay	0.25	0.22	0.12	7	2
pH				7.1	7.5
Electrical conductance dS/m	0.16	-	0.74	0.0235	0.211
Area of acquisition	Barrax Tomelloso	El Bonillo	Pedro Muñoz	Venturina	Orciano
Vegetation cover	Dry and irrigated cultivation	Pine forest and matorral	Vineyards and fallow land	Sun flowers	Wheat

The optical properties of pure minerals are very blurred in natural surface materials due to fractionalization, mixing and weathering. What can be seen in the field may be distinctively different from what is measured in the laboratory. Fig. 3.4 shows the spectral

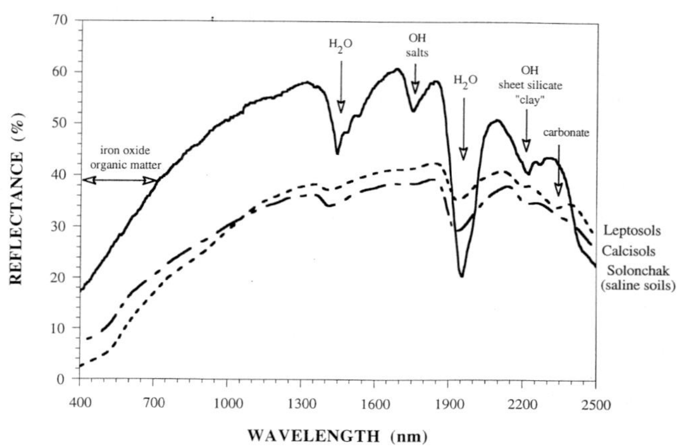

Fig. 3.3. Water absorption bands in the reflectance spectrum of Solonchak and to a lesser extend in Leptosols and Calcisols

measurements of different soils from Tuscany. Photographs of these samples are shown in Fig. 3.5. These soils cover large areas and are exposed after the crops have been harvested so that they may also be identified from space. The first is grey clay from near the village of Orcia, the second is a red soil from the Maremma and the third is a brown marginal agricultural soil from the hilly lands of Colline Metalliferi. The reflectance spans a wide range depending on the solar aspect angle and the degree of smoothness. The very smooth surface of the large piece of grey soil shown in the upper left part of Fig. 3.5 has the highest reflectance if measured against the direction of the sun (268GEQS). The reflectance decreases as the grains become smaller (248G1QS, 248G2QS, and GE&G1579QA). The reflectance of the large flat piece of red soil (268RSQS and upper right corner in Fig. 3.5) is as well high in the near infrared but becomes very low in the visible. Again finer grained soil of this type has a lower reflectance throughout the spectrum (248RMQS, 248RQS, and 248R1QS). The brownish soil with embedded small stones (lower panel of Fig. 3.5) has the lowest reflectance in the near infrared but is similar to the red soil in the visible. From this last soil two probes have been taken which show the potential variability, sample 847DQSSI represents a surface which was smoothed prior to the measurements and the slightly higher one (847DQSS1) is left in its natural state with small stones at the surface (lower left side in Fig. 3.5).

Interaction with Water and Wet Surfaces

Because of its bended structure the water molecule possesses three fundamental vibrations that generate with their overtones a large number of partly overlapping bands. Many of them can be resolved in the water vapour spectrum of the atmosphere (see section 3.3). In the liquid phase these bands grow broader and overlap more strongly. The resulting broad band features are shifted towards longer wavelengths with respect to the water vapour bands of the same transitions as

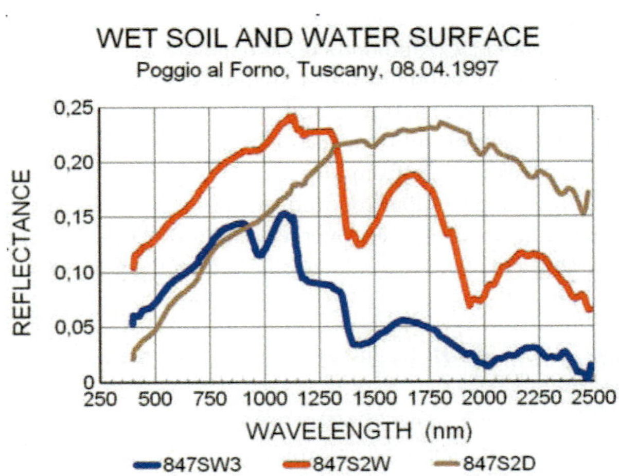

Fig. 3.6. Soil at different stages of wetness: Red - very wet brownish soil partly with sunglint, Blue - thin layer of standing water over brown soil. In addition to the water bands that can be seen in Fig. 3.3 the band at 0.97 μm becomes prominent and the 0.75 μm band starts to appear in the blue spectrum. Brown - spectrum of the dry soil (like Fig. 3.5 bottom) . The higher values at the shortwave end of the wet soils result from sunglint

generally is the case in liquids. The absorption coefficient of pure water has a minimum value around 490 nm. It increases by a factor of 100 towards 200 nm and by a factor of 10^5 towards 2.5 μm with relative maxima at 750 nm, 970 nm, 1200 nm, 1450 nm, and 1920 nm (see Fig. 3.6). Quiet surfaces of pure water reflect solar radiation according to the Fresnel formulae (see above). This has often been confirmed by measurements. The small absorption in the visible part of the spectrum results in high absorption in deep waters and low reflectance values at the surface of 0.020, at high sun. These are determined exclusively by the real part of the index of refraction. As the zenith angle of the sun grows, the reflection increases. Also throughout the middle infrared up to 12 μm the reflectance remains, on the average, of the order of 0.02 though there is anomalous dispersion around the resonance of the fundamental vibrations around 3 μm and 6 μm which makes the reflection vary between 0.005 and 0.05 in narrow spectral intervals (compare Fig. 5.5 and Bolle 1982).

The optical properties of pure water may approximately be observed only at deep calm and clean inland lakes. In all other cases the reflectance is enhanced, especially in the visible part of the spectrum, by (i) reflection of the ground at shallow water bodies, e.g. in coastal zones with bright sands, (ii) the turbidity of the water, and (iii) waves that may cause foam. In nearly all cases, water surfaces have a broadened specular reflectance component ("sunglint") and a diffuse component at much lower intensity. Average broadband albedo values are of the order of 0.06 depending on the wind velocity which causes high reflecting foam (Köpke 1977). These disturbances of the "pure" case are especially strong in the shortwave range which makes it look like that the reflectance decreases from the visible to the near infrared. Ter-Markarayan (1959) computed the dependence of the sea surface broad band albedo of average solar height and cloudiness. At 30° latitude under cloudless skies in winter it increases to 0.095 while in summer it falls to 0.056%. Under overcast skies the value is constant over the year at 7% which demonstrates the influence of the increased irradiance from large zenith angles.

Water bands superimposed to the long wavelength wing of the water vapour bands are also observed over vegetation in the near infrared. The depth of these absorption features indicate whether or not the plants are suffering from water stress.

Interaction With Vegetation Covered Soils

The reflection spectra of plants are dominated in the visible part of the spectrum by the absorption bands of chlorophyll and in the near infrared by the internal micro structure of the leaves and leaf water. Biologically active vegetation has low reflectance in the blue and red part of the spectrum where chlorophyll absorbs. The tiny reflection peak between these bands at 550 nm gives vegetation its green colour (see Fig. 1.23). The magnitude of this peak is responsible for the brightness of the plants. It not only depends on the individual leaf reflectance but also on the structure of the plant which can act more or less strongly as a light trap.

Beyond 700 nm vegetation strongly reflects because the leaves do not absorb anymore but their internal structure scatters the radiation into all directions. Towards

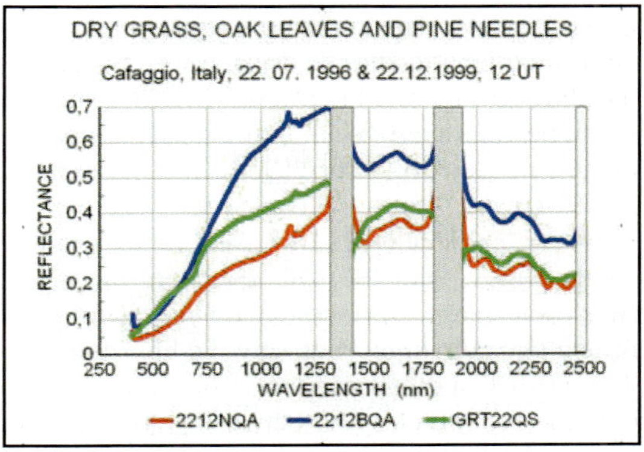

Fig. 3.7. Reflectance spectrum of dry oak leaves (red), dry pine needles (blue), and dry grass (green)

longer near infrared wavelengths reflection decreases again because now the foliage water begins to absorb, first in the water bands at 970 nm and 1200 nm but then from 1400 nm on in the whole domain, because of the strong and broad absorption bands at 1450 nm and 1920 nm.

The intensity of the return signal in this spectral range depends on the optical depth of the canopy or, in other words, the number of scattering and absorbing layers and the gaps between single leaves, which let the radiation penetrate to deeper layers where it can be trapped by multiple scattering. Broadleaf plants like beet, typically, have high reflection values between 700 nm and 1000 nm because there is almost only one scattering layer. On the other hand pine trees with many scattering elements and, in addition, substantial absorbing stem material capture more radiation and reflect less.

If the plant becomes dry, e.g. crops before harvest or deciduous trees in autumn, then chlorophyll does not absorb anymore and the reflectance in the visible part of the spectrum increases because now only the structure matters. For some time in deciduous trees the spectral features of carotene becomes dominant (see Fig. 1.22). At the shortwave end of the near infrared, the reflectance is reduced because of the changes in the leaf structure which occur if the turgor pressure drops. At longer near infrared wavelength the opposite is observed. Without the water absorption the situation is the same as at the shorter wavelengths and the reflectance increases. The reflectance spectrum of dry plants approaches that of dry soils (Fig. 3.7).

The spectral signatures do not exhibit strong characteristic differences from plant to plant which could be used to identify the species. Most conspicuous are differences in the brightness which depends on the intensity of chlorophyll absorption and on the structure of leaves and branches. Well known is the silvery shine of the leaves of olive trees which is caused by a high reflectivity but does not show any specific feature in the spectrum (Fig. 1.23). Apart from relatively small differences in the spectra of species, vegetation introduces an additional dimension of complexity to surface reflectance due to their different vegetation periods, their growth pattern, density, and macro structure.

From the normalized difference between the reflectance in the 700 - 900 nm band and around 650 nm vegetation indices such as the Normalized Difference Vegetation Index (NDVI) can be computed. From its value, based upon absolute reflectances, the Leaf Area Index (LAI) can be deduced with some constraints. This offers an entree into the estimation of vegetation quantities from carefully corrected remote measurements. The inflection wavelength of the "red edge" is correlated with vegetation height, and the dry and wet biomass as long as the plant is growing. But the shift of the red edge can only be measured with high spectral resolution which might become available in future missions (at the expense of spatial resolution). The plant water content is correlated with the depth of the water band at 970 nm (Bach et al. 1994).

If the leaf area index of a green canopy is low or if there is open space between single plants, then the soil or undergrowth properties become important. In the shortwave range, surfaces shadowed from *direct* solar radiation still receive and reflect sky radiation. Thus here the blue end of the reflected spectrum is enhanced. Around 690 nm a minimum is observed because of the strong absorption in the canopy. At wavelengths between 700 nm and 1000 nm the foliage transmits a substantial part of the incident radiation which is reflected at the surface into the atmosphere or back into the canopy. Thus the observed signal is a composite of radiation scattered by the foliage, reflected at the surface (which may be bare soil or soil covered by a different low growing vegetation type), radiation from areas shadowed from direct sun but containing fractions of sky radiation, and radiation transmitted by the canopy, and finally radiation scattered several times between foliage and surface. Within a pixel there may even be contributions from different types of plants.

Because of the mixed information contained in the signal of one pixel, efforts are undertaken to recover the pure components of soil and soil cover. If one knows the reflectances, ρ_{ij}, of the pure components, j, then by solving the linear equation system

$$\rho_i = \sum_{j=1}^{n} \rho_{i,j} f_j \tag{3.26}$$

the fractional coverage, f_j, of each of the pure components can be computed. Because there is no precise solution of such an equation system fed with empirical data, its solution becomes a minimization problem for the residuum. The number of components which can be determined equals the number of available spectral channels.

To improve the relationships between the spectral features and vegetation quantities like *LAI*, biomass, or carbon dioxide fluxes, also other ways have been explored to obtain a more specific measure of vegetation "greenness" than the NDVI offers. As Table 1.2 shows, the NDVI computed for bare soils is not zero but in most cases a number between zero and 0.2. In view of this drawback, vegetation indices, of which some are discussed in section 4.6, have been developed which account for the spectral features of the underlying soil.

The photosynthetic active radiation (PAR) which is absorbed by plants in this shortwave spectral range is one of the limiting factors for their CO_2 uptake and biomass production. It is, therefore, of interest to measure the absorbed fraction of the incident PAR which is often called FPAR. Since radiation of this part of the spectrum also is strongly absorbed by the underlying soil, the determination of the fraction of PAR absorbed by the plants is difficult to estimate if the canopy is not dense. In this case, an accurate determination of FPAR requires estimation of the transmission of the canopy and calculation of the PAR reflected by the underlying soil. The problem is similar as the elimination of the influence of vegetation on remote sensing of soil moisture in the microwave region (see section 4.2.3).

The Albedo Dilemma

An important quantity for the exchange of energy between the surface and the atmosphere is the surface albedo which is defined as the ratio of the upwelling reflected hemispherical radiation flux Φ_r to the incident hemispherical global radiation flux Φ_i of direct solar and diffuse atmospheric radiation:

$$\rho = \frac{\Phi_r^{\uparrow}}{\Phi_i^{\downarrow}} \tag{3.27}$$

This quantity, as fundamental as it is for the energy budget of the surface, cannot be measured from satellites. The downwelling solar radiation has in any case to be estimated applying radiative transfer models or measurements at the ground have to be taken as a guideline. Furthermore there exists no possibility to measure the upwelling flux directly. The reasons are twofold: Firstly, the whole wavelength range is not accessible because of the atmospheric absorption bands, secondly, the full range of observation angles cannot be covered instantaneously. Therefore reflection models have to be constructed for different surfaces which have to be applied to the observation from one direction and observations made in different spectral region have to be combined to cover the whole range from 0.3 to 3 µm wavelength.

In order to relate radiance measurements made from space to albedo measurements made at the ground by sensors with hemispherical fields of view, the term of "technical albedo" was introduced. Here it is assumed that the reflection of the surface is isotropic or "Lambertian" (Lambert 1760). In this case, the flux density per unit area (M) reflected from the surface would be πL if L denotes the radiance perpendicular to a unit surface area. If, furthermore, L_{WLR} is the radiance measured from a non absorbing white Lambertian reflector, then $\pi L_{WLR} = E^{\downarrow}$ where $E^{\downarrow}$ is the irradiance (flux per unit area) at the reflecting surface. The reflection of a surface generally is related to the reflection of a non-absorbing white Lambertian reflector and presented in percent of this reference. This implies that a specular reflectance can be up to π-times higher than the signal from the reference. Sometimes, therefore, the reflection is scaled with the factor π to compare it directly with the incoming flux density.

The experimental approach to determine the spectral albedo can be achieved by one of the following three methods. (i) If the spectral irradiance from each hemisphere is measured near the ground by spectral radiometers of hemispherical (2π) field of view, then the ratio of both readings is the albedo. (ii) If an absolutely calibrated radiometer of limited field of view is used, e.g. from larger distance, by which the radiance into a specific direction is measured, in absolute units, then the downwelling irradiance may be computed or measured as before. Enough measurements have to be available to integrate the reflected light over the lower hemisphere this value can be divided by the computed irradiance to result in the spectral albedo. If only one measurement is available then one might assume that the surface reflects isotropically ("Lambertian") into all directions. The radiance reflected or scattered from the surface into the specific direction has then to be multiplied by π in order to obtain the total return into the hemisphere. (iii) The third possibility is to compare the reflectance of a natural surface with that one of a Lambertian reflector with known albedo (e.g. a MgO_2 or Spectralon surface). If the investigated surface can again be assumed to reflect isotropic as the reference surface does, then the ratio of the (relative) radiance of the surface to that one of the Lambertian reflector (corrected if necessary for its reflectance) gives directly the albedo of the surface. If one has to deal with a specular component this last procedure becomes problematic. One might e.g. measure nearly all of the reflected light in one direction and "albedo" values may be obtained which are larger than one. In this case it is more reasonable to divide the measured value by the value obtained from the Lambertian reference target times π. In addition it has to be considered that the downwelling radiation is a mixture of a directional solar beam and a diffuse (but not isotropic) scattered component.

In some work, the directional reflectance is taken as a surrogate for the albedo. This is acceptable as long as changes are studied at a relative scale and if values taken under equal geometrical conditions are compared. In fact experience showed that the reflectance measured at 90° off the direction of the sun and under about 45° observation angle, for a number of surfaces, comes very close to the albedo.

Net Radiation at Inclined Surfaces

The problem of the radiative energy budget of a surface escalates if inclined surfaces have to be considered. For an inclined, infinite surface the problem can be solved rather easily at least for the direct solar radiation but for hilly terrain where each inclined surface is faced by either an opposite one or at least by a horizontal plain a general solution cannot be offered. A topographic model has to be developed and the radiation which the site at the slope received from surrounding terrestrial targets must be computed.

A single slope is characterized by its inclination i against the horizontal plain and θ_p, the azimuth of the vertical plane which contains the normal to the inclined surface with respect to South. The angle β between the solar beam and the normal of the plane under consideration is determined by

$$\cos\beta = \cos i \sin\gamma_s + \sin i \cos\gamma_s \cos\theta \tag{3.28}$$

where γ_s is the angle between the direction of the solar beam and its projection at the inclined surface (the height of the sun over the inclined plane). θ is the difference of the azimuth of the sun and of the azimuth of the vertical plane which contains the normal to the inclined surface. By combining this relation with the solar geometry the following formula evolves for the irradiance E_p of the inclined plane (Robinson 1966):

$$E_p = S[\cos i(\sin\varphi \sin\delta_s + \cos\varphi\cos\delta_s \cos\omega_s) \\ + \sin i\{\cos\theta_p[tg\varphi(\sin\varphi\sin\delta_s + \cos\varphi\cos\delta_s\cos\omega_s) \\ - \sin\delta_s \sec\varphi] + \sin\theta_p \cos\delta_s \sin\omega_s\}]. \tag{3.29}$$

Here
S = the solar constant corrected for atmospheric effects
φ = geographical latitude
δ_s = solar declination
ω_s = solar hour angle measured from the local meridian.

The reflectance follows the same rules as for horizontal surfaces only that in this case the BDRF function may have to be applied for much larger aspect angles. The definition of albedo becomes a problem. For the energy budget of a site at a slope, the net radiation at the inclined plane is effective. At larger scales the albedo of the horizontal plain matters and meteorological satellites are not capable to resolve small orographic structures. An unconfirmed working hypothesis is that also in hilly terrain the regional scale albedo equals the average local albedos, because each upslope are matched by a downslope, and that the same is true for the net radiation. This of course depends on the scale and may not be the case if north slopes carry different vegetation than south slopes.

3.2.3 Thermal IR Radiation

The Earth surface, as all bodies, emits infrared radiation at wavelengths longer than about 2 µm according to its temperature . For a "black" emitter - such as a cavity of appropriate dimensions - the spectral dependence of this emission is described by Planck´s law. In the wavelength domain, the spectral radiance of a black body is given by:

$$B_\lambda(T) = \frac{M_{B,\lambda}(T)}{\pi} = \frac{c_1/\pi}{\lambda^5}\frac{1}{e^{\frac{c_2}{\lambda T}} - 1} \quad (Wm^{-2}sr^{-1}\mu m^{-1}). \tag{3.30}$$

$M_{B,\lambda}(T)$ denotes the spectral radiation flux density emerging from the black body into the hemisphere. λ is in m, T in K, and the values of the constants are:
$c_1 = 2\pi hc^2 = 3.7418 \cdot 10^{-16}$ Wm^2,
$c_2 = hc/k = 1.4388 \cdot 10^{-2}$ mK,
$h = 6.6262 \cdot 10^{-34}$ Js,
$k = 1.3807 \cdot 10^{-23}$ JK^{-1},
$c = 2.9979 \cdot 10^8$ ms^{-1}.

Most surfaces emit less than the radiance equivalent of their thermodynamic temperature as given by Eq. (3.30). Even carefully blackened or wet surfaces are not ideal black bodies though they may approach that condition. If the radiance measured over such a surface and its temperature is computed by inverting Planck' law without taking into account the emissivity, one obtains a value which is below the thermodynamic temperature of the surface. This is generally known as "radiative", "equivalent" or "brightness" temperatures. The effectively emitted radiance in such a case is

$$L_{\lambda,eff}(T) = \varepsilon(\lambda) B_{\lambda}(T) \qquad (3.31)$$

where $\varepsilon(\lambda)$ is the wavelength dependent spectral emissivity.

If the Planck formula is integrated over all wavelengths the Stefan-Boltzmann law for the flux density results:

$$M_B(T) = \sigma T^4 \quad and \quad M_{B,eff}(T) = \varepsilon\sigma T^4 \qquad (3.32)$$

with $\sigma = (5.67051 \pm 0.0002) \cdot 10^{-8}$ W $^{-2}$ K^{-4}. It should be noted here that, though the same symbols are used, the emissivity, ε, should not be confused with the dielectric constant and the Stefan-Boltzmann constant, σ, not with the electrical conductivity.

The emission of a surface is in principle the result of the same processes as those which produce absorption except that the quantum jump between two energy levels is in the opposite direction. It can be shown that in a state of radiative equilibrium the flux emissivity equals absorptance and (1 - reflectance) which is Kirchhoff's law[1]. Wet surfaces have emissivities close to 1.0 because of the absorption of the water molecule with its permanent dipole momentum. For dry soils emissivity sometimes drops to less than 0.9 and especially in strong absorption bands of minerals "Reststrahlenbanden" (see Fig. 3.1) appear with high reflectance and small emissivity. These bands, however, are not as prominent as for pure crystals because of the fractionation of the surface by which small cavities are generated which act as "black bodies" due to multiple internal reflections. Furthermore the radiances observed from different directions in most cases differ from each other because of the variable efficiency of these cavities and/or the distribution of illuminated and shadowed facets. If radiative temperatures are measured under natural conditions in the open field, the measured effective temperature of a surface may not deviate as much as $\varepsilon^{1/4}$ from the thermodynamic temperature because there is always infrared sky radiation reflected at the surface which partly makes up for the reduced emittance.

The pioneering work of Büttner and Kern (1965) to determine the emissivity of natural surfaces was done on more or less homogeneous material in the laboratory. They found in the 8 - 12 µm region minimum values of 0.71 for agate (quartz). Feldspar, obsidian and granite range between 0.82 and 0.9, and sand around 0.93. In the open the mixture of vegetation, soil surfaces, stones, and shaded fractions contribute to the effective emission which makes the quantity in addition dependent on the angle of observation. The emissivity of a dense canopy is close to one during the growth period due to the water in the leaves. The temperature measured over broken canopies, on the other hand, is ill defined because the infrared radiation flux is composed of contributions from different heights, shadowed and fully irradiated parts. The longer the wavelength the more are also layers below the "skin" involved in the effective emission. Already Lorenz's (1973) measurements from aircraft indicated the angular dependence of the measured effective surface temperature

[1]Radiative equlilibrium ceases in the mesosphere where the path length of photons between two collisions becomes very large

which is due to differences of the reflected radiation stemming from different parts of the sky and shadowed facets of the surface.

Remotely sensed temperatures are difficult to validate. This requires spectral information about the emissivity and surfaces which are homogeneous over an area covered by a pixel. In chapter 4 and 5 some methods in use for this purpose are described. For very uniform fine-grained soils a measurements with fine wire thermocouples mounted at 2 mm depth to minimize radiation effects seem to come close to the thermodynamic surface temperature. In some cases the effective surface temperature is representative for a volume in which a temperature gradient exists. This is especially the case in the microwave region where the radiation emitted from the soil stems from the top few centimetres. But also over vegetation an effective temperature may be measured which is composed of contributions from different leaf levels of different temperature.

The knowledge of the thermodynamic surface temperature is a prerequisite for flux calculations. It also is the basis to calculate the upward thermal infrared flux by using Eq. 3.32. Because of the very limited accessibility of the surface from space within the broad range of the thermal infrared spectrum, the surface temperature can only be measured in narrow spectral bands where emissivity and atmospheric correction are known. The temperature, determined this way, is the thermodynamic surface temperature which can be used in heat flux computations. To obtain the effective radiation flux emerging from the surface, it would, in principle, be necessary to compute, throughout the range of the thermal infrared, for each wavelength the effective spectral emission, and then to sum up over the whole spectrum. However, the spectral emissivity, which is necessary for this procedure, is not known and can not be determined for large portions of the spectrum. Consequently one has to live with an approximation and to use the effective surface temperature, T_{eff}, measured in the accessible atmospheric window without emissivity correction as approximately correct for all wavelengths. Then the flux can be calculated according to Eq. 3.32, right hand part, with $T_{eff} = \varepsilon^{1/4}\ T$. This may be justified if the upward infrared flux is defined as the sum of emitted and reflected radiation because the deviation of the emissivity from the value one is partly (to about 75 %) compensated by the reflected atmospheric radiation.

3.2.4 Microwave Radiation

Spectral Range

The microwave region is subdivided into bands designated by capital letters ranging from 100 MHz (I) to 100 000 MHz (W). The boundaries for these bands are not strictly uniform if different authors are consulted. For remote sensing the following bands are commonly used: 0.8 - 1.1 cm wavelength (40 - 26 GHz frequency, Ka band), 1.1 - 1.7 cm (26.5 - 18.5 GHz, K-band), 2.4 - 3.8 cm (12.5 - 8 GHz, X-band), 3.8 - 7.5 cm (8 - 4 GHz, C-band), 15 - 30 cm (2 - 1 GHz, L-band), and 30 - 100 cm (1 - 0.3 GHz, P-band). The S-band, 7.5 - 15 cm (4 - 2 GHz), is mainly used for communication. A major advantage of lower frequency microwaves lies in the fact that the radiation at these wavelengths is less to negligibly affected by atmospheric gases and clouds, except in some narrow intervals where lines of oxygen and water vapour absorb and in the case of large water droplets which affect the higher of these microwave frequencies (Fig. 3.8).

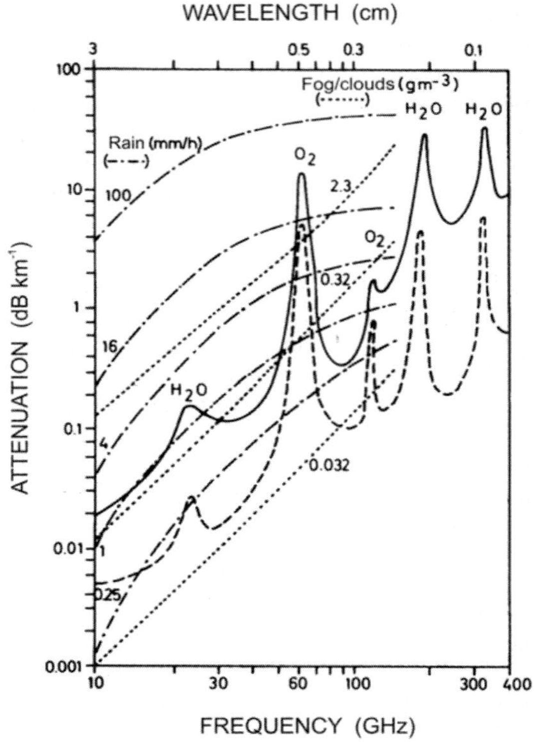

Fig. 3.8. Diminution of mirowave signals by water vapour, oxygen and water droplets. Annonymus (around 1980). The full and dashed lines represent gaseous absorption for a wet and a dry atmosphere, the dotted lines absorption in fog and clounds with different amounts of water mass per cm^3 (0.032 - 2.3 g cm$^{-3)}$, and dashed-dotted lines absorption in rain at different intensities (0.25 - 100 mm/h)

Technology

Concerning the technology and application of microwaves, a clear distinction must be made between passive and active systems. Passive systems use a radiometer to measure the emitted 'thermal microwave radiation'. Active systems (Radar) consist of a source, which produces artificial microwave pulses, and a receiver to measure the radiation which is backscattered from the target (the Earth's surface).

In the passive domain, the emitted radiation has very low intensities. Presently, mainly the C-band and higher frequencies are used for observations from space because longer wavelengths with their low intensities need larger (or synthetic) antennas. The necessity to develop large antennas in space used to be a serious limitation for this technique for a long time. However, the first L-band (1.42 GHz) passive microwave system is planned for launch in 2007. It is the Soil Moisture and Ocean Salinity (SMOS) mission (Kerr et al. 2001), which is an ESA opportunity mission. SMOS is a multi-angle 2-dimensional (imaging) interferometric radiometer which will scan the Earth with a sub-track spatial resolution of ~ 35 km, simultaneously at a large number of incidence angles. This gives the possibility of simultaneous retrieval of a series of surface physical properties (Wigneron et al. 2000, Van de Griend et al. 2002).

In the active domain, the intensity of radiation forms no limitation and, therefore, at lower frequencies, much higher spatial resolutions can be realized compared to the passive domain. The Japanese JERS-1 operates an active L-band instrument with a spatial resolution of 18 m. In contrast, in the passive domain, C-band is the lowest frequency currently operational from space with a spatial resolution of several tens of kilometres. The C-band Scanning Multi-channel Microwave Radiometer (SMMR) operated from space between 1978 and 1987 with a spatial resolution of over 100 km. Currently, the Advanced Microwave Scanning Radiometer-Earth (AMSR-E) observing system launched 4 May 2002 improves the spatial resolution of observations made by the earlier generations of passive microwave instruments (e.g. SMMR and SSM/I) and retrieves information in more frequencies than its predecessors. AMSR-E flies on NASA's Earth Observing System (EOS) Aqua satellite with a C- band radiometer as the lowest frequency.

The potential of C-band (6.6 GHz) microwave signatures for large scale surface moisture monitoring from space has been demonstrated by Owe et al. (1992) and Van de Griend and Owe (1994a, 1994b) on the basis of long-term analysis of SMMR signatures. Higher frequencies (X and K-band) are strongly scattered and absorbed by vegetation and, therefore, can be used for vegetation monitoring. The highest frequencies (Q-band) are strongly influenced by the atmosphere and, therefore, can be used for monitoring atmospheric constituents.

Dielectric Constant

The interaction of microwave radiation follows the same physical rules as that of radiation in the visible and near infrared part of the spectrum. The only difference is that in the microwave domain the movement of free charges in the emitting medium (target) governs the exchange of energy. At low frequencies charged particles can move within their microdomains which creates charge separation (polarization) in addition to the external field. To account for this polarisation, the imaginary part of the relative dielectric constant must be extended in the microwave region to account for the losses due to conductivity by free charges by the term $\sigma/\varepsilon_0\omega$. σ is the electrical conductivity, ε_0 the permittivity (dielectric constant) of free space and ω the circular frequency of the electromagnetic wave.

The effective complex permittivity (or dielectric constant) was defined in Eq. 3.9 as $\varepsilon_{eff} = \varepsilon_0 (\varepsilon_r' - i\varepsilon_r'') = \varepsilon(1 - i\sigma/\omega\varepsilon)$. If the complex dielectric constant is expanded by the term of the conductivity by free charges, the following expression results :

$$\varepsilon = \varepsilon' - i\left(\varepsilon'' + \frac{\sigma}{\varepsilon_0 \omega}\right) \tag{3.33}$$

In dry soil material the imaginary part of the dielectric constant is typically smaler than 0.05 and the real part varies between 2 and 4 according to a bulk formula of Dobson (1985):

$$\varepsilon_{drysoil} = (1.01 + 0.44\rho_b)^2 - 0.062 \tag{3.34}$$

It depends on the bulk density, ρ_b, and is essentially independent of both temperature and frequency. Changes of the dielectric constant are mainly due to the variable water content. The real part of the dielectric constant of sandy loam increases non-linearly from about 2.5 for dry material to 20 - 30 if the volumetric moisture content goes up to 0.4 and the imaginary part increases at the same time from zero to 3 - 10 depending on the frequency. Microwaves therefore are predestinated to measure soil moisture. For the frequencies used in satellite instruments a wet soil surface is characterized by a high dielectric constant which results in a high reflectivity and thus in a low emissivity which typically varies between approximately 0.6 for a wet soil and 0.95 for a completely dry soil. In terms of brightness temperatures, assuming a thermodynamic temperature of 300 K, this means a dynamic range between 180 K for wet soil and 285 K for dry soil. In addition soil type, incidence angle, and surface roughness influence the actual emissivity (Schmugge 1985, Owe et al. 1992).

Emission

Passive systems measure the microwave radiation emitted by the changes of translational movements of charged molecules or free electrons causing polarization and a fluctuating electromagnetic field which depends on temperature.

The energy emitted in a narrow spectral band was defined by Eq. 3.30 for the wavelength domain. In the microwave region radiances and radiation fluxes are generally presented in the frequency domain ($f = c/\lambda$). In both formulations the radiant fluxes (measured in W m^{-2}) must be identical: $B_f\, df = B_\lambda\, d\lambda$. The following relationship evolves from this identity between B_λ and B_f:

$$B_f = B_\lambda \frac{d\lambda}{df} = B_\lambda \frac{d\lambda}{d(c/\lambda)} = B_\lambda \frac{\lambda^2}{c} \tag{3.35}$$

Applied to Eq. (3.30) one obtaines for B_f:

$$B_f = \frac{c_1 f^3}{\pi c^4} \frac{1}{e^{\frac{c_2 f}{cT}} - 1} \quad W\, m^{-2}\, sr^{-1}\, Hz^{-1} \tag{3.36}$$

For long wavelengths the exponential function e^x can be approximated by 1+x and the relationship between the temperature of the emitting surface and the emitted radiance becomes linear, and the Planck formula approaches the formula of Wien (Ulaby, 1986):

$$B_\lambda(T) \xrightarrow[\lambda \to \infty]{} \frac{c_1}{c_2 \pi \lambda^4} T \tag{3.37}$$

The emitted radiance becomes a linear function of the physical temperature of the emitting body:

$$L_\lambda \approx \varepsilon_\lambda T_{eff} \tag{3.38}$$

where ε_λ is the emissivity which takes care of the deviation of the exitance from that of a black body. Due to the linear relationship between radiance and temperature, the surface emissivity can be defined in terms of the radiometric brightness temperature, $T_{B,p}$, where p denotes the polarization, and the thermodynamic temperature of the emitting body by the following linear relationship:

$$\varepsilon_p(\lambda) = \left(\frac{T_{B,p}}{T_{s,eff}} \right)_{\lambda > 1mm} \tag{3.39}$$

Here, the actually measured effective thermodynamic temperature of the emitting elements of the Earth's surface is indicated as an effective surface temperature $T_{s,eff}$. In contrast to the infrared radiation the exitance is not generated directly at the surface but, because of its larger penetration depth, within a volume below the surface which gets larger as the wavelength becomes longer. For bare soil, the effective thermodynamic temperature at these wavelengths therefore is a weighted average value over the penetration depth at the given frequency (Mätzler, 1994):

$$T_{s,eff} = \int_{-\infty}^{0} T(z)\alpha(z) e^{\int_{z}^{0} \alpha(z')dz'} dz \tag{3.40}$$

$T(z)$ is the soil temperature at depth z and α, the absorption coefficient, is defined by (see Eq. 3.14 and 3.15):

$$\alpha(z) = \frac{4\pi}{\lambda_0} \kappa(z) \tag{3.41}$$

where λ_0 is the wavelength in free space (Van de Griend 1994c).

Although $T_{s,eff}$ is related to the thermodynamical surface temperature, it may differ substantially from it, especially during daytime hours when strong temperature gradients may occur within the thermal sampling depth of the top soil. The thermal sampling depth may be approximated as proposed by Ulaby (1986):

$$\delta_p \cong \frac{\lambda\sqrt{\varepsilon'}}{2\pi\varepsilon''} \tag{3.42}$$

During summer, these differences may be up to ~ 10 K for the C-band at daytime and 5 K at nighttime (Van de Griend 2001). If the surface is completely vegetation covered, the effective temperature approaches the canopy temperature.

Beside the temperature structure within the uppermost soil layer the surface roughness controls the emission of the surface. Choudhury et al. (1979) proposed a simple empirical single parameter model which includes a roughness parameter in the formulation of the emissivity. If Kirchhoff's law is applied the emissivity can be expressed by (1 - reflectance). Choudhury et al. (1979) added to this relation an exponential factor:

$$\varepsilon_p(\theta) = 1 - \rho_p^{sp}(\theta) e^{-h\mu^2} \tag{3.43}$$

where ρ_p^{sp} is the soil reflectance, h is an empirical roughness parameter and μ is the cosine of the incidence angle θ. Only for surfaces of which the exponential factor is close to 1.0 and vertical soil moisture distribution is uniform, the polarization reflectivity can be calculated by the Fresnel equations (see Eq. 3.16 and 3.17).

Effect of Vegetation

Vegetation may have a dominating effect on surface emissivity. The effect increases with increasing microwave frequency and with increasing biomass. Therefore, the effect of vegetation must be incorporated in any inversion algorithm to interpret microwave brightness temperatures in terms of surface soil moisture and vegetation biomass. It can be parameterized using two parameters, the single scattering albedo, $\tilde{\omega}$, and the optical depth, δ. The optical depth of the canopy, δ_c, is related to the canopy transmissivity (τ_c) by the relationship: $\tau_c = \exp(-\delta_c/\mu)$. The single scattering albedo characterizes the reflectivity of the leaf surface in the microwave domain. If volume scatering within the canopy is neglected, both ω and τ_c determine the emissivity and the reflectivity of a canopy. According to Van de Griend et al. (1996) the canopy radiation properties can in this case be approximated by the following quantities:

Leaf surface reflectivity (single scattering albedo):	$\tilde{\omega}$
Canopy transmissivity:	τ_c
Canopy reflectivity:	$\tilde{\omega}(1 - \tau_c)$
Canopy emissivity:	$(1 - \tilde{\omega})(1 - \tau_c)$

Making use of the linear relationship between temperature and radiance the three components of the radiation flux emerging from the top of the canopy can be expressed in terms of temperatures:

$T_1 = T_{s,eff}\,\tau_c\,[1 + \tilde{\omega}(1 - \tau_c)(1 - \varepsilon_s)]$ - emission of the surface which penetrates the canopy directly (τ_c) and after one reflection each by the canopy $\omega(1 - \tau_c)$ and the underlying soil $[\rho_s = (1 - \varepsilon_s)]$.

$T_2 = T_c\,(1 - \tilde{\omega})(1 - \tau_c)$ - upward emission of the canopy,

$T_3 = T_c\,(1 - \tilde{\omega})(1 - \tau_c)(1 - \varepsilon_s)\tau_c$ - downward emission of the canopy reflected by the surface and transmitted through the canopy.

The combined signal then becomes in terms of temperatures (Mo et al. 1982) the sum of the three terms:

$$T_{eff} = T_{s,eff}\,\tau_c[1 + \tilde{\omega}(1 - \tau_c)(1 - \varepsilon_s)] + T_c(1 - \tilde{\omega})(1 - \tau_c)[(1 + (1 - \varepsilon_s)\tau_c] \quad (3.44)$$

A more complete formulation would have to take into account also multiple scattering. This is especially necessary if the wavelength is smaller than the individual elements of the canopy (Wigneron et al. 1995a and b).

Several previous works analysed the relationships between surface temperature and microwave emission in quite some depth (e.g. Choudhury et al. 1995, Njoku 1995). Application and validation of these methods under 'real world conditions' with archived satellite data remains still a research task since concurrent information on physical and thermal properties of the land surface is often lacking. This is especially a drawback at the scales of the Nimbus/SMMR microwave radiometer data, although the potential of satellite based measurements at 6.6 GHz for large-scale soil moisture monitoring was demonstrated by Owe et al. (1992) and Van de Griend and Owe (1994a; 1994b).

The vegetation parameters $\tilde{\omega}$ and τ_c are difficult to estimate. Single scattering albedo varies with vegetation type (Mo et al. 1982, Brunfeldt and Ulaby 1984, Jackson and O'Neill 1990) and can be determined in the field using a microwave radiometer (Van de Griend et al. 1996). The method, however, requires repetitive measurements before and after harvesting and is therefore quite labour intensive. For practical purposes and for global monitoring one needs to find other ways of determining the canopy parameters, either by correlating canopy parameters to other satellite derived signatures (e.g. the vegetation indices MPDI or NDVI, see Chapter 4) or by inverse modelling, assuming a certain behaviour of the parameters for the different polarizations (Van de Griend and Owe, 1994b).

Reflection and Scattering

In the microwave region, reflection measurements are only feasible with active systems since the power of the solar radiation is too low at these wavelengths. Advanced active instruments mostly operate at two polarization directions, horizontal (H) and vertical (V). The optical properties of a soil surface are generally described by the backscattering coefficient

$$\sigma_{rt}^{0}\left(\theta_s,\phi_s;\theta,\phi\right) = \frac{4\pi R^2\left|E_r^s\right|^2}{A\left|E_t^i\right|^2} \tag{3.45}$$

where r and t stand for the parallel respectively vertical polarization, θ, φ for the illumination (index s) respectively observation geometry, R is the distance (satellite height) and A the illuminated area. E_t^i is the electrical field vector of the incident wave and E_r^s that one of the scattered wave. The scattering coefficient contains both the effect of the surface roughness and of its dielectric constant. Extensive laboratory studies have shown that the effect of roughness often dominates the effect of changing soil moisture (Wüthrich and Treubig, 1998). Durden et al. (1992) developed a simple scheme to decompose a signal into three components: Single reflectance at the surface, volume scattering in the canopy, and twofold reflection at the ground. This scheme enabled Zink and Schmullius (1994) to attribute high C-band values to volume scattering which decreases as the wavelength increases. One result also was that conifers have higher backscatter coefficients than deciduous forests.

The canopy volume scattering coefficient can be defined by:

$$\sigma_v^{0,i} = \gamma_v^i\left(\frac{1-e^{-\frac{2\delta_i}{\cos\theta}}}{\frac{2\delta_i}{\cos\theta}}\right), \qquad i = HH\,respectively\,VV. \tag{3.46}$$

δ_i is the optical thickness of the canopy, γ_v^i the vegetation volume scattering factor, θ the illumination and backscatter angle. The number two results from the wave travelling twice through the canopy. The term

$$\sigma_{s-v}^{0,i} = \gamma_{s-v}^i\left|R_g^i\right|^2 e^{-\frac{2\delta}{\cos\theta}} \tag{3.47}$$

represents the interaction between the canopy and the soil surface. R is the Fresnel reflection coefficient of the ground surface and γ_{s-v} the surface volume scattering factors.

The contribution of the vegetation covered soil is given by

$$\sigma_{soil}^{0,i} = \sigma_{s}^{0,i} e^{-\frac{2\delta}{\cos\theta}} . \tag{3.48}$$

Here σ_s is the surface backscattering coefficient for a rough surface. In the case of bare soil the exponential function would be 1.

In the case of the active systems the treatment of the scattering at rough surfaces requires special attention. Only a rough bare surface would scatter some radiation back into the direction of the radar. A bare flat surface would just reflect the wave under the reflection angle away from the sensor which would not receive any energy. In the description of the surface roughness the r.m.s. height (σ), the autocorrelation lengths in the x, y directions, $l_x = 2\pi/k_x$, $l_y = 2\pi/k_y$, and the r.m.s. surface slope $m = 2^{1/2}\sigma/l$ play a role. The autocorrelation length can be derived from the normalized power spectral density function at the points where this function falls off to the value 1/e. Three models have been derived:

- the small perturbation model (Ulaby et al. 1986),
- the physical optics model, and
- the geometrical optics model, that uses only normal incidence and therefore provides no expressions for the cross-polarized backscattering.

The small perturbation theory provides estimates of the cross-polarized backscatter coefficient if evaluated up to the second order (Valenzuela 1967). In the first order solution of the physical optics model the depolarizes backscatter coefficients are zero.

3.3 Atmospheric Radiative Transfer

3.3.1 Processes

To remove the influence of the atmosphere from remotely measured signals requires the inversion solution of the equation of radiative transfer or an equivalent approach. There exists a number of mathematical methods, called "radiation codes", to deal with multiple scattering and line structures of absorption bands in solving the equation of radiative transfer. The correction schemes range from very simple approaches which consider only single scattering and bulk absorption coefficients of atmospheric bands, to very elaborate computational techniques, which include multiple scattering, "false" signals introduced by photons reflected from areas beside the target but scattered into the line of view by atmospheric molecules and aerosols, and by correctly taking into account the line structure of atmospheric absorption and emission bands. Atmospheric corrections for the longer wavelength, by which surface temperatures are determined, differ in so far from those applied for the short wavelength in which the surface emissivity plays a role and scattering is mostly negligible. Only few of the more commonly used codes will be described below.

The accuracy of the correction scheme depends on both the quality of the mathematical codes and the quality of the atmospheric parameters by which these codes have to be adjusted to the specific instantaneous conditions. These input parameters, in principle, are vertical profiles of all atmospheric constituents. Especially the highly variable water vapour profiles and temperature profiles are of importance. Carbon dioxide profiles can be taken as constant, though the total amount changes. In a number of cases the total atmospheric ozone content may be sufficient for corrections. To characterize aerosol, its single scattering albedo and number density profiles are required. The trace gases which absorb in the thermal infrared as mentioned above must also be considered. These parameters are often not completely known and must be replaced by "climatological" data or intelligent estimates.

Details regarding various methods of atmospheric correction for estimating land-surface temperature (LST) can be found in the recent comprehensive review article by Dash *et al.* (2002).

Atmospheric constituents interact with the electromagnetic radiation by absorption, emission, and scattering. To determine the effect of the atmosphere on remotely sensed objects requires the solution of the equation of radiative transfer. Assuming an incremental path *ds* within the atmosphere, the attenuation of the electromagnetic wave is proportional to the incident radiance $L(s)$ and the extinction coefficient $\sigma_{ext} = \sigma_{scat} + \sigma_{abs}$, where σ_{abs} is the absorption coefficient and σ_{scat} the scattering coefficient, each at unit pathlength:

$$dL(s) = -\sigma_{ext} L(s)ds \tag{3.49}$$

The geometrical path can be positioned in any direction described by the azimuth angle φ and the zenith angle ζ. It is related to the vertical coordinate z by $ds = dz/\cos\zeta$ if the upward direction is taken as positive. Instead of the vertical height the dimensionless *optical depth* of the atmosphere can be introduced:

$$d\delta = -\sigma_{ext} dz = -\sigma_{ext} \cos\zeta\, ds = -\sigma_{ext}\mu\, ds, \tag{3.50}$$

where $\cos\zeta$ is abbreviated by μ. If a transmission under an inclined zenith angle is considered, it has to be realized that the $\cos\zeta$ approximation only holds for angles $\leq 80°$ because of the curvature of the Earth. Beyond that angle the formula given in section 3.4 has to be applied.

At the same time that the radiance is attenuated by scattering and absorption it is enhanced by radiation scattered out of the radiances from all other directions s' into the direction s and by radiation emitted along the path ds. The radiance scattered into the direction $s(\varphi,\zeta)$ is $dL_{scat} = \hat{J}(\delta;\varphi,\zeta;\varphi',\zeta')ds = \hat{J}(\delta;\varphi,\zeta;\varphi',\zeta')d\delta/\mu\sigma_{ext}$. $\hat{J}$ is the *source function* for the scattered radiation. In analogy the radiance emitted by the volume into the direction s is $dL_{em} = \hat{I}(\delta;\varphi,\zeta)d\delta/\mu\sigma_{ext}$. The change of the radiance due to the interaction with the matter in the volume around *ds* consequently can be described by

$$dL(\delta;\varphi,\mu) = -L(\delta;\varphi,\mu)\frac{d\delta}{\mu} + \hat{J}(\delta;\varphi,\mu;\varphi',\mu')\frac{d\delta}{\mu\sigma_{ext}} + \hat{I}(\delta;\varphi,\zeta)\frac{d\delta}{\mu\sigma_{ext}} \tag{3.51}$$

Division by $d\delta/\mu$ and generalization by writing $(1/\sigma_{ext})(\hat{J}+\hat{I}) = J$ yields

$$\mu\frac{dL(\delta;\varphi,\mu)}{d\delta} = -L(\delta;\varphi,\mu) + J(\delta;\varphi,\zeta;\varphi',\zeta'), \tag{3.52}$$

where J is the general source function. The correction of remotely sensed signals is based upon the solution of this equation of radiative transfer. Fortunately, in the range of the solar spectrum the source function of emission is zero and in the thermal infrared range scattering is small and often can be neglected. Thus it is possible to consider these two cases separately.

Absorption takes place in gases as well as aerosols. The most prominent absorbers are water vapour, carbon dioxide and ozone. But also methane and its derivatives, nitrous oxide, carbon oxide and oxygen can readily be seen in the absorption spectrum of the sun. The strongest absorptions are caused by transitions from a molecular ground state to an exited state. The upper state is characterized by numbers that indicate the overtones of the fundamental vibrations. In the case of water vapour these are three fundamentals (ν_1, ν_2, ν_3) at 2.734 (100), 2.662 (001) and 6.27 (010) μm. The resulting overtone bands in the solar spectrum are the relatively weak ones at 698.38 nm (103), 723.02 nm (301), 795.86 nm (013, vw), 822.96 nm (211), 906.42 nm (003, vw), and the stronger ones that are designated by Greek letters: 913.2 nm - ρ(003), 942.8 nm - σ(201), 970.3 nm - τ(121), 1135.5/1194.2 nm - Φ(121/031), 1379.3/1454.8 nm - Ψ(101/021, w), 1875.5/1910.2 nm - Ω(011/110). The ρ, σ, and τ bands strongly overlap and are often quoted as one ρστ-band. The letter w stands for "weak", vw for "very weak".

Oxygen bands are discernible at 762.1 nm (A-band) and 688.4 nm (B-band). Ozone beside the strong ultraviolett band has a broad absorption feature between 450 and 740 nm(Chappuis bands) which has to be accounted for in remote sensing applications. Its peak absorption by traversing through two air masses is about 7%. Carbon dioxide possesses absorption features in the near infrared consisting of overlapping overtone bands between 1.38 - 1.5 μm (1.434 μm "band"), 1.52 μm - 1.67 μm (1.6 μm "band"), and 1.92 - 2.10 μm (2.0 μm "band"). Most of the various bands in the solar spectrum can be seen in Fig. 3.9 that shows the ratio of the global radiation flux density at the surface to the extraterrestrial solar irradiance for a station in the Republic of Niger when the sun is near to the zenith. Spectra of the reflected solar near infrared solar spectrum observed from the high altitude station Jungfraujoch in Switzerland are presented in Appendix 6.

The thermal infrared is governed by the fundamental vibration bands of water vapour at 6.27 μm (ν_1), 2.73μm (ν_2), and 2.66μm (ν_3), ozone (9.6 μm), and carbon dioxide (3.3 μm and 15 μm) and the pure water vapour rotational spectrum beyond 10 μm. In addition bands of nitrous oxide (7.78 and 8.56 μm), methane (7.66 μm) and chloro-fluor-methanes distributed over the atmospheric "window" in the thermal infrared play a role. In the thermal infrared spectral region the atmospheric gases emit radiation according to their temperature and the pressure and absorb radiation emitted by the ground or by themselves.

Atmospheric molecules scatter shortwave radiation which causes the blue colour of the clear sky. The larger aerosol particles scatter also longer electromagnetic waves and cause the more whitish hazy light. Some components of the aerosols also absorb radiation. Ice crystals scatter also longwave infrared radiation.

In the microwave region some lines of water vapour and oxygen absorb radiation and large water droplets scatter the waves. The efficiency of these scattering processes decreases strongly with increasing wavelength (Fig. 3.8).

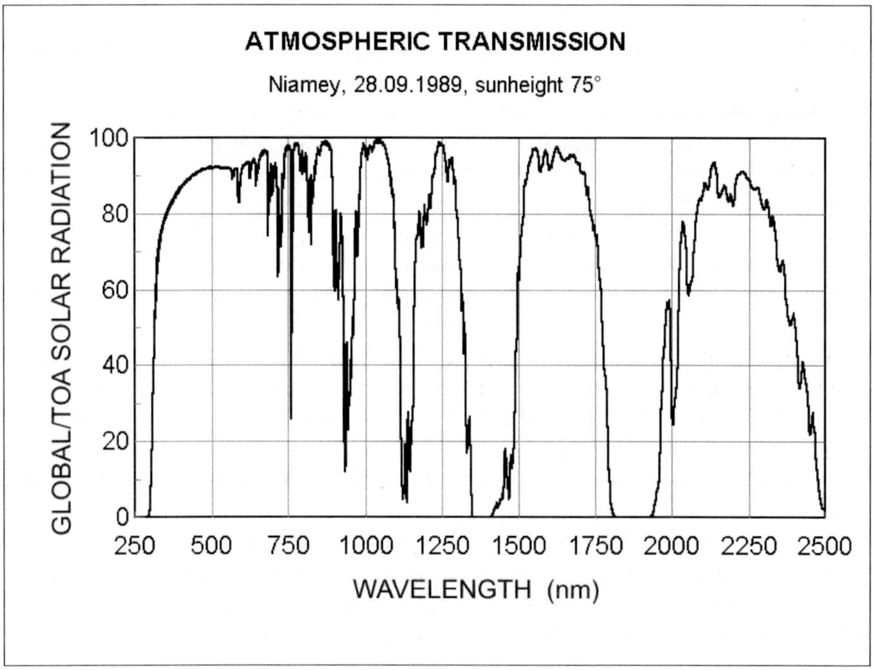

Fig. 3.9. Ratio of global radiation to extraterrestrial solar flux density computed from radiosonde data for Niamey

3.3.2 Radiative Transfer in Scattering Atmospheres

The scattering by molecules, called Rayleigh (Lord Rayleigh 1899, 1902) scattering, can be treated relative easily because of the symmetry of the Rayleigh scattering phase function:

$$p_R(\vartheta) = \frac{3}{4}(1 + \cos^2 \vartheta) \tag{3.53}$$

and the simplicity of the Rayleigh scattering coefficient which depends on the fourth power of the wavelength:

$$\sigma_R(\lambda) = \frac{32\pi^3[n(\lambda) - 1]^2 f}{3N\lambda^4} m^{-1} \tag{3.54}$$

where N = $2.6867 \cdot 10^{25} m^{-3}$ = number of molecules per unit volume (Loschmidt number) and f = 1.054 = Cabannes factor (Cabannes 1929) which corrects for the non-sphericality of the molecules. Because the index of refraction of the air depends slightly on the wavelength (see section 3.4), the overall spectral dependence of the Rayleigh scattering coefficient is proportional $\lambda^{-4.09}$ ($\sigma_R \cong 1.06 \cdot 10^{-8} \lambda^{-4.09}$).

The treatment of the scattering by aerosol particles, called Mie-scattering (Mie 1908), is much more complicated because each aerosol type has an individual scattering function and single scattering albedo (WMO 1986). Aerosol particles cannot be treated as small, approximately spherical spheres. The larger they are in comparison to the wavelength the larger becomes the forward scattering. In addition multiple scattering has to be included in accurate computations. Many algorithms have been developed to deal with the combined atmospheric Rayleigh and Mie scattering as well as ground reflectance processes (a comprehensive summary was given by Lenoble 1985). For most remote sensing applications these techniques are too elaborate and the aerosol input parameters are not always available to be applied individually to each image pixel. In remote sensing applications therefore often simple approximations are used to correct for atmospheric effects.

The principle of the extraction of surface reflectances from the mixed signal is illustrated in Fig. 3.10. Relative to the solar irradiance at the top of the atmosphere the fraction $\tau = \tau(\zeta)$, ζ = solar zenith angle, reaches the surface as direct solar radiation. The transmittance τ contains absorbing components (gases and aerosols), the molecular or Rayleigh scattering and the aerosol or Mie scattering. All these components have different vertical distributions. From the direct solar radiation the fraction (1 - S - A) is taken out of the beam, where S is the scattered flux and A the absorbed energy. To obtain these two components an integration through the whole atmosphere would be necessary. Approximately one might treat the two components separately by taking out the absorbed component first which leaves the fraction (1-A) for the scattering processes by molecules and aerosols. The fraction absorbed by the

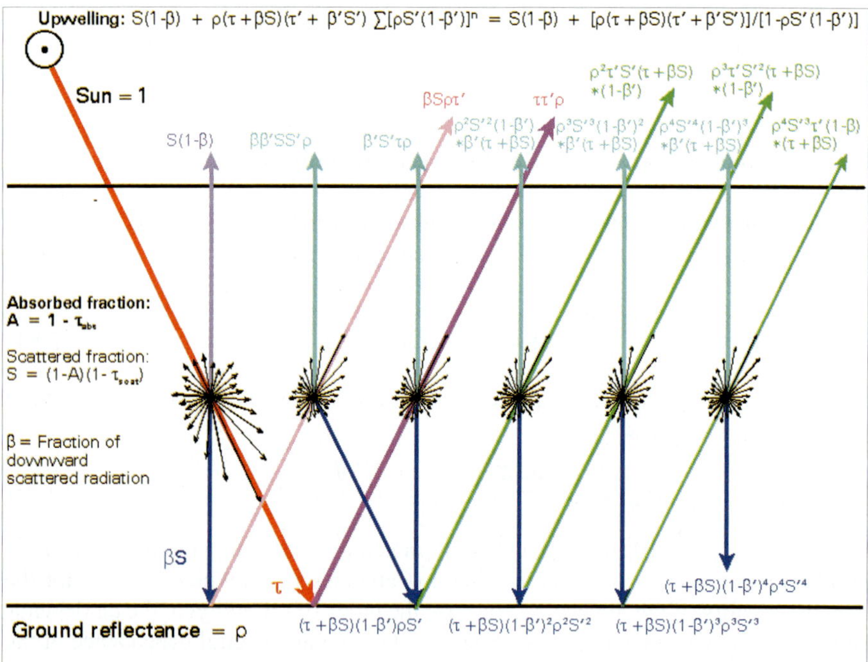

Fig. 3.10. Schematic radiative transfer in the atmosphere-surface system neglecting multiple scattering within the atmosphere

aerosols at the way through the atmosphere lies between 0.1 and 0.2. The molecular fraction of the scattered radiation is nearly equally partitioned into downward and upward fluxes but the single scattering by aerosols is mostly in the forward direction. The fraction β of the radiation is scattered in the direction of the earth surface, the fraction 1-β back to space. $\beta = \beta(\zeta)$ depends on the solar zenith angle and the scattering function of the aerosol. The direct beam reaching the surface is absorbed, scattered or reflected at the surface from where the fraction ρ returns to the atmosphere and the fraction $\rho\tau'$ to space. $\tau'=\tau(\zeta)$ is the transmission on the way through the atmosphere towards the satellite. In addition the fraction ρ of the diffuse sky radiation is reflected back into the atmosphere. The upward reflected fluxes again generates downward scattered fluxes which in successive orders are partly reflected at the surface and *in toto* generate the outgoing flux which defines the albedo at the top of the atmosphere. This simple model neglects the bidirectional reflection features of the surface and the multiple scattering in the atmosphere and does not distinguish between Rayleigh and Mie scattering. It results in the following expression for the upward directed radiance at the TOA:

$$L_{TOA,up}(\vartheta) = \frac{S_0(\bar{d}/d)^2 \cos\zeta}{\pi}\left[S(\zeta)\{1-\beta(\zeta)\} + \rho\frac{\{\tau(\zeta)+\beta(\zeta)S(\zeta)\}+\{\tau(\vartheta)+\beta(\vartheta)S(\vartheta)\}}{1-\rho S(\vartheta)\{1-\beta(\vartheta)\}}\right] \quad (3.55)$$

with $S = \tau_{abs}$ $(1-\tau_{scat})$ and ζ indicating the direction towards the satellite. Eq. 3.55 demonstrates the non-linear relationship between $\rho_{surface} = \rho$ and ρ_{TOA}. To evaluate these and other more sophisticated algorithms a number of parameters have to be calculated for which some guidelines are given in section 3.4.

3.3.3 Radiative Transfer in Absorbing and Emitting Atmospheres

The solar radiation does not play a role in the thermal infrared part of the spectrum. As will be seen in Fig. 3.12 the atmosphere radiates about 10 % of a black body of 300 K at a wavelength of 10 μm from the zenith. Though the sun is much hotter than the Earth (about 5770 K against 300 K) and its radiance is about 500 times as much as the atmosphere radiates in the average from the hemisphere at 10 μm, it covers only $6.8 * 10^5$ sr of the hemisphere.

Atmospheric radiation is generated in bands of atmospheric gases, cöods an aerosols. The bands of gases are characterized by absorption coefficients which result from the overlapping of hundreds and thousands of individual lines. Of almost all of these lines the line strength and the halfwidth as well as their temperature and pressure dependencies are known. For every spectral interval therefore a temperature and pressure dependent absorption coefficient $a(T,p)$ can be computed. The emission of an inclined path of an infinitesimal slab dz of the atmosphere is proportional to the absorbing mass $\rho(z)$ dz/μ, the absorption coefficient $a(T,p)$ and the Planck function $B(T)$:

$$dL(z,\Delta z) = a(p,T)\rho(z)B[T(z)]\frac{dz}{\mu} \quad (3.56)$$

The fraction

$$dL(z,\Delta z)_{z=0} = dL(z,\Delta z)\tau(\mu;z,0)$$

$$= a(z)\rho(z)B[T(z)]\exp\left(-\int_0^z a(z)\rho(z)\frac{dz}{\mu}\right)\frac{dz}{\mu} \tag{3.57}$$

arrives at the surface. If the transmittance τ and the optical depth δ,

$$\tau(z) = \exp\left(-\frac{\delta(z,0)}{\mu}\right), \quad \delta(z,0) = \int_0^z a(z)\rho(z)dz \tag{3.58}$$

are introduced, Eq. (3.57) simplifies to

$$-(dL)_{z=0} = B[T(z)]\frac{d\tau(\mu;z,0)}{dz} \tag{3.59}$$

and the downward directed emitted radiance of the whole atmosphere becomes

$$L(\mu;0,\infty) = -\int_0^\infty B[T(z)]\frac{d\tau(\mu;z,0)}{dz}dz \tag{3.60}$$

The downward looking satellite instrument in addition sees the emission of the surface attenuated by the atmosphere:

$$L(\mu;\infty,0) = \varepsilon B[T(0)]\tau(\mu;0,\infty) - \int_\infty^0 B[T(z)]\frac{d\tau(\mu;z,0)}{dz}dz. \tag{3.61}$$

In this case the integration over the atmosphere has to be inverted which leads to a different atmospheric emission than as seen from below. This immediately becomes clear if one looks into a strong absorption feature: Standing at the ground

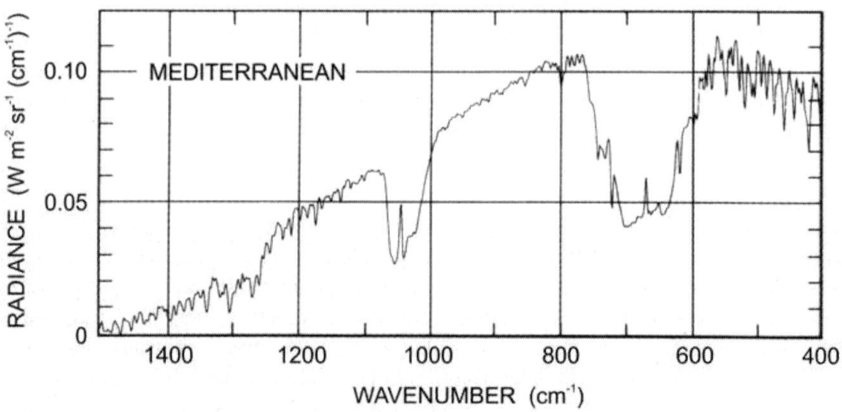

Fig. 3.11. Emission spectrum 400 - 1500 cm^{-1} (6.7 - 20 μm) at the top of the atmosphere measured over the Mediterranean sea on 10 April 1970 after Hanel (1971). To compare with Fig. 3.11 it has to be noted that the spectrum goes beyond 15 μm and is plotted over a linear wavenumber rather than wavelength scale: $L_\lambda = L_\nu\lambda^{-2} = L_\nu\nu^2$

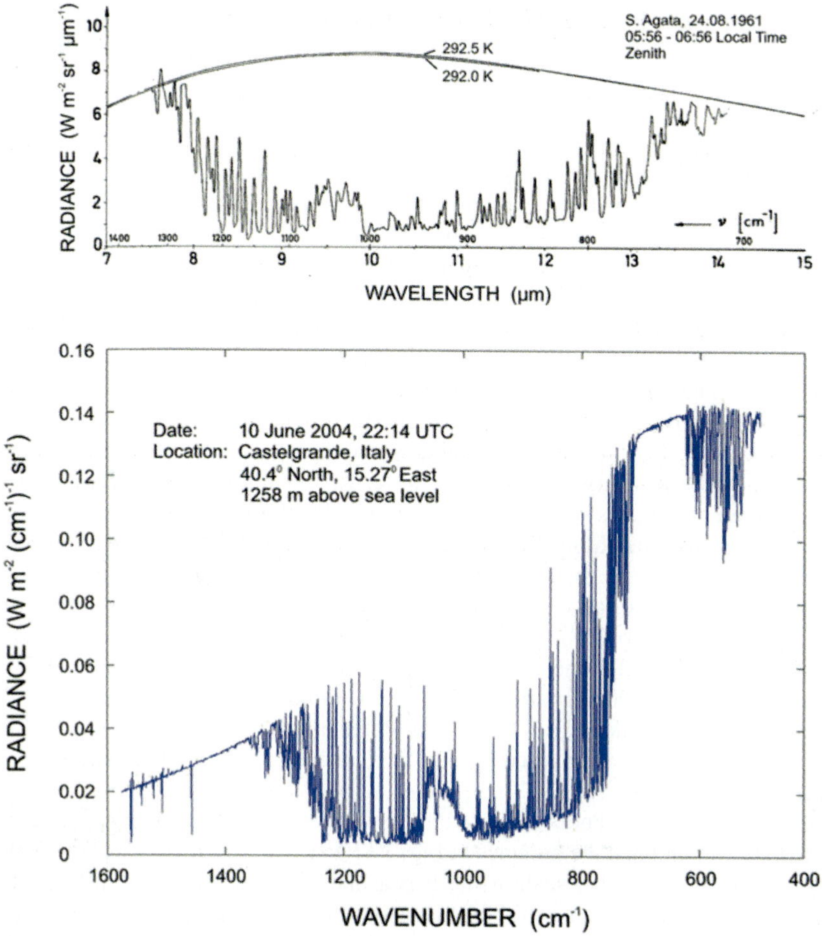

Fig. 3.12. Comparison of thermal infrared sky emission spectra measured with a grating spectrometer in 1961 and modern interferometric technique in 2004. The upper Zenith emission spectrum ranging from 7 - 15 µm wavelength (700 - 1400 cm^{-1} wavenumber) was measured at the peninsula of Sorrento, Italy, altitude 387 m, on 24 August 1961. Note that the Chlorofluoromethane bands were not present at the time. The lower spectrum ranging from 6.25 to 20 µm (500 to 1600 cm^{-1}) was recorded with an interferometer only 105 km apart to the east at Castelgrande, altitude 1258 m, but on 10 June 2004 (courtesy Vincenzo Cuomo, Istituto di Metodologie Per l'Analisi Ambientale del CNR, Tito Scalo, Potenza). The different slope of the spectra results from the fact that the radiance in the upper one is displayed per wavelength, in the lower one per wavenumber (compare text)

one sees an emission close to the emission of a black body at surface temperature and looking from above the emission represents a temperature at higher levels in the troposphere or even the stratosphere. Mathematically speaking, the same increment of the optical depth is multiplied by a different black body emissivity depending on whether one looks from the bottom or the top of the atmosphere, because it is positioned in different heights of the atmosphere. Figs. 3.11 and 3.12 illustrate this fact.

The spectrum shown in Fig. 3.12 was taken with an interferometer from satellite over the Mediterranean sea (Hanel 1971). The emission in the water vapour and carbon dioxide bands is lower than in the window because this is an emission from cooler levels high up in the troposphere.

The spectrum in Fig. 3.12, upper panel, was measured from the peninsula of Sorrento in Italy and shows an early morning zenith spectrum of the atmospheric radiation in the 7 - 14 μm range. In the region of the strong water vapour and carbon dioxide bands the emission is approaching that of a black body at air temperature.The spectrum also shows that the so-called "atmospheric window", the region between 8 and 13 μm, is opaque due to water vapour and ozone emission lines respectively bands and an underlying water vapour continuum. This continuum is caused by the wings of distant strong water vapour lines and the emission from water vapour aggregates $(H_2O)_n$. The spectrum is insofar of historical interest since it was taken when the chlorofluoromethan bands were still not detected. Nowadays the measurements are made more efficiently and with much higher spectral resolution by means of an improved interferometer technique. The lower spectrum ranging from 6.25 to 20 μm (500 to 1600 cm^{-1}) was recorded with a Fourier transform interferometer only 105 km apart to the east at Castelgrande, altitude 1258 m, but on 10 June 2004 (courtesy Vincenzo Cuomo, Istituto di Metodologie Per l'Analisi Ambientale del CNR, Tito Scalo, Potenza). Additional spectra also of the region around 20 μm as covered by Fig. 3.11 and Fig. 3.12 (lower panel) can be found in Appendix 6.

To compare the radiances observed in the three cases of Figs. 3.11 and 3.12 one has to keep in mind that the radiance in Fig. 3.11 and Fig. 3.12 lower panel is plotted over a linear wavenumber scale while it is a linear wavelength scale in Fig. 3.12, upper panel. This results in a shift of the maximum of the spectral radiance because the energy within a given spectral range must be the same in both cases. For a given target the spectral flux density, M_λ, within a specific spectral range from $\lambda_1 = \nu_1^{-1}$ to $\lambda_2 = \nu_2^{-1}$ must be independent of the chosen spectral scale: $M_\lambda \delta\lambda = M_\nu \delta\nu$, where $\delta\lambda = \lambda_2 - \lambda_1$ and $\nu = 1/\lambda$. From this identity follows for narrow spectral intervals $M_\lambda = M_\nu(\partial\nu/\partial\lambda) = M_\nu[\partial(1/\lambda)/\partial\lambda]$ which leads to $M_\lambda = M_\nu \lambda^{-2} = M_\nu \nu^2$ (compare Eq. 3.30, 3.35, and 3.36).

3.3.4 The Overlapping Region Around 3 μm

In the previous sections the spectral intervals for scattering and emitting atmospheres could be treated separately because the solar energy flux density is small as compared to the terrestrial emission at longer infrared wavelengths (the solar radiation comes from a very narrow solid angle while the solid angle of the atmospheric emission is 2π). This is not the case in a small spectral band between 3 and 5 μm where the solar flux density at daytime is comparable to that of the atmosphere. This spectral band nevertheless is of interest for remote sensing because it includes an atmospheric window between 3.4 and 4.0 μm where observations of the Earth surface are possible and in fact are done by the NOAA-AVHRR instrument. This window is not completely transparent - which is true for the other windows as well - it rather can be considered as opaque because the absorption bands of a number of trace gases are positioned here. Fig. 3.13 shows a nighttime emission spectrum measured from the ground in which the bands of the trace gases are identified.

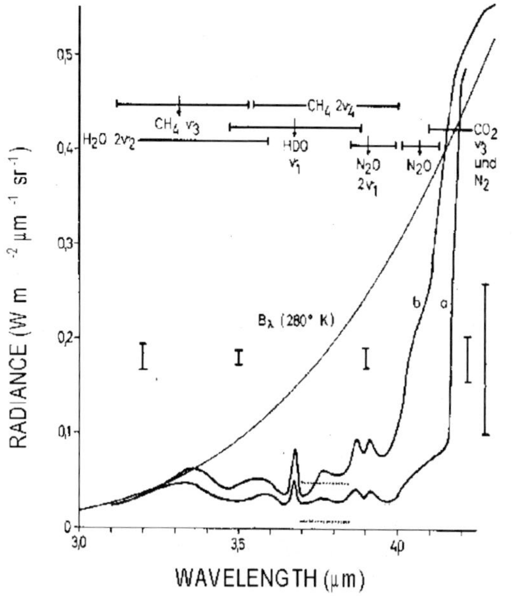

Fig. 3.13. Emission of the night sky between 3 and 4.5μm wavelength, Maisach near Munich, 04. 06. 1966. a = mean of two zenith measurements (01:34 and 02:12 MET), b = 80° zenith angle towards W (01:52 MET). Air temperature 6.5°C, meteorological range 7 km. The dotted lines represent radiances computed on the basis of a radiosonde ascent at 01:00 European Mean Time, the vertical lines the possible error range. After Bertram (1967)

3.4 Radiation Codes Suited for Atmospheric Corrections

3.4.1 A Landsat-TM scheme

As an example for the correction in the visible and near infrared part of the spectrum the code of Gilabert *et al.* (1994) is described here that has been applied to Landsat-TM data. The authors reduce the scheme of Fig. 3.10 to two components: the contribution from single scattering of the direct solar beam by atmospheric constituents and the fraction of the global radiation that is diffusely reflected only once by the earth's surface. In this approximation the signal measured at the satellite consists of two terms, the global radiation E_G reflected at the ground according to its albedo ρ and transmitted through the atmosphere to the satellite position (τ') and the radiation scattered back by the atmosphere, L_{atm}:

$$L_{sat} = E_G \rho \tau' + L_{atm} \tag{3.62}$$

To solve the equation for ρ requires the determination of the global radiation at the surface and the atmospheric radiation into the direction of the satellite which consists of a molecular and an aerosol scattering component.

To obtain information about the scattering properties of the atmosphere the signals measured over dark areas (e.g. deep lakes) in two LANDSAT-TM channels (one and three) are taken as resulting exclusively from single scattering and ozone absorption in the atmosphere. Using climatological ozone data and the absorption coefficients of ozone after Vigroux (1953) the signal is corrected for ozone absorption and the term $(1-\beta')(1-\tau_{scat}')$ is obtained, which represents the radiance scattered back into the direction of the satellite. For Landsat, the direction marked by $'$ is the zenith. As long as the optical depth remains small, $\tau_{scat}' \cong 1 - \delta_{scat}$ with $\delta_{scat} = \delta_R + \delta_M$. β' is taken constant for the two wavelength. The optical depth can be expressed by a function $\delta_0\lambda^m$, where for pure Rayleigh scattering m $\cong$ 4.09, depending slightly on the wavelength ($\delta_{0,R} \cong 1.07$). Then it may be written:

$$\frac{\pi L_{\rho=0}}{S_0 \cos\zeta_0 \left(\bar{d}/d\right)^2 (1-\beta')} = 1 - \tau_{scat} \cong \delta_{scat} = \delta_{scat,0}\lambda^m \tag{3.63}$$

In principle now Linke´s turbidity factor $T = \delta_{scat}/\delta_R$ has to be determined for the backscattered radiation in order to compute the actual mixing of the molecular and the aerosol components. It equals the number of Rayleigh atmospheres that would lead to the same δ_{scat} as molecular and aerosol scattering together. The fraction F_M of aerosol scattering would become $F_M = (T-1)\delta_R/\delta_{scat}$. δ_{scat} is measured and δ_R can be computed. This procedure nevertheless is only an approximation because the multiple scattering in the atmosphere is included in the measured signal but not in the computations. Additional information about the mixing of the components is obtained from the exponent m. The sum of the slopes of the molecular scattering (4.09) and (T-1)-times the aerosol component must result in the slope m_{scat}: $[4.09 + (T-1)m_M]/T = m_{scat}$. The exponent m for total scattering between the two centre wavelengths of TM channel one and three (485 and 660 nm) can be estimated from

$$m = \frac{\ln \partial_i(\lambda_1) - \ln \partial_i(\lambda_3)}{\ln \lambda_1 - \ln \lambda_3}. \tag{3.64}$$

where i stands for $scat$, R and M respectively.

Rayleigh scattering and ozone absorption alone would result in a return signal $L_{R,O3}(\lambda)$ which only depends on the solar (ζ_0) and view (ζ) zenith angles:

$$L_{R,O_3}(\lambda) = \frac{E_0(\lambda)\cos\zeta_0 P_R(\zeta_0 - \zeta)}{4\pi(\cos\zeta_0 + \cos\zeta)}\left[1 - e^{-\delta_R(\lambda)\left(\frac{1}{\cos\zeta_0} + \frac{1}{\cos\zeta}\right)}\right]\tau_{oz}^{\uparrow}(\lambda)\tau_{oz}^{\downarrow}(\lambda) \tag{3.65}$$

where P_R is the Rayleigh phase function $(3/4)[1+\cos^2(\zeta - \zeta_0)]$.

In a next step, the phase functions for aerosol scattering of Rayleigh (P_R) and Mie (P_M) scattering are mixed according to $P_{scat} = F_M P_M + (1-F_M) P_R$. For aerosol scattering the Two-Term Henyey-Greenstein (TTHG) phase function is applied:

$$P_M = \frac{\alpha(1-g_1^2)}{1+g_1^2-2g_1\cos\theta_0)^{3/2}} + \frac{(1-\alpha)(1-g_2^2)}{1+g_2^2-2g_2\cos\theta_0)^{3/2}} \quad (3.66)$$

For the parameters α and g standard values are applied. For continental aerosol: α = 0.978, g_1 = 0.884, g_2 = -0.749 and for maritime aerosol α = 0.935, g_1 = 0.795, g_2 = -0.568. The aerosol single scattering albedo, $\tilde{\omega}_0$, is obtained from $\tilde{\omega}_0 = 0.9\, F_M + (1-F_M)$, assuming 10% absorption in the aerosol. Introducing in Eq. (3.65) the mixed phase function P_{scat} instead of the Rayleigh phase function and replacing the ozone transmission by the combined ozone and Rayleigh transmission the path radiance due to aerosol scattering is obtained. Now also the aerosol optical depth for the downward radiation can be computed and the atmospheric beam transmittances are obtained from:

$\tau\uparrow(\lambda) = \exp\{-\delta_{oz}(\lambda) - \delta_R(\lambda) - \delta_M(\lambda)\}$ (nadir observation),

$\tau\downarrow(\lambda) = \exp\{-[\delta_{oz}(\lambda) + \delta_R(\lambda) + \delta_M(\lambda)]/\cos\zeta\}$.

The global solar irradiance, $E_G(\lambda)$, the sum of direct (beam) irradiance and diffuse irradiance becomes:

$$E_G(\lambda) =$$

$$E_0\cos\zeta_0\left[\tau(\lambda) + \frac{1}{2}\left\{1-\tau_R^{\downarrow}(\lambda)\right\}\tau_M^{\downarrow}(\lambda)\tau_{oz}^{\downarrow}(\lambda) + \beta(\zeta_0)\left\{1-\tau_M^{\downarrow}(\lambda)\right\}\tau_{oz}^{\downarrow}(\lambda)\overline{\omega}_0\right] \quad (3.67)$$

The first term at the l.h.s. represents the direct solar beam, the second term the Rayleigh scattering of which 50% are scattered downward, and the third term the Mie scattering for which β has to be taken from radiative transfer computations (Iqbal 1983).

With this value of the global radiation and the knowledge of the upward transmission into the direction of the satellite, the radiance at the top of the atmosphere can be calculated and Eq. (3.62) be solved for ρ.

3.4.2 A One Parameter Meteosat and NOAA-AVHRR Scheme

The equivalent TOA (brightness) temperature is transformed to the Apparent Radiative Surface Temperature (ARST) by the one parameter model (Peters, 1995):

$$\sigma T_{arst}^4 = c_T^{-1} T_{TOA}^4 . \quad (3.68)$$

The parameter c_T is determined by intercomparing the TOA Meteosat temperature with ground measurements. At te EFEDA field experiment in Castilla - La Mancha for this purpose two cable car systems have been istalled in the plain near Tomelloso over grapevine that measured the surface temperature over a length of 30 m in the 10-11 and 11-12 μm spectral bands. One system measured at 6 m height and the other one at 3 m height. The surface emissivity was assumed to be one for both bands. The atmospheric correction function c_T can be expressed by the formula:

$$c_T = c_1 + c_2 \overline{R}_{81} + c_3 f_c . \tag{3.69}$$

$\overline{R}_{81}$ is the mean broadband radiance determined from 81 Meteosat pixels in the 9x9 pixel window centered at the Tomelloso study site, where the cable car system was installed. f_c is a "compensation function" that allows to compensate for the diurnal variation of the mean value of the observed c_T values. At noon c_T is approximately 0.9, during nighttime about 1.0. The coefficients have the values $c_1 = -\ 1.17184$, $c_2 = 0.001012$, and $c_3 = 1.80107$. Because of the fixed observation angle of Meteosat these coefficients are valid for the EFEDA experimental area exclusively.

3.4.3 Split-window Technique (SWT)

The split-window technique can be applied to the thermal infrared part of the spectrum if two adjacent spectral channels are available like in the AVHRR instrument. It is based upon a linearization of the radiative transfer equation in the form

$$L_{TOA} = \left[\varepsilon B(T_s) + \rho(1-\tau) B(T_a^{\downarrow}) \right] \tau + B(T_a^{\uparrow})(1-\tau) \tag{3.70}$$

The terms have the following meaning: $B(T_s)$ is the Planck function of the surface temperature T_s which is to be determined, ε the emissivity and ρ the reflectance of the surface. $B(T_a^{\downarrow})$ is the downward directed radiance of a black body of the effective emission temperature of the atmosphere and $B(T_a^{\uparrow})$ the equivalent for the upward directed radiance. The equivalent emission temperatures are slightly different because of the temperature profile and water vapour distribution in the atmosphere. $B(T_a^{\downarrow})(1-\tau)$ and $B(T_a^{\uparrow})(1-\tau)$ are the respective radiances from the atmosphere, where τ is the transmission of the atmosphere between the target and the satellite. The signal L_{TOA} at the satellite can be expressed in terms of an effective emission temperature T_{TOA} of the system surface and atmosphere by $B(T_{TOA})$.

The Planck functions can be linearized by developing the function around the temperature T_{TOA}:

$$B(T) = B(T_{TOA}) + (T - T_{TOA}) \frac{\partial B(T)}{\partial T} \tag{3.71}$$

Eq. (3.66) can be transformed into this linearized form provided that the difference between the effective atmospheric temperatures for downward and upward radiation can be neglected. It results:

$$T_s - T_{TOA} = \frac{1}{\varepsilon\rho}\left\{\frac{B(T_{TOA})}{\frac{\partial B(T_{TOA})}{\partial T}}\left[1-\varepsilon\tau-(1-\tau)(1+\rho\tau)\right]-(1-\tau)(1+\rho\tau)(T_a - T_{TOA})\right\} \tag{3.72}$$

If two adjacent channels are available which are distinguished by the indices j and k and if the emission of the atmosphere may be considered as approximately equal in both channels (though the transmissions and the reflectances differ), then one obtains for the surface temperature

$$T_s - T_{TOA,j} = \frac{1}{a_j b_k - a_k b_j}\left[a_j b_k - a_k b_j\right] + \left[b_k \frac{B(T_j)}{\frac{\partial B(T_{TOA,j})}{\partial T}}(1-a_i-b_i) - b_j \frac{B(T_k)}{\frac{\partial B(T_{TOA,k})}{\partial T}}(1-a_k-b_{ik})\right] \tag{3.73}$$

where
$a_i = \varepsilon_i\,\tau_i$ and $b = (1\text{-}\tau_i)(1\text{+}\rho_i\,\tau_i)$, $i = j$ or k. How the surface reflectance $\rho = 1 - \varepsilon$ can be determined is described in Chapter 4. For the transmission Abdullah et al. (1994) have provided the formula

$$\tau = t_i e^{-t_2 w(\vartheta=0)/\cos\vartheta} \tag{3.74}$$

and computed the parameters t_i for different model atmospheres and the two AVHRR channels 4 and 5. $w(\vartheta=0)$ is the amount of water vapour in a vertical path. For a midlatitude atmosphere $t_{4,1}$ ranges around 0.95, $t_{4,2}$ around 0.1 and the respective numbers for channel 5 are 0.91 and 0.15.

A different approximation was introduced by McMillin (1975) for use with a split window in the 10.5 - 12.5 µm atmospheric window. He neglects the reflected sky radiation in eq. (3.70) which is equivalent to setting $\varepsilon = 1$ (as is true for water) and obtains:

$$B_\lambda(T_B) = \tau_\lambda B_\lambda(T_s) + B_\lambda(T_a)\left[1-\tau_\lambda\right] \tag{3.75}$$

with

$$B_\lambda(T_a) = \frac{1}{1-\tau_\lambda}\int_0^1 B_\lambda(T)d\tau \tag{3.76}$$

Here T_a is the mean atmospheric temperature and T_B the measured brightness temperature.

For the two SWT-channels it is assumed that (1) their contribution functions are similar, hence, T_A is equivalent, and (2) differences in transmissivity result from differential absorption in the channels and are mainly due to water vapour. Re-writing equation (3.75) in terms of the two SWT-channels and applying linearization and simplification as before yields analog to eq. (3.73) the simplified split-window equation:

$$T_s = T_{B1} + a(T_{B1} - T_{B2}) + b. \tag{3.77}$$

Equation 3.73 sufficiently accounts for the atmospheric perturbations and the effect of *a priori* emissivity can be included in one of the coefficients. However, these semi-empirical relationships do not always hold and are strictly valid only for the data sets used to derive them; strongly deviating conditions of emissivity and water vapour can lead to substantial error in derived LST. Studies have also shown that non-linear quadratic formulations, which account for emissivity and water vapour, are more efficient than linear formulations (François and Ottlé 1996).

As global *in situ* LST data at the resolution of remote sensing systems are unavailable, the performance of split-window formulations are examined comparatively. Among the various SWTs for LST estimation, some of the most popular are by Price (1984), Becker and Li (1990b), Sobrino et al. (1991), Kerr et al. (1992), Coll et al. (1994), Becker and Li (1995), and Wan and Dozier (1996). These SWTs are formulated by simplifying the atmospheric influence obtained from numerical simulations with radiative transfer models. The SWT of Price (1984) and Becker and Li (1990b) account only for emissivity variations and the coefficients to account for atmosphere are set to fixed values. The coefficients given by Sobrino et al. (1991) and Becker and Li (1995) also depend on the state of the atmosphere. Considering the effects of mixed pixels, Kerr et al. (1992) proposed a method based on the proportion of vegetation, which they obtained from NDVI values. The method classifies the surface as bare soil or as vegetated area and combines the proportional effects in each pixel to obtain effective LST. The formulation of Coll et al. (1994) additionally accounts for the effect of viewing angle. In order to generalize the SWT to the global scale, the coefficients must be able to account for changes in the atmospheric conditions due to varying viewing angle and atmospheric water vapour. The generalized split-window technique (Wan and Dozier 1996) is formulated by optimising a view-angle dependent SWT for separate sub-ranges of atmospheric water vapour, lower boundary temperature, and surface temperature. During application, the actual values of the coefficients are looked up in a LUT, which is accessed via the scan angle and atmospheric variables. Information from atmospheric sounders can be used to determine the optimum range of coefficients for the generalized linear split-window equation. Despite the increased generality of the SWTs, determination of channel emissivity remains a separate and critical issue.

3.4.4 Multi-angle Methods

Multi-angle methods also exploit differential absorption in two measurements, but the difference is due to different atmospheric path-lengths obtained for viewing the same object under different angles. The multi-angle method requires only one channel and it is assumed that the atmospheric column is spatially uniform. Its application involves measurements from a single satellite or simultaneously from different satellites, e.g. Meteosat and TIROS-N (Chedin et al. 1982).

A satellite instrument that allows the use of the multi-angle method is ATSR onboard ERS-1, which provides observations at near-nadir (0°-22°) and at forward view under 55°. Various formulations are given for ATSR assuming that angular variation of emissivity is negligible ($q < 60°$), e.g., Prata (1993), and Sobrino et al. (1996).

Multi-angle techniques also aim at correlating LST with TOA brightness temperatures via derived coefficients, which compensate for the effect of the atmosphere. However, apart from assuming spatial uniformity of the atmospheric column, the algorithm is prone to instability if the observations do not have a significant path difference (Prata 1993). Additionally, the directional variation of emissivity has to be known.

3.4.5 The HITRAN, MODTRAN, and LOWTRAN Family

Starting with the pioneering experimental investigations of Fowle (1917) and the development of the detailed theory of molecular spectra by Herzberg (1945) the understanding of the line spectra of atmospheric gases was achieved in the second half of the 20th century and lead to a detailed knowledge of the structure and properties of the atmospheric absorption and emission bands. In parallel the aerosol scattering processes where studied in detail when the general characteristics of the aerosol had been resolved by Junge (1963). In a third line the theory of radiative transfer in absorbing and scattering media was developed after Lord Rayleigh had layed the basis for the molecular scattering in 1871. The knowledge compiled in thousands of specialized papers was finally amalgamated into an atmospheric radiative transfer scheme that serves practical application in remote sensing. The basic idea is rather simple and straight forward. The radiative transfer equation is solved for each wavelength and then smoothing filters are applied and averaged functions defined to reduce the necessary computations to a reasonable amount. The difficulty lies in the treatment of the combined scattering-emission source function (Eq. 3.50) that varies throughout the atmosphere.

Each spectral line is defined by three parameters. Firstly, its position in the spectrum denoted by the central wave number v_0. Secondly, its strength S that depends on the temperature. With increasing temperature the higher exitation levels of the molecules become more active and therefore the lines resulting from these transitions become stronger. And, thirdly, its halfwidth α that is a function on both pressure and temperature because with increasing pressure and temperature the movements of the molecules are accelerated. This leads to more collisions which broaden the energy levels. From the theory of a damped oscillator follows the most basic formula for a line,

$$a_\nu = S(T)\frac{\alpha}{(\nu - \nu_0)^2 + \alpha^2} \tag{3.78}$$

which is applicable where collision broadening is the overriding process. It is called natural, collision or Lorentz profile (Lorentz 1906, see also Goody 1964). In the upper atmosphere where Doppler broadening influences the line shape, a more complicated formula, the Voigt (1912) profile (see also Unsöld 1955 and Penner 1959), has to be used:

$$a_\nu^{L\&D} = \frac{S(T)}{\pi}\frac{\alpha_L}{\alpha_D}\int_{-\infty}^{+\infty}\frac{\exp\left(-\frac{\ln 2}{\alpha_D^2}(\nu'-\nu_0)^2\right)}{(\nu-\nu')^2+\alpha_L^2}d\nu' \tag{3.79}$$

In this equation α_L denotes the Lorentz half width, α_D the Doppler half width, and $a_\nu^{L\&D}$ is the combined Lorentz and Doppler absorption coefficient.

For an aerosol scattering function a formula like the Henyey-Greenstein function (Eq. 3.66) can be applied.

The line data for all important atmospheric gases are compiled in the *HITRAN* data bank (McClatchey et al. 1971, see also HITRAN Special Issue JQSRT 82 no. 1-4, 15 November - 15 December 2003). These data can be used together with appropriate line shapes to compute atmospheric spectra at any desirable spectral resolution. Figs. 3.9, A.6.2.1, and A.6.2.2 show examples of such computations. Based upon this data bank codes have been developed to compute more rapidly and including the effect of aerosols spectra of degraded resolution such as the Moderate Resolution Transmittance code *MODTRAN* and the low resolution code *LOWTRAN* (the codes are commercially available from ONTAR Corporation, 9 Village Way, North Andover, MA 01845-2000 USA, sales@ontar.com).

Considering their application for the generation of atmospheric corrections to infer e.g. land-surface temperatures from measurements in the thermal infrared, these codes belong to the category of single-channel methods. They perform atmospheric corrections from measurements in one window-channel (Price 1983, Susskind et al. 1984). In order to obtain the single-channel atmospheric correction, it is necessary to accurately know the vertical and horizontal distribution of temperature and water vapour in the atmosphere. Such atmospheric information is obtained from radiosoundings, vertical sounding

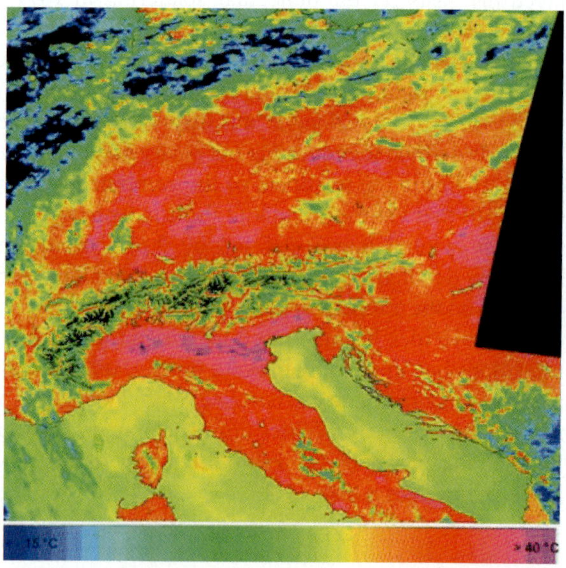

Fig. 3.14. LST map for parts of central Europe (5 - 20.36 E, 40.5 - 55 N) derived with the single-channel method (assumed emissivity: 0.98) from NOAA-14 AVHRR data and ECMWF atmospheric profiles. Date: 20 April 1996, NOAA daytime pass. Black areas denote no data, elevation exceeds a threshold (> 2.4 km), or clouded areas. A separate cloud-processing was not performed.

instruments on board of satellites (e.g., TOVS, VAS), or from numerical weather prediction models (e.g., ECMWF analyses). The method is based on simulations of satellite measurements over a range of surface-related variables (temperature, elevation and emissivity) and solar- and satellite-related variables (solar zenith angle and scan angle) for a given atmosphere. The corrections are then easily obtained from the simulated situations (atmospheric profiles and surface variables) and the given surface variables.

Atmospheric profile data are discrepant in space and time; therefore, LST and atmospheric corrections are usually obtained by interpolating forward calculations (Schroedter et al. 2001). As the method simulates the actual conditions, it is most reliable; however, the corrections will be only as good as the available atmospheric information. Furthermore, surface emissivity must accurately be known. Only for highly homogeneous areas the assumption of a constant emissivity is sufficiently accurate, e.g., in mid-latitude vegetated areas with an assumed constant emissivity of 0.975, an emissivity error of ±0.025 introduces an LST error of ±2 K (Schädlich et al. 2001).

Fig. 3.14 shows the map of both LST and SST for 20 April 1996 over central Europe estimated from NOAA-14 AVHRR channel-5 measurements. The data have been corrected for topography and atmospheric effects using atmospheric profiles from ECMWF re-analyses and employing the single-channel method for an emissivity of 0.98.

3.4.6 The "Simulation of the Satellite Signal in the Solar Spectrum" Algorithm

The 6S (Second Simulation of the Satellite Signal in the Solar Spectrum) algorithm (Vermote et al. 1994) is an improved version of the 5S model developed by Tanre et. al. (1997). It is designed to treat non-homogeneous and non-Lambertian surfaces and considers

- the photons scattered back from the atmosphere not reaching the surface,
- the first order directional reflectance from the target,
- the radiation scattered in the atmosphere and contribute to the illumination of the target surface,
- the average angular reflectance of the target, and
- the contribution of the environment at which photons are reflected and directed into the field of view by scattering processes in the atmosphere.

The radiation signal at the top of the atmosphere is expressed in terms of the reflectance $\rho^* = \rho^*(M;\mu_0,\mu)$ of a small area M at a specific geographical position. It depends on the cosine μ_0 of the solar zenith angle and of the observation angle (μ). ρ_a stands for the contribution of radiation scattered by the atmosphere without touching the surface.

The non-Lambertian behaviour of the surface is accounted for by decomposing the contribution of the target to the signal at the top of the atmosphere into the sum of four terms representing (i) the photons transmitted directly from the sun to the target and back to the sensor, (ii) the photons scattered by the atmosphere, then reflected by the target and from there directly transmitted to the sensor, (iii) the photons directly transmitted to the target but then scattered by the atmosphere on their way to the sensor, and (iv) the photons having two or more interactions with the atmosphere.

In a first approach the surface may be considered as homogeneous. It is illuminated by a direct solar component and a contribution due to scattered radiation. The direct illumination is determined by the transmission through the atmosphere, $\tau_S = e^{-\frac{\delta}{\mu_0}}$, the diffuse component by a diffuse transmission factor $\tau_d(\zeta_0)$. The term $\rho e^{-\frac{\delta}{\mu_0}} e^{-\frac{\delta}{\mu}}$ is the normalized direct reflectance from the target surface with the reflectance ρ. The photons travel twice but under different zenith angles through the atmosphere - from the sun to the surface and back to the satellite.

The normalized diffuse solar radiation $[E(\mu_0)]$ is reflected by an angularly averaged reflectance $\bar{\rho}$ defined by

$$\bar{\rho} = \frac{1}{\mu_0 S_0 + E(\mu_0)} \int_0^{2\pi} \int_0^{-1} \rho(\varphi, \lambda; \mu', \mu, \zeta', -\zeta) L_d^{\downarrow} \mu' d\mu' d\zeta' . \quad (3.80)$$

where $L_d^{\downarrow}$ is the total downwelling radiance that generally also is angle dependent.

If multiple scattering between the atmosphere and the ground is taken into account the total illumination normalized to the extraterrestrial solar radiation is $T(\zeta_0)/[1-\rho r]$ where r is the spherical albedo of the atmosphere and $T(\zeta_0) = e^{-\delta/\mu_0} + \tau_d(\zeta_0)$. In the computation of the transmittances the absorption by atmospheric gases is included. To the incoming radiation that part of the reflected radiation has to be added which is scattered back by the atmosphere to the target (see Fig. 3.9). These two components sum up to $\frac{\rho}{1-\rho r} T(\zeta_0) T(\zeta)$. For the transmission in the return path the same assumptions can be made as for the incoming radiation. We have a direct and a diffuse component and the diffuse transmittance is identical to the former one due to the reciprocity principle of optics. The combined transmittance for the return path therefore is $T(\zeta) = e^{-\delta/\mu} + \tau_d(\zeta)$.

In addition radiation reaches the sensor that is reflected from a surrounding area, the environment of the target pixel, that does not equal the target surface. The target surface is considered as a small area M with the reflectance ρ surrounded by an environment with the reflectance ρ_e. In this case the multiple reflectance of the target surface can be neglected. Only the environment contributes to the diffuse component and the reflectance at the surface as seen by the satellite sensor can now be written

$$\frac{T(\theta_S)}{1-\rho_e S} \left[\rho(M) e^{-\delta/\mu_V} + \rho_e \tau_d(\theta_V) \right] \quad (3.81)$$

If a patchy structure is considered, ρ_e must be replaced by a spatially averaged reflectance $<\rho>$ of all pixels. Adding also the pure atmospheric component the signal at the satellite level can now be written:

$$\rho^* = \rho_a + \frac{T(\zeta)_0}{1 - <\rho(M)> r}\left[\rho(M)e^{-\delta/\mu} + <\rho(M)> \tau_d(\zeta)\right]. \quad (3.82)$$

Actually the required spatial average must be weighted according to the distance from the target pixel because the scattering effect of the atmosphere becomes smaller as the distance from the target increases. This is taken care of by a distance function t(x,y):

$$<\rho> = \frac{1}{E(\mu)} \int_{-\infty}^{+\infty}\int_{-\infty}^{+\infty} \bar{\rho}(\varphi, \lambda) t(\mu; x, y) dx dy \quad (3.83)$$

Finally the sum of all considered components can be written:

$$\begin{aligned} \rho^* = {} & \rho_a + \rho e^{-\frac{\delta}{\mu_0}} e^{-\frac{\delta}{\mu}} \\ & + \bar{\rho} e^{-\frac{\delta}{\mu}} E(\mu_0) + <\rho> \left[E_0(\mu_0 +) + e^{-\frac{\delta}{\mu_0}} \right] E(\mu) \\ & + <\rho> r \left[\frac{\bar{\rho} e^{-\frac{\delta}{\mu}} + \rho E(\mu)}{1 - <\rho> r} \right] \left[E(\mu_0) + e^{-\frac{\delta}{\mu_0}} \right] \end{aligned} \quad (3.84)$$

The third term, $\bar{\rho} e^{-\frac{\delta}{\mu}} E(\mu_0)$, stands for the contribution of the inhomogeneity of the surface target. The fourth term, $\langle\rho\rangle \left[E(\mu_0) + e^{-\frac{\delta}{\mu}} \right] E(\mu_0)$, accounts for the average reflection of the surrounding and can be combined with the fifth term, $<\rho> r \left[\frac{\bar{\rho} e^{-\frac{\delta}{\mu}} + \rho E(\mu)}{1 - <\rho>} \right] \left[E(\mu_0) + e^{-\frac{\delta}{\mu}} \right]$, that summarizes the multi-scattering processes by which radiation from the surrounding is scattered into the field of view of the sensor.

3.4.7 Autonomous Atmospheric Compensation (AAC)

Recently, techniques to obtain at-surface radiances only from TOA spatial hyperspectral IR measurements (without any atmospheric information) have been developed. An example is the Autonomous Atmospheric Compensation (AAC) method (Gu et al. 2000).

The assumptions underlying AAC are: (a) over small areas, atmospheric variables are more uniform than land-surface variables, and (b) radiance spectra are significantly smoother at the surface than atmospheric radiation and absorption. Hence, for two hyperspectral channels, one at a strong absorption band (e.g., 11.73 μm water vapour absorption band) and the other just out of it, upward directed radiances at the surface are similar for the two channels. The difference between the radiances at the sensor in space is mainly due to the atmosphere. This fact is used to obtain the transmittance ratio and the path radiance difference, from which the atmospheric transmission and path radiance can be derived using an empirical relationship. The method was validated using data from the airborne Spatially Enhanced Broadband Array Spectrograph System (SEBASS). However, the problem of mathematical under-determination associated with the extraction of LST and emissivity from radiance at the surface poses a separate problem. The AAC appears potentially reliable at the outset, but further improvements are required.

3.4.8 A Four-stream Atmospheric Correction Procedure for Broadband Albedo

The correction for atmospheric effects often is hampered by the lack of collateral data. Consequently it was attempted to reduce the amount of information needed for this task. An interesting approach therefore is to base the correction on easily measurable radiation fluxes such as the global radiation flux density. The procedure is based on a four-stream radiative model to simulate atmospheric effects suggested by Verhoef (1985). The function describing the atmospheric effect considers the radiation scattered by the atmosphere, the directly reflected radiation, and multiple reflection between the surface and the atmosphere:

$$\rho_p = \rho_a + \rho_0 \frac{\tau_{sun} \tau_{sat}}{1 - \rho_{sa} \rho_0} \tag{3.85}$$

where

ρ_0 = surface albedo

ρ_a = atmospheric albedo

ρ_{sa} = spherical albedo of the atmosphere

$\tau_{sun}\tau_{sat}$ = two way transmittance.

First the planetary broadband albedo ρ_p is derived from the top of the atmosphere reflectances in the AVHRR bands 1 and 2 by the conversion from Brest and Goward (1987) using the NDVI = $(\rho_2 - \rho_1)/(\rho_2 + \rho_1)$ (see chapter 4) at the top of the atmosphere as an additional parameter:

$$\rho_p = 0.526\rho_1 + (0.485 - 0.112 NDVI_{TOA})\rho_2 \tag{3.86}$$

For bare soil the NDVI measured at the top of atmosphere is assumed to be < 0.1 and for closed vegetation > 0.6.

Michels (1989) related the quantities ρ_a, ρ_{sa} and $\tau_{sun}\tau_{sat}$ to the "clearness" of the atmosphere and the solar zenith angle. The "clearness" C of the atmosphere is defined as:

$$C = \frac{E_G^{\downarrow} / S_0}{(E_G^{\downarrow} / S_0)_{ref}} . \tag{3.87}$$

$E_G^{\downarrow}$ = global radiation flux density at the surface (W m^{-2}),
S_0 = exo-atmospheric solar flux density (W m^{-2}),
ref indicates reference atmospheric condition (VIS = 80 km).

With data measured in the western desert of Egypt during field campaigns in the months 7 and 8/1986, 2, 3 and 11/1987, and 6/88, Michels obtained the following relationships which, however, are exclusively strictly valid for the area where they have been determined:

$$\rho_a = 0.01882 + \frac{0.00858}{C \cos\zeta} \tag{3.88a}$$

$$\rho_{sa} = 0.06741 + \frac{0.00398}{C \cos\zeta} \tag{3.88b}$$

$$\tau_{sun} \tau_{sat} = 0.92685 - \frac{0.02854}{C \cos\zeta} . \tag{3.88c}$$

Nevertheless, these relationships can be used as guidelines for similar climatic conditions in the Mediterranean area.

3.5 From Basic Theory to Application

In this chapter some theoretical background was presented with the purpose to refresh the reader's critical judgement about the quality of the information which can be derived from remote sensing data. Many of the methods described here will be applied in the following chapters to gather information about processes at the Mediterranean land surfaces. Radiation processes and radiation transfer belong to the best explored and mathematically formulated physical phenomena. However, the rigorous application of their theoretical foundation meets limits if one has to deal with a manifold of natural conditions. Often these methods and process descriptions will therefore later on appear in simplified versions because basic physical reasoning can not be applied to billions of image pixels and the spatial resolution of satellite data smooths out micro-details. It also must be kept in mind that the application of radiation codes require input data which can only be determined empirically and which are affected by measuring errors. It is therefore imperative that the reader of the following chapters and the user of these remote sensing data knows what kind of reasoning is behind the simplified or "parametrized" versions of the retrieval methods. Some methodological weaknesses, about which further research is necessary, have already been indicated in this chapter.

Chapter 4 Primary Level Products

4.1 Introduction

This section deals with products derived relatively straight forward from spectral measurements made by instruments on board of satellites. These products can be both standardized relative signals or calibrated and geographically corrected spectral radiances which primarily represent the situation at the top of the atmosphere (TOA). If atmospheric corrections are applied, the same type of information is generated for the bottom of the atmosphere (BOA).

From raw data, expressed in terms of counts or grey values, immediately structural information can be inferred, e.g. geometrical patterns, land cover "classes" or vegetation indices. From "classes", identified by their relative spectral characteristics, land cover types can be derived provided that some known training areas are included in the scene. Such products, derived from raw data of individual satellites, generally can hardly be used in comparisons with data obtained from different satellites and other locations or for the generation of longer time series because sensor responses are different and change in time. They also do not provide physical quantities such as reflectance and temperature. To obtain platform independent physical quantities, such as reflectance and temperature, the measured values have to be converted into absolute radiance values by means of calibration and normalization to specific observation geometries.

For clear skies remote sensing from space is limited by atmospheric absorption bands and the scattering and absorption of radiation by aerosols (see chapter 3). This interaction is weak to negligible for microwaves but strong at shorter wavelengths. With the exception of broadband sensors developed to measure the total flux outgoing at the top of the atmosphere, most satellite radiometers primarily measure in accurately defined spectral bands. Passive sensors measure reflected, emitted, or a mixture of both radiances, depending on the spectral region. In addition, active microwave remote sensing methods are applied to derive backscattering coefficients. Primary level products are restricted to spectral reflectances and effective emission temperatures ("brightness temperatures"). Specific indices can be derived by combining two or more of the spectral signals. In some cases their polarization provides additional information.

It is always desirable to control the accuracy of the information derived from these measurements. In the case of TOA data this is only possible by comparing measurements made by different instruments carried by satellites or high flying aircraft. For most measurements no absolute standards exist, one has to rely on the performed calibrations. Such comparisons are not easy. They have to be made at the same time or at least within a short time interval with identical observation and illumination geometries. Such possibilities exist e.g. when a polar orbiting satellite underflies a geostationary satellite. In this context a scaling problem arises as the fields of view of the instruments may differ considerably. Even if data are accurately geo-referenced there still remains an uncertainty in the positioning of a pixel of about 30% of its size. Therefore signals of single pixels measured from different instruments are difficult to compare. A homogeneous scene has to be selected and the measurements have to be averaged over a number of pixels to obtain a compatible result. If a measurement from Meteosat with a resolution of, say 5 km, is compared with a measurement of NOAA-AVHRR, a minimum of $3*3$ Meteosat pixels is used into which the corresponding $15*15$ AVHRR pixels can be fitted. In the same way $120*120$ SPOT pixel fit into a matrix of $3*3$ AVHRR pixels. In most cases the measurements are made under different observation angles. Therefore, bi-directional reflection functions as well as the variable pixel size and geographical orientation of scanning instruments has to be accounted for.

Proceeding to the atmospherically corrected data then allows a comparison with ground measurements. Such *in situ* measurements are only available at specific sites which usually are spots of sub-pixel size. These measurements generally are representative for areas of less than $100*100$ m^2. In SPOT or Landsat scenes these sites can be identified but it is not possible to make a direct comparison with an AVHRR or Meteosat pixel. A specific scaling technique was developed to use measurements made during field experiments or at fixed sites, so-called "Anchor Stations", to test the accuracy of the measurements made from space and the corrections applied for atmospheric effects.

For this purpose, the ground measurements - radiometric surface temperature, directional reflectance, spectral albedo, or spectral indices - are compared with a matrix of $3*3$ pixels of measurements made from SPOT or Landsat. A single pixel may not be accurate enough because of superimposed noise or small surface inhomogeneities. If agreement between ground and satellite measurements exists, the high resolution measurements are aggregated and compared with a $3*3$ pixels matrix of NOAA-AVHRR. The AVHRR data then can be aggregated to the size of a Meteosat matrix. At each step the different spectral characteristics and observation geometry of the instruments have to be accounted for. A serious difficulty exists: simultaneous measurements are very rare. Therefore such comparisons are only possible for surface characteristics which can be assumed to be stable for the time of consecutive overpasses such as the reflectance or the vegetation indices observed at specific times of the day during a period without precipitation. Different instruments have been used at the ground to measure quantities comparable to the signals obtained from space. These will be referred to in the validation section.

Notwithstanding these comparison efforts, it has to be stressed that no absolute standard exists to determine the accuracy of the information inferred from observations made from space and that the possibilities for accuracy checks are limited. For example, no ground based instrument will ever measure reflectance from and temperature of a $20*20$ m^2 spot the same way as a satellite instrument. The distribution of shadows will be different even in the simple case of desert ripples. In the case of vegetation, where the emerging radiation stems from a volume rather than a surface, the distribution of objects may even vary from metre to metre. It is therefore questionable, whether hundreds of measurements made at leaves and stems and put into a model provide a more realistic view of the environment than a satellite instrument which integrates radiances over a certain area. Satellite data will therefore remain a measuring system of its own value which provides, if carefully applied, environmental data of high precision. Using these data, changes and trends can be detected and models can be adapted to the information provided by remote sensing. It has, however, to be kept in mind that the remote sensing products never will be identical with any measurement made at the ground. The measurements made from space can only be *related* to but not identified with quantities measured at the ground by such comparisons.

4.2 From Radiance to Reflectance

4.2.1 Top of Atmosphere Spectral Reflectance

As primary TOA product in the range of the solar spectrum, only the bi-directional reflectance distribution function (Eq. 3,21) $\rho(d\Omega, d\Omega')$ at the top of the atmosphere can be obtained in each spectral channel, *i*, of the various satellite instruments. It depends on the instantaneous solar zenith angle, ζ, and azimuth angle, φ_{sun}, as well as on the instantaneous nadir angle, ϑ_{obs}, and azimuth angle, φ_{obs}, of observation. $d\Omega$ and $d\Omega'$ can therefore be substituted by:

$$\rho_i(\zeta,\varphi_{sun};\vartheta_{obs},\varphi_{obs}) = \frac{L_i^{\uparrow}(\vartheta_{obs},\varphi_{obs}-\varphi_{sun})}{M_i^{\downarrow}(\zeta)} \quad sr^{-1}. \tag{4.1}$$

$L_i\uparrow$ is the radiance measured by the satellite sensor and M_i the solar flux density at a horizontal plane which is independent of the solar azimuth. If $L\uparrow$ is multiplied by π, then $\pi L = M$ can be interpreted as the upwelling flux density into the upper hemisphere from a Lambertian Earth. Therefore sometimes a *technical (spectral) albedo* is defined which is the directional reflectance factor multiplied by the factor of $100*\pi$ and the expression

$$\rho_{i,tech}(\lambda_i,\Delta\lambda_i) = 100\frac{\pi L_i}{S_{0,i}(\lambda_i,\Delta\lambda_i,F_i)} \tag{4.2}$$

is given in %. $S_{0,i}$ is the spectral solar irradiance (spectral "solar constant") of channel *i* at the top of atmosphere with the sun in the zenith. The spectral solar flux has to be computed from the spectral distribution of the extraterrestrial solar radiation (Appendix 5) taking into account the filter function, *F*, of the measuring instrument:

$$S_{0,i}(\lambda,\Delta\lambda) = \frac{1}{W_i}\int_0^{\infty} F_i(\lambda) S_{0,\lambda}\, d\lambda \tag{4.3}$$

where

$$W_i = \int_0^\infty F_i(\lambda)d\lambda \tag{4.4}$$

is the *equivalent width* and $F_i(\lambda)$ the filter function of the spectral channel i.

Accounting for the variable distance between Sun and Earth by introducing the instantaneous distance d and the mean distance d_0 (= 1 Astronomical Unit, A.U.), one can use the expression

$$\rho_{TOA,i} = \left(\frac{d}{d_0}\right)^2 \frac{\rho_{tech,i}}{\cos\zeta} \tag{4.5}$$

as a substitute of the spectral TOA albedo in the channel i. Only in the case of an isotropic (Lambertian) reflector the expression Eq. (4.5) would equal the true TOA albedo.

4.2.2 Normalization of TOA-data for Illumination and Observation Geometry

Methods

Satellite instruments designed for Earth surface observations are looking straight down (nadir view), such as the Landsat instruments, or they observe at small angles off nadir such as the pointable telescope of the SPOT satellite. For a given location and time of overpass and depending on geographical latitude the solar zenith angle varies considerably during the course of the year. Consequently, the reflectance at a given location can vary even if the Earth surface does not change: Size and distribution of shadows change, a surface cover may have an angular dependence of its reflectance, possibly with a specular component, and the path through the atmosphere increases with solar zenith angle.

Polar orbiting meteorological satellites carry scanning instruments, which observe the surface under different angles for consecutive revolutions. The spatial resolution of the AVHRR instrument is determined by the projection of the instantaneous field of view (IFOV) at the Earth's surface. It varies with scan angle from 1 km^2 at nadir to 12.4 km^2 near the horizon. The structure of the surface together with the illumination by the sun and the viewing conditions of the instrument result in varying amounts of shadows in the IFOV and cause differences in directional spectral reflectances of up to 50%. Apart from mountains, the main structural element of the surface is the variety of vegetation types with their diverse states of development throughout the year.

To make observations comparable the data have to be normalized e.g. to nadir observation and mean solar zenith and azimuth angles. For some applications it would even be desirable to normalize the directionally measured signal to 2π spectral albedo. To account for the anisotropy of the surface reflectance the Bidirectional Reflectance Distribution Function (BRDF) $\rho(d\Omega, d\Omega'; \lambda_i)$ at wavelength λ_i has to be known to convert the measured radiances to hemispheric fluxes and thus directional reflectances to albedo. The BRDF varies with changing vegetation cover but also with surface moisture. Approaches to describe and to remove the influence of the BRDF can be divided into empirical methods and physical models.

Once the radiometric calibration problem is solved, the normalization is a question of combining atmospheric and geometrical (or directional) effects. Several studies focussed on one or more specific aspects of the problem such as view angle effects (Holben et al. 1986), sun and view angle effects (Leroy and Roujean 1994, Frulla et al. 1995), atmospheric effects (Rahman 1996, de Haan et al. 1991) and radiometric effects (Markham et al. 1992). Interesting work has also been done by Duggin and Piwinski (1984), Royer et al. (1985), Gerstl and Simmer (1986), Lee and Kaufman (1986), and Koslowsky (1998) in assessing the impact of several of the perturbing factors, geometrical (solar and observation aspect angles) as well as atmospheric effects.

The approaches to obtain information needed for the normalization of the satellite signal differ from one study to another. Most often a multi-temporal AVHRR (Roujean et al. 1992, Leroy and Roujean 1994, Duggin and Piwinski 1984, Koslowsky 1998) or airborne (Royer et al. 1985) data set is used to study the dependence of the signal on the various parameters. Another approach is to simulate satellite measurements. One possibility to obtain simulated satellite data is to combine ground measurements of the BRDF with a radiative transfer model (Holben et al. 1986, Rahman 1996, Lee and Kaufman 1986, Gerstl and Simmer 1986). Another method is to combine a BRDF model with an orbital model (Meyer et al. 1995).

Results of Model Calculations

The information content of satellite measurements can be studied by simulating the signals with various input parameters. One approach is to combine radiative transfer models with measured quantities, another one is the use of individual models for the components (atmosphere, vegetation, topography), which have to be fully coupled (MC-FUME approach). The latter modelling approach is used to study and to quantify perturbing effects of geometry and atmosphere. The results allow to determine the relative importance of the use of actual satellite data (reflectances or vegetation indices) for quantitative and qualitative applications. This method was applied *inter alia* as part of a broader study assessing the unresolved issue of temporal compositing of reflective channels for sensors such as AVHRR and VEGETATION. The outcome is a software package dedicated to sensitivity analyses for the optimisation of temporal compositing of remotely sensed data called SATCO (Sensitivity Analysis Tool for Compositing Optimisation). It is a combination of several models and allows to automatically simulate Top-Of-Atmosphere (TOA) and Top-Of-Canopy (TOC) reflectance factors for a large number of different surfaces and atmospheric conditions.

TOA reflectance factors are computed with the 6S (Second Simulation of the Satellite Signal in the Solar Spectrum) code (Vermote et al. 1995, see also chapter 3). This code is a well established atmospheric radiation transfer model designed to simulate the atmospheric contribution to the satellite signal. It takes into account gaseous absorption, Rayleigh and aerosol scattering, and partly the non-Lambertian, bi-directional and environmental effects of the reflecting surface (see section 3).

The surface bi-directional reflectance (BRF) is characterised by the Coupled Surface-Atmosphere Reflectance Model (CSAR) by Rahman et al. (1993b). The CSAR model is a semi-empirical parametric model requiring three input parameters: Illumination and viewing geometry, atmospheric parameters and surface characteristics. It accounts for the hot spot effect and has been tested in direct and inverse mode against several data sets (satellite images, airborne data, field and laboratory data) under a wide variety of conditions (Pinty and Verstraete, personal communication, Rahman 1993a and 1993b, Engelsen et al. 1996).

Realistic observation geometries for compositing simulations can be obtained for two sensors, namely the VEGETATION sensor on SPOT4 and the NOAA AVHRR sensor on NOAA-9, -10, -11 or -14, which are all platforms for which an orbital model was implemented. The range of sun (ζ) and view zenith (ϑ) angles considered in the simulations was determined by the valid range of the 6S model (i.e. $\zeta < 60°$ and $\vartheta < 50°$). Simulations are done for three spectral bands, namely the BLUE (0.43 - 0.47μm), RED (0.61 - 0.68μm) and NIR (0.78 - 0.89μm).

Albedo calculations or BRDF shape identification can be performed by sampling the BRDF at a large number of points. In this case, simulations cover the BRDF with a resolution of 5° for the view zenith angles and 10° for the view azimuth angles. This sampling is a trade-off between calculation time on the one hand and the detail required to visualise the BRDF on the other hand. The highest variability in reflectance occurs in the principal plane. For this reason and for the sake of clarity, BRDF illustrations in this section will be 2D cross-sections along the principal plane instead of 3D visualization of the BRDF.

Various standard atmospheric profiles are available in the 6S model. However, to include the effect of a changing atmosphere on TOA reflectance, the values for total ozone (O_3), aerosol optical depth (δ_{aer}) and water vapour content (H_2O) were user-defined. Typical ranges for these constituents are: $0 < H_2O < 5$ g/cm², $0 < \delta_{aer} < 0.5$ and $0.1 < O_3 < 0.5$ atm-cm.

Information on surface reflectance characteristics is contained in the three CSAR model parameters (triplets). These triplets result from inversion of the CSAR model for sets of bi-directional reflectance data characterising various surface types. Available data sets were either measured directly in the field (Kimes 1983, Kimes et al. 1985, 1986, 1987, Deering et al. 1995, Engelsen 1995, Rahman et al. 1993b), or model-calculated for simulated, 'generic' scenes (Myneni and Asrar 1993), which are physically as close as possible to natural surfaces. The 'synthetic' biome used for illustration is called Myneni-grass. The corresponding triplets can be considered as representative for radiative properties of grassland vegetation.

The geometrical effects can be quantified by taking the range (maximum - minimum) of BRDF for each illumination direction normalised by the albedo of a Lambertian surface that would exhibit the same amount of reflected radiation as the Myneni-grass BRDF. The albedo of the Myneni-grass biome was calculated through integration of the bi-directional reflectance over all viewing zenith and azimuth angles:

$$\rho(\zeta) = \frac{1}{\pi} \int_0^{2\pi} d\varphi \int_0^{\frac{\pi}{2}} \rho(\zeta; \vartheta, \varphi) \cos\vartheta \, d\vartheta \qquad (4.6)$$

where

ρ : albedo
ζ : solar zenith angle
ϑ : view zenith angle
Φ : relative azimuth angle
$\rho(\zeta, \vartheta, f)$: CSAR model of Rahman et al. (1993)
f : atmospheric parameters and surface characteristics

The TOA BRDF of a Lambertian surface with the same top of canopy (TOC) albedo as the Myneni-grass was calculated through numerical integration of the bi-directional reflectance over the view zenith angle range from -50° to 50° and azimuth angle range from 0 to 180°.

Quantification of atmospheric effects is based on the comparison between TOC and TOA signals for identical acquisition geometries. For this purpose, the ratio of TOA over TOC reflectance was systematically calculated for the different geometrical configurations. Maximum and minimum ratio values indicate the dependence of atmospheric perturbation on acquisition geometry.

Fig. 4.1 a and b show the TOC and TOA reflectances for the red channel in the principal plane. The hot spot phenomenon (reflectance peak in the backscatter direction) can be observed clearly in the TOC curves as well as the influence of an increasing solar zenith angle, which progressively shifts the hot spot towards the position of the solar zenith angle in the backscatter direction. This results in an asymmetrical BRDF with an eccentric hot spot and has to be distinguished from the sun glint effect over water which results in a high reflectance in foreward direction. The geometrical effect is clearly different when considering TOA curves, even though it is a relatively clear atmosphere. Reflectance values increase with increasing *view zenith angles*, the effect being more pronounced when increasing the solar zenith angle. An

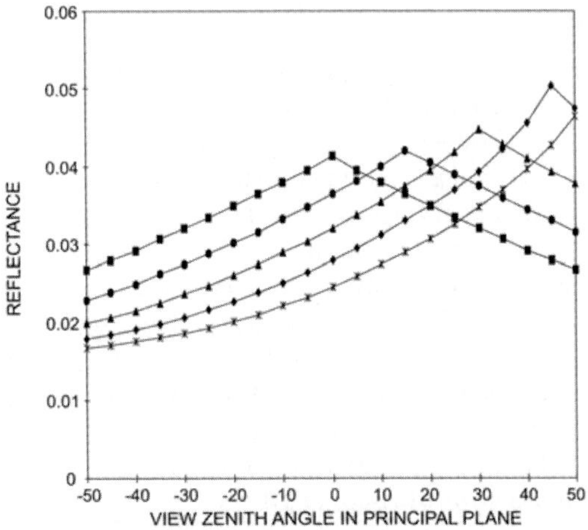

Fig. 4.1a Simulated TOC BRDF in the principal plane for the red channel (Myneni-grasses), for different solar zenith angles (■ = 0°; ● = 15°; ▲ = 30°; ◆ = 45°; ✶ = 60°) and for a relatively clear atmosphere (H_2O= 1 g/cm^2; δ_{aer} = 0.07; O_3 = 0.2 atm-cm). Negative angles indicate forward scattering, positive backward

important difference between TOC and TOA BRDF is that the hot spot effect (or its shift towards the backscatter direction) is smoothened due to atmospheric disturbance. The observed difference between the TOC and the TOA BRDF suggests that geometrical and atmospheric effects are fully coupled and have to be treated in a coupled mode, in simulation models or experiments (Veroustraete et al. 1998).

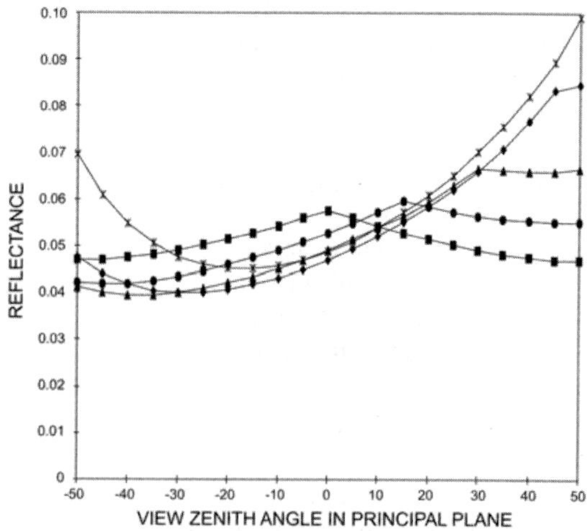

Fig. 4.1b. Simulated TOA BRDF in the principal plane for the red channel (Myneni-grasses), for different solar zenith angles (■ = 0°; ● = 15°; ▲ = 30°; ◆ = 45°; ✶ = 60°) and for a relatively clear atmosphere (H_2O= 1 g/cm^2; δ_{aer} = 0.07; O_3 = 0.2 atm-cm).

The geometrical effect is more important in the visible channels than in the Near Infrared. TOA red and blue are relatively similar and follow the same trend, an increase of the geometrical effect with increasing solar zenith angle. In the NIR channel, the geometrical effect is less pronounced than in the visible channels, although the same trend (increase with increasing θ_s) is followed. TOC and TOA NIR show the same behaviour. These observations can be explained by the presence of the atmosphere, which has a comparable influence in the two visible channels and less of an effect in the NIR channel.

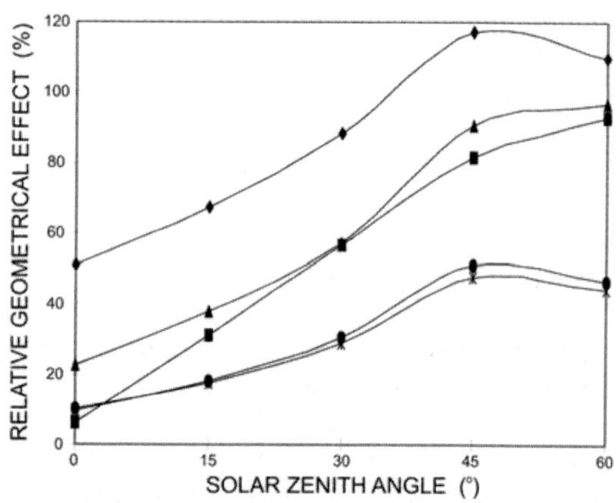

Fig. 4.2a. Relative geometrical effects [(max-min)/albedo$_{lam}$] for individual channels. ◆ Blue/red TOC, ■ blue TOA, ▲ red TOA, ● NIR TOC, ∗ NIR TOA

Results obtained by simulation of geometrical effects are summarised in Fig. 4.2a and b. For the separate reflective channels, it appears that the difference between maximum and minimum reflectance over a range of view zenith angles from –50° to +50° can vary from 6% up to 117% of the albedo. The effect is most pronounced for the TOC in the visible bands, but also for visible TOA it can rise up to 96% of the albedo. Even for the NIR band, the relative effect can rise up to about 51%.

Fig. 4.2b shows results for different vegetation indices that are currently used: NDVI (Normalised Difference Vegetation Index), ARVI (Atmospherically Resistant Vegetation Index), GEMI (Global Environment Monitoring Index) and SAVI (Soil Adjusted Vegetation Index). The vegetation indices were calculated from TOC and TOA reflectance values. Significant geometrical effects can be observed in these cases too. The difference between maximum and minimum TOC SAVI for a certain illumination direction can vary from 5% up to more than 22% of the reference SAVI. The reference index is always calculated with the albedo of a Lambertian surface which would have the same reflection as the Myneni-grass BRDF. For TOA NDVI, the relative effect varies from 4% up to more than 10% of the reference NDVI. Also other indices such as ARVI and GEMI appear to be sensitive to changing viewing and illumination conditions.

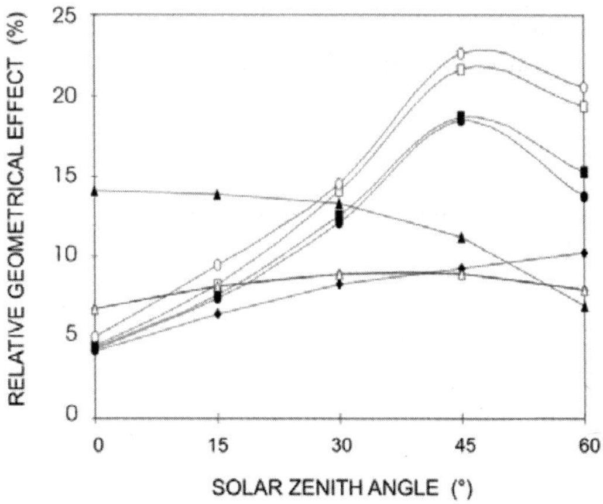

Fig. 4.2b. Relative geometrical effects [(max-min)/albedo$_{lam}$] for different vegetation indices. ♦ TOA NDVI, ● TOA SAVI, ○ TOC SAVI, ■ TOA GEMI, □ TOC GEMI, ▲ TOA ARVI, △ TOC ARVI

Compared to the geometrical effect in the separate reflective channels, vegetation indices reduce the geometrical effect partly, but not completely. This confirms the findings of several authors: bi-directional effects can be substantially reduced through the use of vegetation indices (Lee and Kaufman 1986), although residual effects may remain (Cihlar et al. 1994, Roujean et al. 1992). None of the indices mentioned here was designed to eliminate the bi-directional effects of the surface. Some of them were designed especially to reduce the effects of atmospheric perturbation (Pinty and Verstraete 1992a, Kaufman and Tanré 1992) or to reduce the influence of the soil brightness when dealing with partially vegetated canopies (Huete 1988). Most indices show an increase in geometrical effect with increasing solar zenith angle. SAVI and GEMI appear to have the strongest geometrical effects in the case of a low sun. TOA ARVI shows the opposite effect: the lowest relative effect can be found for a solar zenith angle of 60^0.

To evaluate the perturbing role of the atmosphere on TOA signals, the representative cover type (Myneni-grass) which was used to study geometric effects is coupled to three realistic atmospheres. Emphasis was put on realism so that TOA reflectance could be interpreted close to real cases. The first atmosphere (ATM1) has a low content of perturbing constituents (H_2O = 1.05 g/cm^3, O_3 = 0.2 cm.atm, τ_{aer} =

Table 4.1. Variability of atmospheric effects in the blue, red and NIR bands for three realistic atmospheres due to different acquisition geometries (θ_s = 0°, 60°; θ_v = −50° to +50°)

ζ	Spectral band	Quantification parameter	ATM1	ATM2	ATM3
0°	BLUE	Min TOA/TOC	2.79	3.20	3.46
		Max TOA/TOC	4.39	5.19	5.72
	RED	Min TOA/TOC	1.40	1.60	1.76
		Max TOA/TOC	1.78	2.22	2.58
	NIR	Min TOA/TOC	0.94	0.82	0.83
		Max TOA/TOC	0.94	0.84	0.85
60°	BLUE	Min TOA/TOC	5.15	5.99	6.47
		Max TOA/TOC	8.66	12.22	14.14
	RED	Min TOA/TOC	1.98	2.48	2.84
		Max TOA/TOC	3.22	5.99	7.76
	NIR	Min TOA/TOC	0.89	0.71	0.69
		Max TOA/TOC	0.94	0.85	0.88

0.08 (VIS) and 0.06 (NIR). The second (ATM2) is heavily loaded with water vapour (5.88g/cm^2), medium with ozone (0.25 cm.atm) and aerosol optical thickness is around 0.3. Finally, ATM3 contains a lot of aerosol (0.5), medium water vapour (3 g/cm^2) and medium ozone (0.3 cm.atm). Two different illumination conditions are considered to document the coupling of atmospheric and geometric effects: Θ_S is set to 0° and to 60°. This choice allows to quantify atmospheric effects for minimised and maximised atmospheric paths respectively.

Quantification of the atmospheric effect is done by means of the TOA/TOC ratio, as shown in Table 4.1. It is clear from this table that atmospheric effects in the blue band are much more pronounced (TOA/TOC for ζ = 0° ranges from 2.79 to 5.72) than in the red (TOA/TOC from 1.40 to 2.58) and the NIR band (TOA/TOC ranges from 0.94 to 0.82). For vegetated surfaces, atmospheric effects tend to increase the reflectance in the visible channels. In the NIR, a decrease of reflectance with increasing atmospheric perturbation is observed. The increasing perturbation from ATM1, over ATM2 to ATM3 results from the progressive increase of the aerosol optical thickness from a clear to a polluted atmosphere. In the NIR band, water vapour absorption is an important factor as well. This can be seen by comparing TOA/TOC ratios for ATM2 and ATM3. Thus different combinations of aerosol optical thickness and water vapour absorption (δ_{aer} = 0.3 and H_2O = 5.88 g/cm^2 or δ_{aer} = 0.5 and H_2O = 3g/cm^2) can result in identical signal perturbation. Atmospheric effects can be ranked by the results of a sensitivity analysis: $O_3 < H_2O <$ aerosol. Ozone is the least perturbing factor and is considered negligible with respect to the other perturbations. Aerosol is the most perturbing factor in all three channels and water vapour has medium influence in the red, and a more pronounced one in the NIR channel due to the specific water vapour absorption bands in that spectral range.

The influence of Rayleigh scattering is of importance only in the blue channel.

When discussing Figs. 4.1 and 4.2, clear indications on the coupled occurrence of geometrical and atmospheric effects can be seen by looking at the difference between TOC and TOA curves. Furthermore, Table 4.1 shows that atmospheric perturbation increases significantly with the sun-target path. In the blue band, for instance, TOA/TOC ratios increase by a factor 2 to 3 when increasing the solar zenith angle from 0^0 to 60^0. Furthermore, atmospheric perturbations appear to be larger for off-nadir viewing conditions, i.e. when the atmospheric path target-to-sensor is longer. With the sun in zenith position, this effect is identical for both extreme forward and backward scattering conditions whereas increasing solar zenith angles tend to progressively favour atmospheric perturbations in the forward scattering direction. Thus, TOA measurements depend on the coupling between surface bi-directional properties and atmospheric effects.

These results illustrate the importance of viewing as well as illumination geometry on spectral reflectance and vegetation indices. These findings are of prime importance for multi-temporal analyses of satellite images and for compositing images into temporal syntheses. Another important issue is the coupling between geometrical and atmospheric effects. In summary, standard atmospheric corrections are not sufficient to obtain accurate TOC reflectances.

Simulations have shown that the magnitude of the atmospheric effect is affected by surface type and geometry of acquisition. Consequently, the interdependent nature of the various factors that perturb the satellite signal does not allow to quantify the individual, absolute contribution or perturbation of TOA signals. Nevertheless, relative impacts are identified by the simulations.

Quantification of atmospheric effects stresses the need for reliable and detailed in situ information on tropospheric and stratospheric aerosol distributions (because of strong disturbance in all channels), as well as accurate information on water vapour concentration (especially for the NIR). Instruments such as POLDER, AVIRIS and SSM/I might provide solutions to this problem.

Over short periods of time the illumination angle will change only slightly for sun-synchronous orbiting satellites. However, for the interpretation of images covering several seasons changing illumination conditions have to be accounted for. This is especially true for regions in the mid-latitudes. Polar regions are illuminated only in summer and under extreme solar zenith angles.

When synthesizing satellite images temporally over a certain period of time, attention must be paid to the signal perturbations described above. Although surface changes may be minimal over a 10-day time period, viewing geometry can differ from day to day. Even when comparing different pixels (e.g. for classification) within a single scene of a low spatial resolution sensor such as AVHRR or VEGETATION, the broad scan angle range (between 0^0 and 50^0) can induce an important signal difference. Combination of these anisotropy effects with different atmospheric loading can significantly complicate signal interpretation. Another side effect of compositing is the occurrence of speckle. New techniques for temporal compositing are developed in the framework of the VEGETATION programme.

Empirical Determination of the BRDF

An empirical method to correct long term data series of reflectances for the main observation geometries is to analyse the data themselves, provided that they are measured in the course of the year under different illumination and observation angles. The AVHRR instrument of the NOAA satellites with its large scan angle of ±55° is a source for such data sets. The sun-synchronous orbit of this satellite results in periodic variations of the local observation conditions with a period of about 9 days. During this period the radiation geometry for a specific location changes from off-nadir backscattering over nadir to off-nadir forward scattering conditions. If no clouds are present, it is possible to obtain samples of cross-sections of the BRDF. For sun synchronous orbits the local observation time and accordingly sun height and azimuth are directly related to the sample number and therefore the local zenith distance of the satellite and the day of the year. The orbital drift towards later observation time from year to year change these conditions only slowly which is illustrated in Fig. 4.3.

For each year of operation daily 3-dimensional plots can be generated with the date as the x-axis, the nadir distance of the satellite as the y-axis, and the reflectances as z-axis (Fig. 4.4). Missing values due to data loss or cloudy conditions can be interpolated by a Delauny-triangulation combined with a low pass filtering of the data. In case of a complete data set for one year a cross section through such a distribution at nadir results in a nadir corrected course of the reflectance data for the whole year. A cross section at a specific date provides the actual reflection function, and its reciprocal can be used as a Nadir Correction Function (NCF) to estimate for nadir conditions for a given satellite zenith distance.

Fig. 4.4 shows examples of such 3-dimensional plots of reflectances of CH1,

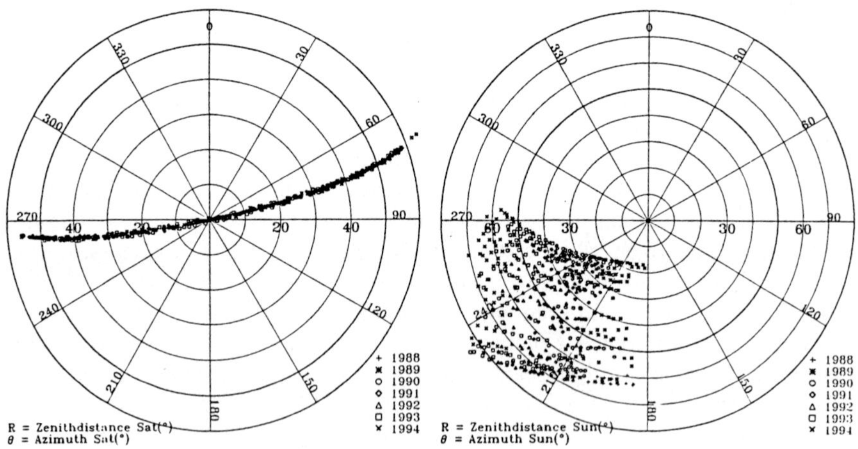

Fig. 4.3 Zenith distance and azimuth variations of satellite and sun during the operational phase of NOAA 11 for a location of Spain

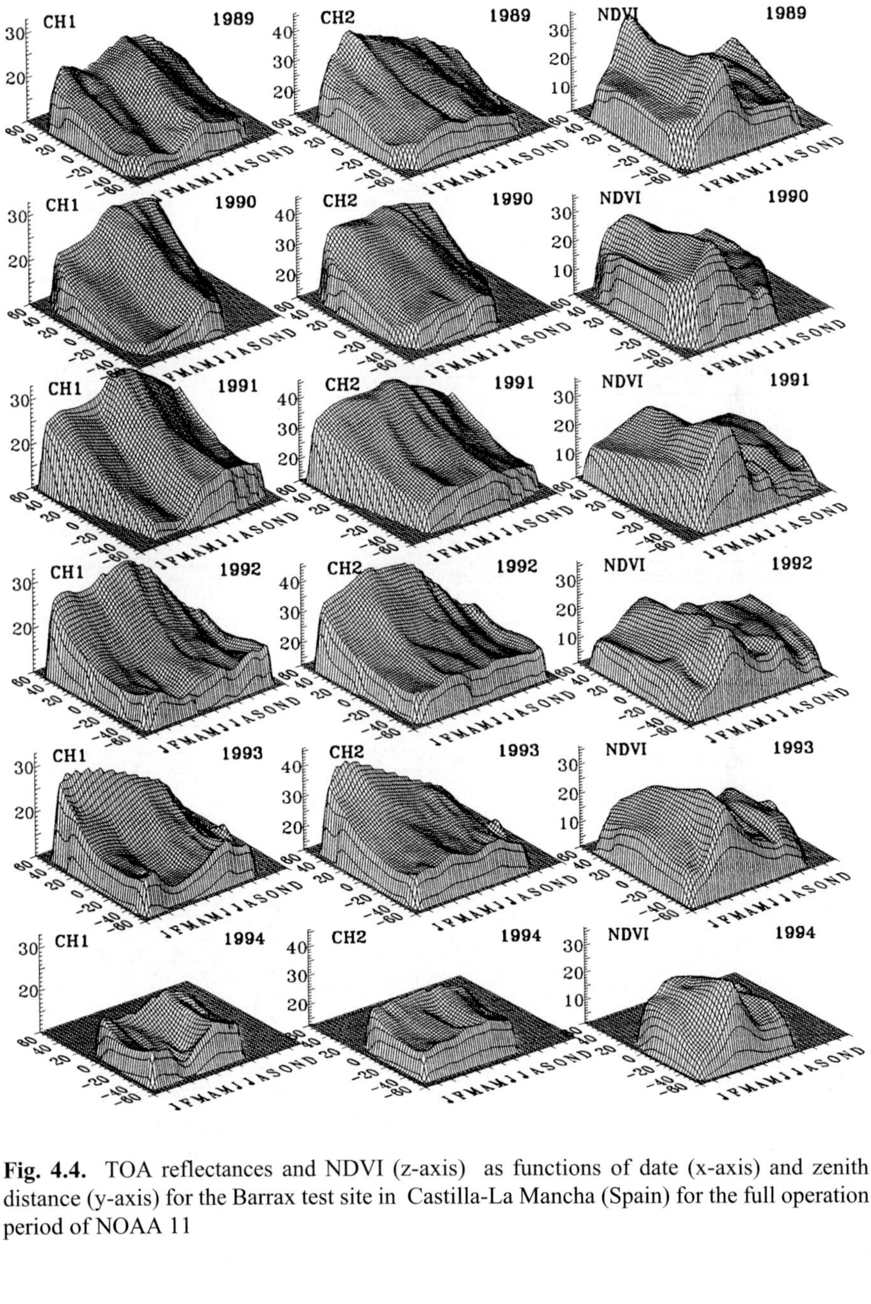

Fig. 4.4. TOA reflectances and NDVI (z-axis) as functions of date (x-axis) and zenith distance (y-axis) for the Barrax test site in Castilla-La Mancha (Spain) for the full operation period of NOAA 11

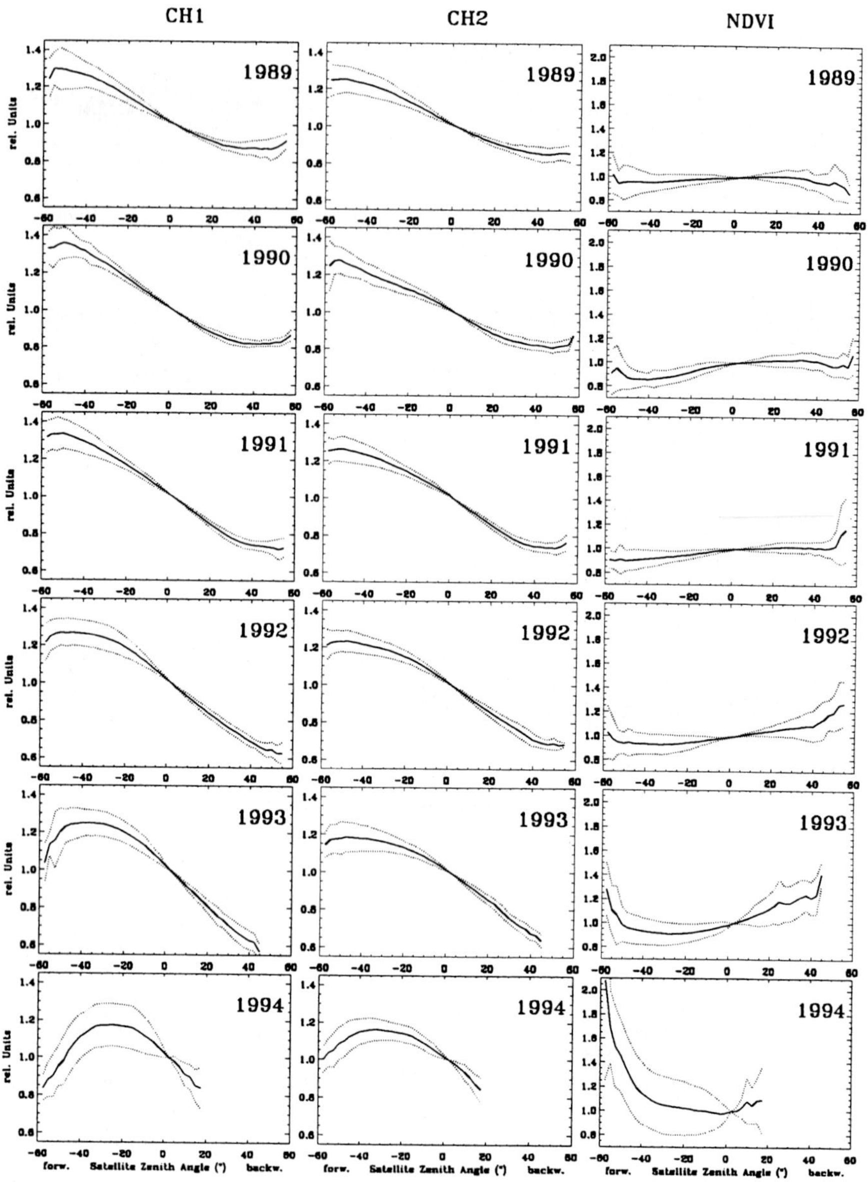

Fig. 4.5. Mean NCF and RMS deviation for the Iberian Peninsula for the month June for 1989 to 1994

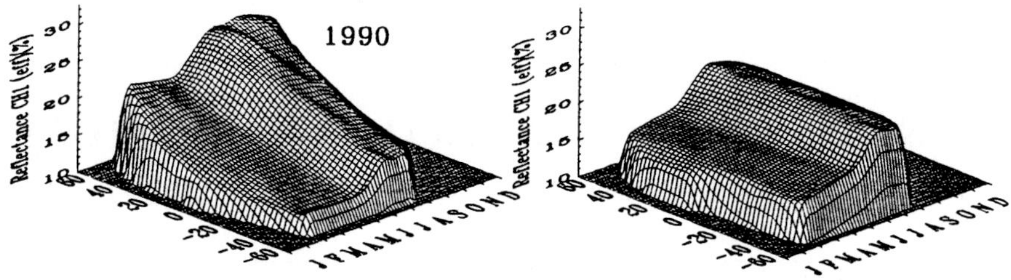

Fig. 4.6. NCF applied to data of a sample half year (1990) for AVHRR-Channel 1 in Castilla-La Mancha, Spain. Left the original data, right the normalized data

CH2 and NDVI respectively over the EFEDA test site Barrax in Castilla-La Mancha, Spain. The pronounced anisotropy of the surface can be seen as well as its variation not only in the course of the year, but also from year to year, mainly caused by the different illumination conditions due to the shift in local observation time.

Even though the NCFs are only valid for a specific land surface type and vegetation cover for which the distribution was generated, averages for the whole Iberian Pensinsula (Fig. 4.6) were applied successfully as a first order correction for surface anisotropy.

If this normalization procedure is applied to TOA data, the NCF corrects over the short period of 10 days for the viewing and illumination geometry as well as for the variable pathlengths through the atmosphere. To infer from these TOA data the surface reflectances it is only necessary to perform radiance transfer calculations for the nadir viewing conditions with atmospheric quantities referring to the mean conditions for such a period. This represents the actual situation better than climatological data. As long as no network of anchor stations is installed in the region of interest, the optical depth is measured operationally only at few stations and during field experiments. Therefore the uncertainty in the distribution of aerosol, ozone and water vapour which affect the shortwave channels of the AVHRR is probably higher than the error due to the sampling period for the NCF. Methods are required to determine the optical depth directly from remote sensing data as promised for future missions. The growing density of the AERONET stations in the Mediterranean as well as the EARLINET network of lidar stations is an important step into this direction.

Mean conditions for the whole Iberian Peninsula including RMS values of the NCF are shown in Fig. 4.5 for the month of June. Even the use of such mean correction functions considerably reduces the scattering caused by the changing viewing geometry of the daily data sets. The resulting corrected annual courses of reflectances and NDVI are better suited for evaluation of characteristic differences from year to year than time series only smoothed by statistical methods. Fig. 4.6 gives one example. The reflectance data of AVHRR channel 1 are plotted from February to July of the year 1990 at the left side without correction and at the right side NCF corrected.

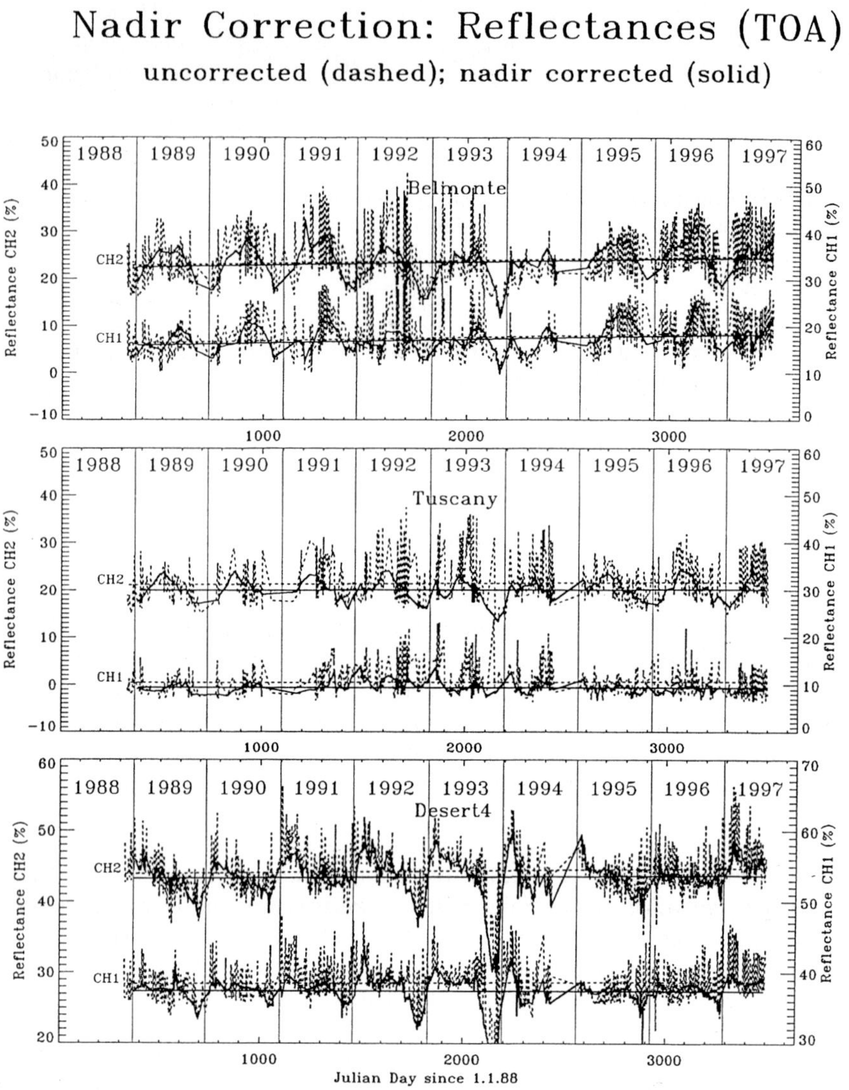

Fig. 4.7. Effect of nadir correction on time series of TOA reflectances derived from AVHRR chann el 1 and 2 measurements for three areas: Belmonte in Castilla - La Mancha, Spain, Tuscany, Italy, and north African desert. The dashed lines represent the uncorrected values and the full lines the corrected data. The horizontal dashed and full lines are the mean values

Fig. 4.7 shows a nine years data set of daily uncorrected and of nadir corrected reflectances for three test sites without filtering. The time series of NOAA 11 ended in September 1994 and is continued by NOAA 14 data since January 1995. The data of these two satellites are carefully inter-calibrated by a RMS fit of the nadir corrected data (see section 2.4; Koslowsky 1997). The regression lines for the desert target show no slopes. This indicates, that the influences of sensor degradation and changing observation time are corrected with sufficient accuracy.

4.2.3 Some Immediate Conclusions

When generating spectral or broadband reflectance maps of the whole Mediterranean region the following items have to be observed:
i) Compositing and mosaicing of spectral or broadband reflectances from AVHRR data does not lead to satisfactory results for such a large area without normalization with respect to the differing observation geometries.
ii) Correcting the combined influence of surface reflectance anisotropy, which is described by the bidirectional reflectance distribution function (BRDF), and of atmospheric effects requires knowledge about the aerosol load of the atmosphere.

In the case of NOAA-AVHRR these problems are aggravated by the missing on-board calibration of the instrument, the degradation of the sensors, and last, but not least, by the precession of the orbital plane, which is responsible for the drift towards later observation times during the lifetime of these missions.

The application of empirically derived mean Nadir Correction Functions (NCF) is capable to remove the main effects of the BRDF for the whole Mediterranean basin. This results in a considerable improvement of reflectance and vegetation index maps, provided that the atmospheric composition shows no considerable long-term trend with respect to atmospheric attenuation.

Often the maximum NDVI composite technique is applied to construct cloudless maps of the vegetation index. In the case of NOAA-AVHRR images this technique favours extreme off-nadir data because of the scene effect (nadir viewing sensors see more soil). If first guess NCF corrections are applied to the reflectances prior to calculating the NDVI, this disadvantage can be compensated to a large extent. Corrections applied to nadir observations not only remove the land surface anisotropy effect but also result in data which are comparable to other sources with higher resolution such as Landsat and SPOT data.

4.2.4 From TOA to Surface Reflectance

In Chapter 3 codes have been discussed to correct TOA radiances. For a large area such as the whole Mediterranean basin it is difficult to obtain the instantaneous information needed to feed radiative transfer calculations with the state of the atmosphere during the satellite overpasses. The gradients of atmospheric water vapour and optical depth are sometimes large. Generally one can apply a correction according to the mean climatological standard atmosphere to get closer to the natural

Table 4.2. Results of the calculation of atmospheric corrections by means of the LOWTRAN 7 code for two days during the EFEDA experiment in Spain.

Location and date	Tomelloso, 39.08°N, 2.98°W 12 June 1991 (day 163)	Barrax 39.06°N, 2.2°W 28 June 1991 (day 179)
Radiosonde	09:32 UT	10:34 UT
Atmospheric model	Model 7 IHAZE 2 Meteorological range 70 km	Model 7 IHAZE 7 Meteorological range 21.8 km
Satellite and solar zenith angle	8.0°/35.7°	8.0°/36.3°
Total amount of water vapour	1.52 g cm^{-2}	1.79 g cm^{-2}
Solar irradiance in the shortwave LANDSAT-TM channels 1-5 and 7 in E-02 Wcm^{-2}	Ch 1 1.149 Ch 2 1.349 Ch 3 0.9964 Ch 4 1.219 Ch 5 0.4438 Ch 7 0.1742	Ch 1 1.146 Ch 2 1.345 Ch 3 0.9936 Ch 4 1.216 Ch 5 0.4425 Ch 7 0.1738
Surface reflectance $\rho_0(i)$ in % as function of the TOA reflectance $\rho(i)$ in % for channel i	LANDSAT-TM Quadratic fit: $\rho_0(3) = -2.33 + 1.14\rho(3) - 3.695\ 10^{-4}\rho(3)^2$ $\rho_0(4) = -1.12 + 1.11\rho(4) - 1.925\ 10^{-4}\rho(4)^2$ Linear fit: $\rho_0(1) = -6.69 + 1.16\rho(1)$ $\rho_0(2) = -3.60 + 1.16\rho(2)$ $\rho_0(3) = -2.08 + 1.11\rho(3)$ $\rho_0(4) = -1.00 + 1.10\rho(4)$ $\rho_0(5) = -0.145 + 1.13\rho(5)$ $\rho_0(7) = -0.0798 + 1.19\rho(7)$	LANDSAT-TM Quadratic fit: $\rho_0(3) = -3.80 + 1.22\rho(3) - 5.043\ 10^{-4}\ \rho(3)^2$ $\rho_0(4) = -2.08 + 1.19\rho(4) - 3.332\ 10^{-4}\rho(4)^2$ Linear fit: $\rho_0(1) = -9.45 + 1.26\rho(1)$ $\rho_0(2) = -5.47 + 1.25\rho(2)$ $\rho_0(3) = -3.45 + 1.18\rho(3)$ $\rho_0(4) = -1.87 + 1.17\rho(4)$ $\rho_0(5) = -0.340 + 1.17\rho(5)$ $\rho_0(7) = -0.187 + 1.22\rho(7)$
	AVHRR ($\varsigma_\odot = 35.08$, $\varsigma_{sat} = 36.58$) $\rho_0(1) = -2.69 + 1.17\rho(1) - 4.096\ 10^{-4}\rho(1)^2$ $\rho_0(2) = -1.36 + 1.23\rho(2) - 2.427\ 10^{-4}\rho(2)^2$	$\rho_0(1) = -4.42 + 1.27\rho(1) - 5.630\ 10^{-4}\ \rho(1)^2$ $\rho_0(2) = -2.39 + 1.34\rho(2) - 4.194\ 10^{-4}\ \rho(2)^2$
	Meteosat:($\varsigma_\odot = 37.32$, $\varsigma_{sat} = 45.42$) $\rho_0(VIS) =$ $-2.62 + 1.20\rho(VIS) - 3.888\ 10^{-4}\ \rho(VIS)^2$	Meteosat:($\varsigma_\odot = 26.07$, $\varsigma_{sat} = 45.3$) $\rho_0(VIS) =$ $-4.19 + 1.31\rho(VIS) - 5.567\ 10^{-4}\ \rho(VIS)^2$

conditions. If higher accuracy is required, one has to rely on the output of weather forecast models, available radiosonde ascents, and a few observations of the optical depth, or, with respect to water vapour, on the measurements of vertical sounders. There is currently no network, which would allow to estimate the distribution of the optical thickness for the whole Mediterranean basin. Furthermore, no satisfying algorithms exist to infer quasi-operationally these quantities over land from satellite based remote sensing measurements. Future satellite missions carrying instruments with more spectral channels and/or lidar may offer the opportunity to obtain information about atmospheric aerosol directly from the satellite data.

The problem of the optical depth determination is especially grave over the Mediterranean area because of the frequent outbreaks of Saharan dust and the pollution of the air due to industry. The meteorological range (horizontal visibility) is only a marginal surrogate of the vertical optical depth since pollution can be caught within a thin atmospheric boundary layer or Saharan dust may be transported in the free troposphere. In the first case meteorological range will overestimate the optical depth and in the second case underestimate it.

For conversion of TOA reflectances into surface reflectances it is convenient to calculate look-up tables using e.g. the LOWTRAN code for varying meteorological ranges, solar and satellite zenith distances, and TOA reflectances. A considerable simplification of the conversion procedure can be achieved if the observed reflectances first are normalized to nadir viewing direction. Corrections with a meteorological range of 30 km account for the main atmospheric attenuation.

Koepke (1989) demonstrated that simple regressions of the form

$$\rho_{surface} = A + B\rho_{TOA} \tag{4.7}$$

represent such calculations well and may be applied to account for the varying aerosol load of the atmosphere. As an example the results of the calculations for two days during the EFEDA experiment in Spain are given in Table 4.2 for two locations at which local radiosondes were available during the time of the field experiment. In this table quadratic correction terms are given as well.

If the surface reflectance is small, the TOA reflectance is larger than the surface reflectance because the scattering of solar radiation adds to the signal. In cases when the surface signal is large the scattering and absorption of the reflected beam is larger than the amount added by the atmosphere and the surface appears darker. The crossover point depends on the wavelength and the aerosol type. It can be computed from the correction formulas presented in Table 4.2.

4.3 Intercomparison and Validation of Reflectance Measurements

4.3.1 Techniques

Validation of reflectances inferred from measurements made in space is not a straight forward procedure since there exists no "ground truth" standard. A number of steps can be made to estimate the accuracy and representativeness of these measurements.

A first step is to verify that the values measured by different instruments agree for the top of the atmosphere. This already implies interpolation between different spectral bands, pixel aggregation to comparable area sizes, and normalization of the geometry of illumination and observation.

The second step is the accurate correction for the atmospheric effects as discussed in section 4.2. Powerful radiative transfer algorithms exist, but it is necessary to check the accuracy of the available atmospheric state variables on which these calculations are based. A relatively simple test of the accuracy of the correction is to compare calculated optical depths with measured ones.

The third step is a comparison of the reflectances obtained after correcting for the atmospheric effects.

A final validation would be, if the inferred reflectances agree with *in situ* measured ones. For such measurements three types of instruments are available: Spectral radiometers, spectral albedometers, and broad band albedometers. None of these instruments can provide the same measuring geometry as measurements from satellites. From satellites the radiation reflected from an area of, say, 625 m^2 is measured which in most cases contain a number of different reflectors. Furthermore the radiance is contaminated by radiation from surrounding areas that is scattered into the sensors field of view. Ground based radiometers measure only the radiation from a small fraction of this area, but the reflectance function can be measured by instruments mounted at a structure that allows multidirectional viewing.

Tests of the accuracy of atmospheric corrections by comparison of remotely inferred and *in situ* measured surface reflectances are difficult to make because it in principle is not possible to measure the reflectance respectively albedo at the ground under the same geometric conditions as observed from space. Nevertheless examples will be presented how ground measurements are related to corrected measurements made from satellites.

4.3.2 Test for Atmospheric Correction

The computation of corrections for atmospheric effects includes the correct simulation of the atmospheric transmission respectively optical depth that can be measured relatively easy from the ground by spectral sun photometers. A meaningful test for radiative transfer computations therefore is, to check how accurate the simulation of the atmospheric transmission respectively optical depth matches the values measured at the ground by spectral sun photometers. The optical depth is obtained if (relative) measurements made at different solar zenith angles are plotted in a logarithmic scale over a linear scale of the relative air mass (for not too large solar zenith angles this equals approximately $1/\cos \zeta$). If the turbidity of the air is not changing during the time the measurements are made, one obtains a straight line, the so-called *Langley-plot*. The slope of the line yields the optical depth.

The LOWTRAN and 6S codes allow to print out the optical depths for specific atmospheric models and wavelengths. These are compared with those computed from the measurements made from the ground. If necessary the aerosol model is iterated. Fig. 4.8 gives an example. With the 6S code the spectral dependence of the optical depth is simulated for an aerosol model which contains 80% dust particles and 20% water soluble material. The density of the aerosol is adjusted by the meteorological range for which three values are adopted in Fig. 4.8 (10, 15 and 20 km). The measured values are plotted in red. The simulation that matches best the measurements made from the ground throughout the spectrum can be considered as

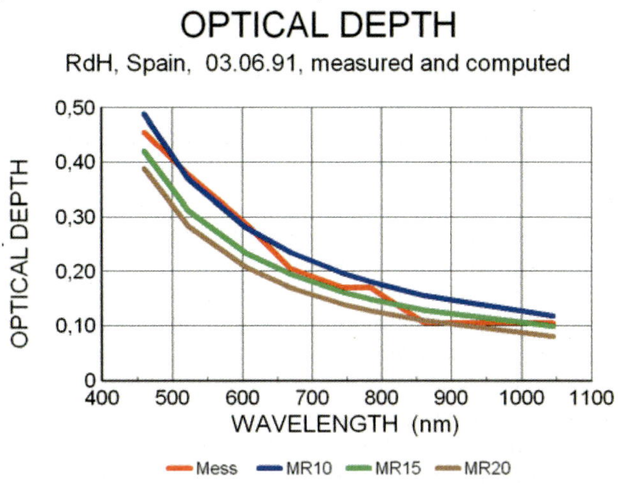

Fig. 4.8. Modelled and measured (red line) optical depth for Rada de Haro, Spain, 03. June 1991. The modelled optical depths are for meteorological ranges of 10, 15, and 20 km

coming closest to the actual conditions. Consequently one would choose this specific model for the radiometric correction of the atmospheric effects, though it may not be the unique solution. It can be seen that in this example the red curve falls in the range of computed values but does not have correctly the same slope. A variation of the aerosol model, e.g. by introducing a multi-modal size distribution, may improve the situation but in view of the local variability of the aerosol and other errors it is often unnecessary to push the accuracy to its limit.

4.3.3 Area Integrated Reflectances of the EFEDA Experimental Areas Derived From Landsat-TM and AVHRR Data

Five cloudless Landsat-TM scenes are available for the period of the EFEDA experiment in Spain, 1991. The following Table 4.3 contains the results of reflectance calculations from both instruments for 10 km x 10 km large areas around the three EFEDA supersites at Rada de Haro, Tomelloso, and Barrax (Fig. 4.9). The coordinates of the Barrax and Tomelloso sites are not identical with those for which the inter-calibration was carried out.

The graphical presentation of the reflectances indicate marginal changes during the observation period. AVHRR Channel 1 data are in relative good agreement with Landsat-TM Channel 3 data for Rada de Haro and Tomelloso, not so for Barrax. AVHRR Channel 2 data are in good agreement with Landsat TM Channel 4 data for all three sites. The absolute values do not differ much though the AVHRR Channel 2 and Landsat TM Channel 4 values are highest in Barrax where in addition AVHRR Channel 1 and Landsat TM Channel 3 values are lower than in Tomelloso. This indicates a slightly higer vegetation index. Barrax is an agricultural area with irrigated fields while the Tomelloso area mainly has dry cultures like grapevine.

4.3.4 Comparative Ground Based Measurements

General Remarks

Surface measurements of local reflectance can be made with three types of instruments. Firstly, there are albedometers that measure fluxes from both hemispheres either simultaneously or intermittently. The ratio of both fluxes yields the albedo of the surface. Albedometers with a black horizontal sensor (mostly a thermocouple) measure the broad band albedo. To obtain spectral albedo values, as needed for comparison with satellite data, filters have to be positioned in front of the sensors, which complicates the measuring geometry. The measurements of an albedometer are approximately representative for a circular area with a radius that equals the height at which the albedometer is mounted above the surface. If this is two metres then the area is about 12.5 m^2. Secondly, spectrometers are used which have a narrow field of view. The measurements generally are representative for a few dm^2 of the surface area. Thirdly, interference filter radiometers are available which measure, similar to spectrometers, the reflected radiance within a narrow field of view. The last two types of instruments do not allow to obtain the reflectance directly but the downwelling illuminating flux can be obtained by measuring the radiance reflected from a nearly white Lambertian reflectance panel of known reflectance. It is difficult to deduce the albedo from these radiometric and spectrometric measurements. For this purpose the instruments have to be moved around in such a way that within a short time interval the same area is looked at from many different directions. This is very demanding with spectrometers and easier to handle with the smaller filter radiometers, but is nevertheless an extensive operation for larger areas. Experience shows that radiometric or spectrometric measurements made 90° off the direction of the sun and at a nadir angle of 45° approximate the albedo values.

Albedometer Measurements

To measure the spectral albedo at the surface one has to use albedometers equipped with interference filters in front of downwards and upwards looking detectors. This combination makes it difficult to build an instrument with a response according to the cosine law which a flat plate broadband albedometer covered by a spherical dome approximately has. This can be done by using integrating spheres (also known as Larché or Ulbricht spheres) in front of the detector system (see Schulze 1970, p.193). Such an instrument becomes bulky and difficult to handle in field. Instead a light instrument was build using diffusor plates in front of the filter-detector system. A system like this does not respond according to the cosine law. Consequently the measured flux has to be corrected for the deviation from the cosine law. The response function can be obtained from measurements in a parallel beam, such as the direct sunlight. This leads, for the upward looking sensor, to a correction function which depends on the response factor $\gamma(\zeta)$ for the direct solar radiation, the response factor κ for the diffuse sky radiation and the ratio α of the diffuse sky

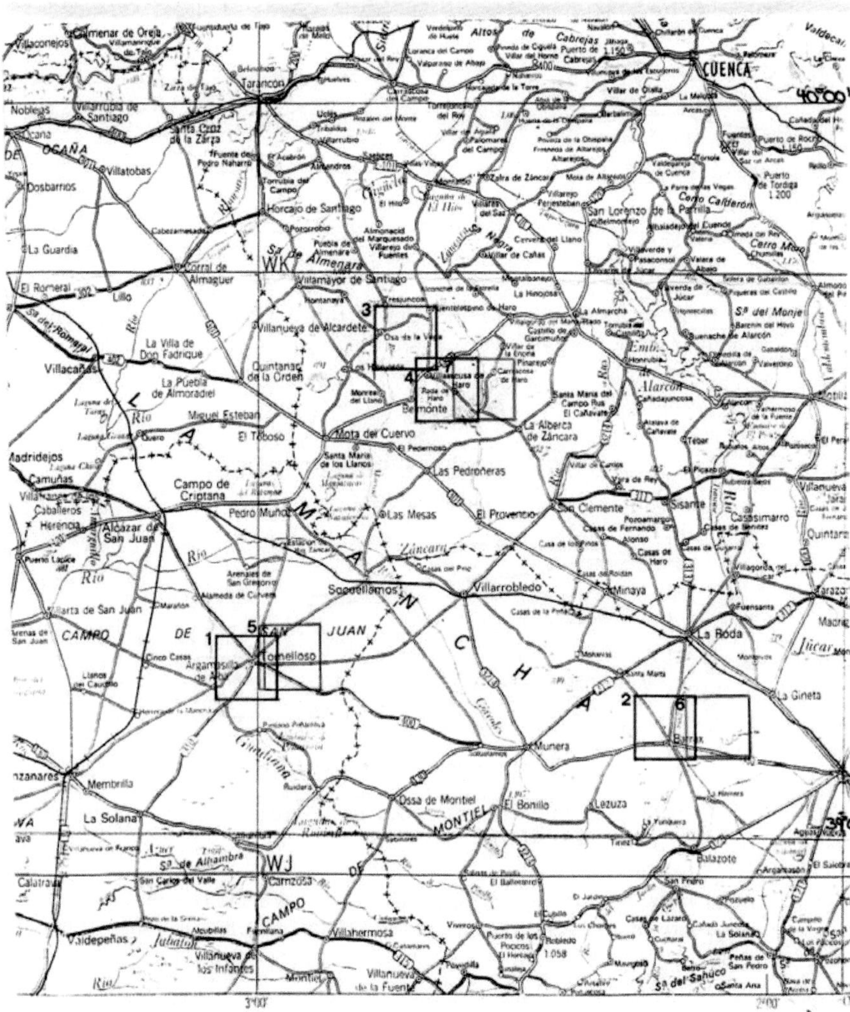

Area 1 (39.15 N/ 3.02 W) (black)
Area 2 (39.06 N/ 2.20 W) (black)
Area 3 (39.65 N/ 2.71 W) (black)
Area 4 (39.65 N/ 2.63 W) (black)
Area 5 (39.08 N/ 2.98 W) (red)
Area 6 (39.03 N/ 2.16 W) (red)
Area 7 (39.15 N/ 2.66 W) (red)

Escala 1:800.000

20.000m. 0 20 40 60 80 100Km.

Fig. 4.9. EFEDA experimental sites 1991

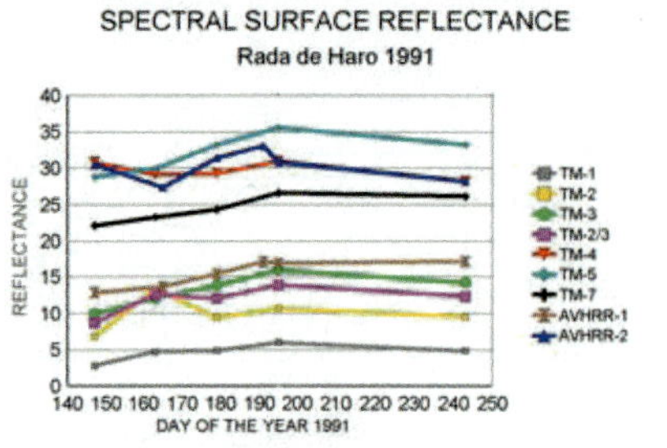

Fig. 4.10a. Time series of the spectral reflectances of the Rada de Haro area derived from AVHRR and Landsat TM data according to Table 4.3

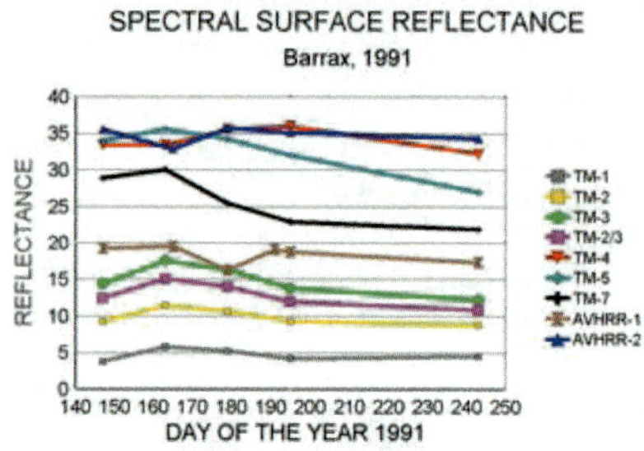

Fig. 4.10b. Time series of the spectral reflectances of the Barrax area derived from AVHRR and Landsat TM data according to Table 4.3

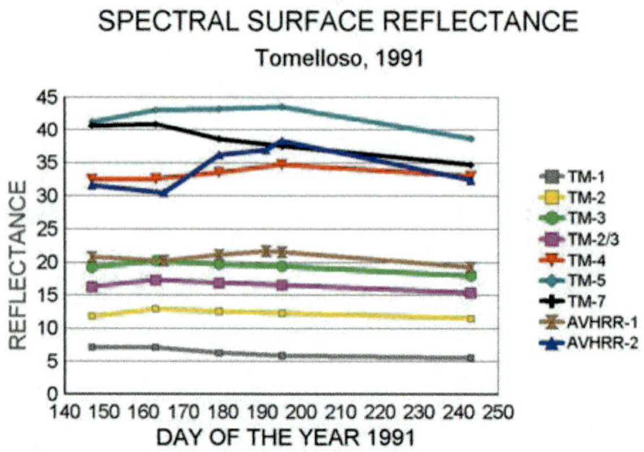

Fig. 4.10c. Time series of the spectral reflectances of the Tomelloso area derived from AVHRR and Landsat TM data according to Table 4.3

Table 4.3. (following pages). Area average of the surface reflectance of 100 km^2 large areas around EFEDA experimental sites corrected for atmospheric effects. The table contains the the raw data (counts), the computed reflectance ρ, and the variance σ. The following radiosonde data have been used in the atmospheric correction algorithm:

LANDSAT-TM	NOAA-AVHRR
12.06. - Tomelloso, 09:32 hours,	27.05. - Madrid, 12:00 hours,
28.06. - Barrax, 10:34 hours,	14.06. - Barrax, 14:15 hours,
10.07., 14.07., and 31.08 - Madrid 12:00 hours.	28.06. - Barrax, 14:55 hours.

(a) Rada de Haro experimental site, central coordinates are 39.15°N, 2.66°W

Channel	Counts Reflectance Variance	Date (1991) 27.5.	12.6.	28.6.	14.7.	31.8.
TM 1	Counts	81.80	88.62	97.45	101.7	87.63
	ρ	**2.82**	**4.72**	**4.89**	**6.00**	**4.86**
	σ	1.02	0.81	1.01	1.13	1.13
TM 2	Counts	42.11	46.30	50.91	54.06	45.72
	ρ	**6.84**	**13.34**	**9.42**	**10.72**	**9.54**
	σ	1.64	1.45	1.67	2.09	2.25
TM 3	Counts	52.88	60.70	69.76	73.82	61.86
	ρ	**9.95**	**12.08**	**13.83**	**15.96**	**14.22**
	σ	2.78	2.61	2.63	3.65	3.51
0.4 TM2 +0.6 TM3	ρ	**8.7**	**12.58**	**12.06**	**13.86**	**12.34**
TM 4	Counts	94.47	90.43	89.03	91.98	76.86
	ρ	**30.83**	**29.11**	**29.28**	**30.97**	**28.22**
	σ	4.36	3.82	3.57	4.36	3.58
TM 5	Counts	129.4	133.2	141.8	152.5	128.0
	ρ	28.79	29.98	33.19	35.58	**33.21**
	σ	3.60	3.46	3.18	3.63	3.89
TM 7 GW	Counts	63.19	65.11	66.36	71.76	63.89
	ρ	**22.10**	**23.28**	**24.32**	**26.59**	**26.11**
	σ	4.35	4.03	3.36	3.65	4.23

NOAA 11 - AVHRR calibration after RAO et al. (1994).

Channel	ρ σ	Date (1991) 27.5.	14.6.	28.6.	10.7.	14.7.	31.8.
1	ρ	12.85	**13.64**	**15.42**	**17.05**	**16.86**	**17.14**
	σ	2.22	1.83	1.49	3.29	3.08	6.58
2	ρ	**30.46**3.	**27.36**	**31.42**	**33.06**4.	**30.82**	**28.22**
	σ	02	2.65	1.99	04	3.96	7.58

(b) Barrax experimental site, central coordinates are 39.03°N, 2.16°W

Channel	Counts Reflectance Variance	Date (1991) 27.5.	12.6.	28.6.	14.7.	31.8.
TM 1	Counts	88.53	96.22	100.1	89.80	85.74
	ρ	**3.83**	**5.82**	**5.28**	**4.24**	**4.55**
	σ	7.00	1.78	8.80	1.88	1.66
TM 2	Counts	49.96	56.45	55.03	49.03	43.64
	ρ	**9.28**	**11.46**	**10.69**	**9.31**	**8.80**
	σ	2.14	3.23	2.46	3.30	2.49
TM 3	Counts	70.08	82.53	76.89	5.91	55.45
	ρ	**14.54**	**17.56**	**16.29**	**13.82**	**12.25**
	σ	3.40	5.54	4.36	5.80	4.17
0.4 TM2 +0.6 TM3	ρ	12.44	15.12	14.05	12.01	10.87
TM 4	Counts	101.7	102.6	104.8	105.4	86.63
	ρ	**33.45**	**33.40**	**35.53**	**35.99**	**32.21**
	σ	2.96	4.17	2.44	2.48	2.35
TM 5	Counts	151.5	156.7	146.0	136.8	105.3
	ρ	**34.01**	**35.55**	**34.23**	**32.01**	**26.96**
	σ	4.88	2.99	5.04	6.89	6.25
TM 7	Counts	80.92	82.73	68.48	62.73	54.43
	ρ	**28.86**	**30.05**	**25.44**	**22.97**	**21.89**
	σ	6.42	3.08	5.88	7.96	7.48

NOAA 11 - AVHRR calibration after RAO et al. (1994).

Channel	ρ σ	Date (1991) 27.5.	14.6.	28.6.	10.7.	14.7.	31.8.
1	ρ	**19.24**	**19.49**	**16.26**	**19.10**	**18.76**	**17.34**
	σ	1.94	1.89	2.07	4.90	3.08	4.10
2	ρ	**35.52**	**32.88**	**35.72**	**41.68**	**35.16**	**34.27**
	σ	1.39	0.95	1.37	1.53	1.09	

(c) Tomelloso experimental site, central coordinates are 39.08°N, 2.98°W

Channel	Counts Reflectance Variance	Date (1991) 27.5.	12.6.	28.6.	14.7.	31.8.
TM 1	Counts	110.3	105.6	106.9	100.8	91.60
	ρ	**7.10**	**7.09**	**6.28**	**5.87**	**5.53**
	σ	1.27	1.51	0.84	0.96	0.85
TM 2	Counts	58.06	61.51	61.02	58.58	51.23
	ρ	**11.80**	**12.91**	**12.55**	**12.29**	**11.49**
	σ	1.80	1.63	1.13	1.22	1.17
TM 3	Counts	87.83	93.21	89.78	86.69	74.44
	ρ	**19.28**	**20.24**	**19.75**	**19.45**	**17.96**
	σ	2.60	2.09	1.43	1.62	1.73
0.4 TM2 +0.6 TM3	ρ	16.28	17.3	16.87	16.58	15.37
TM 4	Counts	99.11	100.2	100.5	102.0	88.45
	ρ	**32.51**	**32.56**	**33.52**	**34.72**	**32.97**
	σ	1.76	1.95	1.77	1.95	1.82
TM 5	Counts	186.9	188.1	182.3	183.5	147.8
	ρ	**41.24**	**42.99**	**43.20**	**43.54**	**38.65**
	σ	2.96	2.51	1.59	1.65	1.86
TM 7	Counts	111.7	110.3	101.2	99.29	83.05
	ρ	**40.65**	**40.85**	**38.6**	**37.53**	**34.68**
	σ	4.43	3.66	2.41	2.17	2.54

NOAA 11 - AVHRR calibration after RAO et al. (1994).

Channel	ρ σ	Date (1991) 27.5.	14.6.	28.6.	10.7.	14.7.	31.8.
1	ρ	**20.82**	**20.29**	**21.17**	**21.73**	**21.60**	**19.32**
	σ	1.88	2.02	0.96	1.40	1.75	1.63
2	ρ	**31.57**	**30.47**	**36.19**	**36.95**	**38.32**	**32.38**
	σ	1.34	2.33	1.26	1.95	1.94	1.74

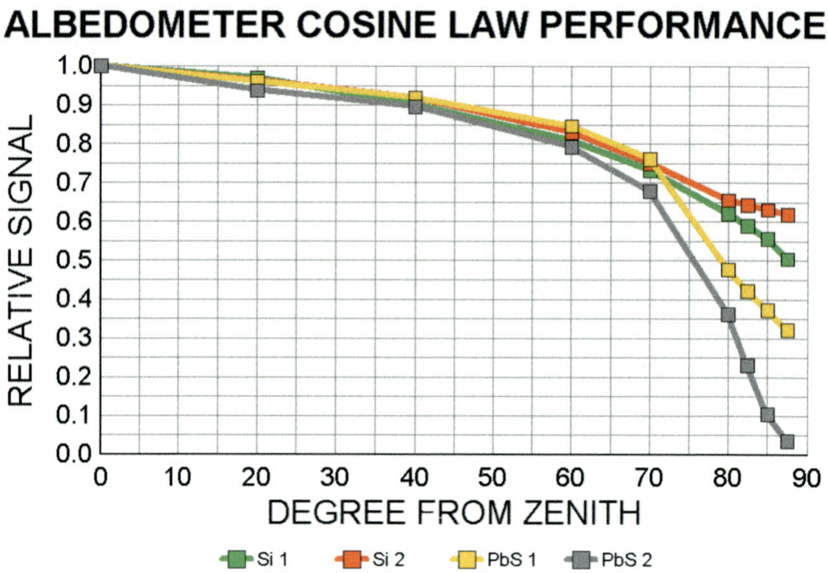

Fig. 4.11. Albedometer correction factor. The curves belong to the four filter-diffusor combinations used in the instrument, two with silicon detectors (green and red), two with lead-sulfide detectors (yellow and gray)

radiation to global radiation. For the downward looking sensor a response factor for the diffusely reflected radiation has to be applied. Because highly sensitive detectors (Si and PbS) have a limited spectral bandwidth, the instruments was equipped with two pairs of detectors discerned by colours (red, green, grey and yellow) the individual directional response factors of which are shown in Fig. 4.11. For simplification the correction for the diffuse radiation was assumed to be the same for the downward and upward reflected radiation and computed for an isotropic radiation flux. Details can be found in Appendix 6.

Reflectance Spectra

Different spectro-radiometers have been used during the EFEDA field experiment to determine the reflectance of specific objects. Some have already been presented in Figs. 1.22 and 1.23. Other examples are given in Fig. 4.16 and Appendix 4. Spectrometers typically have a narrow field of view and, depending on their distance from the target, cover only an extremely small section of the area which is in the field of view of a satellite instrument. Only if the surface is very uniform, a spectrum taken at the ground may be representative for an area of the size of a pixel in a satellite image. In an inhomogeneous area therefore many spectra have to be taken to obtain a representative overview. This can be very time consuming. Small filter radiometers are more flexible in this respect but as can be seen in the following figures spectra are necessary to relate the single points of filter radiometer measurements correctly to each other.

Radiometric Measurements

Interference-filter radiometers often are more flexible and handy than spectro-radiometers with which they have in common a narrow field of view. To obtain reflectances, reference signals from a Lambertian white plate have to be taken as in the case of the spectro-radiometers. In the sometimes hot Mediterranean environment care has to be taken that the temperature dependence of the measurement is minimized. Detectors have to be selected with small temperature sensitivity and the temperature of the instruments has to be kept constant, e.g. near 37 °C by thermostat regulated heating. This reduces the sensitivity of the instrument but keeps it stable and guarantees that the wavelengths defined by the interference filters remains constant. At higher temperatures this constancy is no longer given. Measurements and reference measurements reiterated within short time periods minimize the effect of sensitivity changes but cannot avoid wavelength shifts.

Filter radiometers are especially useful if the position of the instrument has to be changed quickly. This is the case, if angle dependent measurements are to be made.

In a number of comparative field measurements it has been found that measurements made with filter radiometers or spectral radiometers can be used as a surrogate for albedo measurements if the measurement is made under 45° nadir angle and perpendicular to the direction of the solar beam (Appendix 6).

Comparison with Satellite Data

In the following figures the data from Rada de Haro and Tomelloso of Table 4.3 are presented as spectral plots and compared with measurements made at the ground in the areas of the experimental sites shown in Fig. 4.9.

The area of Rada de Haro is characterized by a large variety of land cover. Consequently the scatter of the spectral reflectance can be expected to be large. Fig. 4.12 shows the wide spread of individual spectra measured at different points within the test area. Table 4.4 gives the legend for the identification numbers of the spectra shown in the figures. The satellite data of two dates, 27 May 1991 and 31 August 1991 showing the development from early to late summer as seen from space are given as green respectively red lines which connect the measurements made in the six Landsat-TM channels. It should be mentioned that rain fell at the first of June and the end of June. As can be seen, the NOAA-AVHRR measurements marked by crosses and blue dots fit nicely into the Landsat-TM data. In Fig. 4.13 the mean of all satellite data presented in Table 4.3 is shown as red line and compared with albedo measurements made over different land covers. It can be seen that the governing land cover seems to be low matorral (macchia) represented by measurements over herbs (green line). Dark matorral (dark green line) and holm oak (light blue) compensate for bright soils underneath of grapevine and wheat (which was already ripe at the time of the measurements).

The scatter in the satellite data is shown in Fig. 4.14 where the reflectances ρ inferred from the satellite measurements (Fig. 4.12) are given together with their respective standard deviation $\rho \pm \sigma$. These satellite data are compared with two different mean values computed from spectra shown in Fig. 4.12 and described in Table 4.3. It can be seen that these artificial mean values fall into the range of the satellite data at the wavelength of the Landsat-TM instrument (even in between the points fixed by the TM channels). The scatter of the NOAA-AVHRR data which is not shown in the figure interestingly is larger than that of the TM data.

The Tomelloso area is much more uniform than the Rada de Haro site and though the grapevine was growing during summer the area averaged reflectance did not change much (Fig. 4.10c). Fig. 4.15 shows the spectral reflectance again at the two dates of 27 May and 31 August 1991, which marks the main growing grapevine, and

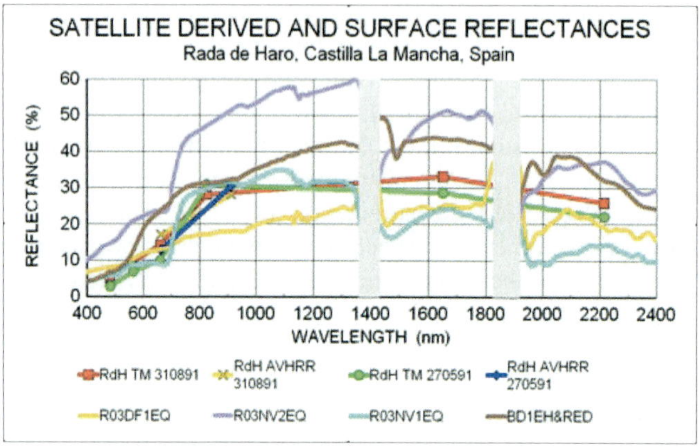

Fig. 4.12. Surface reflectance inferred from Landsat-TM and NOAA-AVHRR measurements, Rada de Haro area, 27. 05. and 31. 08. 1991 (Table 4.3) and selected surface spectra (Table 4.4). Spectral ranges of water vapour absorption are shaded

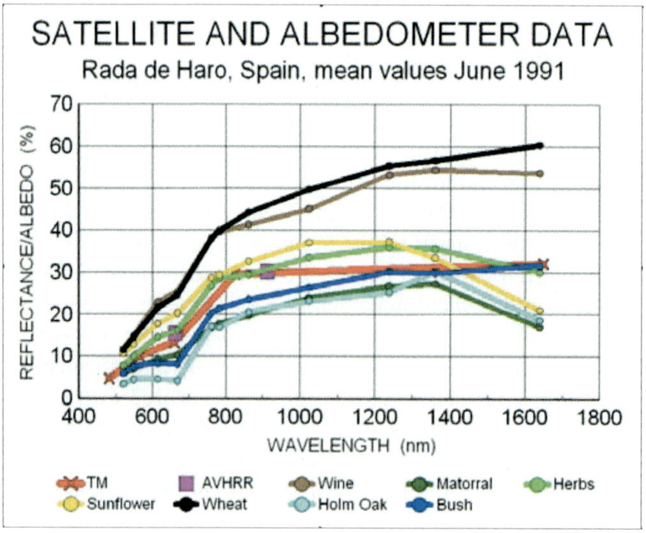

Fig. 4.13. Mean value of the surface reflectance inferred from Landsat-TM and NOAA-AVHRR measurements of the Rada de Haro test area (see Table 4.3) and albedo measurements made over different land cover types at the ground as explained in Table 4.4

two albedo measurements made on 30 April over bare soil. The green and red lines again are the mean values for the test areas indicated in Fig. 4.9. The averaged STD σ is smaller than in the case of Rada de Haro. It can be seen that the satellite measurements represent almost bare soil because the coverage with grapevine and therefore its contribution to the signal is relatively small. As for Rada de Haro the NOAA-AVHRR data agree with the Landsat-TM values.

Table 4.4. Legend of spectra presented in Figs. 4.12 - 4.15 (May - August 1991)

Fig.	ID	Spectrum ID	Location	Date	Time	Description
4.12	A	R03DF1EQ	Rada de Haro	03.06.	07:10 UT	dark soil without vegetation
4.12	B	R03NV1EQ		03.06.	08:47 UT	low macchia, herbs
4.12	C	R03NV2EQ		03.06.	09:18 UT	soil with thin low vegetation
4.12	D	BD1EH&RED				(E + F)/2
	E	R21BD1EH		21.06.	11:40 UT	soil below irrigated alfalfa
	F	RED				mean of three measurements of red soil
4.12	G	RdH TM/ AVHRR		Landsat-TM/NOAA-AVHRR spectra of the Rada de Haro test site on 27. 05. 1991 and 31. 08. 1991		
4.13	H	vine	Rada de Haro	29. 08.	10:09-15:25 40.5° - 29.5° - 51.8°	Average of 6 albedometer measurements above and between vine rows
4.13	I	herbs		01. 05	11:12; 27.6°	Thyme, broom, lavender
4.13	J	short Alfalfa		21. 06.	11:00; 22.5°	Mowed alfalfa, irrigated
4.13	K	bright soil		12. 06.	07:26; 61.3° 09:26; 38.5° 11:20;19.6°	Field without vegetation; mean of three measurements
4.13	L	holm oak		29. 04.	09:09; 46.7°	Albedo measurements over 2m high holm oak bushes
4.13	M	matorral		09. 06. 26.06	12:00; 16.8° 12:14; 16.2°	Low vegetation over reddish soil
4.13	N	TM AVHRR		May - August	mean TM and AVHRR values from Table 4.3	
4.14	O	Mean 1	Rada de Haro			(3∗B+C+4∗A+F+E)/10
4.14	P	Mean 2				(B+E+A/2)/2.5
4.14	Q	Mean 3				(3∗B+4∗A+3∗F+C+E)/12
4.15	R	Tom300491	Tomelloso	30.04.	Albedo measurement over bare soil of wheat field	
4.15	S	Tom310891		31.08.	Albedo measurement over ploughed crop field	

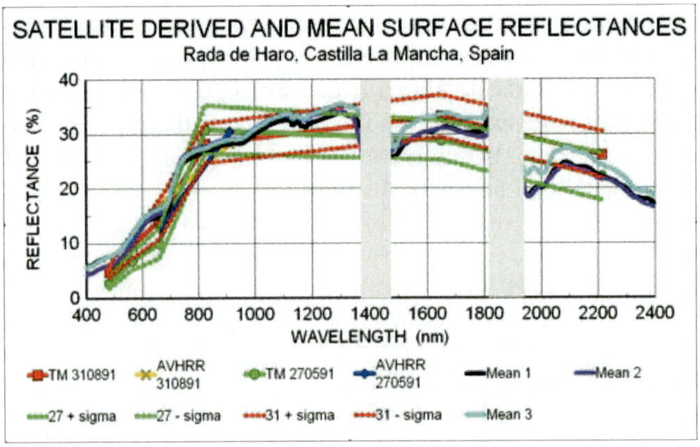

Fig. 4.14. Surface reflectance of the Rada de Haro test site derived from NOAA-AVHRR and Landsat-TM measurements (27. 05. And 31. 08. 1991) compared with two synthetic "mean" spectra derived from surface measurements as described in Table 4.4

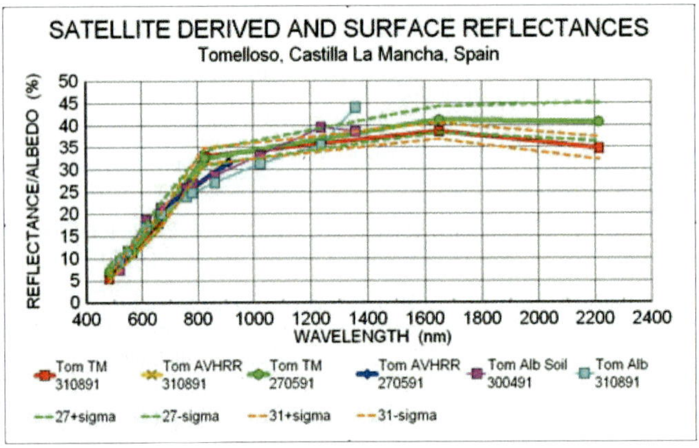

Fig. 4.15. Surface reflectance and its standard deviation inferred from Landsat-TM and NOAA-AVHRR measurements (27. 05. and 31. 08. 1991, see Table 4.3) compared with two albedo measurements over soil (see Table 4.4).

4.4 Radiometric Temperatures

4.4.1 Infrared TOA Radiometric Temperatures

In infrared channels in the range from 3μm to 16 μm (also known as thermal IR, terrestrial IR or far IR) satellite instruments measure a TOA signal that consists of the following components: the upwelling radiance from the ground, consisting of the infrared emission of the surface plus the reflected downwelling radiance, and the radiance coming directly from the atmosphere. The upwelling radiance from the ground is partly attenuated by the atmosphere. The atmospheric contribution to the TOA signal is determined by the emission of the atmospheric gases and aerosols and, if present, by clouds. In the atmospheric window regions, i.e. around 3.7 μm and around 10 μm, for details see Chapter 3, the radiance from the surface dominates the signal, while in the absorption bands - mainly of CO_2 - the atmospheric contribution is dominant. Carefully processed time series of measured radiances or “brightness temperatures” derived formally by inverting the Planck equation provide information about changes in the Earth’s atmosphere-surface system. Interpreted with the aid of models as upwelling infrared fluxes, brightness temperatures constitute an important component of Earth’s radiation budget as they describe the loss of energy in the form of infrared radiation into space. However, the heterogeneity of the land surface makes the use of models to interpret data from larger areas difficult. Furthermore, the scanning instruments of meteorological satellites measure radiances at widely varying times of the day. For an AVHRR instrument the measurements between the western and the eastern end of a scan cover about 3000 km which corresponds to a difference in time of 1.5 h hours. Additionally, at a given location the local solar time of observation varies for consecutive passes up to nearly two hours and the observation geometry differs from pass to pass. The temporal resolution of polar orbiters does not allow to study the temperature dynamics of the surface. From geostationary orbit, i.e. Meteosat for Europe and Africa, the geometry is constant and the time resolution of 0.5 h allows studies of the diurnal temperature cycle; on the other hand, the radiometric and spatial resolution is limited.

4.4.2 Radiometric Surface Temperatures

Radiances can be converted to temperatures using the inverted Planck’s function. Due to a variety of possible corrections (e.g., from atmospheric influence, surface emissivity, and the anisotropy of the radiance field) the nomenclature of the temperatures resulting from such an inversion is not consistent in literature. A clear definition can be made based on the physical description of radiance measurements and the underlying assumptions. The following nomenclature is used in this section for radiance to temperature conversion: TOA radiances result in TOA brightness temperatures, near-surface upwelling radiances in surface apparent brightness temperatures. For known atmospheric profiles and surface emissivity, the radiometric surface temperature can be retrieved by applying the following corrections: the surface apparent brightness temperature must be corrected for the reflected downwelling atmospheric irradiance and in the inverse Planck equation the emissivity has to be accounted for. In addition, TOA brightness temperatures must be corrected for the atmospheric attenuation of the radiances. All radiometric measurements discussed here are measurements of directional radiances. For the case of an isotropic radiance field the radiometric surface temperature equals the thermodynamic temperature.

The radiative transfer equation in the IR range follows the assumptions that: (1) The atmosphere is at local thermodynamic equilibrium, which is approximately valid up to about 50-70 km (2) no scattering occurs, which means only cloud free and non-hazy conditions are considered, and (3) Earth’s surface is a Lambertian reflector (Dash et al. 2002).

In reality, the Earth’s surface behaves in between a Lambertian and a specular reflector. As the atmosphere is not completely transparent even in the atmospheric windows, the radiance going out from the Earth’s surface is influenced by the composition and thermal structure of the atmosphere. The radiance measured by a sensor is:

$$L_i^{sat} = \varepsilon_i \tau_i(\theta) B_i(T_s) + L_i^{\uparrow}(\theta) + \left(1 - \varepsilon_i\right) \tau_i(\theta) L_i^{\downarrow} \tag{4.9}$$

where L_i^{sat} is the TOA channel radiance, i the spectral channel in the IR, $B_i(T_s)$ Planck's function for channel i, $L_i^{\uparrow}$ the atmospheric upwelling radiance, q the zenith angle, $L_i^{\downarrow}$ the downwelling atmospheric radiance (hemispherical irradiance divided by p), t_i the atmospheric transmissivity, and e_i the surface emissivity for channel i.

The first term in Eq. 4.9 represents the emission from the surface, which during daytime is usually warmer than the atmosphere. The second term describes the contribution of the atmosphere, which varies with its vertical structure: warm/moist layers increase its contribution. The third term describes the reflection of downwelling radiance. If the measurements are made in the 3-4 µm range and during daytime, then additionally the reflection and scattering of the solar irradiance, and diffused solar irradiance have to be taken into account (see chapter 3).

Radiometric surface temperature can be retrieved by radiative transfer calculations for known atmospheric situations and known surface emissivity employing a radiative transfer model, e.g. MODTRAN. However, these calculations are very costly in terms of computing power (time-consuming). A significant speed-up can be achieved by using trained neural nets instead of a physical radiative transfer model (Göttsche and Olesen 2002). It was shown that a factor of 10,000 over MODTRAN is possible without significant loss of accuracy.

4.4.3 Sea Surface Temperatures (SST)

Though SST does not belong to land-surface process studies it is addressed here because of its potentially strong influence on land-surface change processes. Mediterranean SST is important for land-surface processes in two respects. Firstly, the difference between land and sea surface temperatures and the related pressure fields drives the land-sea circulation, which is very important for the temperature and precipitation regime of coastal zones. Secondly, the SST is an important factor for evaporation from the sea and thus for the hydrological cycle. High SSTs tend to make the air on top unstable and thus cause severe storms and precipitation events. Therefore it is of interest to know if the Mediterranean SST tends to increase and how “hot spots” are geographically distributed during the seasons. In this section references to external data sources are given.

Mediterranean mean SST’s have been analysed from 1985 through 1995 in the context of the U.S. Pathfinder AVHRR Project and were processed by Ryan et al. (1996) using an objective analysis package developed by Mariano and Brown (1992). Together with wind data these SSTs are available as monthly averaged maps in the internet (http://www.rsmas.miami.edu/~geoff/med.html), from where they can be requested for any geographical location. From these data movies of the sea level anomalies and sea surface temperatures were produced for the time 1992 - 1997 (with the exception of the year 1994) by Bouzinac, ESA-ESTEC, which can be found at http://www.estec.esa.nl/vrwww/SLV_WES_MED/presentation.html. One image distributed by ESA was shown in Chapter 1 (Fig. 1.2). A weekly SST image database of the Mediterranean Sea and Black Sea has been build up by Lascaratos (http://www.oc.phys.uoa.gr/SST/readme.html; http://www.ciesm.org.). The SST maps are based on data from DLR (http://isis.dlr.de) and are provided with a time lag of two to three weeks in JPEG format. Lascarotos points out that the SSTs derived from satellite data are "skin temperatures" representative for the uppermost millimetres only but that in many cases these temperatures are valid for thicker layers as well depending on windspeed.

Millan et al. (1995) reports that high SSTs in the western Mediterranean sea are definitely related to heavy rains respectively thunderstorms. Because these peak temperatures may exist only for short time periods especially if only the uppermost layer is hot (the following cloud cover and rain immediately cools the surface again), the Fundación Centro de Estudios Ambientales del Mediterráneo - CEAM (http://www.gva.es/ceam) in Valencia publishes daily SST images together with other satellite data for the western part of the Mediterranean. Daily data can also be acquired from the DLR archive and the MEDOKADS data archive of the Free University of Berlin.

It can consequently be assumed that an extensive data base of the Mediterranean SST exists for future studies of the Mediterranean water cycle and its relationship with land-surface processes. Only recently this field of research started to draw higher attention in the Mediterranean research communities.

4.4.4 Analytic Representation of the Diurnal Temperature Cycle

For a number of reasons it is desirable to analyse the diurnal cycle of the surface temperature respectively of the TOA "brightness" temperature. Such an analysis is only possible with data from geostationary satellites from which data are available every half hour and in the near future even every quarter of an hour.

Göttsche and Olesen (2001) investigated the possibility to describe the diurnal surface temperature cycle (DTC) by an analytic function, which allows to determine the diurnal cycle with a minimum of observations. The effect of the sun and the decrease of the surface temperature at night is modelled using a harmonic and an exponential term:

$$\begin{aligned} T_{t<t_s} &= T_0 + \Delta T \cos\left(\pi \frac{t - t_m}{t_d}\right) && 0 \le t \le t_s \\ T_{t>t_s} &= (T_0 + \delta T) + \left[\Delta T \cos\left(\pi \frac{t_s - t_m}{t_d}\right) - \delta T\right] e^{-\frac{t-t_s}{k}} && t > t_s \end{aligned} \tag{4.10}$$

The parameters in equation 4.10 are given in Table 4.1 and are illustrated in Fig. 4.16. The choice of the harmonic term is based on the solution of the equation of thermal diffusion (Carslaw and Jaeger 1959, Lin 1980, Price 1989), whereas the exponential term describes typical natural decay-processes. Furthermore, the model assumes clear-sky conditions (cloud-screened pixels) without significant changes in

Table 4.5. Definition of the parameters used in eq. 4.10

Parameter	Dimension	Meaning
T_o	°C	Residual temperature (~ sunrise)
ΔT	°C	Temperature amplitude measured from T_0
t_d	hh:mm	Width over ±p/2 (period/2) of the cosine-term
t_m	solar time	Time of the maximum temperature
t_s	solar time	Start of the attenuation function
k	hh:mm	Attenuation constant
δT	°C	$T_o - T(t\rightarrow\infty)$, where t is the time

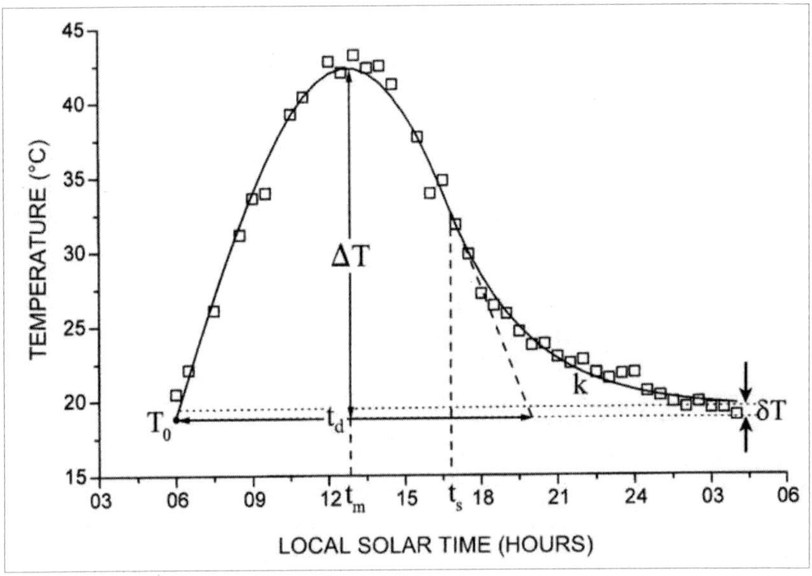

Fig. 4.16. Median brightness temperatures (BT) of Meteosat (boxes) for August 1996, Algeria. The solid line is the fit to the data using the model described by equation 4.10. The parameters are explained in Table 4.5

wind-speed, that only on sunrise takes place, and that the temperature decays freely after the "thermal sunset" defined by t_s. Therefore, time t is limited to the interval between two successive sunrises (48 Meteosat slots). In practice only 44 slots are used (from 6 hours to 4 hours of the next day) to avoid influence of the previous cooling phase and the next sunrise on the fits. The residual temperature T_0 is an extrapolated value: it describes the lowest early morning temperature, which lies outside the modelled interval.

Fitting of the model to measured DTCs yields a set of parameters, which can be used to interpolate missing data, e.g. due to brief cloud cover. Ideally, the thermal behaviour of each pixel is characterised by its respective parameters; since these depend on all the temperatures measured for the pixel, they are hardly influenced by outliers. This improves the determination of the minimum / maximum temperature of the DTC, which is important for the calculation of thermal inertia (Idso et al. 1976, Price 1977, Price 1989).

The model given by Eq. 4.10 is non-linear, which means that the corresponding normal equations cannot be solved explicitly. Therefore, a Levenberg-Marquardt scheme (Press et al. 1990) is utilised to fit the model to the time-series of BTs. The approach works on a pixel basis and allows the automatic calculation of parameters for data of arbitrary size, i.e. time-series of complete Meteosat images.

Errors and convergence problems necessitated to improve the initial model. The width t_d in the cosine terms in Eq. 4.10 was replaced by the hours of daylight (diurnal energy input by the sun). This quantity is computed for a specific location at a given latitude f and with a declination d to the sun according to Duffie and Beckman (1980). Furthermore, in order to account for the observed 'smoothness' of undisturbed (cloudless) DTCs, it is required that the two parts of the model $T_{1,2}$ are differentiable at $t = t_s$. The attenuation constant k can then be calculated from this condition as:

$$k = \frac{t_d}{\pi}\left[\tan^{-1}\left(\pi\,\frac{t_s - t_m}{t_d}\right)\right] - \frac{\delta T}{\Delta T}\sin^{-1}\left(\pi\,\frac{t_s - t_m}{t_d}\right) \qquad (4.11)$$

The dT/ΔT–term in equation 4.11 corrects k for changes of the residual temperature T_0 (e.g., for days with net gain or loss of heat). Finally, because least-squares estimators are very susceptible to outliers e.g. caused by undetected cloud contamination, a so-called 'robust estimator' (Press et al. 1990, Schlittgen and Streitberg 1994) is used. With these changes stable results for all DTCs with a sufficient number of cloudfree Meteosat measurements were obtained. The full details of the model are given in Göttsche and Olesen (2001).

Figure 4.17 shows three fits to DTCs. The corresponding parameters are given in Table 4.6. The DTCs are modelled well and the fit-errors of T_o, T_a, and dT are of the order of METEOSAT's radiometric resolution of 0.5 Kelvin. The amplitude of the broad-leaved forest is the smallest because of evapotranspiration and shading effects.

Table 4.6. The parameters determined for the three diurnal temperature courses shown in Fig. 4.17

parameter		cont. urban fabric	Non-irr. arable land	broad-leaved forest
T_o	°C	6.6 ± 0.6	3.7 ± 0.5	3.9 ± 0.6
ΔT_a	°C	20.8 ± 0.7	22.9 ± 0.7	18.0 ±0.8
t_d	hh:mm	13:56	13:54	13:53
t_m	solar time	12:53 ± 00:04	12:39 ± 00:03	12:13 ± 00:04
t_s	solar time	17:07 ± 00:18	17:04 ± 00:15	17:02 ± 00:22
k	hh:mm	03:21	02:38	01:44
δT	°C	-0.8 ± 0.7	0.9 ± 0.6	2.0±0.6

The scatter after 20 hours, which might be due to navigation error or undetccted clouds (fog), leads to a higher dT for this curve. T_o, the delay of t_m, and k are highest for the town of Milan. This is to be expected because it has the highest thermal inertia of the three examples (urban heat island effect).

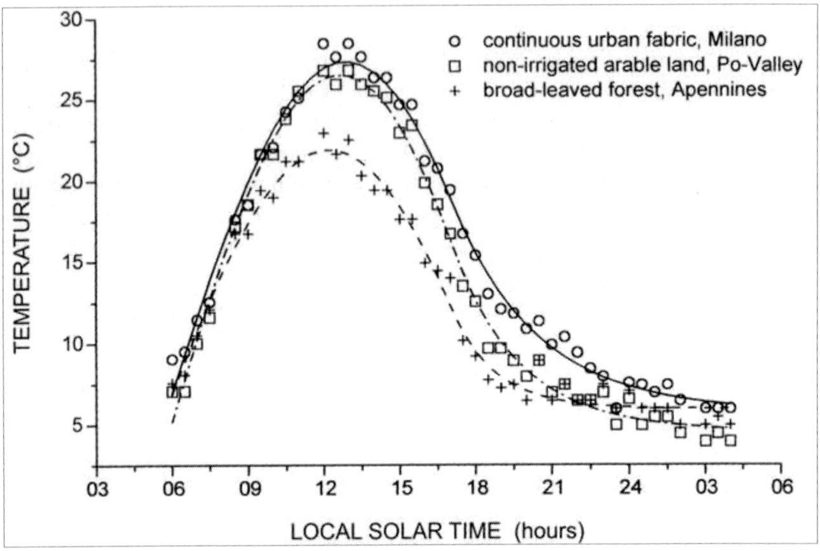

Fig. 4.17. Fits of the final model to Meteosat BTs for April 19.-20., 1996. The land-use information was taken from the CORINE data set. The fits yield the parameters given in Table 4.6 (see text for discussion)

4.4.5 Microwave Brightness Temperatures

Microwave brightness temperatures (for theoretical background see 3.2.4) are a measure of the emitted thermal radiation in the centimetre to decimetre wavelength range. In this range, the emitted radiation is linearly related to the thermodynamic or physical temperature (Rayleigh-Jeans approximation; see: Ulaby 1986) and may, therefore, be expressed in terms of a temperature (the brightness temperature, T_B), defined as:

$$T_B = \varepsilon T \tag{4.12}$$

where T is the thermodynamic temperature and ε is the emissivity.

The emissivity is highly sensitive to variations in soil moisture content and vegetation water content. It varies typically between approximately 0.6 for wet soil and close to 1 for dry soil. However, if soil moisture is not uniformly distributed in the vertical profile of the top soil, the emissivity may vary significantly with frequency because of the fact that the skin depth and the penetration depth both increase with decreasing frequency.

Vegetation absorbs part of the microwave radiation emitted from the soil surface and emits microwave radiation itself. A full canopy, i.e. a canopy with a zero transmissivity (and thus an infinite optical depth) has an emissivity close to 1 and thus emits like a dry soil. Microwave radiation gathered by a radiometer from space, therefore, is composed of fractional components from the top soil on the one hand and the vegetation cover on the other.

The transmissivity of canopies increases with increasing wavelength (decreasing frequency). Therefore, the microwave bands available from the Nimbus/SMMR instrument and used for soil moisture monitoring within the framework of these desertification studies are:

(i) 6.6 GHz, i.e., the lowest frequency to penetrate sparse vegetation and with the largest soil penetration depth, and

(ii) 37 GHz, to determine vegetation water content.

Fig. 4.18 shows the long-term brightness temperatures gathered by Nimbus/SMMR over the La Mancha area in central Spain at 6.6 GHz and 37 GHz in both horizontal and vertical polarization. All time series show a pronounced annual course which is dominated by the thermodynamic temperature. Fluctuations due to variations in soil moisture and vegetation water content are superimposed on the annual course and are difficult to interpret without a model that describes the brightness temperature as a function of T_s, l, w, t and soil moisture (see 3.2.4).

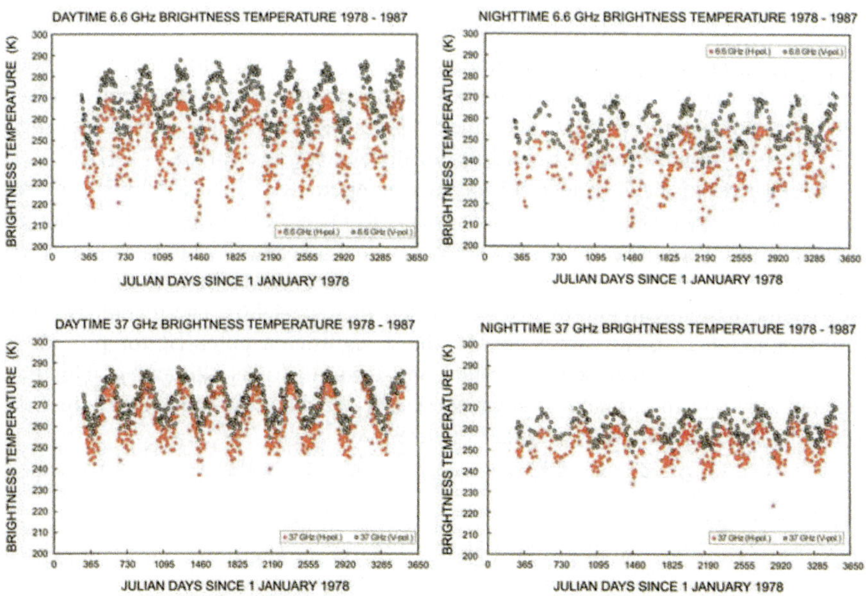

Fig. 4.18. Long term brightness temperatures gathered over the La Mancha area in central Spain from 1978 to 1987 at 6.6 GHz and 37 GHz in H (red) and V (black) polarization

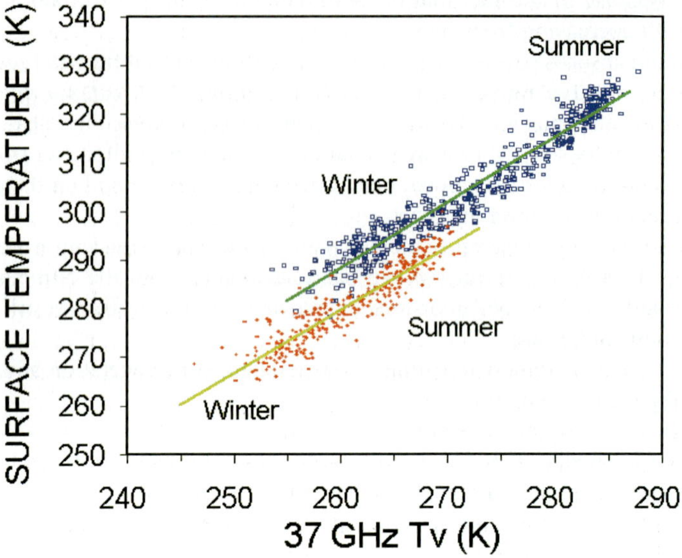

Fig. 4.19. Scattergrams of surface temperatures versus 37 GHz Nimbus/SMMR vertical brightness temperatures for the daytime (blue) at 12.00 UTC (R^2=0.926 with N = 498) and for the nighttime (red) at 0 UTC (R^2= 0.845 with N = 320).

The course of the different signatures, however, is in line with theory. For both frequencies, the V-polarization brightness temperatures are systematically higher than the H-pol. signatures. Moreover, the 6.6 GHz data show a larger dynamic range which results from the higher dominance of the vegetation cover at 37 GHz.

Determination of the (effective) physical temperature of the complex emitting surface in conjunction with microwave observations is not an easy task. Although thermal infrared remote sensing (TIR) is an invaluable means for large-scale surface temperature measurements from space, there are some serious disadvantages for its use in combination with microwave remote sensing (see 4.4.6). Thermal microwave observations at 37 GHz, however, also have significant potential for providing reliable estimates of surface temperature and form a serious alternative. This has been demonstrated by Van de Griend (2001) who calculated the following regression equations from the data presented in Fig. 4.19:

$$T_{s(12\,UT)} = 1.352\, T_{V(37\,GHz)} - 663.4 \quad (K) \tag{4.13}$$

with R^2 = 0.926 (N = 498), and

$$T_{s(00\,UT)} = 1.293\, T_{V(37)} - 56.6 \quad (K) \tag{4.14}$$

with R^2 = 0.845 (N = 320). These equations result in Root Mean Square Errors (RMSE) of T_s of ~3K for both nighttime and daytime observations.

These relationships are based on a statistical analysis of $T_{V(37)}$ brightness temperatures with long-term daytime and nighttime surface temperature estimates (1981-1987), derived from daily maximum and minimum air temperatures. Surface temperature estimates were also compared with METEOSAT derived surface temperatures and field measurements carried out during the EFEDA experiment, in central-Spain. The difference between the nighttime and the daytime relationships is clearly illustrated in Fig. 4.19. The relationships turned out to have approximately the same slope, whereas the scattergrams show only a very slight non-linearity. The difference between T_s and $T_{V(37)}$ is ~30K (daytime) and ~ 20K (nighttime) and increases slightly with increasing T_s in both cases. There are two aspects to be considered in explaining the observed difference of approximately 10K between the daytime and the nighttime, i.e. (i) atmospheric effects at 37 GHz and (ii) effects of sampling depth at 37 GHz. When comparing the daytime and nighttime relationships of Fig. 4.19 it can be seen that daytime surface temperatures in winter correspond to comparable nighttime surface temperatures in summer. This explains why the nighttime $T_{V(37)}$ is ~ 7 K higher than the daytime $T_{V(37)}$ for the same T_s (e.g. T_s = 290 K), because atmospheric effects are smaller in summer as a result of a lower absolute and relative atmospheric humidity compared to the winter season in central Spain. However, also the effect of the thermal sampling depth may contribute to the difference between the daytime and night-time relationships. Although the effect at 37 GHz with a sampling depth < 1 cm may be small, it works in the same direction in the sense that, ignoring atmospheric effects, the daytime $T_{V(37)}$ will be lower than the night-time $T_{V(37)}$ for a given T_s.

The relationships between 6.6 GHz H-pol. brightness temperatures and physical surface temperature derived using Eqs. 2.13 and 2.14 during the period 1978-1987 for the La Mancha area, is shown in Fig. 4.20 for daytime and Fig. 4.21 for night-time. The figures clearly illustrate the stronger variability of T_B during the winter period with low surface temperature compared to the summer period with high temperatures. Obviously, the variations in soil moisture are strongest in winter. Also the difference between the night-time and the daytime illustrates that the soil moisture in the topsoil is more dynamic during night-time, which was already noticed in the Botswana studies (Van de Griend and Owe 1994b).

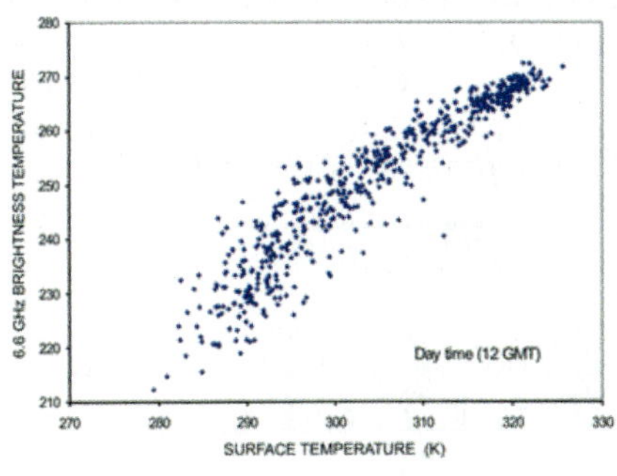

Fig. 4.20. Relationships between 6.6 GHz H-pol. brightness temperatures and physical surface temperature during the period 1978-1987 for the La Mancha area, for daytime

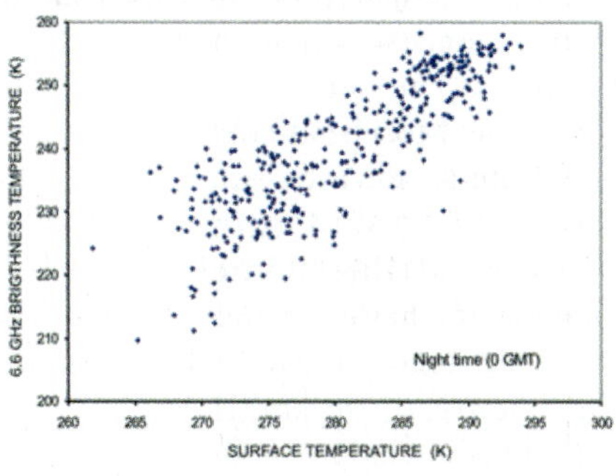

Fig. 4.21. Relationships between 6.6 GHz H-pol. brightness temperatures and physical surface temperature during the period 1978-1987 for the La Mancha area, for night-time

4.4.6 Atmospheric Effects on Microwave Signals

At low frequencies (< 10 GHz) atmospheric effects are relatively small and may be neglected (Ulaby et al., 1986). If we also neglect cosmic background radiation, then the brightness temperature T_P is proportional to ε_p and T_{eff} as given in Eq. 4.12. At 37 GHz the atmospheric effects may be important but seem to be relatively small, especially in semi-arid regions, where atmospheric absorption and emission are more or less balanced (Prigent et al. 1997). In humid environments, however, atmospheric effects at 37 GHz may be as large as several tens of Kelvin, depending on atmospheric conditions (Mätzler, 1994).

4.4.7 Validation of Radiometric Temperatures

Thermal Infrared Radiometric Temperatures

Validation of thermal infrared radiometric temperatures, gathered from space, is not an easy task. It requires spatially distributed, near-simultaneous measurements of the surface physical skin temperature or, alternatively, the radiometric temperature using fields radiometers. Measurement of the surface temperature is usually complex and imprecise when made by conventional thermometric methods such as mercury thermometers, thermistors, etc. Errors in spatial averaging often result from inadequate sampling, while measurement errors are also quite common. Proper sensor placement for measuring the surface skin temperature is difficult on bare soil surfaces and, even more so, for vegetation covered surfaces. Improper air circulation and incoming solar radiation may affect sensors located just below the surface. Consequently, the representativeness or accuracy of large-area mean temperatures estimated from point measurements using conventional methods is often uncertain, which makes this tool not ideal for proper validation. These difficulties are not only obvious for heterogeneous pixels as observed form space, but also count for 'quasi' homogeneous pixels.

As an alternative, one may use hand-held or fixed thermal radiometers, preferably in the same wavelength band as the satellite concerned. Most hand-held radiometers, however, are broad-band (8-14 μm) whereas satellite borne radiometers have smaller bands between 10 and 12 μm, centred in small atmospheric windows with the lowest atmospheric absorption. Moreover, thermal radiometers operating in the field have

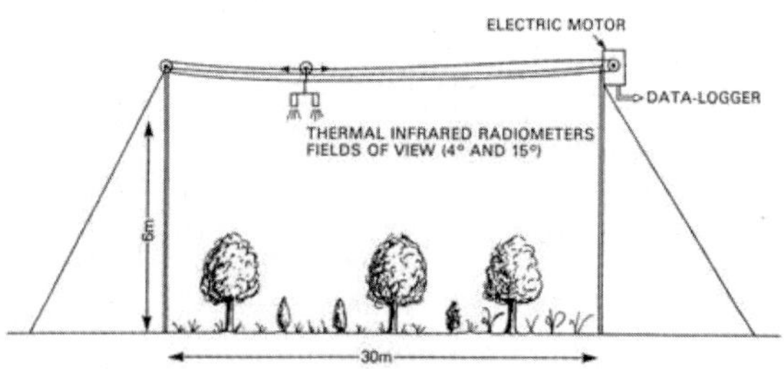

Fig. 4.22. Thermal infrared cable car system for semi-continuous measurement of radiative surface temperatures along a 30 m transect

a limited field of view and, therefore, proper spatial averaging would require a series of radiometers representatively distributed over the study area. To overcome the field of view problem and the problem of small-scale spatial variability, Van de Griend et al. (1989) developed a "cable car system" (Fig. 4.22) to move thermal infrared radiometers over a distance of 30 m. This system is computer operated and is driven by an electric motor.

In order to measure the radiative temperature of a typical vineyard two cable systems were installed in the experimental site near Tomelloso. The cable systems were operated continuously during a six weeks period to cover different stages of growth. Surface radiative temperatures were measured continuously from a height of 6 m. Each cable system contained two radiometers with 4° and 15° fields of view (FOV) respectively. These measurements gave a thorough insight into the small-scale spatial variability and allowed a reasonable estimate of the physical temperature of the soil, the individual wine plants and the integrated radiative temperature of the complex surface (Fig. 4.23).

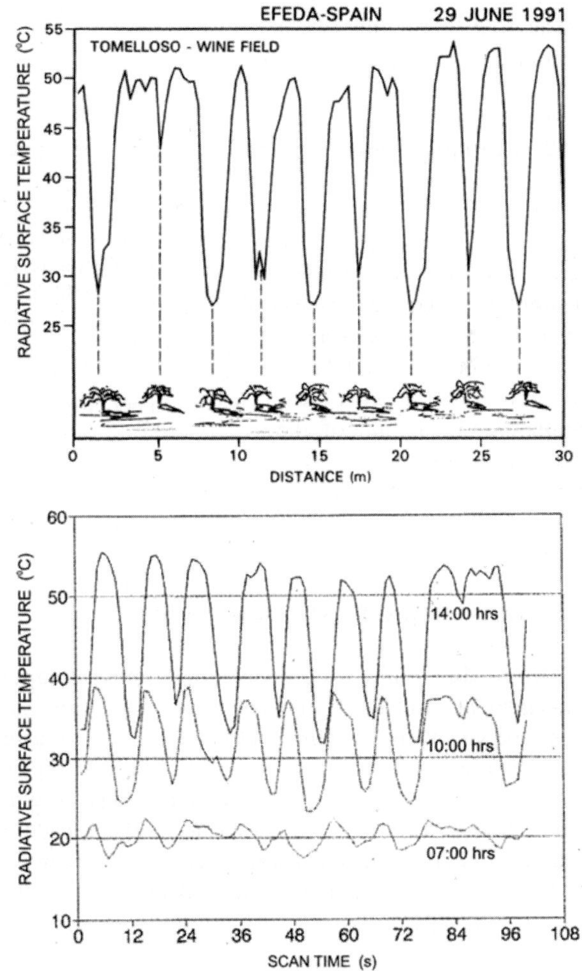

Fig. 4.23 Radiative surface temperature measured along a 30 m transect over the wine field in Tomelloso using the cable system (upper part of the figure). The lower part of the figure shows the development of the temperatures in time (7, 10 and 14 hours) over a similar path as the upper figure. Here the horizontal scale indicates the time in seconds for the trolley to run the transect. Phase shifts between the curves are due to the shift of shadows

The transect integral is shown in Fig. 4.24 for one day in June 1991. Even with such a detailed data set, the validation of the radiometric measurements from space is not easy. *First*, the measurements were done within the broad 8-14 μm band, while Meteosat and AVHRR have smaller band radiometers. *Second*, calculation of the integrated thermal radiation at the top of the atmosphere, as gathered by the satellite, requires the individual surface temperatures of the contributing surfaces (soil and plants) and their respective emissivities in the corresponding wavelength band of the satellite. *Third*, atmospheric corrections have to be carried out separately for the contributing surfaces. Similarly, calculation of the effective surface temperature from satellite thermal observations requires, apart form atmospheric corrections, an 'effective emissivity' of the complex surface of bare soil and plants.

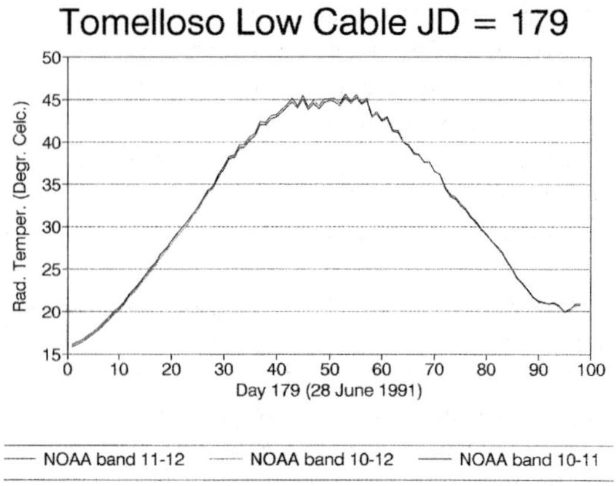

Fig. 4.24. The diurnal course of the radiative surface brightness temperature measured with the cable system over the wine field in Tomelloso during day 179 in 1991. The horizontal scale is in intervals of 15 minutes. Measurements have been made simultaneously in three spectral channels (10-11 μm, 11-12 μm, and 10-12 μm which can not be distinguished in the figure

Microwave Brightness Temperatures

In situ validation of microwave brightness temperatures by ground based measurements is no real option because of the extremely large pixels observed from space and the inherent heterogeneity with respect to surface physical properties such as vegetation cover, surface roughness and soil dielectric properties. The usual validation procedure is to use a reference surface with a known emissivity and surface temperature such as a sea surface. This procedure has been used to check for instrument degradation and instrument drift of Nimbus/SMMR over its lifetime from 1978-1987, from which no significant degradation could be discerned.

4.5 Active Microwave Products

4.5.1 Primary Products

The primary products obtained from active microwave instruments are maps of the backscattering coefficients, which contain information on the state of vegetation and soils and serve, as primary products, the classification of the land surfaces. The C-band (5.3 GHz) instruments on board of ERS-1 and -2 are vv polarization radars with approximately 25 m spatial resolution and a swath width of 100 km. The second instrument which was not particularly designed for land-surface observation is the Windscatterometer. It has a spatial resolution of 45 km and transmits and receives vv polarized radiation.

One of the products is the 16 bit single look data complex from which phase, correlation, and intensity images can be deduced with a spatial resolution of about 10 m. Fig. 4.25 shows a phase image of the Barrax area composed from pairs of overflights from 24 and 25 May 1996 (left side) and from 4 June and 13 August (right side). For the first pair the coherence is high and the phase image shows clear fringes. For the second pair the coherence is much smaller and the phase difference shows a statistical appearance. The reason is that during the long period between the first and the second overflight the vegetation grew, respectively was harvested, and the irrigation changed which in turn changed the dielectric constant. In the left side image the coherence is so high, because during the 24 hours period only the irrigation could have changed. This is readily to be seen in the small almost circular features imbedded into the regular interference fringes. The phase difference to the external dry land is almost π.

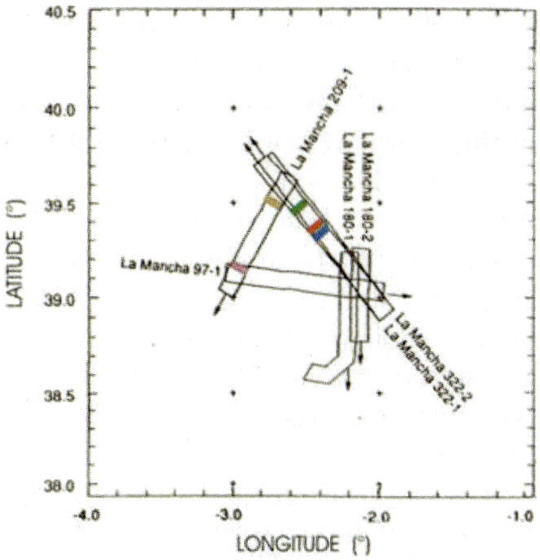

Fig. 4.26. Flight tracks

During the EFEDA experiment 1991 in Castilla - La Mancha the potential of

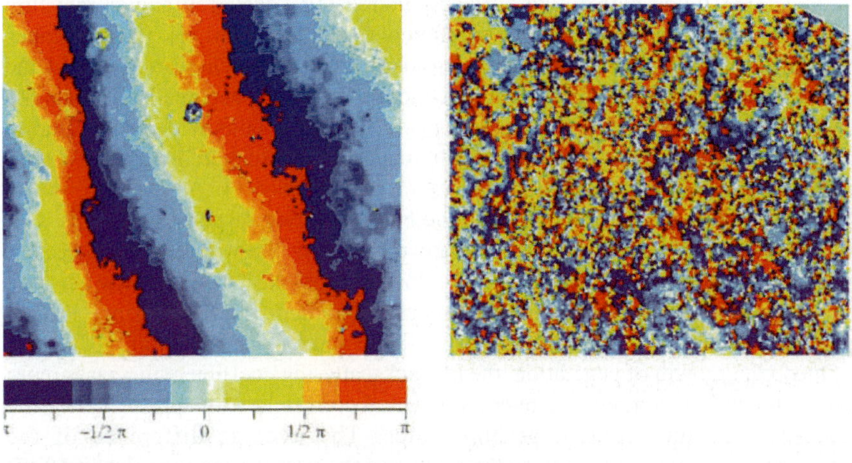

Fig. 4.25. Phase differences between pairs of images obtained from ERS-1 and ERS-2 on 24 and 25 May 1996 (left) and 4 June and 13 August (right)

active microwave remote sensing was explored by an airborne multispectral radar. The Jet Propulsion Laboratory flew the NASA DC-8 aircraft equipped with a multisensor package among which was the Air Synthetic Aperture Radar (ASAR) system (Saatchi et al. 1993). The system operates at three wavelengths, 5.7 cm (C-band), 24 cm (L-band), and 68 cm (P-band) of which the longer ones are not yet operational from satellites. The swath width is 8.5 km over which the angle of observation increases from 15 to 60 degree. The images observed have been analysed jointly with a group of the Agricultural University of Wageningen (Bastiaanssen et al. 1993). The flight tracks are shown in Fig. 4.26 (for geographical location compare Fig. 4.9). The coloured areas in blue, red, green, gold, and brown correspond to the scenes shown in Figs. 4.27 to 4.31 respectively. The scenes 4.27 and 4.28 overlap by about 30%.

The JPL Airsar is a polarimeter that permits measurements of the full polarization signature of the target including the depolarized component measured as cross-polarized returns. This technique allows to synthesise the total returned power rather than only those of the polarized components. The emitted waves are horizontally (h) and vertically (v) polarized with respect to the Earth surface. The first letter in a two letter combination (hh, hv, vh, vv) indicates the polarization direction of the illuminating wave, the second one that of the received wave. In Figs. 4.27 - 4.33 composites of the three bands are reproduced illustrating the wealth of information that can be derived to classify the land surfaces. The figures display the total backscattered power in the three bands as red (P-band), blue (L-band), and green (C-band). Villages appear white (equal backscatter in all bands), yellow areas classify areas with high L- and P-band backscatter (forests), and turquoise is caused by low backscatter in the P-band (low soil moisture). Two images of the Barrax site taken at nearly one month interval (19 June and 14 July) show the changes that occurred during this period. For comparison a vegetation index image is added in Fig. 4.34 which was obtained between the two radar images on 29 June 1991.

In general, areas with strong HH and VV but small HV returns indicate bare or low vegetated fields with rough surfaces. Increasing vegetation and/or moisture content increases the backscattering coefficient for all polarizations. This continues until the vegetation exceeds a certain biomass level at which scattering and the extinction become comparable and the extinction may even become larger than the scattering so that the return signal remains constant or even may decrease. Due to vegetation growth from 19 June to 14 July, the total backscattered power of a corn field increased by approximately 6 dB.

In terms of wavelength dependence the backscattering responses of fields with vegetation or rough surfaces and moisture increase as the wavelength decreases. Areas with large P-band responses (red colour in Figs. 4.27-4.33) usually are bare fields with dry surface and increasing moisture gradient in the soil since P-band signals penetrate deeper into the soil than C-band and L-band radiation.

Due to the growth of vegetation, the hv backscattering coefficient increases more than the hh and vv components over an irrigated corn field with approximately the same soil moisture content at both dates. The average difference of the backscattering coefficient between the early and the later season was almost 10 dB for hv and approximately 5 dB for hh and vv for frequency bands C and L. This

Fig. 4.27. Multispectral microwave image, Castilla - La Mancha, 19 June 1991, $12^h56'44''$ GMT. Central coordinates: 39°22.4' N, 02°19.4'W

4.28. Multispectral microwave image, Castilla - La Mancha, 19 June 1991, $12^h57'19''$ GMT. Central coordinates: 39°25.5' N, 02°22.8'E

4.29. Multispectral microwave image, Castilla - La Mancha, 19 June 1991, $12^h58'12''$ GMT. Central coordinates: 39°30.3' N, 02°27.8'E

Fig. 4.30. Multispectral microwave image, Castilla - La Mancha, 19 June 1991, 13h08'32" GMT. Central Coordinates: 39°29.1' N, 2°45.5' W

Fig. 4.31. Multispectral microwave image, Castilla - La Mancha, 19 June 1991, 13^h11'33" GMT. Central Coordinates: 39°10.4' N, 02°58.6' W

Fig. 4.32. Microwave image of the area of Barax, Spain, showing the irrigated areas on 19 June 1991

Fig. 4.33. Same as Fig. 4.32 but one month later on 27 July 1991

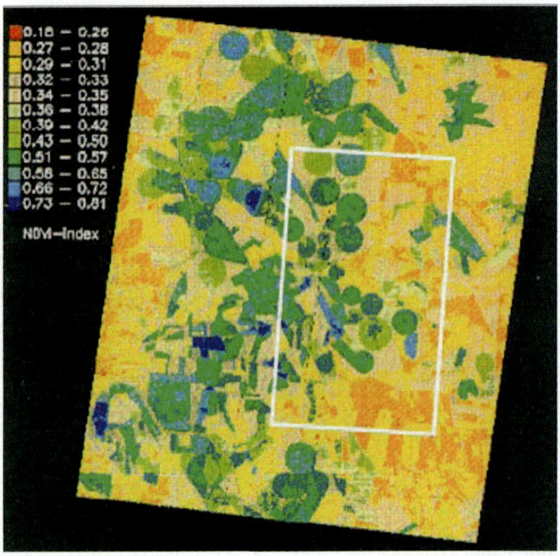

Fig. 4.34. Vegetation index map of the area near Barrax, 29 June 1991. The white rectangle indicates the area shown in Figs. 4.32 and 4.33 (turned by 90°)

corresponds to about 2.5 kg m^{-2} increase of canopy water content.

Another quantity classifying the surface is the phase shift between the hh and vv polarization return signal. Every scattering respectively reflection event adds a 180° phase shift between the vertically and the horizontally polarized electrical fields of the scattered wave. The resulting phase shift of the ensemble is small for short and more horizontally oriented vegetation and bare soil. For corn the phase shift increases with the incidence angle because the vertically oriented wet vegetation stalks with a high dielectric constant act as small antennas. The phase shift was around zero in June when the plants were still small with horizontal leaves near the ground and around 90° in July.

The unsupervised classification scheme of Van Zyl (1989) distinguishes three scattering mechanism:

- scattering at a rough surface or very dense canopy,
- scattering at vertical structures or sparse canopies which leads to a double bounce from the structure to the ground and back to the radar,
- diffuse scattering at a dense canopy.

In the first case the hh and vv signals are returning in phase. The second case can be described as an even number of reflections. The phase difference will be 180°. In the third case the hh and vv return signals are uncorrelated.

4.5.2 Validation of Active Microwave Data

The system is pre-calibrated during system tests to obtain relative response parameters. The absolute calibration is performed by analysing the signals from corner reflectors installed in the target area. The front sides of the trihedral corner reflectors measure 2.4 m at one side. 15 of these reflectors were distributed over the Airsar calibration site in California. Four of them had been deployed in Castilla - La Mancha at the test site near Barrax to ensure the stability of the calibration of the system. Two of these retro-reflectors are shown in Figs. 4.35 and 4.36.

The land use classification was validated by visual inspection and mapping of the land cover. In addition the results obtained with the active microwave systems have

Fig. 4.35. Trihedral retro-reflector for microwave calibration

Fig. 4.36. Retro-reflector for microwave calibration

been compared with classifications and vegetation indices (such as shown in Fig. 4.34) derived from the measurements made in the shortwave range either from aircraft or from satellites (Landsat and SPOT).

4.6 Vegetation Indices

4.6.1 Spectral Shortwave Indices

The spectral reflectances of sugar beet, barley and wheat measured at different times during the vegetation period are given in Fig. 4.37. Beside the minimum around 0.65 µm they show impressively the high reflectances in the near infrared at the maximum of the photosynthetic activity. The filter functions of the shortwave channels of the AVHRR and of TM as well are given below in the same scale. The characteristic difference between bare soil and active vegetation is obvious. While especially in the red channel the reflectances are diminished with increasing vegetation density, a steep increase can be found for the near infrared channels. Therefore, the comparison of red and near infrared reflectances can be used as a measure for the amount of photosynthetically active vegetation and forms the basis for most of the vegetation indices. The use of indices rather than absolute values allows to remove the main effects of atmospheric absorption at different observation angles. These effects are due to larger pathlengths through the atmosphere and the influence of different illumination conditions due to sun height variations. More than 40 vegetation indices have been defined so far and were reviewed by Bannari (1995).

The most widely used vegetation index is the *Normalized Difference Vegetation Index* (NDVI). The difference between near infrared and red reflectances, often also called *Difference Vegetation Index* (DVI) is divided by their sum as normalizing standard:

$$NDVI = \frac{\rho_{NIR} - \rho_{RED}}{\rho_{NIR} + \rho_{RED}} \tag{4.15}$$

The *Simple Ratio Vegetation Index* (SR or RVI) (Tucker 1979) is given by the

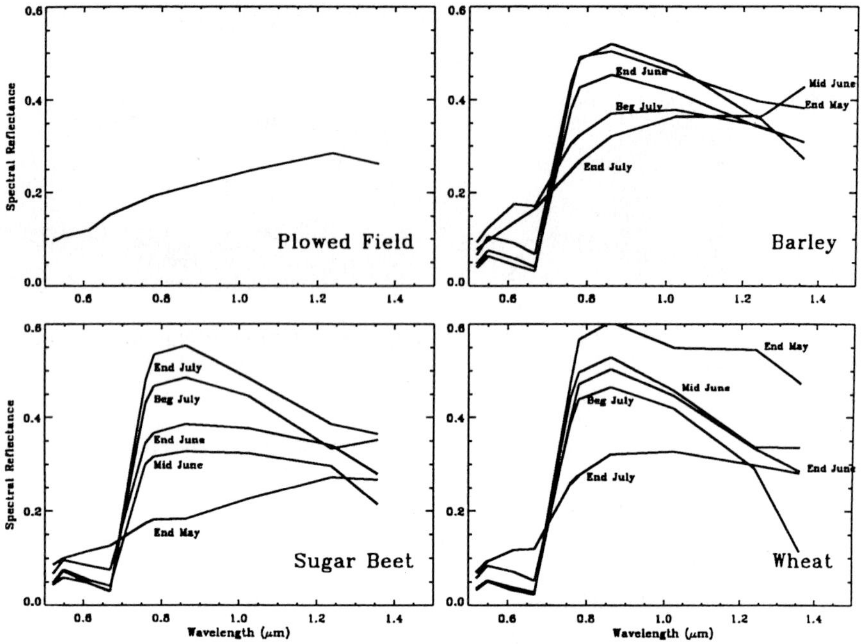

Fig. 4.37. Spectral reflectance of bare soil, barley, sugar beet and wheat at different states of vegetation development

ratio of near infrared and red reflectances:

$$RVI = \frac{\rho_{NIR}}{\rho_{RED}} \tag{4.16}$$

Both indices are related to each other by

$$NDVI = \frac{RVI - 1}{RVI + 1} \tag{4.17}$$

The *NOAA Global Vegetation Index* GVI was introduced 1982 by Tarpley et al. (1984). It uses the digital counts of the AVHRR instrument rather than reflectances. Different calibration coefficients as well as differences in degradation of the sensor response may introduce artificial trends to GVI time series.

If the soil is not totally covered by vegetation the underlying soil modifies the vegetation indices depending on soil type and soil moisture. Therefore attempts were made to reduce this disturbing factor. If the reflectances of the soils are known, a "soil line" can be defined as

$$\rho_{s(NIR)} = a\rho_{s(RED)} + b \tag{4.18}$$

and a soil independent *Perpendicular Vegetation Index* (PVI) (Richardson and Wiegand 1977) can be defined as

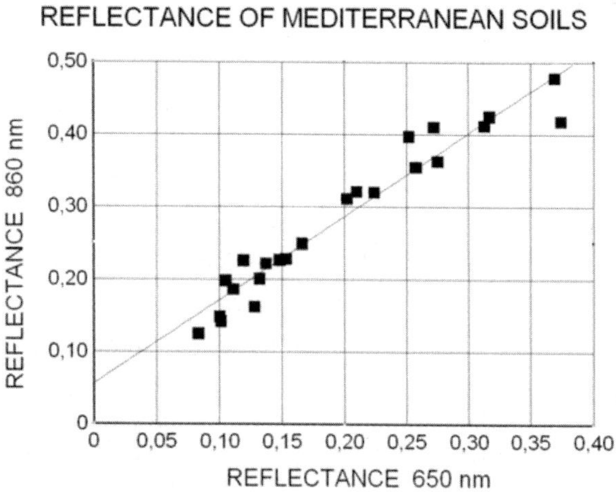

Fig. 4.38. Relationship between the reflectances at 860 nm and 650 nm for the soils listed in Table 1.2

$$PVI = \frac{\rho_{NIR} - a\rho_{RED} - b}{\sqrt{a^2 + 1}} \tag{4.19}$$

In this case a is the slope of the soil line and b its intercept with the NIR axis. Taking the data of Table 1.2 (Fig. 4.38) the following values are obtained: a =1.15, b = 0.06 (r = 0.97). With appropriate scaling factors the same effect can be achieved by the *Weighted Difference Vegetation Index* (WDVI) defined by Richardson and Wiegand (1977) as:

$$WDVI = \rho_{NIR} - a\rho_{RED} \tag{4.20}$$

Another attempt to overcome the soil influence is given by the *Soil Adjusted Vegetation Index* (SAVI) of Huete (1988):

$$SAVI = \frac{\rho_{NIR} - \rho_{RED}}{\rho_{NIR} + \rho_{RED} + L}(1 + L) \tag{4.21}$$

or by the *Transformed SAVI* (TSAVI) which makes use of the soil line rather than the adjustment factor L (Baret and Guyot 1991):

$$TSAVI = \frac{a(\rho_{NIR} - a\rho_{RED} - b)}{a\rho_{NIR} + \rho_{RED} - ab + X(1 + a^2)} \tag{4.22}$$

To overcome the shortcomings of a constant soil adjustment factor L the *Modified Soil Adjusted Vegetation Index (MSAVI)* (Qi et al. 1994) was suggested as

$$MSAVI = \frac{2\rho_{NIR} + 1 - \sqrt{(2\rho_{NIR} + 1)^2 - 8(\rho_{NIR} - \rho_{RED})}}{2} \tag{4.23}$$

The *Atmospherically Resistant Vegetation Index (ARVI)* makes use of the blue channel of the TM instrument to account for atmospheric attenuation (Kaufman and Tanre 1992):

$$ARVI = \frac{\rho_{NIR} - \rho_{rb}}{\rho_{NIR} + \rho_{rb}} \tag{4.24}$$

where

$$\rho_{rb} = \rho_{RED} - \gamma(\rho_{BLUE} - \rho_{RED}). \tag{4.25}$$

The factor γ depends on the aerosol type. This index is not applicable for AVHRR because there is no blue channel.

The *Global Environment Monitoring Index (GEMI)* (Pinty and Verstraete 1992b) has proved to be less affected by atmospheric influences than the NDVI and seems to be the better choice, but is not commonly used (see Govaerts et al. 1999):

$$GEMI = \eta + (1 - 0.25\eta) - \frac{\rho_{RED} - 0.125}{1 - \rho_{RED}} \tag{4.26}$$

with

$$\eta = \frac{2(\rho_{NIR}^2 - \rho_{RED}^2) - 1.5\rho_{NIR} + 0.5\rho_{RED}}{\rho_{NIR} + \rho_{RED} + 0.5} \tag{4.27}$$

For evaluation of MERIS data Gobron et al. (1999) developed an optimized MERIS Global Vegetation Index (MGVI).

4.6.2 Microwave Indices

Microwave Polarization Difference Index

The difference between V- and H-polarization brightness temperatures increases with increasing dielectric constant of the emitting body, and thus increases with increasing surface moisture. It also increases with decreasing biomass, an effect which is stronger at higher frequencies. This biomass effect forms the basis of the microwave based vegetation index $MPDI_{37}$, i.e. the Microwave Polarization Difference Index (Becker and Choudhury 1988, Ferrazzoli et al. 1992), which is defined as:

$$MPDI_{37} = 20\frac{T_{V(37)} - T_{H(37)}}{T_{V(37)} + T_{H(37)}} \tag{4.28}$$

where $T_{V(37)}$ and $T_{H(37)}$ are the vertical and horizontal brightness temperatures at the frequency of 37 GHz. The multiplication with 20 scales the index between zero and one ($0 < MPDI_{37} < 1$). If the vegetation is dominant, the difference between the MPDI between midnight and noon should be small. However, if the vegetation is sparse, a difference in surface moisture (i.e. close to the surface) should lead to a higher MPDI during night-time, because, on average, the surface will dry during daytime and regain some moisture during night-time. The MPDI at 6.6 GHz can be defined in analogy by replacing the index 37 by 6.6 in Eq. 4.28. The difference between daytime and night-time $MPDI_{6.6}$ is expected to be lower than for $MPDI_{37}$ because of the larger soil penetration depth at 6.6 GHz (Ulaby et al. 1986, Owe and Van de Griend 1998) and because diurnal variations in surface moisture will be stronger close to the surface.

Figs. 4.39 and 4.40 show the night-time and daytime courses of $MPDI_{37}$ and $MPDI_{6.6}$ for the 10-year period of 1978-1987, i.e. 3650 days. The graphs clearly show the annual courses with low values in summer and high values in winter. The low values in summer result from the combined effect of drier soils and more vegetation, which both lead to reduced values of MPDI. The graphs also illustrate the above described phenomenon of very similar daytime and night-time values at 6.6 GHz, and the higher values during the night-time at 37 GHz, which is ascribed to the diurnal course in surface moisture.

MPDI and NDVI

The best known vegetation index is the Normalized Difference Vegetation Index (NDVI). It is based on visible and near-infrared reflectances observed by the Advanced Very High Resolution Radiometer on board of NOAA (Tucker 1979, Tucker et al. 1985a and b, Los 1998). For retrospective studies, the use of NDVI is most practical because of the existence of a global database of NDVI data, known as GAC data (Kidwell 1990, 1995). However, the original GAC data may be cloud contaminated, and therefore people tend to use monthly or 10-day maximum NDVI-data, which tend to filter out cloudy pixels with low NDVI-values (Holben 1986). The global data set also suffers from errors resulting from variations in view angle, instrument degradation (Los 1993), and atmospheric absorption (Koslowsky 1996, Los 1998) and therefore should be used with great care. For global climate studies, the original GAC-data were corrected for instrument degradation, view-angle and atmospheric effects (Los, 1993, Los et al., 1994) which resulted in a new global data-set of coarse resolution NDVI data (James and Kalluri 1994, Los 1998). The course of the original uncorrected NDVI GAC-data and the course of the corrected data are shown in Fig. 4.41 for the period 1981-1987.

In order to test the consistency of MPDI, the microwave index is compared in Fig. 4.42 with the NDVI of the La Mancha area in central Spain. In central Spain, vegetation biomass reaches a minimum during the winter period as shown by the annual courses of both NDVI and MPDI in Fig. 4.41. The MPDI is presented as (1 - MPDI) because MPDI is inversely related to wet vegetation biomass. The NDVI maximum precedes the maximum of (1 - MPDI) and the course of the MPDI seems to be more regular than that of the NDVI.

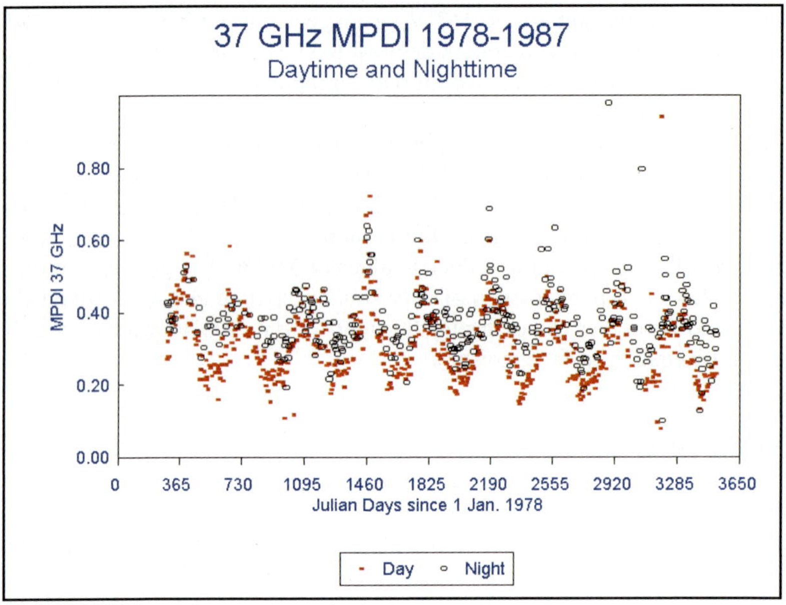

Fig. 4.39. Course of MPDI at 37 GHz for the 10-year period of 1978-1987, derived form NIMBUS/SMMR observations over the EFEDA study area Castilla - La Mancha in central Spain

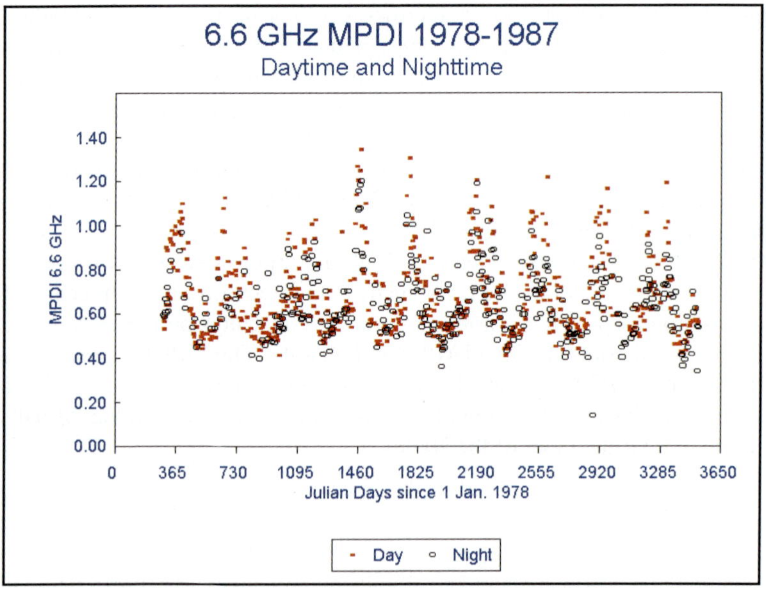

Fig. 4.40. Course of MPDI at 6.6 GHz for the 10-year period of 1978-1987, derived form NIMBUS/SMMR observations over the EFEDA study area Castilla - La Mancha in central Spain

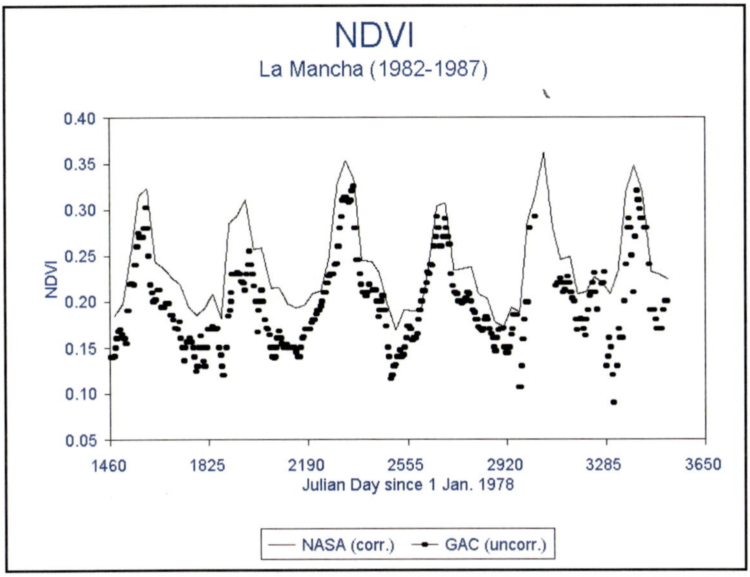

Fig. 4.41. Course of NDVI, as extracted from the original uncorrected GAC data-base and after correction for view-angle, atmospheric constituents and instrument degradation (Los, 1998).

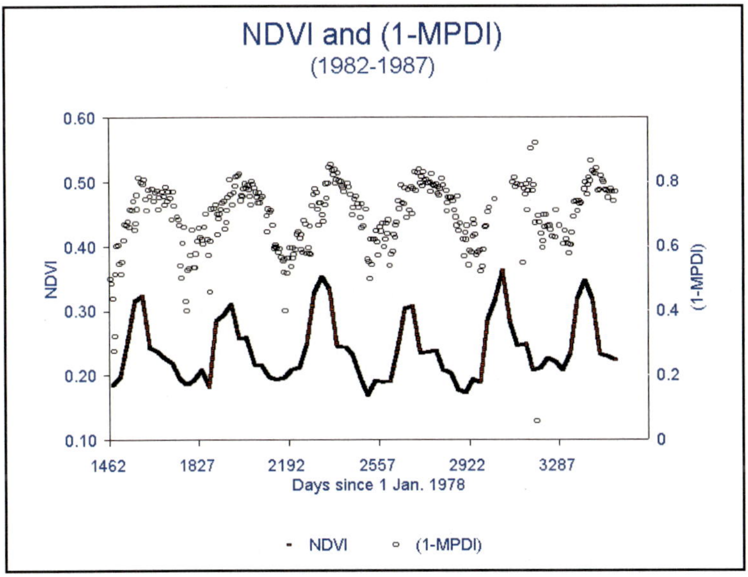

Fig. 4.42. Course of the Normalized Difference Vegetation Index (NDVI), empirically corrected for incidence-angle, sun-angle and atmospheric effects (Los, 1998) and the 37 GHz Microwave Polarization Index, shown as (1-MPDI)

Chapter 5 Higher Level Variables and Their Validation

5.1 Introduction

The ultimate goal of satellite remote sensing of land-surfaces in the context of "Global Change" is to contribute to the understanding of inter-annual variations and long term trends of changes with emphasis on physical and biological processes. At the primary level of satellite data evaluation (chapter 4) one obtains geometric patterns of land-use, spectral radiances, and simple algebraic combinations of these radiances which *inter alia* can be used to characterize the state of vegetation. Time series of these *indicators* are useful to assess *whether* land cover responds to climate variability and to the pressure of man. To understand *why*, *to what extent, and with which consequences* these changes occur requires more, namely a *quantitative* assessment of processes at the land surface. The latter requires the analysis of simultaneous measurements of multiple land-surface properties. This can be accomplished by combining primary radiometric data (chapter 4), and the indices derived from them, with collateral information. While the quality of the primary level data exclusively depends on the accuracy of the sensors, geo-referencing, and eventually atmospheric corrections, the quality of secondary level data depends in addition on the accuracy of the algorithms applied to infer the secondary level variables. These algorithms have to be validated by comparison with ground based measurements.

To assess land-surface processes one needs first to determine the primary forcing factor which is the net radiation and then the mechanism through which land surface dissipates absorbed radiant energy. In this energy conversion process the influence of the hydrological conditions has to be taken into account. The interplay of these processes is revealed by observing the response of vegetation and soils through indicators.

Quantities which in principle can be assessed with remote sensing techniques have been listed in Table 1.1. This chapter follows the logic of the development of the more complex quantities which govern the interaction of the land surfaces with the atmosphere and thus with the climate system. This leads from the assessment of state variables such as temperature, emissivity, and hemispherical reflectance to energy and water fluxes. The chapter concentrates on the methodology to assess and to validate these quantities. The application of these tools in various fields will be the topic of Chapter 6.

The two sections following next deal with the state variables relevant for the computation of the net radiation flux density at the surface:

Section 5.2 describes how the emissivity of the land surfaces is inferred which *per se* is an indicator for the state of the land-surface and is a prerequisite for the determination of the thermodynamic surface temperature which is needed to compute heat fluxes from the surface into the atmosphere. It has also to be investigated to which degree the effective radiative temperature measured in the 10 μm region is representative for the thermal infrared emission of the surface which is needed to compute the long-wave net radiation flux density.

Section 5.3 is dedicated to the computation of the broad band reflection from spectral data from which the hemispherical broad band reflection in the solar spectral range, the *albedo*, is estimated. The albedo is a key quantity in the determination of the shortwave net radiation flux density.

The chapter continues with the estimation of fluxes:

Section 5.4 deals with the net radiation flux at the surface which equals the radiative energy absorbed at the surface and which is available to drive other surface processes.

Section 5.5 discusses the possibilities to derive the heat fluxes into the soil and into the atmosphere into which most of the absorbed radiative energy is transferred. The small amount of the radiation energy which is consumed for photosynthesis is delegated to the last section. Depending on the availability of water in the soil and the vegetation a more or less large fraction of the total heat flux will be released as water vapour into the atmosphere.

The partitioning of the heat fluxes into sensible and latent heat depends on the availability of water. Therefore, the treatment of the fluxes has to include the water budget at the surface, which is the topic of the following two sections:

Section 5.6 takes up the question of the inference of soil moisture by remote sensing techniques, primarily by means of microwaves.

Section 5.7 is a short outlook to the contribution remote sensing techniques can make to assess precipitation that feeds the land surface with water.

The final section of this chapter, section 5.8, addresses the relationships between the physical processes in the surface - atmosphere system and vegetation. It deals with photosynthesis and vegetation characteristics which allow to quantify the state of vegetation and its role in the water cycle. The small amount of solar energy absorbed by the vegetation is of prominent importance for life on Earth. On the other hand, the state of vegetation is an effective indicator for environmental conditions.

5.2 Radiative Properties of Land Surfaces I: Emissivity and Thermodynamic Temperature

5.2.1 TISI Algorithm

Thermodynamic temperature is related to the emitted infrared radiation flux density by

$$M = \sigma \varepsilon T_{therm}^{4} \tag{5.1}$$

with $\sigma = 5.6703$ E-8 W m^{-2} K^{-4} (see chapter 3). Satellite sensors obtain a signal that is limited by the bandwidth of the instruments and affected by the atmosphere. Neclecting the solar radiation one obtains for monochromatic radiation the radiance:

$$L_{\lambda} = L_{\lambda,atm}^{\uparrow} + \left[M_{\lambda,atm}^{\downarrow}(1-\varepsilon_{\lambda}) / \pi + \varepsilon_{\lambda} B_{\lambda}(T_{therm}) \right] \tau \tag{5.2}$$

The measured radiance is composed of three components. Firstly, the upward directed emission from the atmosphere. Secondly, the radiance resulting from the hemispherical flux of the downward directed atmospheric radiation which is reflected by the surface, and, finally, the emission of the surface itself. Both the reflected and emitted components from the surface are transmitted through the atmosphere (multiplication by the transmittance τ) and it is assumed here that the surface acts Lambertian.

To obtain thermodynamic surface temperature two issues have to be addressed simultaneously:

- temperature – emissivity separation;
- atmospheric absorption and emission.

A semi-operational method based on Becker and Li (1990a, 1995) was applied to determine the spectral emissivity with AVHRR Ch.3, 4 and 5 observations. It was found that vegetated surfaces are Lambertian to a good approximation. Bare surfaces in arid zones are not and, therefore, the angular behaviour of the bi-directional reflectance must be taken into account (Bruckert et al. 1995).

Calibrated NOAA-AVHRR satellite data are corrected for atmospheric effects using a meteorological data base, such as provided by Météo-France, called AVISO. For the correction of angular effects only the direct solar radiation and the reflected beam into the direction of the satellite are considered. The resulting correction comes close to the bidirectional reflectance distribution function (BRDF) for the principal and the perpendicular reflection plane.

A Temperature Independent Spectral Index (TISI) is introduced to characterize the surface. This Index is the product of the emissivities determined in two or three spectral AVHRR bands after an individual exponent has been applied to each of the emissivities which makes the product temperature independent.

The temperature dependence results from the approximation method applied to the Planck-function. Because of the strong temperature dependence near 3 µm it is not appropriate to apply a linear approximation to the Planck-function. Instead, for each spectral channel an approximation of the form

$$B(T) = \alpha T^{n} \tag{5.3}$$

is used for the radiance of a black body. For the three thermal infrared channels of the AVHRR instrument the exponents have the following values: $n_3 = 12.876$, $n_4 = 4.591$, and $n_5 = 4.152$. For simplification all radiances are expressed in terms of a Planck-function of the corresponding effective black body temperature. In general, the radiance leaving the surface can be expressed by

$$\begin{aligned} L_s &= B(T_{s,eff}) + \rho S = \varepsilon B(T_s) + (1-\varepsilon) B(T^{\downarrow}_{a,eff}) + \rho S \\ &= \varepsilon \alpha T_s^n + (1-\varepsilon)\alpha (T^{\downarrow}_{a,eff})^n + \rho S \\ &= \varepsilon \alpha T_s^n \left[1 + \frac{1-\varepsilon}{\varepsilon} \frac{(T^{\downarrow}_{a,eff})^n}{T_s^n} \right] + \rho S = \varepsilon \alpha T_s^n C + \rho S \end{aligned} \tag{5.4}$$

The following notations are used:

$T_{s,\,eff}$ = effective emission temperature of the earth surface,

$T^{\downarrow}_{a,eff}$ = effective radiation temperature of the atmosphere for downward directed radiation,

ε = emisivity of the surface,

ρ = surface reflectance distribution function (sr^{-1}),

S = solar irradiance at the surface.

The longwave surface reflectance may differ from (1-ε) because of the angular dependence: the reflectance function primarily is a directional-directional value while (1-ε) is a hemispherical-directional value (see section 3.2.1).

The *C*-value as defined by Eq. (5.4) and α may be different for day- and nighttime which will be marked by the indices D and N where relationships between day- and nighttime values of the longwave radiances are deduced.

In a next step a temperature independent function of the ratio of the day- and nighttime non-solar components of the radiances at the surface are constructed for channel 3 and channel 4 (or 5). For this purpose first an exponent is applied to the effective radiation which is 1 for channel 3 and $a_4 = n_3/n_4$ for channel 4. The ratio

$$\frac{L_{s,N,3}}{L^{a_4}_{s,N,4}} = \frac{B(T_{s,eff,N,3})}{\left[B(T_{s,eff,N,4})\right]^{a_4}} = \frac{\varepsilon_3 \alpha_{N,3} T^{n_3}_{s,N} C_{N,3}}{\varepsilon_4^{a_4} \alpha^{a_4}_{N,4} T^{n_4 a_4}_{s,N} C^{a_4}_{N,4}} = \frac{\varepsilon_3}{\varepsilon_4^{a_4}} \alpha_N C_N \tag{5.5}$$

is independent of temperature ($C = C_3/C_4^{a4}$, $\alpha = \alpha_3/\alpha_4^{a4}$). The same is true for the non-solar component of the daytime radiance at the ground. In order to get rid of the emissivities one has to assume that the emissivity during daytime equals the emissivity during nighttime. This requires pairs of nighttime and daytime measurements which are close together. Then, the ratio between day and night measurements eliminates the emissivities and it results

$$\frac{B(T_{s,eff,D,3})}{B(T_{s,eff,N,3})} = \left[\frac{B(T_{s,eff,D,4})}{B(T_{s,eff,N,4})}\right]^{a_4} \frac{C_D}{C_N} \frac{\alpha_D}{\alpha_N} \tag{5.6}$$

The reflectance in AVHRR channel 3 depends on the solar zenith angle and the observation zenith angle. For a specific situation it can be obtained from the daytime measurement via the first line of Eq. (5.4) as

$$\rho_3 = \frac{L_{s,D,3} - B(T_{s,eff,D,3})}{S_3} = \frac{1}{S_3}\left[L_{s,D,3} - B(T_{s,eff,D,3})\left(\frac{B(T_{s,eff,D,4})}{B(T_{s,eff,N,4})}\right)^{a_4} \frac{C_D \alpha_D}{C_N \alpha_N}\right] \tag{5.7}$$

ε_3 is computed from

$$\varepsilon_3 = 1 - \rho_{3,hem} \tag{5.8}$$

where $\rho_{3,hem}$ is the albedo. To obtain the albedo the reflectance has to be integrated over all combinations of illumination and observation angles. Observations of the reflectance at different but limited combinations of the illumination and measuring geometry are possible for a specific area due to the daily shift of the satellite orbit. These data are then used to determine the coefficients of a semi-empirical reflection model. Subsequent integration of this model over the hemisphere provides the albedo for which the formula results

$$\rho_{3,hem} = \frac{2\pi}{k+1}\rho_0 \cos^{k-1}(\theta) \tag{5.9}$$

and, if the surface can be treated like a Lambertian reflector, simply $\varepsilon = 1 - \pi\rho$.

The emissivities of the other two thermal infrared AVHRR channels 4 and 5 can be estimated from the nighttime measurements. In this case for each channel the following expression will be valid for the thermodynamic surface temperature:

$$\varepsilon_i T_s^{n_i} = \alpha_i T_{s,eff}^{n_i} - (1-\varepsilon)\alpha_i T_{a,i}^{n_i\downarrow} \tag{5.10}$$

T_s has the same value for all channels. If one assumes that the atmospheric contributions to the effective emission in all channels do not differ much and that variations compensate each other, it follows with j = 4 respectively 5:

$$\varepsilon_j = \varepsilon_3^{n_j/n_3}\left(\frac{T_{s,eff,j}}{T_{s,eff,3}}\right)^{n_j} \tag{5.11}$$

The TISI is directly derived from the night time ground radiance B_s.

$$TISI(3,4,5) = \prod_{i=3}^{5}\left(\frac{\alpha_i \varepsilon_i T_s^{n_i}}{\alpha_i}\right)^{a_i} = \prod_{i=3}^{5}(\varepsilon_i T_s^{n_i})^{a_i} = T_s^{\sum_i a_i n_i}\prod_{i=3}^{5}\varepsilon_i^{a_i} \tag{5.12}$$

The condition to make the expression temperature-independent is

$$\sum_{i=3}^{5} a_i n_i = 0 \tag{5.13}$$

If one chooses $a_3 = 1$, then the following coefficients result from the relationships $a_4 = - n_3/n_4$ and $a_5 = - n_3/n_5$ and for the exponents n_i according to the approximation of the Planck-function, Eq. (5.3):

n_3=12.876, n_4 = 4.591, n_5 = 4.152; a_3 = 1, a_4= -2.804, a_5= -3.101.

Following Becker and Li (1990a) the *TISI* of channels 4 and 5 can be used for an alternative computation of the emissivities in channels 4 and 5:

$$\varepsilon_4 = \left(\frac{\varepsilon_3}{TISI(4,5)TISI(3N)} \right)^{1/(n45+n4)}$$
$$\varepsilon_5 = \varepsilon_4^{n45} TISI(4,5N) \tag{5.14}$$

where

$$TISI(4,5) = \frac{2.383 B_5(T_{g5})}{B_4(T_{g4})^{n45}} \tag{5.15}$$

N indicates TISI-night, **B** = radiance observed at the ground, and $n45 = 0.913$.

The steps involved in the procedure are summarized in Fig. 5.1. Step 1 is the application of radiometric corrections for atmospheric effects. Step 2 includes the calculations of the TISIs. Step 3 is the calculation of the emissivity in channel 3 from daytime and nighttime TISIs and of the emissivities in channels 4 and 5 by combining the TISI of one of the channels with the TISI constructed for channel 4 and 5 [Eq. (5.16)]. Once the emissivities have been derived for the AVHRR channels 3, 4, and 5, the surface temperature can be computed in step 4 from the satellite radiances using a split window technique that gives the surface temperature T_s as a linear combination of the brightness temperatures T_4 and T_5 measured in the adjacent thermal infrared channels 4 and 5 (Becker and Li 1990b):

$$T_s = A_0 + P\frac{T_4 + T_5}{2} + Q\frac{T_4 - T_5}{2}$$

with:

$$A_0 = 1.274$$

$$P = 1 + 0.15616\frac{1-\varepsilon}{\varepsilon} - 0.482\frac{\Delta\varepsilon}{\varepsilon^2} \tag{5.16}$$

$$Q = 6.26 + 3.98\frac{1-\varepsilon}{\varepsilon} + 38.33\frac{\Delta\varepsilon}{\varepsilon^2}$$

$$\varepsilon = \frac{\varepsilon_4 + \varepsilon_5}{2}, \quad \Delta\varepsilon = \frac{\varepsilon_4 - \varepsilon_5}{2}$$

A number of parameters evolve which can be used to characterize the surface. The variability of the TISI, e.g. TISI(3,4), correspond over vegetated areas with the NDVI. Over North Africa, where the NDVI remains below 0.1, it shows a large variability (between 0.1 and 0.8) the causes of which are to date unexplained.Obviously the chemical composition of the soil respectively sand plays a role. This is reflected in the strong variability of the reflectance of various materials of which examples are given in Appendix 6.

From a combination of the AVHRR channel 4 and 5 data also the atmospheric water vapor content can be estimated which is necessary for atmospheric corrections.

Sobrino et al. (1994) used for this purpose variance and co-variance, Choudhury et al. (1995) the difference between the Ch.4 and Ch5 brightness temperature.

At-surface radiances can be obtained from at-satellite radiances using a radiative transfer model, e.g. MODTRAN. According to Becker and Li (1993) as well as Li and Becker(1993) it should not be necessary to know the vertical atmospheric profiles of air temperature and humidity with ultimate accuracy.

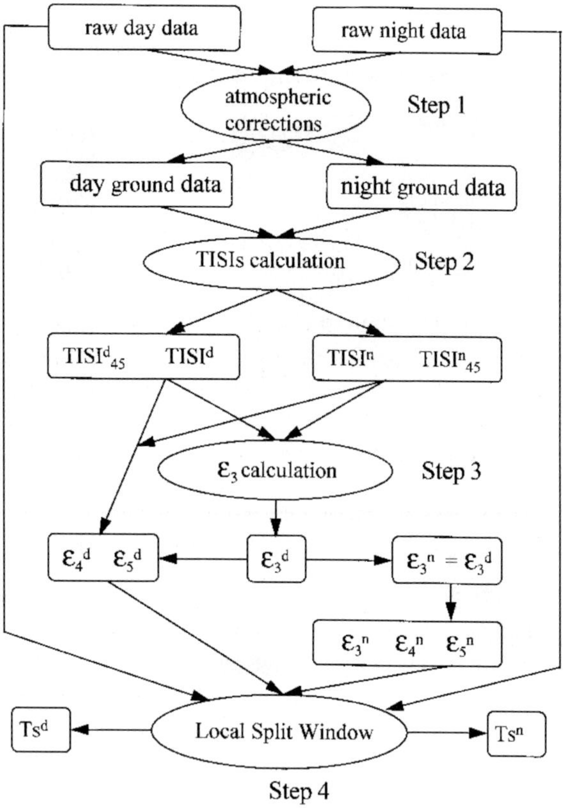

Fig. 5.1. Flow diagram of the TISI procedure

This statement was tested during the EFEDA experiment with data from Tomelloso. The atmospheric correction was performed with a radiosounding from nearby Barrax as well as with two standard atmospheres. The MODTRAN algorithm computed 1.55 g cm^{-2} precipitable water for the radiosonde ascent at 16:00 hours (12 June 1991), for the US Standard Atmosphere 1.42 g cm^{-2} and for the mid-latitude summer atmosphere 2.92 g cm^{-2}. The differences of the computed emissivities between the data corrected with the radiosonde ascent and the US Standard atmosphere were between 0% and 0.3% (Fig. 5.2). This results in an error for the temperature of about 1 K. The comparison between the data corrected with the radiosonde ascent and the mid-latitude summer atmosphere, however, resulted in temperature errors up to 3.5 K (Fig. 5.3), which is not acceptable. It must be concluded, therefore, that the method is robust but knowledge of local atmospheric conditions is necessary to choose the appropriate reference atmosphere.

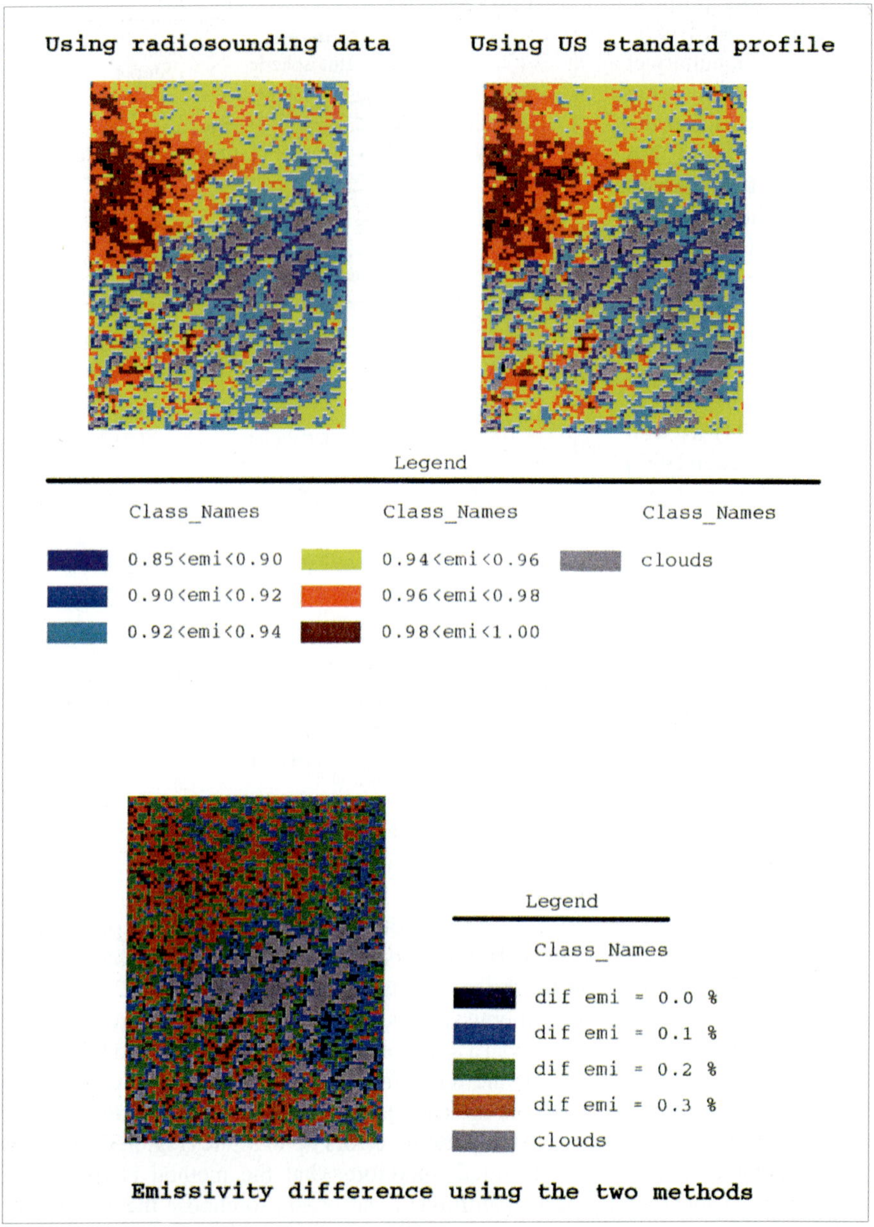

Fig. 5.2. Comparison between the emissivity retrieved in AVHRR-channel 5 at the EFEDA experimental area (see Fig. 5.7) using 12 June 1991 radiosounding respectively US-Standard Atmosphere vertical profiles for the atmospheric correction

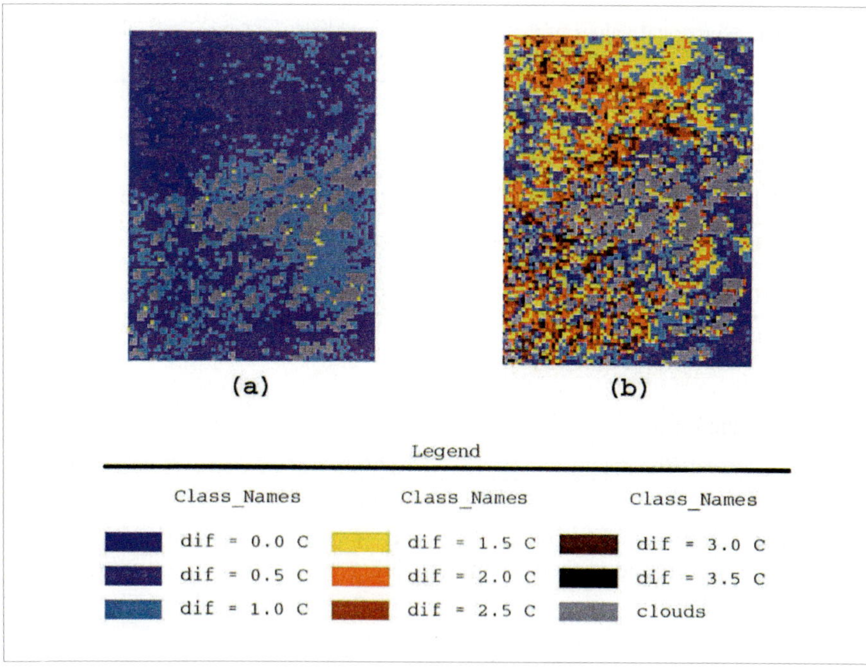

Fig. 5.3 Differences of surface temperatures inferred with atmospheric corrections of different profiles. (a) Barrax radiosounding minus US-Standard Atmosphere, (b) Barrax radiosounding minus mid latitude summer atmosphere

5.2.2 Relation Between Emissivity and NDVI

The operational application of the method to retrieve emissivities as described in 5.1.1 is limited by the availability of consecutive cloudless day and night images. To bypass this limitation relationships between the emissivity and the Normalized Difference Vegetation Index (NDVI) have been studied by Van de Griend and Owe (1993) and Nerry (1997). If it were possible to establish a relationship between emissivity and NDVI, obtained in an independent way, the knowledge of NDVI would permit the determination of emissivity for days where pairs of day/night images are not available. The general relationship that conveys the fact that the emissivity increases with the density of vegetation is

$$\varepsilon = a + b\log(NDVI)\,. \tag{5.17}$$

It seems to fit the experimental data quite well within a central range of the NDVI and with the values for a and b in Table 5.1.

If information about the NDVI for soil and vegetation is available, the approach of Valor and Caselles (1996) can be applied. They define fractional vegetation

Table 5.1. Numerical values of the coefficients a and b of Eq. 5.17

Author	Coefficient *a*	Coefficient *b*	Correlation Coefficient *r*
Van de Griend and Owe	1.0094	0.047	0.941
Nerry 12.06.91	1.0087	0.0314	0.631
Nerry 28.06.91	0.9958	0.0252	0.52

cover P_v by the ratio

$$P_v = \frac{NDVI - NDVI_s}{NDVI_v - NDVI_s} \tag{5.18}$$

where the subscript s stands for bare soil, v for full vegetation cover. The emissivityof the ground is then computed with a "vegetation structure parameter" $\langle d\varepsilon \rangle$ according to

$$\varepsilon_0 = \varepsilon_v P_v + \varepsilon_s (1 - P_v) + 4\langle d\varepsilon \rangle P_v (1 - P_v) \tag{5.19}$$

If no other information is available the following numbers can be used for the coefficients in the 8-14μm band: $\varepsilon_s = 0.91$, $\varepsilon_v = 0.99$, $\langle d\varepsilon \rangle = 0.02$.

5.2.3 Validation

Ground Based Measurements

Emissivities determined by means of space techniques can be validated in two different ways: by direct measurements of the emissivity of surface materials either in the field or in the laboratory or by validation of the inferred thermodynamic temperatures.

Emissivity can be measured in the field by the so-called box method originally proposed by Buettner and Kern (1965) and modified by Becker et al. (1986), see Fig. 5.4. Inside the box is lined with highly reflecting aluminium. The top consists of a heated black body (honeycomb type) radiation source. The bottom of the box is open and is placed on the examined surface. A thermal infrared radiometer equipped with filters similar to those used in the space instrument looks down through a small hole at the top of the box at this surface. Below the infrared radiation source sliding plates are located which are also highly reflecting and block the radiation from the black body source. In this configuration the radiance measured by the radiometer on top of the box after thermal equilibrium is adjusted is given by

$$L_{1,\lambda} = \varepsilon_\lambda L_\lambda (T_s) + (1 - \varepsilon_\lambda) L_\lambda (T_s). \tag{5.20}$$

If the reflecting slides are opened for a short time so that the surface temperature is not affected, the radiance at the radiometer is given by

$$L_{2,\lambda} = \varepsilon_\lambda L_\lambda(T_s) + (1-\varepsilon_\lambda) L_\lambda(T_B), \tag{5.21}$$

where T_B is the black body temperature. Finally a measurement is made with the box closed at the bottom as well by a reflecting surface. In this configuration the radiometer measures only the black body radiation:

$$L_{3,\lambda} = L_\lambda(T_B). \tag{5.22}$$

From the three equations $L_\lambda(T_s)$ and $L_\lambda(T_B)$ can be eliminated and it results:

$$\varepsilon_\lambda = \frac{L_{2,\lambda} - L_{3,\lambda}}{L_{1,\lambda} - L_{3,\lambda}}. \tag{5.23}$$

With this improved technique various emissivities have been determined as shown in Table 5.2. The data originally obtained by Buettner and Kern (1965) are compiled for comparison in Table 5.3. These data give some information as well about the differences between polished and rough material. Melting snow is the "blackest" material observed with an emissivity of 0.99 (Falckenberg 1928). The emissivity of ice is very high below 10 μm but has a reflection band at 13 μm.

Reflection measurements can be regarded as as a measure of (1 - emissivity). If made in the field, the separation of the reflected atmospheric radiation causes a problem. Reflection spectra from plain water surfaces taken in the laboratory are reproduced in Fig. 5.5. Laboratory measurements of the reflectance of solid natural surfaces at thermal infrared wavelengths also offer the possibility to obtain information about the angle dependence of emissivity. In Appendix 6 some examples are given of spectral bidirectional reflectances of natural surfaces which clearly show the angular behaviour of reflectance in the 1 - 12 μm spectral region. The reflected radiation can also be polarized as shown for water in Fig. 5.5 at the right side. In addition these measurements show that one has to be very careful in interpreting broad band emissivity measurements because of the existing reflection bands. Very strong is the 10 μm *restrahlen* band of quartz (see Fig. 3.1).

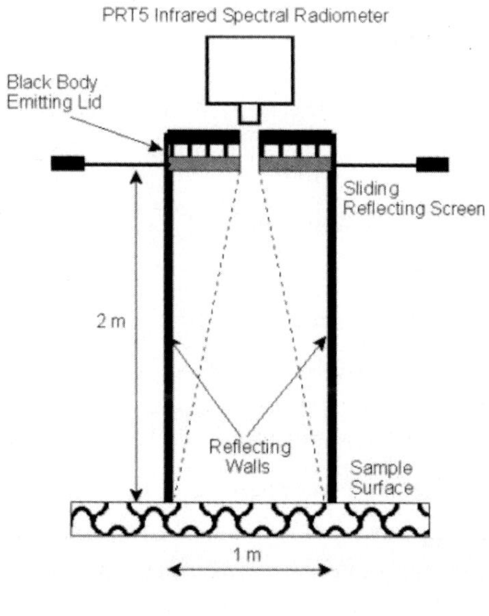

Fig. 5.4. "Black Box" for emissivity measurements

Table 5.2. Emissivity of different surfaces measured with the emissivity box in the 8 - 14 μm spectral region

Botswana savanna environment	ε	σ	NDVI	σ
Loamy sand	0.914	0.011	0.157	0
Sorgum - covering 10% of soil	0.94	0.01	0.246	0.01
Graminea (grass) - partly covering soil	0.956	0.013	0.28	0
Graminea (grass) - almost complete cover	0.958	0.01	0.294	0
Graminea (grass) - open, partly covering soil	0.949	0.01	0.278	0.01
Long grass(35 cm) - complete cover	0.958	0	0.276	0.011
Solanaceae (cultivated weed) - almost complete coverage	0.985	0.01	0.506	0.01
Rhygozum brevispinosum (shrub) - partial cover	0.952	0.01	0.367	0.01
Euclea undulata (shrub) - partial coverage	0.976	0.01	0.476	0.019
Euclea undulata (shrub) - complete coverage	0.986	0.01	0.727	0.015
Barrax, Tomelloso, Rada de Haro and other locations in Spain			$\varepsilon_{8\text{-}14\ \mu m}$	$\varepsilon_{10\text{-}12\ \mu m}$
Bare soil adjacent to corn (Barrax)			0.977	0.983
Bare soil and corn			0.964	0.983
Bare soil and dry vegetation			0.979	0.961
Ploughed bare soil			0.978	0.983
Irrigated barley			0.996	0.995
Dry barley			0.99	0.989
Grapevine			0.993	0.99
Bare soil adjacent to grapevine			0.974	0.98
Bare soil and sun flowers			0.977	0.983
Sand (after Nerry, Labed, Stoll, 1988)				0.947
Loess (after Nerry, Labed, Stoll, 1988)				0.964
Orange tree (modified box approach after Caselles, Sobrino, Becker, 1988)			0.985 ± 0.006	
Ground underneath orange trees			0.957 ± 0.006	

Table 5.3. Vertically measured emissivity of various materials determined from the Barnes IT-2 Infrared Thermometer (roughly 8 - 12 μm, the precise filter function is not known) and the Emissivity Box of Buettner and Kern (1965). Values in brackets are integrated spectral values 8 - 12 μm at 293 K. Material for which the chemical composition is annotated a destinct temperature dependence of the emissivity was measured. Granite = SiO_2-$KAlSi_3O_8$-$NaAlSi_3O_8$, Lavrotite = $CaMgSi_2O_6$+ Vanadium, Obsidian = SiO_2-$NaAlSi_3O_4$-$CaAl_2Si_2O_8$.

Material	ε	Material	ε
Quartz (Agate), SiO_2	0.712 (0.672)	Dolomite rough	0.958
Granite	0.815	Dolomite gravel (5 mm)	0.959
Granite, rough side	0.898	Silicon glass plate	0.865
Feldspar, $KAlSi_3O_8$	0.870 (0.819)	Parson's black, plain	0.988
Obsidian	0.837 (0.828)	Plywood	0.962
Obsidian, rough side	0.837	Human skin	0.98
Basalt, $CaMg(SiO_3)_2$	0.904 (0.906)	Quartz sand grains	0.914
Basalt, rough side	0.934	Quartz sand wet	0.936
Dunite	0.856 (0.861)	Sand, Monterey	0.928
Dunite, rough side	0.892	Concrete, dry walkway	0.966
Lavrotite	-0.812	Asphalt paving	0.956
Marble, $CaMg(CO_3)_2$	-0.942	Water, pure	0.993
Silicon sandstone polished	0.909	Water plus thin film of petroleum	0.972
Silicon sandstone rough	0.935	Water plus thin film of corn oil	0.961
Dolomite	0.929	Water plus thin sheet of polyethylene	0.961

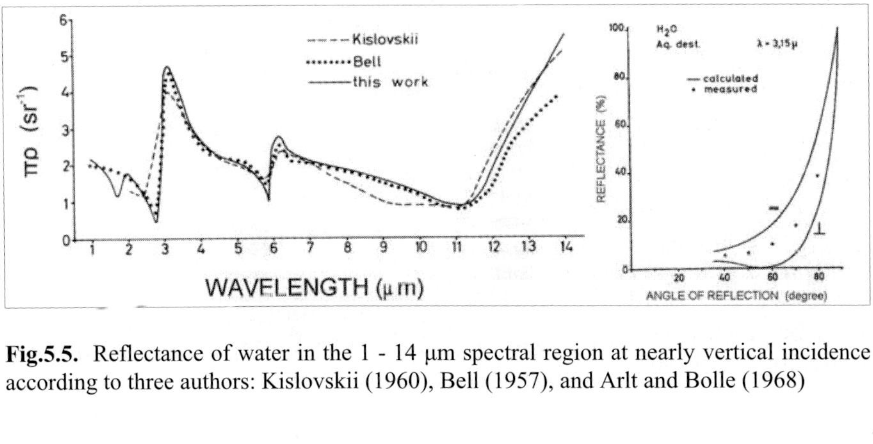

Fig.5.5. Reflectance of water in the 1 - 14 μm spectral region at nearly vertical incidence according to three authors: Kislovskii (1960), Bell (1957), and Arlt and Bolle (1968)

Results Obtained with Satellite Data

Tests have been performed in the EFEDA experimental area. Specifically investigated was the relationship between emissivity and NDVI. The positions of the three EFEDA super-sites are superimposed to an emissivity map in Fig. 5.6. Fig. 5.7 shows the NDVI from the EFEDA data base (Valencia University) and the emissivity in channel 4 of June 13, 1991. The Tomelloso area has a low NDVI but a high emissivity. Distributions of the emissivity in channel 4 (obtained by the TISI method) in dependence of the NDVI are shown in Fig. 5.8 for the 12th and 28th of June 1991. For areas with a mixture of bare soil and vegetation an increase of the emissivity with increasing NDVI is expected. This is observed in both figures, although they show significant differences at the low NDVI end. On June 12, instead of having a smaller emissivity value which could describe a bare soil associated with low

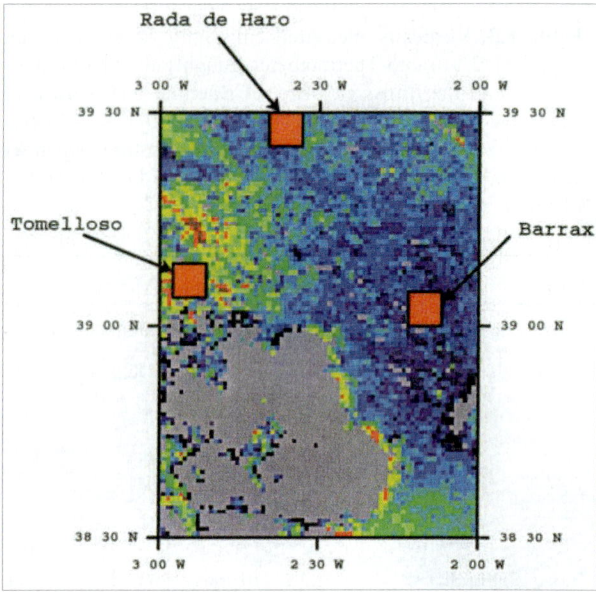

Fig. 5.6. The three EFEDA test sites superimposed to a AVHRR channel 3 emissivity map of 13 June 1991

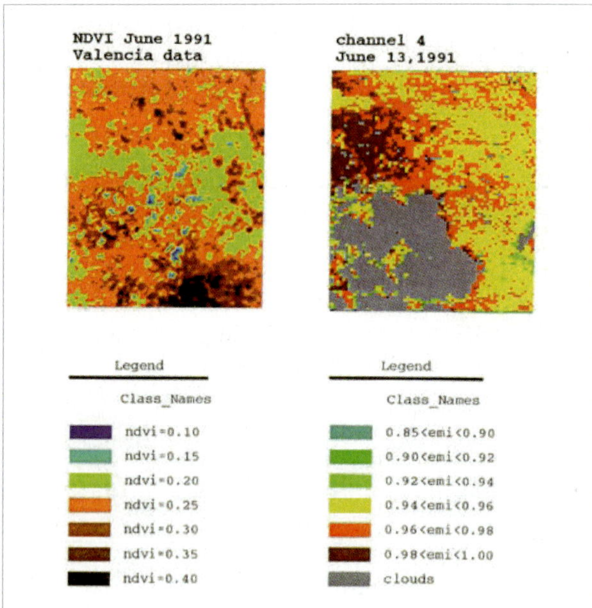

Fig. 5.7. Comparison between NDVI and emissivity in AVHRR-channel 4 for June 1991.

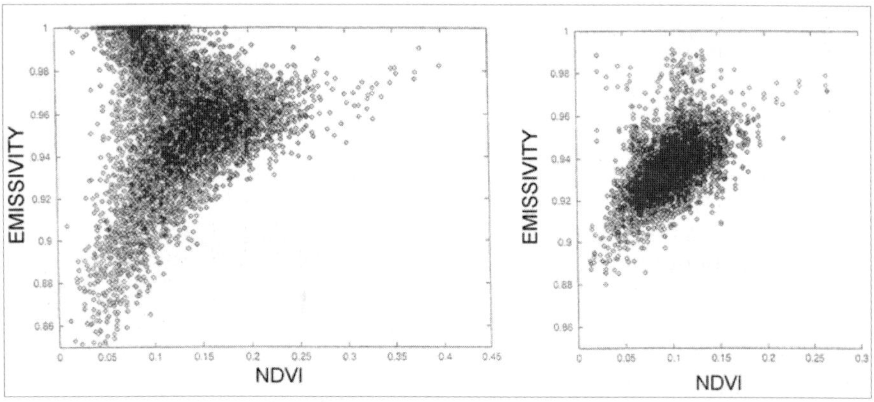

Fig. 5.8. Scatterogram of emissivity in AVHRR channel 4 against NDVI for 12 June 1991 (left) and 28 June 1991 (right).

NDVI, a large number of pixels is observed having high emissivities close to one. The difference between June 12th and June 28th might be due to the fact that the vegetation was growing between the two dates. Another likely explanation is that "anomalous" points correspond to a bare soil with low NDVI but in wet condition (high emissivity). This could be due to the intensive irrigation made during this period where the vegetation was only at its first stage of growing which means that a large part of the surface was still bare soil. Another cause for anomalous points could be the fact that also bare soil can have intrinsically high emissivity values (up to 0.98). Field and laboratory measurements on samples coming from the EFEDA area (EFEDA, Final Report, 1993) reinforce this argument. On June 28 furthermore part of the area was cloud covered and this area is not taken into account in the scatterogramm.

Average values of the emissivities have been derived for 10 x 10 km^2 squares of the three EFEDA test areas indicated in Fig. 5.6. The results of three days are presented in Table 5.4. Though no one-to-one comparison with ground measurements is possible, the values are in reasonable agreement with those given in Table 5.2. These emissivities allow to specify the coefficients P and Q of Eq. 5.16 for the split window method which are presented in Table 5.5. Examples of emissivity maps are given in section 6.7.

A rough estimate of the *a* and *b* coefficients in Eq. 5.17 is performed by eliminating the anomalous points for June 28th, 1991. The Nemani and Running (1989) algorithm, based on the negative correlation between NDVI and surface temperature, is used. Table 5.2 shows the parameters a and b and correlation coefficient found by Van de Griend and Owe (1993b) and those from the EFEDA data. The coefficients *a* and *b* obtained here are quite similar for the two dates and not too far from those determined by Van de Griend and Owe. The conclusion might be that a trend could be observed, although, on a physical basis, a relationship between NDVI and ε is difficult to establish.

Table 5.4. Emissivity values of three EFEDA test areas averaged over 10 x 10 km^2. Bold typed values are retrieved in cases with less than 10% cloud cover, italic typed values refer to cases contaminated with clouds

Date	Barrax	Tomelloso	Rada de Haro
12 June 1991	ε_3 = *0.759 ±0.034* ε_4 = *0.914 ±0.017* ε_5 = *0.920 ±0.015*	ε_3 = **0.914 ±0.024** ε_4 = **0.980 ±0.011** ε_5 = **0.984 ±0.010**	ε_3 = **0.833 ±0.016** ε_4 = **0.945 ±0.006** ε_5 = **0.949 ±0.006**
13 June 1991	ε_3 = **0.838 ±0.025** ε_4 = **0.951 ±0.007** ε_5 = **0.957 ±0.006**	ε_3 = **0.915 ±0.036** ε_4 = **0.980 ±0.014** ε_5 = **0.985 ±0.013**	ε_3 = **0.870 ±0.015** ε_4 = **0.956 ±0.005** ε_5 = **0,959 ±0.005**
28 June 1991	ε_3 = *0.774 ±0.016* ε_4 = *0.924 ±0.012* ε_5 = *0.922 ±0.001*	ε_3 = *0.774 ±0.016* ε_4 = *0.924 ±0.012* ε_5 = *0.922 ±0.001*	clouds

Table 5.5. Coefficients for the split window technique Eq. 5.16

Coeff.	Barrax		Tomelloso		Rada de Haro	
	12.06.91	13.06.91	12.06.91	13.06.91	12.06.91	13.06.91
P	1.0158	1.0091	1.0038	1.004	1.0098	1.0077
Q	6.4836	6.3257	6.2535	6.2351	6.3973	6.374

Observations of surface temperature determined from AVHRR and Thematic Mapper observations were also compared with ground measurements (Table 5.6). The latter apply to specific points at the surface while the satellite values are obtained from mean radiance over an area of 10 km x 10 km. In the case of Tomelloso a radiometer was mounted on a rail (see Fig. 4.22) and the temperature given is the mean value over a distance of 30 m. Since surface temperature changes rapidly with time, validation of estimates obtained from radiometric data collected from space requires simultaneous measurements over rather large areas. Spatial variability of surface temperature is significant as documented by the ground measurements collected in Barrax.

The main problem encountered with TM data to retrieve surface temperature maps at a higher spatial resolution (120m in channel 6) is the single IR channel which does not allow a Split Window method to solve for surface temperature and emissivity simultaneously. A possible way to overcome this difficulty is to apply the NDVI - ε relationship found using the AVHRR data. However, comparison of

NDVIs obtained from AVHRR and TM can show significant differences. This means that it will not be possible to apply the reported NDVI - ε relationship found with the AVHRR data to the TM data. The difference in NDVI computed from the two sensors might have various origins:

- mismatch of spectral channels;
- difference in the level of heterogeneity between NOAA and TM data because of the pixel size;
- view angle and anisotropy effects.

Obtaining reliable ground measurements for proper validation of satellite observations remains a challenging problem. If fine grained soil is present it is possible to approach surface temperature using thin wire thermocouples or resistances. These devices must be in thermodynamic equilibrium with the soil if attached to the surface and a radiation effect must be avoided. Reasonable results are obtained if they are placed at 2 mm depth although also in this case an absolute reference is not available. If vegetation is present, green leaves will approximately be at canopy air temperature.

Table 5.6. Comparison of surface temperatures inferred from satellite measurements with those measured at the surface

Site	Time	12. June 1991			28. June 1991		
		Sensor	Temp. (°C)	Ground Temp. (°C)	Sensor	Temp. (°C)	Ground Temp. (°C)
Barrax	0.427	TM	42.2	46.5-31.2	TM	40	39.1-32.5
	0.611	AVHRR	37.2	58.8-36.5	AVHRR	48.6	51.2-42.4
Rada de Haro	0.427	TM	42	40.7	TM	40.7	39.6
	0.611	AVHRR	47.2	46.3	AVHRR	48.5	47.2
Tom-elloso	0.427	TM	49.1	43.4	TM	40.5	36.8
	0.611	AVHRR	49.1	43.4	AVHRR	50.7	42.8

5.3 Radiative Properties of Land Surfaces II: Broad-band Hemispherical Reflectance

5.3.1 Narrow-band to Broad-band Conversion

Hemispherical broad band reflection of solar radiation or (meteorological) albedo is defined as the ratio of the hemispherical fluxes from the surface and from the sky.

The combined direct and scattered solar irradiance is also called "global radiation". Albedo plays an important role in the computation of the surface energy budget. From space this quantity is not directly accessible because of the atmospheric absorption bands and the impossibility to measure the flux emerging from a pixel of the surface into the upper hemisphere. Only approximative methods can be applied based on the measurements that can be made in between the atmospheric absorption bands or within the bands sorted out by the filters in satellite instruments. The basic equation to obtain broad-band albedo $\overline{\rho}$ from spectral measurements is

$$\overline{\rho} = \frac{\int_{300nm}^{3\mu m} \rho(\lambda)\, M_{SW}^{\downarrow}(\lambda) d\lambda}{\int_{300nm}^{3\mu m} M_{SW}^{\downarrow}(\lambda) d\lambda} = \frac{\sum_{i=1}^{N} \int_{\lambda_i}^{\lambda_{i+1}} \rho(\lambda)\, M_{SW}^{\downarrow}(\lambda) d\lambda}{\int_{300nm}^{3\mu m} M_{SW}^{\downarrow}(\lambda) d\lambda} \tag{5.24}$$

$M_{SW}^{\downarrow}(\lambda)$ is the spectral shortwave flux density measured in W m^{-2} (unit wavelength)$^{-1}$ at wavelength λ.

The integrals under the sum can only be solved approximately because the spectral bands are limited by the filter functions $F(\lambda)$ and because of the large bands in between the spectral channels that are blocked by water vapour absorption. There are different possibilities to handle this equation. One is to take the values within the spectral channels as representative for the larger bands reaching to the half distances to the neighbour channels:

$$\begin{aligned} \lambda_i &= \frac{\lambda\left(F_{i,\max}\right) + \lambda\left(F_{i+1,\max}\right)}{2} \qquad i = 1, \ldots N \\ \lambda_0 &= \frac{3}{2}\lambda\left(F_{1,\max}\right) - \frac{1}{2}\lambda\left(F_{2,\max}\right) \\ \lambda_{N+1} &= \frac{3}{2}\lambda\left(F_{N,\max}\right) - \frac{1}{2}\lambda\left(F_{N-1,\max}\right) \end{aligned} \tag{5.25}$$

where $\lambda\left(F_{i,\max}\right)$ is the wavelength where filter i has its maximum. In view of the small contributions of the shortwave and longwave ends of the spectral interval, λ_0 and λ_N here are somewhat arbitrarily defined, one could also set $\lambda_0 = 300$ nm and $\lambda_N = 3\mu$m, depending on the position of the available filters. For each band the albedo can now be calculated for the (narrower) bandwidth of the channel:

$$\rho_i = \frac{\int_{\lambda_{i-1}}^{\lambda_i} F(\lambda)\rho(\lambda) M_{SW}^{\downarrow}(\lambda) d\lambda}{\int_{\lambda_{i-1}}^{\lambda_i} F(\lambda) M_{SW}^{\downarrow}(\lambda) d\lambda} \tag{5.26}$$

The broadband albedo can then be written:

$$\overline{\rho} = \sum_{i=1}^{N} \left(\rho_i \frac{\int_{\lambda_{i-1}}^{\lambda_i} M_{SW}^{\downarrow}(\lambda) d\lambda}{\int_{300nm}^{3\mu m} M_{SW}^{\downarrow}(\lambda) d\lambda} \right) = \sum_{i=1}^{N} g_i \rho_i \tag{5.27}$$

where the weights g_i for each channel are

$$g_i = \frac{\int_{\lambda_{i-1}}^{\lambda_i} M_{SW}^{\downarrow}(\lambda) d\lambda}{\int^{3\mu m} M_{SW}^{\downarrow}(\lambda) d\lambda} \approx \frac{\overline{M_{SW}^{\downarrow}(\lambda_{i-1}, \lambda_i)}(\lambda_i - \lambda_{i-1})}{\int_{300nm}^{3\mu m} M_{SW}^{\downarrow}(\lambda) d\lambda}$$
$$\approx \frac{M_{SW}^{\downarrow}\left[\lambda\left(F_{i,\max}\right)\right](\lambda_i - \lambda_{i-1})}{\int_{300nm}^{3\mu m} M_{SW}^{\downarrow}(\lambda) d\lambda} \tag{5.28}$$

A simple consideration based on the spectral distribution of the extraterrestrial solar irradiance with a Solar Constant of 1368 W m^{-2} is presented in Table 5.7. First the original TM spectral bands are considered which represent 43% of the solar spectrum and in a second approach these spectral bands are expanded to represent 83% of the solar spectrum. Only the strongest water vapour bands are left out here because in these bands no radiative transfer occurs at the surface. It can be seen that especially the assumptions made for TM channel 4 are very critical. For other channels there is more consistency with the results of other authors.

Looking at the spectral distribution of the global radiation at the surface (Fig. 5.9) one can see that the described approach may not be optimal because the central filter wavelength radiance may not be representative for the broader spectral range.

Another possibility to determine the weighting factors is to measure broadband

Table 5.7. Conversion of Landsat-TM band reflectance to broad band albedo based upon the extraterrestrial solar spectrum. S_λ = Solar irradiance at the central wavelength of the respective bands (W m^{-2} µm^{-1})

TM channel (µm)	S_λ	$S_\lambda\,\Delta\lambda$	g_i	Expanded $\Delta\lambda$	S_λ	$S_\lambda\,\Delta\lambda$	g_i
1: 0.45-0.52	196	136.99	0.23	0.400-0.520	203	243.6	0.213
2: 0.52-0.60	183	146.32	0.25	0.520-0.615	185	175.3	0.1534
3: 0.63-0.69	156	93.42	0.16	0.615-0.725	154	168.96	0.1478
4: 0.76-0.90	105	146.58	0.25	0.725-1.350	69	432.5	0.3784
5: 1.55-1.75	21.9	43.86	0.1	1.450-1.775	25	81.25	0.0711
7: 2.08-2.35	7.46	20.131	0	1.950-2.450	8.3	41.35	0.0362
Represented solar energy		587.3 (43%)				1142.96 (83%)	

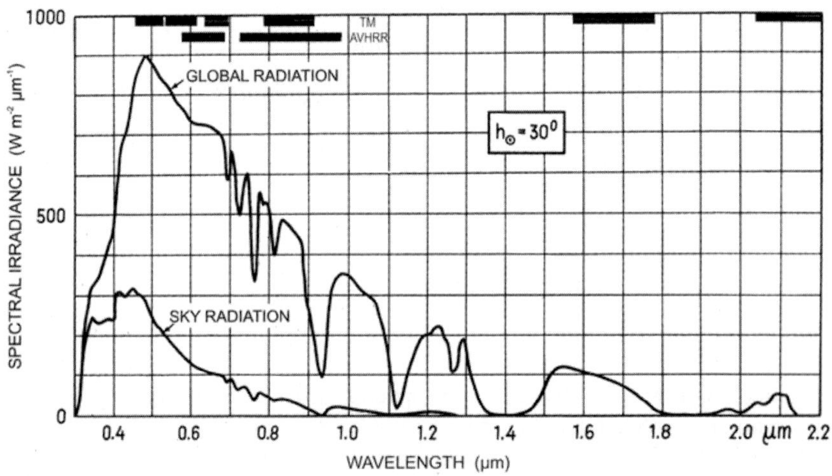

Fig. 5.9. Spectra of global and sky radiation after Dogniaux (1940) with indication of the position of spectral Landsat-TM and NOAA-AVHRR channels

albedo as well as the spectral reflectances over a number of surfaces with different albedo and to apply a multivariate correlation analysis to the data. Li and Leighton (1992) used autocorrelated reflectance measurements to determine narrow- to broadband conversion. Valiente et al. (1995) found from such a statistical analysis the following relations (compare Table 5.8, pos. 11):

$$\begin{aligned} \bar{\rho} &= 0.974\ \rho_{Meteosat} \\ \bar{\rho} &= 0.545\ \rho_{AVHRR-Ch1} + 0.320\ \rho_{AVHRR-Ch2} + 0.035 \end{aligned} \tag{5.29}$$

with standard errors of 7.3% and 4.4% and correlation coefficients of $r^2 = 0.938$ and 0.982. These authors also related the albedo obtained with AVHRR data (ρ_1, ρ_2) to those obtained with Meteosat data. For TOA they found $\rho_{Met} = 0.440\ \rho_1 + 0.529\ \rho_2$, or, with about 1% better agreement, $\rho_{Met} = 0.425\ \rho_1 + 0.474\ \rho_2 + 0.018$, and for data corrected for atmospheric effects: $\rho_{Met} = 0.459\ \rho_1 + 0.481\ \rho_2$. Tristan (1997) introduced the dependency on the solar zenith angle:

$$\begin{aligned} \bar{\rho} &= a(\zeta)\rho_{Meteosat} + b(\zeta) \\ a(\zeta) &= 1.09 \\ b(\zeta) &= -3.67\cdot10^{-4} + 1.23\cdot10^{-4}\zeta + 5.55\cdot10^{-3}\sin x + 2.18\cdot10^{-3}\cos x \\ x &= 2.32\cdot10^{-2}\zeta + 2.53 \end{aligned} \tag{5.30}$$

This leads to the generalized equation:

$$\bar{\rho} = \sum_{i=1}^{N} g_i \rho_i + d\ . \tag{5.31}$$

Depending on how the spectral intervals have actually been defined, various authors found different weighting factors. A number of AVHRR weighting factors are summarized in Table 5.8.

Large differences in the weighting factors listed in Table 5.8 partly result from the use of different targets. If the broadband albedo is determined over vegetated areas the weight of the high reflectance measured in AVHRR channel 2 is too high for the longer wavelength where, outside AVHRR channel 2, reflectance decreases while over bare soil in most cases it increases with increasing wavelength. Few authors therefore tried to compensate for this effect by parameterizing the weight with the NDVI which leads to the following approximations:

$$\bar{\rho} = 0.526\rho_{AVHRR-Ch1} + \left(0.485 - 0.112\ NDVI\right)\rho_{AVHRR-Ch2} \tag{5.32}$$

after Brest and Goward (1987) and

$$\begin{aligned} g_1 &= 0.494\ NDVI^2 - 0.329\ NDVI + 0.372 \\ g_2 &= -1.439\ NDVI^2 + 1.209\ NDVI + 0.587 \end{aligned} \tag{5.33}$$

after Song and Gao (1999).

In this context also the compatibility of spectral channels has to be discussed. A very accurate comparison between two systems would require adjustments with respect to the filter functions. Fortunately, the AVHRR channels 1 and 2 correspond fairly well with the Landsat-TM channels 3 and 4 (and the DAEDALUS scanner channels 5 and 7, an airborne instrument). Over bare light soil, however, the AVHRR channel 1, because of its larger halfwidth, eventually yields lower values than TM channel 3 and over dense vegetation canopies the opposite may occur. But as Table 4.4 shows, TM channel 3 values are closer to AVHRR-Ch1 values than those of the weighted average 0.4 $\rho_{TM\text{-}Ch2}$ + 0.6 $\rho_{TM\text{-}Ch3}$. In the near infrared AVHRR channel 2 the reflectance is slightly underestimated over vegetation in comparison to TM channel 4.

Jagdhuhn (1994) computed weighting factors in four different ways for Landsat-TM as well as AVHRR. The applied methods are specified in Table 5.9. After testing these methods (see section 5.3.2) he concluded that only the results of methods 3 and 4 give acceptable results. Best results are obtained for Landsat-TM data with method 3 and for AVHRR data with method 4. An alternative set of weights proposed by Jagdhuhn (1994) is the following:
$g_{TM\text{-}1} = 0.271$, $g_{TM\text{-}2} = 0.149$, $g_{TM\text{-}3} = 0.146$, $g_{TM\text{-}4} = 0.305$, $g_{TM\text{-}5} = 0.097$, $g_{TM\text{-}7} = 0.032$, $g_{AVHRR\text{-}1} = 0.507$, $g_{AVHRR\text{-}2} = 0.493$.

Table 5.8. Weighting factors to convert AVHRR spectral reflectance measurements to broadband albedo according to Eq. 5.31

Id	Authors	Year	$g_{AVHRR\text{-}Ch1}$	$g_{AVHRR\text{-}Ch2}$	d
1	Brest and Goward, vegetation	1987	0.526	0.418	0
2	Brest and Goward, bare soil	1987	0.526	0.474	0
3	Wydick et al.	1987	0.36	0.73	-0.7
4	He et al.	1987	0.332	0.678	0
5	Saunders	1990	0.502	0.408	0
6	Potdar and Narayana	1993	0.798	0.188	0.05
7	Fell	1993	0.664	0.336	0
8	Jagdhuhn	1994	0.456	0.544	0
9			0.507	0.493	
10	Hucek and Jacobowitz	1995	0.347	0.65	0.746
11	Valiente et al.	1995	0.545	0.32	0.04
12	Russel et al.	1997	0.441	0.67	0.04

Table 5.9. Spectral to broad band conversion according to Jagdhuhn (1994)

	Method	Weights	Remarks
1	$$g_i = \frac{\frac{1}{\Delta\lambda}\int_{Ch\,i} F(\lambda)\, M_{SW}^{\downarrow}(\lambda)d\lambda \cdot}{\sum_{1=1}^{N}\left[\frac{1}{\Delta\lambda}\int_{Ch\,i} F(\lambda)\, M_{SW}^{\downarrow}(\lambda)d\lambda\right]}$$	$g_{TM-1} = 0.262$ $g_{TM-2} = 0.267$ $g_{TM-3} = 0.269$ $g_{TM-4} = 0.157$ $g_{TM-5} = 0.035$ $g_{TM-7} = 0.010$ $g_{AVHRR-1} = 0.664$ $g_{AVHRR-2} = 0.336$	Method developed by Fell (1993)
2	$$g_i = \frac{\int_{Ch\,i} F(\lambda)\, M_{SW}^{\downarrow}(\lambda)d\lambda}{\sum_{i=1}^{N}\int_{Ch\,i} F(\lambda)\, M_{SW}^{\downarrow}(\lambda)d\lambda}$$	$g_{TM-1} = 0.211$ $g_{TM-2} = 0.247$ $g_{TM-3} = 0.186$ $g_{TM-4} = 0.235$ $g_{TM-5} = 0.085$ $g_{TM-7} = 0.036$ $g_{AVHRR-1} = 0.456$ $g_{AVHRR-2} = 0.544$	
3	$$g_i = \frac{\int_{\lambda_i}^{\lambda_{i+1}} M_{SW}^{\downarrow}(\lambda)d\lambda}{\int_{350nm}^{3\mu m} M_{SW}^{\downarrow}(\lambda)d\lambda}$$	$g_{TM-1} = 0.238$ $g_{TM-2} = 0.151$ $g_{TM-3} = 0.151$ $g_{TM-4} = 0.323$ $g_{TM-5} = 0.103$ $g_{TM-7} = 0.043$ $g_{AVHRR-1} = 0.502$ $g_{AVHRR-2} = 0.498$	$\lambda_i \ldots \lambda_{i+1}$ nm 350-525 525-620 620-735 735-1240 1240-1940 1940-3000 350-700 700-3000
4	The weighting functions found by Brest and Goward (Brest 1987) for Landsat MSS are adopted for the TM channels 2, 4, and 5. Model computations with 5S result in similar values (0.535, 0.358, 0.107). For AVHRR-1 the value of TM-2 is taken. The AVHRR-2 weighting factor is the sum of the TM-4 and TM-5 weighting factors but in the case of vegetated surfaces $g_{TM-4} + 0.5\, g_{TM-5}$ is taken (value marked with *).	$g_{TM-1} = 0$ $g_{TM-2} = 0.526$ $g_{TM-3} = 0$ $g_{TM-4} = 0.362$ $g_{TM-5} = 0.112$ $g_{TM-7} = 0$ $g_{AVHRR-1} = 0.526$ $g_{AVHRR-2} = 0.474$ 0.418*	

5.3.2 Validation

Validation of albedo derived from narrowband measurements from space can only be tested over large uniform land surface units because the ground based measurements are only representative for small areas. A number of albedo measurements have been made at the three instrumented sites of the EFEDA experimental area as indicated in Fig. 4.9. The measured values are shown in Table 5.10. The spectral satellite data compiled in Table 4.4 were converted to broadband albedo with the weighting factors of Table 5.8. The results are presented in Table 5.11.

Inspection of Table 5.11 shows that the methods No. 3, 4, 10, and 12 are less consistent with the rest. This is confirmed by comparison with measured values. Of the three sites the Tomelloso area is the most homogeneous one with a mixture of soil 15% grapevine coverage in June 1991 and a small contribution of vetch[1]. Albedo values around 0.26 should therefore be expected, a minimum value would probably

Table 5.10. Measured broadband albedo in Castilla - La Mancha

Area	Date	Type of surface	Albedo			
Tomelloso	11.06.91	Vine	range: 0.252 - 0.276, mean: 0.266 (Min.) 06:00: 0.311, 08:00: 0.282; 10:00: 0.268; 12:00: 0.266; 14:00: 0.268; 16:00: 0.279; 18:00: 0.298			
	28.06.91		range: 0.260 - 0.268, mean: 0.265 (Min.)			
	29.08.91		0.253			
Tomelloso	28.06.91	Vetch	0.228 - 0.241 (Min.), mean: 0.235			
Tomelloso	26.04.91	Wheat	0.202 (noon, min)			
Tomelloso	28.06.91	Bare soil	0.24 (noon)			
	31.08.91		0.233			
Rada de Haro,	11.06.91	Matorral	0.177	27.06.91	Vine	0.276
	23.06.91		0.182	28.06.91		0.277
	27.06.91		0.188	29.06.91		0.275
	28.06.91		0.188			
	29.06.91		0.191			
Barrax,	11.06.91	Irrigated	0.256	29.06.91	Irrigated	0.163
		not irrigated	0.267		not irrigated	0.277

[1]Vetch denotes here sparse vicia crop, type of a legume used for animal feed, 20 - 30 cm high, mixed with some wild flowers and deeper rooting weed with low transpiration.

be 0.235, the albedo of the soil. Methods No. 1, 2, 5, 7, and 11 fall into this category and for the beginning of the measuring period also 8 and 9. The Rada de Haro area is primarily determined by matorral vegetation but there is also some mostly dry agriculture. Albedo values around 0.19 would reflect this situation which are found with methods 1,3, 5, 7, 11. Difficult to be validated are the albedo values of the Barrax area because of the changing irrigation patterns. The values should lie well above 0.16 and not be larger than 0.27 which is the case for methods No. 1, 2, 3, 5, 6, 7, 8(?), 9, and 11. In summary it seems that the method of Valiente et al. will fit the observations best, but also the double pack of Brest and Goward gives good results. Also the methods of Saunders and Fell seem to be reasonable, which is astonishing in the case of Fell because, according to Jagdhuhn, this method is less reliable in the case of Landsat-TM.

Jagdhuhn tested the methods of Table 5.9 for Landsat-TM and NOAA-AVHRR scenes in Tuscany, Italy. The test area near the mouth of the river Cornia is shown in Fig. 5.10 as a false colour composite of a Landsat-TM scene. Several test areas are indicated for which the broadband albedo was calculated.

Table 5.11. Broadband albedo computed for test areas in Castilla - La Mancha, Spain, from spectral reflectance (Table 4.) with the narrow-to-broadband conversion formulas of Table 5.7. Tom, RdH, and B are respectively Tomelloso, Rada de Haro, and Barrax. The numbers refer to the dates in June 1991.

ID	Tom14	Tom28	RdH14	RdH28	B14	B28
1	0.234	0.263	0.186	0.212	0.240	0.235
2	0.251	0.283	0.201	0.230	0.258	0.255
3	0.225	0.270	0.179	0.215	0.240	0.249
4	0.274	0.316	0.231	0.264	0.288	0.296
5	0.226	0.254	0.180	0.206	0.232	0.227
6	0.270	0.288	0.211	0.233	0.268	0.248
7	0.237	0.262	0.183	0.208	0.240	0.228
8	0.258	0.293	0.211	0.241	0.268	0.268
9	0.253	0.286	0.204	0.233	0.261	0.259
10	0.343	0.383	0.300	0.332	0.356	0.363
11	0.243	0.266	0.197	0.220	0.246	0.238
12	0.338	0.380	0.287	0.323	0.350	0.355

A first test was made for spectral data over the sea surface target no. 1. In the 5S radiative transfer algorithm a sea surface model is available which can be used as a reference. In Table 5.12 the forward computations made with this sea surface reflectance model are compared with radiances measured in the Landsat-TM channels. Three model calculations have been carried out. One with the 5S code (a predecessor of the 6S code) and two with the LOWTRAN code, one with the standard summer mid-latitudes atmosphere and one with radiosonde data. The spectral reflectance values inferred from the Landsat-TM measurements scene were considerably smaller than those computed with the models. An effect of the radiometric correction for atmospheric effects could be excluded because the computations using the 5S and the LOTRAN codes were all very close with the exception of the TM-5 and TM-7 values, where the signal is nearly zero. The inferred albedo values which are shown in Table 5.12 agree much better with the model of Köpke (1977) than with the reflection model in the 5S code. Furthermore model computations performed with a 20% surface albedo in the 5S code resulted in higher values than LOTRAN computations.

In the target area over land two measuring stations were operated at the western and eastern slope of a hill covered with olive trees and macchia. The position of the sites, which are representative for a larger area in the Colline Metalliferi, could be identified in the Landsat-TM image. At both sites the albedo measured at the time of the Landsat overpass was 0.175. At the time of the NOAA-10 overpass (07:30 LST) the values have been 0.195 at the eastern slope and 0.125 at the western slope. Because one AVHRR pixel covers a much larger area the mean value of 0.16 seems

Table 5.12. Measured spectral radiances in W m^{-2} sr^{-1} at the satellite over the sea test area 1 of Fig. 5.10 and computed spectral radiances using the sea surface model of the 5S code in the 5S and LOWTRAN 7 (L7) models.

Channel	Measured	5S	L7 (Standard Atm.)	L7 (Radiosonde)	SS Model Reflectance (%)
TM-1	3.7	3.1	3.4	3.4	4.2
TM-2	2.6	2.9	3.1	3.2	5.5
TM-3	1.3	1.5	1.5	1.6	3.9
TM-4	1.5	0.5	0.5	0.5	0
TM-5	0.1	0.07	0.03	0.03	0
TM-7	0.05	0.02	0.01	0.01	0
AVHRR-1	2.2	2	2.2	-	4.7
AVHRR-2	2	0.6	0.8	-	0.2

Table 5.13. Broadband albedo for seven target areas near the mouth of the Cornia river computed with the four methods given in Table 5.9 after Jagdhuhn.

Target	Method 1		Method 2		Method 3		Method 4	
	TM	AV HRR	TM	AV HRR	TM	AV HRR	TM	AV HRR
Forest 1	0.1	0.13	0.12	0.18	0.14	0.17	0.16	0.149
Forest 2	0.12	0.16	0.15	0.22	0.17	0.21	0.19	0.179
Forest 3	0.1	0.13	0.12	0.18	0.14	0.17	0.16	0.145
Agricultural land 1	0.18	0.19	0.2	0.23	0.22	0.22	0.23	0.216
Agricultural land 2	0.17	0.18	0.19	0.21	0.2	0.2	0.21	0.201
Large area A	0.13	0.16	0.15	0.2	0.17	0.19	0.19	0.166* 0.182
Large area B	0.12	0.14	0.14	0.18	0.16	0.17	0.17	0.152* 0.166
Test site	0.15	0.16	0.17	0.2	0.19	0.19	0.2	0.173* 0.190
Sea 1	0	0.1	0	0	0	0	0	0.05
Sea 2	0	0.1	0	0.1	0	0.1	0	0.05
Area incl. quarry	0.2	0.2	0.22	0.24	0.24	0.23	0.26	0.223
Quarry (TM only)	0.32 - 0.44		0.35 - 0.48		0.37 - 0.50		0.38 - 0.51	

to be representative for a larger area but a value closer to 0.195 can be expected if the pixel is more biased by eastern slopes than western slopes.

Broad band reflectances computed with the four methods from the Landsat-TM data are given in Table 5.13. It can be seen that the albedo values computed with methods 2 and 3 from Landsat-TM and in the case of the AVHRR data with method 1 and method 4, the latter one with the factor for vegetation, come closest to the mean albedo measured at the ground. If, however, the AVHRR pixel is more representative for the eastern slope, then the methods 3 and 4 would yield the best approximation.

Field measurements were performed 1996 in Italy *inter alia* in the area where Jagdhuhn made his investigations in 1990. Though the times are not the same, some albedo values may provide a further guideline for the quality assessment of the spectral-to-broadband conversion. The experiment was carried out close to the "agricultural area 1" in April/May. Albedo values between 0.20 and 0.25 were found in two fields at noon, the average of which agrees fairly well with the results obtained with method 4.

With the exception of method 4 the AVHRR values are higher than the TM values. This is very reasonable because of the earlier cross over time of NOAA-10 (07:30 LST). The albedo of vegetation covered surfaces normally increases at low sun. The TM value obtained with method 4 therefore must be considered as being too high.

5.4 Radiation Fluxes

5.4.1 Nomenclature

Net radiation at the surface determines its energy budget and drives land-surface processes, such as evaporation. The net radiative flux is balanced by the fluxes into or out of soil and canopy, where heat intermediately is stored, and the turbulent heat fluxes from the ground into the atmosphere. In the following text fluxes directed toward the surface are counted positive and those leaving the surface are counted negative.

Net radiative flux is composed of a shortwave, sun originated, and a thermal infrared component originating from the atmosphere and the surface respectively. Following the definition that fluxes directed (downward) towards the surface are counted positive, the net surface radiation flux is positive during daytime and zero at night. The net thermal infrared flux (downward directed atmospheric radiation minus upward directed emission from the surface) generally is negative. Only in exceptional cases it may become zero or positive, e.g. when the effective (or radiative) sky temperature exceeds the ground temperature. This could be the case, if a warm cloudy airmass shifts over a surface that cooled down under a former cloudless sky.

The term "flux" in this context is used as a shorthand for flux (Φ, measured in W) per unit area (W m^{-2}) for which the symbol E is used if an irradiance, a flux incident onto an area element, is meant and the symbol M if it is a radiant exitance, a flux emerging from an area. For free floating fluxes the symbols M is preferably been used but also E can be taken. If an irradiance is integrated over some specified time interval, one talks about radiant exposure measured in $J\,m^{-2}$ per exposure time.

5.4.2 Net Shortwave Radiation Flux Density

Only some instruments on board of satellites like ERBE have the capacity to infer the planetary albedo at the top of the atmosphere. Most satellite instruments measure the outgoing spectral radiance at the top of the atmosphere. To obtain the surface net flux density $M^*_{SW,surface}$ it is necessary to apply radiative transfer models to the filter bands, $\Delta\lambda$, and to integrate over the whole spectrum in a similar way as explained for the broad-band albedo. This integration can be performed using empirical relationships between the spectral fluxes and the whole solar spectrum. In principle one can say, that within an atmospheric window one receives in the first approximation radiation (neglecting multiple scattering) scattered back from the atmospheric gases and aerosols and radiation reflected from the surface. The flux density, M, may be approximated by multiplying the radiance L by π and if the symbol S_{date} is introduced for the instantaneous solar irradiance at a horizontal plane at the top of the atmosphere at a specific date (which takes care of the solar aspect angle and the sun - earth distance) it may be written according to chapter 3:

$$M^{\uparrow}_{SW,sat}(\lambda,\Delta\lambda) \approx \pi L_{sat}(\lambda,\Delta\lambda) = S_{date}(\lambda,\Delta\lambda)\cdot \left[\{1-\tau(\lambda,\Delta\lambda)\}a + \{(1-\tau(\lambda,\Delta\lambda))b + \tau(\lambda,\Delta\lambda)\}\tau(\lambda,\Delta\lambda)\rho_{surface}(\lambda,\Delta\lambda)\right] \tag{5.34}$$

a denotes the fraction of scattered solar radiation into the direction of the satellite sensor, b the fraction scattered towards the surface, $\rho_{surface}$ the surface albedo, and τ the flux transmissivity, which depends on the optical path through the atmosphere. All quantities are defined for the spectral interval $\Delta\lambda$ around the centre wavelength λ of the respective spectral channel. Eq. 5.34 can be solved for the surface albedo. If, for the sake of shortness, the indication for the wavelength is dropped, it can be written

$$\rho_{surface} = \frac{\dfrac{M^{\uparrow}_{SW,sat}}{S_{date}} - (1-\tau)a}{\left[(1-\tau)b+\tau\right]\tau} \tag{5.35}$$

The spectral net flux at the surface is

$$M^{*}_{SW,surface}(\lambda,\Delta\lambda) = S_{date}(\lambda,\Delta\lambda)\cdot \left[(1-\tau(\lambda,\Delta\lambda))b + \tau(\lambda,\Delta\lambda)\right]\left[1-\rho_{surface}(\lambda,\Delta\lambda)\right] \tag{5.36}$$

which can be transformed with Eq. 5.35 and $M^{\uparrow}_{SW,sat} / S_{date} = \rho_p$, the planetary albedo or reflectance, and $a + b = 1$ to

$$\frac{M^{*}_{SW,surface}}{S_{date}} = \rho_p + \tau + (1-\tau)b - \frac{1-\tau}{\tau}a = \rho_p + 1 + \frac{\tau^2-1}{\tau}a \tag{5.37}$$

or

$$M^{*}_{SW,surface} = M^{\uparrow}_{SW,sat} + C(\tau,a,b,S_{date}) \tag{5.38}$$

This equation suggests that a linear relationship exist between the (normalized) net flux at the surface and the planetary albedo. C can be computed for model atmospheres or determined empirically by comparison with measurements at the ground.

Equation 5.37 allows also to compute τ if the surface albedo is determined independently according to section 5.3. In this case τ can be computed by solving the quadratic equation

$$\tau^2 + \frac{a-b\rho_{surface}}{(b-1)\rho_{surface}}\tau + \frac{a-\rho_p}{(b-1)\rho_{surface}} = 0. \tag{5.39}$$

$a + b = 1$ and in the case of a Rayleigh atmosphere a = b = 0.5. For a turbid atmosphere these values have to be computed with an appropriate aerosol code.

Eq. 5.38 can be generalized after Li et al. (1993) as

$$a = \alpha(\mu, w; C, D) - \beta(\mu, w; A, B)\rho_p \tag{5.40}$$

where a is the fraction of the incident irradiance at the top of the atmosphere that is absorbed at the surface $\left(M^*_{surface} / S_{date}\right)$, ρ_p is the planetary albedo, μ the cosine of the solar zenith angle, and w is the amount of preciptable water in the atmosphere. The explicit expressions for the coefficients α and β are:

$$\alpha(\mu, w; C, D) = \alpha_0(\mu; C, D) + \Delta\alpha(\mu, w)$$

$$\alpha_0(\mu; C, D) = 1 - \left(\frac{C}{\mu} + \frac{D}{\sqrt{\mu}}\right) \tag{5.41}$$

$$\Delta\alpha(\mu, w) = \frac{1}{\mu}\left(1 - e^{-\mu}\right)\left(0.0699 - 0.0683\sqrt{w}\right)$$

and

$$\beta(\mu, w; A, B) = \beta_0(\mu; A, B) + \Delta\beta(\mu, w)$$

$$\beta_0(\mu; A, B) = 1 + (A + B \ln \mu) \tag{5.42}$$

$$\Delta\beta(\mu, w) = -0.0273 + 0.0216\sqrt{w}$$

The original values of the coefficients A, B, C, and D are given in Table 5.14. In addition the intercept α and the slope β are presented in dependence of the air mass respectively μ in Fig. 5.11 disregarding the precipitable water content term. Since the

Table 5.14. Coefficients of Eq. 5.40. The mean value of water clouds was deduced from Fig. 5.11

Cloud model	A	B	C	D
Clear sky	0.0815	0.0139	-0.01124	0.1487
St II	0.1356	0.1045	-0.0062	0.1415
Sc II	0.1766	0.0863	-0.0077	0.1399
Cu	0.1838	0.082	-0.008	0.1397
Ci	0.1591	0.2516	0.00255	0.1334
Mean of water clouds	0.1609	0.0958	-0.007	0.1404

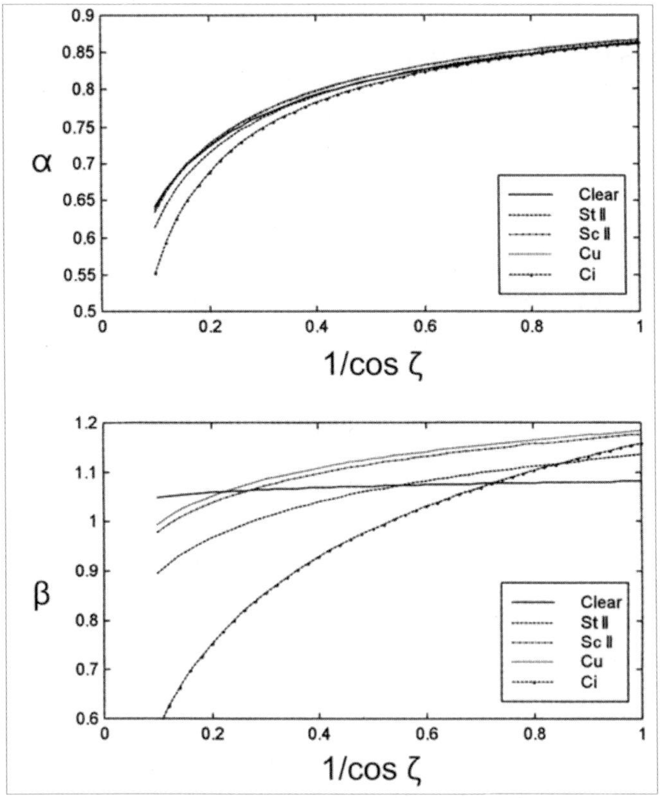

Fig. 5.11. Dependence of intercept (top) and slope (bottom) of the function Eq. 5.41 and 5.42 on airmass

behaviour of the functions is quite similar for water clouds (see also last line of Table 5.14), it seems to be justified to simplify the problem by only introducing three types of atmospheres: clear, water clouds, and cirrus.

Based on experience gained for the Iberian peninsula Fortea and Lopez-Baeza (1997) used for a cloudy atmosphere the formula:

$$M^*_{SW,surface,cloudy} = M^*_{SW,surface,clear\ sky}(1-0.33C_L)(1-0.54C_M)(1-0.05C_H) \tag{5.43}$$

where C_L, C_M, and C_H are the fractional covers with low, middle and high clouds respectively. For a rapid evaluation of a large number of Meteosat scenes, which becomes necessary if one wants to determine the diurnal course of the flux, Melia et al. (1999) developed a quasi-operational scheme of which a simplified flow diagram is shown in Fig. 5.12. The amount of water vapour in the atmosphere is derived from the ECMWF analysis and re-scaled into the METEOSAT format. An example is shown in Fig. 5.13 (see also Femenia et al. 1995).

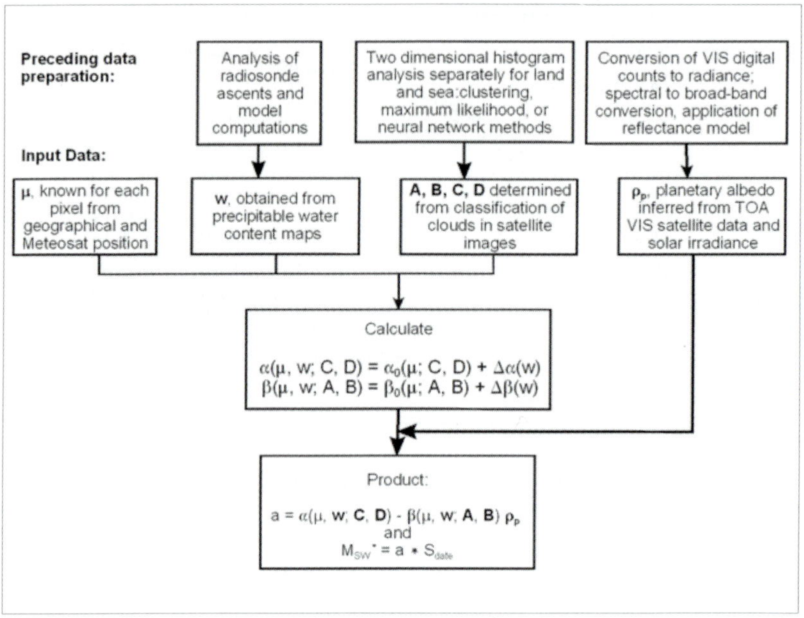

Fig. 5.12. Scheme for quasi-operational estimates of the shortwave net flux at the surface

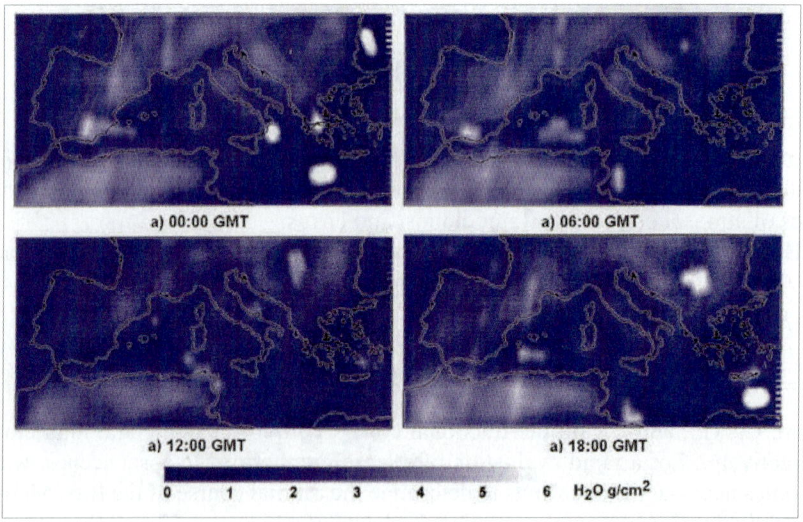

Fig. 5.13. Evolution of the atmospheric precipitable water content of the atmosphere on 11 June 1996. The water vapour column density was derived from the ECMWF analysis and resampled into the Meteosat scale

5.4.3 Net Longwave Radiation Flux Density

To determine the net infrared radiative energy flux its downwelling and upwelling components have to be determined. The clear sky downwelling thermal radiation can be calculated if temperature and water vapour profiles of the atmosphere are known. In radiative transfer codes, such as LOWTRAN, the radiative transfer equation is solved for a beam. This result has to be integrated over all incident angles of the hemisphere to calculate the downwelling flux. For cloudless atmospheres it can be shown (e.g. Bolle, 1982, p 206) that the radiance at an incidence angle of 53° multiplied by π is representative for the hemispherical radiation flux. Hurley and Boers (1996) presented coefficients to adjust the clear sky net longwave radiance according to the amount of high, middle, and low clouds. For overcast situations the downwelling radiation is determined by the temperature of the cloud basis. Graphical estimates of the atmospheric thermal infrared radiation are made by radiation diagrams with radiosonde data input (Möller 1943, Yamamoto 1952, Kondratiev and Niilisk 1960). Approximation formulas use temperature and/or moisture to estimate the thermal sky radiation. Some of these formulas are listed in Table 5.15.

Atmospheric thermal infrared radiation and the emission of the surface can be combined to obtain the longwave net flux. To do this with approximation formulae some authors used in the formulation of the atmospheric emission already $\varepsilon\sigma T^4$ instead of σT^4 which simplifies the combination. Two approximation formulae which neglect the surface emissivity are added in Table 5.15.

The upwelling flux is determined by the surface emissivity and temperature. Under clear skies the upwelling infrared flux respectively the radiative surface temperature can be inferred as described in chapter 4. As the satellite instruments measure the thermal radiance in W m^{-2} sr^{-1} μm^{-1} at the top of the atmosphere in the small instantaneous field of view of the sensor these spectral measurements in principle have to be converted

- from top of atmosphere (TOA) to bottom of atmosphere (BOA) radiance by applying radiometric corrections for the effect of the atmosphere,
- from spectral radiative temperature to thermodynamic surface temperature using the emissivity within the spectral band,
- from spectral radiances of the satellite sensor (e.g. 10 - 12 μm), which is measured in W m^{-2} sr^{-1} μm^{-1}, to total thermal radiance within the spectral range of 3 - 300 μm measured in W m^{-2} sr^{-1}, which needs the application of spectral emissivities throughout the spectrum or the use of an average, weighted emissivity,
- from directional radiance to hemispherically integrated radiative flux measured in W m^{-2} accounting for the spectral and directional properties of the surface emissivity.
- Finally the reflected atmospheric radiation has to be added to the pure emission.

As long as the determination of the surface emissivities has not reached the status that for each pixel of the Mediterranean basin as well spectral as broadband, angle and time dependent emissivities are available, the relationship between surface flux and TOA radiance must be estimated by regression analysis with simultaneous ground based measurements at anchor stations.

Table 5.15. Approximation formulae for atmospheric thermal infrared radiation

Author	M_{LW} (W m^{-2})	Parameter
Ångström (1916) Bolz and Falkenberg (1949)	$\sigma T_0^4 \left(0.82 - 0.25 \cdot 10^{-0.12 e_0}\right)$	T_0 = surface temperature e_0 = water vapour pressure (hPa)
	cloudless: 0.82 σT_H^4	T_H = 2 m air temperature
	cloud covered: 0.94 σT_H^4	
Brunt (1932)	$\sigma T_0^4 \left(a + b\sqrt{e_0}\right)$ originally: a = 0.256, b = 0.065	Medians of evaluations by various authors: a = 0.605, b = 0.048
Swinbank (1963)	$\sigma T_a^4 \cdot 0.398 \cdot 10^{-5}\ T_a^{2.148}$ $\approx 0.92 \cdot 10^{-5}\ \sigma T_a^6$	T_H = 2 m air temperature
Yamamoto (1954)	$\sigma T_0^4 \left(1 - 0.76 \cdot 10^{-0.353\, m \sqrt{w}}\right)$ $\sigma T_0^4 \left(0.51 + 0.066\sqrt{e_0}\right)$	w = precipitable amount of water vapour (cm) e_0 = water vapour pressure at the surface (hPa)
Brutsaert (1984)	$1.24\ \sigma T_a^4 \left(\frac{e_a}{T_a}\right)^{\frac{1}{7}}$ $0.552\ \sigma T_a^4\ e_a^{1/7}$	T_a = air temperature at shelter level. In the parentheses it can be approximated by the surface air temperature of the standard atmosphere, 288 K
Thermal infrared net flux:		
Feußner (1953)	$\sigma T_0^4 \cdot 10^{-0.534}\ w^{0.108}$, $\sigma T_0^4 \cdot 10^{-0.456}\ e_o^{0.108}$	
Lönnquist (1950/51 and 1954)	$0.01\, \sigma T_0^4 \left[47.8 + 0.06t + (0.18 + 0.12w)\gamma_m - 10 \ln w a_0\right]$ γ_m = mean temperature gradient, w = humidity factor, a_0 = absolute humidity at the surface	

For cloudy situations the ground temperature cannot be determined from satellite measurements in the thermal infrared window around 10 - 12 µm. However, the ground temperature may be inferred from high frequency microwave measurements, e.g. from the currently available SSM/I instrument, as shown by Owe and Van de Griend (2001) and Van de Griend (2001) on the basis of 37 Ghz V-polarization brightness temperatures with a spatial resolution of about 30 km from Nimbus/SMMR.

Generally, during daytime for dry desert-like surfaces the ground temperature will be lower for overcast conditions than under clear skies. While the surface temperature decreases, the downwelling flux increases leading to a reduction of the net longwave flux. At the same time the cold cloud tops produce low values of the TOA outgoing longwave flux. For broken clouds a correlation between the outgoing and the net longwave fluxes at the surface can be expected. For overcast situations the net fluxes become small and no correlation exist anymore with cloud top temperatures. During field experiments described in section 5.4.5 such correlations have been explored. The following average correlation for cloudless conditions was found for land as well as sea surfaces:

$$M^*_{LW,\,surface} = -1.2\left(\widetilde{M}_{LW,\,TOA} - 275\right) = 330 - 1.2\,\widetilde{M}_{LW,\,TOA} \qquad (5.44)$$

$\widetilde{M}_{LW,TOA} \geq 316\ Wm^{-2}$ is marked by a tilde because it is not the longwave flux that actually is emitted at the TOA but a virtual black body flux corresponding to the effective radiation temperature measured e.g. in the Meteosat IR channel 11 - 12 µm. For cloudy conditions, $\widetilde{M}_{LW,\,TOA} < 316\ Wm^{-2}$, which corresponds to emission temperatures < 0°C, a constant IR flux of - 50 Wm^{-2} was obtained for the conditions met during field campaigns in the Mediterranean area.

5.4.4 Radiant Exposure

Instantaneous radiation fluxes are of limited value for energy budget studies . Daily and longer term net radiation sums are necessary to determine, e.g., evaporation over a certain time period. Polar orbiting satellites assess each point of the Earth only once per day and once per night. In some cases an integration is possible, if one instantaneous value is available from a satellite and if the relative diurnal cycle of the studied quantity is known. This may be feasible during a sequence of cloudless days - like in the case of the diurnal temperature wave as discussed in chapter 4. A real measurement of the diurnal cycle is only possible from geostationary satellites. In the case of the Mediterranean this is Meteosat. These data become available every 15 - 30 minutes. An example for the shortwave radiation exposure is given in Fig. 5.14.

The data from the net Visible and net IR exposure are added to yield the daily exposure as illustrated in Fig. 5.15. Table 5.16 presents the contributions from the solar and IR-cooling to the total radiative energy balance for 4 different areas and three months.

Over the open water, the net visible energy input reaches its largest values of 27

MJ m^{-2} day^{-1} in June due to the low albedo of water surfaces. The infrared exposure, averaged over the whole decade, day and night, reaches values up to 15 MJ $m^{-2}day^{-1}$ over Lybia and Spain. Over the North African Atlas Mouintains, these energy losses amount to only 50% of these largest values. The available net radiative energy as well reaches its largest values over open water. In March, 12 MJ $m^{-2}day^{-1}$ are reached, which increases to values of 16 MJ m^{-2} day^{-1} in June. If the whole radiative energy would be converted into latent heat, this would be equivalent to an evaporation of 6.6 l m^{-2} (1 MJ m^{-2} ≅ 0.4 mm evaporation).

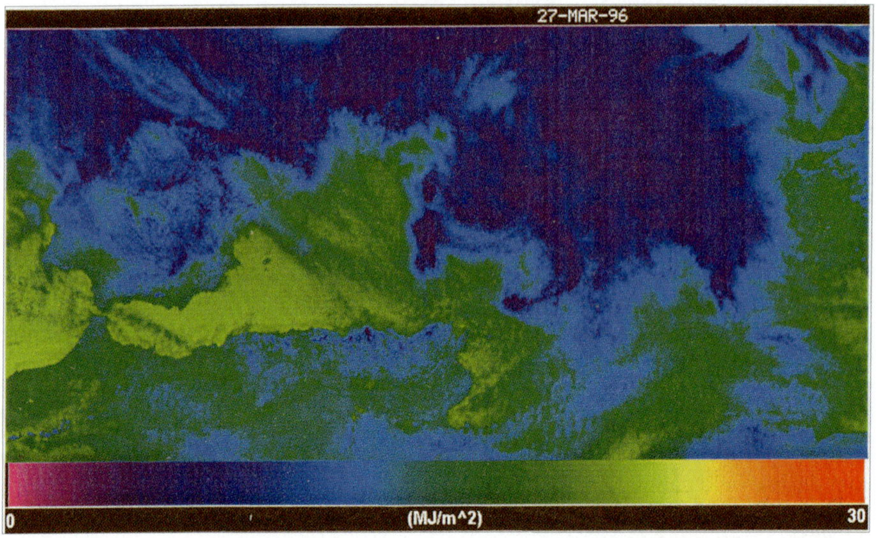

Fig. 5.14. Net surface shortwave radiation exposure in the Mediterranean basin between 06:00 and 17:00 hours GMT on 27 March 1996. The scale is from 0 to 30 MJ $^{-2}$. At the left side the contours of the Iberian Peninsula and North Africa can be seen with the Strait of Gibraltar in between

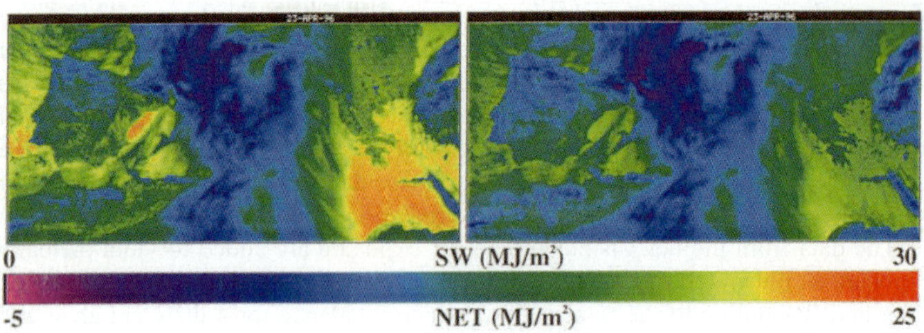

Fig. 5.15. Shortwave and total exposure in the Mediterranean on 27 April 1996

Table 5.16. Mean daily radiative energy (MJ m^{-2} day^{-1})

Month	March 1996			April 1996			June 1996		
Area	VIS	IR	tot	VIS	IR	tot	VIS	IR	tot
Spain	13	-7	6	15	-6	9	21	-14	7
Med-sea	21	-9	12	23	-9	14	27	-12	15
Bulgaria	10	-5	5	18	-8	10	18	-9	9.5
Sahara	14	-8	6	16	-6	10	17	-8	9

5.4.5 Validation

Instrumentation

Validation of radiation fluxes derived from measurements made in space has one major source of potential error: A measurement at the ground is only representative for an area of about 10 m^2 while e.g. one Meteosat data point covers an area of several km^2. If not carefully selected the area covered by a Meteosat pixel may contain various different land cover types not assessed by the ground measurement

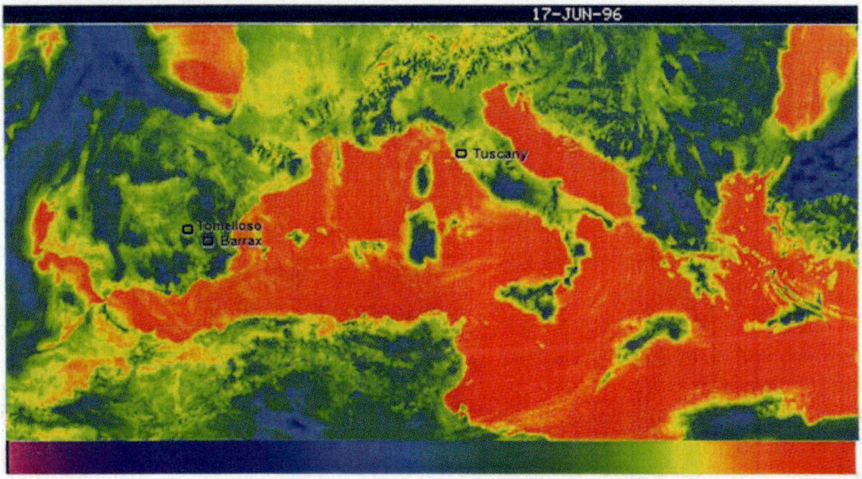

Fig. 5.16. Map of surface net shortwave radiation exposure on 17 June 1996 with the position of Anchor stations in Spain and Tuscany. The scale ranges from 0 to 30 MJ m^{-2}

equipment. Since it is impossible to localize observation sites at the ground within a Meteosat image pixel the effect of clouds on the ground measurement can differ very much from its effect in the image pixel. It therefore is not astonishing that one is confronted with a wide scatter of data points when both types of measurements are compared with each other. Only with large ensembles of comparisons it is possible to establish reliable correlations.

During the EFEDA experiment 1991 of the order of 20 sites were in operation with various radiation flux measuring instruments. In 1996 surface measurements of radiation fluxes were available at anchor stations in Tuscany and Castilla - La Mancha, which are shown in Fig. 5.16 together with a map of the net surface shortwave radiation exposure for 17 June 1996. Five periods during different seasons throughout 1996 could be used for tests. The types of instruments used for comparative measurements at the ground are listed in Table 5.17. Some comparative measurements have also been made in Crete.

A combination of a pyrradiometer type Schulze-Lange and an albedometer type Kipp&Zonen allows to calculate the individual long wave radiation fluxes by subtraction. The first instrument measures radiation in the wavelength range 0.3 μm to 30 μm and the second one in the range 0.3 to 3 μm. During daytime, the large signal of the shortwave radiation which has to be subtracted can cause large errors of the smaller IR signal. The accuracy of the IR measurement therefore depends critically on the calibration factors (which can be different for shortwave and longwave radiation), the accuracy of the temperature measurement inside the instrument, and the temperature of the covering domes. To reduce the last type of errors the instruments are ventilated keeping all components close to air temperature. An absolute calibration of the instruments during field work is not feasible. To obtain a measure of the precision of the used instruments they usually are compared for a series of days. In the case of the EFEDA experiment this was done at the Tomelloso site (Fig. 5.17) by mounting all sensors side by side over a homogeneous bare soil surface. The range of measurements obtained with different instruments for global radiation and net shortwave fluxes can be seen in Figs. 5.18 and 5.19. For global radiation differences of

Fig. 5.17. Comparison of radiation flux instruments at the Tomelloso site in Spain. Partial view of the arrangement over bare soil in Tomelloso, Spain, 20.06.1994

about ± 20 W m^{-2} or ± 2 % were measured which is not quite within the expectations of the World Climate research Programme but also not too far off. The situation is worse for the longwave radiation fluxes shown in Figs. 5.20 for the sky radiation and Fig. 5.21 for the net flux for which differences up to ± 33 W m^{-2} or ± 13.5 % were found.

The only instrument designed for direct measurement of longwave thermal infrared sky radiation is the Eppley pyrgeometer, in which shortwave radiation is blocked by an interference filter dome that still has an optical thickness of 0.3 in the long wave range. Therefore the temperature of the dome affects the signal. To eliminate this error source the pyrgeometer is ventilated. Measurements made with a four component pyrradiometer (LXG 55 "Schulze" and CM-7 Solarimeter SBM) may lead to higher values of the atmospheric longwave radiation than those measured with the Eppley instrument, probably because of different time constants of the ventilated LXG and the non-ventilated CM-7. Comparisons of this difference

Table 5.17. Typical ground based instruments in radiation flux validation experiments

Type	Wavelength Range (μm)	Construction	Direction of measurement
Pyranometer	0.3 - 3	Flat thermopile covered by glass dome	upward or downward
Pyrradiometer	0.3 - 30	Flat blackened thermopile covered by polyethylene dome (sometimes different sensitivity for short- and longwave radiation	upward and downward
Pyrheliometer (Solarimeter)	0.3 - 3	Pyrradiometer used for global radiation measurements	upward
Pyrgeometer	3 - 30	Pyrradiometer sometimes without cover or detector covered by a filter (e.g. in the shape of a dome) that blocks shortwave radiation	downward at night only upward and downward
Albedometer	0.3 - 3	Two thermopiles (or separate pyrranometers) mounted back-to-back covered separately by glass domes	one sensor up, the other down
Net radiometer	0.3 - 30	Two thermopiles mounted directly back-to-back, covered in both directions by polyethylene domes	one sensor up, the other one down
Four component radiation budget meter	0.3 - 3 0.3 - 30 3-30	Combination of two pyranometers and two pyrradiometers respectively one pyrradiometer and one pyrgeometer	from each pair one sensor up and one down

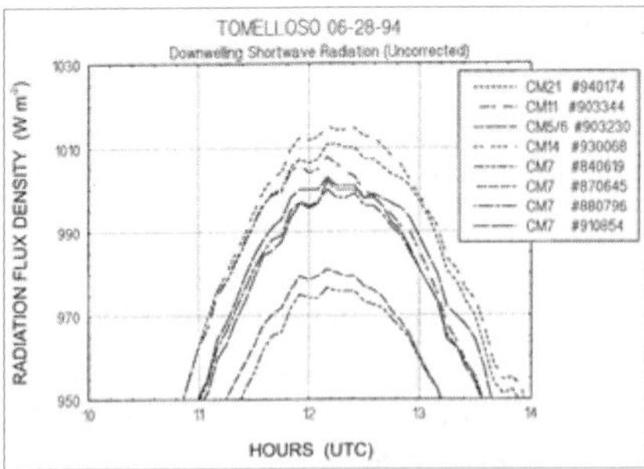

Fig. 5.18. Examples of global radiation measurements with different instruments evaluated with individual calibration constants. The figure shows only the peak values with greatly enhanced scale

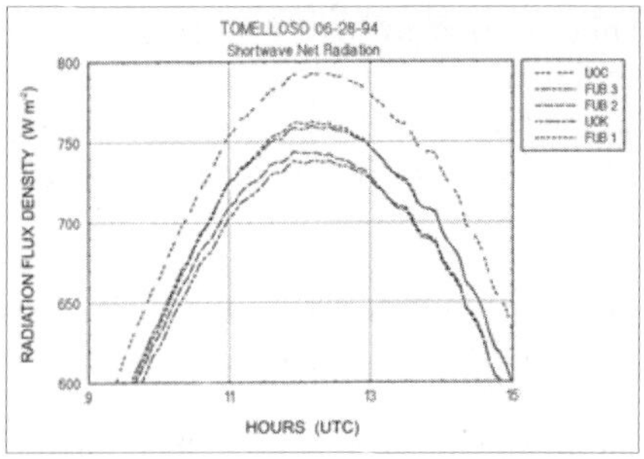

Fig. 5.19. Net shortwave radiation flux densities measured with different sensors and evaluated with individual calibration constants. The figure shows only the peak values at enhanced scale

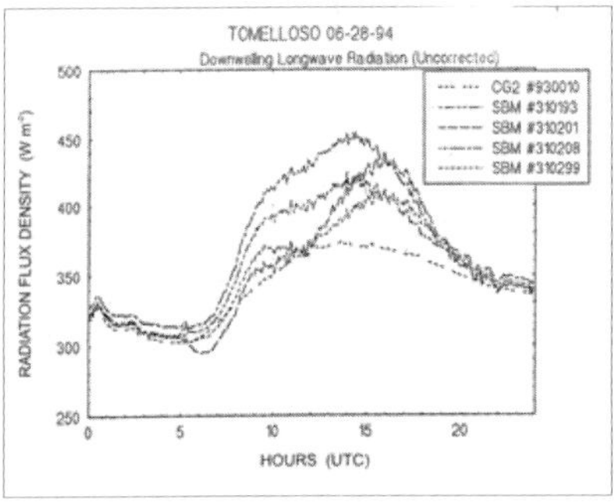

Fig. 5.20. Atmospheric longwave infrared radiation flux densities measured by different instruments

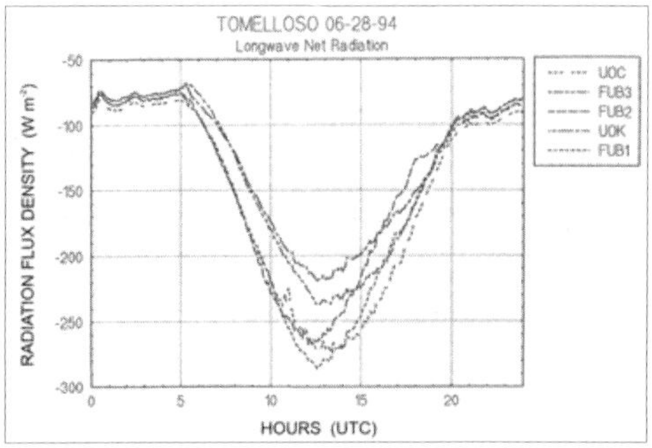

Fig. 5.21. Net longwave radiation flux densities measured with different instruments

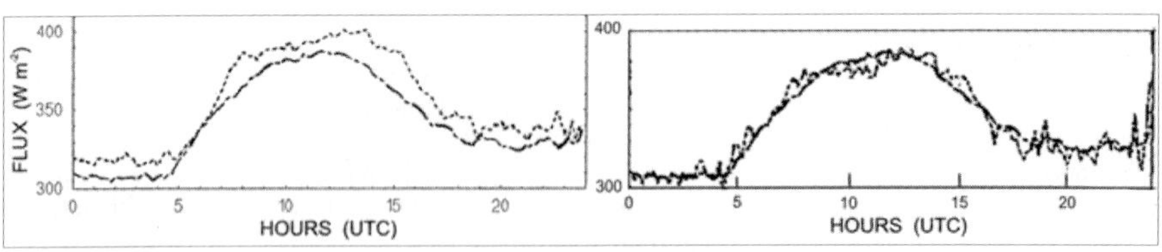

Fig. 5.22. Comparison of diurnal infrared fluxes determined from four component measurement equipment and measured by the Eppley instrument prior (left) and after (right) correction

with both global radiation and air temperature lead to a corrected flux computed according to:

$$M_{corr} = M_{measured} + 14 * \frac{\Delta T}{10\,\text{min}} - 9.6 - 0.008 * M_{global} \tag{5.45}$$

$M_{measured}$ is the flux determined as difference between pyrradiometer and solarimeter, M_{global} the global radiation and ΔT is the temperature change in 10 minutes. If this correction is applied, the diurnal courses get very close as can be seen in Fig. 5.21. Eq. 5.45 certainly will not be applicable universally but gives an indication of the kind of uncertainty to be expected in the ground truth operations.

Fig. 5.23 shows the directly measured total net radiation flux densities for which the spreading of the measurements at noon is about ±37 W m^{-2} or ±7.7 % of the mean.

Correction of the calibration constants according to a generally accepted "standard" instrument brings the data very close together but does not guarantee that the absolute accuracy will be very high because such a reference instrument may have lost its accuracy for various reasons.

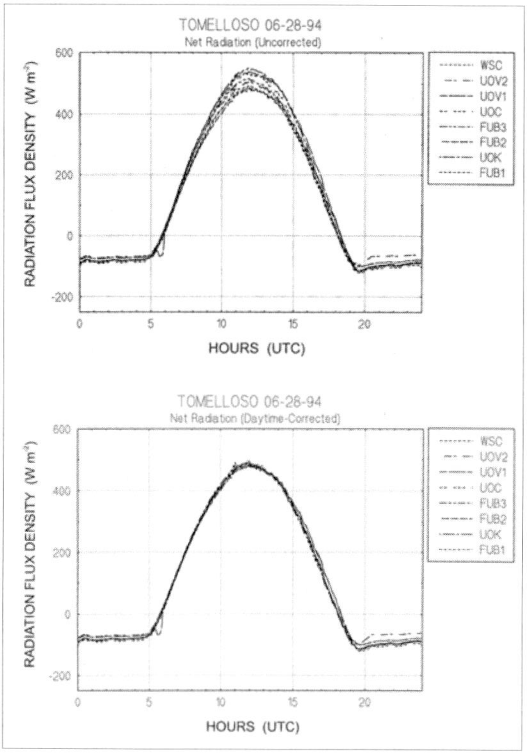

Fig. 5.23. Comparison of total net flux instruments near Tomelloso, June 1994: The upper diagram shows the fluxes computed with individual calibration factors, the lower one with correction factors referred to a reference instrument for daytime conditions

Typical Radiation Fluxes at the Surface

Measurements at the ground can be made with much higher temporal resolution than satellite data can provide. Consequently not only a spatial but also a temporal adjustment is necessary if ground measurements are compared with those made from space. Fig. 5.24 shows the diurnal course of the four fluxes under a cloudless sky which can be integrated to yield the daily radiant exposure which is given in Table 5.18 as average over three cloudless days at the beginning and at the end of June

1991. It can be seen that the global radiation increased only slightly, the thermal infrared fluxes increased because both surface and atmosphere warmed up during June and the net exposure decreased. The daily averaged albedo increased as well by 1% absolute due to the drying of the vegetation and the brighter soil below.

Similar trends have been observed near Tomelloso (Table 5.19) though here the albedo over grapevine decreased, but only marginally, because of the growth of the plants. The first days of June were still disturbed by low albedo because on the last day of May there was rainfall while June remained dry.

It is not astonishing that the global radiation is the same in Rada de Haro and Tomelloso but the albedo is considerably higher near Tomelloso than at the matorral site near Rada de Haro and consequently also the net fluxes are lower in Tomelloso.

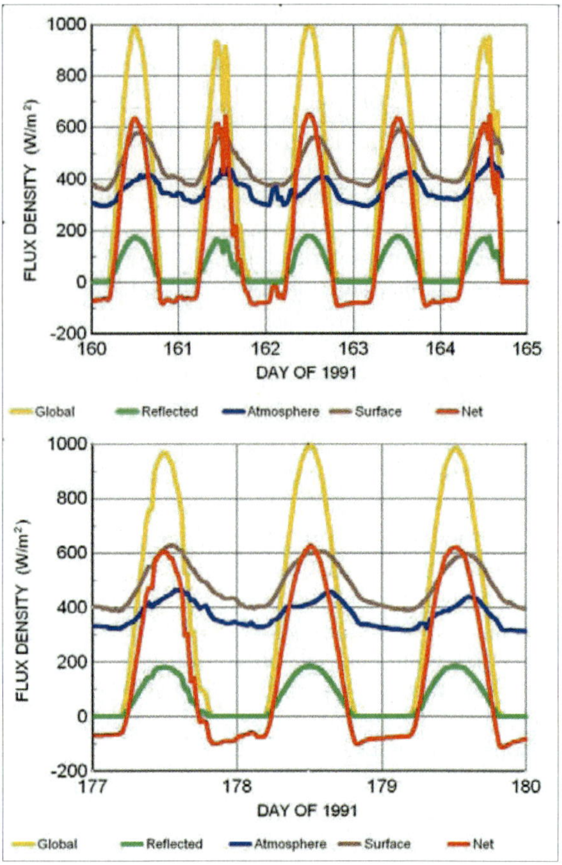

Fig. 5.24. Diurnal course of the net radiation flux density and its components at beginning and end of June 1991 near Rada de Haro, Spain

Table 5.18. Radiation exposure (MJ m^{-2} day^{-1}) at the surface over matorral vegetation near Rada de Haro, Spain, in the beginning and the end of the month June 1991 at nearly cloudless days

Days 1991	Global	Reflected	Atmosph.	Surface	Net	Albedo
160, 162, 163	30.391	5.401	30.508	39.667	15.832	0.178
178, 179, 180	30.475	5.787	31.347	41.097	14.938	0.190

Table 5.19. Trends of radiation exposure over grapevine and vetch near Tomelloso, Spain, during the month of June 1991. The numbers refer to nearly cloudless days if not stated otherwise. D (= 3 ... 30) is the day of June

Flux	Grapevine site	Vetch site
Global radiation	30.43 - 0.0263 (D-1)	30.57 - 0.0397 (D-1)
Reflected shortwave radiation	8.33 - 0.0097 (D-1)	7.03 + 0.0102 (D-1)
Infrared flux from the atmosphere	no measurement	all days: 30 + 0.085 (D-1) ± 3.27
Net shortwave radiation	22.10 - 0.035 (D-1)	23.54 - 0.0499 (D-1)
Albedo (computed)	Begin: 0.274 End: 0.271	Begin: 0.230 End: 0.231
Net total radiation	Max.: 10.78 - 0.0127 (D-1) all days: 10.0	12.0 - 0.05 (D-1) all days: 11.22 - 0.0407 (D-1) ± 1.99

Comparison with Satellite Data

One of the first task was to test, whether or not a comparison of shortwave net fluxes can be performed at the level of one image pixel. Fig. 5.25 demonstrates that the scatter in the data does not increase if one pixel is used instead of a matrix of 3 x 3 pixels. Comparisons were made separately for clear sky and cloudy conditions with data of the Tomelloso anchor station. Though, as expected, the scatter is larger with clouds, the linear relationship between Meteosat measurements and ground based measurements in the average holds (Fig. 5.26). Fig. 5.27 in addition shows how the time evolution of the net shortwave flux is reflected in the Meteosat data as compared to ground measurements. It is clear that differences exist under a cloudy sky but the general trend is reproduced.

To test the hypothesis that observations in the 10-12 μm window are correlated with the longwave infrared net flux at the ground the radiance measurement in the Meteosat IR channel was converted to a fictive flux at the top of the atmosphere. The radiance was first converted to an effective TOA radiative temperature and by application of the σT^4-law a virtual flux is obtained. This flux is related to the net infrared flux measurements at the ground for five observation periods. Fig. 5.28 presents the relationship between the fictive infrared flux at the top of the atmosphere computed from the effective emission temperature measured in the METEOSAT IR channel and the IR net flux measured at the ground for clear sky conditions at the Tomelloso site in Spain, 11. - 15. June 1996. In this case one obtaines an almost ideal linear relationship with very little scatter on which Eq. 5.44 was built. The calculated flux at the top of the atmosphere varies between 320 W m^{-2} before sunrise and 430 W m^{-2} at noon. The rms of the deviation from a linear fit is 13 W m^{-2}. Scatter immediately grows if clouds are present (Fig. 5.29a-c). In these cases the (fictive) flux at the top of the atmosphere may drop to 150 W m^{-2} while the net flux at the surface remains of the order of -50 W m^{-2} where it reaches saturation. A bias of + 31 W m^{-2} results and a standard deviation of 26 W m^{-2}. The measurements have been made during different seasons and at different locations as indicated in the individual figures.

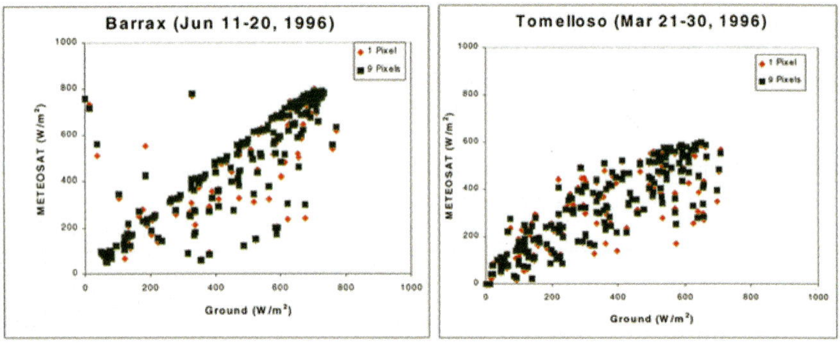

Fig. 5.25. Results of a test whether one Meteosat pixel may be used for comparisons between ground and satellite data or whether a matrix is necessary

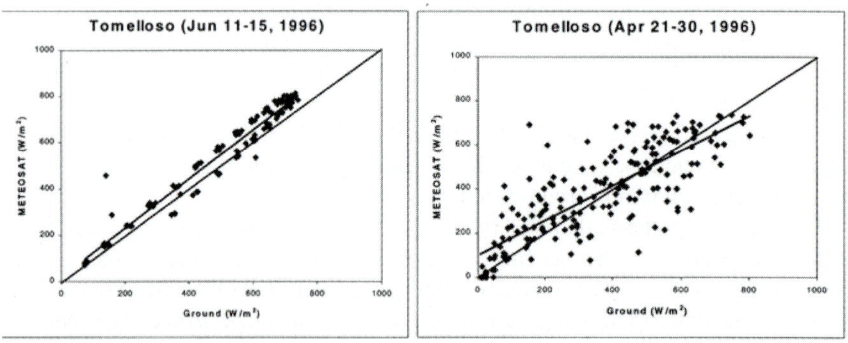

Fig. 5.26. Comparison between surface fluxes derived from Meteosat data and ground measurements

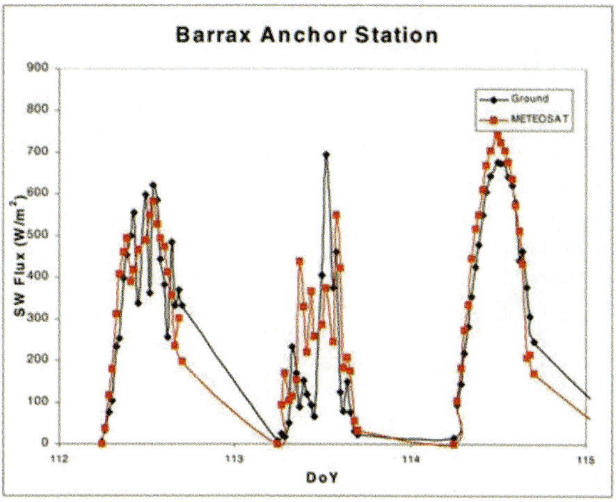

Fig. 5.27. Diurnal development of the global radiation measured at the ground (black) and inferred from Meteosat measurements (red)

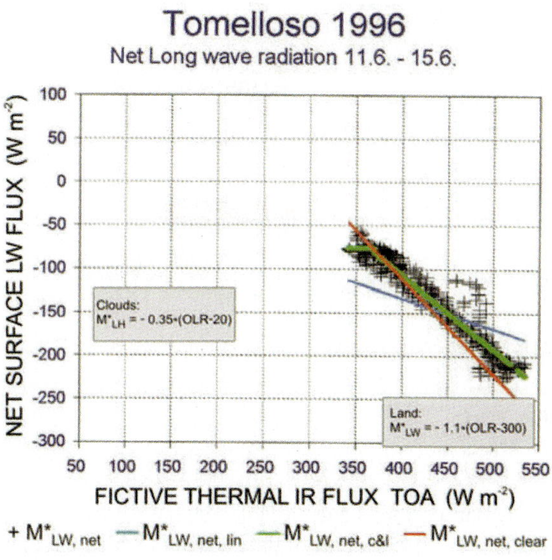

Fig. 5.28. IR net flux at the surface in dependence of the fictive infrared flux at the top of the atmosphere for cloudless skies in the area of Tomelloso, Spain. Explanation see text

The following symbols are used:

+ individual data points
○ flux density calculated for the surface using the Eppley pyrgeometer for the downwelling radiation flux,
—— linear fit of individual data points
—— separate fit for data points in clear and cloudy skies
—— data fit using Eppley data for clear sky atmospheric radiation flux density.

Figs. 5.29a-c are the same diagrams for partly cloudy skies at Tomelloso, Barax, and Radicondoli (Tuscany).

Fig. 5.30 summarizes the results. It can be seen that there are differences in the fit

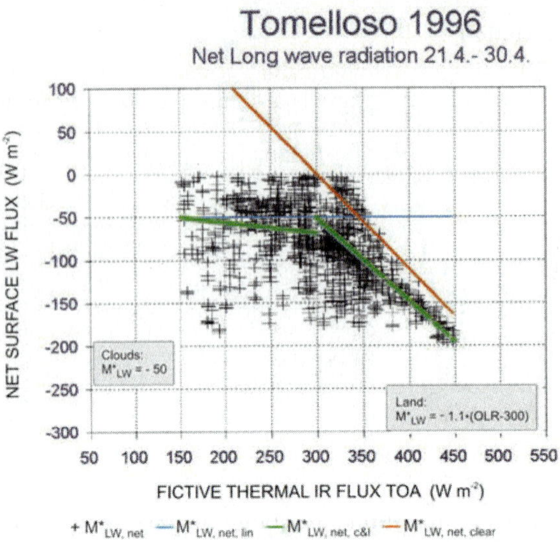

Fig. 5.29a. As Fig. 5.28 but for both cloudy and cloudless conditions

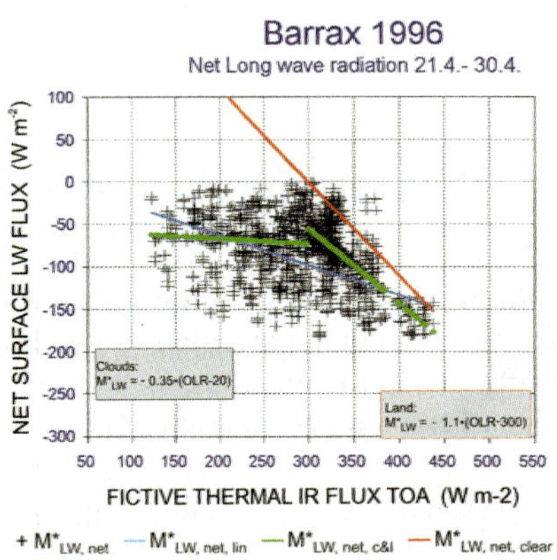

Fig. 5.29b. As Fig. 5.29a but for the Barrax area

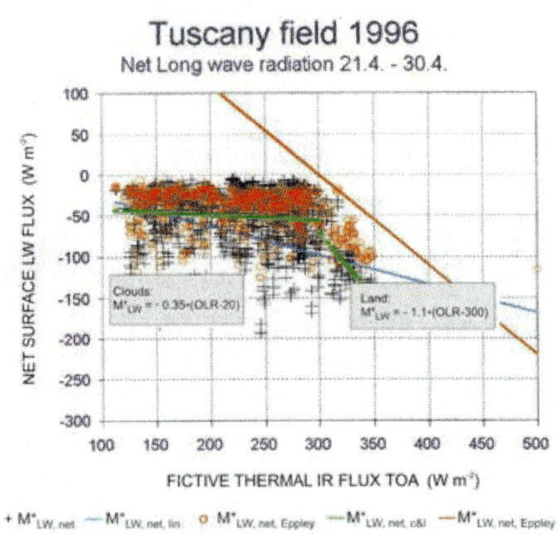

Fig. 5.29c. As Fig. 29b but for Tuscany and a second set of data added in which the Eppley pyrgeometer was used for measurements of the atmospheric infrared radiation flux

parameters obtained under the different conditions. The coefficients used in Eq. 5.44 resulted as a weighted mean value.

For cloudy conditions, the ground measurements are scattered between 0 and –190 Wm^{-2} and show no correlation with cloud top measurements. In Barrax the ground is partly irrigated. Thus the ground temperatures are much more inhomogeneous and show therefore a larger scatter in the net longwave/top of atmosphere diagram.

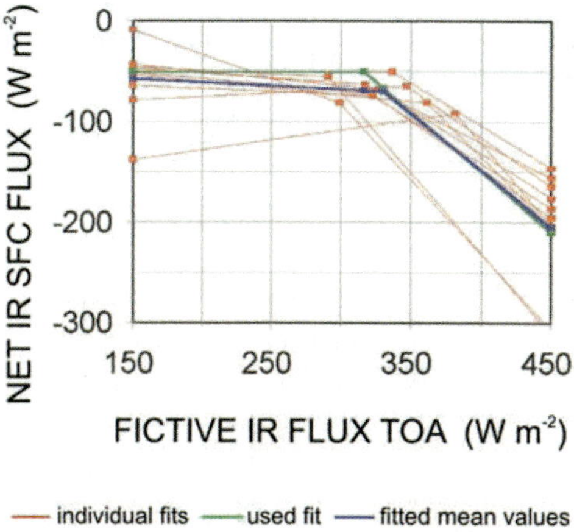

Fig. 5.30. Summary of regressions between longwave net radiation observed at the surface vs. thermal IR flux at the top of the atmosphere inferred from Meteosat data by converting 10 - 12 μm radiances into emission temperature and temperature into (fictive) hemispherical flux. Red lines - individual fits, green line - finally used weighted fit, blue line - fit resulting from mean of all measurements

For five different decades, the fits show a similar behaviour for all three sites. To apply a simple method for a first estimate of net long wave radiative flux, the linear regression for all samples were averaged to be used for the whole Mediterranen area and for all periods between March and August.

Radiation Fluxes at High Spatial Resolution

During a field experiment in Piano di Rosia, Tuscany, Italy (Parlow and Vogt 1998, Parlow et al. 1999), it was important to obtain instantaneous radiation fluxes with higher resolution than obtained on the basis of Meteosat data. A scene of a Landsat-TM overpass was analysed according to a scheme that is presented in Fig. 5.31. After correction of illumination effects induced by terrain topography, the visible and near infrared wavelengths were used to compute albedo. From an elevation model, which was digitised from topographic maps, slope and aspect were calculated for each pixel at a resolution of 30 m. From Landsat-TM channel 6, radiometrically corrected for atmospheric effects, the longwave emission was computed. Atmospheric infrared sky radiation can be estimated using temperature and vapour pressure field data. Finally net radiation flux density is calculated. The influences of land use and terrain on radiation fluxes are considered in all these data layers resulting in significant variations in the spatial domain of up to several hundred W m^{-2}.

Results obtained for the shortwave part of the spectrum are shown in Fig. 5.32. A comparison with the NDVI (Fig. 5.33) shows that the land use at Piano di Rosia is responsible to variations in shortwave net radiation of the order of a few tenths of W m^{-2} (Fig. 5.33). Only in mountainous terrain in the west the different orientation of the slopes cause variations up to 150 W m^{-2}. The net radiation is presented in Fig. 5.34.

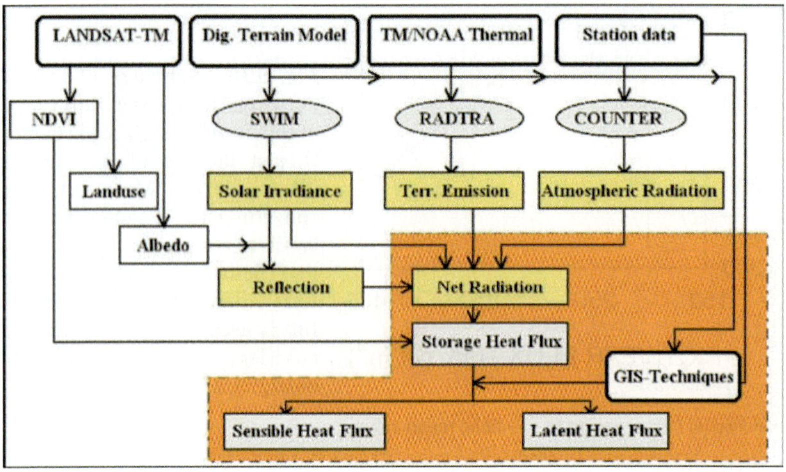

Fig. 5.31. Evaluation scheme to compute fluxes using satellite and supplemental data

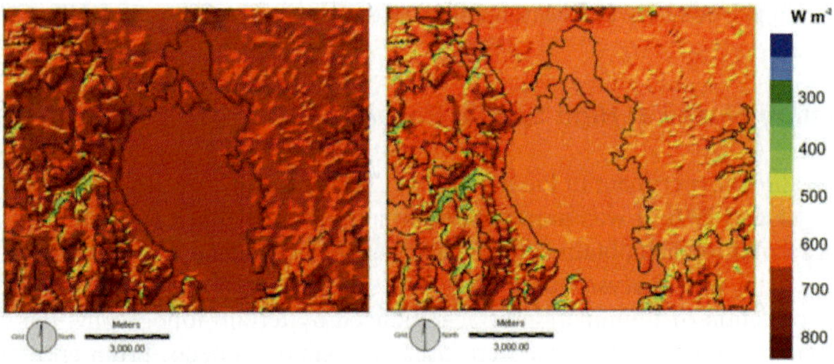

Fig. 5.32. Solar irradiance (left) and shortwave net radiation flux density (right) computed for the time of the Landsat overpass

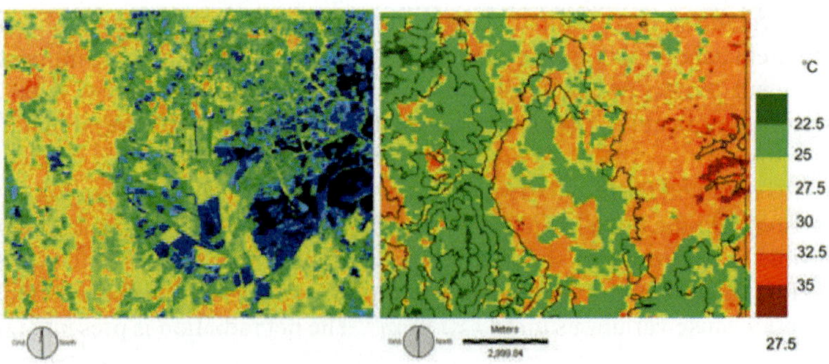

Fig. 5.33. NDVI (left) and surface temperature (right) at the time of the satellite overpass. The colour code of NDVI is relative and ranges from low over medium to high for blue, green, and orange, respectively

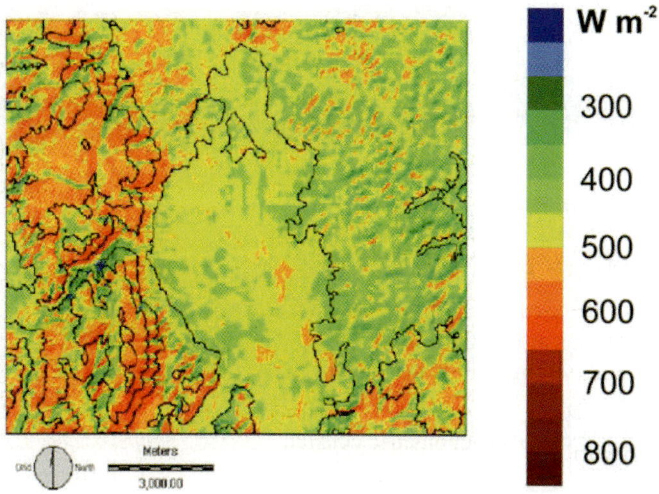

Fig. 5.34. Net radiation flux density at the time of the satellite overpass

5.5 Surface Heat Fluxes

5.5.1 Theoretical Basis

Net radiation energy flux density at the surface is balanced by the sum of the heat flux densities into (respectively from) the ground and the atmosphere:

$$M_R^* - M_{SH} - M_{LH} - M_G = 0 \tag{5.46}$$

If the radiation net flux is known, also the sum of the heat fluxes is known. To obtain the instantaneous sum of the turbulent sensible and latent heat fluxes, $M_{SH} + M_{LH}$, into the atmosphere, the heat flux M_G into soil and vegetation has to be subtracted from the net radiation energy. It should be annotated here that the daily mean values of M_G are small (directed into the soil in spring, out of the soil in autumn). Apart from momentum transfer the coupling between surface and atmosphere critically depends on the distribution of the available energy into latent and sensible heat. Therefore it is of eminent importance to determine the magnitude of the individual turbulent fluxes, respectively their ratio, which generally is expressed as Bowen ratio $\beta = \Phi_{SH}/\Phi_{LH}$. Hydro-meteorological applications of satellite data ultimately aim at the determination of evaporation. A convenient macroscopic formulation for heat transport is that of a flux following the gradient of the concentration of the transported quantity with a magnitude that is reciprocal to the flux resistance over the considered path. If this path is taken from the surface, denoted by subscript 0, to a reference level z in the atmosphere, denoted by subscript *a*, and if the subscript *s* is used in the soil, it can be written:

$$
\begin{aligned}
(a)\quad M_{LH} &= \rho_q\left[q_0 - q(z)\right]/(r_{av} + r_{st}) \\
&= \frac{\rho_a c_p}{\gamma}(e_0 - e_a)/(r_{av} + r_{st}) \\
(b)\quad M_{SH} &= \rho_a c_p\left(T_0 - T_a\right)/r_{ah} \\
(c)\quad M_G &= \rho_s c_s\left(T_0 - T_s\right)/r_{sh}
\end{aligned}
\tag{5.47}
$$

The following notations are used:

$q = \rho_v/\rho_a$ specific humidity = mass of water vapour ρ_v per unit mass of moist air, ρ_a

$$\gamma = \frac{c_p p}{L\frac{m_{H_2O}}{m_{air}}} = \frac{c_p p}{0.622\ L} = \text{psychometric constant}$$

γ = 0.67 mb K^{-1} at STP
c_p = specific heat of air at constant pressure
L = latent heat of vaporization ≈ $2.5 \cdot 10^6$ J kg^{-1}
m_i = molecular weight for water and air respectively
e^* = saturation water vapour pressure in air
e = actual water vapour pressure
r_a = resistance of heat and vapour transfer in air
r_s = resistance of heat and vapour
r_{sv} = resistance for vapour transfer between liquid-moist air interface and the physical surface of the evaporating system
r_{sh} = resistance for heat transfer between liquid-moist air interface and the physical surface of the system
r_{ah} =resistance for sensible heat transport in air
r_{av} = resistance for vapour transport in air
r_{st} = stomata resistance

The resistances are complex functions of the stratification of the atmospheric boundary layer and wind speed respectively of the vegetation type.

Often it is practical to get rid of the boundary condition at the surface because it is not well defined in the presence of vegetation. For this purpose it is recalled that through the Clausius-Clayperon equation saturation water pressure is related to temperature by $\partial e^*/\partial T = s$, the slope of saturated soil water pressure curve in air, which is a non-linear function that takes values between 0.4438 hPa K^{-1} at 0°C and 3.933 hPa K^{-1} at 40°C (20°C: 1.448 hPa K^{-1}). Assuming that one is starting at a wet surface with e_0^* then e_0 - e_a can be replaced by

$$e_0^* - e_a = s(T_s - T_a) \tag{5.48}$$

if the temperature difference is not too high (s should be taken for the mean temperature). The temperature difference furthermore can be expressed by the sensible heat flux (Eq. 5.47) and the sensible heat flux in turn by the energy balance equation. This leads to

$$M_{LH} = \frac{\rho_a c_p}{\gamma(r_{av} + r_{st})}\left[\left(e_a^* - e_a\right) + \frac{s r_{ah}}{\rho_a c_p}\left(M_R^* - M_G - M_{LH}\right)\right] \quad (5.49)$$

Solving this equation for the latent heat flux leads to

$$M_{LH} = \frac{s(M_R^* - M_G) + (e_a^* - e_a)\rho_a c_p / r_{ah}}{\gamma\left(1 + \frac{r_{st}}{r_{ah}}\right) + s} \quad (5.50)$$

The four methods which will be described in the following use these equations in different ways. Three of them called DEMI-SEC, SEBAL, and SEBI attack the problem of areal evaporation from the side of the atmosphere while the fourth one starts from the side of the soil moisture depletion and couples the soil hydrological model with the top soil moisture determined from space by means of microwave remote sensing.

5.5.2 Computation Schemes

DEMI-SEC

The DEMI-SEC Inversion Scheme (Peters 1995) was developed for a savanna type of landscape and is based upon Meteosat data. A central element of this scheme is a set of linear unmixing equations to derive the surface temperature within each pixel separately for bare soil and vegetation covered fractions of the surface. Net radiation, sensible heat flux and soil heat flux are calculated separately for the two fractions within a pixel. Finally, evaporation is computed separately for each fraction from the energy balance equation. It is built of eight computational modules as listed in Table 5.20.

SEBAL

The Surface Energy Balance Algorithm (SEBAL) developed by Bastiaanssen et al. (1994, 1997, 1998a, 1998b and Bastiaanssen 1995) is based upon the energy balance equation (5.46) where

$$M_R^* = (1 - \rho_0) M_{SW}^{\downarrow} + \varepsilon' \sigma T_a^4 - \varepsilon_0 \sigma T_0^4 \quad (5.51)$$

It attempts to map out pixelwise the following quantities which are necessary to compute the components of Eq. 5.46: NDVI, ρ_0, T_0, T_a, z_0, ε_0, M_G/M_R, ε', and $M_{SW}^{\downarrow}$. Examples of the conversion equations used in the SEBAL parametrizations for Tomelloso and Barrax are given in Table 5.21.

The pixel values of Eq. 5.46 are found in the following way whereby the first three quantities can be measured from space:

Table 5.20. Modules of the DEMI-SEC algorithm

	Quantity	Aim	Method of computation
I	Periodical Constants		
	Maximum TIR radiation (relative)	fractionation of land cover type	bare soil fraction = $(TIR_{observed} - TIR_{min})/(TIR_{max} - TIR_{min})$
	Minimum albedo	net SW radiation flux computation	cloud eliminated composite maps (cloud threshold 0.35)
II	Hourly and daily preparatory calculations		
	Cloudiness (%CC)	hourly and daily	from Meteosat images expressed in daily mean percentage
	Optical depth	computation of global radiation	$\delta = 0.3545 \exp(0.01321\ \%CC)$
	9x9 pixel TIR	atmospheric correction	extraction of the average temperature of a 9x9 pixel matrix from the centre of the scene
	Pixel cloud contamination	fractional cloud cover	TIR thresholding
III	Pixel net SW radiation calculation: After this step the program branches. If pixel is cloudless the energy balance approach is used, if cloud contaminated, use empirical linear relationships between M_{LH} and M_{sd} and M_{global} and M_{net}		
	Global radiation	Radiation budget	Eq. 4.
	Surface net SW	Radiation budget	Estimate from global radiation and albedo
IV	Temperature calculations (hourly)		
	Calibrated TIR	Absolute temperature values	Meteosat calibration
	Atmospheric correction	$\sigma T_{s,eff}^4 = \sigma T_{s,TOA}^4 / c_T$ using cable system surface data	$c_T = c_3 + C_4 \overline{M_{R,81}} + c_5 \sin\left(\frac{t}{c_1}\right)\cos\left(\frac{t}{c_2}\right)$
	Linear unmixing (bare soil and vegetation)	Determination of component temperatures	$T_{s,bare} = \dfrac{T_{s,eff} - 12.1686 f_1 - 174.641 f_2}{\dfrac{f_1}{1.044} + \dfrac{f_2}{2.41}}$ $T_{s,veg} = \dfrac{T_{s,eff} + 390.978 f_2}{f_1 + 2.3084 f_2}$

V Fractional net radiation			
	Air temperature		$T_a = 301 + 7 \ln [0.03 (T_s - 284)]$ sorghum $T_a = 301 + 14 \ln [0.03 (T_s - 273)]$ savanna (daily data)
	Mean emissivity	flux-to-temperature conversion	$\varepsilon = 0.9031 + 0.00350\ \theta_s$ sorghum $\varepsilon = 0.9115 + 0.00273\ \theta_s$ savanna θ = volumetric soil moisture fraction
	Atmospheric emissivity		estimated from water wapour pressure related to cloudiness ($w = 0.0103 + 2.846 \cdot 10^{-5}\%CC$ kg m^{-3})
	Net radiation of fractions		$M_{R,i}{}^* = M_{SW,i}{}^{\downarrow}(1-\rho_s) + \varepsilon_i M_{LW}{}^{\downarrow} - \varepsilon_{s,i}\sigma T_{s,i}{}^4$ $i = 1, 2$
VI Fractional sensible heat flux			
	Wind speed	%CC < 50: %CC > 50:	$u_t = (2.5 + 0.06\ \%CC)\ u_{n,t} + 0.5$ m/s $u_t = (5.5 - 0.02\ \%CC)\ u_{n,t} + 0.5$ m/s
	Surface resistance	to compute M_{sh}	calculated from T_s, T_a, u and field determined z_0 for both components
	Sensible heat flux	using aerodynamical resistance r_{ah}	$M_{SH} = \rho c_p (T_s - T_a)/r_{ah}$
VII Fractional soil heat flux			
	Estimation of $M_G/M_R{}^*$	determination of M_G	$\overline{M_G} = A - B\,\overline{M_R^*}$ savanna: A = 16.309, B =0.153 sorghum: A = 27.890, B = 2.2376 bare soil: A 31.094, B = 0.412
VIII Latent heat flux			
	Fractional $M_{LE,i}$		$M_{LE,i}{=}M_{G,i} - M_{SH,i} - M_{R,i}$
	Pixel integration		$M_{LE} = f_1\ M_{LE,1} + f_2\ M_{LE,2}$
	Diurnal latent heat flux	total evaporation	summation over 24 hours
IX Hourly empirical approach for cloud contaminated pixel			
			$M_{LE} = a + b\ M_R{}^*$ a and b empirically to be determined

Table 5.21. Conversion equations used in the SEBAL parametrization for the Tomelloso and Barrax sub-regions acquired with NS001 data on 29 June 1991, at 10:21 GMT

Computation step	Unit	Tomelloso sub-region	Barrax sub-region
8 $r_0(x,y)$	-	$-0.085+1.47r_p(x,y)$	$-0.085+1.47r_p(x,y)$
13 $\varepsilon_0(x,y)$	-	$1.009+0.047\ln[NDVI(x,y)]$	$1.009+0.047\ln[NDVI(x,y)]$
14 $T_0(x,y)$	K	$[T_{sat}(x,y)^4/0.91]^{0.25}$	$[T_{sat}(x,y)^4/0.91]^{0.25}$
15 $K^{\downarrow}(x,y)$	$W\ m^{-2}$	$1358{*}0.66{*}\cos\phi_{su}(x,y)/1.016^2$	$1358{*}0.66{*}\cos\phi_{su}(x,y)/1.016^2$
18 $\Gamma(x,y)$	-	$T_0(x,y)/r_0(x,y)\ [0.32r_0^{avg}(x,y)+0.62r_0^{avg}(x,y)]\ [1-0.98NDVI^4(x,y)]\ 10^{-2}$	
19 $G_0(x,y)$	$W\ m^{-2}$	$49.0+1.23T_0(x,y)$	$-4.1+2.64T_0(x,y)$
20 $z_{0m}(x,y)$	m	$\exp[-5.5+5.8\ NDVI(x,y)]$	$\exp[-5.2+5.3NDVI(x,y)]$
21 $z_{0h}(x,y)$	m	$0.01{*}z_{0m}(x,y)$	$0.01{*}z_{0m}(x,y)$
25 $T_0(x,y)$	K	$292+305r_0-1160r_0^2+1330r_0^3$	$285+107r_0+191r_0^2-940r_0^3$
30 $u_*(x,y)$	$m\ s^{-1}$	$\{(0.41\ 2.5)^{-1}\ [\ln(5.0/z_{0m}(x,y))-2.01]\}^{-1}$	$[(0.41\ 8.5)^{-1}\ \{\ln[5.0/z_{0m}(x,y)]-0.43]\}^{-1}$
30 $L(x,y)$	m	$-[1175\ u_*^3(x,y)\ T_0(x,y)]/[k\ g\ 151]$	$-[1175\ u_*^3(x,y)\ T_0(x,y)]/[k\ g\ 167]$
32 $r_{ah\text{-}sur}(x,y)$	$s\ m^{-1}$	$[0.41\ u_*(x,y)]^{-1}\ \{\ln[5.0/z_{0h}(x,y)]-\psi_h\}$	$[0.41\ u_*(x,y)]^{-1}\ \{\ln[5.0/z_{0h}(x,y)]-\psi_h\}$
33 $\delta T_{a\text{-}sur}(x,y)$	°C	$-19.0+0.73T_0(x,y)$	$-13.3+0.61T_0(x,y)$

- $M_{SW}^{\downarrow}$ is estimated from S_{date} in combination with clear sky transmission factors derived from ground based observations,
- ρ_0 is derived from spectral reflectance measurements (chapter 5),
- $M_G\ /\ M_R^*$ was parametrized based on surface measurements in the following way:

$$M_G = M_R^* \left\{ \frac{T_s}{\rho_0} \left[0.32\rho_0' + 0.62\rho_0'^2 \right] \cdot \left[1 - 0.98NDVI^4 \right] 10^{-2} \right\} \tag{5.52}$$

Here

T_s = surface temperature in °C

$\rho_0{}'$ = average of surface albedo for the period M_G is directed downward (during daytime).

The other quantities in Eq. 5.49/50 are computed as follows:

$$r_{ah} = \left(0.41\,u_*\right)^{-1} \ln\left(\frac{z}{z_0}\right) \tag{5.53}$$

$$z_0 = \exp(-a + b \times NDVI)$$

For a and b the following numbers were found:

Tomelloso: $a = 5.5$, $b = 5.8$

Barrax: $a = 5.2$, $b = 5.3$.

Part of the necessary information can be drawn from the first partial derivative of the energy balance equation with respect to T_0:

$$\frac{1}{M_{SW}^{\downarrow}}\left(\frac{\partial M_{SW}^{*}}{\partial T_0}+\frac{\partial M_{SH}}{\partial T_0}+\frac{\partial M_{G_0}}{\partial T_0}+\frac{\partial M_{LH}}{\partial T_0}\right)-\frac{\partial \rho_0}{\partial T_0}=0 \tag{5.54}$$

The important tool is the relationship between ρ_0 and T_0 (Menenti et al. 1989) which is obtained from satellite data. Under the assumption that $M_{SW}^{\downarrow}$, T_a, M_{LH}, and M_G are constant (e.g. under dry conditions with $M_{LH} = 0$) $\partial\rho_0/\partial T_0$ depends only on the outgoing longwave and the sensible heat flux:

$$\rho_0 = b' - aT_0 = b' - \left[\frac{1}{M_{SW}^{\downarrow}}\left(-4\varepsilon\sigma T_0^3 - \rho_a c_p / r_{ah}\right)\right] T_0 \tag{5.55}$$

The slope in the brackets through these pixel points allows to estimate r_{ah} respectively u_* which is found from $\partial M_{SH}/\partial T_0$. Values of first order derivatives for Tomelloso and Barrax are given in Table 5.22.

If, as is the case in wet spots, $M_{SH} \approx 0$ and $M_{SW}^{\downarrow}$, T_a, and M_G can be assumed to be constant, then a in Eq. 5.54 has to be replaced by

$$a' = a - \frac{1}{M_{SW}^{\downarrow}}\frac{\partial M_{LH}}{\partial T_0} \tag{5.56}$$

And, if furthermore only $M_{SW}^{\downarrow}$ and T_a are kept constant, it follows for the slope:

$$a'' = a' - \frac{1}{M_{SW}^{\downarrow}}\frac{\partial M_{LH}}{\partial T_0} \tag{5.57}$$

The Simplified Surface Energy Balance Index (SEBI) is a simplification of the SEBAL algorithm for quasi-operational computation of the evaporative fraction $M_{LH}/(M_{LH} + M_{SH})$. Provided that the net radiation flux is derived and that M_G can be subtracted it is a matter to determine the partitioning of the available energy between latent and sensible heat. The determination of the evaporative fraction again is based upon the ρ_0 - T_0 relationship. The assumption is that in wet areas the sensible heat flux is negligible and the rerflectance dependent temperature $T_0 \rightarrow T_{LH}$ is minimum. In dry areas the latent heat flux can be neglected and the temperature $T_0 \rightarrow T_{SH}$ tends to maximum values. In a two dimensional ρ_0 - T_0 diagram one has to find the line that tops the field of data points and that one which limits the field of data points at the lower temperature boundary. If one picks out one pixel point with a certain

Table 5.22. Values of the first order derivatives of surface albedo r_0 and the surface flux densities $L^{\uparrow}$, G_0, H, λE with respect to surface temperature on the dry-lands of the Tomelloso and Barrax super-sites on 29 June, 1991, at 10:21 GMT

Slope	Unit	Tomelloso sub-scene	Barrax sub-scene
$\partial r_0/\partial T_0^{dry}$	K^{-1}	-0.0412	-0.0768
$\partial L^{\uparrow}/\partial T_0^{dry}$	$W\ m^{-2}\ K^{-1}$	-6.85	-6.45
$\partial G_0/\partial T_0^{dry}$	$W\ m^{-2}\ K^{-1}$	1.23	2.64
$\partial H/\partial T_0^{dry}$	$W\ m^{-2}\ K^{-1}$	22.4	48.1
$\partial \lambda E/\partial T_0^{dry}$	$W\ m^{-2}\ K^{-1}$	0.0	0.0

reflectance then the line through this point parallel to the y-axis cuts through these boundaries at T_{LH} and T_{SH} and the evaporative fraction is:

$$M_{LH} / (M_{LH} + M_{SH}) = (T_{SH} - T_{measured}) / (T_{LH} + T_{SH}) \qquad (5.58)$$

The range of surface temperature and maximum evaporation can be calculated as a function of surface albedo and PBL conditions as described by Menenti and Choudhury (1993). The equation of the latent heat flux is re-written with r_e = external resistance and r_i = internal resistance as follows (Menenti 1984 p19) with M_G being a negative quantity:

$$M_{LH} = \frac{s\, r_e (M_R^* + M_G) + \rho_a c_p (e^* - e)}{(s+\gamma) r_e + \gamma\, r_i} \qquad (5.59)$$

By substituting both this equation (5.59) and the sensible heat flux

$$M_{SH} = -\rho_a c_p (T_0 - T_a) / r_e \qquad (5.60)$$

into the energy balance equation, one obtaines:

The maximum value of $(T_0 - T_a)$ can be determined by taking the limit of Eq. (5.61) for $r_i \rightarrow \infty$:

$$T_0 - T_a = \frac{\frac{r_i + r_e}{\rho_a c_p}(M_R^* + M_G) - \frac{1}{\gamma}(e^* - e)}{1 + \frac{s}{\gamma} + \frac{r_i}{r_e}} \qquad (5.61)$$

$$(T_0 - T_a)_{upper\ \lim} = \frac{(r_e)_u}{\rho_a c_p}(M_R^* + M_G) \qquad (5.62)$$

And the minimum value for $r_i \rightarrow 0$:

$$(T_0 - T_a)_{lower\ \lim} = \frac{\frac{(r_e)_l}{\rho_a c_p}(M_R^* + M_G) - \frac{1}{\gamma}(e^* - e)}{1 + \frac{s}{\gamma}} \qquad (5.63)$$

$(T_0 - T_a)_{lower\ lim}$ stands for the condition of potential evaporation. The span between zero and potentioal evaporation is given by the difference of Eq. (5.62) and (5.63). Thus the distance between the measured ΔT at constant ρ and the lower limit divided by the span is a measure for the ratio of actual to potential evaporation (Fig. 5.35). If the variability of the resistance along the line between maximum and minimum evaporation is taken into account this ratio can be expressed by:

$$\frac{E}{E_{pot}} = 1 - \frac{\frac{T_0 - T_a}{r_e} - \frac{(T_0 - T_a)_l}{r_{e,l}}}{\frac{(T_0 - T_a)_u}{r_{e,u}} - \frac{(T_0 - T_a)_l}{r_{e,l}}} = 1 - SEBI \qquad (5.64)$$

Eq. (5.64) shifts the problem to the determination of $r_{e,u}$ ande $r_{e,l}$. Furthermore the air temperature may not be the same over dry and wet sites in an inhomogeneous area. To circumvent this problem, the potential air temperature at the top of the atmospheric boundary layer is taken as reference. Now SEBI can be written:

$$SEBI = \frac{(T_0 - T_{a,pot}) - (T_0 - T_{a,pot})_l}{(T_0 - T_{a,pot})_u - (T_0 - T_{a,pot})_l} \tag{5.65}$$

where $T_{a,pot}$ is the potential temperature at the top of the PBL. In this case according to Brutsaert (1982, p78) the resistance is obtained by:

$$r_e = \frac{\ln \dfrac{\delta}{z_{0,h}} - 2\ln\left(\dfrac{1+\sqrt{1-16\mu_i}}{2}\right)}{k\,u_*} \tag{5.66}$$

h_i = observed height of the ABL,
z_{0h} = aerodynamic roughness length for heat transport ($\approx 0.1\ z_{om}$, the roughness length of momentum,
k = 0.41, von Karman constant,
μ_i = $h_i/L \leq 0$ (neutral to unstable cases), L = Monin Obukhov length.

Effective resistances determined for the dry Tomelloso area range around 1500 s/m and for the partially irrigated Barrax area around 500 s/m.

Practical use of the SEBI diagram requires meteorological observations of air temperature, vapour pressure, and net radiation. Moreover, the resistance r_e has to be determined. As indicated above in relation with SVAT models, processes at the land-atmosphere interface cannot be studied independently of the planetary boundary layer. Accordingly, Menenti and Choudhury (1993) proposed to use SEBI with PBL instead of near surface values of air temperature, vapour pressure, friction velocity, and net radiation. Once the upper and lower bounds (Fig. 5.35) have been calculated, values of SEBI are obtained by plotting paired observations of $[T_0, \rho_0]$ into this diagram. The value of E_d/E_{max} is then easily determined as indicated in Fig. 5.35. The use of the PBL variables implies that PBL and radiative forcing is constant over large areas and observations of $[T_0, \rho_0]$ describe how elements of the land surface react to forcing.

Hydrological Approach

Van de Griend et al. (1997) suggested a combination of an evapotranspiration and a soil hydrology model, the coupling of which is governed by the moisture in the top soil layer which is to be derived from microwave satellite data. The Penman-Monteith equation is replaced by the Priestly and Taylor (1972) equation:

$$M_{LE} = \alpha(M_R^* + M_G)\left(\frac{s}{s+\gamma}\right) \tag{5.67}$$

The soil heat flux is estimated by

$$M_G = M_R^*\left[0.266 - 0.197\,e^{-6.39(1-\theta)_{fc}}\right] \quad (5.68)$$

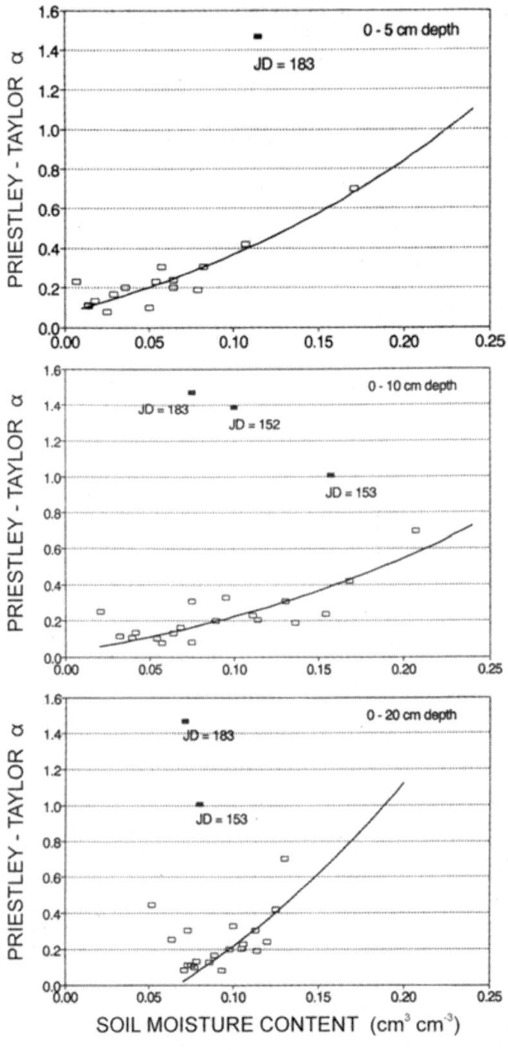

Fig. 5.36. α-θ relationships for the layers 0 - 5 cm (top),0 - 10 cm (middle), and 0 - 20 cm (bottom) at the Tomelloso vetch site. Filled squares represent days with precipitation (Julian days 152, 153, and 183 with 10, 12.5 respectively 3 mm precipitation). The coefficients in Eq. 5.69 have the following values: a = -0.44, -0.33, -1.22; b = 0.51, 0.35, 0.88; c = 4.6, 4.6, 4.9 respectively

where α is an empirical parameter related to surface conditions. According to Owe and Van de Griend (1990) it can be described by

$$\alpha = a + b\exp(c\theta_{fc}) \quad (5.69)$$

θ_{fc} is the moisture content of the surface 20 cm profile expressed as a fraction of the field capacity; a, b, and c are fitted empirically (see Fig. 5.36).

In the soil, the soil moisture changes according to

$$D\frac{\partial \theta_{top}}{\partial t} = -\frac{M_{LH}}{L} + P \quad (5.70)$$

where D is the thickness of the top soil layer and P the precipitation. Richard's (1931) equation was used to compute the flow in the soil:

$$\frac{\partial \theta}{\partial t} = \frac{\partial}{\partial z}\left[K\left(\frac{\partial \psi}{\partial z} - 1\right)\right]. \quad (5.71)$$

The matric potential

$$\psi(\theta) = \psi_s\left(\frac{\theta}{\theta_s}\right)^{-\beta} \quad (5.72)$$

and the hydraulic conductivity

$$K(\theta) = K_s\left(\frac{\theta}{\theta_s}\right)^{2\beta+3} \quad (5.73)$$

were modelled according to Clapp and Hornberger (1978). The index s stands for saturation and the exponent β depends on soil texture.

5.5.3 Results and Validation

Surface Temperatures as Wetness Indicator

During studies performed by the Vrije Universiteit Amsterdam (Peters et al. 1996) it turned out that surface reflectance measured e.g. in the broad shortwave Meteosat channel is an unsuitable measure of vegetation because it varies with soil moisture and in some cases the broad band reflectances of soil and vegetation are not very different. A solution to the problem of vegetation cover mapping was found in the use of thermal infrared observations. It was shown by aircraft measurements that the fractional vegetation cover is linearly related to the NDVI, which in turn is linearly related to the instantaneous surface temperature. It was conjectured that the maximum temperature of a pixel at noon over a number of days is a measurement of the amount of dry bare soil within a pixel.

Under given radiative and boundary layer forcing, observed surface radiometric temperatures of a heterogeneous surface will span a range comprised between a

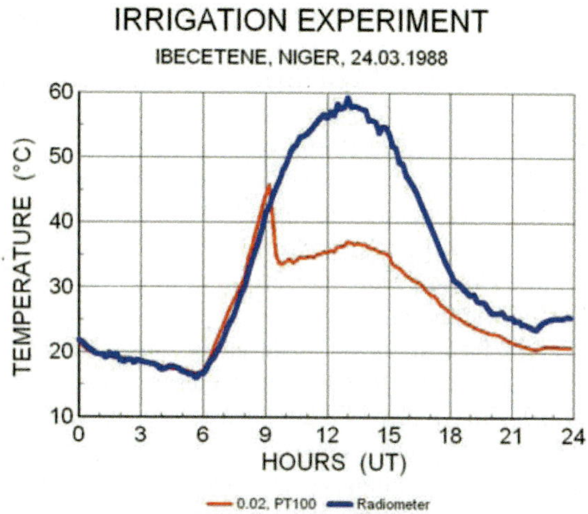

Fig. 5.37. Development of the surface temperature after irrigation at the experimental site of Ibecetene in the Republic of Niger

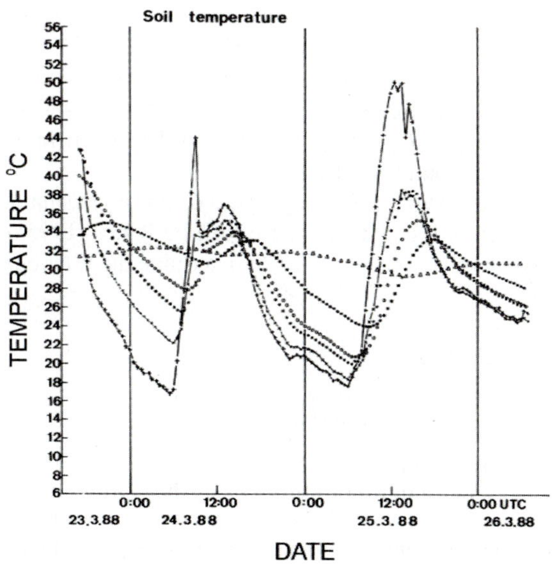

Fig. 5.38. Development of soil temperatures at 0.2, 2, 5, 9.2, 20.2, and 41 cm depth after irrigation on 24 March 1988 at the experimental site of Ibecetene in the Republic of Niger

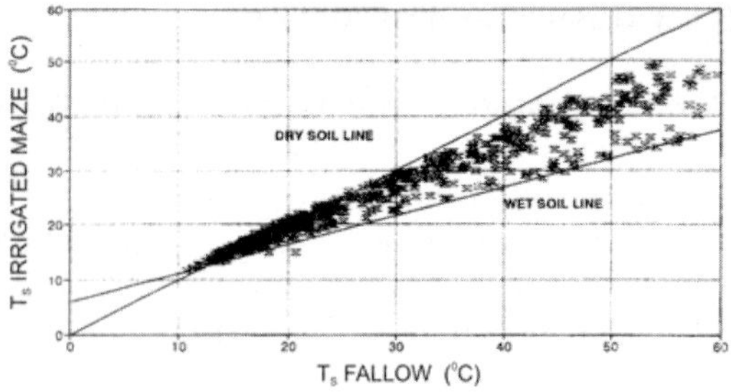

Fig. 5.39. Observed radiometric surface temperatures of fallow land and irrigated maize near Barrax, Castilla - La Mancha

minimum value for a completely wet surface and a maximum value for a completely dry surface. Fig. 5.37 shows this for a bare soil which was irrigated at one day with approximately 25 l m^{-2}. Soil albedo was reduced from 0.267 to 0.217 at 615 nm and from 0.465 to 0.40 at 1024 nm. The irrigation caused an immediate temperature drop of 10 - 12 degrees which increased to little more than 20 degrees after noon.

From measurements made at the surface or with high spatial resolution from space it is possible to determine for a specific ecosystem the relationship between the surface temperature of a dry surface such as dry bare soil and an adjacent irrigated vegetated area such as maize. All measurements made at this area will then fall in between a dry soil line and a wet (respectively vegetated and irrigated) soil line (Fig. 5.38). The position of observed surface radiometric temperatures between the wet and the dry reference line is determined by the fractional vegetation cover (Peters 1995) and by the evaporative fraction (Menenti and Choudhury 1993). If, e.g. the dry soil temperature and the wet surface temperature lines are determined for a specific ecosystem it becomes possible to estimate the surface wetness for a specific pixel. In case, one deals with bare soil exclusively and one measures a temperature T_m below the dry soil line, then the ratio $(T_{dry} - T_m)/(T_{dry} - T_{wet})$ is a measure for the wetness of the soil surface. If the surface is partly vegetated and if the vegetated fraction f_v is known (see section 6.8), then the effective soil temperature can be calculated according to: $T_{soil} = (T_m - T_v f_v)/(1 - f_v)$. This temperature can then be used as T_m in the first equation for the bare soil.

In section 6.8 it is shown how a composing technique based upon the temperature diagram of Fig. 5.39 and described by Peters (1995) can be used to decompose the content of single pixels and to determine the fractional vegetation cover.

Soil Heat Flux

Soil heat flux at the surface is defined by

$$M_G = \lambda \left(\frac{\partial T_s}{\partial z} \right)_{z=0} \tag{5.74}$$

λ is the thermal conductivity. Fig. 5.37 already demonstrated how surface temperature and temperature gradient are affected by precipitation. To allow a determination of the ground heat flux in the context of satellite data evaluation it is necessary to relate the flux to quantities that can be measured from space. This parametrization needs empirical data but $M_G(z = 0)$ cannot be measured directly. Generally, flux measurements consist of a combination of temperature and flux-plate (Thorthwaite plates) measurements at some cm depth in the soil from which the surface conditions have to be extrapolated. Between the surface and the level where the measurement takes place the magnitude of the heat flux changes according to the heat the soil takes up. The change of the soil heat content is proportional to its volumetric heat capacity $C_s = \rho_s c_s$ (ρ_s = density of the soil, c_s = specific heat capacity) and the temperature change in time is given by:

$$\frac{\partial M_G}{\partial z} = \rho_s c_s \frac{\partial T}{\partial t} \tag{5.75}$$

Provided that the thermal properties of the soil are constant over the considered depth range, both equations can be combined to

$$\frac{\partial T}{\partial t} = D \frac{\partial^2 T}{\partial z^2} \tag{5.76}$$

where $D = \lambda / C_s$ is the thermal diffusivity.

The combination of Eq. 5.54 and 5.75 can either be evaluated algebraically or by application of a Fourier analysis to soil temperature measurements. The algebraic formulation for the heat flux at the surface leads to

$$M_G(z=0) = M_G(z) + C_s \frac{\overline{\Delta T}}{\Delta t} z \tag{5.77}$$

where ΔT should be an average value for the layer from 0 to z.

Alternatively the temperature development can be described by a Fourier series

$$T_s = A_0 + \sum_{n=1}^{N} \left[A_n \cos(n\omega t) + B_n \sin(n\omega t) \right] \tag{5.78}$$

where N is the number of terms in the series. Often the diurnal course is coarsely approximated by only the first term. If this approximation is used, the soil heat flux in the depth z can be computed as

$$M_G(z;t) = \Delta T_0 \exp(-\alpha z) I \sin\left(\omega \left[t - t_0 \right] + \frac{\pi}{4} - \alpha z \right) \tag{5.79}$$

ΔT_0 is the amplitude of the temperature at the surface during one day. $\alpha = \sqrt{C_s \omega / 2\lambda} = I / \left(\lambda \sqrt{2} \right) = \sqrt{\omega / 2D}$, $\omega = 2\pi/P$, $P = 86400$ s is the time of one Earth rotation. I is called the thermal inertia. The soil heat flux at the surface ($z = 0$) is maximal if $\omega(t - t_0) = -\pi/4$ which is three hours prior to the time when ΔT has reached its maximum which means at the time when the net radiation at the surface has its steepest increase. This condition may slightly be changed due to the fact that the diurnal temperature wave is not strictly a sine-function. α is the penetration depth, which is the depth at which the amplitude of the wave is damped to 1/e of its initial value. Fig. 5.40 gives an example from a Spanish site.

Soil temperature, T_s, is measured by resistance thermometers or thermocouples at different depths. Flux and temperature gradient can be measured across a few centimetres. It is not advisable to place the measuring devices too close together because then the determination of vertical distances becomes inaccurate especially if measurements are carried out for a longer time period during which the soil, into which the devices are imbedded, may slightly settle down.

Thermal conductivity and heat capacity can be determined in field measurements according to their definition (Eq. 5.74 and 5.75) by an adequate positioning of the measuring devices:

a) λ can be estimated from temperatures in two depths and one flux measurement in between:

$$\lambda = \frac{\Delta z \; M_G(z)}{T\left(z - \frac{\Delta z}{2}\right) - T\left(z + \frac{\Delta z}{2}\right)} \tag{5.80}$$

b) C_s results from flux measurements at two depth and the change in time of one temperature measurement in between:

$$C_s = \frac{M_G(z_1,t) - M_G(z_2,t)}{\dfrac{T(z,t_2) - T(z,t_1)}{t_2 - t_1}}, \quad z = \frac{z_1 + z_2}{2} \quad t = \frac{t_1 + t_2}{2} \tag{5.81}$$

The reliability of the results of such measurements can be controlled by measuring in addition their ratio λ/C_s which is related to the penetration depth α by $C_s / \lambda = (2/\omega)\alpha^2$, using the following methods for the determination of α:

c) measurement of the change of the diurnal temperature amplitude or of the heat

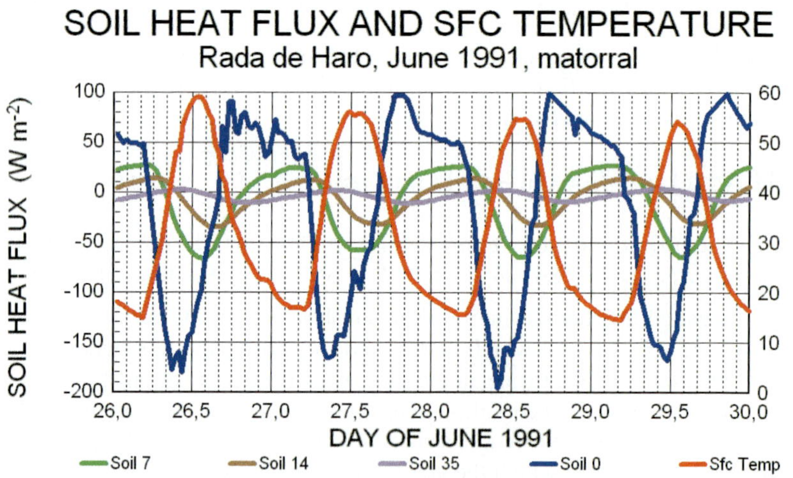

Fig. 5.40. Soil heat fluxes (left hand scale) measured at three depth (7, 14, and 35 cm) and computed for z = 0 (upward direction positive) as well as the surface temperature (red; right hand scale) measured at 2 mm depth for the days 26 - 29 June 1991 showing phase shift and amplitude decline

flux wave with depth as can be seen in Fig. 5.40, or
d) determining α from the shift of the phase of the temperature or flux wave with depth using Eq. 5.79.

The thermal conductivity, varies with soil moisture and takes e.g. values of 0.3 W m^{-1} K^{-1} for dry sand and 2 W m^{-1} K^{-1} for sand at 40% volumetric wetness. Typical values for the volumetric heat capacity are $1.25 \cdot 10^6$ J m^{-3} K^{-1} for dry sand and a value of $2.9 \cdot 10^6$ J m^{-3} K^{-1} for wet sand. It can approximately be computed according to $C = 0.48 f_m + 0.60 f_o + f_w$ if the composition fractions of mineral matter, f_m, organic matter, f_o, and water, f_w, is known. Table 5.23 presents a few measured data. Based on these values the soil heat flux at the surface can be computed for $z = 0$ shown as well in Fig. 5.40.

For remote sensing applications it is necessary to relate the soil heat flux to quantities measurable from space. The physically most obvious connection exist to the net radiation at the surface. A number of attempts have been made to parametrize the soil heat flux this way. One must distinguish here between a climatic approach where long term averages are required and solutions for instantaneous observations. Because of the phase shift between soil heat fluxes and net radiation a single linear relationship can lead to large errors if applied inadequately. Fig. 5.40 shows the relationship between the soil heat flux and net radiation for four cloudless days during a dry period lasting for most of the month of June 1991 in Castilla - La

Table 5.23. Thermal properties of some Mediterranean and African soils

Location (Author)	Depth (cm)	Density (kg m^{-3})	λ (W m^{-1} K^{-1})			C_s (MJ $m^{-3}K^{-1}$)	Water % of dry mass
La Crau, France (Raffy and Dugdale, 1990)	00 - 10	1363	0.278			1.19	0
	10 - 20	1444	0.295			1.10	
	20 - 30	1473	0.291			1.42	
	30 - 40	2600	4.72			2.05	
	00 - 10	1483	0.597			1.69	8.83
	10 - 20	1592	0.689			1.72	10.27
	20 - 30	1642	0.735			2.13	11.48
Rada de Haro Spain (M Berger, 1997)	5		0.725			1.36	
		λ (W m^{-1} K^{-1})					
		Night	0.25	0.5			
Ibecetene Niger (Bolle, 1990)	4.4	0.232	0.223	0.21			D = 1.7 10^{-6}
	7.1	0.398	0.407	0.35		1.02	m^2 s^{-1}
	15.5	0.354	0.448	0.30			α= 11.1 avg

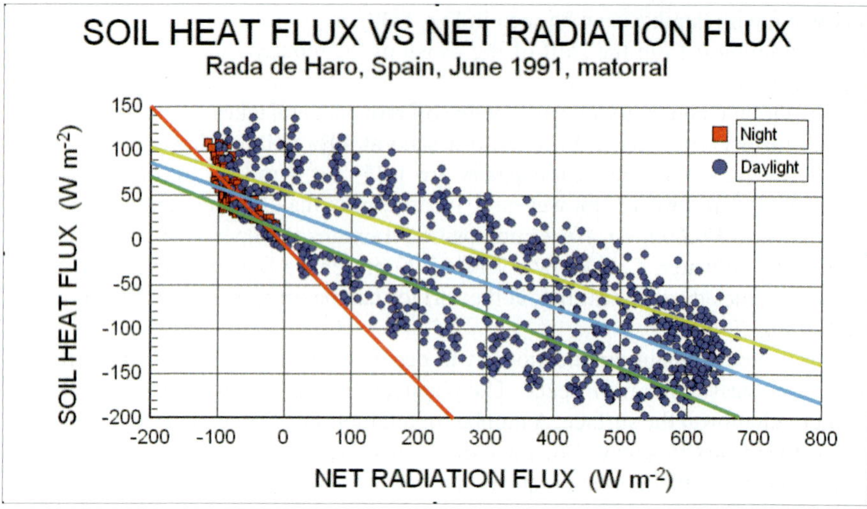

Fig. 5.41. Soil heat flux versus net radiation June 1991. Red for data for zero global radiation, blue for data with positive global radiation (daylight). Linear trends separately for night (red), morning hours (green), afternoon hours (yellow) and all data (light blue)

Table 5.24. Radiation - soil heat flux relationships for sites in Rada de Haro, Spain. Fluxes towards the surface are counted positive

Site	Date	Type of data	Unit	Global rad.	Net rad.	Soil heat	Soil/ Net rad.
RdH	June 91	all data means	W m^{-2}	332.41	171.1	13.6	0.0794
	June 91 average	Daylight mean $M_{SW} > 0$	W m^{-2}		316.7	51.3	0.1618
		Night mean $M_{SW} < 0$	W m^{-2}		74.85	52.4	0.7
	June 27-29 cloud-less	Sums: Day, Net>0	J m^{-2}/day	30.248	18	3.2	0.178
		Night, Net<0		0.227	-3.09	2.54	0.823

Table 5.25. Parametrizations of the soil heat flux (M^* = net radiation flux density). Compare also Eq. 5.52 and Table 5.20/VII

Formula for M_G	Valid for	Author
$-0.27\,M^* + 33.36$ $-0.31\,M^* + 10.15$ (r = 0.94) $-0.32\,M^* + 10.21$ (r = 0.95) $-0.24\,M^* + 55.98$ (r = - 0.90) $-0.26\,M^* + 75.76$ (r = - 0.91) $-0.77\,M^* - 5.57$ (r = - 0.74) $-0.27\,M^* + 35.03$ (r = - 0.77) $-0.1618\,M^*$ (from mean values Table 5.24) $-0.6998\,M^*$ (from mean values Table 5.24) $-0.1780\,M^*$ (from mean values Table 5.24) $-0.8232\,M^*$ (from mean values Table 5.24) $-0.0794\,M^*$ (from monthly mean values)	00 - 24 hour 00 - 12 hour, all days 00 - 12 hour, (cloudless) 12 - 24 hour 12 - 24 hour, (cloudless) night daytime $M_{SW} > 0$ $M_{SW} = 0$ $\sum M^* > 0$ $\sum M^* < 0$	EFEDA experiment Rada de Haro, matorral
$-(0.28\,M^* - 20)$ W m^{-2} $-(0.34\,M^* - 21)$ W m^{-2}	Vine, $\rho = 0.26$ Vetch, $\rho = 0.22$	Oliver and Sene 1992
$0.1\,M^*$	Daytime hours	Clothier et al. 1986
$0.35\,M^*$; $0.31\,M^*$	Daytime; noon	Anderson 1997
$0.19\,M^*$; $0.32\,M^*$	Daytime; night	Mohan and Siddiqui 1998
$0.77(SR)^{-1.23}\exp(K/\cos\zeta)\,M^*$ $(0.4 - 0.33\,NDVI)\,M^*$ $(0.36 - 0.02\,[\rho_{NIR}/\rho_{red}])\,M^*$	K = 0.25	Kustas et al. 1993 Kustas et al. 1994
$0.539\,(SR)^{-0.53}\exp(-K/\cos\zeta)\,M^*$	K = 0.25	Gao et al.

Mancha, Spain. Fluxes into the soil have negative values. In Fig. 5.41 all days of the June period are included and linear trends are indicated which result if night, morning, afternoon or all values are considered. The numerical values of the trend lines are given in Table 5.24 and in Table 5.25 together with relationships reported by other authors. As can be seen, the relationship varies strongly for different hours of the day.

Bastiaanssen (1995) tested the SEBAL algorithm with respect to the soil heat flux in the area of Castilla - La Mancha. Data were taken from Meteosat and from aircraft with the NASA Thematic Mapper Simulator instrument on 29 June 1991 at 10:21

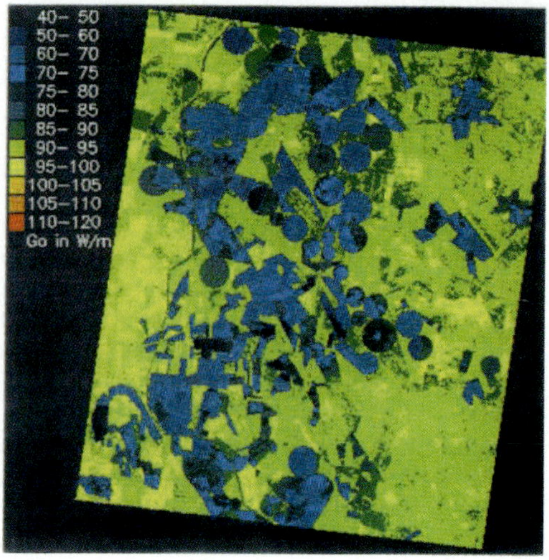

Fig. 5.42. Soil heat flux calculated with TMS001 observations, Barrax

and Landsat-TM data have been used for 12 June 1991. Both were cloudless days. The Simulator soil heat flux map of the site near Barrax can be seen in Fig. 5.42. It shows the partially irrigated area with strong contrasts. In irrigated areas (blue colours) the soil heat flux is well below 80 W m^{-2}. This agrees with the results of the Ibecetene irrigation experiment (Fig. 5.37) where the soil heat flux collapsed to values around 25 W m^{-2} after irrigation until the water was evaporated. In non-irrigated areas the soil heat flux was around 100 W m^{-2}. On average, the flux density here was 83 W m^{-2}. It increased to 105 W m^{-2} at the dry farming area near Tomelloso. From Meteosat data, fluxes of 76 W m^{-2} for Rada de Haro, 81 W m^{-2} for Barrax, and 85 W m^{-2} for Tomelloso were derived. On 12 June the respective values were 65, 66, and 69 W m^{-2}. The measurements were performed at a time when soil heat flux was maximum as can be seen in Fig. 5.40 and seem to be relatively small as compared to ground measurements. The cause may be that the parametrization was done with all day data sets rather than those appropriate for late morning hours (compare Table 5.25). If the all day linear regression is applied to morning hours, the result will be a too small soil heat flux and if applied to the afternoon hours a too large one. In the Rada de Haro area the averaged values include also areas covered by dense vegetation which explains why the values are lower than at the other sites. Here the monthly means lead to a ratio of soil heat flux to net radiation flux of 0.078 (and a ratio of net flux to global radiation of 0.52) over matorral vegetation.

Sensible and Latent Heat Fluxes

To test the performance of algorithms, sensible and latent heat flux densities as well as the evaporative fraction have been computed from aircraft TM-simulator measurements over the Tomelloso and Barrax areas, from Landsat-TM data for the whole experimental area (Fig. 5.43), and from Meteosat data for the Iberian Peninsula. To validate the obtained results a dense network of surface measurements was available at the three sites near Tomelloso, Barrax and Rada de Haro. In addition fluxes have been measured from aircraft at different altitudes. In comparing these data particular attention must be paid to the spatial resolution of the observations in relation to the length-scale of the eddies contributing to turbulent heat transport in the atmosphere (Bastiaanssen et al. 1995). Turbulent heat flux

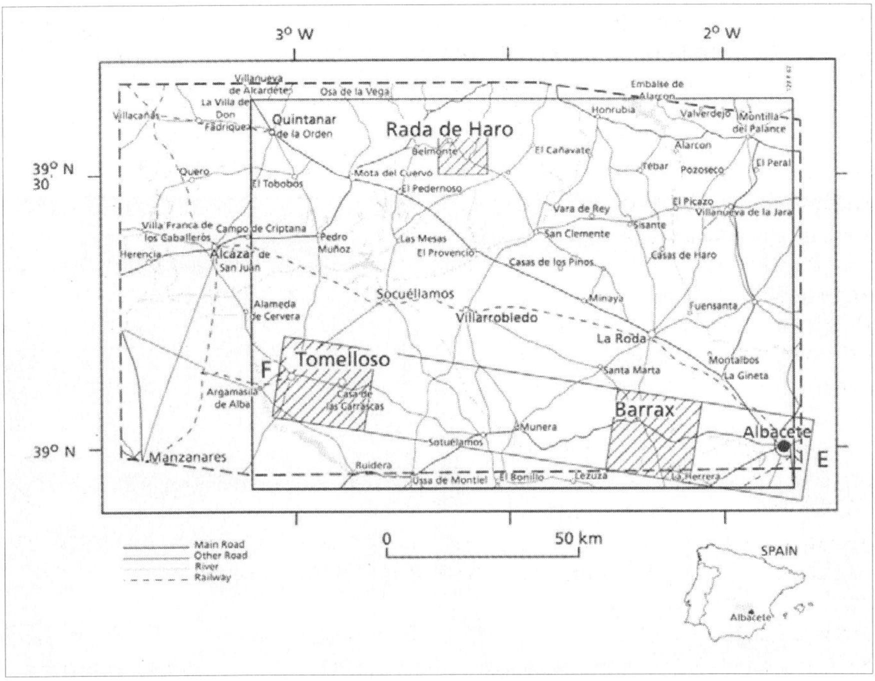

Fig. 5.43. The test sites in the Castilla - La Mancha area in Spain

densities measured at a given elevation and location are determined by the heat released from a relatively large area upwind of the location where the measurements are done (so-called foot-print of the instrument). On the other hand, pixel-wise satellite-based estimates of heat flux densities are local values, i.e. they apply to the exact location of a given pixel. To compare the satellite based estimates with ground measurements, the former have to be integrated over the foot-print. This was done in different ways with the SEBAL estimates described above. The agreement with ground measurements of heat flux densities is poor when using SEBAL estimates at the pixel where the ground measurements are located and it mproves if the SEBAL estimates are averaged over different foot-prints.

For the Tomelloso test site the map of sensible and latent heat fluxes are shown in Figs. 5.44 and 5.45. For the Barrax test site maps of the evaporative fraction, which is the ratio between latent heat flux and net radiation, and of the diurnal evaporation are shown in Fig. 5.46. In the area of Tomelloso maximum values between 100 and 160 W m^{-2} of latent heat flux have been measured over vetch and vine and, in the average, values of the order of 28 W m^{-2} over bare soil. This corresponds well with the values mapped in Fig. 5.45. Only very small irrigated areas with higher evaporative fluxes can be seen which correspond to the values found in such cases in the Barrax area (Fig. 5.46).

Measurements in the irrigated areas of Barrax were difficult to perform because instruments could not be installed within the circles of the rotating irrigation systems. In addition these measurements could not be carried out simultaneously with the TM-simulator overflight because the irrigation is changing day by day. Measurements made at different days nevertheless provide a ground truth of the range of data.

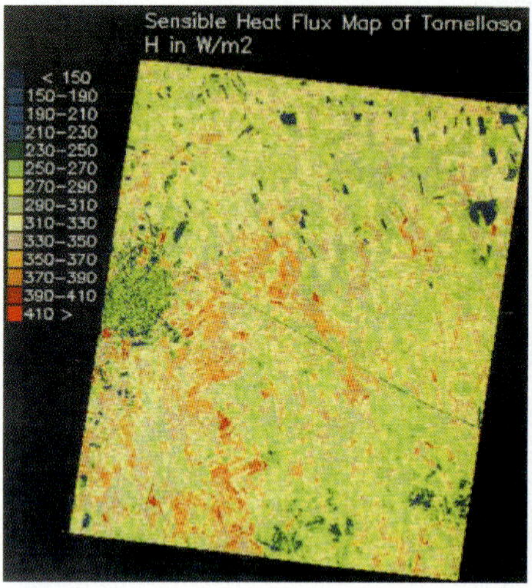

Fig. 5.44. Sensible heat flux over the area east of Tomelloso, Spain, computed with the SEBAL algorithms from TM simulator data

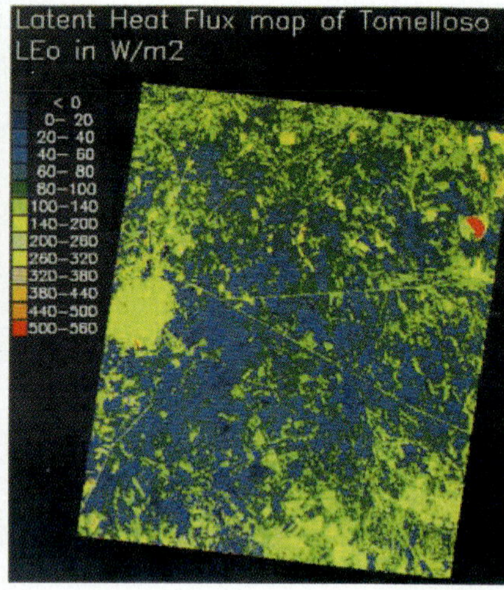

Fig. 5.45. Latent heat flux over the area east of Tomelloso, Spain, computed with the SEBAL algorithms from TM simulator data

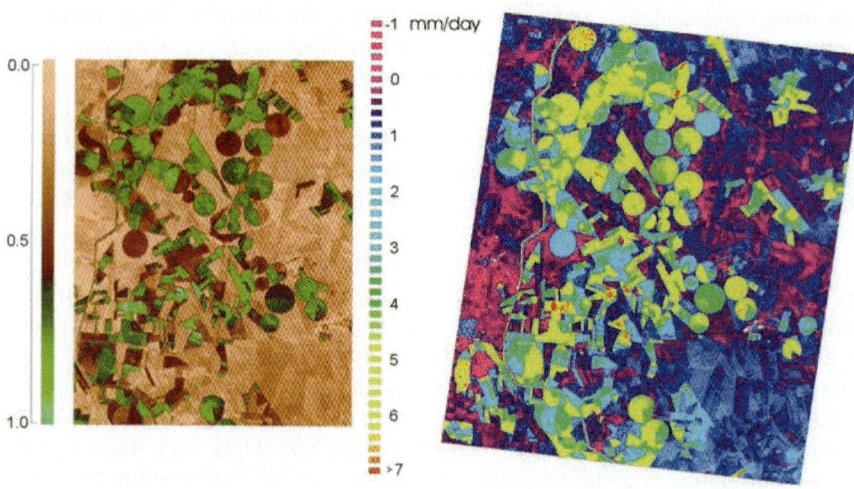

Fig. 5.46. Evaporative fraction (left, scaled from 0.0 - light brown to 1.0 - green) and diurnal evaporation (right, scaled from -1 to +7 mm evaporation per day) of the Barrax test site on 29. 06. 91 derived from TM simulator data by means of the SEBAL algorithm

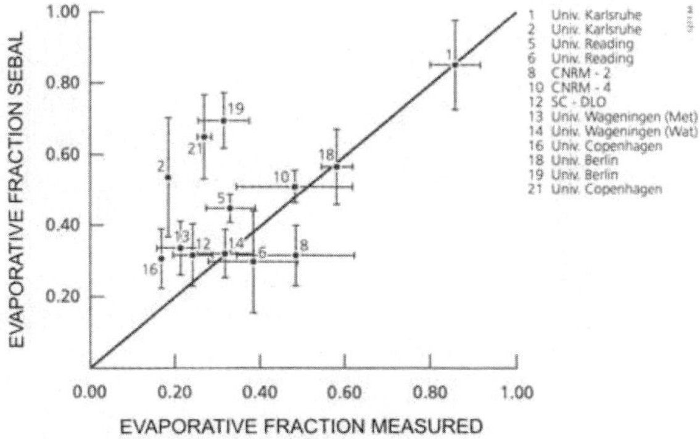

Fig. 5.47. Comparison between evaporative fractions derived from Landsat-TM and measurements from masts at the surface for 12 June 1991, 10:12 GMT (after Bastiaanssen 1995)

Few measurements that have been obtained shortly after irrigation indicate values for the evaporative fraction between 0.52 for irrigated barley and 0.75 for irrigated maize. These correspond to the dark brown and dark green colours in Fig. 5.46. In this figure values up to about 0.9 can be seen which are credible because SVAT model computations initiated with measured data confirmed the existence of Bowen ratios of 0.2 which results in an evaporative fraction of 1/1.2 = 0.8. In one case shortly after irrigation values close to one have been simulated. Between two irrigation spells the ratio quickly drops to values below 0.5. For dry farming areas values of 0.14 to 0.2 have been measured, for bare soils even values around 0.05 which correspond to the light brown areas in Fig. 5.46. The whole span of evaporation lies between 550 and 25 W m^{-2} which is equivalent to 19 and 0.8 mm/day.

The evaporative fraction was computed from the flux measurements made at masts (see Fig. 1.6) at the three experimental sites. Fig. 5.47 shows such a comparison for 12 June, 1991. The three outliers (2, 19, and 21) are from less homogenous areas, either surrounded by matorral vegetation in the hilly area near Rada de Haro (19, 21) or at a partly irrigated area near Barrax (2).

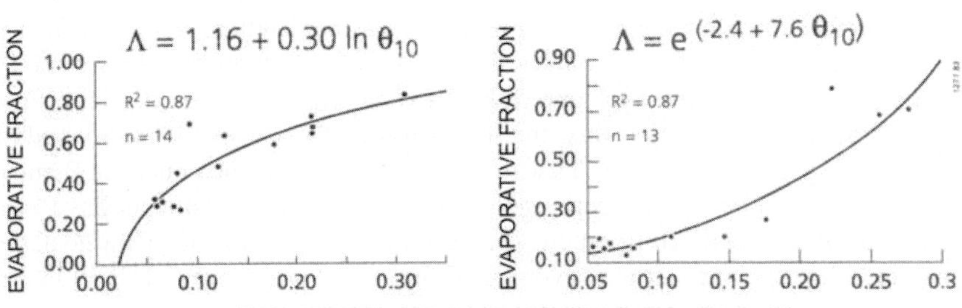

Fig. 5.48. Relationship between soil moisture content at 10 cm depth and evaporative fraction in the Tomelloso and Barrax area enclosing bare soil as well as vineyard, vetch, maize, barley and alfalfa cultures. Left: 12 June 1991, right: 29 June 1991 (after Bastiaanssen1995)

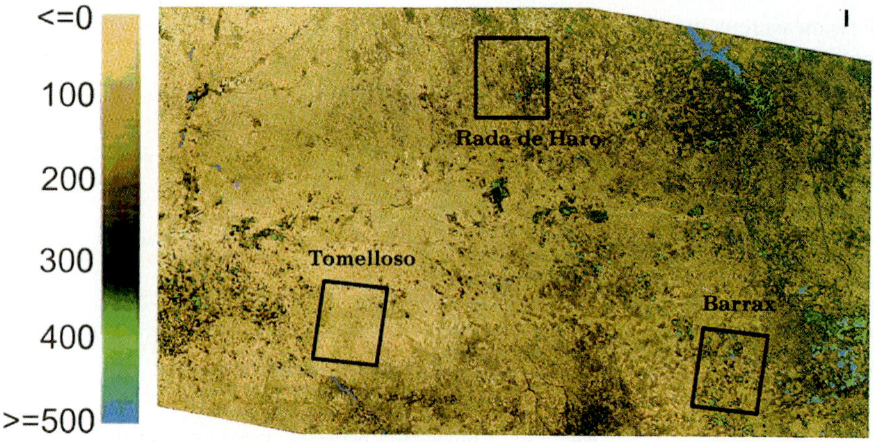

Fig. 5.49. Latent heat flux map of the EFEDA experimental area derived from Landsat-TM data of 12 June1991 (after Bastiaanssen 1995).

The energy partition under arid conditions relates primarily to the available soil water storage. Soil moisture was measured between 5 and 50 cm depths at 46 plots in over 15 fields scattered across the Barrax and Tomelloso sites. These data have been related to the evaporative fraction Λ as shown in Fig. 5. 48. The relationship can be expressed by an exponential function as shown in the diagrams and depends strongly on the season as the difference between the June 12 and 29 data suggest.

An evaporation map of the whole EFEDA experimental area is presented in Fig. 5.49. The bulk evaporative fraction computed from all pixels enclosed in the image are 0.4 for the Tomelloso area, 0.68 for the Barrax area and 0.61 for the whole EFEDA area (Bastiaanssen 1995).

The SEBI algorithm is designed for high as well as medium spatial resolution images. Fig. 5.50 gives an example for its application over an area with mixed agricultural use in Tuscany, Italy. Opposite to the plain areas in Spain this is a hilly to mountainous landscape. Consequently a Digital Elevation Model (DEM) was used to correct for solar incidence angle at the surface and the cooling effect of altitude on temperature. The left hand side shows a large section of Tuscany and the right hand side the test area of Piano di Rosia known from Figs. 5.32 - 5.34. From top to bottom the net radiation flux, the soil heat flux, the sensible heat flux and the latent heat flux are shown for the Tuscan scene at the left and the enhanced Piano di Rosia test area at the right, where comparative ground measurements have been performed. *Inter alia* a scintillometer was used to measure the sensible heat flux across the plain. A five minutes average of the scintillometer measurements (see Appendix 7) at the time of the satellite overpass agrees very well with the SEBI result while eddy correlation and Bowen ratio measurements underestimated systematically the area average of the sensible heat flux by about 20 W m^{-2} (Fig. 5.51). On the other hand the Bowen ratio measurement agreed perfectly with the SEBI latent heat flux estimate. The medium resolution image of the latent heat flux

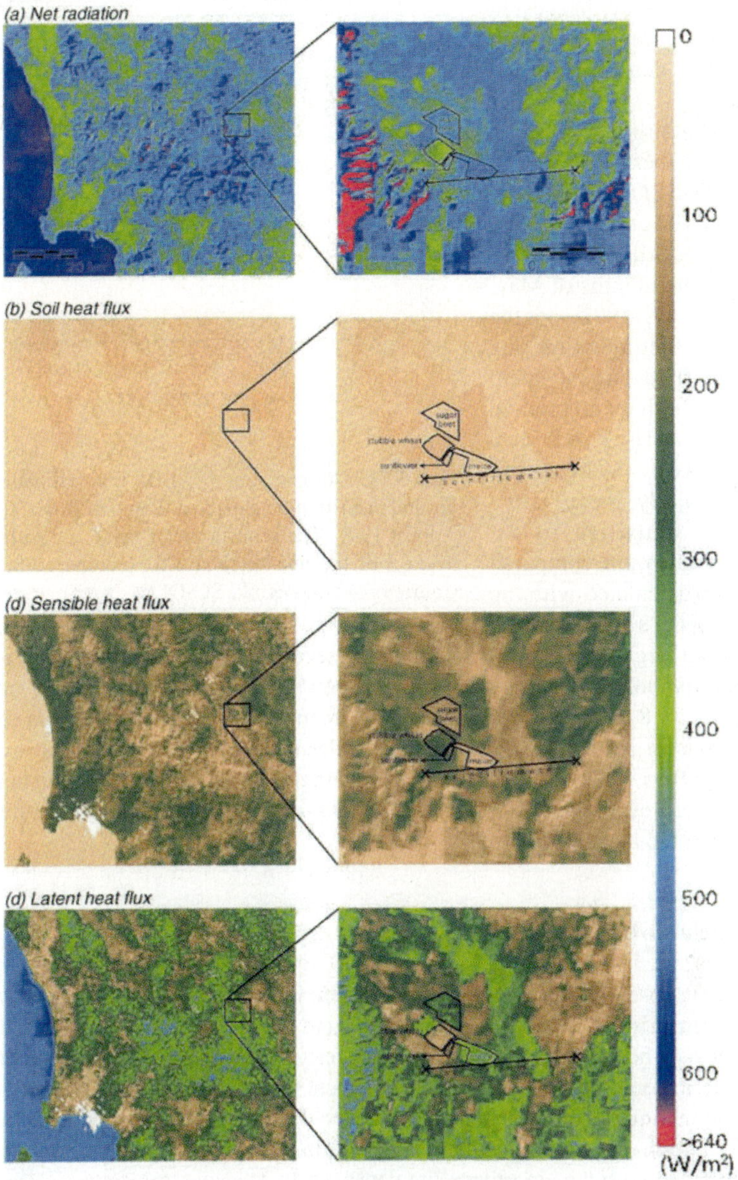

Fig. 5.50. Maps of surface energy fluxes as derived for 23 August 1997, 09:30 GMT, using the SEBI algorithm and Landsat-TM data for Tuscany, Italy. From top to bottom the flux densities of net radiation, soil heat flux, sensible heat, and latent heat can be seen. At the right the test area of Piano di Rosia is enlarged (courtesy of G. Roerink, SC-DLO)

shows small values for the, at that time, mostly harvested agricultural areas and high values over densely forested regions.

For even larger areas, such as the whole Iberian Peninsula, the DEMI-SEC algorithmn, based on Meteosat data, provides results. Fig. 5.52 shows for the months of May, June and July, 1991, on the left side the net radiation flux and on the right side the latent heat flux, both expressed in the equivalent of mm evaporation. When the measured daily mean latent heat flux densities of all EFEDA sites were compared with those calculated with the DEMI-SEC algorithm it resulted in a large scatter. Agreement was better for monthly mean values. For June 1991 the evaporation computed with DEMI-SEC resulted in 38.9 W m^{-2} for Tomelloso, 38.8 W m^{-2} for Barrax, and 39.8 W m^{-2} for the Belmonte - Rada de Haro area. This is in reasonable agreement with the measured data integrated over the month, e.g. for vine near Tomelloso in the average 34.96 W m^{-2} day^{-1} latent heat flux was measured during June 1991. The EFEDA experimental area is indicated as a rectangle in the maps of Fig. 5.52.

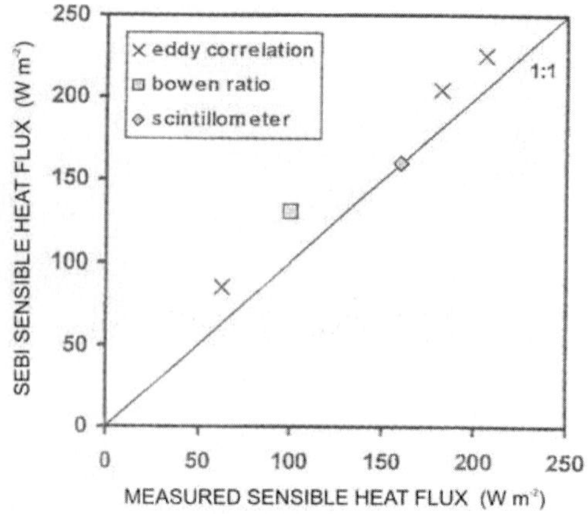

Fig. 5.51. Comparison of SEBI flux estimates with ground measurements made at Piano di Rosia, Italy, 23 August 1997. The scintillometer value is a path-length average of the sensible heat flux (compare Fig. 5.47). Courtesy of G. Roerink, SC-DLO

Diurnal Evolution of Fluxes

Radiometric observations from satellites provide instantaneous estimates of heat flux densities. In principle, daily evolutions of heat fluxes can only be inferred from hourly or half-hourly Meteosat data. However; if the relative diurnal evolution is known or a generalized relationship between instantaneous measurements (e.g. around noon) and the diurnal sum can be derived from ground based measurements, this information may be used to deduce the diurnal evolution from one instantaneous value obtained with the aid of satellite data.

In the case of the soil heat flux the diurnal evolution is shown in Fig. 5.40 for four cloudless days. The average value for the diurnal flux density is 10.48 W m^{-2} (resulting in 0.905 MJ m^{-2} day^{-1}) and the average of the maximum flux density from 09:00 to 12:00 hours is 152.12 W m^{-2}. The relation of diurnal mean flux to average maximum flux therefore is 0.069. For all days during the measuring period of June 1991 the average diurnal soil heat flux density was 13.58 W m^{-2}. This relationship

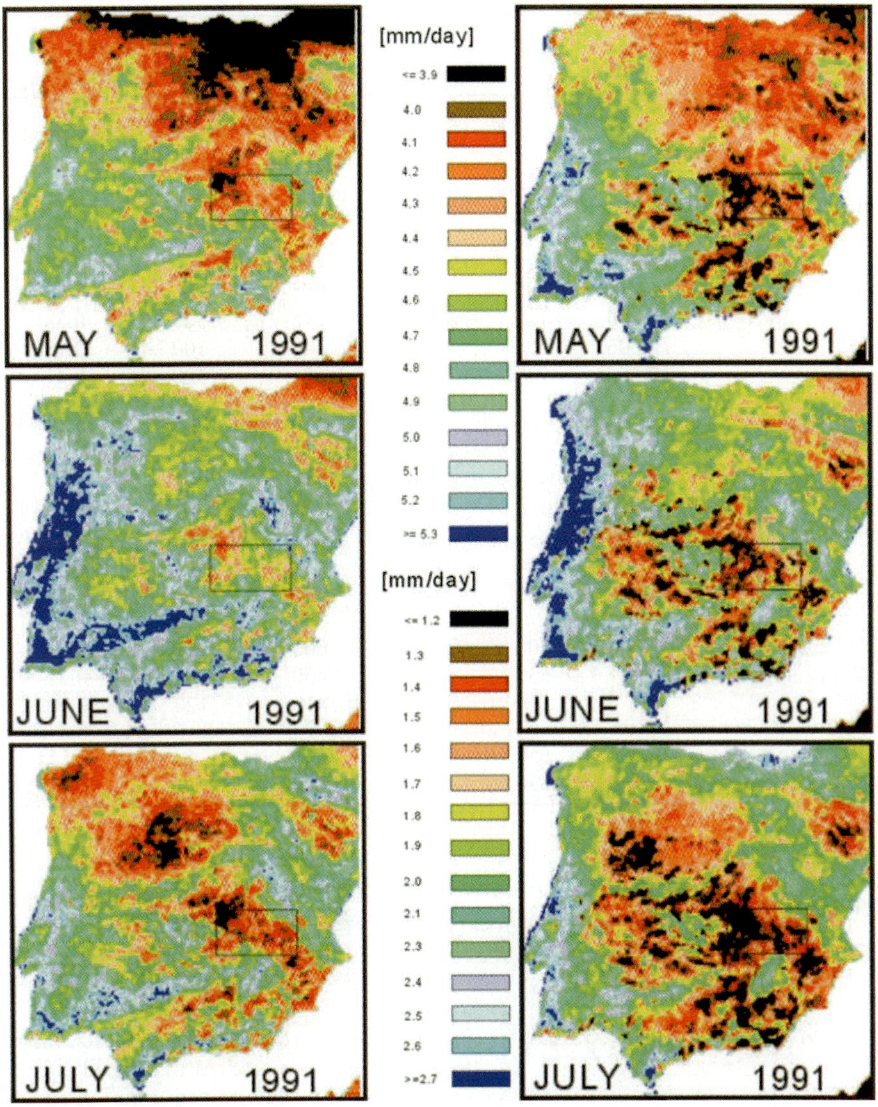

Fig. 5.52. Monthly mean values of the net radiation flux density (left, upper scale) and evaporation (right, lower scale) both expressed in mm evaporation for the months of May, June, and July, 1991 after Peters (1998). The rectangular box indicates the position of the EFEDA experimental area where ground measurements have been performed.

Table 5.26. Diurnal maximum of fluxes measured around noon at the ground and evaporative fraction for diferent locations in the EFEDA experimental area

Type of surface	Date 1991	M_R* (W m^{-2})	M_G (W m^{-2})	M_R*-M_G (W m^{-2})	M_{LH} (W m^{-2})	M_{LH}/ (M_R*-M_G)
Irrigated barley Barrax	11.06	673	100	573	300	0.52
Irrigated maize	11.06.	550	160	390	150	0.31
Barrax	12.06.	740	120	620	460	0.74
	21.06.	590	175	415	150	0.36
	22.06	555	180	375	140	0.37
	23.06.	725	80	645	375	0.58
	28.06.	600	160	440	100	0.15
	29.06.	750	75	675	320	0.47
Dry farming	11.06.	500	170	330	50	0.15
fields	12.06.	460	210	250	50	0.20
Barrax	21.06.	500	175	325	50	0.15
	22.06.	450	175	275	48	0.17
	23.06.	450	160	290	40	0.14
	28.06.	470	150	320	30	0.09
	29.06.	480	130	350	25-30	0.07-0.09
Vineyards	28.06.	700	approx.	600	160	0.27
Tommelloso		550	100	450	100	0.22
		510	100	410	120	0.29
Matorral	11&12	651	125	526	140	0.2151
Rada de Haro	27&28	630	110	520	90	0.1429

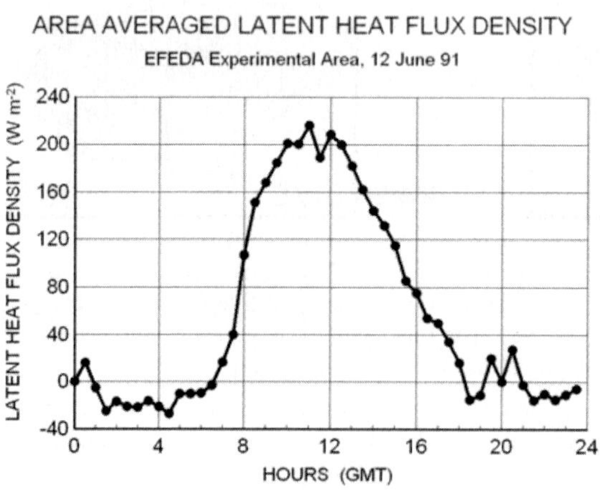

Fig. 5.53. Daily evolution of the area-averaged latent heat flux density for the entire EFEDA grid box determined from in-situ flux measurements combined with the SEBAL-derived frequency distribution of the latent heat flux density and weighting coefficients; June, 12th 1994

certainly is not universal and has to be established for each area and season. Nota bene that the diurnal soil heat flux which is maximum in spring changes sign in autumn.

Pelgrum and Bastiaanssen (1996) reconstructed the area-averaged daily evolution of the latent heat flux density (Fig. 5.53) by using weighting coefficients determined on the basis of the instantaneous spatial patterns computed with SEBAL to combine ground measurements at all available EFEDA sites. In this case the ratio of the diurnal average and the maximum heat flux is 0.269. Thus, if around noon maximum values of 550 W m^{-2} are measured, as was the case just after irrigation, then this would correspond to 5.11 mm evaporation per day. The average values for irrigated areas of Table 5.26 are between 100 and 300 W m^{-2}, say 200 W m^{-2}, corresponding to 1.8 mm. The minimum of measured latent heat fluxes was 25 W m^{-2} resulting in 0.23 mm evaporation. In Fig. 5.46 (right hand side) these values are represented as yellow, light blue, and dark blue to pink colours respectively.

Other SEBAL Products: Resistance to Evaporation

According to Table 5.21 there are several intermediate products until the evaporation emerges from the algorithm. One is the surface resistance to evaporation. The pair of Figs. 5.54 and 5.55 shows this quantity for Tomelloso and Barrax. Tested over these areas the algorithm was applied for the whole EFEDA experimental area (Fig. 5.56) and the cloudless part of the Iberian peninsula (Fig. 5.57). In Fig. 5.56 one can see how the resistance increases from east to west as the vegetation becomes thinner. The blue areas in the east are a water reservoir (embalse) in the north and irrigated fields in the south near Albacete-Barrax. The four figures demonstrate as well the loss of details when the scale is changed.

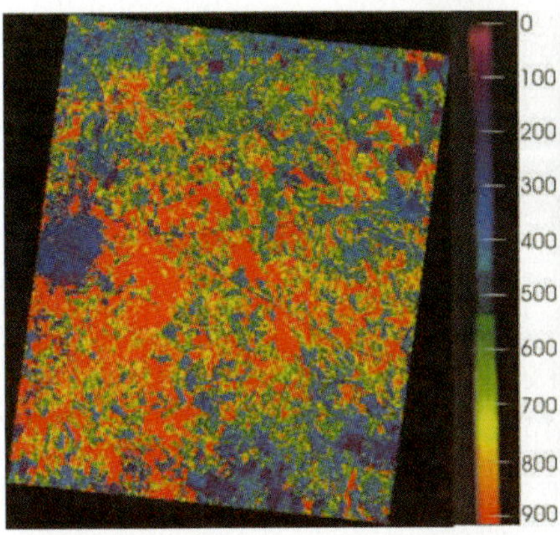

Fig. 5.54. Surface resistance to evaporation for the Tomelloso experimental site in units of s/m derived with the SEBAL algorithm for 29 June 1991

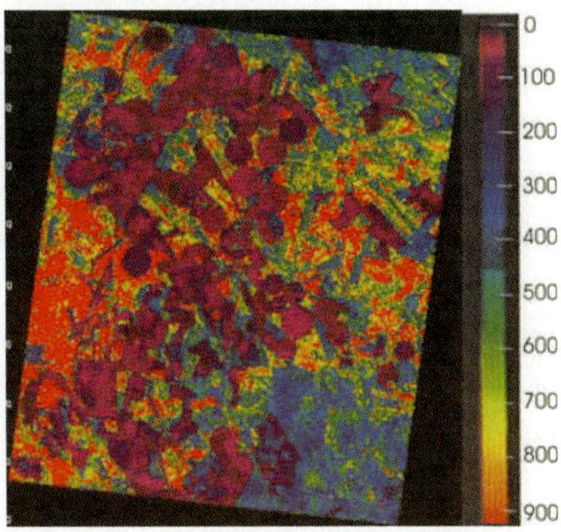

Fig. 5.55. Surface resistance to evaporation for the Barrax experimental site in units of s/m derived with the SEBAL algorithm for 29 June 1991

Fig. 5.56. Surface resistance of evaporation for the EFEDA experimental area (see Fig. 5.44) in units of s/m derived with the SEBAL algorithm for June 1991

Fig. 5.57. Surface resistance of evaporation for the Iberian Peninsula in units of s/m derived with the SEBAL algorithm for 29 June 1991. The scale is the same as in Fig. 5.56

5.6 Hydrological Aspects I: Soil Moisture

5.6.1 Dual-Frequency Approach

Algorithm

Microwave radiation emitted from a complex (vegetation covered) surface can be expressed in terms of the corresponding effective or brightness temperature defined in Eq. 3.44. Accounting for soil surface and vegetation properties, and assuming *equal vegetation and soil temperatures* ($T_s = T_c = T_{sc}$), the effective temperature for polarized radiation, $T_{eff,p}$, can be re-written according to Kirdiashev et al. (1979), Mo et al. (1982), Van de Griend and Owe (1996) as:

$$T_{eff,p} = T_{sc}\left[\left\{1 + (1-\varepsilon_s)\tau_c\right\}(1-\tau)(1-\tilde{\omega}) + \varepsilon_s \tau_c\right] \tag{5.82}$$

Here, τ_c is the transmissivity and $\tilde{\omega}$ the single scattering albedo of the vegetation, ε_s the emissivity of the soil surface, and T_{sc} the thermodynamic temperature weighted over all physical elements of soil (subscript s) and canopy (subscript c) contributing to the microwave radiation received by the antenna. $T_{eff,p}$ stands for the effective (or brightness) temperature at the direction of polarization p.

The *dual frequency approach* uses two frequencies, one at 6.6 GHz for soil moisture ($T_{H(6.6)}$), and one at 37 GHz (i.e. MPDI) to account for vegetation transmittance. It is further assumed that the relationship between the transmittance at 6.6 GHz and MPDI found for Botswana (see Fig. 4.12, Eq. 4.25) holds for Spain as well. Under this assumption Eq. (5.82) can be used to estimate surface moisture.

The assumption of equal soil and foliage temperatures is quite erroneous for daytime observations. For night-time observations, this assumption may be more realistic. During daytime, the difference between canopy and soil temperature may be extremely variable throughout the year and is difficult to correct for on a routine basis. It was found that the daytime effective temperature, averaged over the emitting targets at 6.6 Ghz, including the topsoil, may be up to 12 °C lower than the actual bulk thermodynamic temperature (T_s) of the surface (Van de Griend, 2001). The use of T_{eff}, however, compensates to some extent for the error which results from the assumption of equal foliage and soil temperatures. This, because the daytime foliage temperature will also be substantially lower than the temperature of the soil surface itself.

Solution of Eq. 5.82 for $\tilde{\omega} = 0.09$, and $T_{eff,H} = T_{H(6.6)}$ leads to a time series of the topsoil emissivity ε_s. Surface emissivity is determined by the reflectivity (Kirchoff's Law) which in turn is a function of the soil dielectric properties and surface roughness. Choudhury et al. (1979) developed the empirical roughness model for horizontal polarization

$$\varepsilon_{s,H} = 1 - \rho\, e^{-h \cos^2 \mu} \tag{5.83}$$

where ρ is the soil reflectivity, h is an empirical roughness parameter and μ is the nadir incidence angle. For a flat surface with a uniform vertical soil moisture distribution, the horizontal polarization reflectivity can be calculated from the

Fresnel equation according to

$$R_H = \left| \frac{\cos\mu - \sqrt{\kappa - \sin^2\mu}}{\cos\mu + \sqrt{\kappa - \sin^2\mu}} \right|^2 \quad (5.84)$$

where κ is the complex dielectric constant of the soil with a real part κ' and an imaginary part κ''.

The dielectric properties of a soil are frequency dependent and require a specialized laboratory (such as IROE/CNR in Florence) for their determination. Fig. 5.58 shows the laboratory determined L-band (f = 1.4 GHz) values of κ' and κ'' as a function of soil water content together with a curve fitted through the points. The soil sample (sandy loam with a bulk density of 1.34 g cm^{-3}) was taken from the surroundings of Tomelloso in the centre of the EFEDA experimental area.

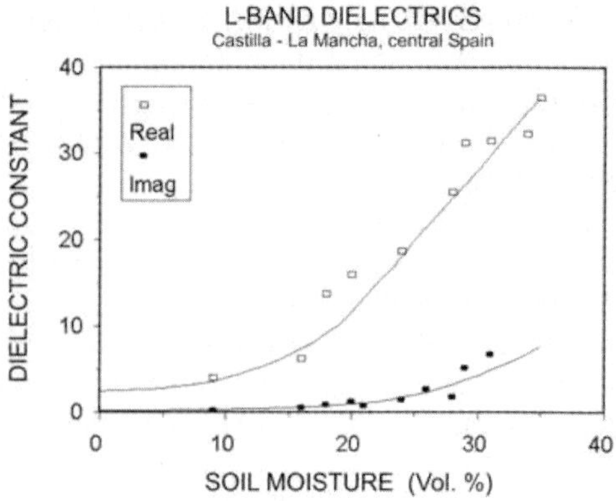

Fig. 5.58 The real and imaginary parts of the dielectric constant of a sandy-loam soil sample taken from the surroundings of Tomelloso (central-Spain), analysed by CNR/IROE in Florence (Italy). For explanation see text

Using the dielectric mixing model by Dobson et al. (1985) to determine differences in dielectric properties between L-band and C-band for a sandy-loam, it was found that the ratio $\kappa'_{(C)}/\kappa'_{(L)} = 0.96$ for a soil moisture content θ of 30% and $\kappa'_{(C)}/\kappa'_{(L)} = 0.93$ for $\theta = 10\%$. For the imaginary part the ratios the following values are found: $\kappa''_{(C)}/\kappa''_{(L)} = 1.17$ for $\theta = 30\%$ and $\kappa''_{(C)}/\kappa''_{(L)} = 1$ for $\theta = 10$ %. These ratios were used to estimate the C-band values of κ' and κ''.

The roughness parameter h is difficult to determine, especially at the scale of SMMR pixels. From a sensitivity study, applied to the Botswana savannas (Van de Griend and Owe 1994b), an optimum was found for zero roughness. Although the surface roughness was measured at field scale in several parts of the EFEDA area, it is completely unknown what the overall roughness might be for the La Mancha area as a whole, with a flat central basin surrounded by hills. Therefore, zero roughness was assumed in simulations for Spain.

Validation

Model calculations were done independently for both daytime and night-time satellite observations. An elaborated example, over an area of 150 km × 150 km centred around the La Mancha area is given in section 6.12. The night- and daytime observations turned out to be quite consistent, realizing *(i)* that daytime and night-time observations are completely independent, *(ii)* that daytime and night-time observations are usually not on the very same day but with a gap of one or more days in between, and *(iii)* that precipitation may have occurred in between the day- and night-time satellite observations.

Absolute calibration and validation of a inversion model, such as the dual-frequency model described above, and its individual parameters (vegetation and surface/soil properties) is impossible without long-term soil moisture data. Therefore, validation was done in a relative sense, by comparing model output with observed time series of precipitation. The use of thermodynamic temperature of the surface to derive microwave emissivity leads, especially for daytime observations, to significant errors as a result of the before mentioned possible difference between the thermodynamic temperature of the soil-surface and the *'effective'* thermodynamic temperature of the microwave emitting soil layer. However, the use of 37 GHz vertical polarization brightness temperature is an effective means to estimate the effective thermodynamic temperature of the emitting soil layer, both for daytime and night-time observations (Owe and Van de Griend 2001).

Notwithstanding the uncertainty of parameters, the soil moisture data determined from space seem to be quite reasonable in the sense that:

- the long-term range over the period 1978-1987 reported in section 6.12 is within reasonable boundaries of (6 % $< \theta <$ 18 %);
- night- and daytime calculations are comparable.

Because of the uncertainty in an absolute sense, the method is especially useful for change detection, which does not require absolute precision but rather long-term consistency. Therefore, the method has been applied for aridification/desertification monitoring over the Iberian Peninsula using all Nimbus/SMMR data gathered during its life-time of 1978-1987. The methodology and results of these studies are described in Chapter 6 section 12.

5.6.2 Estimating Near Surface Soil Moisture with SAR Systems

Introduction

In this section a study is presented on the errors involved in the determination of areal soil water content by means of active microwave remote sensing. Comparisons are made between the use of the polarimetric JPL-AIRSAR (P-, L-, and C-band, polarimetric) and the ERS-1 SAR (C-band, single polarization VV). The simulations and comparisons are limited to C-band (about 5.3 GHz). The heterogeneity of the surface parameters and the speckle are considered to be the source of errors in the soil moisture estimation.

Speckle is that part of the noise within a backscattered signal which is caused by the interference of waves originating from scattering centres illuminated by a coherent source of electromagnetic radiation (for the discussion of speckle see e.g. van Oevelen 2000). Depending on the structure of the scatter centres, speckle can expose a distinct spatial pattern if the centres are arranged geometrically or a "grainy" structure if the centres are randomly distributed. Thus the speckle pattern contains information about structure and geometrical order of the scattering elements. From a homogeneously vegetated surface without speckle a radar image would return which shows uniform pixel values. If each leaf and each branch generates a backscattered wave then all these waves interfere causing randomly distributed brighter and darker picture elements. This speckle noise can be reduced by applying an averaging filter. If the homogeneous surface is large enough this smoothing would result in the expected homogeneous picture. If the surface is structured, e.g. with roads and buildings, this smoothing would also reduce resolution, small objects may vanish.

The influence of vegetation in the detection of soil moisture can be significant and may limit the applicability of the sensors. Through sensitivity analysis of the applied soil moisture retrieval algorithm using the Monte Carlo approach it can be shown what problems may arise and how they may be circumvented. Knowledge of the limitations of soil moisture retrieval by active microwave remote sensing is invaluable in hydrological and climatological modelling since it gives insight into *attainable resolutions* and *accompanying errors*. The theory and applicability of the microwave retrieval algorithms has been described by Van Oevelen and Hoekman (1996).

Method

Microwave backscatter modelling. The description of microwave backscattering is limited to that of *bare soil* surfaces. Various empirical and theoretical approaches have been developed to describe the backscatter from these random rough surfaces e.g. Fung (1994) and Ulaby et al. (1986). Only two of them will be discussed in the following text. The Integral Equation Method (IEM) model of Fung et al. (1992) is a theoretical physically based model that describes the backscatter coefficient (σ^0) as a function of surface roughness, soil dielectric properties (expressed as a complex dielectric constant, $\kappa = \kappa' - i\kappa''$) and incidence angle ($\theta_{inc}$). The IEM model combines three other models which can be looked at as simplifications, the so-called Kirchhoff approximations, that are valid for certain ranges of roughness conditions: (i) The Geometrical Optics (GO) model being valid for 'very rough surfaces', (ii) the Physical Optics (PO) model being valid for 'intermediate rough surfaces', and (iii) the Small Perturbation Model (SPM) being valid for 'smooth surfaces' (Chen and Fung 1988, Oh et al. 1992). Each model has its own range of validity for the surface roughness parameters which can be adequately described for surfaces with normally distributed autocorrelation functions. Surface roughness can be expressed as root mean square (rms) of the height differences ($ks = 2\pi s/\lambda$) relative to the wavelength, the autocorrelation length (kl) relative to the wavelength and the rms slope ($m = 2\ s/l)^{1/2}$ (Borgeaud and Noll, 1993). The wavenumber is defined as $k = 2\pi/\lambda$, with λ as wavelength and s = rms of the surface height fluctuation. The IEM model is applicable to a much wider range of parameter values including those of the component models.

An example of an empirical model is that of Oh, Sarabandi and Ulaby (1992), hereafter referred to as OSU. This model is based upon field measurements over bare soils using an L-, C-, and X-band radar. The model was tested over a large range of soil moisture and surface roughness conditions: $0.09 < \theta < 0.31$, $2.6 < kl < 19.7$ and $0.1 < ks < 6.0$. The model showed also good agreement when tested against data measured for surfaces with parameters outside the above mentioned ranges.

Inverse modeling of radar backscatter. In general, forward modeling of radar backscatter poses a tremendous problem when dealing with an object of a complex nature such as a vegetated surface. The most descriptive theoretical models available require detailed information about the object and are of such nature that an analytical inversion to retrieve parameters of interest such as soil moisture, is virtually impossible. To circumvent these problems one can either simplify the models and their boundary conditions and apply numerical inversion techniques or use empirical approaches linking field data directly to measured radar backscatter or use semi-empirical approaches which combines both previous methods (Dubois et al., 1995, Oh et al. 1992). Here, the first approach is chosen and IEM is inverted to estimate soil moisture for bare soil surfaces. This model has been chosen because it can be applied to single as well as multiple frequency sensors.

The inverse IEM model, hereafter referred to as INVIEM, is based upon the forward solution of the IEM model (Van Oevelen and Hoekman, 1996). A priori knowledge on the surface roughness characteristics (*ks* and *kl*) and the incidence angle can be utilized. The first step is to simulate the radar backscatter for the required frequencies and polarization as a function of soil moisture and incidence angle for a number of distinct roughness states. These simulated data are then tabulated in the form of a look up table (LUT) which shows per frequency and polarization the backscatter coefficient σ^0 as a function of volumetric soil moisture content, θ, for the different roughness states and the various incidence angles. For higher soil water contents θ the relationship $\sigma^0(\theta)$ is less sensitive than at lower θ-values. The relationship of the backscatter coefficient σ^0 with the incidence angle depends strongly upon the surface roughness. Soil mixing models have been used to convert the fraction soil moisture and soil type into a dielectric constant as is required by the IEM model (Dobson et al. 1985, Hallikainen et al. 1985, Wang and Schmugge 1980). Naturally the INVIEM model is valid only where the IEM model is valid of which a discussion can be found in Chapter 6 of Fung (1994). The IEM model has been validated for certain roughness states where the physical optics, geometrical optics and small perturbation approaches are valid (Ulaby et al. 1986). Surfaces periodic roughness characteristics such as plowed fields can be outside the IEM model's validity range depending upon the view angle of the sensor.

The inverted OSU model, hereafter referred to as INVOSU, is an empirical model based upon the knowledge of scattering behaviour in the limiting cases (e.g. large *ks*) using measured backscatter and ground-truth data (Oh et al., 1992). The model needs HH-, VV- and cross polarized radar data along with incidence angle θ_{inc} as input. The model gives the real part of the relative dielectric constant and roughness measure *ks* as output. The mathematical formulation of the model is a non-linear function:

$$\left(2\frac{\theta_{inc}}{\pi}\right)^{\frac{1}{3\rho}}\left[1-\frac{q}{0.23\sqrt{\rho}}\right]+\sqrt{\rho}-1=0 \qquad (5.85)$$

with:

$$p = \frac{\sigma_{hh}^{0}}{\sigma_{vv}^{0}}, \quad q = \frac{\sigma_{hv}^{0}}{\sigma_{vv}^{0}} \tag{5.86}$$

and:

$$\rho = \left(\frac{1 - \sqrt{\varepsilon_r}}{1 + \sqrt{\varepsilon_r}} \right)^2 \tag{5.87}$$

where:

σ_{ij}^{0} = backscatter coefficient with subscripts i and j denoting receiving and sending polarization respectively (dB)

ρ = squared Fresnel coefficient of the surface at nadir (-)

ε_r = real part of relative dielectric constant (-)

θ_{inc} = incidence angle (rad).

The advantage of using INVIEM over INVOSU is that first can be applied on single polarization data as well as polarimetric data. However, for both algorithms polarimetric data is preferable since it simplifies the masking of the vegetation influence and enhances the estimation of soil moisture since more than one polarization can be used. The INVOSU model requires no a-priori knowledge about the surface characteristics accept for the incidence angle and soil type information. As said before, applicability of both models is restricted to bare soils.

Estimation of systematic and random errors. Consider a bare soil field with a field averaged soil moisture content $\langle\theta\rangle$ and field averaged roughness parameters $\langle ks\rangle$ and $\langle kl\rangle$. The field consists of n-pixels, each with a soil moisture content θ_i (i=1,...,n) and roughness parameters, ks_i and kl_i that deviate from the field average values. The distribution of the variables is considered to have a Gaussian distribution with the mean equal to the field averages and a standard deviation determined from field observations. Furthermore each pixel has a incidence angle $(\theta_{inc})_i$ which is not exactly known. The above mentioned variation in surface parameters can be considered as random errors. Another source of errors in radar observations is speckle which statistics are well known (Hoekman 1991). For logarithmically scaled data, i.e. σ^0 in dB, the standard deviation is a constant depending only on the number of looks and the distribution can be approximated by a Gaussian distribution. To simulate as a first attempt the influence of errors and error propagation on the model performance the following method is employed (Bastiaanssen et al. 1995).

Given a random variable X, e.g. the radar backscatter, which is a function of the input parameters x_i , e.g. θ, ks etc., with i = 1,2,...n:

$$X = f(x_1, \quad \ldots, x_i, \quad \ldots, x_n) \tag{5.88}$$

and a set of reference values of x_i, called r_i , e.g. field averages of θ, ks etc., determined from field observations which give as a result the reference value of X hereafter called R:

$$R = f(r_1 \quad \ldots, r_i, \quad \ldots, r_n) \tag{5.89}$$

Thus R is the backscatter coefficient (model value) computed with the field averaged values of θ, *ks* etc.. The input parameters are assumed to have a random error that is normally distributed with the reference value as a mean and the standard deviation set at a percentage of the mean. Using sets of the randomly distributed input parameters to simulate errors in model input parameters, model output X can be generated from which the probability density function or distribution of X can be generated. A large number of realizations of X is necessary to retrieve meaningful statistics. One of the important conditions is that the number of realizations must be large enough for the statistics to become constant. The above described simulation method is generally known as the Monte-Carlo method.

One is usually not interested in the distribution of X but in the distribution of the deviation S_e from the reference value. The error S_e is defined as:

$$S_e = |X - R| l \tag{5.90}$$

The cumulative probability density function (pdf) or distribution of S_e can thus be calculated.

In this study it is assumed that the reference value is the true value for X. However, the reference value for the backscattering coefficient σ° is determined using the reference input parameters, which were considered to be the true values. For the soil moisture retrieval algorithm the areal or field averaged value used to generate the backscattering coefficients is used as the reference value. This implies that only the true value of X is retrieved when a perfect model is used.

Two different situations have been considered. The first situation is the estimation of the soil moisture on a pixel by pixel basis for both pixels with or without speckle (referred to as pixel-based). The second situation is the estimation of soil moisture on a field or object basis using field averaged backscatter values also for fields (number of pixels per field is varied to a maximum of 10^4) with or without speckle (referred to as field-based).

Sar Estimates of Soil Moisture Content During Efeda-spain and Hapex-sahel

Relationship between the electromagnetic variables and surface parameters. In section 3.2.4 the backscattering coefficients have been defined and examples have been given in section 4.5 how polarization, frequency dependence, and phase shift can be used to classify the surface. For most applications the aim is to determine the water content of the vegetation canopy and/or the soil. The step towards these quantitative information required to establish a relationship between the electromagnetic variables and surface parameters.

The optical thickness for the H component can be written as

$$\delta_H = \frac{2\pi\rho}{k} \operatorname{Im} < f_{HH} >, \tag{5.91}$$

where k is the propagation constant, ρ the density of the scattering particles, and $<f_{HH}>$ the complex ensemble average of the forward scattering amplitude of an individual scatterer. Im stands for its imaginary part. It has been shown that in the low frequency quasi static domain the scattering amplitude is proportional to the square of the propagation constant and is also a function of the scatter size, orientation, and dielectric constant. If $\overline{F}_{HH}$ is used for the normalized value of the averaged scattering amplitude it follows after Schmugge and Jackson (1992) and Saatchi et al. (1992):

$$\delta_H = 2\pi k \rho \overline{F}_{HH} = k B_H W, \tag{5.92}$$

where B_H is a frequency independent factor described and measured by Schmugge and Jackson (1992) and W is the vegetation water content. It could also be shown by Saatchi et al. (1992) that the volume scattering term can be approximated by

$$\gamma_v^{HV} = k^{\frac{3}{2}} A_{HV} W \sin\theta \tag{5.93}$$

where A_{HV} is an approximately constant frequency independent factor.

EFEDA – Spain. The INVOSU model is only applied on the JPL-AIRSAR data from Spain, since it needs multipolarized data. In Table 5.27 the results are shown using the model of Wang and Schmugge (1980) to invert the dielectric constant to soil moisture values. Comparison with ground measurements shows that estimates from C-band are in general too high and that there is little variation in the estimates. Since there are only a few measurements at larger depths, comparison with L- and P-band estimates is difficult. However, the L-band estimates seem to be close to the C-band estimates and also show little variation over the various fields. P-band estimates show clearly more variation and higher or wetter estimates.

HAPEX – Sahel. The Normalized Difference Vegetation Index (NDVI) data derived from SPOT imagery is used to mask out the pixels having too much vegetation. For that a NDVI threshold of 0.1 above which the vegetation influence is supposed to be present seemed appropriate. The INVIEM model is than applied on the remaining ERS-1 data which is not masked by the filter. The soil moisture was measured over a depth of 6 cm which is regarded an appropriate depth for comparisons with L-band emission (Wang 1987). Although, measurement ranging depths from 2 to 5 cm are preferable (Chanzy et al. 1995). For comparison with shorter wavelengths such as C-band, the measurement depth of 6 cm is probably larger than the with this wavelength associated sensing depth. In a climate as that of Niger, with short intensive rainfall periods followed by warm and sunny periods, the soil moisture profile dries out towards the surface. The C-band soil moisture estimates are therefore probably lower than the measured values. The results in Table 5.28 confirm this for all the plots and for the whole range of soil moisture estimates.

Only a few measurements are available for comparison but there seems to be a considerable difference between the results from INVOSU and INVIEM. Both model results are not significantly outside the range of measured soil moisture, an exception is P-band. However, the INVIEM data (Fig. 5.59) appear to capture the variance in the field better and show a difference in estimates between the various bands although the ranges do overlap.

If the soil moisture estimates using both models are visualized, the difference

Table 5.27. The AIRSAR C-, L, and P-band INVOSU results for nine fields near Barrax, Spain during the EFEDA experiment 1991. Fields 2 and 5 correspond to the in situ measured bare soil and maize fields. The "ci" indicates a part or slice out of a pivot irrigation circle. The inverse model of Wang and Schmugge (1980) is used to retrieve the volumetric soil water content θ_v from the real part of the dielectric constant ε_r.

Site	No. of pixel	C - Band		L - Band		P - Band	
		ε_r	θ_v	ε_r	θ_v	ε_r	θ_v
1	250	4.67	0.16	4.28	0.156	4.58	0.159
2	122	4.36	0.157	4.52	0.159	17.99	0.318
5	76	4.36	0.16	4.11	0.154	15.9	0.296
8	18	4.96	0.163	8.09	0.2	34.37	0.453
9	91	4.47	0.158	4	0.153	5.41	0.169
10ci	699	4.74	0.161	4.01	0.153	4.07	0.153
11ci	377	4.88	0.162	5.39	0.168	5.91	0.174
12ci	374	4.86	0.162	5.14	0.165	5.09	0.165
13ci	522	4.9	0.163	4.35	0.157	0.421	0.155

Fig. 5.28. INVIEM θ_v estimates (est) using C-band VV polarization of ERS-1 (12 September, 1992) for a number of sub-transects in Niger compared with the averaged TDR measurements over 6 cm depth (me). Given are the lowest, mean, and highest values of both measurements and estimates. For sub-transect 7 no measurements were available because the soil surface was too hard to be penetrated by the TDR rods. For sub-transect 4 no estimates are available because of too much vegetation

Field No.	3		4		5		6		7		8		9		10		11	
est/me	est	me	est	me	est	me	est	me	est	me	est	me	est	me	est	me	est	me
m_v(low)	2.1	7.8	-	5.6	2	6.4	2.6	6.4	5.1	-	1	4.9	0	5.5	1.5	7.8	1	6.2
m_v(med)	6.8	13	-	9.4	6.4	9.3	8.2	12	11	-	4	11	3	9.9	5.7	11	5	16
m_v(high)	12	24		13	11	13	14	16	16	-	8	16	6	14	10	13	10	29

becomes even more apparent. The INVIEM estimates (Fig. 5.59) show a clear spatial variation in soil moisture content due to the irrigation practices. The same irrigation circles can be observed in Fig. 5.60 (INVOSU model) but the soil moisture variation within these circles is not as apparent. Overall, both models give reasonable estimates of soil moisture content as far as can be deduced from the limited ground truth data and visual inspection.

Discussion. The mean value and the standard deviation of model parameters considered as significant error sources were estimated from field data (Van Oevelen et al. 1993). For both ERS-1 and AIRSAR three fields with different roughness situations were chosen, each field considered representative of a certain roughness case. The simulation for 10.000 pixels for these fields or roughness cases will give

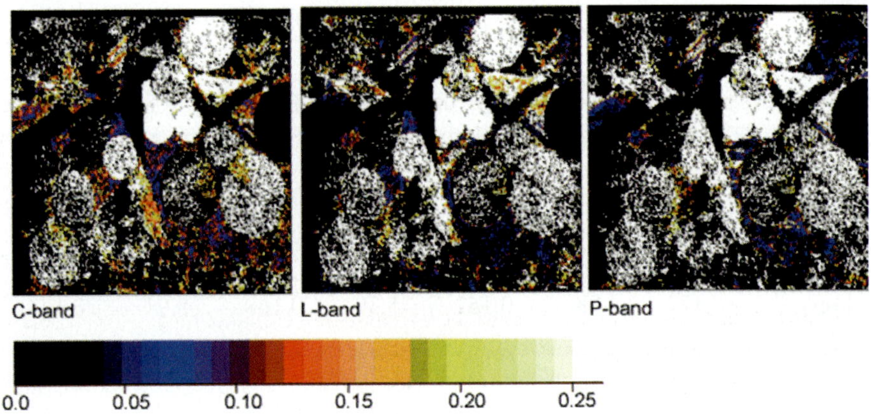

Fig. 5.59. Multiband medium INVIEM soil moisture estimates for Barrax, Spain. From left to right: C-, L-, and P-band

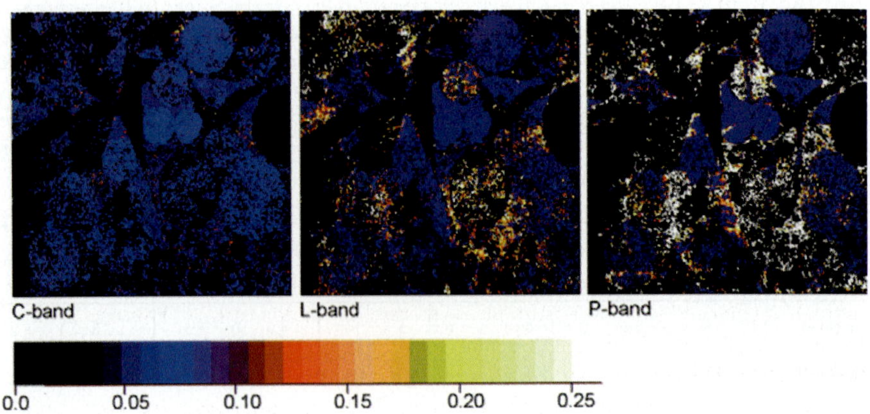

Fig. 5.60. Multiband soil moisture estimates using the medium INVOSU estimates in Barax, Spain, for C-, L-, and P-band

a distribution of the backscatter coefficient σ° for each field or roughness case. INVIEM is applied on these backscatter coefficients to yield a distribution of soil moisture estimates. The speckle is determined from 6 bare soil fields both in Spain for AIRSAR data and in Niger for ERS-1 data and compared with theoretical values from Hoekman (1991). For ERS-1 the standard deviation is about 2.5 dB (3 looks) and for the AIRSAR it is about 1.33 dB (16 looks). Hoekman (1991) gives for 3- and 16-look data, a standard deviation of 2.73 dB and 1.10 dB respectively. Thus the experimental data are in good agreement with the theoretical values.

For the soil moisture content estimates θ_v the results are listed in Table 5.29. Clearly, the average standard errors are in most cases much larger for the individual pixel soil moisture estimates from Niger compared to those of Spain. However, if looked at the field estimates the error decreases significantly even lower than those of the field estimates of Spain. The most likely explanation are the higher standard deviations for the surface parameters for Niger, which give large deviations from the true mean backscatter coefficient for a low number of pixels and thus introduce an error when inverted. The standard error in estimating the soil moisture content for a individual pixel ranged for the Niger data set from 10% to 15% for pixel-based estimates and from 1% to 3% for field-based estimates (Table 5.29). For Spain the errors ranged from 5% to 10% for pixel-based estimates and from 3 to 8% for field-based estimates.

Currently the INVIEM model uses, when available, both VV- and HH-polarizations to obtain soil moisture estimates. Usually, the ranges of soil moisture estimates from VV and HH overlap. In that case the minimum and maximum of the overlap are taken as the boundaries of the soil moisture estimate range. The problem arises when there is no overlap, in that case the inversion is not possible. This has been the case for field 3 for which for a large number of pixels no inversion was possible.

After averaging over a certain number of pixels the average soil moisture estimate becomes almost constant. Each individual estimate, however, can deviate strongly from the reference or true value. When for all separate surface roughness cases or field the estimates are averaged the result generally becomes closer to the reference value (Table 5.29). In the case where all values are averaged over a number of pixels larger then 25 there is only a slight deviation of the soil moisture estimates from the reference values (Table 5.30). For the field-based estimates there is almost no difference between the cases with speckle and without. For the pixel-based estimates even after averaging over a large number of pixels there is a constant difference between the cases with and without speckle.

The result of the error analysis provides a basis to assess the quantitative value of the spatial information provided by detailed remote sensing data. The advantage of remote sensing data, especially if at high resolution is that a number of samples (i.e. pixels) can be used to obtain estimates with smaller random errors. This can be done by averaging over a limited number of pixels, at the price of degrading the spatial resolution. However, to avoid mixing measurements relating to different physical objects, this averaging can be performed over *stationary heterogeneous* targets only.

The large variation of surface roughness makes it a difficult parameter to describe at different resolutions. The roughness characteristics can vary so much within a

Table 5.29. The standard error S_e on soil moisture content θ_v estimation using C-band and three different roughness cases. Given is a comparison between pixel based estimates and field based estimates, i.e. using the average backscatter over all the pixels ($n=10^4$). For case 3 in Spain, the field average backscatter coefficients did not give a solution. For this case the pixel based estimation yielded only a result in 60% of the pixels. In all the other inversions the outfall was less then 15%.

Case	Spain		Niger	
	Pixel	Field	Pixel	Field
1 with speckle	0.09	0.08	0.13	0.03
1 without speckle	0.1	0.08	0.13	0.03
2 with speckle	0.04	0.03	0.15	0.01
2 without speckle	0.05	0.03	0.1	0.01
3 with speckle	0.05	–	0.1	0.03
3 without speckle	0.05	–	0.12	0.03

Table 5.30. The standard error of the soil moisture estimates as function of the number of pixels for a single realisation over which the ERS-1 and JPL-AIRSAR backscatter coefficients are averaged

No. of Pixels	Niger				Spain			
	Case 1		Case 2		Case 1		Case 2	
	Speckle		Speckle		Speckle		Speckle	
	No	Yes	No	Yes	No	Yes	No	Yes
2	0.05	0.09	0.03	0.07	0.11	0	0.01	0
10	0.05	0.02	0.01	0.02	0.1	0.04	0.03	0.05
25	0.03	0.01	0.02	0.01	0.07	0.06	0.03	0.03
50	0.03	0.01	0.01	0.01	0.08	0.07	0.03	0.03
100	0.03	0.02	0.02	0	0.08	0.06	0.03	0.03
250	0.03	0.03	0.01	0	0.08	0.07	0.03	0.04
500	0.03	0.03	0	0	0.08	0.08	0.03	0.04
1000	0.04	0.03	0	0.01	0.08	0.07	0.03	0.03

might be inadequate for a good description. Agricultural fields which are regularly treated with practices such as tilling and plowing have surface roughness characteristics that are difficult to describe. Most models that describe the surface backscatter, assume a certain distribution for the roughness characteristics. This distribution has to be adapted for surfaces such as the agricultural fields as mentioned above.

Conclusions

The approach used in this study is to analyse the effect of spatial heterogeneity on remotely sensed soil moisture estimates. This heterogeneity is treated as a random error source on the backscatter measurements. Here only a specific case is treated, namly the use of INVIEM for given surface object parameters (e.g. surface roughness) and sensor parameters (e.g. frequency, polarization). Generalization of the results is therefore extremely difficult. To extend the validity of this type of error analysis a more general and systematic approach is needed. From this study it is clear that soil moisture estimates using radar data based upon a single pixel have to be treated with caution. The presence of speckle, inherent to radar data, makes this problem more complicated but does not pose a real problem when averaged over a small number of pixels. The error in individual pixel estimates can become unacceptably high. However, averaging over a small number of pixels will already reduce the error significantly.

The soil moisture estimates using INVIEM require knowledge of the roughness characteristics of the surface. So far the ground measurements available of these characteristics have been used. Future developments will include generalization of the INVIEM inversion algorithm in order to estimate surface roughness, or surface soil moisture or both depending upon information available. Furthermore, the model currently uses only HH- and VV-polarizations; inclusion of the multi-polarization data or employing polarization ratios can improve the performance of the model. To extend INVIEM to vegetated surfaces vegetation correction algorithms have to be developed

5.7 Hydrological Aspects II: Precipitation

Since the 1960s several techniques were developed to estimate surface rainfall based on visible and infrared imagery collected by geostationary satellite instruments (Lovejoy et al. 1979, Barett et al. 1986, Adler et al. 1988, Goodman at al. 1994. The relevance of IR imagery is demonstrated by the evidence that intense thunderstorms are characterized by very cold cloud tops and rapid changes in time and space of the cloud structures (Vicente et al. 1996). In more recent work (Hsu et al. 1997), artificial neural networks were employed for off-line and adaptive calibration of satellite estimates (i.e., calibration of the rainfall estimates "on the fly" with the available updated rain gauge and radar observations). In the frame of this technique, the specific environmental conditions were not considered and good performances were obtained only activating the adaptive option, especially in the mid-latitudes. If no adaptive algorithms are used (either due to hard live data access, sparse or insufficient rain gauge or radar data), the assimilation of environmental conditions is needed.

Based upon recent research carried on at the Atmospheric Research and Applications Division (ARAD) at NOAA/NESDIS an automatic technique was developed for the Mediterranean. It performs the automatic reconstruction of rainfall fields over large areas based on thermal infrared geostationary satellite (METEOSAT-7) data and atmospheric limited area model fields. The spatial resolution is 0.2° longitude by 0.2° latitude and the data are accumulated on hourly to daily basis. The core of the technique is, that not only cloud top brightness temperatures are considered but also their time evolution and the cloud top structure are evaluated by means of the hessian/gradient analysis. Information drawn from meteorological model analysis and short term forecasts concern environmental (low to middle troposphere) moisture content. This was shown to be a critical element in the NOAA/NESDIS technique called "Auto-Estimator".

Two artificial neural network algorithms were designed to (i) identify the "rainy" pixels and (ii) quantitatively evaluate the rainfall within rainy pixel areas for the desired time period (30 minutes). The input data for both networks (as for the "Auto-Estimator" technique) were derived from the analysis of satellte imagery and meteorological models. Both neural networks obtain their input data from three satellite scenes. It is necessary to screen all situations the network may face. Most of the rainfall events (i.e. half an hour with rainfall over a specific pixel) occurs with cloudy conditions in all three satellite scenes. Cases with clear sky at least in the last two satellite scenes had zero rainfall in the last 30 minutes. Other combinations involved a small number of cases. Simple linear or exponential regressions were used to derive the probability of rainfall occurrence (screening of "rainy pixels") and of the quantitative rainfall estimates on the basis of the useful satellite and meteorological model data.

The two neural networks employed in the development of the new methodology were designed separately both for the best performance on the independent test data sets and the stability. The neural networks belong to the multi-layer perceptron (MLP) global neural network class (Lawrence et al. 1996) and uses the back-propagation tool. This is an extremely effective learning tool that can be applied to a wide variety of problems. Back-propagation related paradigms require supervised training. This means, they must be taught using a set of training data where known solutions are supplied. Their performance was evaluated against independent data sets in a broad range of events. The analysis of the potential predictors for rainfall has been performed by means of the comparison against spatialised rain gauge data in Tuscany, Italy. The technique is now routinely running and the results are available on Internet (www.lamma.rete.toscana.it).

Meteosat infrared scenes, meteorological model and ground rainfall data were collected during more than twenty storm events that occurred in 1997 and 1998. The Meteosat images were resampled to the choosen resolution by means of an averaging procedure. Geo-referencing assured accurate correspondence between satellite and ground-based data (the errors remain less than the original Meteosat infrared pixel size).

Time series of brightness temperatures and cloud top structures (convexity, concavity, or saddle) were produced for each series of three consecutive Meteosat IR scenes on a pixel-by-pixel basis. Time series of simultaneously derived meteorological model data were produced, namely the product of precipitable water (PW) from the surface to 500 hPa layer and the average relative humidity (RH) in the same layer, hereafter called the PWRH factor. In addition, time series of 30 minutes cumulated rainfall were collected from the rain gauges and averaged over the selected pixel areas, by means of a simple arithmetic procedure. The whole data set consists of 1460 records of satellite infrared, meteorological model and raingauge interpolated data.

The exploratory data analysis aimed at the identification of clear relationships between cloud top features and ground rainfall. The relative frequency of rain occurrence was compared on a pixel-by-pixel basis to cloud top temperatures in the last Meteosat IR scene at the end of the 30 minutes rain period. It was also compared to the cloud top growth rate from the first to the second scene, just before the start of the rainy period. The same comparison was performed for the PWRH factor. Both PW and RH are computed from short time forecast of atmospheric fields from a limited area model (DALAM) nested into the global circulation mode of the European Centre for Medium-range Weather Forecast (ECMWF) and running over southern Europe and the Mediterranean basin (Buzzi 1994, Perrini et al. 1995). The atmospheric fields are available at a time resolution of three hours and the quantities PW and RH are interpolated in time to get the mean value for the time interval between the first and the last of the three Meteosat scenes (one hour). It revealed a larger relative frequency of rainy events in the lowest cloud top temperature range

Table 5.31. Performance indices for the rainfall estimation technique. Symbols mean: POD, the Probability Of Detection of rainy pixels; FAR, the False Alarm Ratio in the screening of rainy pixels; COR, the linear correlation for the rainfall in 30 minutes over actually rainy pixels; RMSE, the Root Mean Square Error for the rainfall in 30 minutes over actually rainy pixels; BIAS, the bias for the rainfall in 30 minutes over actually rainy pixels. The number of test cases was: 450 for the "screening" network, 200 for the quant*i*tative network.

Statistics	POD	FAR	COR	RMSE (mm)	BIAS (mm)
Screening of rainy pixel	0.65 ± 0.04	0.46 ± 0.06			
Rainfall estimate over rainy pixel			0.28 ± 0.01	1.2 ± 0.2	0.02 ± 0.17

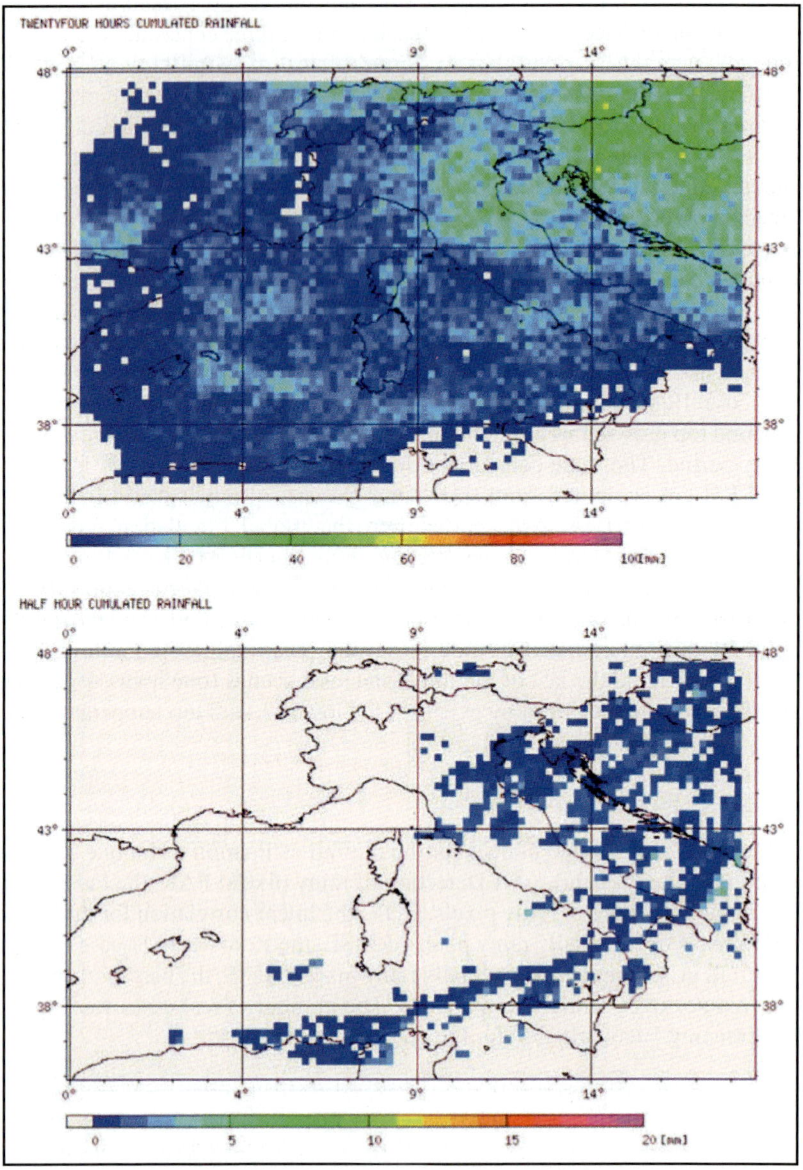

Fig. 5.61. Automatic estimated rainfall maps of 11 November 1998 processed by Neural Rain Estimator developed by LaMMA (Laboratory for Meteorology and Environmantal Modeling. Latitude - longitude geographical reference system with 0.2° × 0.2° spatial resolution. Cumulations over twenty four hours (top) and 30 minutes (bottom).

(< 210 K), in the lowest cloud top growth range, i.e. in conditions of overshooting cloud tops, and in the highest range of the PWRH values (2 mm%) in agreement with the results by Vincente et al. (1966).

The best performance and stability of the network trained over the binary data set (rain / no rain) were obtained with the configuration set at one hidden layer and three hidden nodes. The best performance and stability of the network trained over the rainfall amounts computed for the pixels where actually rainfall occurred were obtained with the configuration set at two hidden layers, the first one with five hidden nodes and the second one with three hidden nodes.

The performances of the two neural networks were evaluated by means of several statistical indices (Table 5.31): the Probability Of Detection (POD) and the False Alarm Ratio (FAR) for the screening of rainy pixels, the linear correlation (COR), the Root Mean Square Error (RMSE) and the bias (BIAS) for the estimation of 30minutes rainfall over actually rainy pixels.

The ability of the neural network approach to identify heavy rainfall situations was tested too. With the rainfall threshold fixed at 1 mm in half an hour the correlation coefficient over the five runs grows to 0.36±0.2, the RMSE lowers to 0.7±0.4 mm and the bias slightly increases in absolute value to –1.5±0.3 mm.

Once the quantitative network was checked to be sufficiently stable with regard to whatever test set was employed, it was re-calibrated over the whole data set to allow the most complete representation of the whole range of conditions (especially heavy rain) and likely provide somewhat more accurate results than the networks employed in the cross-validation phase.

Rainfall maps are now routinely produced and concern Italy and the central-western Mediterranean. An example of a half-hour cumulated rainfall map and a twenty-four hours cumulated rainfall map is shown in Fig. 5.61.

5.8 Indicators of Vegetation Conditions

5.8.1 Photosynthetic Active Radiation, Vegetation Indices and LAI

Definitions

Biomass is produced in the presence of water by absorption of *Photosynthetic Active Radiation (PAR)*. The biological efficiency of the shortwave solar radiation has been shown in Fig. 1.22. The *Absorbed* Photosynthetic Active Radiation (*APAR*) is computed as the difference between the incoming $PAR^{\downarrow}$ and the radiation reflected by the vegetation-soil system, $PAR^{\uparrow}$, diminished by that fraction of the radiative energy which is absorbed at the soil. This latter fraction is the difference between the radiation flux transmitted through the canopy, $PAR_t^{\downarrow}$, and the radiation reflected by the underlying surface, $PAR_s^{\uparrow}$ (= $\rho_s PAR_t^{\downarrow}$). With this notation *APAR* can be expressed by:

$$APAR = PAR^{\downarrow} - PAR^{\uparrow} - PAR_t^{\downarrow}(1-\rho_s) \tag{5.94}$$

The *Fractional* Absorbed Photosynthetic Active Radiation (f_{APAR}) characterizes the photosynthetic efficiency of the vegetation canopy

$$f_{APAR} = APAR / PAR^{\downarrow} \tag{5.95}$$

To obtain the *Daily Total* Fractional Absorbed Photosynthetic Active Radiation (DTf_{APAR}) the instantaneous values have to be integrated from sunrise t_0 to sunset t_n:

$$DTf_{APAR} = \int_{t_0}^{t_n} APAR\, dt \Bigg/ \int_{t_0}^{t_n} PAR^{\downarrow}\, dt \tag{5.96}$$

The filter function of the AVHRR channel 1 covers about half of the *PAR* fraction of the incoming solar radiation, ranging from 0.40 to 0.70 µm while Landsat-TM offers three channels between 0.45 and 0.7 µm. The second of these channels coincides with the dip in the chlorophyll absorption spectrum which causes the small reflection maximum around 0.55 µm.

The main factors affecting interception, scattering and emission of radiation at the vegetated surface are leaf optical properties, leaf size and shape, leaf location in the three-dimensional space (vertical and horizontal distribution of leaves and degree of 'clumping'), leaf orientation in three-dimensional space (leaf inclination and azimuth angle distributions), reflectance of the undergrowth vegetation, soil or other background, geometry of illumination and view, and atmospheric conditions (Myneni and Ross 1991, Danson 1995, Walter-Shea and Bieffl 1990).

The perhaps most commonly used canopy structure parameter is the *Leaf Area Index* (*LAI*) which is frequently considered as useful characteristic of vegetation for quantitative analyses of interest for ecology (Botkin and Running 1984). It is one of the main input parameters for Surface-Vegetation-Atmophere-Transport (SVAT) models or for crop yield models and is thus related to the biomass production by photosynthesis during the growing period. This is discussed in more detail in section 6.2. The leaf area index is the sum of the one sided area of the leaves covering one unit ground area. Other definitions, which are sometimes used like double sided leaf area or needle surface, have explicitly to be defined.

Leaf Area - Vegetation Index Relationship

Many efforts went into the determination of relationships between *LAI* and quantities that can be observed from space. Candidates are vegetation indices which measure the "greenness" of the vegetation. Spectral bands of reflected radiation that have most value in designing vegetation indices are wavelength in the red and near-infrared part of the spectrum. Reflected red radiation is negatively correlated with chlorophyll concentration and thus to green leaf area. Reflected near-infrared radiation is positively correlated with the amount of multiple scattering at the interfaces between cells and the air, and therefore also to leaf area (Knipling 1970). Vegetation indices computed from these two bands have been related to green *LAI* (Price 1992, Wiegand et al. 1990) and either through this relationship or directly to other vegetation canopy properties (Fig. 5.62, see also section 6.2), canopy biomass (Anderson and Hanson 1992), absorbed photosynthetically active radiation

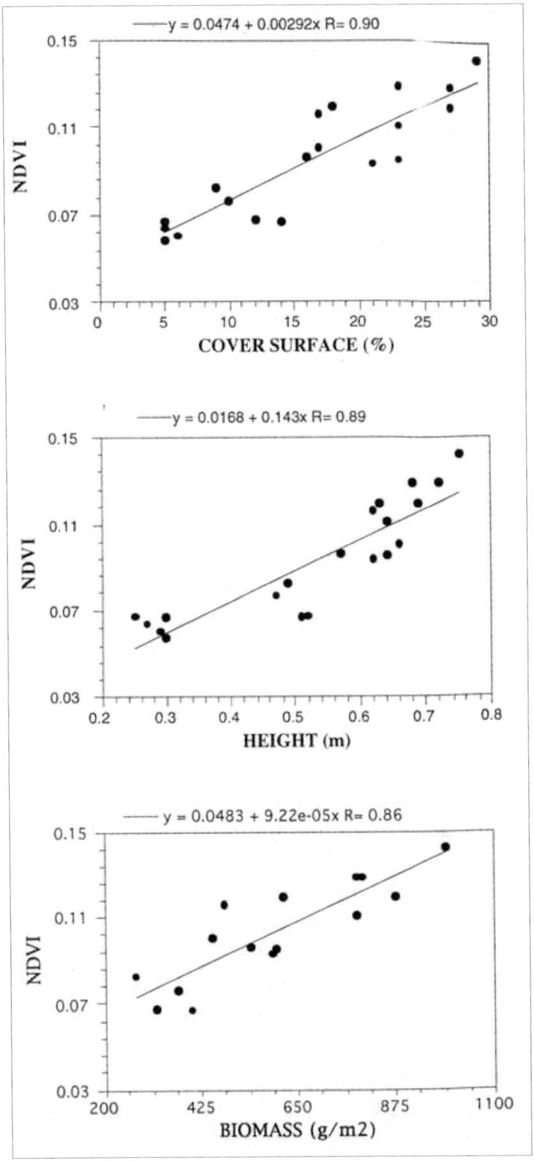

Fig. 5.62. Relationship between NDVI computed from Landsat-TM data and physiological parameters of grapevine (Airen) measured by the University of Castilla - La Mancha from May to July in 1991, 1994, and 1996 near Tomelloso, Castilla - La Mancha, Spain, after Lanjeri (1996). Correlation between *NDVI* and *LAI* from1996 (not shown here) was *NDVI* = 0.021 + 0.044 *LAI*, R = 0.92

(Daughtry et al. 1992), grain yield (Ashcroft et al. 1990) and total nitrogen content (Ashcroft et al. 1990, Jensen et al. 1990).

The relationship between canopy *LAI* and remotely sensed data has been examined at a range of spatial and temporal scales. Baret and Guyot (1991) proposed the following generic relationship between a vegetation index (*VI*) and the canopy *LAI*:

$$VI = VI_{\infty} + (VI_s - VI_{\infty})\, e^{-k\, LAI} = a + b \exp(-c\ LAI) \tag{5.97}$$

where VI_{∞} is the asymptotic vegetation index value reached when *LAI* tends towards infinity, VI_s is the vegetation index value for bare soil, and k is an attenuation coefficient. Once this relationship is established (and the coefficients are known), the canopy LAI can be estimated by inverting Eq. (5.97), resulting in

$$LAI = -\frac{1}{k} \ln\left(\frac{VI - VI_{\infty}}{VI_s - VI_{\infty}}\right) \tag{5.98}$$

Although some authors have proposed a set of coefficients to estimate *LAI* from Eq. 5.98 for different applications (Hill 1996), it is noteworthy that they are strongly dependent on the soil background and crop type (Kanemasu et al. 1990). Therefore, they can not be extrapolated or adapted and should be determined for each specific study.

The measurement of the vegetation index is mostly based on AVHRR measurements because they offer one observation per day within a time window defined by the variation of the revolution time of 102 minutes and long-term trends due to orbital drift. Without the knowledge of the diurnal cloudiness, till now only derivable from low resolution data of the geostationary satellites, it is not possible to compute the diurnally integrated quantities.

In semiarid regions and during the start and end of the vegetation period the instantaneous field of view of the AVHRR is only partly filled by vegetation cover. For instance grape vine cultivation in the Castilla-La Mancha (Spain) at maximum plant development shows only a fractional vegetation cover (shielding factor) between 13% and 18 % (Fig. 5.63). Thus the observable signal is strongly affected by the radiometric properties of the underlying and surrounding soil (Bausch 1993).

It is obvious, that these indices which stand for the instantaneous ability to produce biomass can only be used in combination with *PAR* to estimate the actual biomass production. Under the assumption that the mean daily variation of the PAR over a period of a decade can be treated as constant, a single sample of the vegetation index, e.g. the NDVI (see section 4.6), obtained from a NOAA overpass can be linearly related to the daily integral. This was verified by observations that show very high correlations between the DTf_{APAR}, defined in Eq. 5.96, and NDVI (e.g. Goward and Huemmrich 1992). But the investigations also show that the NDVIquickly runs into saturation for $LAI > 2$ depending on the fractional vegetation cover.

Some other variables such as background spectral properties or illumination conditions may be considered as "confounding factors" (Baret, 1995), whose influence is minimised when using vegetation indices instead of reflectance values in single bands to estimate the canopy leaf area index.

Furthermore the total biomass production or the crop yield of a growing cycle can not be related linearly to the NDVI, because the production rate of biomass is increasing till the maximum green vegetation development, followed by a plateau phase with nearly constant rate and the final maturing phase with decreasing rates. The errors in estimating biomass production by NDVI measurement with coarse time resolution will be highly dependent on the relationship of growing conditions (water availability, temperature) and biomass development. Coniferous forest will show a completely different development than sugar beet, wheat or deciduous forests. If a linear relationship with a high correlation coefficient can be found, it is only an indicator that the inter-annual variability for this type of vegetation is low. In either case the numbers that have to be compared are the values integrated over the whole growing cycle.

Fig. 5.63. View from an observation tower at an experimental site near Tomelloso, Castilla - La Mancha, Spain, showing the relatively thin vegetation cover and the absolutely "clean" soil background. The picture was taken mid June and is thus not representing the final development state of the grapevine.

"Red Edge" Shift and Chlorophyll Content

Considerable improvements and new opportunities to find characteristic spectral features correlated to the crop status may be expected from the extension of the spectral resolution down to bandwidths of a few nanometers. New techniques such as derivative spectroscopy, an established method in analytical chemistry, can be used to define spectral parameters which correlate with biophysical data. In this context, the 'red edge' is important, i.e. the sharp change in vegetation reflectance between approximately 650 and 800 μm, where the vegetation reflectance changes from very low in the chlorophyll red absorption wavelength region to very high in the *NIR*. The relation between leaf chlorophyll content and the wavelength of maximum slope on the red edge is a well known application of derivative spectroscopy. Horler et al. (1983) defined a red-edge parameter, λ_{re}, as the wavelength of maximum slope and found that it depends on chlorophyll concentration. Red-edge shifts associated with phenological crop development were demonstrated by sampling flag leaves of winter wheat and spring barley on several occasions. As chlorophyll concentration increases, λ_{re} moves to longer wavelengths, and then reverts as senescence begin.

The applicability of derivative techniques to obtain the λ_{re} and some other parameters from the reflectance spectrum is well documented by Baret et al. (1992), Demetriades-Shah et al. (1990), Gernmell and Colls (1993), Li et al. (1993), and Miller et al. (1991). A general result of these studies is the high sensitivity of such spectral parameters as slopes or wavelengths of inflection points to leaf area index and chlorophyll content. By all these reasons, the parameters associated to the red edge can be considered also as vegetation index.

5.8.2 Field Radiometry and Data Processing

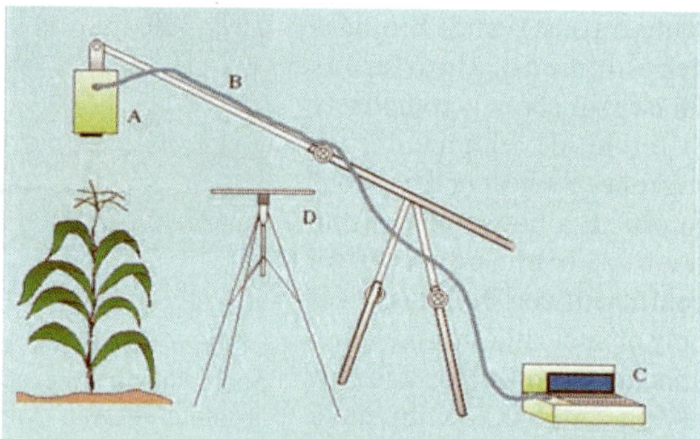

Fig. 5.64. Experimental design for field spectral radiometry. The optical head of the GER SIRIS spectroradiometer (A) is mounted on a boom (B) which eventually is carried by a pick-up truck. The instrument is operated from a PC (C) and the reference panel is mounted at a tripod at approximately equal level with the canopy

The reflectance of natural surfaces is fully specified by its "bidirectional distribution function" (Nicodemus et al. 1977). This quantity can not or at least not perfectly be measured directly. It is more usual in remote sensing to measure the "bidirectional reflectance factor" (Silva, 1978). As defined in Chapter 3 this is the ratio of the radiance reflected from a target and from a reference panel into a specific direction (ϑ_r, φ_r) whereas both target and reference are identically irradiated out of a direction (ϑ_i, φ_i). Since the irradiance is the same, the reflectance factor can be expressed by the ratio of the radiances measured for target, dL (ϑ_r, φ_r), and reference, $dL^*(\vartheta_r, \varphi_r)$, which should be an ideal, completely and perfectly diffuse reflecting "Lambertian"surface:

$$\rho(\vartheta_i, \varphi_i; \vartheta_r, \varphi_r) = \kappa(\vartheta_i, \varphi_i; \vartheta_r, \varphi_r) \frac{dL(\vartheta_r, \varphi_r)}{dL^*(\vartheta_r, \varphi_r)}. \qquad (5.99)$$

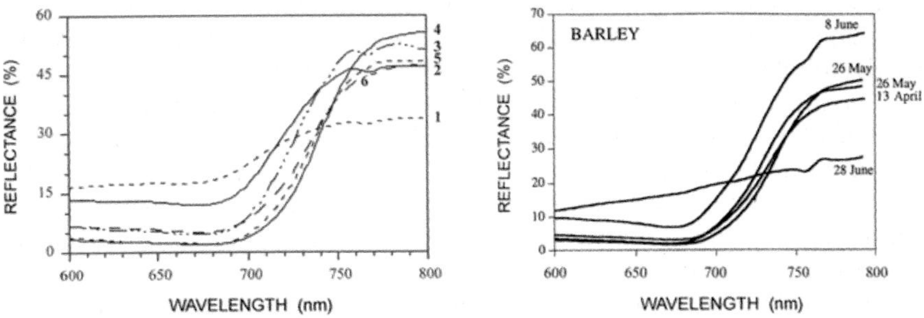

Fig. 5.65. Maize (left) and barley (right) canopy reflectance spectra between 600 and 800 nm for all available dates between 13 April and 15 September 1992. 1 - 19 June, 2 - 1 July, 3 - 14 July, 4 - 6 August, 5 - 15 August, 6 - 15 September

κ is the absolute calibration factor of the reference panel that generally is not strictly Lambertian.

Spectral radiance measurements from 400 to 2500 nm generally are acquired at approximately 2-weeks intervals from middle of April till the middle of September in case of the vegetation period of barley respectively maize. The experimental about 1 m vertically on top of the canopy on an adjustable frame and measurements are made along rows in which the plants are planted. A layout of the design is schematically shown in Fig. 5.64.

In the experiments described below, a Spectralon-2 panel was used as diffuse reference surface. Reference panel measurements were collected immediately before the radiance measurements from the targets. Each pair of measurements took about four minutes. Field radiometry measurements are also affected by errors due to non-simultaneous sampling of target and reference radiance. To minimize effects due to changing irradiance the measurements were taken within 1 hour either side of solar noon at cloudless days exclusively. To ensure comparability among data recorded on different dates, the radiance data were acquired for a longer period at two dates (1 July and the 6 August, representing different vegetation density and development stages) to capture a wider range of possible solar zenith

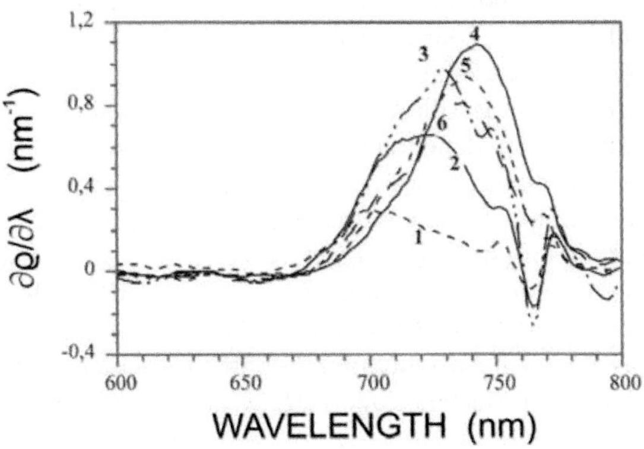

Fig. 5.66. Derivatives of the canopy reflectance spectra of maize between 600 and 800 nm wavelength for the analyzed dates of the year 1992: 19 June (1), 1 July (2), 14 July (3), 6 August (4), 15 August (5), and 15 September (5)

angles. The reflectance of the bare soil was also measured, in dry conditions. Fig. 5.65 shows the canopy reflectance spectra in the 600-800 mn domain for all the dates analyzed in the study.

Two different treatments were applied to the reflectance spectra. For the broad band analysis, filters reproducing the relative spectral response of the Thematic Mapper bands (Markham and Barker 1985) have been applied to each spectrum, in order to reproduce the reflectance in TM3 (red) and TM4 (near infrared) bands. These values have been used to calculate the normalized difference vegetation index, NDVI, and the soil adjusted vegetation index, SAVI (see chapter 4).

For the high resolution analysis, the wavelength corresponding to the red edge was determined throughout the phenological cycle of the corn. The reflectance spectra were first smoothed with respect to random noise. Then, the first derivative curves were calculated (Fig, 5.66). Although the smoothing technique could affect the maxima of the derivative spectra, Rollin and Milton (1994) showed that the level of smoothing does not result in a significant shift in the wavelength position of the peaks in the visible and in the near-infrared regions, probably due to the high spectral resolution of the raw data in this wavelength region. The spectral zones between 600 and 800 mn in the derivative spectra were analyzed in terms of the wavelength corresponding to the slope maximum (Horler et al. 1983), λ_{re}, and the value of the derivative at this point, $(d\rho/d\lambda)_{re}$.

5.8.3 A Case Study on Corn and Barley

Experimental Site and Ground Measurements

Ground measurements have been performed at the EFEDA experimental site in Barrax, Spain, where 60% of the area is occupied by dry farming, mainly cereals such as winter wheat, and corn occupies about 60% of the remaining irrigated area. Field measurements have been carried out in a corn canopy irrigated with a centre pivot sprinkler. The corn (Zea mays L.) was planted in 70cm wide eastwest rows on 5 June 1992 and thinned out to 7.6 plants/m^2. The crop reached the 10-leaf stage on 20 July and the silk stage[2] about 20 August. The end of the physiological grain ripening was reached in the first week of October.

The canopy was grown on calcisols, which are the most common soils in the area since the calcic and/or petrocalcic horizons are always at depths of no more than 125 cm. This soil, with a loam texture, had a high calcium carbonate content (between 25 and 30% at the surface and between 40 and 65% in the calcic horizon) and a level of organic matter content around 1.5%, which decreased at a larger depths (according to the Munsell classification the soil had a colour code of 5YR 5/4).

Canopy reflectance and *LAI* measurements were performed on a regular basis

---

[2] In the phenology of maize the following phases/stages can be distinguished: tips of stigmata visible, stigmata fully emerged, stigmata drying and stigmata completely dry. Silk stage refers to the first phase (tips of stigmata visible) when silk threads appear to form grains.

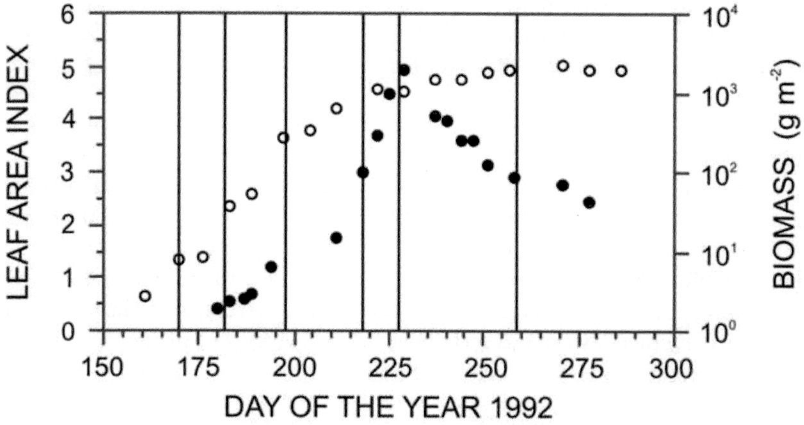

Fig. 5.67. Measured LAI (dots) and biomass (circles) of corn versus the day of the year. Vertical lines indicate the days when Landsat-TM data were available

Table 5.32. Date and phenological stage of the corn when reflectance measurements were acquired

Date	Day of the year	Phenological Stage
19. Juni	170	Four leaves out
1. Juli	182	Six leaves out
14. Juli	195	Eight leaves out
6. August	218	Fourteen leaves out
15. August	227	Sixteen leaves out (spike tip showing)
15. September	258	Pasty grain

from June to October during the 1992 growing season. *LAI* measurements were acquired approximately every seven days throughout the growing season using a LI-COR LAI 2000 Plant Canopy 2 instrument. This instrument estimates the *LAI* through the radiation transmittance values obtained by rationing the radiation on the top and on the bottom of the canopy (LI-COR 1992). To obtain a representative value for the *LAI* in each date, different transects were carried out on the canopy, each one scanning the row and the furrow (in three different places). Afterwards the mean value was calculated. *LAI* measurements with this instrument were slightly below *LAI* data obtained in the laboratory with an area meter (Model LI-30002, LI-

COR), with a coefficient of determination 0.92 (Martin de Santa Olalla, personal communication). The performance of this instrument is discussed by Brenner et al. (1995), who compared indirect estimates of *LAI* with directly measures of individual bushes.

Fig. 5.67 shows *LAI* values (closed symbols) versus the day of the year. Biomass values (open symbols) are also represented to better illustrate the crop phenology. The figure shows that *LAI* reaches a maximum around the middle of August, corresponding to maximum photosynthetic activity. After this period *LAI* decreases as due to the senescence of the leaves. Standard deviations of *LAI* values are compressed between 5-15%, reaching the lowest value around the end of July and the beginning of August. Vertical lines in the graph represent the dates where spectral measurements were taken (Gilabert et al. 1996). The phenological stage of the crop at these dates is described in Table 5.32.

Results of Broad Band Analysis

As a preliminary analysis (Gilabert et al., 1996), spectral data corresponding to the two longer series (1st July and 6th August) were plotted as a function of the solar zenith angle, in order to study the dispersion among the reflectance values due to the spatial variation (along a row) and to the temporal variation (time of day when the measurements were taken). As a result, it was found that: (i) For a given solar zenith angle, dispersion among measurements relates to the gaps in the canopy, which are more marked in July (when the corn plants were small) than in August (when canopy was almost fully developed). Since heterogeneity is intrinsic in the observed scene, dispersion appears in all the spectral bands and vegetation indices. (ii) The diurnal variation in reflectance observed at different solar angles within the same day shows a decrease of reflectance in the red and near-infrared regions near solar noon (Gilabert and Meliá 1993). This behavior, observed in both dates, is partly removed by the vegetation indices which present an almost flat distribution of values. Table 5.33 shows averaged NDVI and SAVI values for each measurement date. For the last one, an optimized value for a constant *L* for each *LAI* value has been used as suggested by Bausch (1993) in a study carried out in a corn canopy.

The evolution of the vegetation index values as a function of the day (N) of the year can be approximated by a quadratic equation (Eq. 5.100 and 5.101), which shows the increase of the values until they reach a maximum (around the middle of

Table 5.33. Mean value and standard deviation (in parenthesis) of NDVI and SAVI values measured 1992 for each date

Date	19 June	1 July	14 July	6 August	15 August	15 September
NDVI	0.30 (0.11)	0.57 (0.14)	0.80 (0.06)	0.92 (0.03)	0.93 (0.04)	0.80 (0.04)
SAVI	0.23 (0.08)	0.45 (0.11)	0.68 (0.07)	0.84 (0.03)	0.79 (0.03)	0.72 (0.02)

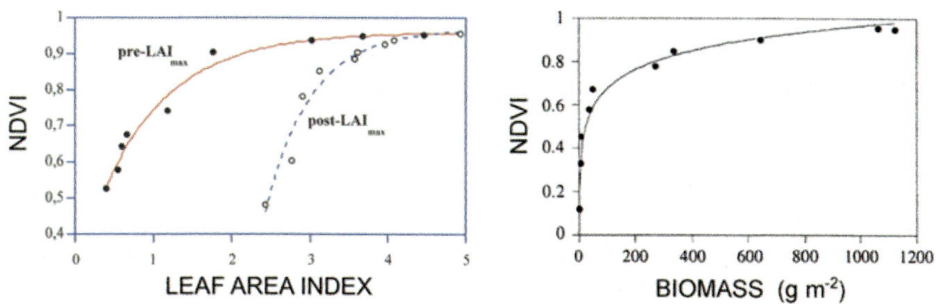

Fig. 5.68. NDVI values versus LAI and biomass of maize for the 1992 vegetation period. Data which correspond to the development and period of maximum activity of the plants are represented by closed symbols (pre-LAI_{max} data), whereas the others, corresponding to the senescence of vegetation, are shown as open circles (post-LAI_{max} data). In this case, the values resulting from extrapolation (beyond day 258) are not represented, although all of them habe been used in the data fit. The curves resulting from the adjustment of the values are shown as continuous and dotted lines respectively).

August) and a subsequent decrease in the senescence stage of the crop:

$$NDVI = -8.80 + 8.57*10^{-2}\, N - 1.88*10^{-4}\, N^2 \tag{5.100}$$

$$SAVI = -8.08 + 7.77*10^{-2}\, N - 1.69*10^{-4}\, N^2 \tag{5.101}$$

In both cases $r^2 = 0.98$.

Eq. 5.97 was used to estimate the *NDVI* values which would correspond to the dates when *LAI* measurements were available. Fig. 5.68 (l.h.s.) shows these values versus measured *LAI* data. Closed symbols represent values which correspond to the development and maximum photosynthetic activity. Open symbols correspond to the senescence phase of the vegetation. For this second stage , the *NDVI* values are more affected by errors since only spectral measurements on 15 August and 15 September were available to fit the *NDVI* as a function of the day of the year. Extrapolated data beyond the last date of observations are not shown.

The observed asymptotic behavior showing that the relation is of most value for small amounts of vegetation was also reported e.g. by Baret et al. (1989), Sellers (1989), and Wiegand et al. (1991).

The coefficients resulting for Eq. 5.97 are: a = 0.960, b = −0.673, and c = 1.13 with r^2= 0.98 and r^2 = 0.96 for the first and the second phase, respectively. Extrapolation to LAI = 0 results in a virtual soil vegetation index of 0.29. Direct measurement of the soil reflectance spectrum suggest rather a soil NDVI of 0.14. If this value is used in the numerical approximation, b becomes −0.82. The respective values for barley are a = 0.96, b = − 0.85, and c = 0.96. The main problem with the relationship between NDVI and LAI is the lack of sensitivity to LAI greater than 3 because of saturation.

A similar study was carried out 1994 on irrigated barley and maize. Fig. 5.69 shows measured NDVI values for each canopy along the day of year and the curves obtained from a quadratic law fit the data. The data of 1994 confirm the development

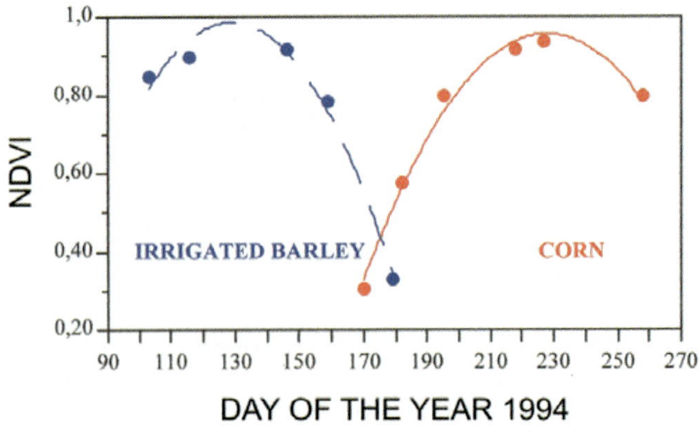

Fig. 5.69. Development of NDVI measured (points) and calculated (lines) from quadratic law f nd synthetic Landsat-TM channel 3 and 4 reflectances for barley and maize in the 1994 experiment

of the NDVI throughout the respective vegetation periods as measured in 1992. The data for barley in this case show a larger scatter because of the lack of measurements around the maximum of the photosynthetic activity of this crop.

A similar procedure as described above for the corn was applied to obtain the relationship between *NVDI* and *LAI* for the barley canopy. In this case, the soil background correction mentioned earlier was not applied, since the barley canopy is more homogeneous and the sensor field-of-view allows to obtain reflectance measurements representative of the whole canopy.

As shown in Fig. 5.70, the *NDVI* - *LAI* relationships for both canopies are rather similar, although the slope for the corn is more pronounced. Towards the end of the phenological cycles, both canopies tend to reach a similar *NDVI* value.

Table 5.34 summarizes the coefficients obtained for *NDVI*/*LAI* relationships for both canopies, as following Eq. 5.97. As the growth periods for both canopies do not overlap, as seen in Fig. 5.69, the sets of coefficients could be used to assess *LAI* according to Eq. 5.98. Further research is required to establish the coefficient values for satellite images since reflectance variations are observed when studying the canopies at lower spatial resolution.

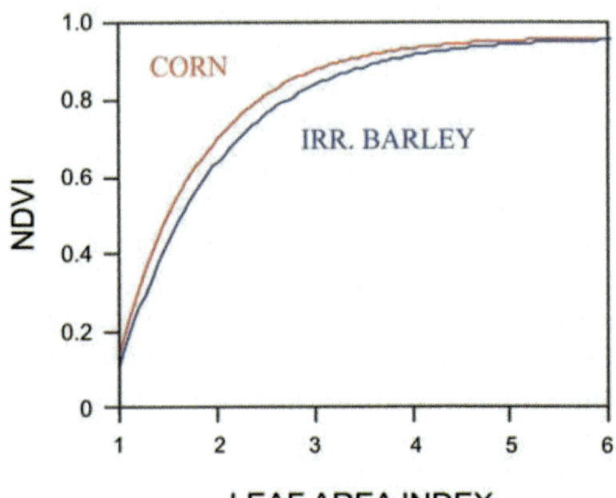

Fig. 5.70. *NDVI* and *LAI* values for corn and barley canopies corresponding to the pre-maximum *LAI* period

Table 5.34. Summary of the coefficients of the adjustments *NVDI* versus *LAI* (see Eq. 5.97) for the corn and barley canopies (Barrax site), corresponding to the pre-maximum LAI period: $NDVI = NDVI_{\infty} + \left(NDVI_{s} - NDVI_{\infty} \right) e^{-c*LAI}$

Coefficient	$NDVI_{\infty}$	$NDVI_s$ - $NDVI_{\infty}$	k
Corn	0.96	-0.82	1.13
Barley	0.96	-0.85	0.96

Results of the High Spectral Resolution Analysis

The red edge is not correctly defined for gaps of the canopy where the soil contribution is dominant. To investigate this parameter for the whole vegetation period therefore an average of all spectra obtained at each date shown in Fig. 5.65 has been calculated. Fig. 5.66 shows the derivatives of the reflectances of the maize spectra. The red edge analysis was performed in the spectral domain 600 - 800 by determining the wavelength corresponding to the maximum derivative, λ_{re}, and the value of the derivative at this point, $(\partial\rho/\partial\lambda)_{re}$. This latter parameter turned out to be highly correlated with the broad band vegetation index ($r = 0.98$). Table 5.35 shows the λ_{re}, values obtained for each date from the derivatives of the corresponding spectra.

As in the case of the vegetation indices the relationship between the wavelength of maximum slope and the day of the year can be expressed by a quadratic equation:

$$\lambda_{re} = 76.5 + 5.69\, N - 1.21{*}10^{-2}\, N^{2}, \quad r^{2} = 0.9 \qquad (102)$$

The quadratic equation (Eq. 5.102) fitted to the temporal evolution of the maize culture allows to determine by interpolation the wavelength of the red edge at the dates when biophysical measurements are available. In the investigated case λ_{re}, reaches its maximum about one week later than the *NDVI*.

An analysis similar to that described in the broad band analysis section was carried out for the correlations between the wavelength of the red edge, λ_{re}, and *LAI* values (Fig. 5.71).

Table 5.35. The wavelength of the maximum first derivative, λ_{re}, values obtained for each date in 1992

Date	38156	38168	38181	6 Aug.	15 Aug.	15 Sept.
λ_{re} (nm)	691	714	719	744	739	736

In Fig. 5.71, λ_{re} moves to longer wavelengths as *LAI* increases and reverses its direction as senescence begin. λ_{re} also tends to a plateau. This temporary saturation, which disappears with the subsequent blue-shift of the maximum inflection, remains for a longer period of time than for the *NDVI*. λ_{re} seems to present more inertia to return to pre-maximum than *LAI* values. Therefore, hysteresis is less evident in this case than it was for *NDVI*. The mathematical coefficients obtained for the pre-maximum *LAI* fit, according to model shown in Equ. 5.97, are: $a = 742$, $b = -52.0$ and $c = 0.99$, with a coefficient of determination of $r^2 = 0.98$ and, for the second stage, $r^2 = 0.94$.

5.8.4 Comments

Both the *NDVI* and the red-edge position, λ_{re}, were found to be useful to describe phenological events for an irrigated corn canopy, since they are both significantly correlated with the leave area index. All relationships analysed in the study resulted in coefficients of determination (r^2) which ranged between 0.94 and 0.98.

It is noteworthy that reflectance measurements were centred over the row of corn and did not include inter-row spacings. Therefore, experimental spectral indices of the development of corn are not quite as representative for the whole canopy as they should be because the contribution to the signal of some soil background was excluded. Nevertheless, LAI was estimated by averaging data samples along a row representative of the canopy.

If one considers the exponential relationship between *NDVI* and *LAI* (Equ. 5.97), the value of b = −0.673 implies a *NDVI* value for soil background of 0.29. Although reflectance of the soil background is difficult to specify for a crop under field conditions because of periodic irrigation, rainfall, etc., it was found experimentally that a representative *NDVI* value for the soil background was 0.14 (corresponding to dry conditions). If the hypotheses is imposed that: (i) this *NDVI* value is appropriate during the growing season, and (ii) the extinction-related coefficient c (in Equ. 5.97) is the same for the row and for the whole canopy, one could obtain a general functional relationship between *NDVI* and *LAI* for the corn canopy. It would correspond to an exponential form as shown in Eq. 5.97 with the following coefficients: $a = 0.96$ (the same as before since this saturation value is practically unaffected by soil background), $b = -0.82$ (obtained taking into account the measured value for soil background *NDVI*), and $c = 1.13$. Coefficients a and b are about the same order of magnitude as those shown by Wiegand et al. (1990) and by Daughtry et al. (1992). Wiegand et al. (1990), who combined data from four locations, found that $a = 0.834$, $b = -0.816$ and $c = 0.860$ for the pre-maximum *LAI* growth period. Daughtry et al. (1992) derived equations for the fraction of absorbed *PAR* as a function of *LAI* and as a function of *NDVI*, also for a corn canopy. From those two equations, the dependence of *NDVI* with *LAI* can be estimated by an exponential law with coefficients: $a = 0.942$, $b = -0.778$ and $c = 0.463$.

Fig.5.72 shows this "predicted" curve, which corresponds to the canopy, and the "measured" curve obtained by scanning over a corn row. *NDVI* values for soil background (*LAI* = 0) have been also included: 0.14 for the estimated curve and 0.29

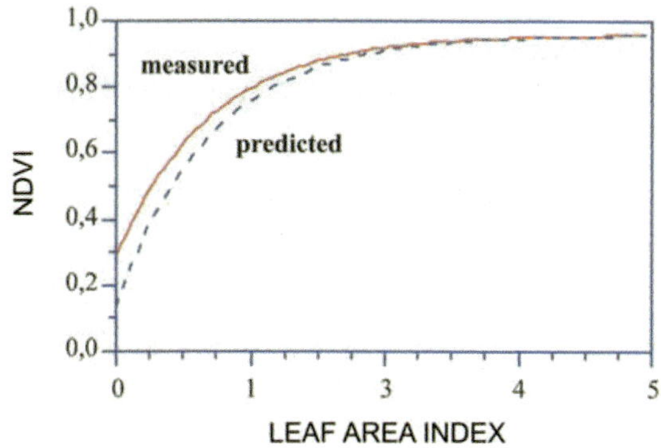

Fig. 5.72. NDVI versus *LAI* values for the first period of the phenological cycle of corn (pre-LAI_{max} data) corresponding to the measurements obtained for a row of the canopy (continuous line) and the estimated data for the whole canopy (dotted line) by using the true soil background *NDVI* (see comments in the text).

for the experimental case. As expected, the difference between the two curves diminishes as *LAI* increases since soil influence diminishes with increasing vegetative cover. The relationship between *NDVI* and *LAI* obtained by following this procedure could be recommended for general use in remote sensing when studying corn canopies.

Since the leaf area index is closely related to other plant physiological quantities the NDVI can as well be related to vegetation coverage, plant height or biomass as shown in Fig. 5.62 for grapevine. These relationships are canopy dependent and have to be determined for each of the investigated vegetation types. As was shown in Fig. 5.67, the *NDVI-LAI* relationships for corn and barley are similar, but the LAI-biomass/coverage/height relationships are not the same.

All figures in this section show higher *NDVI* values as one would obtain with NOAA-AVHRR data. The reason is that these values are computed from the narrower and more specific spectral bands of Landsdat-TM for which the effect of the atmosphere is almost zero.

The obtained results suggest that the leaf area index (*LAI*) can be assessed from *NDVI* by inverting the exponential equation (Eq. 5.97). This is an easier way than the determination of λ_{re}. Furthermore it can be concluded from a comparison of Fig. 5.65 and 5.68 that λ_{re} does not have any significative advantage with respect to the determination of the *NDVI*.

5.8.5 Canopy Water Content Derived from Optical Data

Algorithm

Advanced instruments build for satellites, like MERIS, with high spectral resolution allow to infer additional indicators for vegetation conditions such as the canopy water content. The estimation of this quantity requires high spectral resolution measurements because the liquid water absorption bands in the near infrared have to be distinguished from the partly overlapping atmospheric water vapour bands (Fig. 5.73). During EFEDA, such measurements have been made from aircraft. To make use of this new hyperspectral technology - which means remote sensing in many narrow spectral channels such as provided by a spectrometer like shown in Fig. 5.65 - requires new or modified methods of analysis. Algorithms used for this purpose which are based on radiative transfer require:

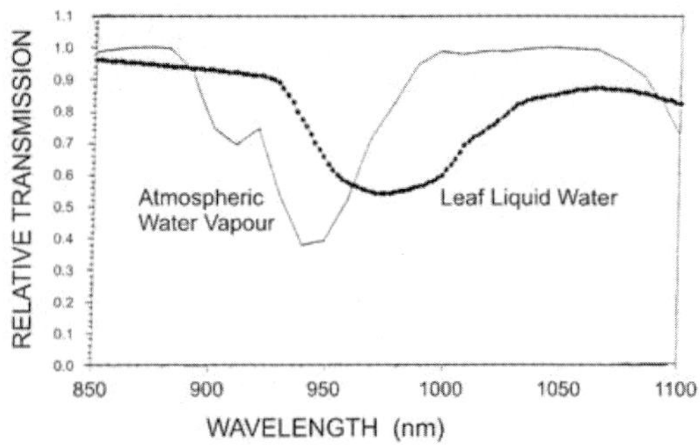

Fig. 5.73. Superposition of the atmospheric water vapour and canopy water absorption bands near 970 nm wavelength

- Adequate spectral resolution to really see the absorption bands.
- Accurate calibration to allow inversion of physically-based models.
- Spectral stability, since spectral shifts are critical for accurate retrievals.

A full algorithm was developed using model inversion techniques (Moreno and Green, 1996) to retrieve information about both canopy water content and atmospheric water vapour content from directional optical hyper-spectral radiometric measurements. Special attention was given not only to the derivation of the model itself but as well to proper inversion techniques which must be robust with respect to noise in the data and inadequacies of the model. The problem of non-linear effects has been addressed in detail, as it has been demonstrated to be a major source of error in the numerical values retrieved by more simple, linear-based approaches. Non-linear effects are especially critical for algorithms based on the use of data within absorption bands, but a proper model can handle such non-linear effects which is especially important over vegetated surfaces.

As the model is intended for application mostly over vegetated surfaces, more emphasis is put on the modelling of vegetation components than on soil variability. The reflectance model has been developed by combining some independent modules: a model for the spectral reflectance and transmittance of leaves, a model for the reflectance of the soil background and a model for the canopy structure consisting of leaves over this background. For the reflectance and transmittance of the leaves, an adaptation of the 'prospect' model (Jacquemoud and Baret 1990) is used. The main advantage of this parametrization is given by the fact that only three parameters (leaf specific biomass, leaf chlorophyll concentration and leaf liquid water content) determine the spectral reflectance and transmittance of the leaves over the full range from 0.4 to 2.5 μm with reasonable accuracy, as an optimum compromise between accuracy requirements and minimum parametrization. For the reflectance of bare soil, the model starts from the same assumptions as the 'soilspect' model (Jacquemoud et al., 1992); that is, separability between macroscopic morphological structure of the soil (giving angular dependencies), assumed to be wavelength-independent, and the microscopic optical properties (single scattering albedo), assumed to be wavelength-dependent. Both effects are parameterized in a different way, and the phase function and single scattering albedo are also modelled in a more general way, by properly taking into account particle size distributions as a function of wavelength. In any case, the difficulties in modelling angular and, specially, spectral behaviour of bare soil reflectances in a general way are well known and a pragmatic modelling with simple assumptions is all that can be expected for realistic approaches. Trying to cover the most general situations as good as possible, the canopy model developed in this case uses 8 parameters: LAI, ground vegetation cover, canopy height, two parameters determining leaf distribution and three more canopy structural parameters. Some structural and hot-spot effects are only included in first-order scattering contributions, while multiple scattering contributions are calculated by using a discrete-ordinates code (applied to a simplified canopy model to save computation time, as architectural effects have less for multiple-scattering contributions). To account for the effects of direct/diffuse irradiance ratio and to model directional irradiance, the surface reflectance model is coupled with an atmospheric model, which is actually a modification of a part of the 6S code (see section 3.4.6. The atmospheric model provides the irradiance field over the scene as a function of wavelength, as the illumination conditions are highly dependent on wavelength.

Once the surface and atmospheric models are coupled, measured radiances can be inverted to fit the model and to give the full set of required surface and atmospheric parameters to explain such radiance values. The model runs with a spectral resolution of 2.5 nm and full bidirectional effects are considered for each single 2.5 nm channel. Finally reflectance is calculated for each 2.5 nm channel. The AVIRIS bands are simulated by using a Gaussian filter for each band, with the specifications for the AVIRIS data being used in each case. The final accuracy depends essentially on the accuracy by which the central band positions are known. For the AVIRIS data available over the EFEDA Castilla-La Mancha area the uncertainty in band centre positions was so high, that a fine-tuning by allowing an additional degree of freedom in the position of the channels (by assuming a constant shift across each spectrometer) was required.

After a theoretical model is available, the second step is the development of an appropriate inversion technique in order to retrieve information from the measured data (Jacquemoud 1993). Inversion technique is a critical issue (compare section 6.8). Three main aspects have to be considered. The first one is that most probably the model will not be able to fit the data perfectly, and then some degree of freedom must be allowed in the inversion method. In this case, a multi-resolution constrained method is used to isolate pixels to which the model does not apply and to obtain more robust estimated from those pixels where the model applies by means of such multi-resolution estimates. Constrained linear mixtures of model components are used when no convergence is obtained. The second aspect to be taken into account is the problem of noise in the data. In the case of AVIRIS, two types of noise have to be taken into account: the spatial noise and the problem of knowing the exact spectral position of each channel. The spectral stability being a critical issue, the inversion technique must allow for some kind of fine-tuning in the centre positions of channels (or equivalent re-calibration in the measured radiances) so that the model can fit properly the measured data. The third aspect to be considered is computational efficiency. Because of the large number of channels and parameters in the model, any optimization becomes critical. The use of first-guesses as accurate as posible to start the iterative inversion is also critical to decrease the number of iterations needed. In this case, semi-empirical relationships are used which are obtained by running the model over typical situations and fitting the results to polynomials, so that first-guesses can be easily estimated for initialization of model parameters before inversion. The method used for numerical inversion of the reflectance model is the downhill simplex method due to noise robustness, with two limiting conditions: maximum error and maximum number of iterations allowed. Numerical optimization techniques are used to allow computations on a pixel-by-pixel basis.

Results

Three instruments were flown on board of aircraft over the experimental site of Castilla - La Mancha, AVIRIS, the Digital Airborne Imaging Spectrometer (DAIS), and AIRSAR.

Examples of the analysis of these data presented here are based on the following data sets:
(i) AVIRIS data from the MAC-91 campaign for the Barrax study area in Spain, plus corresponding atmospheric and ground truth information (Fig. 5.74a and b), and
(ii) data obtained during DAISEX-1998 on 11 August 1998 (Fig. 5.75a and b), in the same area of Barrax in Spain, plus the available corresponding atmospheric and ground truth data, also acquired under the framework of DAISEX-1998.
radiosoundings that are available as part of the intensive field campaign in the EFEDA'91 experiment (Fig. 5.76). A problem in this intercomparison is the altitude of the radiosoundings. Extrapolation to the highest atmospheric levels needs some modelling by using a standard atmosphere, as the altitude of AVIRIS (20 km) ranges over the availability of the direct measurements. In any case, the correspondence of values is of the order of the experimental errors. An inferred distributions of the column integrated values over the Barrax area is presented in Fig. 5.77.

The atmospheric water vapour content is needed for the separation of the water vapour signal and the signal of the leaf (canopy) water content, which is the main parameter of interest as part of the general program to provide inputs to surface energy balance models by using remote sensing data.

A simulation of the reflected signal on the canopy water content and LAI is shown in Fig. 5.78. The figure gives an impression of the accuracy which can be expected from the retrieval.

A map of relative canopy liquid water contents in the Barrax area retrieved from AVIRIS data (by fitting the shape of the absorption bands, Fig. 5.7) is shown in Fig.5.79. The DAIS observations have lower spectral resolution than AVIRIS in the mid and short-wave infrared. Notwithstanding this draw-back, estimated leaf water content confirms the actual possibilities to derive such important information from hyperspectral optical data when compared with field observations (Table 5.36). Quantitative data obtained from DAIS measurements can be seen in Figs. 5.80a & b.

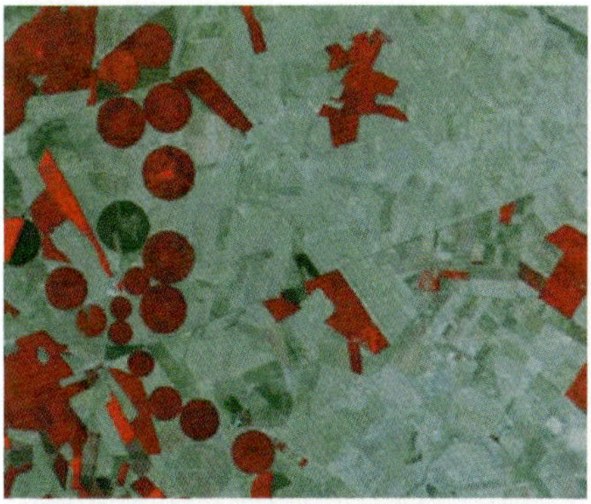

Fig. 5.74a. AVIRIS data available for the area of Barrax, Spain, of 29 June 1991

Fig. 5.74b. AVIRIS data available for the area of Barrax for 15 July 1991

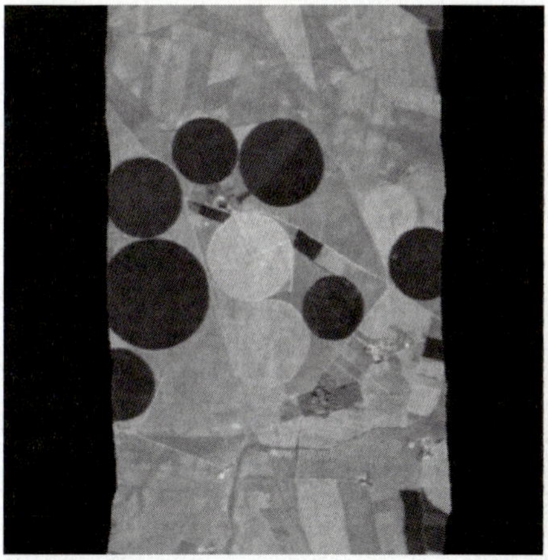

Fig. 5.75a. DAIS data available for the area of Barrax, Spain, of 11 August 1998: North-south track

Fig. 5.75b. DAIS data available for the area of Barrax, Spain, of 11 August 1998: East - west track

They agree reasonably well with ground measurements in the case of low LAI (or low vegetation cover), even when a linear model is used for the retrieval of the amount of water in the leaves. As the LAI increases, the non-linear effects due to multiple scattering contributions and canopy geometry effects start to become significant (Fig. 5.78). In order to avoid such anomalous behaviour the fully non-linear method described above is used to retrieve simultaneously both LAI and leaf water content as separate contributions to the measured reflectance. Although the advantages of linear approaches is well known, and even if such linear approaches can provide good results (specially for low LAI values), the importance of non-linear effects is in any case not negligible.

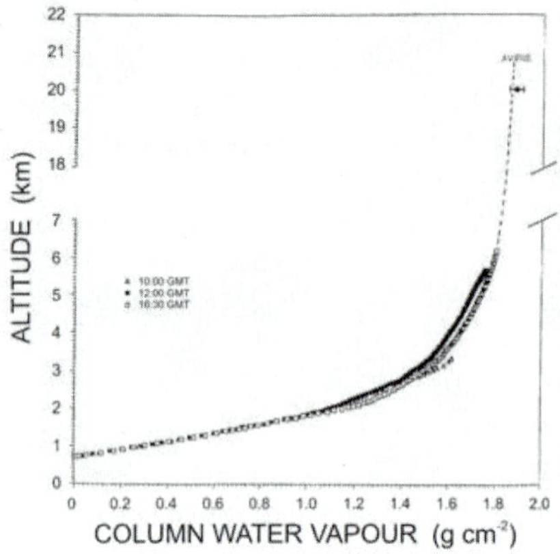

Fig. 5.76. Comparison of an AVIRIS derived water vapour profile with three radiosoundings made in the area of Barrax during EFEDA 1991. The horizontal bar in the AVIRIS-derived profile indicates the standard deviation within the full scene

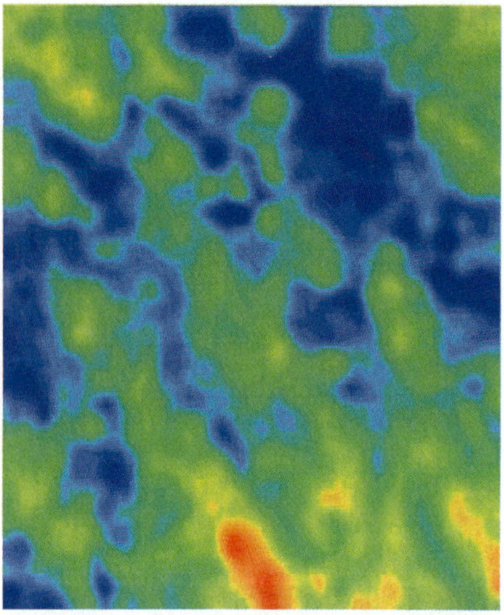

Fig. 5.77. From AVIRIS data derived column integrated amounts of atmospheric water vapour (relative units)

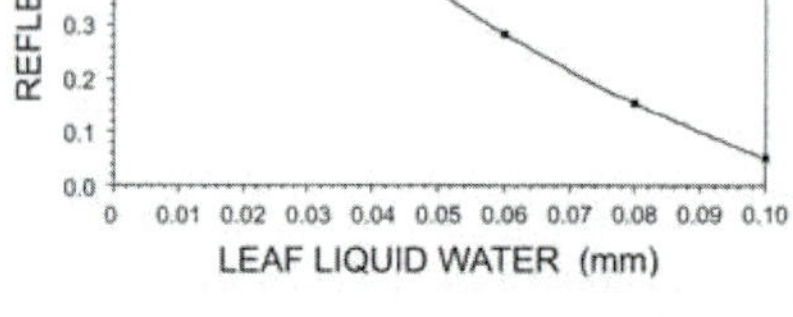

Fig. 5.78. Simulated reflectance ratio in dependence of leaf liquid water content for two values of LAI

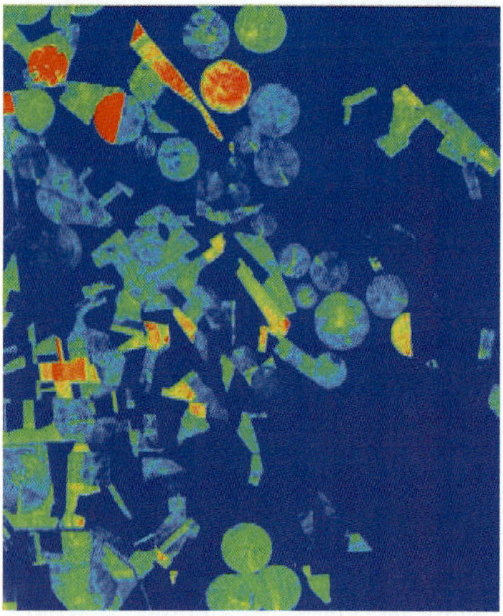

Fig. 5.79. Canopy liquid water content (relative scale) derived from AVIRIS data for Barrax. The figure shows the same area as Fig. 5.74b

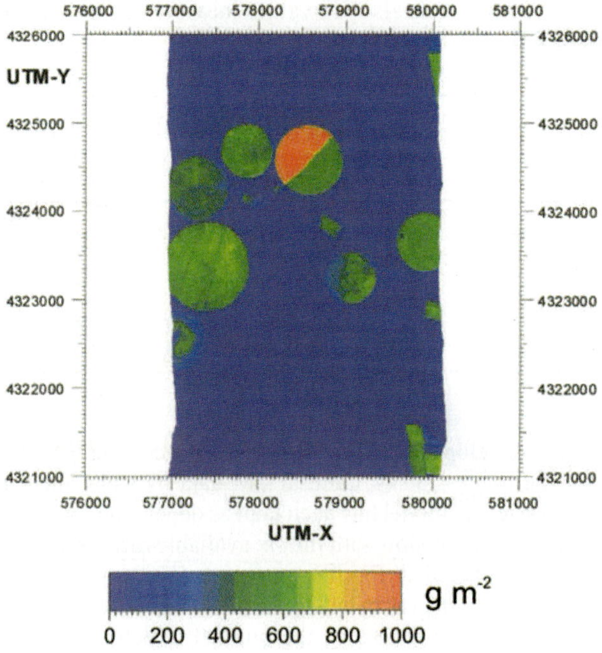

Fig. 5.80a. Leaf water content expressed in g m^{-2}, derived from DAIS data for the North-South track. Barrax, 11 August 1998.

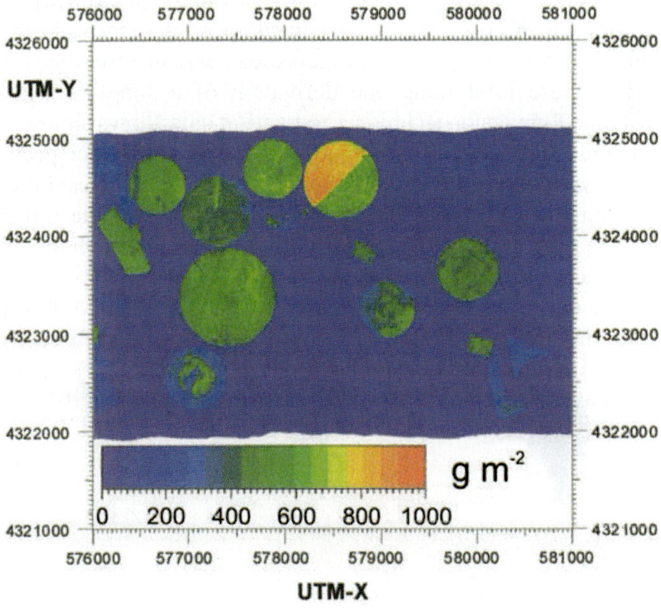

Fig. 5.80b. As Fig. 5.80a but for the East - West track.

Table 5.36. Comparison of DAIS retrievals with values measured at the ground (pixel numbers in brackets [ ]).

Crop Type	Ground Value (g m^{-2})	DAIS Retrievals (g m^{-2})	
		N-S Track	E-W Track
Alfalfa	255.1 - 304.9	443 ± 53 [268, 602]	442 ± 51 [280, 608]
Corn	460.3 - 683.5	613 ± 62 [424, 776]	579 ± 60 [424, 772]

The algorithm originally developed for AVIRIS data processing has been simplified and adapted for the processing of Landsat TM data by using the same modeling approaches. In this case, the model has been kept as depending on only six parameters in order to allow model inversion with the six available measurements in Landsat TM channels (thermal excluded for this purpose). The capabilities of Landsat data to provide the major parameters by using the model-inversion technique are really surprising, but explainable because of the optimal selection of the six channels and the radiometric stability of broad-band channels as compared to hyperspectral data. However, multispectral data (like Landsat) have limitations for separating LAI and ground cover effects. Also, the retrievals of canopy water vapour from Landsat data give a strong underestimation (although explicable due to the small penetration capabilities at these wavelengths), while only a global estimation of pigment absorption (chlorophyll plus others together) can be derived from Landsat data. However, these limitations, the derivation of a simplified algorithm for application of model-inversion techniques to Landsat data allows almost 'operational' and multitemporal studies, which would not be possible with hyperspectral data, for which we only have very limited data for selected dates. Application of the same type of algorithms to SPOT data is not possible because of the limited spectral information. The capabilities provided by multi-directional data in the same type of modeling approaches is being pursued.

In summary, it can be said that the account for all coupled absorptions and multiple scattering effects is a major improvement in comparison with other algorithms. The algorithm is quite robust. Empirical relationships between canopy water content and spectral indices are easier to use, but case-specific.

5.9 Climate Aspects

The impact of climate variability on ecosystems is strongest in regions with a fragile balance between climate and ecosystem, like in the Sahel or parts of the Mediterranean. Both the growing season as the total amount of vegetation, which is probably the most sensitive part of the ecosystem, are strongly affected by climate changes. Precipitation and net radiation are the prominent quantities that affect the vegetation. Its variability can be assessed for large areas by remotely sensed vegetation indices. To understand the relationship between the vegetation index and these climate parameters coherent data series have to be constructed. Precipitation and net radiation can be obtained from operational climatological measurements but the network of stations at which net radiation is measured is not very dense. Another way to obtain these data is from numerical weather prediction models. Here the Regional Atmospheric Climate Model (RACMO) was used which is operated by the Royal Dutch Meteorological Institute (Christensen and Van Meijgaard 1992). The test area was confined to most of western and central Europe and parts of North Africa 32°N - 55°N, 10°W - 20°E).

Precipitation, P, and net radiation R_n can be combined to a "Climate Indicator"(CI), P/Rn, which is the inverse of the Budyko dryness index R_n/P (Budyko 1974). To relate the two quantities to each other the precipitation measured in mm/day have to be converted into W m^{-2} according to the amount of energy necessary to evaporate it. The CI represent the ratio of the annually averaged P and R_n values which range from zero in Northern Africa and some parts of Spain to values between 0.5 and 1 in western and central Europe and to values larger than 2 in mountainous areas. Fig. 5.81 is a map of the CI for the year 1996. The spatial resolution of the CI is 0.5°.

For the derivation of the vegetation index NDVI 10 days maximum values from NOAA-14 AVHRR were available for the years 1996 and 1997 with a resolution of 0.01°. In order to extract the characteristics of the vegetation cycle from this data set, a Fourier analysis has to be performed to obtain start, length and magnitude of the cycle. In Chapter 2, section 2.6.5, it was shown that time series can be reconstructed from measurements with temporary gaps by means of the Harmonic Analysis of Time Series (HANTS) algorithm (Verhoef 1996). This allows to construct remote sensing data series which are compatible with the operational or modelled meteorological/climatolog

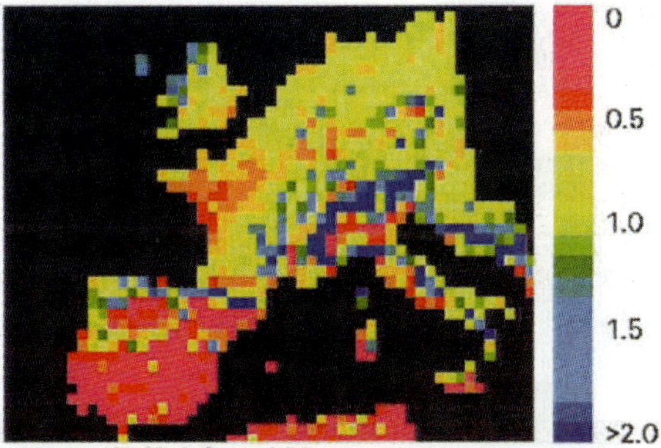

Fig. 5.81. Map of the Climate Indicator (P/R_n) for 1996 derived from RACMO simulations

ical information. This algorithm considers only the most significant frequencies expected to be present in the time series (0 - mean, 1 - annual cycle, and 2 - 6 months cycle) and applies a least squares curve fitting based on harmonic components (sines and cosines). For each frequency amplitude and phase of the cosine function is determined during an iterative procedure. Input data points, which have a large positive or negative deviation from the current curve (like cloudy and missing pixels), are removed by assigning a weight of zero to them. After recalculation of the coefficients on the basis of the remaining points, the procedure is repeated until the maximum error is acceptable or if the number of remaining points has become too small. Fig. 5.82 gives examples of the result of this procedure for the classes 5, 10 and 19 (from top to bottom) of the following Fig. 5.84. The mean NDVI values of these three classes are 0.23, 0.36, and 0.18 respectively. At the left side the amplitudes are given for the 12 (green), 6 (blue) and 3 (brown) months periods. The scale (at the left) ranges from 0 - 0.2. The red bar represents the mean NDVI value for which the right side scale is relevant which ranges from 0 to 0.5. The centre contains the modelled decadal values of the NDVI for the decades 1 - 35. At the right side the temperatures are platted for the mean temperature (red), the 12 months amplitude (green), and the 6 months amplitude (blue). The scale ranges from 0 to 30 °C. A composite of such an amplitude analysis of NDVI data is shown in Fig. 5.83 where the three amplitudes are displayed in red (annual mean value), green (annual amplitude) and blue (6 months amplitude).

Here evergreen forest with nearly constant average NDVI are shown in red, areas with a strong annual amplitude in green and areas with a six month period like those with spring and autumn vegetation and a dry summer in blue. Desert areas are

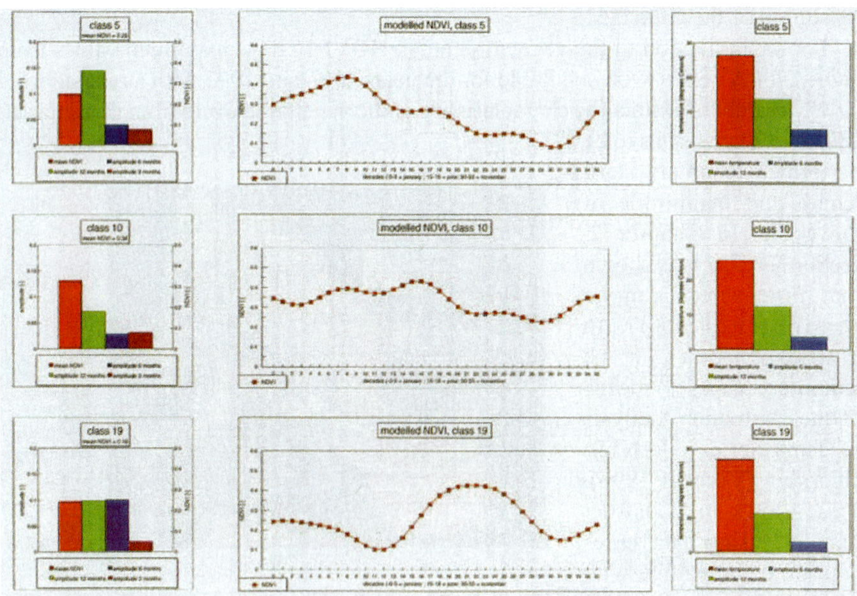

Fig. 5.82. Examples of the result of the Fourier analysis (details see text)

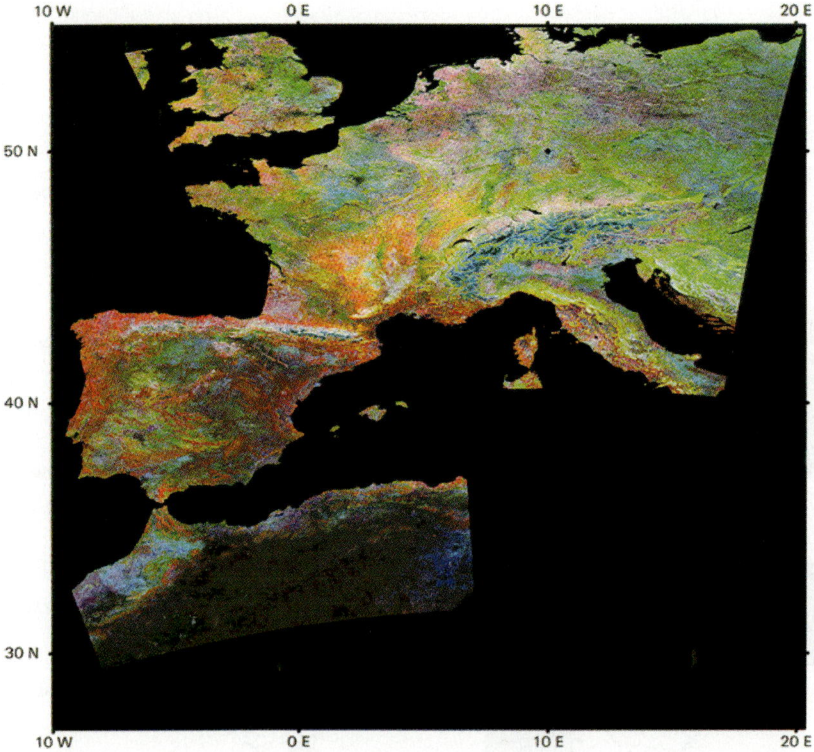

Fig. 5.83. Composite of the average NDVI (red), the amplitude of one year (green), and the amplitude of six months (blue)

black because of the small noisy signals.

By combining the amplitude information with that of the phases (compare Fig. 6.9.7), each pixel of the NDVI map gets labelled with the three amplitudes and three phases which makes it possible classify a large number of "isogrowth zones". Each of these classes includes crops and "natural" vegetation types which may not be of the same species but have the same growth pattern. 25 of theses classes could be discerned as can be seen in Fig.5.84. The map derived from the NDVI values have been validated with CORINNE (CEC 1993) land cover classes.

To assess the relationship between the information derived from the Fourier analysis of NDVI data and the CI, the NDVI map has to be rescaled into the CI map. If one plots the individual pixel values of the mean value or one years amplitude the scatter is large because not only the climate but also other sources affect the NDVI. If the NDVI values are averaged for each value of the CI the curve shown in Fig. 5.85 emerge which show a clear relationship between the two data sets. This becomes even more evident if the derivatives for space and time of the three NDVI data sets are plotted against the CI (Fig. 5.86). The spatial relationship is computed

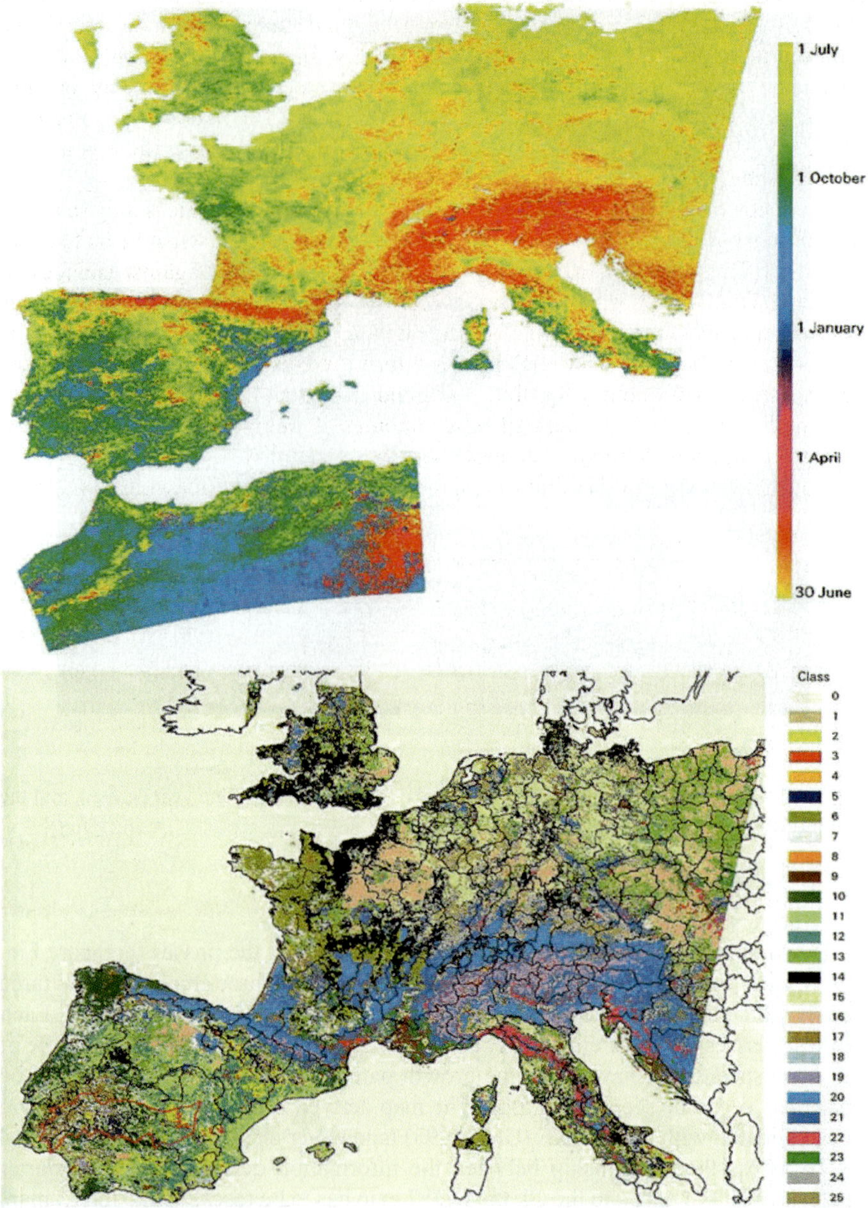

Fig. 5.84. Maps of the phase of the one year component of the FFT algorithm applied to the NDVI time series (top) and of the vegetation-soil-climate complexes with regional boundaries overlaid in black, and the boundaries of the Guadiana watershed area overlaid in red (bottom)

according to Eq. 5.104. The temporal relationship is that of the years 1996 and 1997 (Eq. 5.105). Especially the mean NDVI value and the annual amplitude are suitable parameters for the detection of climate variability. Both the relations in time and space show clear trends, which reveals that the vegetation variability is most sensitive to P/R_n changes in the dryer areas where the P/R_n values are very low. From $P/R_n = 0$, where the vegetation response to P/R_n is highest, there is a trend towards saturation where there is no response anymore.

There is, however, an important difference: The relationship in time is only sensitive for P/R_n changes from 0 to 0.3 while the relationship in space is sensitive up to about $P/R_n = 1$. This suggests that vegetation has a natural buffer against changes of climate factors for short time periods, e.g. one year. Only in dry environments vegetation is sensitive for changes of the climate parameters within one year. It is also observed that the mean NDVI takes a turn towards a negative trend for P/R_n values larger than about 1.8, while the trend of the annual amplitude is further upward directed. This indicates that larger amounts of precipitation become a burden for the average vegetation but increases seasonal variability.

The numerical analysis of these relationships result in the following formula:

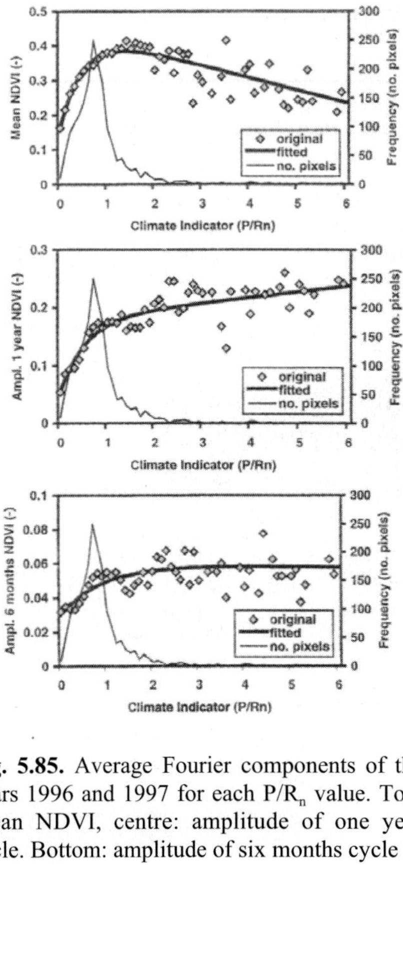

Fig. 5.85. Average Fourier components of the years 1996 and 1997 for each P/R_n value. Top: Mean NDVI, centre: amplitude of one year cycle. Bottom: amplitude of six months cycle

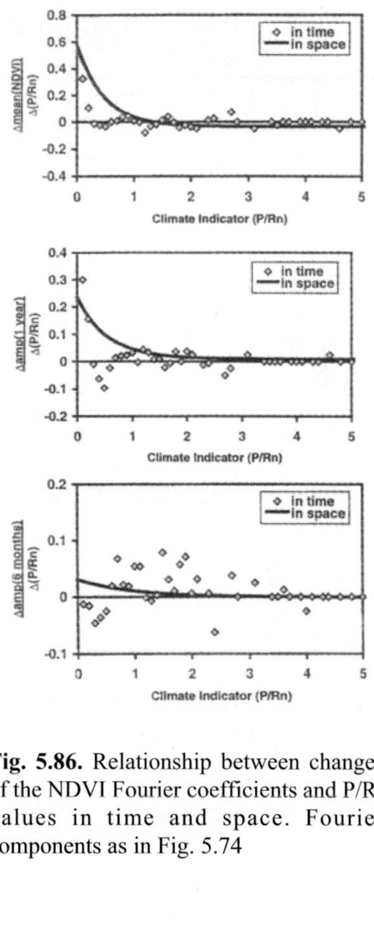

Fig. 5.86. Relationship between changes of the NDVI Fourier coefficients and P/R_n values in time and space. Fourier components as in Fig. 5.74

$$NDVI\ Fourier\ component = a + b\left(\frac{P}{R_n}\right) + cd^{P/R_n} \tag{5.103}$$

The first derivative of Equ. 5.103 is:

$$\frac{\partial(NDVI\ Fourier\ components)}{\partial(P/R_n)} = b + c(\ln d)d^{P/R_n}\ . \tag{5.104}$$

The temporal relationship is derived as:

$$\frac{\Delta(NDVI\ Fourier\ components)}{\Delta(P/R_n)} = \frac{(NDVIFour.comp.)_{1997} - (NDVI\ Four.comp.)_{1996}}{(P/R_n)_{1997} - (P/R_n)_{1996}}\ . \tag{5.105}$$

The regression coefficients which have been found are given in Table 5.37.

Table 5.37. Regression coefficients found for Eq. 5.103

Fourier Component	a	b	c	d	R^2
Mean NDVI (frequency =0)					
1996	0.430	-0.030	-0.257	0.111	0.85
1997	0.443	-0.033	-0.296	0.090	0.82
both years	0.449	-0.035	-0.287	0.119	0.90
Amplitude 1 year (frequency = 1)					
1996	0.188	0.0066	-0.137	0.123	0.79
1997	0.213	0.0008	-0.147	0.358	0.84
both years	0.179	0.0098	-0.123	0.160	0.93
Amplitude 6 months (frequency = 2)					
1996	0.060	-0.0009	-0.032	0.243	
1997	0.051	0.0001	-0.023	0.125	
both years	0.060	-0.0003	-0.029	0.344	

Chapter 6 From Research to Application

6.1 Introduction

In previous chapters it has been shown that the methodology is in place to derive a variety of quantities which either play a role in the mechanisms that drive land-surface processes or are indicators for changes occurring at the land surfaces. It was also emphasised that the combination of remote sensing data with measurements carried out at the ground by a network of stations greatly can enhance the quality of information obtained from either system. Until a basin-wide network of such "anchor stations" is established, specific ground based data become only available during short term field experiments and in some areas by operational networks installed at the scale of watersheds. Application projects preferably have been carried out at these smaller scales that are easier to overlook and in most cases more supplemental information can be raised. Size and heterogeneity of the Mediterranean basin complicates extrapolation from water-sheds scales to the entire basin.

Supplemental information generally is congregated into Geo-Information-Systems (GIS) and stored in a way that it may easily be mapped into the satellite images formats. Basic supplemental information needed for most interpretation tasks are topographical maps, information on land-use/land-cover and vegetation, and on soils (see also Appendix 1). In addition information about aquifers may be needed and data about water consume (e.g. for irrigation), population density, etc.

Applications of satellite data at the larger, basin-wide, scale are of different nature than the detailed studies controlled by measurements made at the surface. They are mainly based on the simpler approaches of the *level one products* described in chapter 4.

To assess the role, remote sensing methods can play in land-surface studies, some examples are presented in this chapter to demonstrate the use of some of the methods discussed in chapters 4 and 5. One purpose to acquire such information is to use it in conjunction with land-surface models, both mesoscale and global ones. Specifically, it is shown that, for different parameters and at different scales, one can obtain information regarding temporal changes of the state of the land cover and the state of vegetation for several years in a row which is a first step towards diagnosing the risk of desertification in the Mediterranean basin.

An overview and definitions of key variables needed for the modelling at different scales of biospheric processes is given in the following section 6.2. It is followed by a discussion of the characterization of local vegetation development (section 6.3). In section 6.4 methods are reviewed which are applied to detect vegetation changes from the measurements made in space. Knowledge about the spatial land-surface characteristics is necessary to interpret correctly the measurements made with instruments of different field of views which integrate over topographic structures of various sizes. In section 6.5 examples of these structures are presented for different test areas. To follow changes that occur at the land surfaces the repetition rate of the observations have to be appropriate. Also this question is raised in section 6.5 in the context with changes experienced in the course of field experiments and the evaluation of longer term operational observations. Examples of aggregation are given which is used to compare high with medium spatial resolution satellite images.

Small scale changes developing in adjoining areas in opposite directions may not be captured by medium to low resolution instruments. In section 6.6 it is described how rare high resolution data are combined with frequently available medium resolution data to detect these changes and to study the dynamics of functional ecosystem types.

The characterization of such land-cover units is the theme of the following section 6.7. First, multispectral classification methods are discussed using spectral bands in the visible and near infrared part of the spectrum. Then the use is described of the Temperature Independent Spectral Indices (TISI) derived from the signals received in the AVHRR channels 3, 4 and 5 to characterize the land surfaces by their emissivity. Finally the application of microwaves to quantify the density of the vegetation canopy by their microwave transmissivity is reported.

Going further into details contained in single pixels requires the decomposition of their signatures into "end members". This is the topic of section 6.8. First the use of shortwave bands and then that of thermal infrared bands is described.

Seasonal and interannual variability as seen from space is the theme of section 6.9 in which images are discussed that are derived from the NOAA-AVHRR instrument. The 1 km resolution MEDOKADS data set introduced in chapter 2 is used to map radiometric temperatures of the atmosphere - surface system for cloudless conditions for decadal, seasonal and annual periods in relation to climate variability. Using the shortwave bands allows to map for the whole Mediterranean the variability of the vegetation cover for any chosen time interval. Combination of thermal infrared and shortwave data allows to produce secondary level data sets which are useful for vegetation and desertification studies.

In the following two sections the results of change detection studies are reported. NDVI technique was applied to investigate the effect of land-use change to the health of vineyards in the Castilla la Mancha region of Spain (section 6.10). The NDVI differencing image combined with thresholding technique on one hand and the NDVI change vector analysis on other, has proven to be effective in detecting and categorizing inter-annual changes in the vineyard crop during 1991-1996 period. In the other study (section 6.11) wet and dry years are identified by their NDVI-values due to the relationship between the state and density of vegetation to climate factors. NOAA/AVHRR data are used to investigate the shift of the Israeli transition zone between Mediterranean vegetation and semiarid regions. The NDVI database used in this example was developed from atmospherically and radiometrically corrected NDVI composites from observations spanning three years (March 1995-April 1998). The NDVI of the transition zone was found to be very sensitive to rainfall, a phenomenon which can be used as an indicator of environmental changes in this region.

A technique to estimate directly moisture in the top layer of soils uses passive microwave remote sensing data and is applied to monitor desertification respectively aridification. This is described in 6.12 in application to the Iberian Peninsula. The first example of the application project is for the whole Iberian Peninsula in which Nimbus/SMMR passive microwave signatures have been used to determine the spatial distribution and temporal variations in surface soil moisture for the whole lifetime of the Nimbus satellite (1978-1987).

In the following study (section 6.13) transpiration of a Mediterranean forest ecosystems is simulated by a model which was originally created for temperate areas (FOREST-BGC) and adapted for the Mediterranean. The model was calibrated and applied on two deciduous forest stands in Tuscany by using as inputs information from different sources about the two stands, meteorological data extrapolated from a nearby station, transpiration values obtained by a sap flow method, and LAI data derived from satellite NDVI values.

The use of GAC NDVI data for cropland identification and yield forecasting in Mediterranean African countries is the topic of section 6.14. A preliminary analysis showed that relatively high correlations exist between crop yield and mean NDVI values for specific months computed at national level. This work uses a statistical method to estimate the per-pixel fractions of agricultural and non-agricultural vegetation. The methodology was applied operationally.

In another study (section 6.15) the relationship of climatic conditions and natural vegetation was tested (the amount of green vegetation measured by NDVI at a regional scale, using Landsat and SPOT data) to study the climatic influence on the natural post-fire regeneration of the burned area. It could be shown that, in burned areas and dependent on the type of vegetation, the regeneration follows an exponential curve, stronger in wet zones, and with high correlation coefficients.

The use of data from geostationary meteorological satellites to generate soil moisture fields for the initialization of meteorological models is the theme of section 6.16. It is shown how methods described in previous chapters can operationally be applied to estimate soil moisture fields for large areas and thus improve the forecast capacity of the models.

Finally, in section 6.17 conclusions resulting from the work carried out under the EC Framework Programme are summarized.

6.2 Biosphere Processes: Key Variables, Models and Scales

6.2.1 Main Driving Variables of Biosphere Processes

Processes

In the following, the main driving variables of the processes in the biosphere are defined and briefly described (compare section 5.8).

Photosynthesis is the process by which vegetation fixes atmospheric carbon dioxide into biomass, using solar energy. The rate of carbon fixation by the vegetation depends first and foremost on the amount of photosynthetically active radiation (PAR, nominally between 400 and 700 nm), absorbed by the canopy (McCree 1972). The *fraction of absorbed PAR* (*fAPAR*) depends on the architecture of the canopy and on the optical properties of its elements (leaves, stems, trunks, branches, background) (e.g., Sellers 1985). Canopy architecture is traditionally described by the *leaf area index (LAI)*: the total one-sided leaf area of the canopy per unit horizontal ground area), although more sophisticated representations of leaf clustering in plants and plant clumping are emerging. The efficiency with which the PAR energy absorbed by the canopy is actually transformed into biomass strongly depends on species, phenology and the environmental conditions experienced by the vegetation. Atmospheric carbon dioxide is transferred to chloroplasts (where the chlorophyll molecules are located and photosynthesis takes place) through small holes in the leaf surface called stomata. Plants control the opening and closing of their stomata, and thereby regulate the rates at which carbon dioxide molecules enter and water molecules exit these stomata. Since both photosynthesis and evaporation of liquid water in the stomata require significant energy inputs, the carbon, water and energy cycles are intimately linked through the suite of chemical and biological mechanisms associated with the primary productivity of plants. These relationships are biome-dependent, as illustrated in Figure 6.2.1. It is therefore mandatory to describe concurrently the water, energy and carbon exchanges to be able to understand photosynthesis. The rate of photosynthesis also depends on many other factors, including carbon dioxide concentration in the atmosphere, leaf temperature, or mineral deficiencies (in particular nitrogen) in the soil. The nitrogen content of leaves is strongly related to their chlorophyll content (Field et al. 1986).

Autotrophic respiration is the process by which some of the chemical energy stored by photosynthesis is used by the plants themselves to grow and develop. This process is critical to the carbon cycle because it results in the rapid release of a large fraction of the carbon initially stored through photosynthesis back to the atmosphere. Autotrophic respiration depends on foliage temperature, growth rates, total biomass, as well as the biochemical composition of the products formed in the plants.

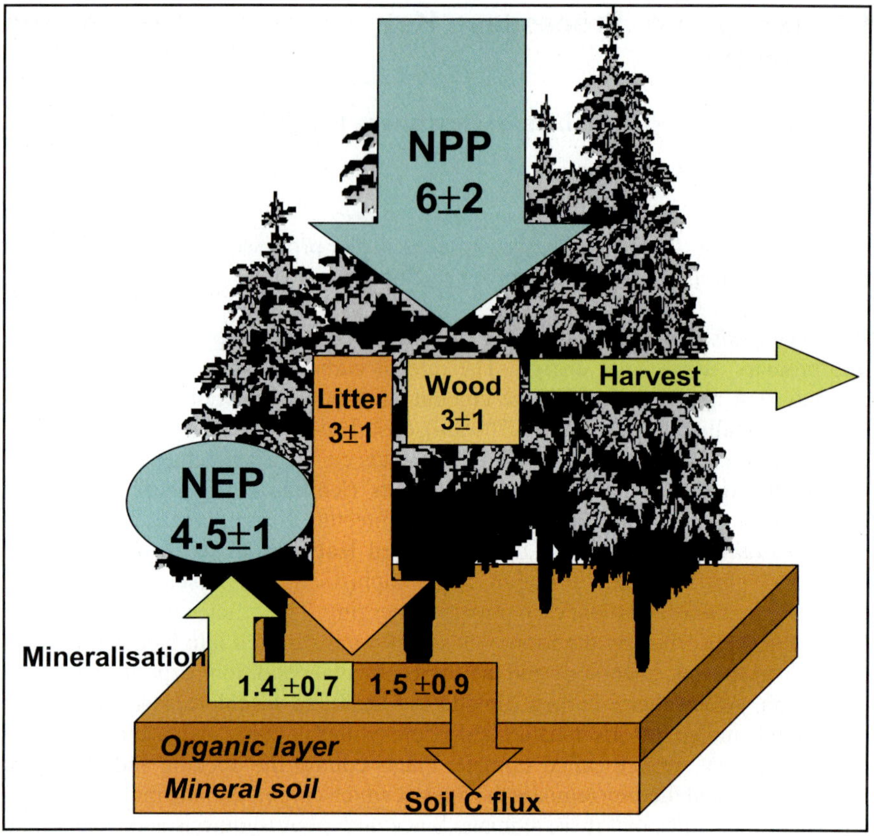

Fig. 6.2.1. Typical values of carbon fluxes (t C ha^{-1} yr^{-1}) observed over forest ecosystems, and their associated uncertainties. (Adapted from Valentini et al., 2000)

Heterotrophic respiration corresponds to a similar process in soils. The soil carbon reservoir can be very large as compared to the aboveground biomass. Understanding the fluxes of carbon to (senescence, mortality) and from (respiration or mineralization) this soil reservoir becomes a major issue when closing the carbon cycle at the local scale. Heterotrophic respiration is very dependent on soil temperature, and the availability of water and nutrients, particularly nitrogen. Apart from nitrogen fertilisation or deposition, symbiotic fixation of atmospheric nitrogen, and leakage or volatilisation, the nitrogen cycle is intimately linked to the carbon cycle within the soil via the biotic activity. Evaluation of heterotrophic respiration is a major challenge in the description and modelling of NEP.

Water and energy both play a critical role in plant physiology, as seen above. Water is used by plants to transport nutrients from the soil through the roots, to the various plant organs, to redistribute assimilates within and between these organs, to control structural stability and optimise solar radiation interception (through turgescence) and to ensure the thermal regulation of plant tissues. The soil constitutes the main reservoir of water for plants. This reservoir is mainly replenished by rainfall, flooding, irrigation or capillary rise from a lower water table. Conversely, water losses are due to evaporation from the soil, extraction of water by the roots to support plant transpiration, run-off and drainage. Evaporation and transpiration are often combined into evapotranspiration to describe the total water flux to the atmosphere. And of course the water and energy cycles are closely linked through the latent heat needed to evaporate liquid water in plants or in the soil. The energy balance, in turn, is primarily driven by the available incident solar energy (both direct and diffuse radiation). In the solar domain (300 to 3000 nm), the *bidirectional spectral reflectance* of the soil and canopy control how much radiation is actually available at the surface. The radiative energy absorbed by the plants but not used in photosynthesis, as well as the energy absorbed in soils, is transformed into heat and ultimately released to the environment through conduction, convection (sensible and latent heat fluxes), or thermal emission. The latter is controlled by the relevant temperature and emissivity. The convective fluxes are largely determined by gradients of temperature (sensible heat flux) or humidity (latent heat flux), and by a variety of complex mechanisms collectively represented through empirical coefficients, known as resistances, that depend on leaf area index and vegetation type. (Menenti and Ritchie, 1994).

Ecosystem dynamics. Many of the processes described above are dependent on the type of ecosystem or biome considered. Ecosystems are built from a collection of species that are organised according to complex rules. As a matter of fact, ecosystems tend to adapt themselves to the local soil and climatic conditions, and develop specific strategies for surviving, always subject to the competition of a wide range of species. Whenever external conditions are changing, either due to climatic variability, fire, or human activities, the fragile equilibrium between species may shift, leading to changes in the ecosystem functioning and composition. In the absence of direct human interference by deforestation, fire and agriculture, this process is generally slow and can be observed only over long time periods or along specific pedo-climatic gradients. However, when simulating long-term climate changes, these processes become very important and must be accounted for (Prentice et al. 1989; Steffen et al. 1992).

Models

These processes are explicitly represented in models (Figure 6.2.2) that may be classified as follows:

Land Surface Models (LSM). These models describe physiological and biophysical processes, as well as soil biochemical and physical processes. Exchanges with the atmosphere are described by a Soil Vegetation Atmosphere Transfer (SVAT) module that runs under forced climatic conditions. In these models, soil properties such as temperature, moisture and nitrogen content vary with time, but all other characteristics are supposed to be stable (e.g., Brisson 1998, Spitters 1989

Hoogenboom et al. 1999). Some of these LSMs include a radiative transfer module to simulate the reflectance of the canopy, they can be used to assimilate radiance measurements gathered from space (e.g., Weiss et al. 2001; Schneider 1999).

Dynamic Global Vegetation Models (DGVM). These models typically include a LSM and an additional module that describes the dynamic evolution of the ecosystem as a function of climatic variables. Interactions with soil properties may be included to describe the dynamics of variables other than temperature, moisture and nitrogen content. However, they are still running under forced climatic conditions. Cramer et al. (2001) recently reviewed and compared various DGVMs. They differ significantly in complexity and reliance on parameterisations. Simpler models are more efficient but also more limited to describe particular processes. They are suitable for long-term studies or to investigate feedbacks in coupled Earth System models (Cox et al. 2001). On the other hand, more complex models allow for direct studies of eco-physiological processes and their implications at the global scale. These differences in model complexity and reliance on parameterisation lead to large uncertainties in the terrestrial carbon balance. Detailed remote sensing observations at high resolution will be critical to improve these parameterisations and evaluate the performance of these models.

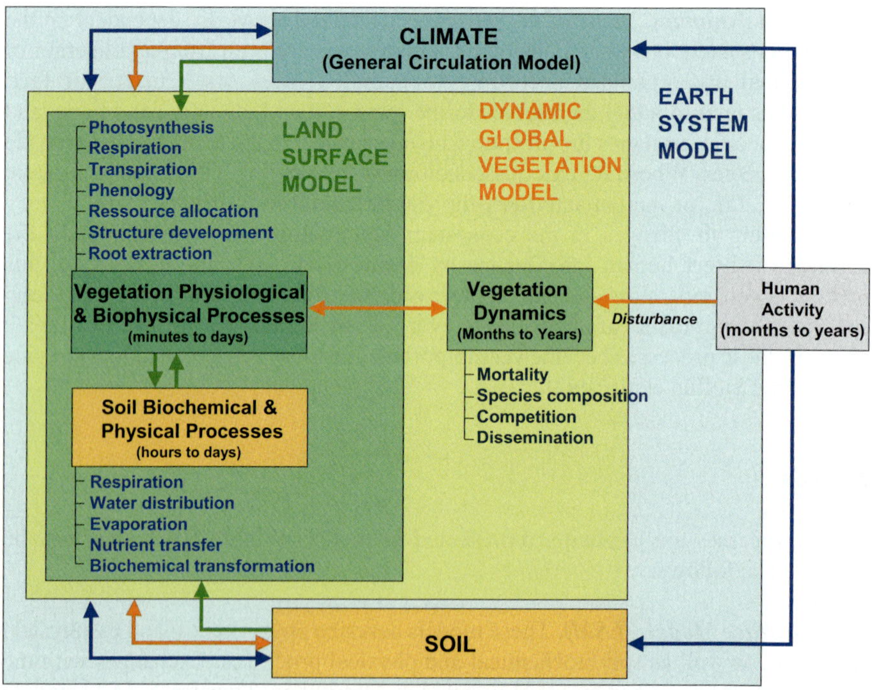

Figure 6.2.2. Schematic representation of the processes determining the interactions between vegetation and the climate system, at various space and time scales, and of the types of models needed to describe them (Courtesy F. Baret).

Earth System Models (ESM). These models attempt to couple all the relevant processes together, with an emphasis on the atmosphere-surface interactions. They focus on the global spatial scale, and are used to simulate the evolution of the entire system over time scales of up to decades. These models necessarily rely on effective parameterisations of small-scale processes; their results are thus very sensitive to the accuracy of these parameterisations.

This discussion has identified a long list of variables relevant to describe the state and evolution of the land surface. Some of them (highlighted in *italic* in the preceding sections) can be retrieved from an analysis of radiometric measurements made in space, and used in combination with these models.

The exploitation of remote sensing observations can follow two different approaches. In the first one, *forcing*, biophysical variables such as those identified above (e.g., fAPAR, LAI, soil temperature, etc.) are determined from radiance measurements and directly input in the models. The second approach consists in the *assimilation* of either geophysical products or radiometric measurements in dynamic models. This is achieved by adjusting one or more model parameters until the model matches the observations.

Each approach has its advantages. Data assimilation guarantees a high degree of internal model consistency, may be less sensitive to gaps in the data sets (such as produced by the obscuration of the surface by clouds) and may lead to the determination of optimal solutions. In the case of forcing, satellite data can be analysed off-line with generally more detailed radiative models to generate products independent from the type and sophistication of the environmental models that may use this information. Data assimilation may lead to the parameterisation of complex biome properties, such as canopy conductance for gas exchange. Observations of leaf water content and leaf dry matter are quite useful for this purpose. Improved parameterisations may be used to forecast long-term trends, even when sparse or no concurrent data are available.

6.2.2 Local Processes and Global Models: Spatial Heterogeneity and Scaling

Spatial heterogeneity is a defining feature that distinguishes terrestrial surfaces from the oceans and the atmosphere. Reliable and accurate fluxes of carbon, water and energy are difficult to estimate at the regional to global scale because of the heterogeneity of the landscapes.

Recent literature demonstrates that forecasts on the long-term evolution of climate depend quantitatively and qualitatively on how the response of terrestrial vegetation is parameterised. For example, Cox et al. (2001) demonstrated that plausible assumptions on the ratio of assimilation to respiration led to substantially different evolutions of the global climate in response to increased CO_2 concentration in the atmosphere. Similarly, Claussen (1997) showed that different representations of biosphere processes may or may not lead to re-growth of vegetation in the Sahara in response to long-term climate trends.

From the point of view of modelling, we may distinguish two main approaches to represent land-atmosphere interactions in global models:

"Frozen biosphere": A map of global biomes (e.g., Fig. 6.2.3) is used to determine abundance of each land cover type for all model grid boxes. The exchanges of energy, water and carbon within each grid box are computed using a Land Surface Model; The properties of each biome are time-independent and are specified, typically in tabular form. This approach may be extended to use time-dependent biome properties, specified in an off-line database.

"Interactive biosphere": The LSMs in this category are far more complex than for models in the previous category. We may further distinguish between models that only describe assimilation and respiration, i.e., where only the amount of vegetation changes, and those that allow these biomes to evolve, for instance in terms of species composition, in response to climate forcing.

Documenting the state and evolution of terrestrial ecosystems requires describing a rather wide range of complex processes operating at the local scale. On the other hand, the LSMs implemented in global Earth System Models relate to entire regions, rather than samples of truly homogeneous biomes characterized by well defined, observable biosphere properties. While all relevant processes may be included in a model of the type depicted in Fig. 6.2.3 such models lump the underlying heterogeneity and non-linearity of terrestrial biosphere processes into parameterisations at the scale of a model grid, say 100 km.

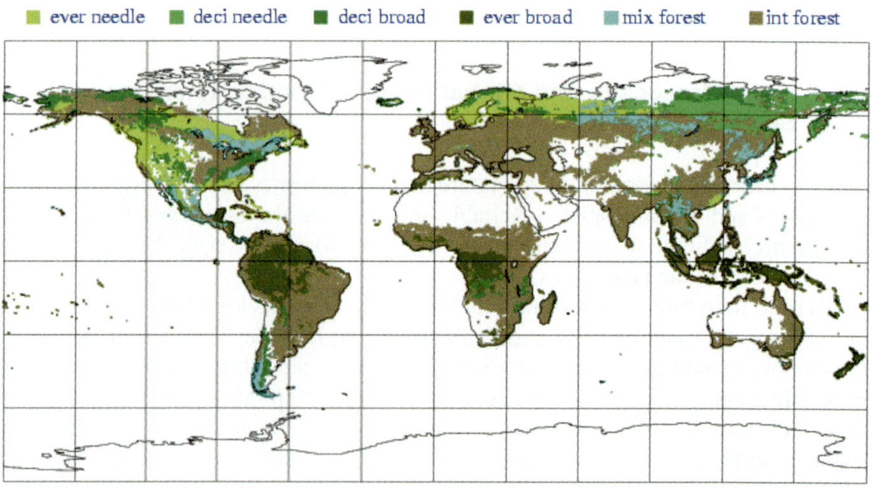

Fig. 6.2.3. Map of tall vegetation types currently used by the global ECMWF NWP model to parameterise the exchanges of energy, water and carbon between the terrestrial biosphere and the atmosphere (Courtesy: ECMWF)

The implications of this modelling approach become evident when comparing the true complexity of heterogeneous landscapes (Figure 6.2.4) with a map of any of its properties as represented in most global models. Indeed, anyone of these landscapes, no matter how heterogeneous or complex, is represented by a small set of properties that are kept constant over the entire area of 100 km × 100 km, as depicted in Figure 6.2.5a. Distributed models of terrestrial landscapes (see e.g. Figure 6.2.6) do a significantly better job at describing heterogeneous landscapes and the mechanisms determining their functioning, and are the subject of intense on-going research.

There is an evident knowledge gap between what is known empirically through local measurements and the idealizations contained in the parameterisations used in global models. *A robust and scientifically sound strategy to fill this gap will require*

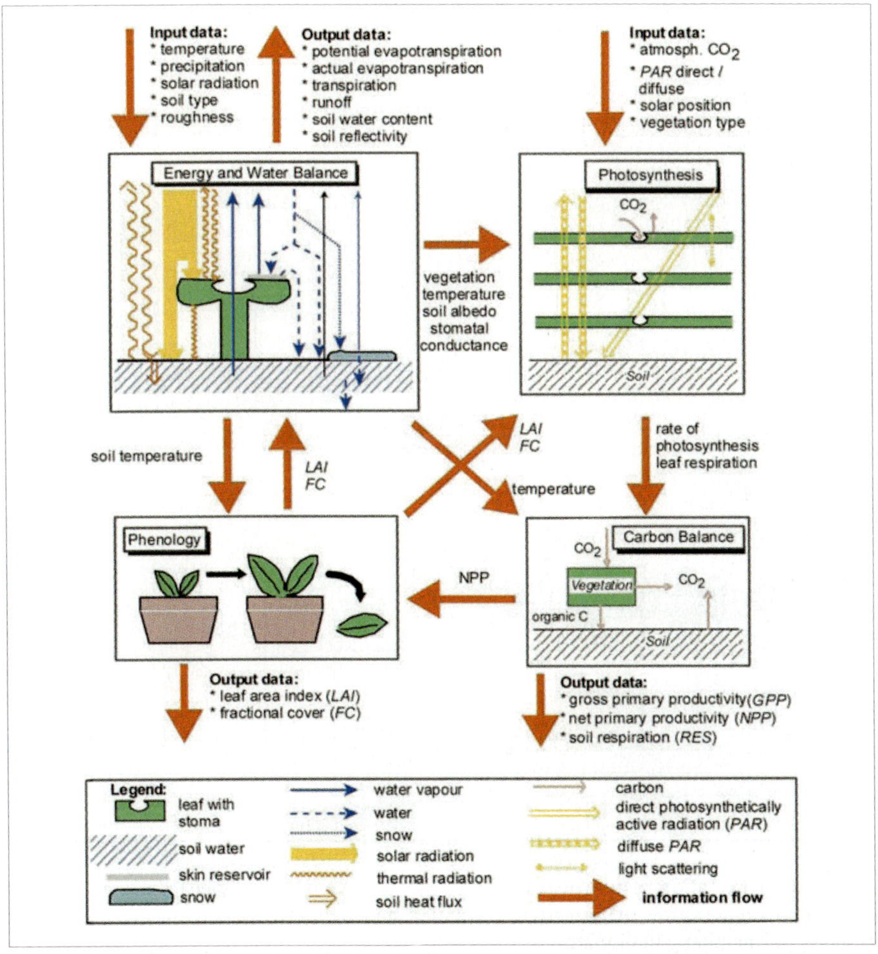

Fig. 6.2.4. Variables and processes determining the response of terrestrial vegetation to environmental forcing in Land Surface Models (Knorr and Heimann 2001)

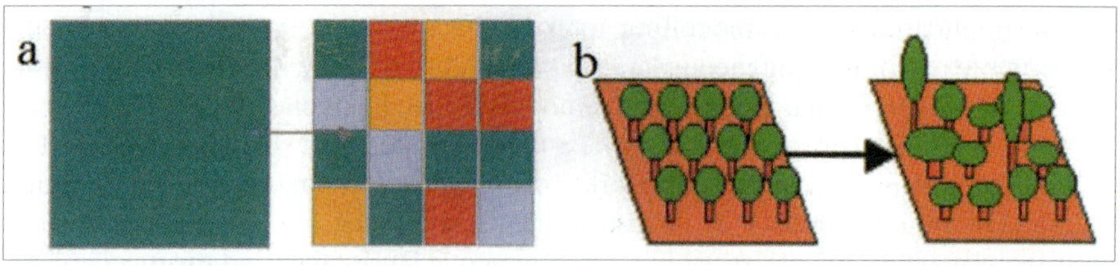

Fig. 6.2.5. Schematic representation of the way global models capture the heterogeneity of terrestrial vegetation. The land-surface parametrization implemented at ECMWF permits to distinguish multiple types of land cover within each GCM grid box (a, small squares), including low and tall vegetation types (b). Courtesy of B. J. J. M. Van den Hurk.

a focused experimental and modelling plan covering an ensemble of regions,

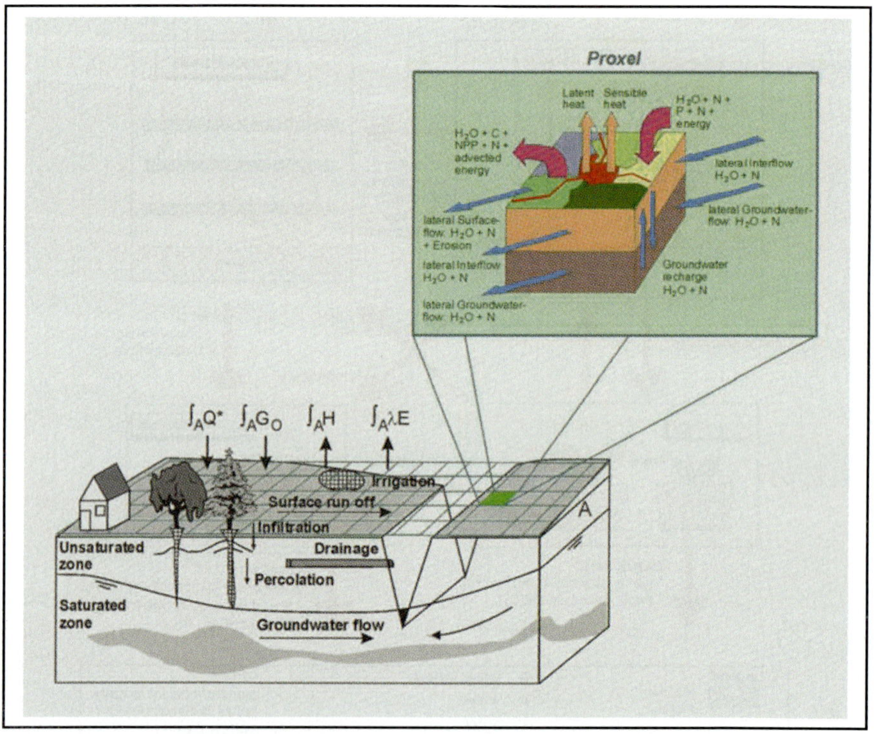

Fig. 6.2.6. Schematic structure of a distributed ecosystem model designed to take into account the spatial organization of terrestrial vegetation and of the processes that determine the response of the biome to climate.

representative of all relevant terrestrial biomes, and the development of suitable parameterisations for each one of them.

To meet this challenge, it will be necessary to study in detail the functioning of terrestrial vegetation over areas as large as the size of a grid box in global models. This will provide a solid basis for grid-box scale parameterisations of biosphere processes. Three steps might be distinguished in this process:

- Observations of the terrestrial biosphere are collected at a spatial resolution sufficient to capture its heterogeneity (Fig.6.2.3).
- A detailed, distributed model of the region is constructed in a way that preserves the spatial organization of the observed landscape (Fig. 6.2.4) while permitting the estimation of overall quantities over the entire region covered by the distributed model.
- A simplified representation of the heterogeneity of the landscape within a model grid box (Fig. 6.2.5) is developed for use in global models. State-of-the-art global models are evolving towards incorporating heterogeneity in two ways: either taking into account sub-grid variability of land cover, i.e., the so-called tiling or *mosaic* approach, where each tile interacts independently with the atmosphere (6.2.5a) or by treating in a different way tall and low vegetation types (6.2.5b) within the entire grid box.

The next stage in this process of building up our understanding of terrestrial vegetation over a range of scales is to determine the relevant properties of these biomes at a spatial resolution sufficient to capture heterogeneity. This can be done by combining observations at high spatial resolution with data assimilation and modelling at the regional scale.

6.2.3 From Local to Global Scales

It is common practice to define spatial scales of biosphere processes as follows: The *local scale* (10-100 m) corresponds to the typical length scale of the internal variability of landscapes (Briggs et al. 1995). Depending on the biome, it also represents the largest scale at which all relevant environmental variables can be exhaustively documented in field studies. It is therefore the native scale of most dynamic vegetation models and corresponds to the typical scale at which human activities take place (agriculture, deforestation, reforestation, afforestation, fires, etc). When observed at the local scale, a landscape appears as the one in Fig. 6.2.7.

The *regional scale* (10-100 km) is defined here as the typical scale of variability of a biome. At this scale, topography may play a dominant role on the transfer and storage of water with strong implications on vegetation functioning. Yet, realistic models (Figure 6.2.6) can be implemented to take into account the heterogeneity of variables and processes. This is also a useful scale to document the long-term evolution of spatial gradients, due to past disturbances, or to derive parameterisations for the DGVMs. Observations at this scale will also directly contribute to meet the national reporting requirements of international conventions such as the Kyoto Protocol.

The *global scale*, the relevant spatial variability is driven by the distribution of land and oceans, as well as by that of biomes over the continents. Global assessments of the carbon cycle must necessarily rely on parameterisations of processes occurring at scales too small to be resolved in global models. These parameterisations may rely on a simplified representation of heterogeneity, e.g., by lumping the area covered by each vegetation type in a homogeneous tile and disaggregating a model grid-box into several tiles (Fig. 6.2.5a). In one case, tall and low vegetation types are treated separately (Fig. 6.2.5b). The heterogeneity of the land surfaces is then taken into account by associating biome-specific parameterizations with a global biome distribution map shown in Fig. 6.2.3.

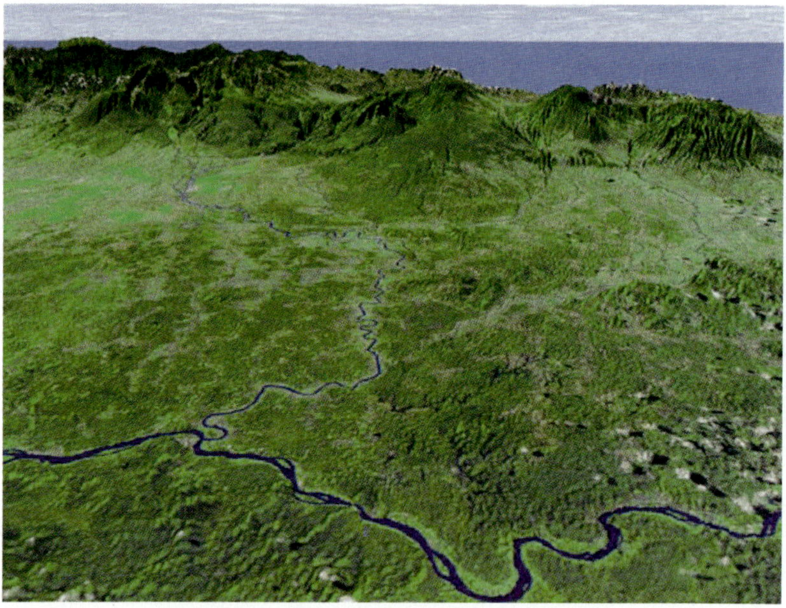

Fig. 6.2.7. This ASTER-derived image, re-projected on a DEM, captures the heterogeneity of complex terrestrial landscapes as observed with a modern high resolution sensor over complex terrestrial landscapes: 3-D view of a region in Costa Rica (Image courtesy of NASA)

6.3 Characterization of Local Vegetation Development

High temporal resolution analysis is extremely important for environmental studies and research. In particular, monitoring vegetation extension in space and activity in time allows to gather valuable information on the main ecosystems (Leith 1974, Benedetti et al. 1994). More specifically, multi-temporal NDVI data can be used to characterize the development of local vegetation, thanks to the direct relationship between NDVI and the Fraction of Photosynthetically Active Radiation Absorbed by vegetation (fAPAR) (Myneni and Williams 1994).

As an example of vegetation characterization at regional scale a land cover map of Tuscany (Fig. 6.3.1) is used to analyse multitemporal NDVI images. In particular, a long time series of NDVI data from the Telespazio NOAA/AVHRR DataBase (1985-1994), was used to compute the vegetation index time series for the main natural vegetation classes defined in the CORINE land cover classification. These time series were then interpreted in ecological terms compared to the main factors affecting vegetation distribution and development.

The CORINE Land Cover Classification of Tuscany is a remote sensing-based land use classification of the entire Tuscany Region produced at a working scale of 1:100000. The procedure followed is defined as *"computer aided photointerpretation*

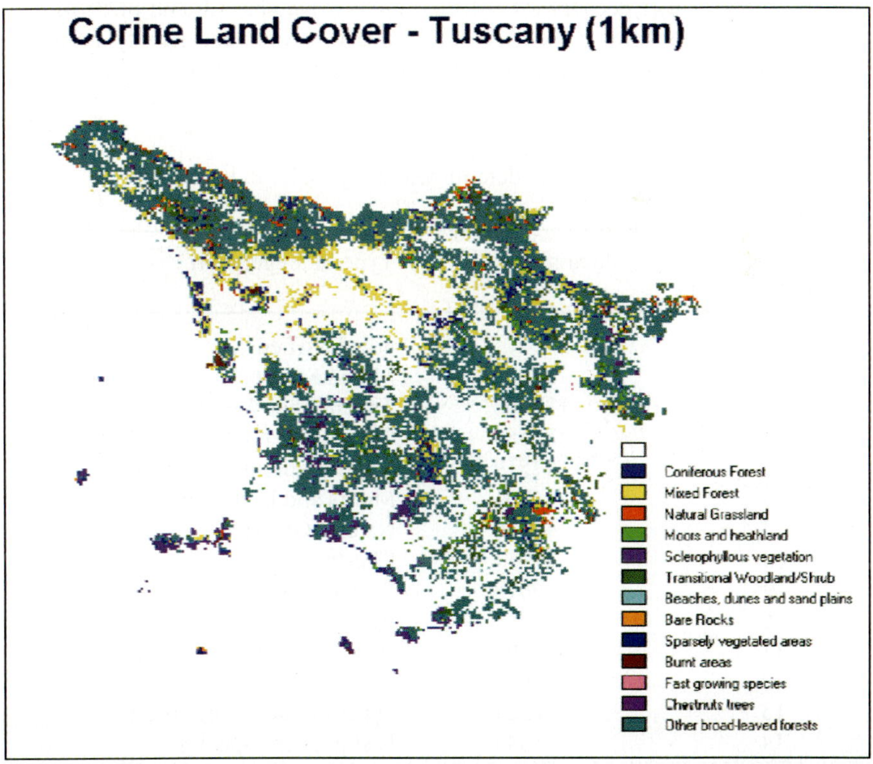

Fig. 6.3.1. CORINE land cover classification of Tuscany with the spatial resolution of 1 km

of false colour images". The satellite data used are Landsat TM-MSS and SPOT HRV XS. However, a number of ancillary data (aerial images, local cartography, etc.) were also used as well. The final land cover nomenclature comprises 3 level with 5, 15 and 44 items respectively. Due to ground resolution of images and ancillary data used, the smallest area detected is about 25 hectare, i.e. 1/4 of a AVHRR pixel.

As a consequence the NOAA pixel is well described by the CORINE database. In order to allow direct comparison between the two data sets, it is possible to rasterize the CORINE images to a 1 x 1 km resolution assigning each pixel to the more frequent class present in the 100 hectares wide square area. The area of each class is reported in Table 6.3.1 in hectares. The average annual NDVI course for the main natural vegetation classes computed over the nine study years are shown in Fig. 6.3.2 and the maximal values in Fig. 6.3.3. Ten-day sampling is reported on the abscissa of Fig. 6.3.2 starting from the 19th decade (1-10 June) till the 18th decade of the following year (21-31 May). NDVI values are reported on the Y axis. Approximately a NDVI value of 0.2 or less has been assumed as indication of absence of vegetation activity (Malingreau 1986). Some conclusions can be drawn on the basis of profile dynamics in the year. All vegetation classes show very similar dynamics with the significant exception of sclerophyllous vegetation. It can be seen that, while all other vegetation classes pass through a period of almost absence of photosynthetic activity, in the winter decades (D33-D06, i.e. 20 November till end of February), Sclerophyllous vegetation presents, even if reduced, a continuos activity during the whole year (Lloyd 1990). This is probably due to two major facts.

Table 6.3.1. Area of each class determined in Fig. 6.3.1 in hectares

Class (from Fig. 6.3.1)	Area in ha	Type of Vegetation
1		not classified
2	756813	Mixed Forest
3	126496	Transitional Woodland/Shrub
4	81383	Coniferous Forest
5	47519	Sclerophyllous vegetation
6	44861	Moors and heathland
7	40180	Natural Grassland
8	21931	Chestnuts trees
9	9098	Sparsely vegetated areas
10	8939	Burnt areas
11	5190	Beaches, dunes and sand plains
12	2959	Bare Rocks
13	2229	Other broad-leaved forests
14	1343	Fast growing species

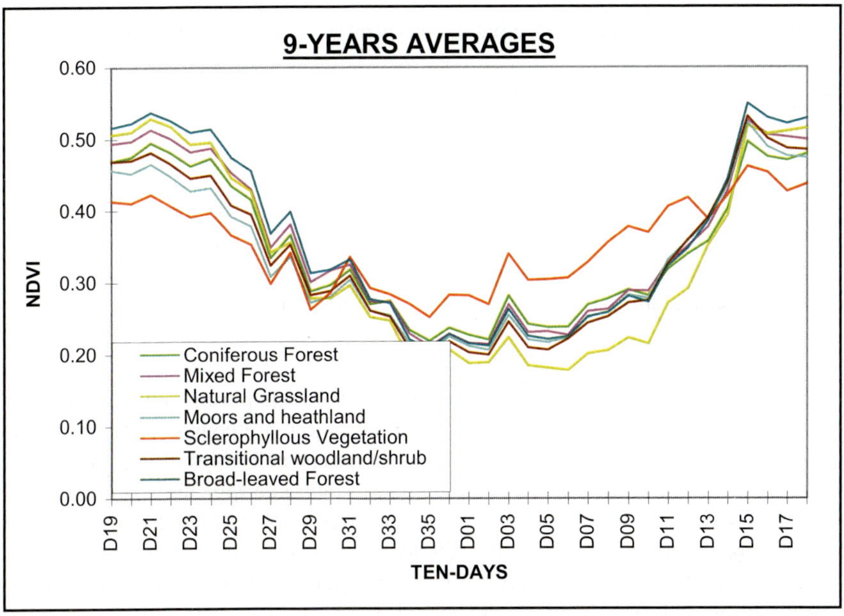

Fig. 6.3.2. Average annual course of the NDVI for the CORINE vegetation classes in Tuscany from the 19th decade (1-10 June) till the 18th decade (21-31 May) for the years 1985 - 1994 based on 10-day mean values

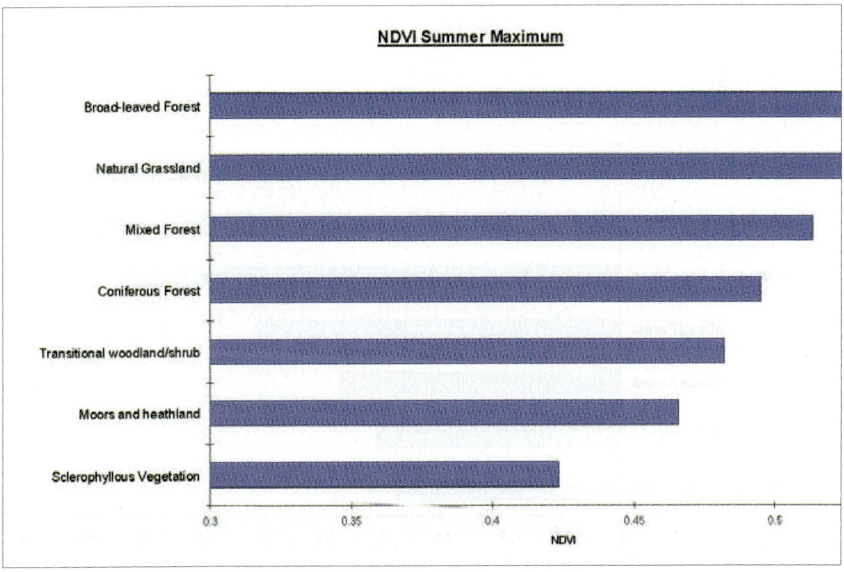

Fig. 6.3.3. NDVI summer maximum for the CORINE vegetation classes

Table 6.3.2. Average CO_2 uptake for different plant groups according to Larcher (1995)

Plant group	CO_2 uptake (mmol m^{-2} s^{-1})
Deciduous trees	10-15
Grass and sedges	5-15
Sclerophyllous	4-10
Dwarf shrubs of heath deciduous	3-6
Coniferous evergreen	3-6

Sclerophyllae are evergreen plants and preserve their photosynthetic apparatus also in winter time. Secondly they live generally in low hill coastal zones where the presence of sea allows a mild winter time when very cold days are almost absent. However their main constraint is represented by water shortage and excessively hot summer that do not allow them to have an elevated photosynthetic rate in summer months. Consequently, their profile is the lowest in late spring and summer months.

By inter-comparing the profiles of the other vegetation classes it is possible to see a gradient in activity from broad-leaved forest to Sclerophyllous vegetation especially in summer months. In the following figure the maximum activity rate in summer months is plotted for each vegetation classes.

This reflects the decreasing primary production rate of the different canopies and confirm that broad-leaved forests in good health and with an adequate water content in soils produce the highest activity rate of the region. This is in good accordance with the average maximum values for net photosynthesis for different plant group reported in the literature and measured with traditional instruments. Under optimal

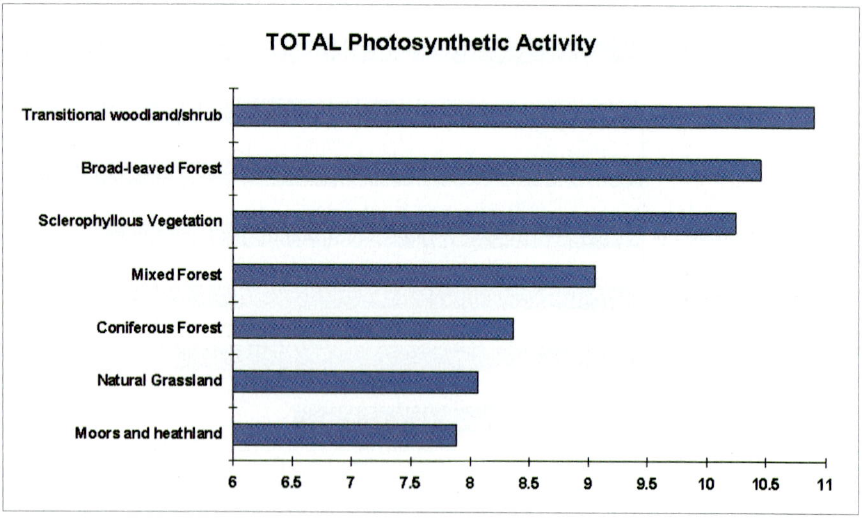

Fig. 6.3.4. Total annual NDVI sum for the CORINE vegetation classes

conditions (natural CO_2 availability, saturating light intensity, optimal temperature and adequate water supply) a number of works showed that the CO_2 uptake shows a wide range of values for different plant groups (see Table 6.3.2, extracted from Larcher 1995).

Furthermore, if we assume that the sum of NDVI decade values over 0.3 is directly correlated with the total photosynthetic activity performed during the whole year, one obtains Fig. 6.3.4.

The order of vegetation classes is altered and Sclerophillae are located in the third decreasing position. Since this parameter is probably linked to the total primary production assessed by each canopy (Prince 1991), one can conclude that as expected, the intensity of activity is not necessarily correlated to total green biomass produced and that the possibility of low but continuous photosynthetic activity during the whole year allows a considerable quantity of total primary production.

6.4 Change Detection Methods

6.4.1 Mapping and Monitoring of Land-surface Changes

Mapping of land-surface changes is an important tool for controlling and planning the correct and sustainable exploitation of environmental resources. Change detection is useful in applications such as land use change analysis, monitoring of shifting cultivation, assessment of deforestation, study of changes in vegetation phenology, seasonal changes in pasture production, damage assessment, crop stress identification, disaster monitoring, and hydrological information. Monitoring of land-surface change is essential, in particular, for developing countries, where they are usually extremely fast, dramatic in amplitude and uncontrolled.

It is a common experience that land surface-changes occur at extremely diversified temporal and spatial scales. If one look at such changes from an environmental perspective, these scales correspond to those which characterize the temporal and spatial variability of land surface-processes (i.e. variations in land cover and conditions, water and carbon balances, primary productivity, etc.). From this viewpoint, the spatial scales of interest can range from those affecting individuals (organisms, plants) to plots (fields, stands), basins, regions, and continents; while temporal scales range from instantaneous to daily, seasonal, multi-year and historical.

The origin of such changes may be both natural and human-induced, but more often the two sources of changes are intermingled in a complex way, which leads to multiple-scale processes in both spatial and temporal dimensions. At the field level, for example, changes in vegetation cover and conditions are determined mostly by variations of plant water status and phenology during the growing season, while inter-year land cover changes are determined by agricultural practices and changes on the longer period are affected by both anthropic actions and climate variability.

This implies that the monitoring of land-surface changes can not be carried out without a clear comprehension of their environmental origin and properties, which is essential in deciding the appropriate characteristics of the instruments and methodologies to apply (Waring and Running 1998). This is true for methods based both on conventional ground measurements and remotely sensed data, but, while the collection and processing of the former are often expensive and time consuming, the latter offer great advantages thanks to their simplicity and relatively low costs and human time requirements. Change detection and interpretation have therefore become major applications of satellite remotely sensed data, because of repetitive coverage at short intervals and consistent image quality (Sing 1989).

When using remote sensing data for change assessment, the properties to consider for each application concern the spectral, spatial and temporal features of the sensor which acquires the data, as well as the processing methods which will be used. In practice, no sensor is capable of providing the optimal resolutions for following all kinds of environmental phenomena, so that different data types must be used separately or jointly in each specific case. This often originates the need for integrating data with different spectral, spatial and temporal features. The current section will deal with the spatial and temporal scales of land surface changes as can be viewed by remote sensing techniques, and particularly by the use of different satellite systems.

6.4.2 Change Detection Techniques

The digital nature of satellite imagery makes them suitable for computer-aided analysis of land cover changes. Several automatic or semi-automatic change analysis techniques have therefore been developed in the last few years. The basic premise of all these techniques is that changes in land cover must result in change in radiance values which are higher than changes due to external, disturbing factors (differences in illumination conditions, soil moisture, vegetation phenology, etc.). Since these factors are sometimes very influential on radiance values, some of the methodologies developed are specifically aimed at their reduction, which is generally contemporaneous to an enhancement of the useful, interesting signal about land cover changes.

The main methods for change detection using satellite data include image differencing/ratioing, vegetation index differencing, image regression, principal component analysis, change vector analysis and post-classification comparisons. All these methods have some advantages and some drawbacks, which highlights the importance of a proper selection of the most efficient method for each environmental situation and data set used. A brief review will be therefore presented, which will serve to justify the selection of the specific methodology applied.

As a premise to the description of these methods, it must be mentioned that accurate spatial registration of the two images is essential for most change detection techniques. If this phase is not accomplished well, this may hamper the efficiency of the all procedure and produce results which are inaccurate or even completely useless.

In univariate image differencing two images taken at time t_1 and t_2 are subtracted pixel by pixel to produce a further image which represent the change between the two times. The input data can be raw or corrected images, and this often causes differences in the results obtainable. In general, a threshold is applied after the differencing to identify areas of changes, and the selection of the suitable threshold is the most crucial step of the whole procedure. Vegetation index differencing is practically a variation of this method, where a vegetation index is used instead of the orignal bands. This latter method is obviously specifically suited to identifying changes in vegetation cover, which, however, may be affected by the disturbing factors previously mentioned.

A similar method to highlight areas of possible changes is image ratioing, where two registered images from different dates are rationed on a pixel by pixel basis. In the image which is obtained the changed areas will display values significantly different from 1, which can be isolated again by the use of suitable thresholds. Similarly, in image regression, pixels from time t_1 are assumed to be a linear function of the time t_2 pixels. The residuals of the regression between the two images can therefore be interpreted as change values. Also these methods, however, suffer from the same problems already mentioned, especially in complex areas where several factors can influence the spectral responses of the surfaces in a way difficult to predict.

Principal Component Analysis is generally used to reduce the number of spectral image channels to fewer, uncorrelated components, but it has also been used successfully for change analysis studies. When applied to multitemporal/ multispectral scenes, principal component analysis results in fact in gross variations associated with overall radiation and atmospheric changes appearing in the major components, and statistical minor changes associated with local changes in land cover appearing in the minor component images. The method is particularly efficient when applied to only two images (or image transforms, such as vegetation indices) taken at times t_1 and t_2. In these case in fact, the second principal component, PC2, contains only information related to changes (Fig. 6.4.1). Thanks to these features, this method has been applied with good success in several operational cases (Singh 1989).

Change vector analysis is based on the concept that each pixel of a multispectral scene can be described as a vector in the n-dimensional feature space. When considering two scenes from different dates, the direction and magnitude of change from the first to the second date is represented by a change vector. The decision that a change has occurred is made if the magnitude of the computed spectral change vector exceeds a specified threshold criterion. The direction of the change vector gives information about the type of the change. The main advantage of the method is clearly that it fully exploits the multispectral information contained in the scenes, but it may suffer from the same problems of all threshold-based procedures.

Post classification comparison is the most obvious method of change detection which requires the comparison of independently produced classified images. By properly coding the classification results for times t_1, and t_2, the analyst can produce change maps which show a complete matrix of changes. In addition, selective grouping of classification results allows the analyst to observe any subset of changes which may be of interest. Post classification comparison holds promise because data from two dates are separately classified, thereby minimizing the problem of normalizing for atmospheric and sensor differences between two dates. The method also bypasses the problem of getting accurate registration of multi-date images. However, if one considers the land cover classification generated from a single date of Landsat data, it is not difficult to see that the change map product of two Landsat classifications is likely to exhibit accuracies similar to the product of the accuracies of each individual classification. Hence it can produce a large

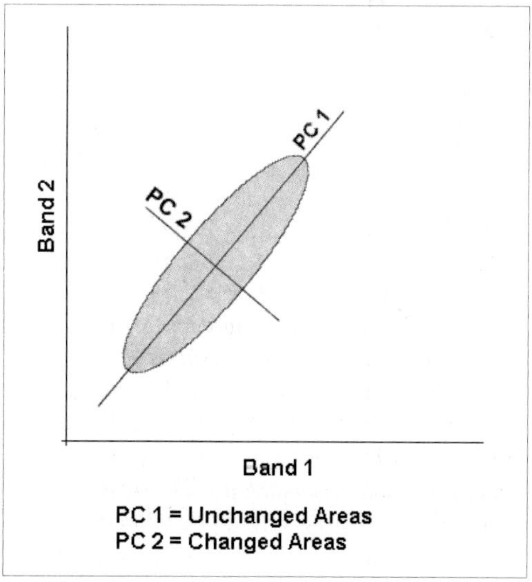

Fig. 6.4.1. Use of PCA to identify changed areas

number of erroneous change indications since an error on either date gives a false indication of change. For example, two images classified with 80 per cent accuracy might have only a 0.80 x 0.80 x 100 = 64 per cent correct joint classification rate.

Fisher and Pathirana (1993) have argued that among the causes of inaccuracy in land cover classification is the inability of conventional classification approaches to manage the problem of mixed pixels. They and others have therefore proposed the use of a fuzzy classifier (Wang 1990), or use of all the probabilities derived from the maximum likelihood classifier as a means of deriving information about which cover types contribute to mixed pixels. This approach was actually successful in reducing erroneous change detection rates, but requires a more careful interpretation of the results, since fuzzy change grades are more difficult to interpret than usual hard change maps.

6.5 Scales of Land-surface Variability

6.5.1 Spatial Pattern

In this section it is discussed how the structure of the Mediterranean landscape may limit the assessment of changes if the observational scale grows. Examples will be given of the spatial variability of soil characteristics in Castilla-La Mancha and the structure of agricultural respectively forested areas in the investigated landscapes. Since changes may occur in time at closely adjoining areas (fires!) in opposite directions an observation at a larger scale may not capture such changes.

Mediterranean Topographic Structures

In the Mediterranean area one deals with the wide range of structures from small agricultural fields to large nearly uniform woods or agricultural monocultures and deserts. Geological processes created a large variety of soil formations that exist in close neighbourhood. One example, mentioned in connection with Fig. 3.5, is the co-existence of grey and red soils in Tuscany. Fig. 6.5.1 shows a picture of one of these areas in Tuscany which is a composite of SPOT channels 1, 2, and 4. East of the Autostrada which can be seen in the centre most of the fields consist of bright greyish soil while in the west of it and in the south reddish soil is predominant. In channel 1 the measured values over the bright field are about 30% higher than the reddish ones. In channel 2 the relation is slightly smaller and drops to 16 % in channel 3 and nearly zero in channel 4.

Fig. 6.5.1. Valley of Collesalvetti in Tuscany with the coexistence of two types of soil. Composite of SPOT channel 1, 2, and 4 values of 07. August 1998 after most of the fields (mainly wheat) have been harvested

In agricultural areas the average size of quasi-homogeneous fields varies considerably and even within a field soil quality and vegetation growth can be variable. Crop types change from field to field, during the course of the year, and from year to year. To an increasing extent fields are covered by plastic greenhouses that introduce a new element to the land cover (Fig. 6.5.2).

The following figures present examples of different land use pattern: the vine area east of

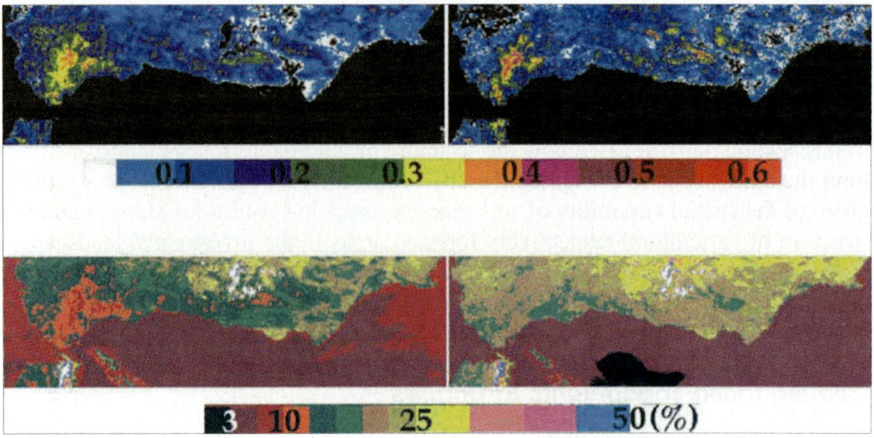

Fig. 6.5.2. The figure shows four views at a section of the southern coast of Spain. In the upper part the NDVI for 12 June 1991 (left) and 25 August 1994 (right) is to be seen, in the lower part the reflectances computed for AVHRR channel 1 (left) and channel 2 (right). The upper scale is for NDVI, the lower one for reflectance (TOA). While the peninsula of Almeria stands out clearly in the lower pictures (yellow colour), it is erased in the NDVI due to the equal reflectance in both AVHRR channels of the plastic greenhouses that cover the whole peninsula

Tomelloso and irrigation pattern west of Albacete in Spain, an agricultural area near the mouth of the Cornia river in Tuscany, Italy, and a section of the Messara valley of Crete. Castilla-La Mancha had been selected as experimental site of the EFEDA pilot study because it is plain and looks very homogeneous with monocultures of grapevine over large areas. If one looks closer even here a considerable variability is observed. 6.5.3 reproduces a false colour image of the area (SPOT channels1, 2, and 3) with four profiles of which three are expanding radially from the centre of the town. The values of channel 3 counts and of the NDVI derived from SPOT channels 2 and 3 are plotted in Fig. 6.5.4. In the case of the NDVI it must be observed that the values derived from SPOT measurements are not identical with those derived from AVHRR measurements, but here only scaling effects are discussed. It can be seen that the correlation between the channel 3 signal and the NDVI is low. The vegetation index remains fairly constant at a very low level while the variability of channel 3 signals (and consequently also those of channel 2) are primarily determined by the variability of the soil type and surface structure. The strongly oscillating Ch-3 record is not markedly different in the three directions.

Figure 6.5.5 shows the Messara valley on Crete with two nearly parallel profiles (Fig. 6.5.6). In this case the vegetation index shows more correlation with the variability of the Ch-3 record. Its average value is higher. In the centre of the records a strong difference can be seen since the more southern traverse goes through less cultivated land (compare Fig. 6.5.5).

A closer view at the structure of the surface is obtained if the full resolution of the SPOT images is visualized. This is done in the following figures in which profiles of

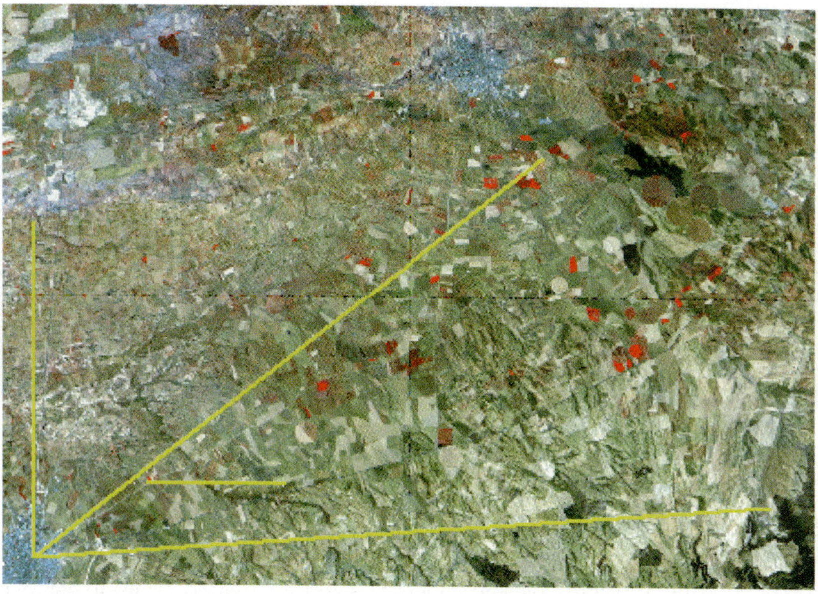

Fig. 6.5.3. Area east of Tomelloso, Castilla - La Mancha, Spain with profiles. SPOT image of 07.07.1994

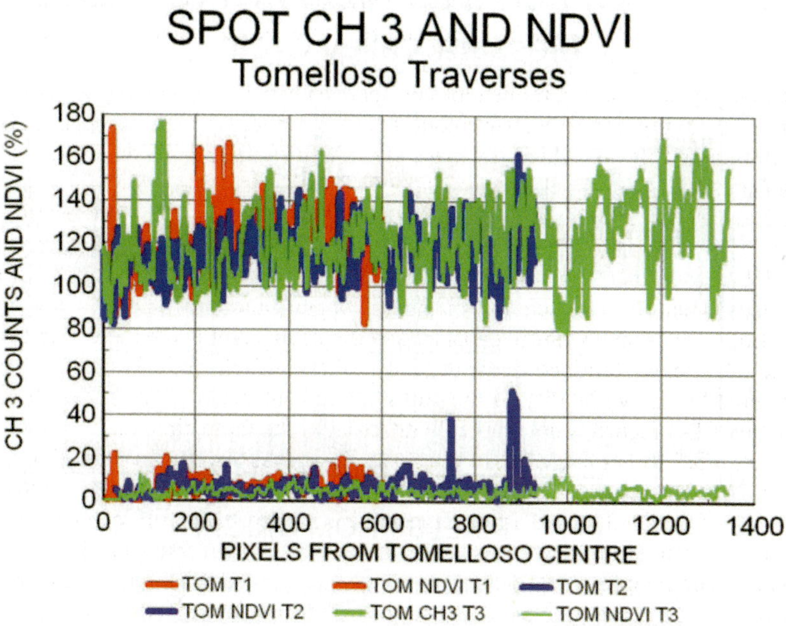

Fig. 6.5.4. SPOT channel 3 counts and NDVI for profiles 1 (N–S) and 2 (inclined) of Fig. 6.5.3.

Fig. 6.5.5. Messara valley, Crete, with two profiles, The location of the greenhouses of Fig. 1.14 is at the left hand side about one third from the top and near the coast

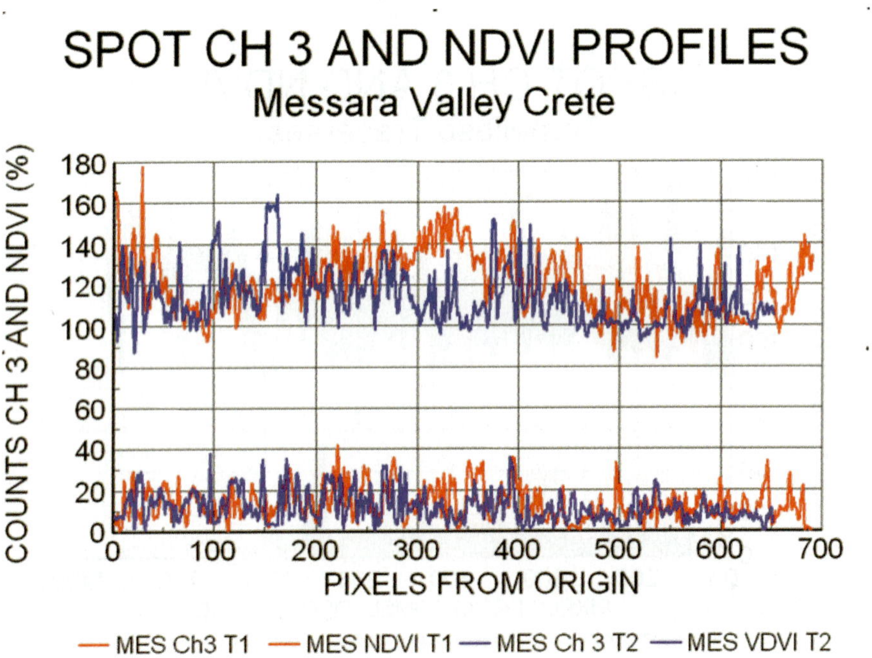

Fig. 6.5.6. SPOT channel 3 counts and NDVI for two profiles through the western part of the Messara valley as shown in Fig. 6.5.7

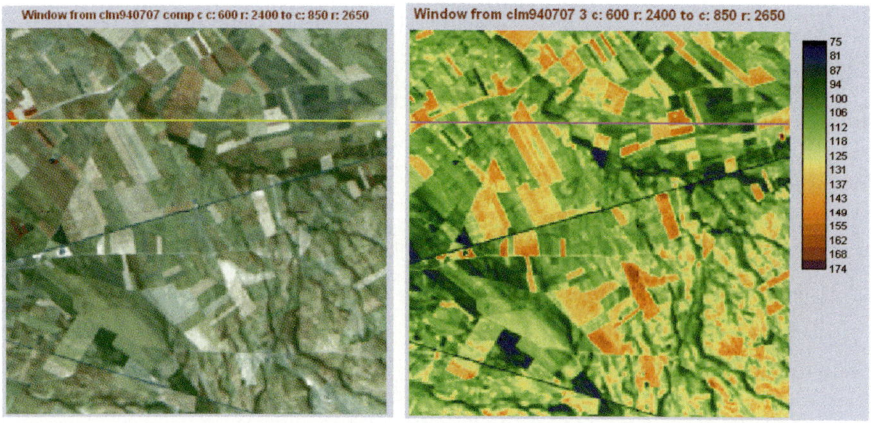

Fig. 6.5.7. High resolution view at the short profile shown in Fig. 6.5.3. The full resolution SPOT image (to the left a colour composite, to the right channel 3 reflectances) reveals the fine structure of the landscape: The area is streaked with rills and erosion marks stemming from the rainy period. Even within single fields different colours can be detected

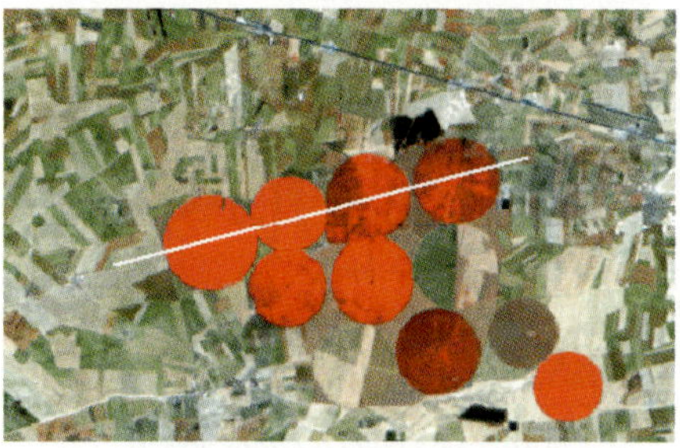

Fig. 6.5.8. Profile through four irrigation circles near Barrax in Castilla-La Mancha

Fig. 6.5.9. The figure shows how soil structures can be seen through thin vegetation canopies in irrigation circles

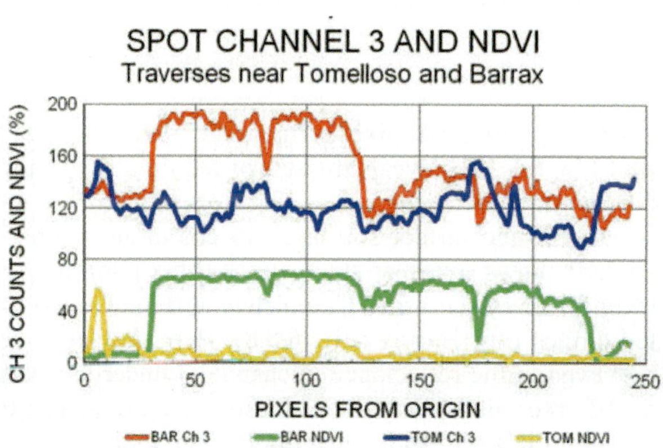

Fig. 6.5.10. Cross section of SPOT channel 3 and NDVI values through the 250 wide profiles shown in Figs. 6.5.7 and 6.5.8. The figure shows strong structures in the channel 3 track not or to a much lesser degree reflected in the NDVI

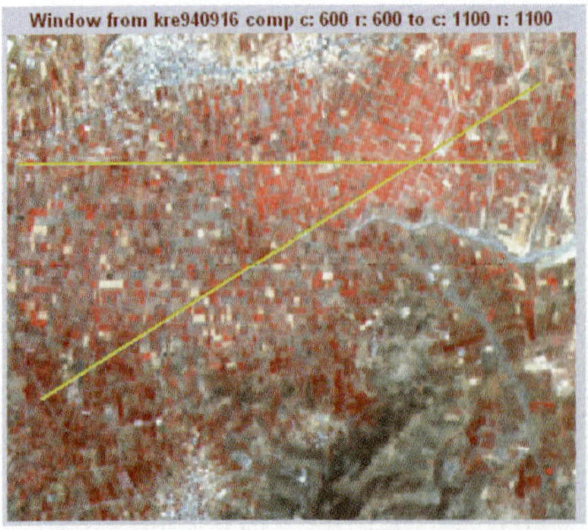

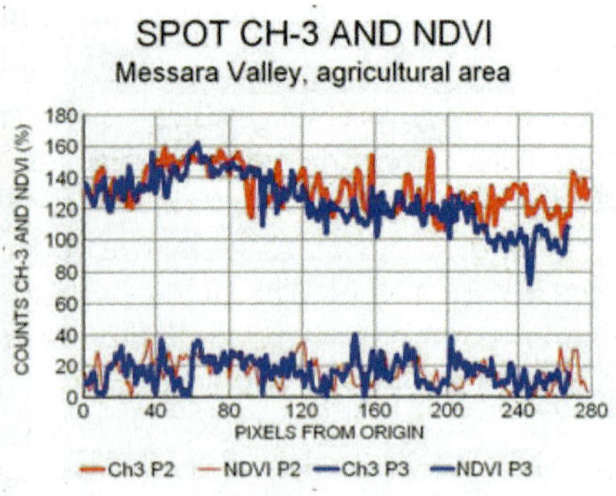

Fig. 6.5.11. Section of the Messara Valley presented as false colour image combined of the SPOT channels 1, 2, and 3

Fig. 6.5.12. SPOT channel 3 counts and NDVI for 16 September 1994. Profile 3 (blue) is the more inclined one

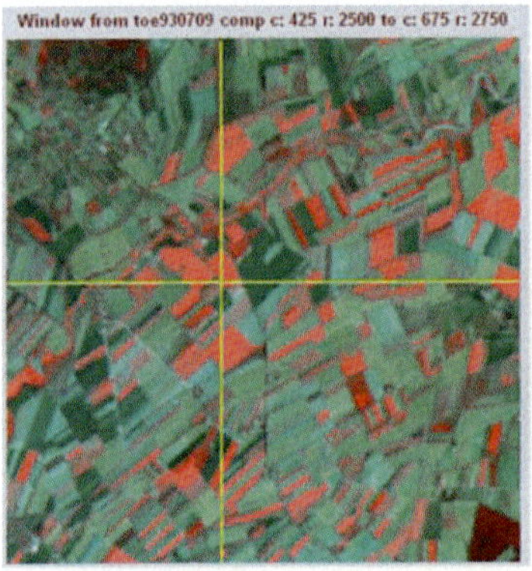

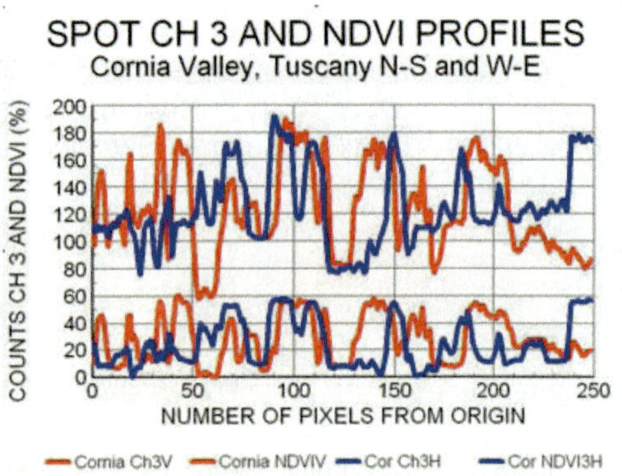

Fig. 6.5.13. Profiles through a colour composite of an agricultural area near the mouth of the river Cornia in Tuscany, Italy

Fig. 6.5.14. Cross sections through the area shown in Fig. 6.5.13. V (red) stands for the vertical line, H (blue) for the horizontal one

250 SPOT pixels are analysed. The 250 x 250 pixel images 6.5.7 show the area of the short horizontal profile of Fig. 6.5.3. It reveals the size of the fields and shows that erosion features like rills run through the area stemming from the rainy periods which have changed the structure of the surface soil layer, its colour and probably also its composition. Fig. 6.5.10 shows irregular fluctuations (blue line) and low vegetation indices (yellow line) for this profile. Completely different looks the profile through the irrigated circles near Barrax (Fig. 6.5.8). Here the vegetation index remains at a high level even in the reflectance in channel 3 undergoes large variations (Fig. 6.5.10 red and green lines). Fig. 6.5.9 illustrates that even in the irrigated areas features of the top soil layer can be seen through the canopy.

In the closer view of Fig. 6.5.11 at the Messara valley a large number of small rectangular fields (Fig. 6.5.11) can be recognized. Also in the higher resolution profiles of Fig. 6.5.11 a rather smooth track for both the channel 3 counts and the NDVI can be seen. Obviously the small field size smooth out the variability between the fields and the small pathes between them.

Completely different is the structure of an agricultural area of Tuscany shown in Figs. 6.5.13 and 14. The fields are large and the variability between vegetated and ploughed fields is strong. Both the vegetation index and the reflectance measured in channel 3 have large amplitudes.

The question now arises what effect scaling may have on the quantitative assessment of radiometric properties and the vegetation index. To investigate this issue, four areas of Tuscany indicated by the rectangles in Figure 6.5.15 were used. These are a large terrestrial area (mostly Provincia Livorno) containing nearly 6 million pixel (large black rectangle), a mainly agricultural sub-section of this image with some forests (small black rectangle) near the mouth of the river Cornia (0.576 million pixels), a large coastal area (coded L, 2.83 million pixel) containing a strip of the sea (white rectangle) and a small coastal area (coded S, 0.384 million pixel) with a strong gradient of reflectances from forest over an agricultural area to the sea (blue rectangle).

The small black rectangle can be seen in the upper parts of the Figs. 6.5.16 and 17 for 12 March and 7 August 1998 in original resolution for SPOT channel 3 and the SPOT NDVI. Below are the re-sampled images to the resolution of an AVHRR

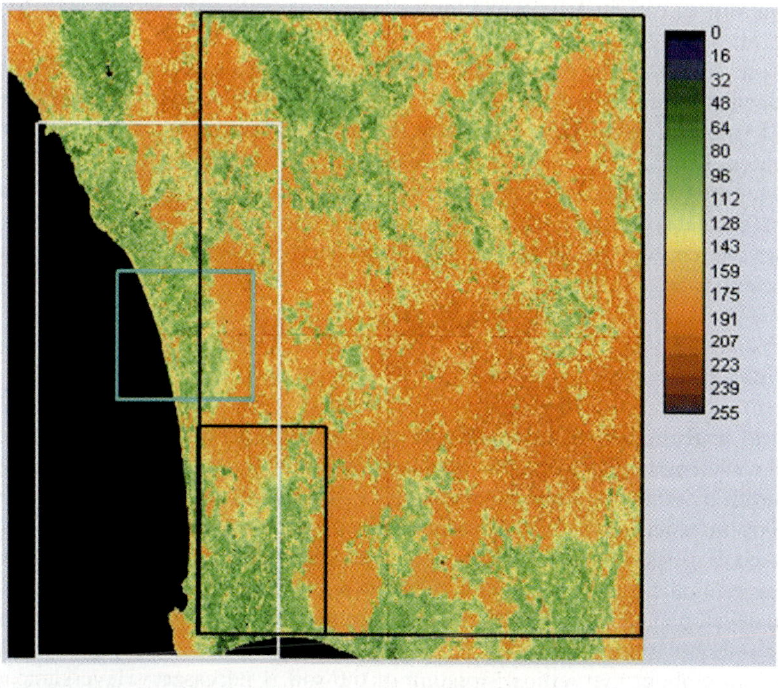

Fig. 6.5.15. SPOT NDVI image of 07.08.1998 with test areas

image. By comparing the images the following has to be kept in mind. Fig. 6.5.15 is a NDVI image in a relative scale which was stretched to avoid the occurrence of negative values and to enhance contrasts. Fig. 6.5.16 was derived from the original data but the colour scales were adjusted to cover just the whole range of values. Therefore the same colours represent different values. The original SPOT values have been averaged over 40 x 40 pixels to construct an image similar to the ones produced by AVHRR with a resolution of about 1 km. Each "AVHRR" pixel represents the average of 1600 SPOT pixels and the areal averages in both the original and the aggregated images are identical provided that one applies pixel accumulation (and not line omitting) in the averaging process and selects an area which results, when degraded, in an image of an integer number of AVHRR pixels.

The re-sampling process narrows the distributions because extreme values are mixed with less extreme ones. Comparing the data in the upper and lower parts of Figs. 6.5.16 and 17 it can be seen that the range of the scale is shrinking for the aggregated data and consequently the colour code gets a different meaning. As can be seen in Fig. 6.5.16 and 17 the averaging process creates a different spatial distribution of reflectances and NDVI values, the histograms look quite different. Furthermore the colour scale of the NDVI is inverted with respect to those used for chanel 3.

Table 6.5.1 presents an overview of the effect of scaling from SPOT to AVHRR resolution. It shows the original SPOT values of channels 2 and 3 and the NDVI computed from them averaged over the four areas previously defined. The data show the strong decrease of the land-NDVI from values of 52% for the agricultural area in May to 16% in August and for the larger area with more woods from 61% to 27%.

If one computes the mean NDVI from the mean values of channel 2 and 3 data, one obtains a value which differs from the average of the pixel by pixel determined NDVI values. The difference between the two numbers grows as the gradient of individual numbers becomes large across the image (highly bimodal distribution). Thus for the two images which contain only land the difference is only of the order of 1% (absolute) or below. This error grows considerably for the areas which include the sea (which, like the dry soil, in this uncorrected data set has negative vegetation index values).

Vertical Structures

Vertical structures of canopies and soils can hardly be assessed by remote sensing. At all wavelength the penetration depth into solid and liquid material is very limited. However, by combining measurements of surface quantities with models and other *a priori* information, some information about processes in deeper layers can be assessed. E. g. some information about soil moisture in the root zone can be obtained by observing the state of vegetation which is fed by this moisture. Techniques have been developed to determine soil moisture profiles by combining microwave measurements with hydrological models. To apply this technique properly, information about the vertical structure of the soil is necessary. Haverkamp (see: Vauclin et al. 1992) pointed out the importance of the layering of the soil for

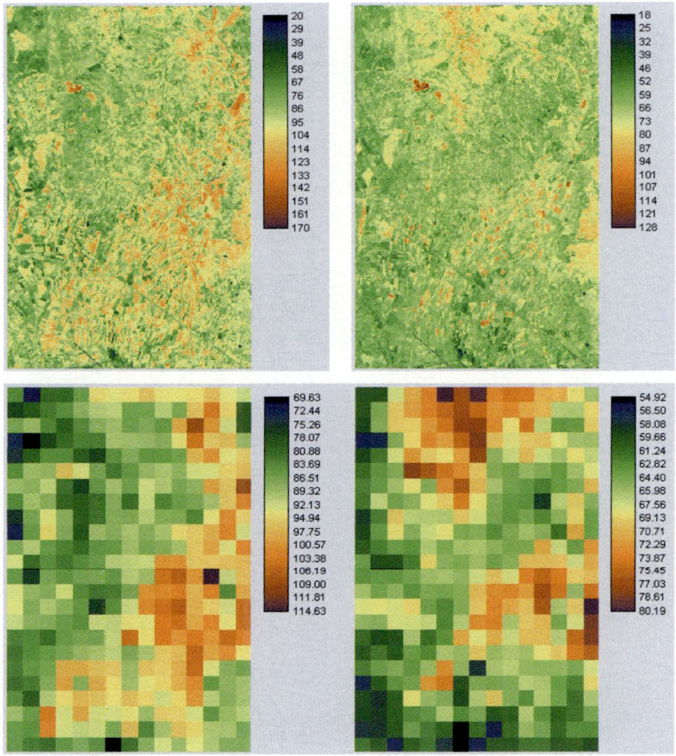

Fig. 6.5.16. SPOT channel 3 image of Cornia valley. Left 12.05.1998, right 07.08.98, top original resolution, bottom degraded to AVHRR resolution (averaged over 40x40 SPOT pixels)

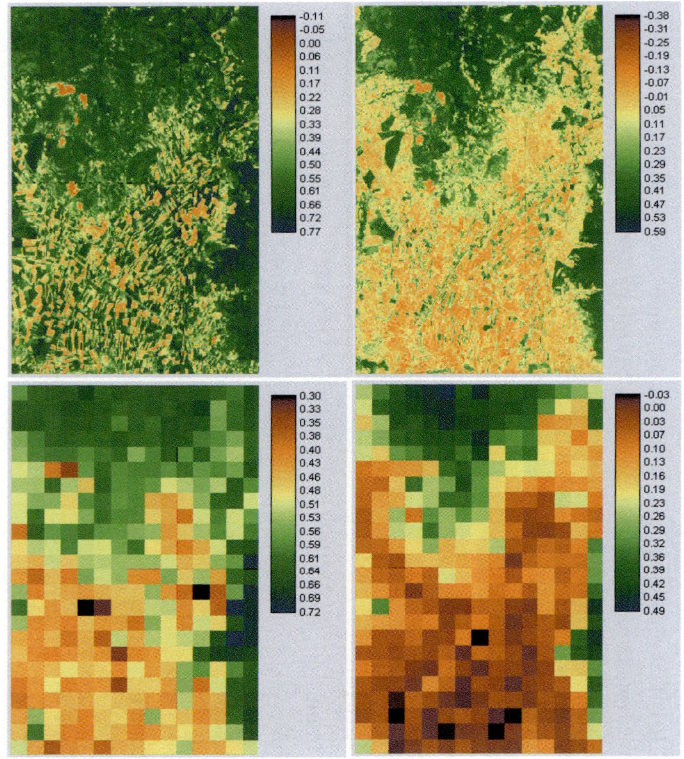

Fig. 6.5.17. SPOT NDVI image of Cornia valley. Left 12.05.1998, right 07.08.98, top original resolution, bottom degraded to AVHRR resolution (average over 40x40 pixels

Table 6.5.1. Effect of aggregation on area mean values. In the first of the NDVI columns the vegetation index was computed pixel by pixel from the original SPOPT data and then averaged over 40 x 40 pixels. In the second column it was computed from SPOT channel 3 and 4 reflectance values averaged over 40 x 40 pixel (~1 km^2). The five line "profiles" refer to those in Fig. 6.5.3 and 6.5.8. Here the line mean values are given

Area and no. of pixel	Date	SPOT Ch 2	SPOT Ch3	NDVI	
				SPOT	Comp.
Livorno Land 5990400	12.05.1998	23.241	96.836	0.6105	0.6129
	07.08.1998	43.043	73.714	0.2759	0.2627
Cornia Land 576000	12.05.1998	28.893	90.506	0.5185	0.516
	07.08.1998	49.134	67.188	0.1691	0.1552
Coastal Zone L 2830400	07.08.1998	34.613	40.372	-0.08	0.077
Coastal Zone S 384000	07.08.1998	38.536	50.997	0.039	0.1392
Profile 1 601	07.07.1994	107.48	121.25	0.061	0.06
Profile 2 929	07.07.1994	101.76	115.35	0.065	0.063
Profile 3 1340	07.07.1994	112.98	122.43	0.041	0.04
Profile 4 250	07.07.1994	104.32	120.06	0.072	0.07
Profile 5 Barrax 250	07.07.1994	52.868	150.66	0.4885	0.4805

agriculture apart from its state of physical, chemical, and/or biological degradation. In the Tomelloso area a spongy calcite crust layer underlies the arable top soil layer of 50 to 80 cm depth and forms a water supply reservoir. It acts as a buffer for winter rain water that is released to the plants in summer. The extremely dry arable top layer helps to maintain this sponge effect by sealing of the more humid crust soil from evaporation. Only few plants like vine and olive trees can survive under these conditions without irrigation. In the Barrax area the soil is irrigated and the arable top soil layer of about one metre depth is the horizon for plant growth, mainly annual crops.

Very dry surface layers are also faced in Italy where, as an example, in summer

4.3 % moisture have been measured near 5 cm depth, 9.8% at 10 cm depth and 8.9% at 20 cm depth (Campiglia M.ma, 31.07.1990, 11:30 MEST). These soil structures have to be taken into account if soil moisture profiles are inferred from remotely sensed top soil layer moisture and hydrological models (see section 5.6).

6.5.2 Classification and Aggregation

In previous chapters it was shown, which quantities can be derived from measurements made in space and how the algorithms used to infer them can be tested by simultaneous measurements made at the ground. These ground measuring sites generally are representative for not more than 100 to 10^4 m^2 while it is expected that applications of the information inferred from measurements made by satellite sensors cover much larger areas. The question to be discussed here is the relationship between the variability within a target area and the spatial resolution achieved by the satellite sensors.

It was mentioned that land-surface changes can occur at spatial scales that can range from individual elements up to the entire continents. The analysis of these scales in the present context must therefore start from the consideration of how remote sensing data are capable of observing and reporting information on these objects and the processes that change them. The use of digital remote sensing data, in particular, implies that reality is represented with a fixed and discrete grid of spectral observations, corresponding to the image pixels. Thus, the identification and study of surface objects is only possible if the pixel resolution is sufficiently high. On the other hand, if this resolution is too high, intra-object spatial variability is retained in the images which can complicate their interpretation (Woodcock and Strahler 1987). This trade-off has been investigated in several studies, which have proposed a series of solutions to the problem. In general, since the resolution of a certain data type is fixed while the surface objects have different sizes, the preferred strategy is to acquire data with the highest possible resolution. This is however expensive and requires the use of subsequent spatial processing aimed at reducing the above mentioned intra-object variability. Several of these processing methods are based on spatial smoothing or aggregation operations, whose success is critically dependent on the size of the area to consider in each case.

The existence of different optimal aggregation levels for the identification of different cover types has been experimentally confirmed by Gong and Howarth (1992a, 1992b) at the urban fringe of a Canadian town and by Conese and Maselli (1996) for a sub-tropical landscape in Senegal (Western Africa). In these cases, variable sizes of a moving window were found to produce optimum frequency histograms for the classification of different land cover classes. The size of the moving window was mainly dependent on the relationship between the pixel size and the mean size of the objects examined (plots, fields, buildings, etc.), but other complicating factors could intervene connected with within-class spatial variability. Unfortunately, the existence of different optimal window sizes for the identification of the ground elements in the same landscape is an obstacle to the use of simple filtering techniques. The application of a single moving window to an image tends in fact to facilitate the identification of some objects at the expense of others with different spatial structures. As already mentioned, the problem is especially relevant to the classification of spatially variable, complex environments, where several cover types are present with varying characteristics.

This calls for a multi-scale approach to the classification of complex landscapes. The problem has been dealt with by Marceau et al. (1994a, 1994b), who proposed the use of intra and inter-class variances to identify the optimum aggregation level of remotely sensed data with respect to the study of ground elements. A more efficient solution, however, should be inherent in the specific objective of the study, in the current case image classification. Since several aggregation levels are contemporaneously optimal for the classification of a given area, a method for merging the interesting spatial details of images with different spatial structures would be desirable. Actually, as the optimal scale for the identification of scene elements varies not only between cover categories but also within each class, the issue should be addressed by a procedure capable of identifying this parameter for each pixel. The proposal recently put forward by Maselli et al. (1996) is that such an identification can be carried out using a measure of classification reliability based on the full exploitation of the class membership information produced during the pixel attribution process.

Several publications (Wang 1990a, 1990b, Foody 1992, 1994) demonstrated that multivariate discrimination procedures can yield much more information than is conventionally utilized for the production of "hard" thematic maps. In particular, the maximum likelihood membership grades obtained following the theory of "fuzzy" sets are highly informative on the reliability of the attribution of a pixel to the cover classes. A fuzzy membership grade close to 1 indicates a confident assignment, while low membership grades are indicative of unreliable assignment (Maselli et al. 1994). This property can be utilized in the present context to identify the optimum resolution for the attribution of each pixel in a remotely sensed scene. In practice, if various versions of the same scene differently filtered are available, the optimum filtering size for the identification of each pixel can be found by maximizing the relevant fuzzy membership grades. The rationale for this is that the spatial scale for the most confident attribution of a pixel to a class should correspond to the highest relative membership grade. In this way the process is calibrated with respect to the characteristics of every single pixel. The original spatial resolution of the images is not completely lost as in the conventional filtering techniques, since the highest resolution details are retained when corresponding to confident attributions. On the other hand, when the highest resolution is not sufficient for reliable attributions, new, larger aggregation levels are evaluated. Finally, the classification is accomplished based on the most confident attributions among those examined in the multi-scale data set. In alternative, the maximised fuzzy membership grades can be retained for further analysis, which could be useful for probabilistic change assessment.

6.5.3 Temporal Scales

As in the case of spatial scales, the use of remotely sensed data must correspond to the response time of the land surfaces to the environmental processes which determine the changes. This knowledge is essential in deciding which is the appropriate revisiting time to monitor each process. An obvious consideration is that the image acquisition frequency should be higher than the typical frequency of variation of the observed process. In this regard it must be kept in mind that the actual acquisition frequency is often much lower than the theoretical one, especially for optical or thermal infrared data which are affected by atmospheric perturbations. These can reduce the real acquisition frequency to a point which *a priori* is often undeterminable, depending on the average cloudiness of each area/season (Holben 1986).

Land-surface processes occur in a large span of time scales of which only a small section can be monitored by remote sensing from space. High frequency changes such as changes due to cloud fields and short term precipitation events eventually can be followed in intervals of ten minutes from geostationary satellites at moderate spatial resolution. The main capacity of remote sensing methods lies in the assessment of decadal to interannual changes.

Figs. 6.5.18, 5.37 and 5.38 give an impression how quickly land-surface reflectance and temperature recover after a rain event of short duration provided it is followed by clear sky conditions at daytime. The typical time constant is one day. After strong rainfall and under cloudy weather conditions the surface layer may remain wet for a couple of days. Vegetation may change considerably within a period of ten days and its annual cycle is clearly reflected in ten days averages of the

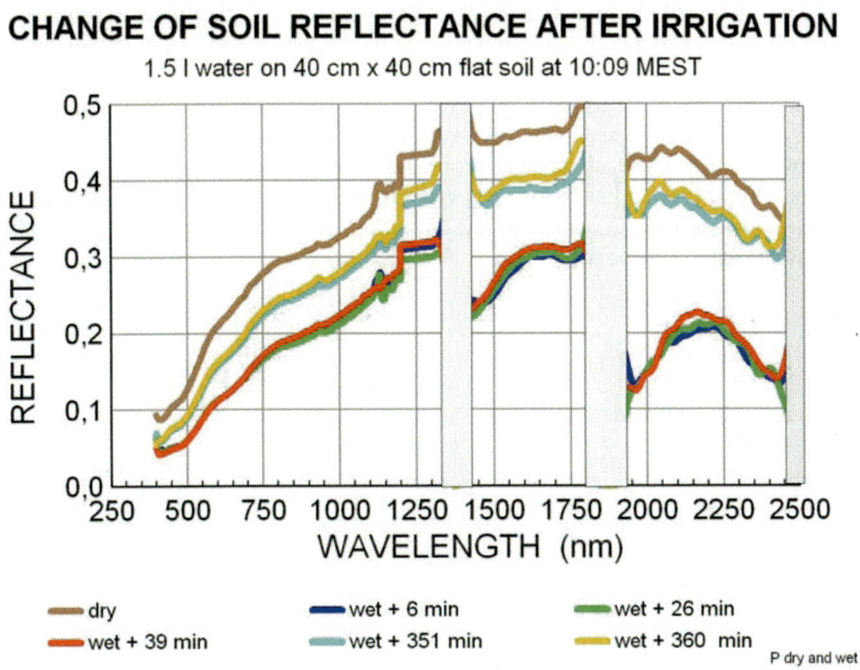

Fig. 6.5.18. Reflectance change of the soil shown in Fig. 3.5 lower panel after irrigation at 10:09 UT

vegetation index. Land use changes from year to year as far as crops and vegetables are concerned.

When studying changes in vegetated areas, two processes must be taken into consideration, shifts of biosynthetic activity due to anomalous weather situations and slowly progressing land-use changes. Depending in the interest in one or the other processes different observation sequences have to be taken. For example, the monitoring of crop water status during a growing season ideally requires daily data, which are generally difficult to obtain with suitable spatial resolutions. When instead studying land use changes for agricultural or forest inventories, intra-seasonal variations are often considered as noise, and the preferred strategy is to use multi-year images from a season in which vegetation is rather stable to minimize the effect of differences in vegetation phenology.

After a fire, vegetation may recover within five to seven years sometimes with different species but may also stay bare for longer periods, depending on the quality of the soil. Fig. 1.5 is an impressive example how long an anthropogenic impact remains visible in a matorral land. The same time it takes for a weathering process of soft stones as can be seen in Fig. 3.2. The freshly broken stone was lying in its position for about ten years. Other processes such as the degradation respectively decomposition of soils (salinization, mineralization, humification, argilization, carbonization) develop at longer time scales. Sanchez Diaz et al. (1992, 1994, 1996) not only mapped the soil types in the Castilla - La Mancha test area but also determined and mapped the stage of soil degradation that can be categorized as physical, chemical, biological, and total. This degradation takes place essentially at three time scales of which two are well beyond the present observation capacity of satellites.

A critical task is the assessment of trends. In view of what has been said before, trends can only be assessed with time series which are much longer than the typical relaxation times of the underlying processes and the quasi-periodicity of weather events. One therefore has to be very careful in the interpretation of the available satellite time series with respect to trends that may continue in the future. Examples will be given in Appendix 8.

6.6 Combination of Satellite Data of Different Provenience

6.6.1 Integration of Data with Different Spatial and Temporal Features

The above mentioned existence of mutiple-scale processes in both the temporal and spatial domains limits the utility of current systems for operational monitoring of land-surface changes, due to problems of under sampling in space and time (Richards 1993, Bolle 1996). While in fact high spatial resolution sensors such as Landsat Thematic Mapper (TM) or SPOT High Resolution Visible (HRV) have an adequate capacity for collecting information with high spatial detail, their actual temporal frequency is too low to follow rapidly changing phenomena connected to vegetation phenology or responses to environmental changes (Allan 1990). On the other hand, sensors onboard meteorological satellites such as the NOAA Advanced Very High Resolution Radiometer (AVHRR) are able to collect information very frequently but with poor spatial resolution. Recently launched satellite sensors, such as Terra-MODIS and Envisat-MERIS, can only partially alleviate this problem, since their temporal and spatial resolutions, which are intermediate between those of the mentioned systems (a few days and 250-300 m, respectively), are the same sub-optimal for most environmental monitoring applications. The use of data from very high resolution sensors, such as IKONOS and Quickbird, is finally unfeasible for large area monitoring, due the their high costs and processing requirements.

The efficient integration of the useful spatial and temporal features of different data types has therefore become a major challenge for remote sensing experts interested in environmental applications (Puyou-Lascassies et al. 1994, Oleson et al. 1995). The paramount importance of the subject is testified by the numerous investigations performed in recent years, which have achieved varying degrees of success. A common and straightforward strategy is to assume a linear combination of the spectral signal to compare high and low resolution data by regression analysis techniques (Puyou-Lascassies et al. 1994, Kerdiles and Grondona, 1995, Oleson et al. 1995). This strategy was followed in a recent paper (Maselli et al. 1998b), which presented a procedure for integrating NDVI data from TM and AVHRR sensors by classification and regression analysis techniques. By this integration, NDVI estimates were produced with the spatial resolution of TM images and the temporal frequency of AVHRR data.

6.6.2 Merging NOAA-AVHRR and Landsat-TM Vegetation Indices

The concept of functional vegetation classes is strictly linked to the ecological role that terrestrial plants exert on the land-surface in terms of productivity and bio-geo-chemical fluxes. In practice, such classes represent plant assemblages which perform relatively homogeneous eco-physiological processes, particularly transpiration and photosynthesis. As these processes are extremely variable in space and time, the mapping of functional vegetation classes is a complex task to accomplish by conventional ground surveys. Alternatively, an efficient method for identifying and mapping functional vegetation types is by integrating diverse satellite and collateral data through advanced processing methodologies which take into account the spatial and temporal variability of the observed vegetation surfaces.

In particular, recent studies have shown that these problems can be tackled by the statistical integration of data from different sources (Puyou-Lascassies et al. 1994, Oleson et al. 1995). Several methods have therefore been developed for the integration of remotely sensed images from different sensors and with ancillary data. These methods are specifically effective in complex terrain, where the spectral information provided by one data source is usually insufficient for the discrimination of functional vegetation classes (Maselli et al. 1998).

An example is presented of the integration of data sets with different spatial and temporal resolutions relying on pixel unmixing and fuzzy classification procedures. The case study concerned the Radicondoli study area in Tuscany, where Landsat TM and NOAA AVHRR data were processed by soft classification procedures to evaluate vegetation properties and conditions with high spatial and temporal details.

Land cover is classified first at high spatial resolution using TM data. Temporal profiles of NDVI are then constructed using AVHRR observations of the same spatial domain. For each pixel the NDVI profile is decomposed in temporal "endmembers" using the high resolution land cover classification. Abundance maps of each class at the AVHRR spatial resolution are obtained using a hard classification of the TM multispetctral data and class membership values obtained using fuzzy classifier of the same TM data.

The high spatial resolution satellite data consisted of three Landsat-5 TM frames (192/30) taken on 1st June, 4th August and 8th November 1990. From a visual inspection the three frames were found to be cloud free and practically unaffected by other atmospheric perturbations all over the Radicondoli area. A scene of 512x512 pixels was extracted from each of the three frames and georeferenced by a nearest neighbour resampling algorithm. TM bands 4 (NIR) and 3 (red) were then converted into apparent reflectance values (Gilabert et al. 1994), in order to compute corrected NDVI images.

The AVHRR data, already in the form of NDVI, were provided by Nuova Telespazio. The standard procedure for the production of these data comprised the georeferencing of the original images by a cubic convolution algorithm, the radiometric calibration of the first two bands to derive apparent reflectances, and the computation of NDVIs to finally obtain maximum value composites, MVCs, on a monthly basis. The final products were therefore 12 monthly NDVI MVCs covering all 1990, which were used for the current analysis.

The data processing chain consisted of three main steps, which are: (i) generation of a high resolution reference map and its degradation to produce abundance images with AHVRR pixel size; (ii) multivariate regression of these abundance images on AVHRR NDVI data to identify class end-members; (iii) integration of the high resolution map and NDVI end-members to produce multitemporal NDVI maps.

The generation of a reference land use map was carried out by a supervised procedure based on a modified maximum likelihood classifier, MLC (Maselli et al. 1992). The land use references were directly identified on the TM images with the aid of air photos taken in 1991 and personal knowledge from direct ground surveys. Four land use classes were deemed sufficient to cover almost all the variability in vegetation features of the study area (coniferous forest, deciduous forest, spring crops and pasture). To avoid redundancy of the spectral data, only bands TM3, TM4 and TM5 from each scene were retained for the classifications, as suggested by Horler and Ahern (1986). After finding the training statistics of each class (means, variance-covariance matrix and frequency histograms, Maselli et al. 1992) the modified MLC was applied both in "hard" and "fuzzy" ways, in order to evaluate the possible differences deriving from the use of the relevant outputs (Wang 1990). When utilized in a fuzzy mode, the classifier yielded the "a posteriori" probabilities of the study classes as output membership grades, thus retaining all information produced during the classification process (Wang 1990). By applying the hard classification, a conventional thematic map was thus produced, while the output of the fuzzy procedure were four membership grade images.

Each hard map was first transformed into four dichotomous membership grade images (i.e. images with pixels having 0 values for no attribution, 1 for full attribution). All membership grade images from both the hard and fuzzy classifications (dichotomous or continuous, respectively) were then degraded to produce abundance images with the AVHRR pixel size. The degradation was done in two ways: (i) By simple average (mean filtering), as suggested by Kerdiles and Grondona (1995), and (ii) by applying a two-dimensional Gaussian filter to simulate the AVHRR point spread function, following Moreno and Meliá (1994). The results of these operations were abundance images which reported the proportion of each land use class in every AVHRR pixel as derived from the high spatial resolution classifications.

Uni- and multivariate linear regression analyses were then performed between the abundance images and the monthly NDVIs to retrieve the pure class profiles (Maselli et al. 1998). In the uni-variate case, the NDVI values of each monthly composite were regressed against the proportion of each class and the "pure NDVI" was estimated by extrapolating the proportion to 1 (Kerdiles and Grondona 1995). In the multivariate case, the same NDVI values were regressed contemporaneously on all four classes; the pure NDVI of each class was then found by extrapolating the proportion of that class to 1 and those of the other classes to 0 (Oleson et al. 1995). Globally, eight complete NDVI profiles of all classes were obtained from the classifications with the multi-temporal data set (every combination from hard and fuzzy classifications, mean and Gaussian filtering, uni- and multivariate regressions). The accuracy of these profiles was evaluated by comparison with the mean NDVI values of the four classes derived from the apparent reflectance TM data.

The most accurate NDVI profiles found in the previous phase were finally used to produce NDVI images of all 12 months with high spatial resolution. For this purpose a procedure based on the use of the high resolution classification outputs was employed. For each month, the NDVI estimated for every TM pixel ($NDVI_{est}$) was computed as the summation of the average NDVI values of all classes weighted for the relevant membership grade (Fig. 6.6.1). The method of course yielded discontinuous or continuous outputs depending on the use of hard or fuzzy classification representations. The 12 high resolution NDVI images were finally subjected to a procedure for re-calibrating the estimated NDVI values on the AVHRR index. In practice, the NDVI value of each high spatial resolution pixel for each month was adjusted so that the average of the relevant low resolution window would match that of the monthly AVHRR NDVI. The comparison of the TM and

AVHRR NDVI values gave interesting results for all the treatments examined. As regards the effect of the representation of the classification outputs, the two classification procedures produced rather similar estimates of the NDVI profiles (Table 6.6.1), but slightly better results were obtained by the fuzzy representation. Concerning the effects of the degradation methods compared, they produced substantially similar results. Again from Table 6.6.1 it can be seen that the major effect on the NDVI estimation process was attributable to the regression method. The univariate procedure in fact gave relatively poor estimates in all cases, while the multivariate method yielded results which were in absolute the best obtained in the study.

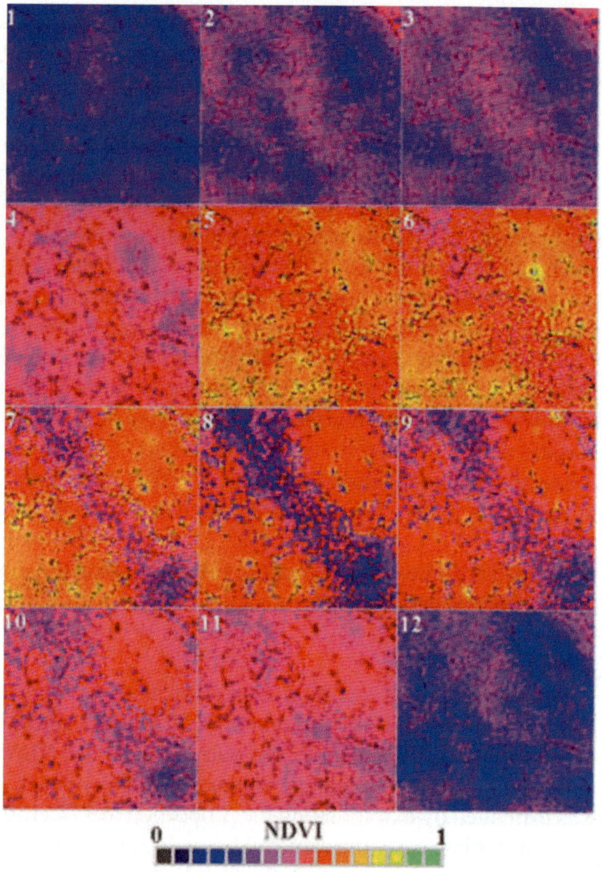

Fig. 6.6.1. Estimated monthly NDVI images at TM spatial resolution; Radicondoli 1990.

The NDVI profiles found by this best procedure were also quite realistic for all classes. Coniferous forests, which mainly cover the upper mountain zones, showed an NDVI maximum in July-August, with moderate variations during the year. Deciduous forest had a higher contrast from winter to spring-summer NDVI values, with a maximum in June. The profile of class 3 was typical of spring crops in Mediterranean areas, with a clear maximum in April and two minima in winter and summer due to the cold and arid seasons, respectively. The profile of pastures finally showed moderate variations throughout the year, with a smoother maximum from May to July and the same minima as in the previous case.

From a visual examination of the 12 final NDVI images (Fig. 6.6.1) obtained by the optimal procedure developed (fuzzy classification, mean filtering, multivariate regression) it can be seen that they are informative on the spatial distribution of the index in the study area all over the year. Thanks to the re-calibration procedure applied, the mean NDVI value in each low spatial resolution window is the same as that of the AVHRR NDVI, so that the images produced can be considered unbiased estimates with respect to the latter data.

Table 6.6.1 Determination coefficients (r^2) and root mean square error (RMSE) for the correlations between TM and interpolated AVHRR NDVI data obtained by different classification, degradation and regression procedures (all correlations are highly significant, $P < 0.01$; the results of the best estimation method are bold underlined).

Classification and Degradation Procedures	Regression Procedure	
	Univariate	Multivariate
Hard and Mean Filter	r = 0.768 RMSE = 0.108	r = 0.847 RMSE = 0.066
Hard and Gaussian Filter	r = 0.764 RMSE = 0.114	r = 0.853 RMSE = 0.064
Fuzzy and Mean Filter	r = 0.781 RMSE = 0.119	**r^2 = 0.864** **RMSE = 0.062**
Fuzzy and Gaussian Filter	r = 0.769 RMSE = 0.122	r = 0.857 RMSE = 0.063

6.7 Multispectral Classification of Land-surface Types

6.7.1 Methodology

The classical approach to classify a land-surface type is its characterization by the spectral signals emerging from the surface. For this procedure exist many algorithms in every Geographical Information System (GIS). The simplest way to obtain a qualitative overview of the variability of the land-surface cover is to generate a false colour composite of three spectral images. More accurate, physically based and reproducible methods make use of calibrated and even atmospherically corrected radiances or reflectances. It is to distinguish between unsupervised and supervised methods. Given specific thresholds and the number of types which should be recognized the unsupervised methods automatically find out groups of pixel with approximately identical spectral signatures, without informing the user about the nature of the land cover. There are two groups of supervised methods. The first one needs *a priory* knowledge about the spectral signatures of the classes into which the land-surface types should be grouped. The second one requires test sites of known land-use types for the classes into which the land surface should be sorted. In this second case the known test sites first are analysed with respect to their spectral signatures and establish the groups one is looking for.

6.7.2 Example of a supervised classification

For a classification of the EFEDA experimental area in Castilla - La Mancha, Spain, the supervised maximum likelihood method was used. The selection of the images and the adequate bands were carried out with a detailed analysis of the reflectance information at the different dates with the objective to enhancing discrimination between classes. Five spectral classes were considered in the land use classification; cereal, fallow (barbecho), forest area, olive parcels and vineyards. Principal component analysis and unsupervised classification (for TM images acquired on 27 May 1991) were used for discrimination of the cereal class from the others. The selection of the adequate TM images for the discrimination between the considered classes depends on the timing of the maximum photosynthetic activity of the plant canopies. TM3, TM4 and TM5 of the same image were used for classification of the forest class, while TM3, TM4, TM5 and TM7 of the same date were used for discrimination of olive parcels (as shown in Fig. 6.7.1). TM3, TM4 and TM5 of 14 July 1991 images were used for discrimination of the fallow and vineyard classes. The land use classification image combines the sum of all the classified images.

In the soil type classification, Landsat-5 TM images of 9 April 1991 were used for the haplic rodic calcisol and petric calcisol classes. TM images of 27 May 1991 were used for discrimination of leptosol soils. The lithic and haplic calcisol were discriminated using TM4, TM5 and TM7 of 28 June 1991 images. The reflectance properties of the selected images and classes are shown in Fig. 6.7.2.

Validation of the classifications was achieved by respectively comparing the results obtained with the land use maps offered by the University of Castilla - La

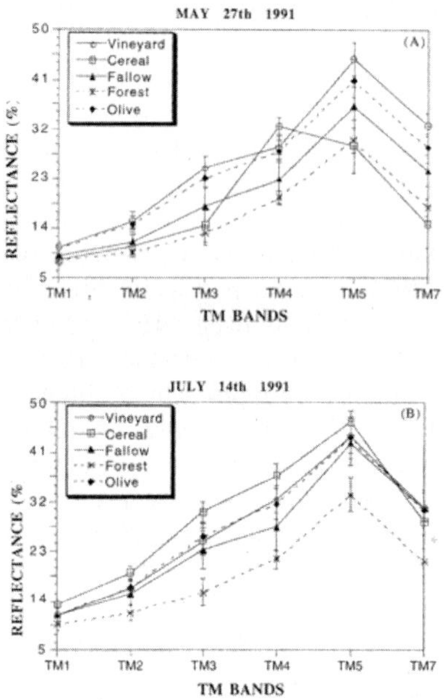

Fig. 6.7.1. Spectral signatures of land-use classes at two dates in 1991.
(A): Differentiation between cereal, olive, and forest classes by successively applying principal component analysis to the TL channels 3, 4, 5, and 7 and TM 3, 4, and 5 of the image of 27 May, 1991. (B): Differentiation between vineyards and fallow land classes was achieved by using the TM channels 3, 4, and 5 of the image of 14 July 1991

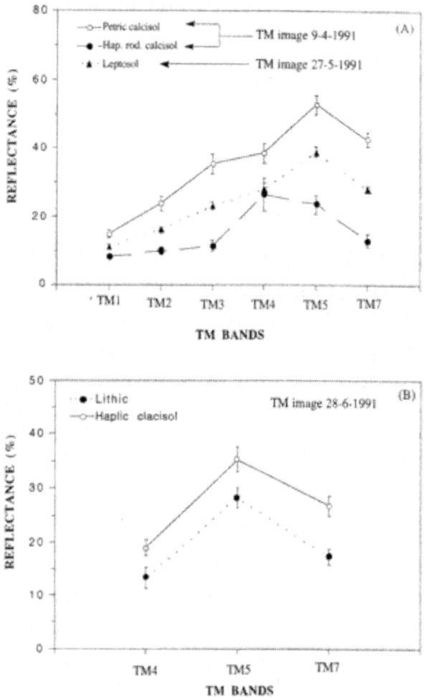

Fig. 6.7.2. Spectral signatures of soil type classes in multitemporal LANDSAT-5 TM images.
(A): Differentiation between petric and haplic rodic calsol in the image of 9 April 1991 and of the leptosol class in the image of 27 May 1991. (B): Differentiation between the lithic and haplic calcisol classes by using the TM channels 4, 5, and 7 in the image of 28 June 1991

Mancha and the soil maps from the Dept. Biologia Vegetal. The comparison includes spatial distribution and proportion of the classes considered.

The classification result presented in Fig. 6.7.3 shows a high degree of correspondence with the land use and soil type maps. The estimated percentages of the considered classes is given in Table 6.7.1.

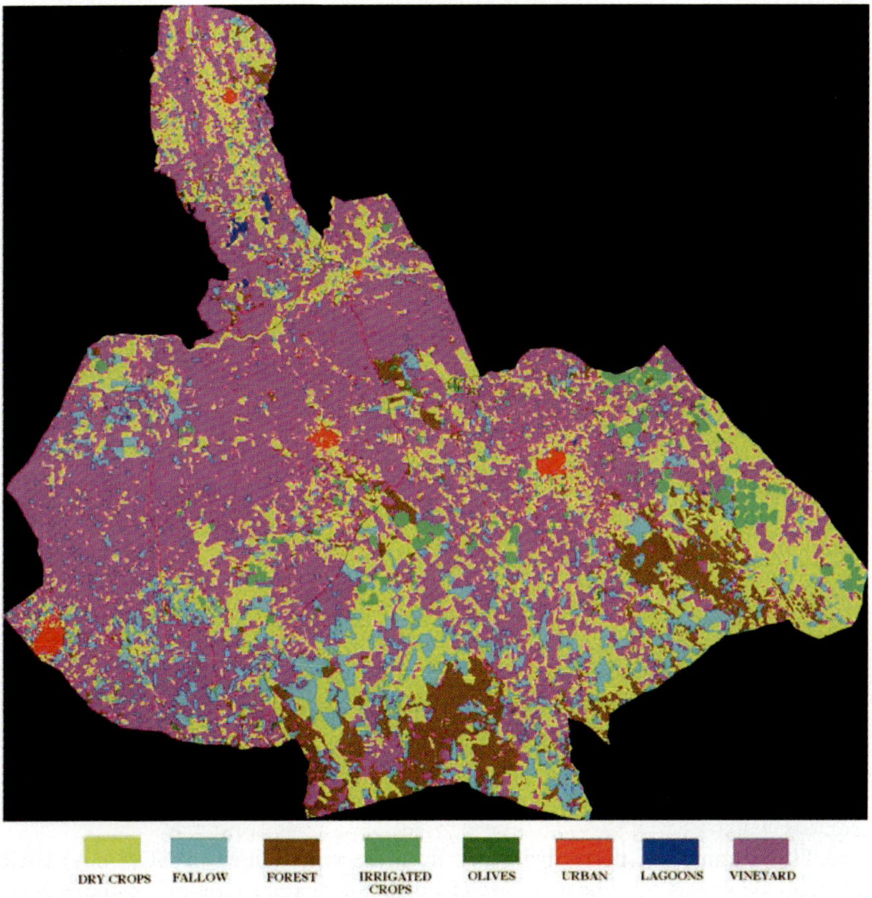

Fig. 6.7.3. Result of the classification of vegetation, urban areas and lagoons in the Tomelloso area of Castilla-La Mancha, Spain

Table 6.7.1. Estimated percentages of the classes considered in the classified images and the result of ground surveys in the Tomelloso area

Class	Classification (%)	Ground survey (%)
	Vegetation	
Vineyards	62	76
Cereals and fallow land	29	20
Forests	5	1
Olive plantation	4	3
	Soil types	
Lithic	1	1
Haplic rodic calcisol	4	4
Haplic calcisol	9	12
Petric calcisol	80	75
Leptosol	6	8

6.7.3 Surface Discrimination by Application of the Temperature Independent Spectral Indices (TISI)

Introduction

The TISI parameters, derived directly from the emissivities in NOAA-AVHRR channels 3, 4 and 5, appear to be adequate for characterization of the natural surfaces.

- For vegetated areas, these indices present a dynamic equivalent to the NDVI parameter.
- For arid zones, where NDVI is no more consistent, TISI depict a high dynamic representing well the different types of surface encountered.

In this paper, the TISI and NDVI parameters are conjointly presented and studied on two types of region: The first one, sized 1000 x 1400 km^2 includes the Iberian peninsula which presents a cover rather vegetated. The second one (800 x 900 km^2) is an arid zone located in North Africa.

Methodology of Parameters Retrieval

The TISI are derived from emissivity in at least two thermal infrared channels using the following formula (Becker and Li 1990a):

$$TISI_{1,n} = \prod_{i=1}^{n} \varepsilon_i^{a_i} \tag{6.7.1}$$

where the a_i are chosen in a way that this product is temperature independent. Extraction of emissivities is achieved following the methodology from by determining first the bi-directional reflectivity in AVHHR channel 3 as described in (Nerry et al. 1998). The use of a set of about thirty day and night pairs of AVHRR images allows the complete determination of the bi-directional reflectivity, which is then modelled and integrated over the upper hemisphere leading to the emissivity in the AVHRR channel 3. This allows to resolve the so called "missing equation" and therefore to determine the emissivities in AVHRR channels 4 and 5.

NDVI values used for this study are derived from reflectivity in channel 1 and 2 of AVHRR corrected from atmospheric perturbations and from angular variations (Petitcolin 1997). The angular correction is performed by the means of a model which allows to retrieve the reflectance for a nadir view and for a normal incidence angular configuration.

Data Used

AVHRR images, calibrated, navigated and atmospherically corrected, covering the Iberian Peninsula and a North African region, have been used for this study. For the Iberian Peninsula window, the month of March 1997 has been selected for this study due to the continuous clear sky condition during this period. For the region in Northern Africa, summer and winter images of the year 1995 have been selected. Fig. 6.7.4 presents emissivities maps in the three TIR AVHRR channels and for the two regions.

Results

$TISI_{3,4}$ and $TISI_{3,5}$ are then derived from the emissivity values and maps of these parameters are plotted in Fig. 6.7.5 together with the NDVI.

One can see on these maps that NDVI the TISI parameters seem to be correlated with NVDI for the Spanish region : a high value of NDVI corresponds to a high value of TISI and vice versa. For the case of the Northern African region, TISI parameters appear to be more sensitive to the surface characteristics presenting interesting features.

Two transects, one in each region, of about 300 pixels (kilometers), have been extracted from these images. They are materialized by the two black lines on the top image of Fig.6.7.5. NDVI and $TISI_{3,4}$ are then represented together along these transects in Fig. 6.7.6. The figure on the top corresponds to the Iberian Peninsula

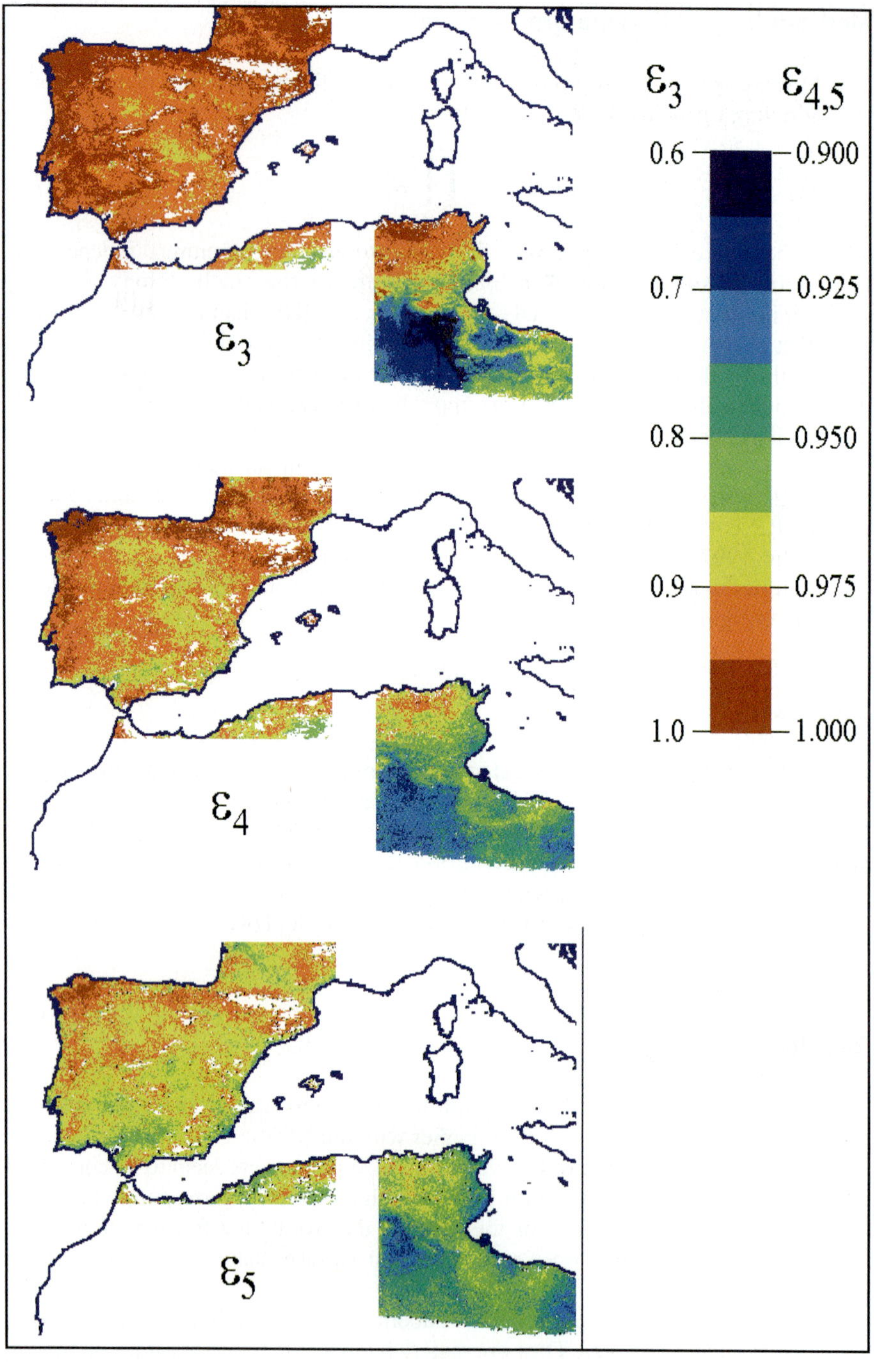

Fig. 6.7.4. TIR emissivities in the three AVHRR channels

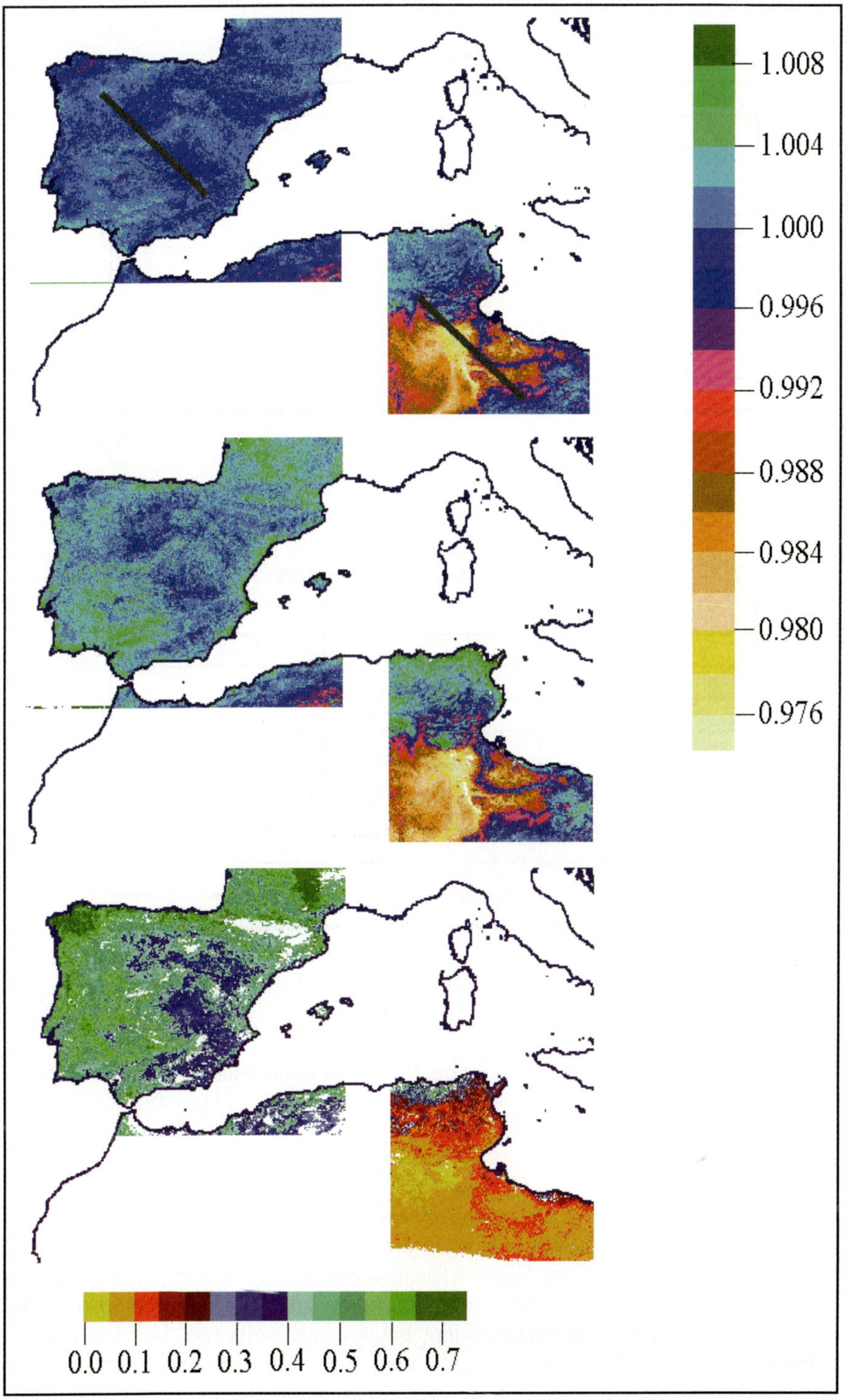

Fig. 6.7.5. $TISI_{3,4}$ (top) and $TISI_{3,5}$ (middle) and NDVI (bottom) maps for the two regions

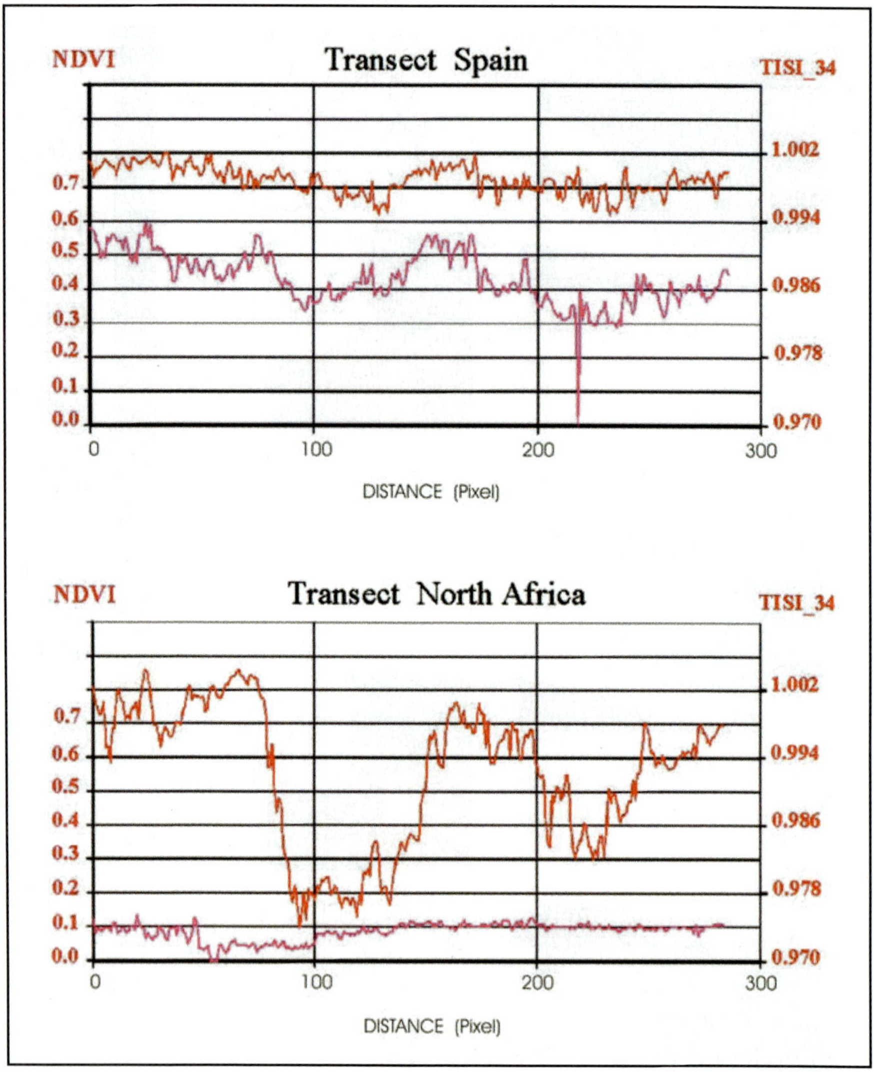

Fig. 6.7.6. $TISI_{3,4}$ and NDVI transects for the Iberian Peninsula window (top) and the North Africa window (bottom)

window and the one on the bottom part to the North Africa window. The transects for NDVI and $TISI_{3,4}$ in the Spain window are quite similar, varying as a function of the vegetation cover. For the North African window, NDVI transect is rather flat whereas $TISI_{3,4}$ exhibits a huge dynamic. That shows that this index could be useful not only for vegetation characterization, but for bare soils too. Validation using external data such as land use, geological maps have still to be performed.

Conclusions

The TISI parameters, derived from emissivity values, appear to be a very promising index for the characterization of the vegetation cover and for the soils discrimination. Compared with NDVI, they exhibit the same type of dynamic for the vegetated areas but a higher one in the case of bare soil. Results still need to be validated against ancillary data.

6.7.4 Vegetation Canopy Characterization by Microwave Transmittance

The relationship between the microwave vegetation transmissivity at 6.6 Ghz, $\tau_{veg,6.6}$, and NDVI (uncorrected GAC) has been derived for the Botswana savannas from satellite observations and modelled soil emissivity using a database of large-scale surface soil moisture (Van de Griend and Owe 1994a). The relationship is shown in Fig. 6.7.7 and is valid for a single scattering albedo $\tilde{\omega}$ = 0.09, which is a representative value determined for the Botswana savannas (Van de Griend and Owe 1993a). Since the value of $\tilde{\omega}$ is vegetation-specific, it is somewhat dangerous to apply the relationship found for Botswana to the Spain data. Moreover, the vegetation of central Spain is dominated by vineyards with an approximate partial vegetation cover of less than 10% at maximum. Therefore, the bare soil fraction will have a dominant effect on NDVI. The Botswana savanna is predominantly covered by grasses, shrubs and isolated trees. Therefore, NDVI is probably a better index for vegetation wet biomass (and thus for the microwave transmissivity) in Botswana than for Spain.

In section 4.6.2 examples have been shown of the annual courses of the NDVI and the MPDI. The regression lines between NDVI (uncorrected GAC) and MPDI found for Botswana and Spain are very similar. They are:

for Botswana:

$$NDVI = 0.411 - 0.473 * MPDI \tag{6.7.2}$$

for Spain:

$$NDVI = 0.371 - 0.471 * MPDI \tag{6.7.3}$$

For sparse vegetation the MPDI at 37 GHz is probably a better index than the NDVI (especially in terms of microwave radiative properties) because MPDI is less effected by clouds and bare soil - unless the very topsoil (~ 0.5 cm) is wet, which probably occurs only during or just after a storm. Moreover, the probability of an entire SMMR pixel (150 x 100 km^2) to be wet at the very surface is also very small. Another reason why MPDI is preferred above NDVI for routine soil moisture monitoring is that 37 GHz (vegetation frequency) and 6.6 GHz (soil moisture frequency) observe footprints with the same centre and at the same time. Therefore, it is assumed that the relationship between $\tau_{veg,6.6}$ and $MPDI_{37}$ found for Botswana, given in Fig. 6.7.8, gives a first order estimate for the situation in Spain. The relationship determined for Botswana for $\tilde{\omega}$ = 0.09 is of the form:

$$\tau_{veg,6.6} = 0.477 + 0.289 * MPDI\ . \tag{6.7.4}$$

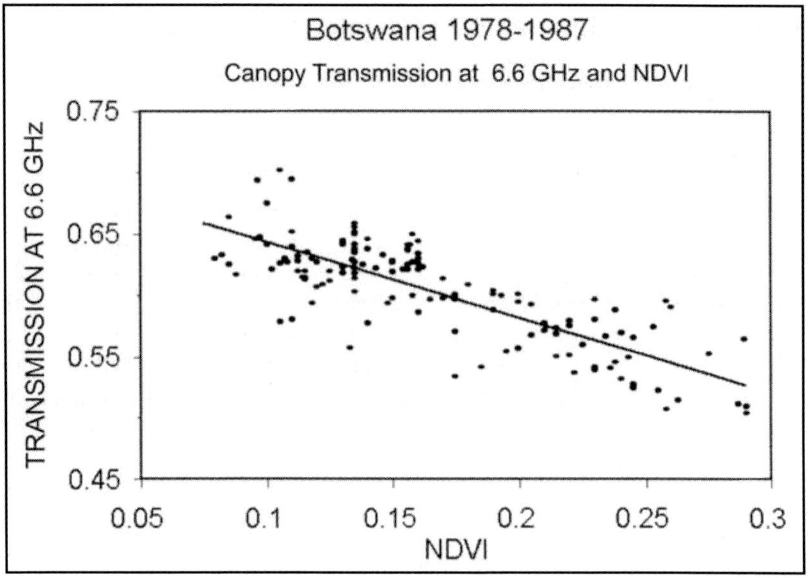

Fig. 6.7.7. The relationship between $\tau_{6.6\ GHz}$ and NDVI derived for the Botswana savannas for $\tilde{\omega} = 0.09$ (From: Van de Griend and Owe 1993a)

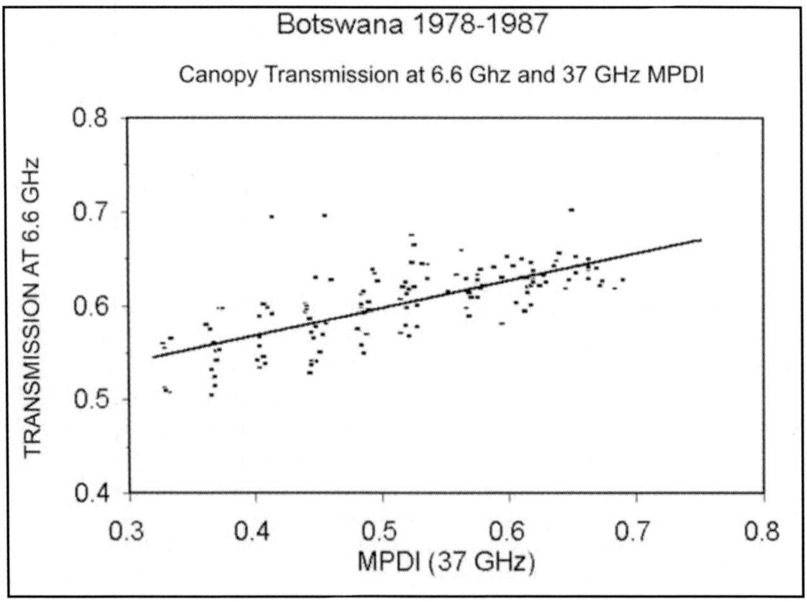

Fig. 6.7.8. Relationship between $\tau_{6.6GHz}$ and $MPDI_{37}$ derived for the Botswana savannas for $\tilde{\omega} = 0.09$ (From: Van de Griend and Owe, 1993a)

6.8 Decomposition of Pixel Contents

6.8.1 Spectral Decomposition

Prerequisits

Spectral decomposition or unmixing of measured reflectances was not specifically a task of this EFEDA since it is dealt with in other EC sponsored projects. Spectral decomposition primarely is done to determine the components filling one pixel. To assess as an example changes of the vegetation if its coverage is sparse, it is necessary to know the spectral response of the underlying soil. In order to validate remote sensing data it was necessary to measure e.g. the reflectance of an area that can be resolved in satellite measurements. In the case of Landsat this is about 30 m x 30 m and in the case of NOAA-AVHRR 1 km^2. Since spectrometers or spectral radiometers operated at the ground provide a much higher resolution it was necessary to determine the spectral characteristics of surfaces that are homogenous over a larger area. In the course of these collateral measurements the spectral characteristics of various pure vegetation species and soils have been measured as well and are partly presented in the respective following chapters.

These measurements have been carried out with either spectrometers or interference filter radiometers that have a narrow field of view. This is necessary for the rationing of the measured reflected radiance to white targets that are available only at small sizes of a few square decimeters. These white targets are usually oriented horizontally to represent by its reflected radiance the total irradiance incident on a horizontal surface. These white plates are not 100% white but spectral calibration data are supplied with these plates that can be applied as a correction. The horizontal orientation of the plates is very convenient if soils or very low vegetation are investigated. In the case of bushes and trees it is sometimes difficult to apply the same technique if not an adequate elevated platform is available to view the target as well as the reference plate under approximately the same angle. Sometimes therefore the measurement at high vegetation was performed by placing the reference plate vertically to the direction of the measurement in front of the plant. Such a measurement provides realistic reflectance data but not necessarely the same as a measurement from nadir would provide.

As a preliminary stage to analyse the usefulness of remotely sensed data - such as those from the Landsat 5 TM sensor - to study the soils in the EFEDA area, laboratory measurements of the reflectance of some selected soil samples were carried out by the University of Valencia to obtain relationships between the shapes of the spectral reflectance observed and the soil composition (mainly clay, organic matter and salt contents). Certainly, spectrometry enables a more quantitative approach of the relationships between spectral and physical properties.

In a first approach, the spectral characterization of a given soil should be performed by measuring the spectral reflectance of its different horizons. Nevertheless, when dealing with applications for remote sensing, we will focus on the first horizon, since the upper layer of the surface is the only one which can be observed from space.

Soil spectral reflectance curves have been reported in several publications; a good review can be found in Escadafal (1994). The first striking feature of soil reflectance is the high brightness variability. For example, a low reflectance in the visible-near infrared range is typical of organic matter-rich soils. But soils are not only dark or bright, their reflectance curves also vary in shape due to the presence of some absorption features occurring where specific wavelengths of light are absorbed. In the visible domain, reflectance is determined by soil colour due to the presence of iron oxides such as hematite and goethite that produce absorption features caused by electronic processes (Kruse and Lefkoff 1993). In the 0.9-2.5 μm region, vibrational absorption bands occur as overtones and combination tones of fundamental vibration bands due to anion groups such as OH^- and CO_3^{2}. Additionally, vibrational absorption bands may also result from overtone and combinations of the fundamental modes of molecules such as water trapped in the crystal structure. Therefore, the shape of reflectance curves in this region is due to clay content (sheet silicates and carbonate minerals content) and salt content. The strong water absorption bands of wet soils are the same as those of the atmosphere and consequently, they are not usable in remote sensing.

In this work, laboratory measurements on soil samples were carried to obtain the reflectance spectrum from 400 to 2500 nm, with a spectral resolution of 2-4 nm, depending on the wavelength, by means of a GER SIRIS spectroradiometer. The study was twofolded:

(i) a broad band analysis of the data was performed to reproduce reflectance values in the TM bands to check their suitability to monitoring soil physical properties and soil mapping.

(ii) a high spectral resolution analysis was applied to the data by using 'derivative techniques' which enable detection of subtle spectral features and normalization of spectra from illumination variations. The first derivative shows a maximum in the wavelength corresponding to an inflection point in the original spectrum, meanwhile the second derivative shows maxima in the positions of absorption features which produced minima in the reflectance spectra, whereas it shows zero values in the original inflection points.

The study was carried out by the Remote Sensing Group (spectral properties) in collaboration with the Dept. Biologia Vegetal (physical properties), both from the University of Valencia.

Theory

In principle the spatial resolution achieved by a satellite-borne sensor is defined by the size of the detector, the optics in front of the detector and the observation geometry which define the size of the pixel at the Earth's surface. However, it is possible to discern land-cover components even within a pixel by the *spectral decomposition* method provided that enough spectral information is available about the components contained in one pixel.

A pixel normally contains n components (e. g. bare soil, stones, various kinds of vegetation shadowed soil and/or differently inclined and insolated surfaces). The radiometric signal s_{ik} from a surface mixture k measured in a wave band i is the sum of the responses r_{ij} of the n components j in the wave bands i times the fractional coverage f_{jk} of the components:

$$s_{ik} = \sum_{j=1}^{n} r_{ij} f_{jk} \qquad (6.8.1)$$

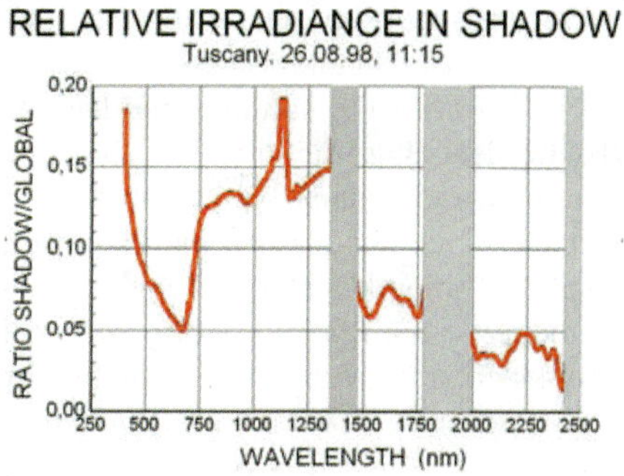

Fig. 6.8.1. Ratio of irradiance under olive trees to global radiation

If a single mixture k with j components is investigated, f_{jk} is a vector. In matrix notation Eq. (6.1) can then be written:

$$\mathbf{s} = \mathbf{Rf} \qquad (6.8.2)$$

The task is now, to determine first the number of components needed to approximate the condition and then to determine their fractional coverage which means to solve Eq. 6.8.2 for **f**. This is a classical inversion problem: to find a minimum number of components with spectral features r_{ij} and relative area coverage f_{jk} the combination of which approximates the measured signal s_{ik} to a pre-specified degree of accuracy.

The method to solve the problem was developed in the 1960s in the field of optics (Simonds 1963). A basic discussion of the mathematics of inversion in remote sensing was given by Twomey (1977) followed by applications in chemistry (Knorr and Futrell 1979, Gillete et al. 1983, Malinowsky and Hoery 1984, Windig and Meuzelaar 1984). It was used by Huete (1986) to single out cotton from a cotton-soil mixture in remote sensing data. With increasing number of components the method quickly runs into severe mathematical problems. Prior to applying information gained with satellite instrumentation therefore some *a priori* information about the area not only is helpful but occasionally also essential.

Eq. 6.8.1 is only correct for radiances. If one deals with higher level information the linear relationship may not exist. Already in the case of vegetation indices different results are obtained if one sums up indices computed pixel by pixel or if one computes indices from averaged radiances. These differences may be small as shown above but could become large if extreme values are present in neighboured fractions of the investigated area.

In the simplest case one has to deal with three components, a specific type of vegetation, a sunlit soil and the same soil in the shadow of the plant. Within pixel sizes of Meteosat sometimes already two to three types of soils in principle would have to be regarded. On the other hand, soil areas in the shadow of vegetation often can be neglected, especially if the sun is high. In the visible part of the spectrum shadows may be approximated as black areas but the fraction of radiation reflected from soils shadowed by green plants increases in the near infrared. Fig. 6.8.1 shows the irradiance under olive trees. High values at the shortwave end result from illumination by diffuse sky radiation. Beyond 700 nm the irradiance results from solar radiation transmitted by the canopy.

The formal solution $\mathbf{f} = \mathbf{R}^{-1}\,\mathbf{s}$ of Eq. 6.8.2 leads back precisely to the initial **f**. *Nota bene,* in this case the reflectance matrix has to be square which means there must be as many reflectance classes as spectral channels. If, however, the radiance or signal vector **s** is superimposed by noise, as it is the case in real measurements, the solution becomes unstable and useless.

To stabilize the solution one has to introduce a constraint. This could be the requirement that the square of the residual of a confined solution **f**″ becomes a minimum and remains below a small number of the order of the measurement errors. But this does not lead to improvement. An additional constraint is to make the solution "smooth" which is equivalent to minimize the differences between the elements of the **f** vector and to suppress large oscillations. The differences of the components of a vector **f** are obtained by multiplication of **f** with a matrix **K** of the type used in Eq. 6.8.3 for a three components vector **f**:

$$\mathbf{Kf} = \begin{pmatrix} 0 & 0 & 0 \\ 1 & -1 & 0 \\ 0 & 1 & -1 \end{pmatrix} \begin{pmatrix} f_1 \\ f_2 \\ f_3 \end{pmatrix} = \begin{bmatrix} 0 \\ f_1 - f_2 \\ f_2 - f_3 \end{bmatrix} \tag{6.8.3}$$

Minimizing the differences between the **f**-components requires that the quadratic form $\Sigma(f_{j-1}-f_j)^2$ becomes a minimum. The quadratic form is computed by multiplying **Kf** with its transposed matrix (**Kf**)*. By this operation a symmetric matrix **H** with only near diagonal elements is obtained for the constrained solution which may be written **f**″:

$$(\mathbf{Kf}'')^*\mathbf{Kf}'' = \mathbf{f}''^*\mathbf{K}^*\mathbf{Kf}'' = \mathbf{f}''\mathbf{H}^*\mathbf{f}'' = \mathbf{f}''^* \begin{bmatrix} 1 & -1 & 0 \\ -1 & 2 & -1 \\ 0 & -1 & 1 \end{bmatrix} \mathbf{f}'' \tag{6.8.4}$$

The extremum problem now can be formulated in the following way: The quadratic form $|\mathbf{Rf} - \mathbf{s}^2| = (\mathbf{Rf} - \mathbf{s})^*(\mathbf{Rf} - \mathbf{s}) \le \varepsilon^2$ should be held constant below a specified level ($= e^2 \le \varepsilon^2$) and at the same time **f*Hf** should be minimum. It is therefore required that (omitting the ″ for the constraint solution and as well the terms not containing **f**):

$$\frac{\partial}{\partial f_k}\left\{\mathbf{f}^*\mathbf{R}^*\mathbf{Rf} - \mathbf{s}^*\mathbf{Rf} - \mathbf{f}^*\mathbf{R}^*\mathbf{s} + \gamma\mathbf{f}^*\mathbf{Hf}\right\} = 0 \tag{6.8.5}$$

Here an undetermined Lagrangian multiplier γ is introduced that can be chosen arbitrarily to force **f*Rf** to a minimum. Differentiation leads to

$$\mathbf{e}_k^*(\mathbf{R}^*\mathbf{Rf} - \mathbf{R}^*\mathbf{s} + \gamma\mathbf{Hf}) + (\mathbf{f}^*\mathbf{A}^*\mathbf{A} - \mathbf{s}^*\mathbf{A} + \gamma\mathbf{f}^*\mathbf{H})\mathbf{e}_k = 0 \tag{6.8.6}$$

The second term is the transpose of the first one. If the sum is zero both terms have to vanish separately and if one takes the first term it results:

$$\mathbf{f} = (\mathbf{R}^*\mathbf{R} + \gamma\mathbf{H})^{-1}\mathbf{R}^*\mathbf{s} \tag{6.8.7}$$

This is the constrained solution $\mathbf{f}''$ which can be computed for several γ and the residual $|\mathbf{R}*\mathbf{f}'' - \mathbf{s}|$ can be determined. If the residual turns out to be larger than the overall estimated error in **s**, then γ is too large. If the residual is considerably smaller than the error, then one has an underconstrained solution, γ is too small. If $\gamma = 0$ one falls back to the solution without the smoothness constraint.

Example

To consider an example, a mixture is chosen of olive trees, soil, and soil shadowed by the olive trees sensed by the LANDSAT-TM sensor with six channels in the visible and near infrared. The reflectances are taken from Table 6.8.2. The columns stand for olive, soil, and shadowed soil:

$$\mathbf{R} = \begin{bmatrix} 0.039 & 0.078 & 0.0072 \\ 0.082 & 0.122 & 0.005 \\ 0.048 & 0.150 & 0.0049 \\ 0.586 & 0.216 & 0.165 \\ 0.195 & 0.318 & 0.0136 \\ 0.079 & 0.268 & 0.0062 \end{bmatrix} \tag{6.8.6}$$

It is assumed further that the fractional coverage of olive trees is 30%. The coverage of the 70% soil by shadow depends on the solar zenith angle. One may assume the values 10, 30, 50, 70 and 90%. In this case the vector **f** expands into the following matrix F:

$$\mathbf{F} = \begin{bmatrix} 0.3 & 0.3 & 0.3 & 0.3 & 0.3 \\ 0.07 & 0.21 & 0.35 & 0.49 & 0.63 \\ 0.63 & 0.49 & 0.35 & 0.21 & 0.07 \end{bmatrix} \tag{6.8.7}$$

The signal matrix **S** then becomes:

$$\begin{vmatrix} 0.021696 & 0.031608 & 0.04152 & 0.051432 & 0.061344 \\ 0.038621 & 0.054483 & 0.070345 & 0.086207 & 0.102069 \\ 0.029751 & 0.049673 & 0.069595 & 0.089517 & 0.109439 \\ 0.211458 & 0.244134 & 0.27681 & 0.309486 & 0.342162 \\ 0.094809 & 0.136207 & 0.177605 & 0.219003 & 0.260401 \\ 0.050587 & 0.086301 & 0.122015 & 0.157729 & 0.193443 \end{vmatrix}$$

Since it is dealt in this example with three components, it is adequate to consider three spectral channels. We select the channels 3, 4, and 5. The direct inversion of

Table 6.8.2. Reflectances of different ground cover measured within the six visible and near infrared bands of the Landsat-TM sensor.

Channel	1 482 nm	2 565 nm	3 660 nm	4 825 nm	5 1.65 µm	7 2.215 µm
Olive tree	39	82	48	586	195	79
Dry soil	78	122	150	216	318	268
Wet soil	44	70	96	129	194	129
Dry grass	103	163	208	349	424	282
Ratio of irradiances in shadow and clear sky	92	71	51	128	7	48
Shadow at bare dry soil	72	87	77	276	223	135
Shadow at gras covered soil	95	116	106	447	297	135

Eq. 6.8.2 with these data naturally reproduces the values given in Eq. 6.8.7. But if errors are introduced to the signals this is no longer true. To test this, the numbers of the signal matrix are rounded to the following values:

$$\begin{vmatrix} 0.0297 & 0.05 & 0.07 & 0.089 & 0.11 \\ 0.21 & 0.24 & 0.28 & 0.31 & 0.34 \\ 0.095 & 0.14 & 0.18 & 0.22 & 0.26 \end{vmatrix}$$

If this matrix is used in the direct inversion of Eq. 6.8.2 the following solution is obtained for the matrix **F**:

$$\begin{vmatrix} 0.29259 & 0.25703 & 0.29584 & 0.28732 & 0.30228 \\ 0.06353 & 0.16489 & 0.33751 & 0.46248 & 0.6485 \\ 0.79559 & 1.67911 & 0.67186 & 0.75801 & -0.23174 \end{vmatrix}$$

It can be seen that this leads to unrealistic, even negative, values. The constrained solution with $\gamma = 1$ delivers an even worse result:

$$\begin{vmatrix} 0.0898 & 0.11448 & 0.13917 & 0.16385 & 0.18852 \\ -0.08162 & -0.10433 & -0.12703 & -0.14973 & -0.17243 \\ -0.0082 & -0.01016 & -0.01214 & -0.01411 & -0.01609 \end{vmatrix}$$

Already with a $\gamma = 0.0001$ the solution becomes more realistic:

$$\begin{vmatrix} 0.31205 & 0.30605 & 0.30006 & 0.29406 & 0.28806 \\ 0.09842 & 0.22412 & 0.34981 & 0.47551 & 0.60121 \\ 0.10548 & 0.22833 & 0.35118 & 0.47403 & 0.59688 \end{vmatrix}$$

Satisfactory values result for γ = 1E-7 or 1E-8 and an excellent result is obtained with γ = 1E-10:

$$\begin{vmatrix} 0.3 & 0.3 & 0.3 & 0.3 & 0.3 \\ 0.07 & 0.21 & 0.35 & 0.49 & 0.63 \\ 0.62998 & 0.48999 & 0.35 & 0.21001 & 0.07002 \end{vmatrix}$$

In practise such an outstanding result will be difficult to achieve because in addition to the errors in the radiometer measurements also the reflectances of the components are only approximately known.

6.8.2 Fractional Vegetation Cover Determination by Unmixing Surface Temperature

From measurements made at the surface or with high spatial resolution from space it is possible to determine, for a specific ecosystem, the relationship between the surface temperature of a dry surface such as dry bare soil and an adjacent irrigated vegetated area such as maize. All measurements made within this area will then fall in between a dry soil line and a wet (respectively vegetated and irrigated) soil line (Fig. 6.8.4). If, e.g. the dry soil temperature is measured then the wet surface

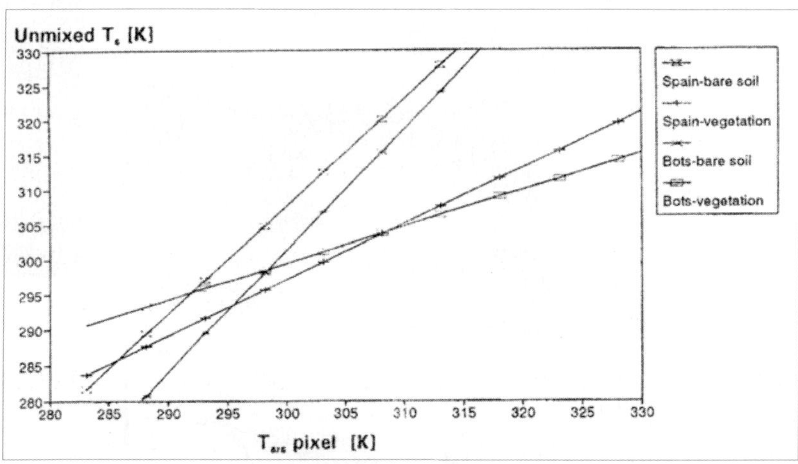

Fig. 6.8.4. The unmixed physical surface temperature for fallow land and irrigated maize (Spain) and bare soil and savanna vegetation (Botswana) as a function of the satellite observed mean area (pixel) apparent radiative surface temperature

Fig. 6.8.5. Maps of the relative fractions of bare soil and healthy vegetation derived from noon thermal infrared Meteosat images. White color indicates the fraction of bare soil = 1

temperature can be estimated, and if the instrument measures a temperature below the dry soil value, the fractional land cover can be determined or, *vice versa*, if the fractional land cover is known the dry and wet surface temperatures can be estimated (Peters1995).

During studies performed by the Vrije Universiteit Amsterdam it turned out that the surface reflectance as measured in the broad shortwave Meteosat channel is an unsuitable measure of vegetation because it varies with soil moisture and because, in some cases, the broad band reflectances of soil and vegetation are not very different. A solution to the problem of vegetation cover mapping was found in the use of thermal infrared observations. It was shown by aircraft measurements that the fractional vegetation cover was linearly related to the NDVI, which in turn is linearly related to the instantaneous surface temperature (Fig. 6.8.4). It was conjectured that the maximum temperature of a pixel at noon, over a number of days, is a measure for the amount of dry bare soil within a pixel. A composing technique was derived and described by Peters (1995). Vegetation mapping using thermal information has the additional advantage that the state of the vegetation is implicitly included. This is of importance for the temperature unmixing method, since it attempts to separately treat healthy vegetation and bare soil.

The technique was applied to noon thermal images of Spain over the three months period May to July 1991 (Fig. 6.8.5). The figures represent the vegetation cover at the end of each month. A striking feature of these images is their consistency. Disturbing factors are the influence of topographical elevation on the temperature causing an overestimation of the fractional vegetation cover. From May to July the amount of bare soil increases, especially in central areas of Spain. The vegetation zone in the north remains constant.

6.9 Seasonal and Interannual Variability as Seen in NOAA-AVHRR Images

6.9.1 Thermal Infrared Data Series

The MEDOKADS data set, described in Chapter 2, allows to construct temporal averages, trends and differences for the basic AVHRR channel data and their algebraic combinations. Because of the accurate georeferencing the construction of such data sets can be accomplished for almost each individual pixel though, because of the inherent jitter in the geographical allocation of the data, it is advisable to interpret the data finally for larger complexes such as 2 × 2 or 3 × 3 pixel matrices. Temporal averages can be constructed for any period larger than about 10 days and smaller than 16 years. The lower limit is determined by the necessity to eliminate pixels which contain clouds by compositing weekly or decadal means. The upper limit can be expanded as the data series grows.

The main purpose of this sub-chapter is to illustrate the different products which can be derived from the NOAA-AVHRR data set and how they can be used in Mediterranean land-surface process studies. In Chapter 1 (Fig. 1.1) already use has been made of thermal infrared AVHRR data converted to surface temperatures to illustrate the average temperature gradient across the Mediterranean sea. Here this data set will be presented in more detail for the years 1989 to 2004 to document seasonal and interannual variability.

In Fig. 6.9.1 and 6.9.2 seasonally averaged temperatures are plotted in longitude-latitude coordinates for cloudless conditions in the years 1989 - 1998 and 1999 - 2000 at the time of the NOAA satellite overpass (nominal 15:50 UT for NOAA-14 which was the same for NOAA-11 at launch but by March 1995 it had drifted to 17:33 UT). The colours indicate temperature in steps of five degrees ranging from light blue (-10 °C) to red (+60 °C). The figures are composites of cloudless pixels from two to three overpasses per day throughout the specified season. Over the sea, they represent the daytime sea surface temperature (SST). Over land, they represent a close to maximum daytime land-surface temperature (LST). But one has to remember, that, because of the large swath angle, the measurements are made within a time interval of plus/minus one hour. Only because the overpass time changes from day to day by about ten minutes and repeats itself approximately after eight and a half days, the composite approximates a mean value for the early afternoon and because this geometry repeats every year, the composites allow comparisons from one year to the other. Unfortunately a shift of the orbit of polar orbiting satellites as observed for NOAA 11 complicates interannual comparisons of land-surface temperature distributions but has no serious effect on the sea surface temperature (SST).

At the first view the sequence of images reveals a high degree of regularity from year to year. But, in spring 1996 in the western Mediterranean cooler temperatures are observed as compared to spring 1995 and 1997 and lower autumn temperatures in Greece and Turkey in 1997 as compared to 1996 and 1998. Again in spring 2000 it was generally cooler that in 1999. Also remarkable is the from year to year changing SST in the Adriatic Sea.

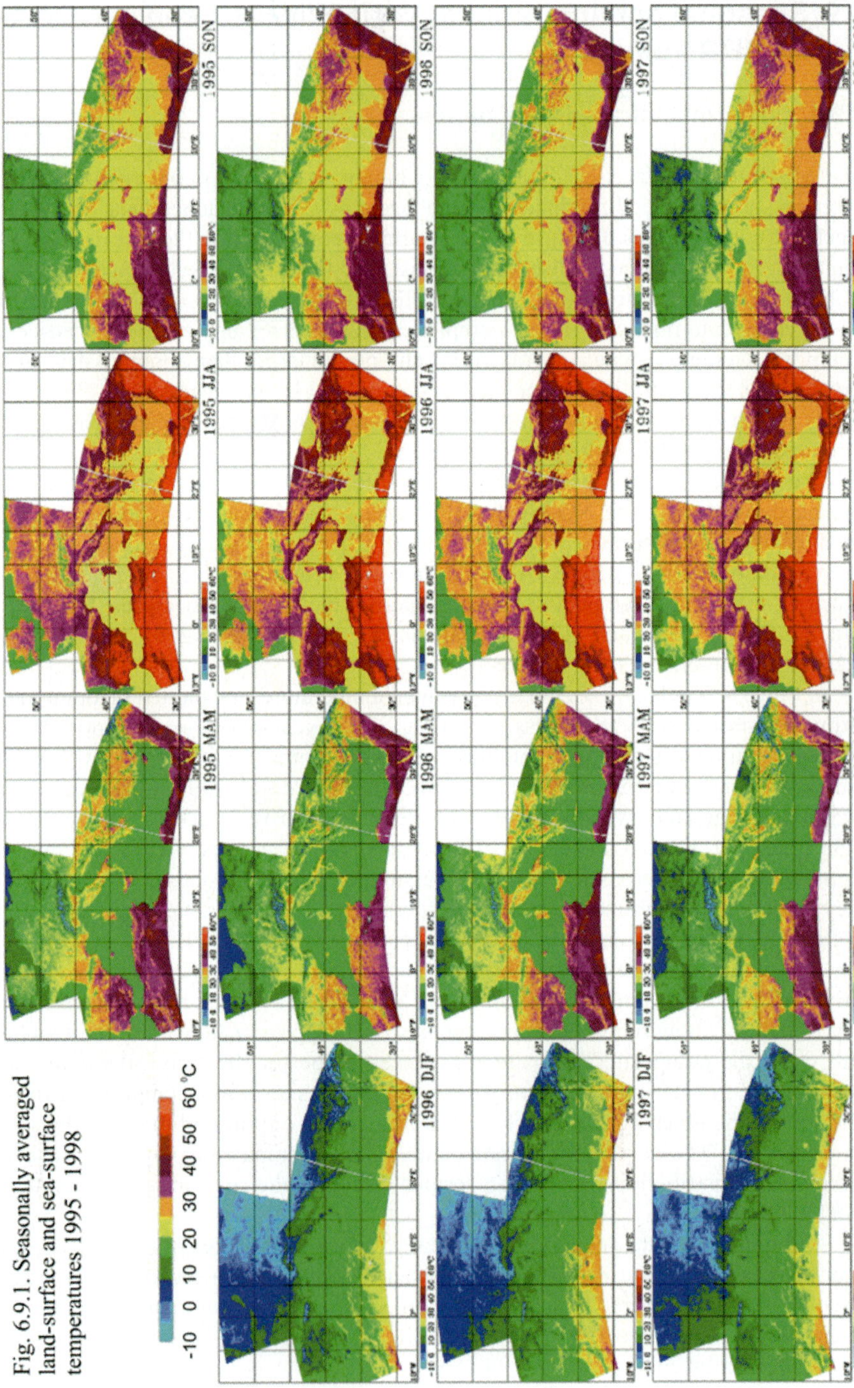

Fig. 6.9.1. Seasonally averaged land-surface and sea-surface temperatures 1995 - 1998

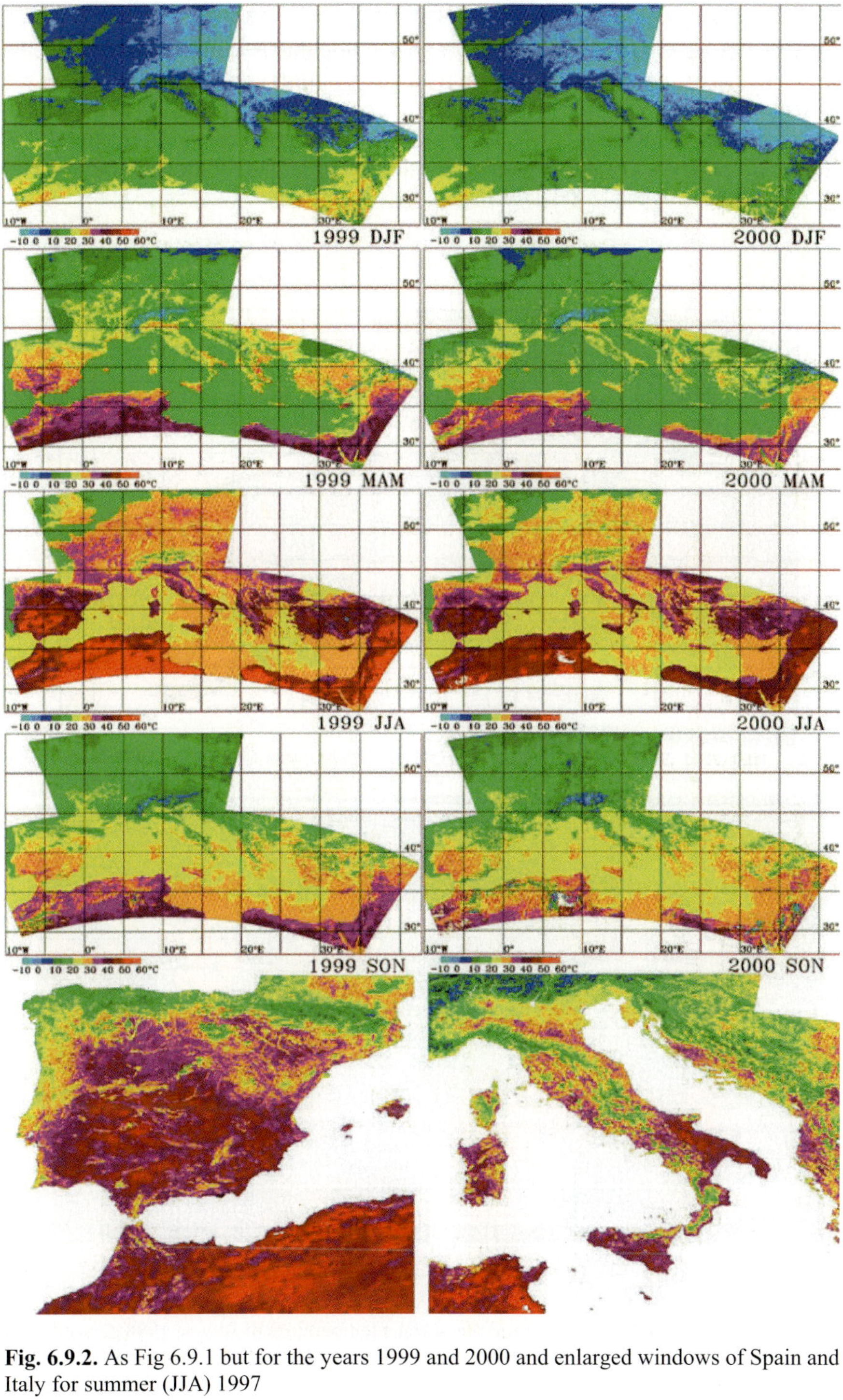

Fig. 6.9.2. As Fig 6.9.1 but for the years 1999 and 2000 and enlarged windows of Spain and Italy for summer (JJA) 1997

For Spain and Italy enlarged August 1996 images are added at the bottom of Fig. 6.9.2 to give an impression of the spatial resolution that is achieved. Here it can be seen how the *Sistema Central* in Spain with temperatures around 35 °C separates the hotter plains north-west and south of it. From the Castilla - La Mancha area down to Andalucia there is a number of "hot spots" with temperatures like in the deserts of North Africa, areas that are sensitive to desertification. In Italy the Po-valley warms up in spring as soon as the Basilicata and Sicily.

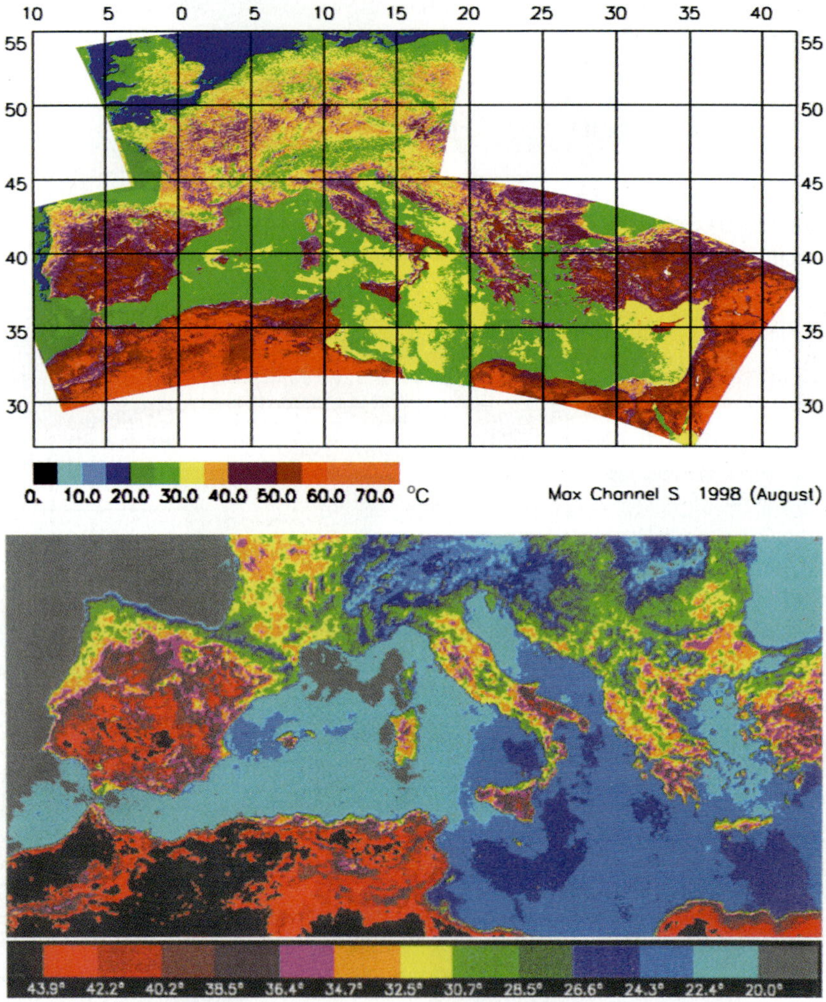

Fig. 6.9.3. Maximum land-surface and sea-surface temperature of August 1998 derived from AVHRR data (top) and maximum TOA temperature for decade 22 in August 1998 derived from Meteosat data (bottom)

In Fig. 6.9.3 the Maximum temperature reached in August 1996 picked out of the data set of which examples are shown in Figs. 6.9.1 and 2. It is of interest to compare these temperature data with the mean temperatures of Fig. 1.1 and with the maximum temperatures inferred from Meteosat data of which one example (for August 1998) is shown in the lower part of Fig. 6.9.3. Though the numbers at the colour codes are slightly different the general picture is very similar. Even small details show up like the hot spot in Corsica (green colour) and the cooler area in northern Spain. The Meteosat image in addition shows the distribution of the cold water upwelling from the gulf of Lyon.

The evaluation of long-term data series (1989 - 2004) will be discussed in Appendix 8.

6.9.2 Short-wave Channels Data Series

The Role of Vegetation as Change Indicator

In the thermal infrared time series presented in section 6.9.1 changes have been detected which are related to climate variability and the state of vegetation. The reason for changes of the state of vegetation in turn can partially be sought in climate variability but can have other reasons as well. A measure for the state of the vegetation is its "greenness" which can be monitored from space in the shortwave range of the spectrum. A convenient measure of this quantity are vegetation indices. During dry years the "greenness" of vegetation is lower than in wet years and/or the length of the vegetation period is shorter. This can be seen in the mean annual vegetation index, its annual sum, and the development of the index within each year. The general conclusion, that high vegetation indices indicate wet years and that low ones indicate dry years is only valid if larger areas show the same signal. At smaller scales changes from year to year may also be caused by changed land use. As an example, the reduction of the vegetation index can result from wilting of otherwise unchanged vegetation due to drought or from a change of land-use e.g. by converting macchia into an orchard with young trees or irrigating the desert (Fig. 1.25). Vineyards can look like thin vegetation if grown on bare soil (see Fig. 5.60) or it can look richly vegetated if grass grows in between the rows of the grapevine. The vegetation index can also increase if a forest is replaced by grassland but the annual cycle will be different. A momentary view can lead to wrong conclusions. Long term time series are generally needed for a correct interpretation of processes occurring at the land surfaces. The amplitude of the vegetation index and its development in the course of years contains useful information about the vegetation type. To be on the save side with the interpretation of these data in addition local inspections are advisable such as expected from the surroundings of anchor stations.

Turning now to the short wave bands of the AVHRR instrument the development of the vegetation often is characterized by the NDVI (see Chapter 2). Though any vegetation index may be constructed from the primary data set, only the NDVI is considered here because, despite some deficits, this vegetation index derived from AVHRR channel 1 and 2 is widely used to assess changes of the vegetation and therefore allows comparisons with other studies. As has been mentioned earlier the

NDVI composed of the Landsat-TM bands 3 and 4 give similar results. It is shown in section 4.6.2 that also from microwave signals vegetation indices can be constructed.

In the following reproductions of the data for the whole Mediterranean basin many details contained in the original data set are lost as will be seen by comparing figures which show the full Mediterranean basin with those presenting windows cut out of the complete MEDOKADS data set.

Basin-wide Data Representation

In Fig. 6.9.4 an overview is given of the whole Mediterranean basin showing the reflectance measured in AVHRR channel 1 and 2 (ρ_{Ch1}, ρ_{Ch2}) and the NDVI composed of the reflectances measured in these two channels. Figures of the type of data shown in Fig. 6.9.4 can also be used to compute (see Chapter 3) the broad band reflectance for the whole Mediterranean area. One of these results is reproduced in Fig. 6.9.5. Data of this type are important as input for atmospheric circulation models. It is desired to supply these models with actual and periodically updated broad band reflection or albedo maps. The albedo of Fig. 6.9.5 is computed by application of the formula of Brest and Goward (1987, compare Chapter 4):

$$\rho = 0.526\rho_{Ch1} + (0.485 - 0.112\,NDVI)\rho_{Ch2} \qquad (6.9.1)$$

It is obvious that the higher albedo values in Fig. 6.9.5 correspond to the lower NDVI values of Fig. 6.9.4 and the higher temperature values in Fig. 6.9.3 and 4.

An example is given in Fig. 6.9.6 for the change of the NDVI from April to August. It shows at the top the NDVI for the first decade of April, in the middle that of August and at the bottom the difference between both images. The vegetation does not shift in parallel with the surface temperature since it depends on the water availability as well. Its phase differs from place to place. This can be seen in Fig. 6.9.7 where the decade of the NDVI maximum is displayed. It shows a wide range by travelling from southern countries with summer drought towards the north where the growth of vegetation depends more on sunshine than water. In some places the NDVI has a double peak (spring and autumn) which does not show up in this graph. Only the main maximum between January and October is displayed. November/December maxima do not show up, because only the decades 1 - 26 are used for this figure. A comparison with the bottom part of Fig. 6.9.4 shows the relationship to the average NDVI. The lower part of Fig. 6.9.7 indicates the temperatures at the time when the NDVI maximum occurs. Inland of northern Africa the (small) maximum of the NDVI occurs early in the year at already high temperatures. The late maxima in the near coastal zones are related to topography. In Spain one can find closely adjacent areas with early and late development of the vegetation at similar temperatures.

This leads to the question of vegetation changes which can be studied by looking at the development of the vegetation index over the years. Fig. 6.9.8 shows such a development from the year 1995 to the year 2000 for the whole Mediterranean basin and parts of central Europe. As ill be discussed in Appendix 8, such short term

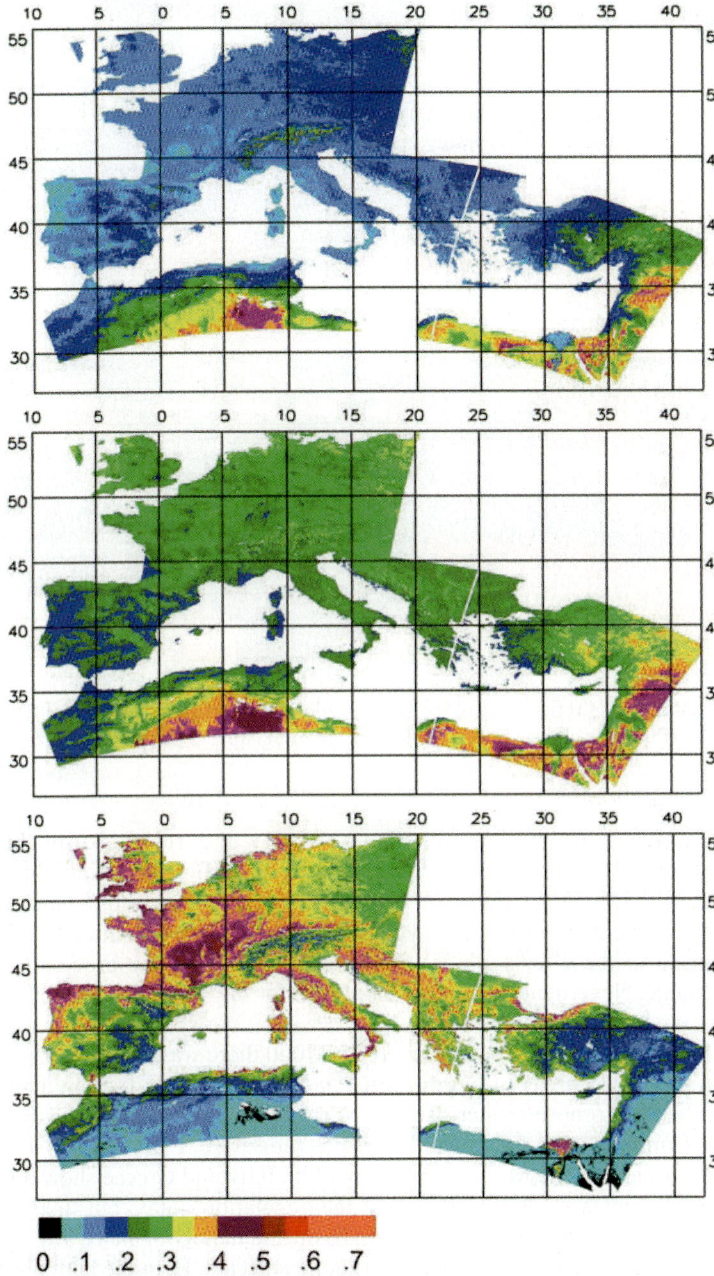

Fig. 6.9.5. Mean value composite of AVHRR channels 1 (top), channel 2 (centre), and NDVI (bottom) for the time period 1995 - 2000

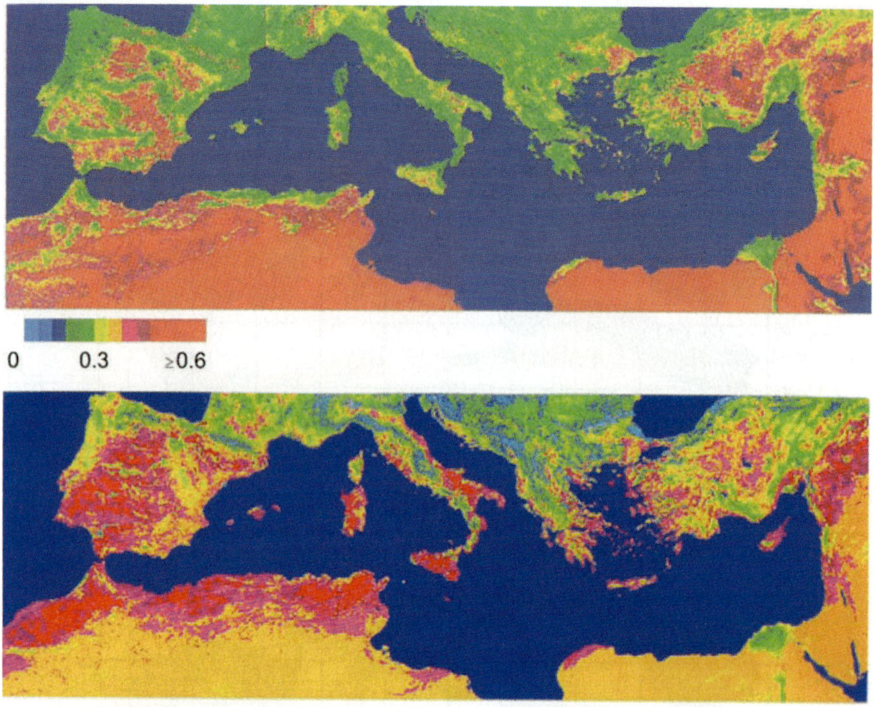

Fig. 6.9.5. Broad band reflectance of the first decade in August 1996 (top) and differ ence of broadband rerflectance change minus NDVI change from April to August (bottom, details see text)

"trend" analyses are not appropriate to study the relationships between desertification and global climate change because in different periods the "trend" may change its sign. Vegetation reacts as one sees in these data rather quick to short term climate anomalies of a few years duration which constitute the "noise" superimposed to the effects which might result from the slow process of global climate change. Nevertheless, they are indicative for the sensibility of landscapes to climate change.

During the first 26 decades of years 1995 - 2000 there was, as to be expected, no change in the desert areas of North Africa and the Near East (yellow colour). In Spain and North Morocco an increase of the vegetation index is observed that shows recovery from the drought period 1994/5 (compare Figs. 6.9.1, 6.9.9 and 6.9.10).Other parts of the Mediterranean area like Italy and Greece show a mixture of small scale increases and decreases of the vegetation index, but there are also areas distributed all over the Mediterranean basin and mainly in the Near East where the reduction of vegetation was severe during these years. Both the high positive as well as high negative values point to areas which react sensible already to short duration climate anomalies and would therefore be vulnerable for longer term climate changes.

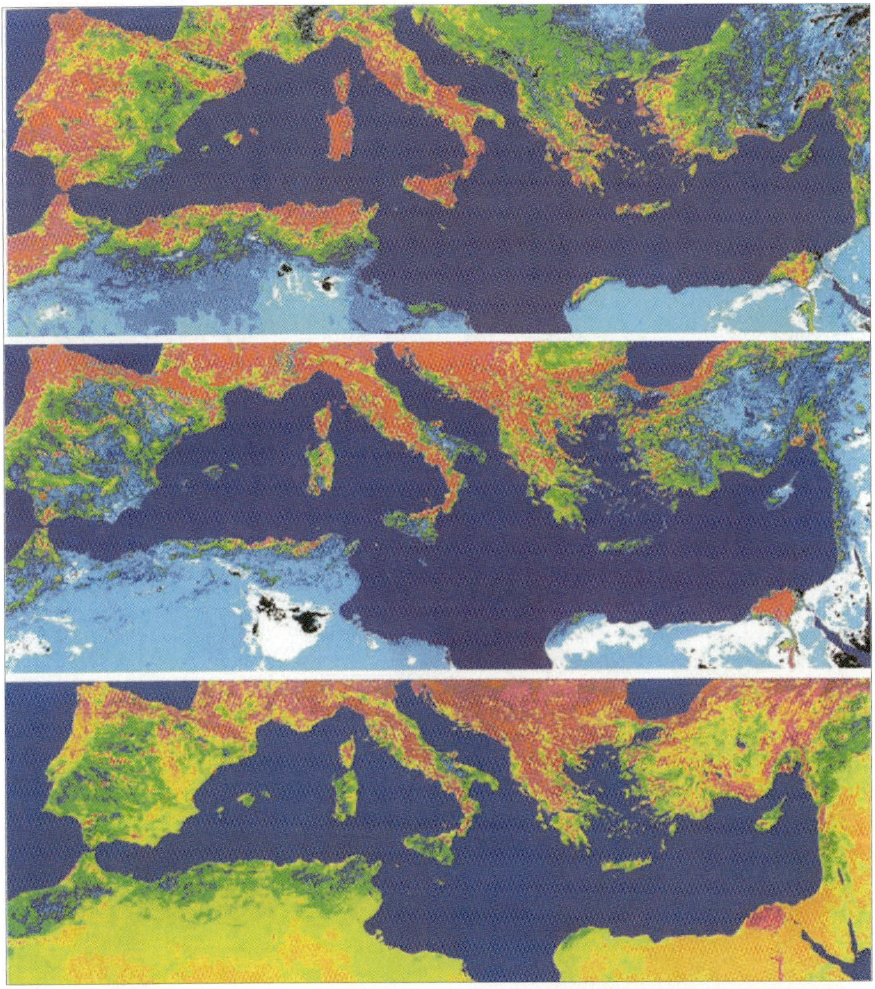

Fig. 6.9.6. Ten days maximum NDVI composite of the Mediterranean basin in polar stereographic projection for the second decade of the month of April 1996 (top) and of the first decade of the month of August (centre) showing the areas of maximum summer drought. The spatial resolution of the original data is 1.2 km. The colour scale ranges from 0 to 0.8. Blue stands fror low values, red for high values. The scene at the bottom is the difference between the August and the April values of the NDVI. Yellow to blue colours indicate increasing negative changes and yellow to red colours increasing positive changes

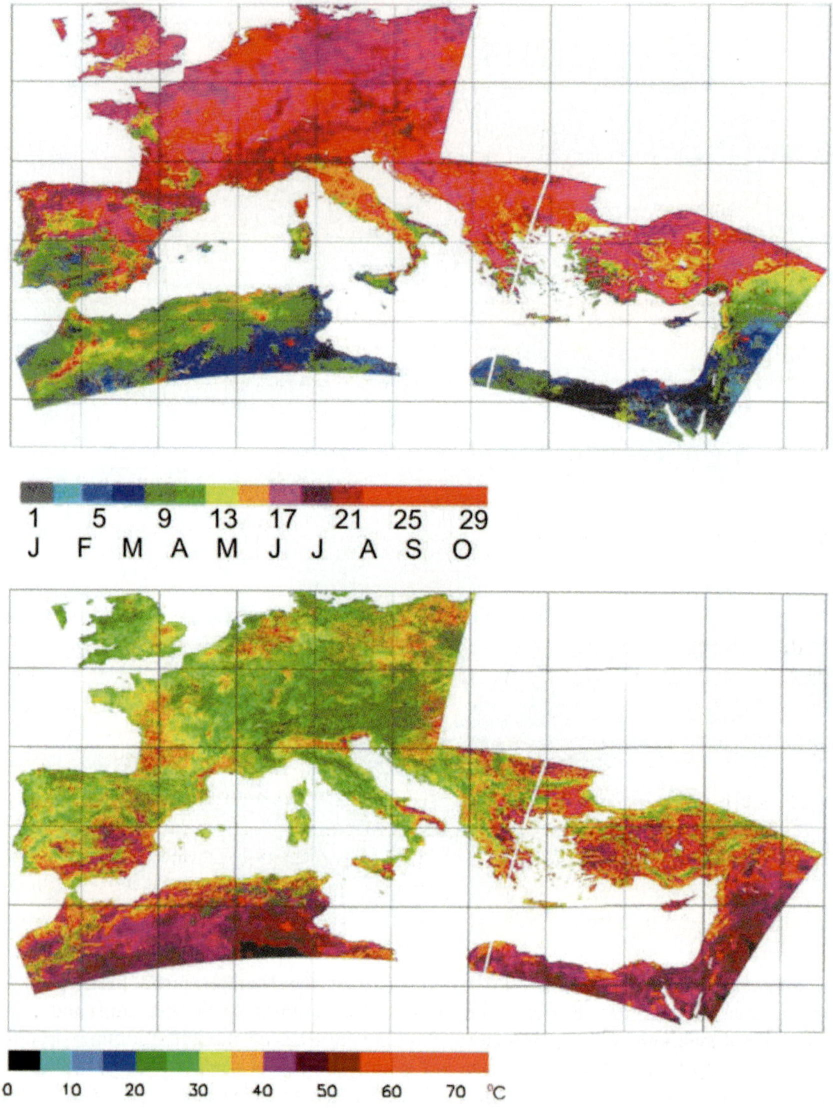

Fig. 6.9.7. Decade in which the maximum of the vegetation index occurred in 1996 for each pixel between January (decade 1) and October (decade 29) (top) and temperature at the time of the occurrence of the maximum NDVI (bottom)

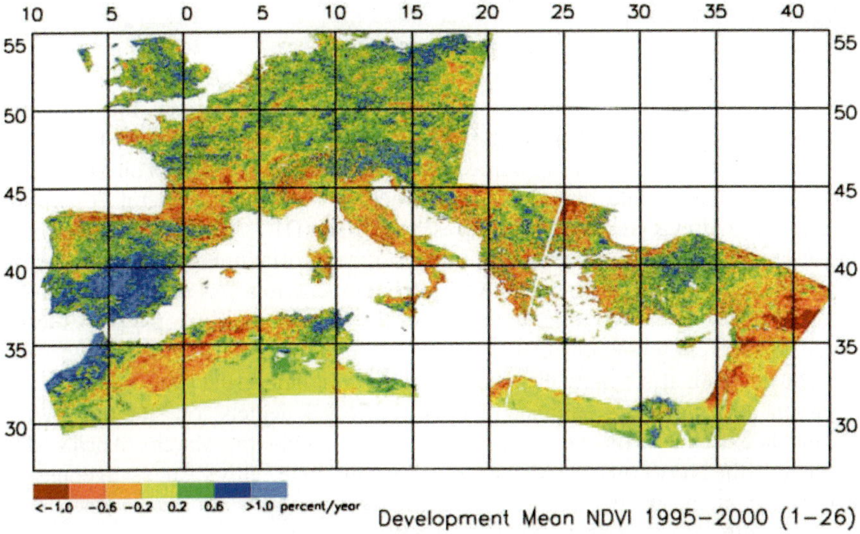

Fig. 6.9.8. Development of the NDVI for the decades 1 to 26 from 1995 to 2000

Another tool to identify areas which are sensible to desertification is shown in the lower part of Fig. 6.9.5. In this map the difference between the broad band albedo change and the NDVI change between August and April (see Fig. 6.9.4, bottom) is plotted for the Mediterranean area: $\Delta\rho$ - ΔNDVI = $(\rho_{August} - \rho_{April})$ - $(NDVI_{August} - NDVI_{April})$. Blue colours represent negative values. In this case the NDVI increase is larger than a possible albedo increase since it is not likely that the albedo decreases in summer. This signals a stable situation. Orange colours indicate zero change. Either both quantities do not change at all or their changes compensate each other, this is an indication for either a non-vegetated surface with stable reflectance (desert) or an indifferent situation (such as vegetation over dark soil). Red colours stand for positive values. This occurs (i) if the albedo increases in summer and the NDVI decreases, (ii) if the albedo increase is larger than the NDVI change, or (iii) if the NDVI decreases to a larger degree than any albedo change. Since a decrease of the vegetation index is mostly accompanied by an increase of the albedo, the red areas are those which would be vulnerable to desertification or in this example sumer drought. The condition for ongoing desertification would be that this index is positive for the majority of pairs of months for a series of years, e.g.:

$$\sum_{n}\sum_{i}\left[\left(\rho_{i,n} - \rho_{i,n+1}\right) - \left(NDVI_{i,n} - NDVI_{i,n+1}\right)\right] > 0 \qquad (6.9.1)$$

Where i = 1, 2, ... 12 is the number of months, n = 1, 2, ... N the years of a sequence and j is a reference year. Alternatively one could take one reference year and compare it with a series of following years. The index has not yet been applied but seems to be promising for the future.

Regional Variability

Only a few examples of regional variability will be discussed here, following the considerations about seasonality in the preceding section. Examples for long term changes are presented in Appendix 8. These data series underscore that taking too short periods would give wrong impressions about trends. From 1990 on to 1995, as an example, one would diagnose for the Iberian Peninsula a general downward "trend" but with varying regional centres. 1992 the NDVI minimum was mainly in the west, 1993 in the southwest, 1995 in the southeast. Maximum temperatures rose with declining NDVI or the season with high temperatures was prolonged in these areas. Starting with 1996 in large regions of the Iberian peninsula the NDVI trend reversed and kept going upwards.

The question arises, how the situation of the first half of the year influences the situation during the dry summer months. Fig. 6.9.9 depicts four for the Iberian peninsula extreme years and shows in four image blocks the March - April - May average of the NDVI (ul), the May average (ur), the June - July - August average (ll) and the annual mean value (lr). The scale reaches in all figures from zero (deep blue) to 0.7 (bright red). The year 1998 (Fig. 6.9.9.b, lower block) starts in spring with a high NDVI which signals wet and warm conditions. In May the NDVI is very high all over the peninsula and in summer some areas show low NDVIs such as Murcia and the Ebro-valley. The annual mean still shows a moderate NDVI. In contrast the year 1995 (Fig. 6.9.9.b upper block) starts already with relatively low NDVIs in the eastern part and these low NDVIs spread in May over half of the peninsula. In summer most of the peninsula is deep blue and also in the annual average most of the peninsula has low NDVIs. This severe drought had precursors in the preceding years. In the year 1992 (Fig. 6.9.9.a, lower block) one sees after a moderate May large areas with low NDVI in summer and in the annual average high NDVI values are missing, even in the northern forested areas. 1993 (not presented here) shows similar features. This is a very unusual aspect caused by the injection of aerosols into the atmosphere by the eruption of the Pinatubo volcano in 1991 (compare Apendix 8). The year 1990 (Fig. 6.9.9.a, upper block) was moderate in all seasons and the typical summer structure which repeats year for year with different intensity is clearly to be seen. At the Iberian peninsula summer NDVIs always show similar features independently of how the year starts but if the NDVI is high in spring and lasts high during May, it can be expected that the summer drought is less severe.

Fig. 6.9.10 supplements this analysis for Spain by a series of maps for Italy and Greece. Here, in contrast to Fig. 6.9.9, the *deviation* of annual mean values from the long term 1989 - 2004 mean are used to demonstrate changes.. The maps on top of Fig. 6.9.10a show these long term mean values. In the following figures the yellow colour stands for zero deviation from the mean. In the maps in the middle of this figure a very uniform negative deviation of about 5 % in the average from the mean values can be seen. This is the map of the year 1992 when the Pinatubo eruption had probably its maximum impact. The corresponding map of Spain (Fig. 6.9.9a) would look very similar if presented as deviation from the mean. Another extreme case is seen in the lower map of Fig. 6.9.10a. This is the year 2004 with maximum positive deviations. As discussed in more detail in Appendix 8, the measurements since the

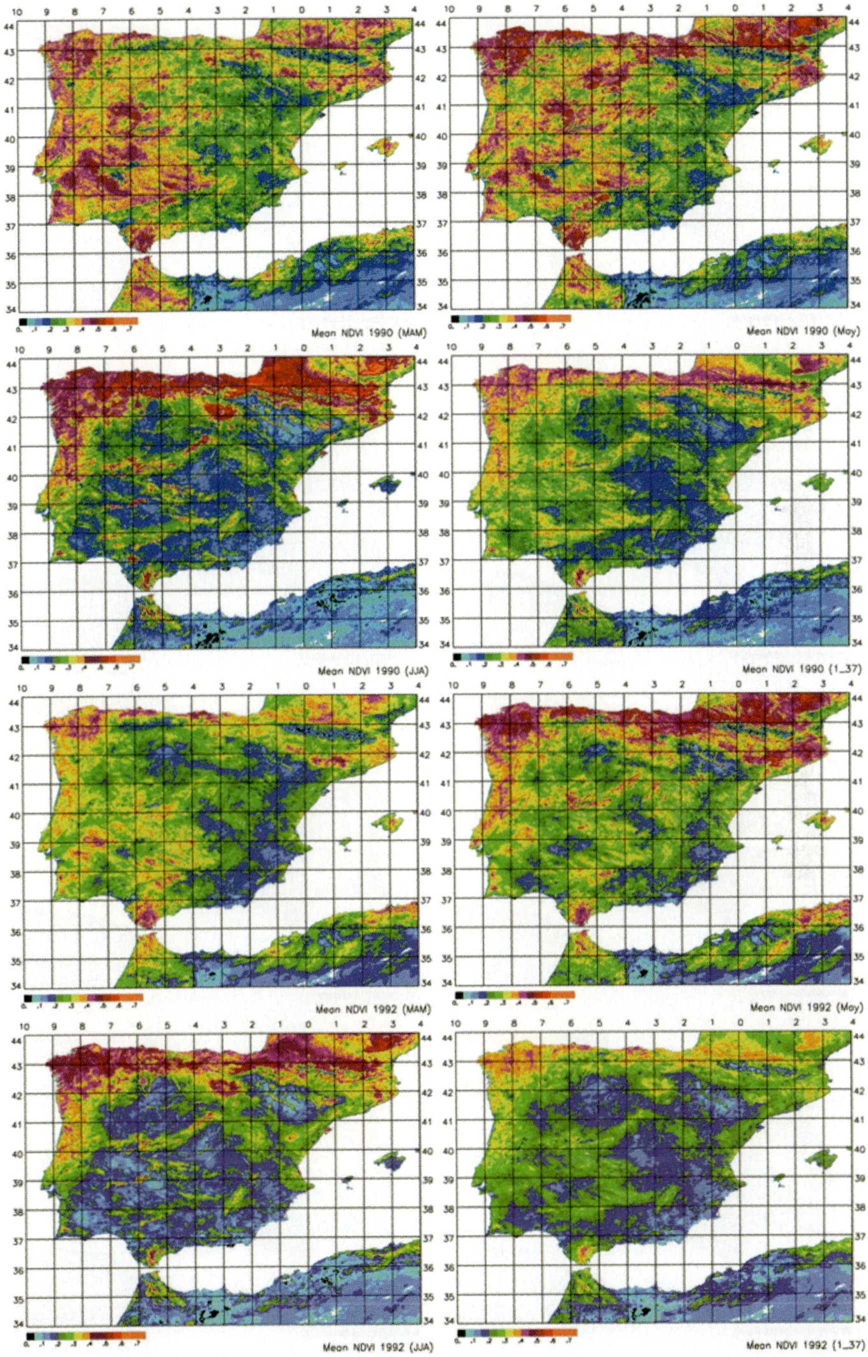

Fig. 6.9.9a. Mean Spring, May, summer and annual NDVI values for Spain, 1990 and 1992

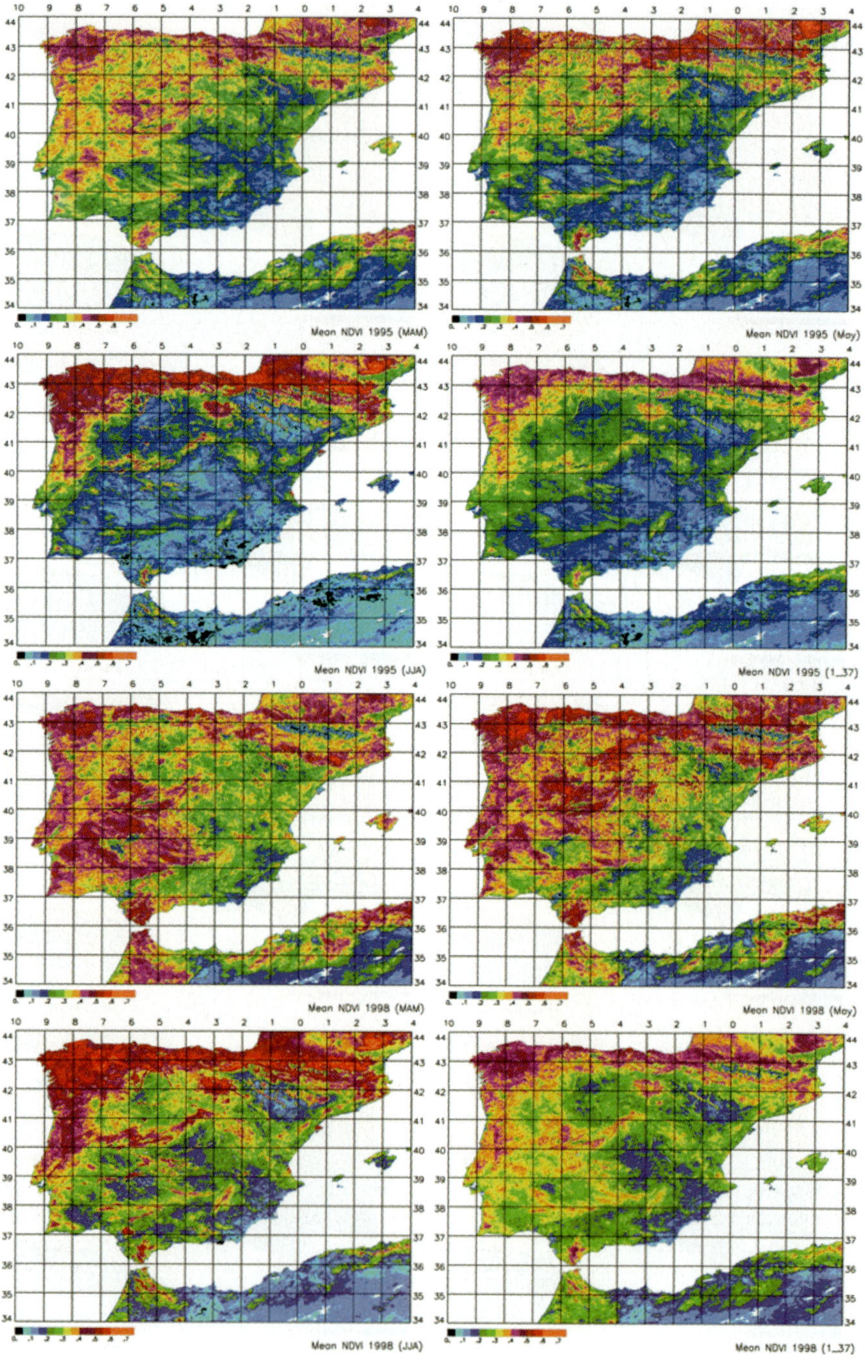

Fig. 6.9.9b. Mean Spring, May, summer and annual NDVI values for Spain, 1995 and 1998

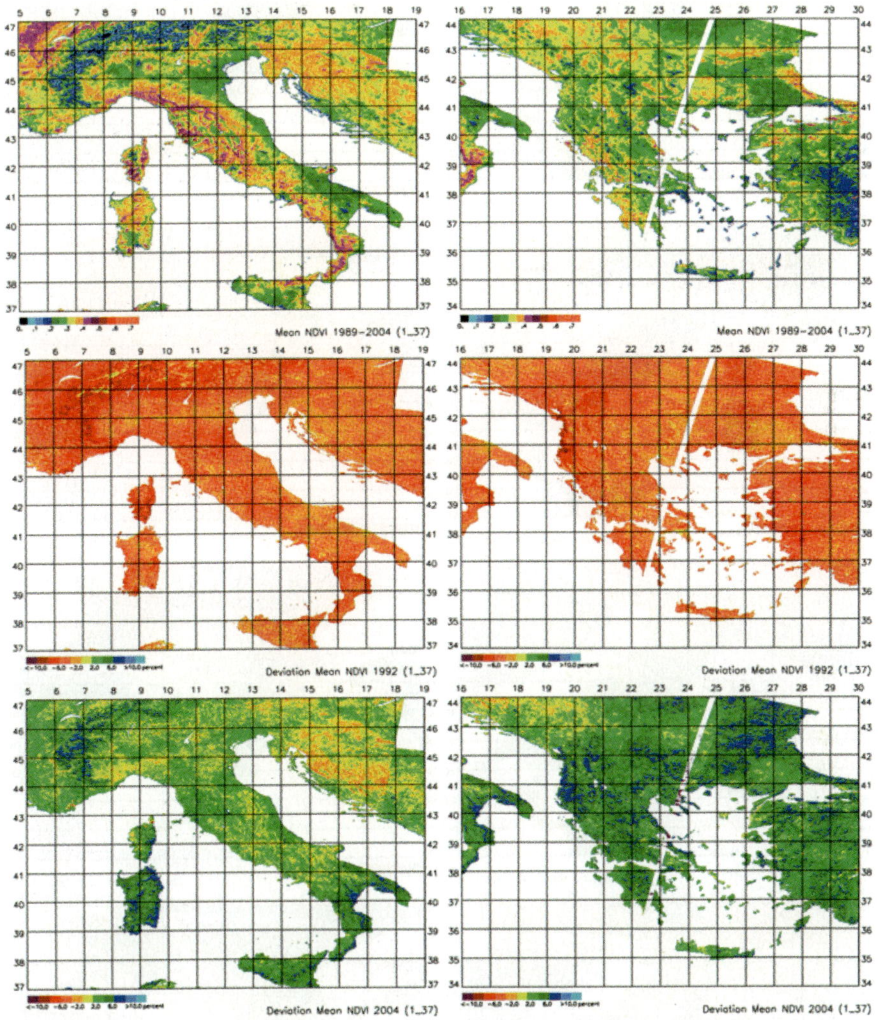

Fig. 6.9.10a. Annual mean NDVI maps for Italy and Greece. The upper maps are annual mean values averaged from 1989 to 2004. The following maps are deviations from these averages. The central maps show the situation of the post Pinatubo eruption year 1992. In the lower part the maps of 2004 are presented which habe been obtaines by the advanced NOAA-16 AVHRR instrument but normalized to the NOAA-9 conditions

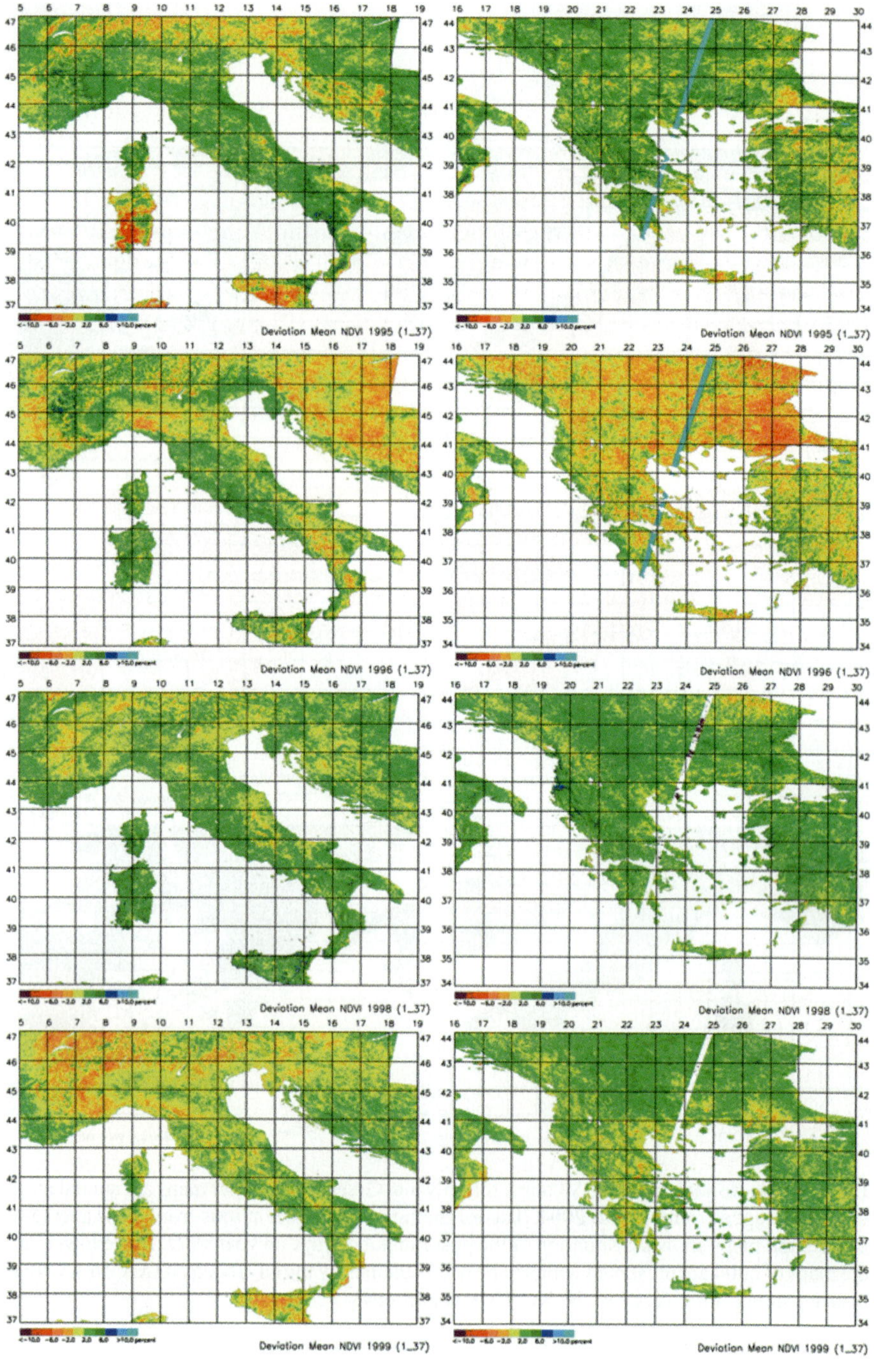

Fig. 6.9.10b, Deviation from the 1989 - 2004 mean NDVI values for Italy and Greece, 1995, 1996, 1998 and 1999

year 2001 with an advanced AVHRR instrument result in higher NDVI values as older instruments produced. Though the data shown here are all normalized to NOAA-9 conditions, it can not be excluded that still a difference remains. Notwithstanding the absolute level of the NDVI, the relative distribution of its values mediate the impression of an elevated NDVI in the Alps and large areas of Greece, the eastern part of the Balkans and Turkey as compared to the Adriatic part of the Balkans. Fig. 6.9.10b gives an overview of more "normal" years. The uppermost map of 1995 corresponds to the drought map of Spain in the upper part of Fig. 6.9.9b. It can be seen that with the exception of parts of the Alps, Sardinia, Sicily, Crete, and the inner parts of Turkey the opposite is true for most of Italy and Greece. A small positive deviation from the mean can be noted (the corresponding map for Spain shows a deep red colour with the exception of the northern coastal areas and the Pyrenees). The following maps, 1996, suggest a drought situation in large areas of the Balkans and the Po valley indicated by negative deviations from the mean. This is again opposite to Spain where the land obviously recovers from the 1995 drought. The years 1998 and 1999 which follow in the lower part of Fig. 6.9.10b show a more average situation but with negative deviations in the western part of the Alps as well as again Sardinia and Sicily and in the average slightly positive deviations for the Balkans.

In the images reproduced in the Figs. 6.9.9 and 6.9.10 regional details tend to get lost because of the restricted resolution. Two scenes are therefore shown in Fig. 6.9.11 at full spatial resolution to show a case of "irregular" seasonality of an area with intensive irrigation. It covers a section of the Albacete - Valencia region in Spain. Until July the area developed "normal" with increasing drought and decreasing NDVI. In the July to September mean values the irrigated areas near the village of Barrax and the natural reserve Elche at the coast south of Valencia pop out while in the surrounding the vegetation index is further reduced. If one looks at individual years the expansion of the irrigated area can be followed.

The trend from 1989 to 1997 in the south-eastern part of Spain is plotted in a special way in Fig. 6.9.12. At the top the changes of the vegetation index and the

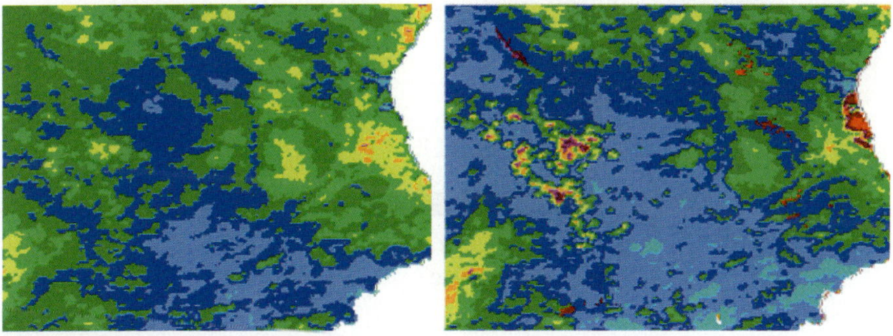

Fig. 6.9.11. Ten years mean (1989 - 1998) of the annual mean NDVI of the Valencia - Albacete region. Left: February to July, right: July to September. The irrigated areas around Barrax and the natural reservation Elche south of Valencia clearly stand out in late summer. The irrigated area grew during this period and its pattern slightly changed from year to year.

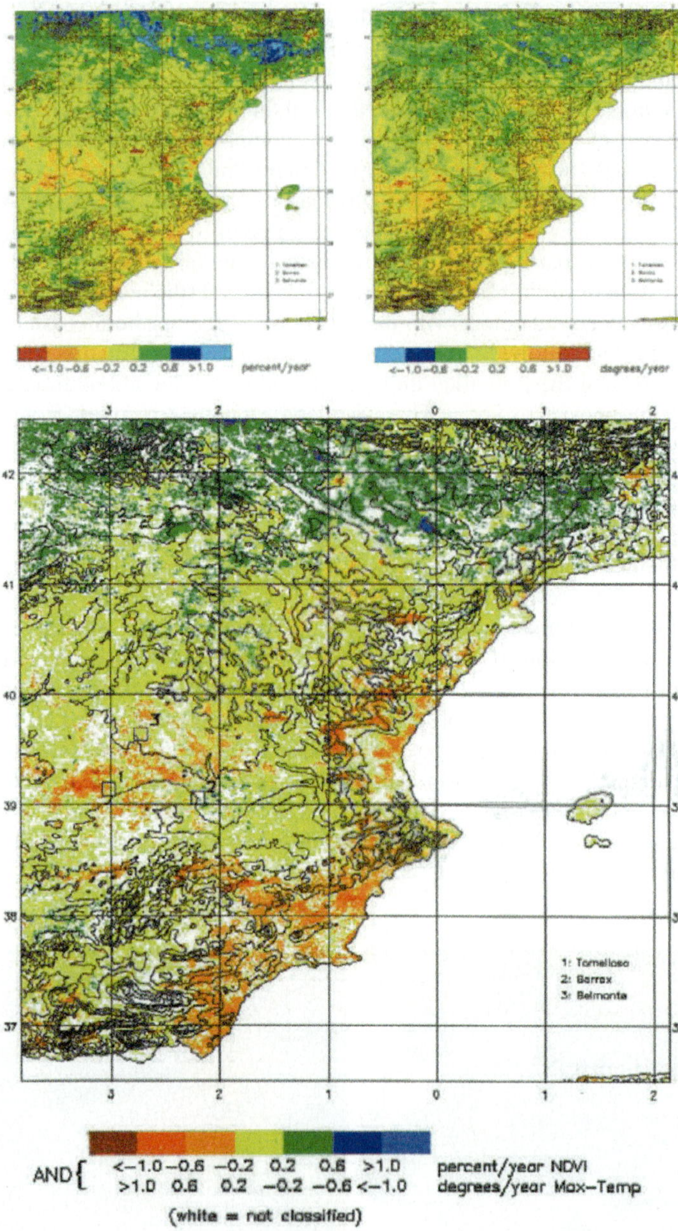

Fig. 6.9.12. Change of the annual NDVI and the maximum surface temperature from 1989 to 1997 in south-eastern Spain. The upper figures show the individual changes which are composed into one image in the lower figure

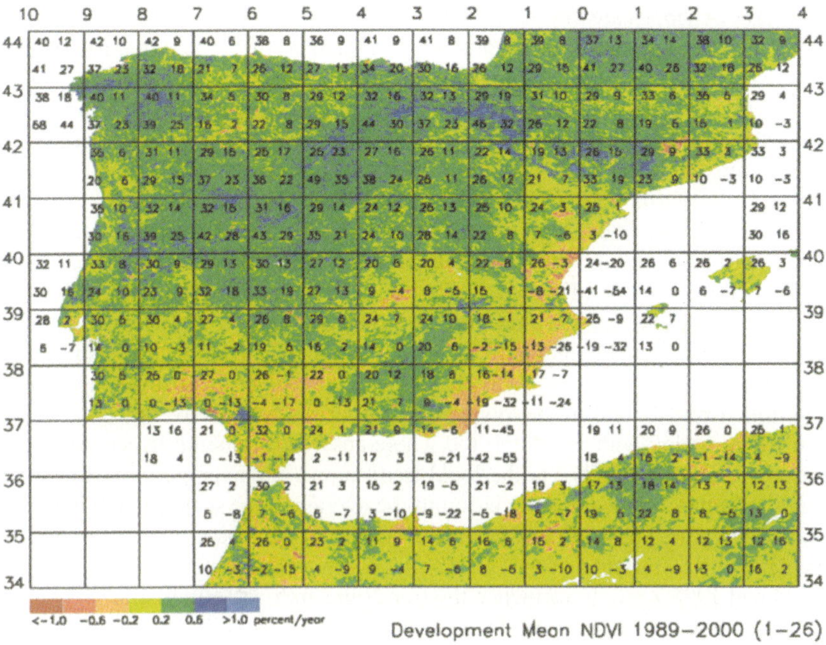

Fig. 6.9.13. Change of the NDVI averaged over the decades 1 - 26 from 1989 to 2000 for the Iberian Peninsula. The numbers in the 1 x 1 degree boxes have the meaning: mean value (top left), change 1989 - 2000 (top right), change in % per year times 10 (bottom left), relative change per year multiplied by 10 with respect to the whole analysed frame

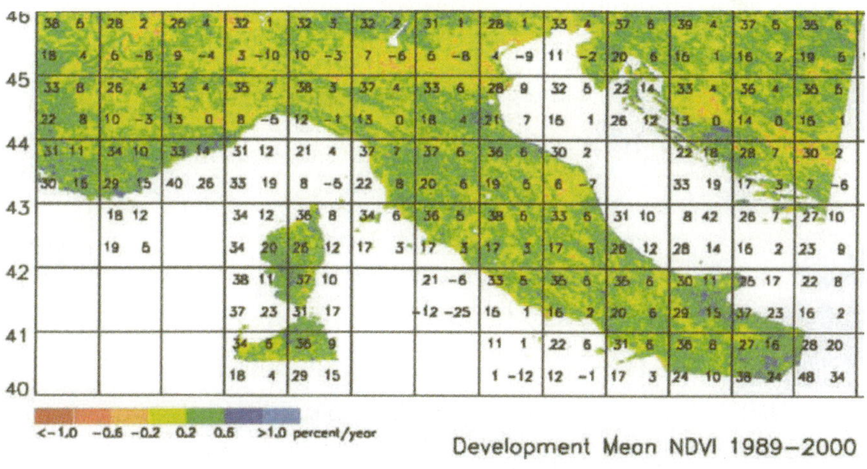

Fig. 6.9.14. Same as Fig. 6.9.14 but for northern and middle Italy

temperature are shown separately. In the lower enhanced scene the changes are compiled into one image after appropriate scaling. It shows clearly that temperature and NDVI changes are closely correlated with opposite signs. A one percent change of the NDVI per year corresponds to -1 °C change of the maximum temperature. Only small areas are left white where the correlation is not obvious.

Trends can also be quantified for selected areas rather than pixelwise. This is shown in Fig. 6.9.13 for the Iberian Peninsula and in Fig. 6.9.14 for northern Italy. The figures represents the change of the mean vegetation index during decades 1 - 26 for the 12 years period from 1989 - 2000. The change is obtained by computing for each pixel the linear regression line for the 12 years. The scene is subdivided in 1 x 1 degree boxes into which four numbers are written which contain the full NDVI information for the land surface in that box. In the upper left corner one finds the mean value of the NDVI within the box multiplied by 100, in the upper right corner the change from the beginning to the end of the period in percent of this mean value, in the lower left corner the local change ("trend") given in absolute units per year multiplied by 10^4, and in the lower right corner the relative average change during the whole period with respect to the whole analysed frame in percent of the mean value. This latter number is the local trend minus the average trend of the whole analysed frame or the local deviation of the average large scale change in percent of the mean value. Some of the areas show land-surface changes due to fire which also explains the restoration of the NDVI after a few years. But also the very dry years 1994 and 1995 strongly influence the trends in south-easterly Spain.

The same analysis was performed for parts of Italy (Fig. 6.9.14) where a slight decrease can be recognized in the Po valley and a gain of vegetation in the south.

6.10 Vineyard Change Detection

6.10.1 Introduction

In the Mediterranean Basin, vine is preferably grown in dry-farming conditions, establishing a strong environmental relationship that overcomes local, regional and national borders. Improper management of dry-farming systems under semiarid conditions enhance erosion and land degradation, waste rainfall resources and give rise to rural depopulation and land abandonment. Dry farming is so far the prevailing system of growing vines in Spain and even more so in Castilla-La Mancha, where it is the most common woody crop. The changing environmental conditions (soil, climate and availability) make it difficult to promote any action to regularise annual yields, which could be a solution in a sector that represents one fifth of the total regional agrarian production, in 1993, spreading over 609.000 ha surface in Castilla-La Mancha (MAPA 1997), figuring 17.4 % lower that in 1983. Nearly 99% of the grape production is devoted to wine.

The process of reformation of the EU Common Agricultural Policy, CAP, is outstandingly affecting the vine and wine sector in Spain (Sotés 1996). The actions proposed are to decrease productions in farms, to root out a large surface of vine stocks, to sustain vine growing in marginal areas, as well as facilitating training, research to improve the added value of vine products. Most of these proposals are affecting negatively vine growing in arid and semiarid regions because financial accompanying measures hardly compensate for the socio-economic and environmental damage caused. That human-induced changes occur at rates which are unprecedented in natural conditions and are likely to increase in importance if not controlled and/or mitigated by policy interventions.

The importance of mapping, quantifying, and monitoring the changes in physical characteristics of land cover has been widely recognised in the scientific community as a key element in the study of global change (e.g IGBP 1990; Henderson-Sellers et al. 1992). Change detection is a powerful application of remote sensing, in that the spectral and temporal resolution of multi-band sensors can be used to advantage in monitoring both significant and subtle land cover changes over time at global scale (Townshend et al. 1991), because of repetitive coverage at short intervals and consistent image quality.

The vineyard crop is considered as magnificent indicator of the desertification process in the RESMEDES/RESYSMED projects. Castilla-La Mancha is a region, changing very rapidly especially regarding the vineyard crop. This decrease is mainly due to economic and social factors. Therefore, the following study of the vineyard changes is required for determining the consequences of this fact in the landscape system, including the desertification process.

The changes occurred on vineyard crop give three different forms of agricultural use of soil: abandoned, bare soil and cereal. During the period of study (1991-1996), most of originally abandoned vineyards were extracted few years later for cereal cultivation and so the soil can be observed as a cereal parcel or like a bare soil one. Abandoned vineyard with natural vegetation and vine coverage is extremely rare for long time period.

The purpose of this paper is to detect changes in an extensive area of vineyard culture, for the period 1991-96. For that we will use two methods: NDVI differencing images and NDVI change vector analysis. Since the ability to detect change is related to its spectral and temporal characteristics, suitable image data from appropriate acquisition dates must be analysed. In this case, the vineyard change detection was performed using Landsat-5 TM images of 1991 and 1996. The two methods permit to detect areas of change and categorise the type of change processes occurring.

The study area situated in the experimental zone, Castilla-La Mancha, of the EFEDA project, is located in central Spain and constituted by five councils, Mota del Cuervo, Las Mesas, Tomelloso, Socuellamos and Villarobledo, with a surface area of about 3.534 km^2. The UTM co-ordinates are X 495096-557110 and Y 4321694-4379098. The study area is described in section 6.7.2 and Appendix 1.

6.10.2 Methodology

The change detection method should be able to detect areas of change and categorise the type of change processes occurring in the vineyard crop during 1991-1996 period. So, it is suitable to make an spectral analysis of the interannual curves of the NDVI for some training areas where we are sure that changes like those commented previously (abandoned, bare soil and cereal) have been occurred. The study of the interannual vineyard crop variability during 1991-1996 period will be carried out within three steps:

1. Applying the multitemporal masking classification method to map the vineyard crop in the study area for 1991.
2. Spectral analysis of the interannual temporal development curves of the vineyard change processes.
3. Using NDVI change vector analysis and NDVI differencing to detect and categorise vineyard change processes in the study area using the temporal resolution satellite data.

6.10.3 Results

In order to carry out this study (step 1), a series of six Landsat-5 TM scenes were acquired for 1991 and other three for 1996. From this multiannual series the images of the following dates have been selected for anylysis: May 27th , and July 14th of 1991, and May 24th and July 11th of 1996. They were analysed and checked for their suitability to the objectives of this study. In section 6.7.2 the followed procedure for image processing and classification is described. For the vineyard change detection purpose, a mask is created to eliminate the remaining land use classes.

Five training areas have been selected where the Castilla-La Mancha group made ground measurements during the 1991 EFEDA field campaign. Fig. 6.10.1 shows the phenology, for 1991 and 1996, of parcels that undergo each of the three changes considered, conversion to abandoned, bare soil, and cereal. Analysing the curves, it is suitable to use methods like NDVI differencing and NDVI change vector analysis in order to take advantage of the differences observed in terms of NDVI between the vineyard crop and three changes categories.

To better adapt the method to the vineyard change, the phenological NDVI evolution in the two study years was investigated. On
this basis the following images have been selected for the evaluation: May 27th (M91) and July 14th (J91) for the year 1991,May 24th (M96) and July 11th (J96) for the year 1996 are selected. A thresholding technique was applied to identify the vineyard surface change. The threshold values, in DC, are derived from the histogram of the two NDVI differencing images. The threshold was chosen as mean value + standard deviation. For the first difference the threshold was 110 and for the second difference 108. The difference between the two NDVI images"M91" and "M96" are used to detect vineyard areas that change to "cereal+abandoned areas". Then the NDVI difference between "J91" and "J96" allows to recognise the vineyard areas which are considered as bare soil in 1996. It results for cereal and abandoned vineyard:

$$NDVI_M = 100 + NDVI_{M91} - NDVI_{M96} \qquad (6.10.1)$$

and for bare soil:

$$NDVI_J = 100 + NDVI_{J91} - NDVI_{J96} \qquad (6.10.2)$$

The image which combines both differences shows that the second difference corresponding to the vineyard zones that change to bare soil have noise, and there is confusion between some snatched vines and the vineyard zones which lose vegetation between the two years of study 1991 and 1996. So it can be tried with the method of the NDVI change vector analysis to analyse extensively that possible confusion.

The change vector compares the difference in time-trajectory of biophysical indicator, such as the normalised difference vegetation index, for two successive time periods (Lambin and Strahler 1994). The change vector is simply the vector difference between successive time-trajectories, each represented a vector in multidimensional measurement space. The length of the change vector indicates the magnitude of the interannual change, while its direction indicates the nature of change. In our case, the multidimensional vector change was reduced to a bidimensional one as shown in Fig. 6.10.2.

The magnitude of change vector, calculated as the Euclidean distance between the two positions is expressed as:

$$\left|C_{NDVI}\right| = \sqrt{\left(x_2 - x_1\right)^2 + \left(y_2 - y_1\right)^2} \qquad (6.10.3)$$

Where x_1, x_2 are values of pixels belonging to the images acquired on May 27th 1991 and May 24th 1996, respectively, and y_1, y_2 are values of pixels belonging to the images acquired on July 14th 1991 and July 11th 1996, respectively.

For better understanding of the vineyard change vector analysis, we have used five vineyard training areas selected by the Castilla-La Mancha group for the 1991 campaign, where changes, as commented previously, occurred on ones of them as confirmed by the ground truth. Thus, we have calculated the magnitude change vector for the five training areas as represented in the Figs. 6.10.2 and 6.10.3. The

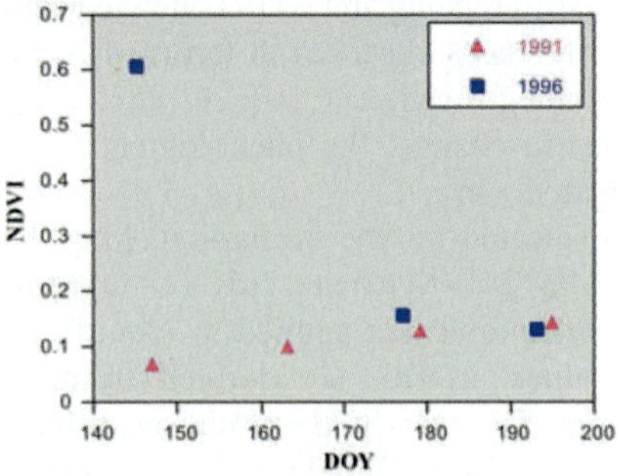

Fig. 6.10.1a. NDVI phenology for change from vinyard to abandoned land

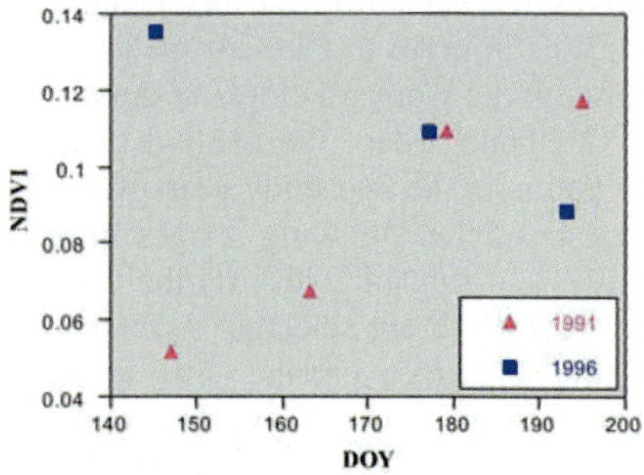

Fig. 6.10.1b. NDVI phenology for change from vineyard to bare soil

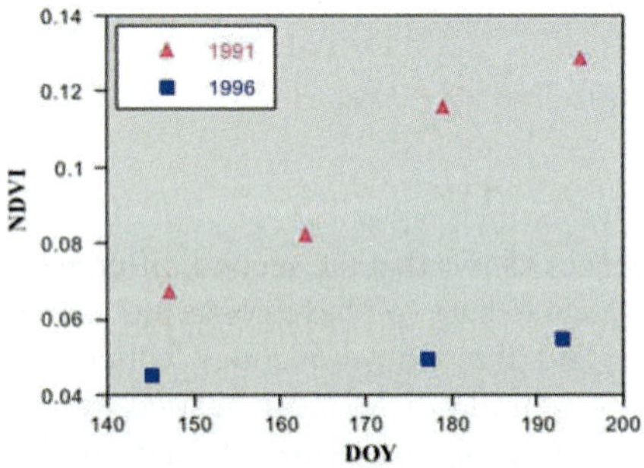

Fig. 6.10.1c. NDVI phenology for change from vineyard to cereal

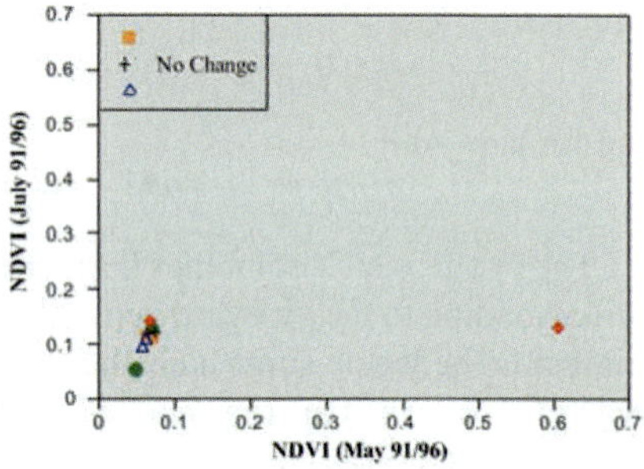

Fig. 6.10.2. Multiannual phenological NDVI change vector from May to July for the years 1991 - 1996

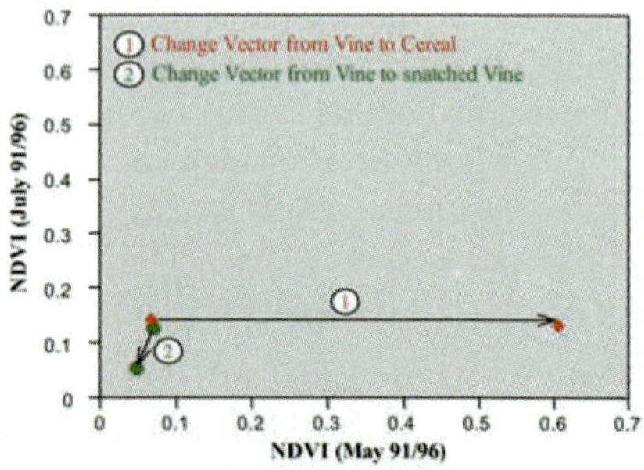

Fig. 6.10.3. Abandoned vineyard vector change. Vector 1 (red) indicates change from vineyard to abandoned and cereal, vector 2 (green) from vineyard to bare soil.

Fig. 6.10.2 represents the multiannual (1991-96) - phenological (May-July) NDVI diagram where the vector change is represented. Evolution (2) identifies changes to bare soil and evolution (1) corresponds to abandoned and cereal, that can not be separated by this procedure.

Analysing both Figs. 6.10.2 and 6.10.3, it can be confirmed that in the magnitude vector image high values will be obtained for changes that occurs from vineyard cover to cereal one or to the intermediate states like abandoned vineyard. While small values are obtained for changes that occur from vineyard cover to snatched vineyard. And finally, values near or equal to zero correspond to no change. However, attention has been paid to the vineyard areas which lose vegetation, in other words, the NDVI values decrease from 1991 to 1996 and can be confused, as commented previously, with the snatched vineyard (bare soil).

The analysis of Fig. 6.10.2 shows that the magnitude of the variability introduced by the changes is small as compared to the phenological variability. However, it had to be verified whether this is true for the whole study area. The analysis of the data was performed in four steps:

1. The magnitude of the interannual change vector was calculated for the whole study area. An image of $|C_{NDVI}|$ values was created and interpreted.
2. Statistical analysis of a window which covers approximately 1000 pixels belonging to the $|C_{NDVI}|$ image.
3. With the magnitude change vector image and two images acquired on 24 May and 11 July, 1996, a R-G-B (Red-Green-Blue) colour composite is constructed of $|C_{NDVI}|$, M96 and J96.
4. Classification of the composite image using the clustering method.

In Fig. 6.10.4, the vineyard changes during the 1991-1996 period are presented as map. The results also are shown in the Table 6.10.1 in terms of the surface (ha) occupied by each category change in each council of the study area.

Table 6.10.1. Change of vineyard surface from 1991 to 1996 into cereal or abandoned land and bare soil

Councils	Total Vineyard surface 1991	Vineyard surface change into cereal and abandoned land		Vineyard surface change into bare soil	
	hectare	hectare	fraction (%)	hectare	fraction (%)
Mota del Cuervo	9640	1363	14.1	704	7.3
Las Mesas	5900	578	9.8	176	3
Socuellamos	25180	4544	18	1875	7.4
Tomelloso	16390	2996	18.3	1822	11.1
Villarobledo	30100	7895	26.23	2856	9.5
Total	87210	17376	19.9	7433	8.5

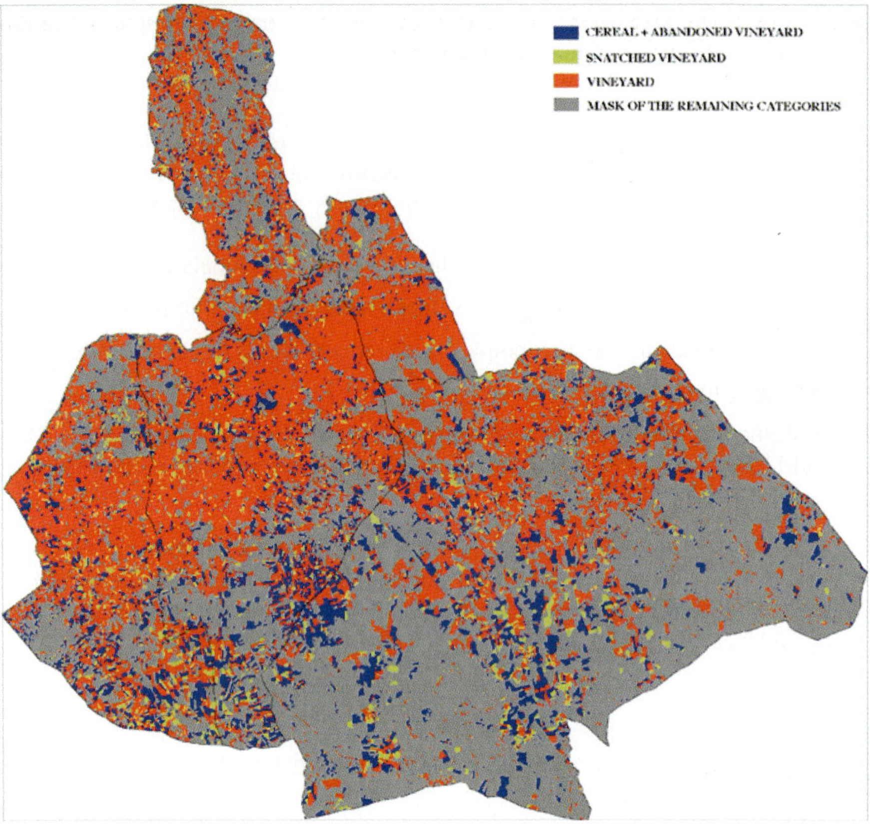

Fig. 6.10.4. Vineyard change in the period 1991 - 1996 in the Tomelloso study area based upon Landsat-5 TM data. The red area is the area of vineyards in 1991 as well as in 1996 (no change). The blue colour indicates the vinyard area of 1991 that changed into cereal and abondoned land. The yellow area gives the corresponding change of vinyards in 1991 into bare soil in1996. The grey area is occupied by other land-surface categories

The validation of the change map results was carried out by comparing to the land use information offered by Consergeria de Agicultura y Medio Ambiente de Castilla-La Mancha for 1998. This comparison includes the surface area of changes estimated by remote sensing from 1991 to 1996 and the funded snatch vineyard during 1992-1996 estimated by the Consegeria de Agricultura y Medio Ambiente de Castilla-La Mancha. Also we have selected some test areas of changes to be checked in the ground in 1998. All those areas have been confirmed affirmatively in the ground.

The NDVI differencing image combined with thresholding technique on one hand and the NDVI change vector analysis on other hand, has proven to be effective in detecting and categorizing interannual changes in the vineyard crop during 1991-1996 period. The interpretation of the results of the NDVI differencing have allowed to detect the changes which occurs from vineyard to cereal and abandoned vineyard.

While, the second method of the NDVI vector analysis have allowed to obtain the first changes and snatched vineyard (bare soil) map.

6.11 Estimation of Weather Impact on Vegetation Cover Along the Israeli Transition Zone Using AVHRR Data

6.11.1 Introduction

This paper discusses the use of NOAA/AVHRR data for monitoring seasonal dynamics of vegetation cover in the Mediterranean vegetation region, the transition zone, and semi-arid region in Israel and estimation of weather impact on the spatial and temporal variation of vegetation cover in this region.

Land cover information is crucial for understanding processes such as global carbon cycling, terrestrial primary productivity, hydrological cycles, and terrestrial energy balance. Remote sensing methods are being used over a number of regions worldwide to monitor vegetation changes (e.g., Tucker et al. 1983, Townshend and Justice 1986, Justice et al. 1986, Maselli et al. 1993, Prince et al. 1995, Bastin et al. 1995, Hobbs 1995). The results of these studies show that remote sensing can be considered to be a useful tool for studying arid and semi-arid ecosystems.

Several sources of satellite data provide for an objective analysis of ecological variables. Among them, the Advanced Very High Resolution Radiometer (AVHRR), with a high temporal frequency, from the National Oceanic Atmospheric Administration (NOAA), plays a significant role (e.g., Gutman 1989). The most important AVHRR-derived variable for ecological applications is the Normalized Difference Vegetation Index (NDVI) developed by Rouse et al. (1974), which shows correlation with photosynthetic activity, vegetation cover, biomass, and Leaf Area Index.

The Vegetation Condition Index (VCI) is a modified version of the NDVI (Kogan 1990). It was shown that the VCI is almost not affected by land heterogeneity caused e.g. by varieties of soils and topography, and thus facilitates the estimation of weather impacts on vegetation and their spatial and temporal changes (Gitelson et al. 1996 and 1998, Kogan 1997).

The use of satellite data gathered by the Advanced Very High Resolution Radiometer (AVHRR) flown on NOAA polar orbiting satellites was studied for monitoring seasonal dynamics of vegetation cover in the Israeli transition zone between the Mediterranean vegetation and semi-arid regions. The NDVI database used was developed from atmospherically- and radiometrically-corrected NDVI composites from observations spanning three years (March 1995-April 1998). The results show that three regions – the Mediterranean vegetation region, the transition zone, and the semi-arid region - can be geographically separated by means of NDVI values. The phenological characteristics for all three vegetation zones were found to be very similar and stable during three years of observations (a relatively dry year and two years with near average rainfall). The NDVI of the Israeli transition zone was found to be very sensitive to rainfall. The difference between maximum and minimum NDVI values in the rainy season in the transition zone was at least three times higher than those in the Mediterranean zone and semi-arid region. This phenomenon can be used as an indicator of environmental changes in this region.

6.11.2 Study Area

Israel is geographically located between 29.5° to 34° N latitude and 34° to 35.5° E longitude. Precipitation is concentrated in the rainy season (November-March), ranging from about 800 mm/year in the north to 100 mm/year in the semi-arid region and to practically zero in the south.

The AVHRR observations took place during three years with different rainfall rates: a year with slightly above the average rainfall for this region - in 1996/97 the rainfall measured at Sede Boker (in the southern semi-arid region) was 116 mm/a, while the average for this region is 100 mm/a; and in a dry year, 1995/96, with rainfall much lower than the average precipitation (50 mm/a). In 1997/98, the measured rainfall had an average value of 96 mm/a. The differences in the vegetation cover are mainly caused by the rainfall gradient from the north to the south. The country is divided into several vegetation zones (Zohary 1962), ranging from typical Mediterranean vegetation in the north, through an Irano-Turanian phytogeographic region, to a Saharo-Arabian region with typical desert vegetation in the south. Besides relatively dense natural vegetation in the north, this region is intensively used for agriculture. The southern Negev Desert has natural vegetation, mainly consisting of shrubs with a density ranging from 20% to less than 5% further south. Therefore, the vegetation cover changes from the north with almost 100% vegetation cover to the south with less than 5% of vegetation cover.

The central part of Israel, including the transition zone, was chosen as the study area. The Israeli transition zone is located between the Mediterranean zone in the north and the semi-arid region in the south and is characterized by a distinctive spatial and seasonal variation of the vegetation cover. In the rainy season, the southern vegetation border moves further south and shifts back to the north during the dry period, showing almost the same features as the southern semi-arid region.

6.11.3 Method

This study is based on NOAA/AVHRR data, which were acquired in High-Resolution Picture Transmission (HPRT) format with a spatial resolution of 1.1 km at nadir in Sede Boker (Israel, the Negev Desert). NOAA-14 AVHRR images of Israel were obtained for the three year time period from June 1995 to April 1998.

The process of data calibration in the visible and the NIR channels is based on radiometric calibration coefficients. Within this study, the results of the post-launch calibration of the NOAA-14 AVHRR channels 1 and 2 by the calibration group at the NOAA/NESDIS Office of Research Application were used (Rao and Chen 1996). Atmospheric correction of the top-of-atmosphere (TOA) reflectances was carried out using the 6S algorithm (Vermote et al., 1997). Estimates of total precipitable water and optical thickness of the atmosphere were obtained from an automatic tracking sunphotometer (CIMEL), installed in Sede Boker. The image data were geometrically corrected to a master image using ground control points. The accuracy of the correction lies at the subpixel level. Only AVHRR images with a satellite zenith angle of less than 30° were analyzed, altogether 67 images.

The NDVI was calculated for all AVHRR images with *corrected* surface reflectance values ρ_1 and ρ_2 calculated from radiances in channels 1 and 2, as NDVI = $(\rho_2 - \rho_1)/(\rho_2 + \rho_1)$. The maximum value compositing technique has been used on a monthly basis to overcome factors that reduce the NDVI: water vapor, aerosol content viewing and illumination geometry (Holben 1986). On a pixel-by-pixel basis, each NDVI value is examined, and only the highest value is retained for each pixel location for a given period of time.

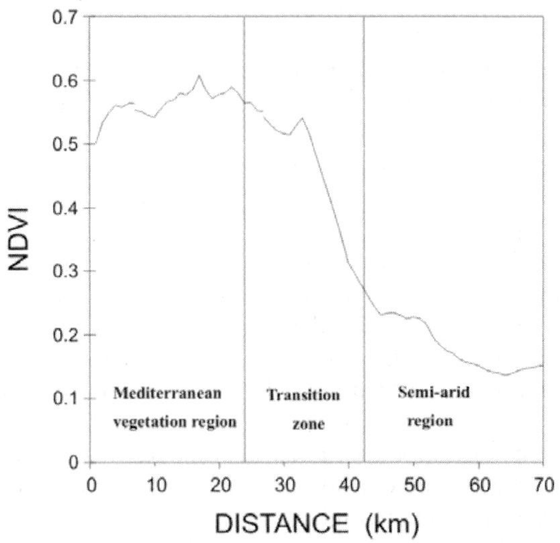

Fig. 6.11.1. Spatial variability of NDVI along a North-South transect. The transect is based on the average of three transects, which are at about 30 km distance from each other. The transect shows a distinguishable difference in the NDVI values, from the northern part with dense vegetation to the semi-arid southern part. The transition zone is located between these vegetation zones. It is represented by a significant drop in the NDVI values.

To estimate the weather-related component of the NDVI and the current state of vegetation, a Vegetation Condition Index (VCI) was used (Kogan 1990) of the form VCI = $(\mathrm{NDVI} - \mathrm{NDVI}_{min})/(\mathrm{NDVI}_{max} - \mathrm{NDVI}_{min})$, where NDVI_{max} and NDVI_{min}, are the monthly maximum and minimum NDVI values for the observation period of three years.

6.11.4 Results and Discussion

The spatial variability of the NDVI from east to west was found to be quite small: less than 0.08-0.1. Therefore, one north-south transect represented vegetation cover in the Israeli transition zone (Fig. 6.11.1). The transect shows a distinguishable difference in the NDVI values between the northern part, the Mediterranean region with quite dense vegetation, and the semi-arid southern region with very low vegetation cover. Between these regions, a narrow (width at about 20 km) transition zone is located; it is characterized by a significant drop (near three-fold) in the NDVI values.

Seasonal variation of vegetation cover in this region was very pronounced: maximal values of NDVI (0.35-0.55) occurred in the rainy season from February until April. The most protracted dry season, from May until October, had very low (about 0.05) and virtually unchanged NDVI values. The following months,

November – January, differed from the dry season with a slight increase of the NDVI values. These variations can be seen clearly in Figs. 6.11.2a and 6.11.2b, showing monthly minimal and maximal NDVI values for each pixel along the transect studied. In the rainy season, the $NDVI_{min}$ in the Mediterranean vegetation region and the transition zone was much higher than that in dry and pre-rainy seasons, where $NDVI_{min}$. values were almost the same. In the semi-arid region, the picture changed: maximal difference in the $NDVI_{min}$ between all three seasons was only 0.05, indicating that in this region weather impact on the $NDVI_{min}$ is minimal. It's not clear whether these small differences are significant or if they represent the margin of error in registration and processing. Further investigation is required to provide support for $NDVI_{min}$ variation between different seasons in the semi-arid zone. In dry years with minimal rainfall, the vegetation cover in the semi-arid region is minimal (less than 20%); the $NDVI_{min}$ is very low (about 0.05) and represents almost bare surfaces. Therefore, in

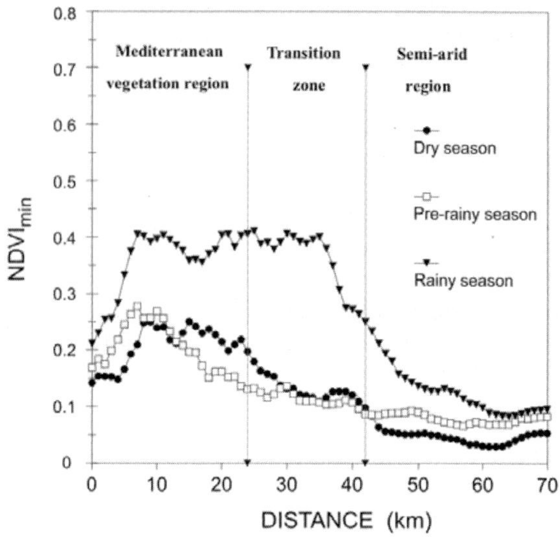

Fig. 6.11.2a. The $NDVI_{min}$ along the North-South transect for three seasons: dry, pre-rainy, and rainy.

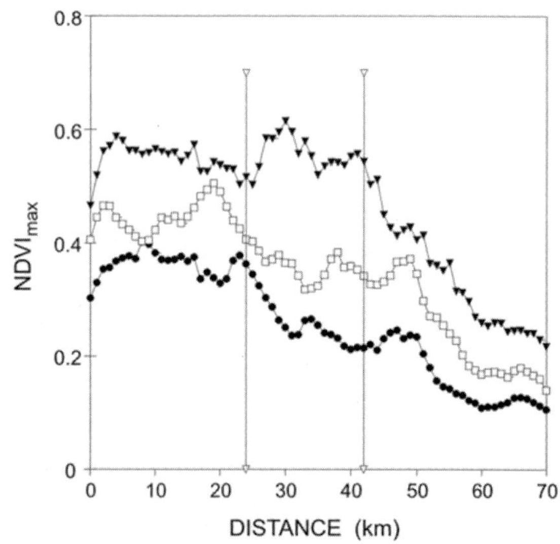

Fig. 6.11.2b. The $NDVI_{max}$ along the North-South transect for three seasons: dry, pre-rainy, and rainy.

semi-arid environments, the information on the soil background can be retrieved from the coarse resolution satellite data as the $NDVI_{min}$.

The $NDVI_{max}$ values for all three seasons were very different (Fig. 6.11.2b). The short rainy season was characterized by the highest $NDVI_{max}$ values in all three climatic regions. During this season, the $NDVI_{max}$ in the Israeli transition zone was almost the same as that in the Mediterranean vegetation region. As a result, the geographical position of a sharp decrease in the $NDVI_{max}$ values shifts 5-7 km towards the south (compare the position of sharp decrease in the $NDVI_{max}$ in Fig. 6.11.2b with that in the $NDVI_{min}$ in Fig. 6.11.2a). In the following dry season, both the $NDVI_{min}$ (Fig. 6.11.2a) and $NDVI_{max}$ (Fig. 6.11.2b) in the Mediterranean vegetation region were quite stable, but dropped sharply in the transition zone and the semi-arid region. At the end of the rainy season and at the beginning of the dry season, the position of sharp decrease in the $NDVI_{max}$ moved back to the north. During the short pre-rainy season, November - January, a significant variability in

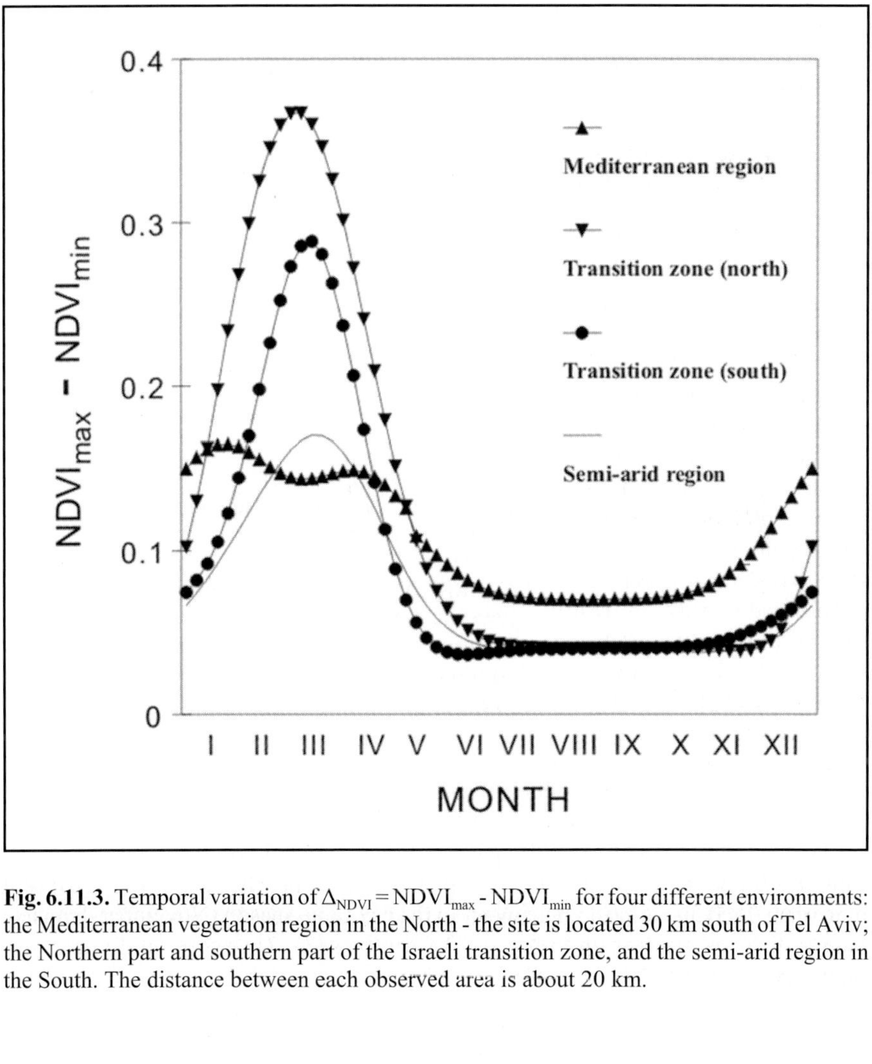

Fig. 6.11.3. Temporal variation of $\Delta_{NDVI} = NDVI_{max} - NDVI_{min}$ for four different environments: the Mediterranean vegetation region in the North - the site is located 30 km south of Tel Aviv; the Northern part and southern part of the Israeli transition zone, and the semi-arid region in the South. The distance between each observed area is about 20 km.

the $NDVI_{max}$ and $NDVI_{min}$ values existed in the Mediterranean vegetation region and the transition zone.

The $NDVI_{max}$ contains the information on both weather impact and ecosystem resources, whereas the $NDVI_{min}$ corresponds to unfavorable weather (minimal rainfall), and, as a result, is mainly an indicator of the ecosystem resources. Thus, the difference between them $\Delta_{NDVI} = NDVI_{max} - NDVI_{min}$ primarily represents weather impact on the NDVI (Kogan 1990, Gitelson et al. 1998). To study temporal variation of the difference Δ_{NDVI}, four areas along the North-South transect were selected. Figure 6.11.3 shows Δ_{NDVI} values extracted from all available AVHRR images of the three years of observation in four areas that represent the Mediterranean vegetation region, the northern and southern transition zones, and the semi-arid region in the south. A remarkable feature of the temporal behavior of the weather-driven component of the NDVI (Δ_{NDVI}) is a pronounced peak in rainy season. In the rainy season, a minimal Δ_{NDVI}. value was found to be in the Mediterranean vegetation zone (0.14-0.16), and the highest values were found in the transition zone (0.36 for north and 0.3 for the south). In the semi-arid region, the Δ_{NDVI} was also significant - 0.17.

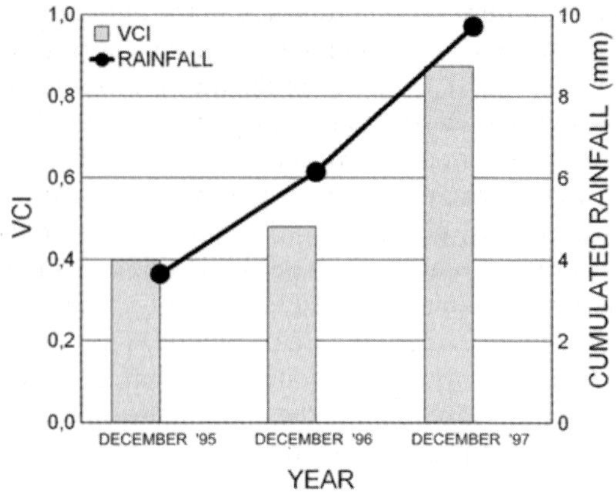

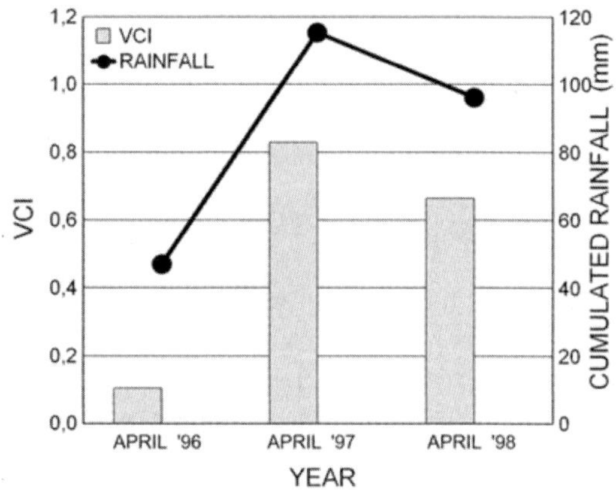

Fig. 6.11.4. Top: Rainfall, cumulated from the beginning of each rainy season, measured in the Sede Boker area, the southern part of semi-arid region, and Vegetation Condition Index (VCI) retrieved from AVHRR data in the pre-rainy season (21 December 1995, 22 December 1996, and 14 December 1997). Bottom: VCI and cumulated rainfall in the rainy season (1 April 1996, 11 April 1997, and 10 April 1998)

During the May to October dry period, the Δ_{NDVI} in the semi-arid region showed a similar temporal behavior as that of the transition zone: the Δ_{NDVI}. was small (not more than 0.05) and almost unchanged. In the pre-rainy season (November-January) only a slight increase of up to Δ_{NDVI}.= 0.1 can be seen. Therefore, in dry and pre-rainy seasons, the NDVI response to weather (in terms of the Δ_{NDVI}) was almost the same in the transition zone and in the semi-arid region.

The most sensitive index to describe spatial variation of the Δ_{NDVI} was found to be $(\Delta_{NDVI})_{rainy}/(\Delta_{NDVI})_{dry}$. For the transition zone, the index ranged from 12 to 17, for the semi-arid region it decreases more than twice up to 6, and in the Mediterranean vegetation region it dropped to 2.7. Thus, this index shows that the transition zone was at least three times more sensitive to weather impact than the other climatic zones (see also Gitelson and Schmidt 1999 and Schmidt and Gitelson 2000).

In Fig. 6.11.4, the VCI values are plotted together with the rainfall, cumulated from the beginning of each rainy season (Fig. 6.11.4a), measured in the Sede Boker area (the southern part of semi-arid region). The cumulated rainfall in Sede Boker is given for 21 December 1995, 22 December 1996, and 14 December 1997. The VCI was calculated from the data obtained on the same dates. The VCI was found to be sufficiently sensitive to reflect even small variations in cumulated rainfall in the pre-rainy season (at about 4 mm against a background of 10 mm). The VCI and the cumulated rainfall in the rainy season (April) is given for 1 April 1996, 11 April 1997, and 10 April 1998 (Fig. 6.11.4b). The VCI was retrieved from AVHRR data obtained on the same dates. The VCI was sensitive to small variations in rainfall (about 20 mm) against a background of 100 mm. As can be seen, the temporal behavior of the VCI is in a good agreement with that of cumulated rainfall. The VCI was even sensitive enough to reflect quite a small decrease in cumulated rainfall (not more than 20 mm) between April'97 and April'98.

6.11.5 Conclusion

The ability of NOAA/AVHRR data to monitor the spatial and temporal distributions of the vegetation cover in the Israeli transition zone was studied. The results demonstrate the strong potential of AVHRR data to distinguish different vegetation zones and their temporal variability along a North-South transect in Israel. The spatial variability of vegetation cover from north to south was found to be very high over a distance of only 100 km, while the variability from east to west along the entire North-South transect was very small.

The highest sensitivity of the NDVI to rainfall was found in the Israeli transition zone (at least three times higher than those in the Mediterranean and the semi-arid regions). The Vegetation Condition Index was very sensitive to cumulated rainfall. It was found that besides the high seasonal variability of vegetation cover in the transition zone and the semi-arid region, the inter-annual change of the vegetation cover has to be considered as well. The change in vegetation cover in the Israeli transition zone, recorded remotely, can be used as a sensitive indicator of environmental changes.

It's evident that three years of data hardly satisfy statistical requirements to formulate reliable conclusions. However, the concept of minimal NDVI suggests the existence the only one year in the study period with extremely small rainfall. Such a winter in the study period appeared in 1995/96, when cumulated rainfall was two times less than average value. Maximal NDVI was selected as a criterion approximating the ecological potential of the area (Kogan, 1997). During study period ecological potential of the Negev Desert probably was not achieved; maximal cumulated rainfall was little bit large than average. Thus, the $\Delta_{NDVI} = NDVI_{max} - NDVI_{min}$ was underestimated, and the VCI was overestimated. When $NDVI_{max}$ will be achieved, the estimates may be more accurate; nevertheless, seasonal and spatial dynamics of vegetation cover in southern Israel, revealed in this study likely remain the same.

6.12 Monitoring of Soil Moisture Fields and Change Detection by Passive Microwave Remote Sensing

6.12.1 Introduction

The use of passive low frequency microwaves for surface moisture and high frequency microwaves for vegetation monitoring from space using polar orbiting satellites has been a focus of scientific research for many years now (Van de Griend and Owe 1994a, Kerr et al. 2001, Pellarin et al. 2003). The validation of theoretical models, however, has largely been restricted to field experiments because of the inherently large footprint of microwave radiometers, determined by the low radiation intensities emitted from natural surfaces. The footprint of Nimbus/SMMR typically varies between approximately 150x100 km^2 (6.6 GHz) and 35x25 km^2 (37 Ghz). Therefore, the availability of adequate ground data at such scales is obviously limited. The data-base of soil moisture and climate data compiled by NASA/GSFC and the Vrije Universiteit Amsterdam for a 150x150 km^2 area in south-eastern Botswana from archived data, seems to have been, until recently, the only one existing extensive and long-term data base. This database has been extensively used by both institutes for testing and validating algorithms for surface moisture monitoring from space. Based on this work, two inversion models have been developed and tested using the Botswana data set, i.e. a synergistic approach (Owe et al. 1992, Van de Griend and Owe 1994c) and a dual-polarisation approach (Van de Griend and Owe 1994b).

In addition, a dual-frequency approach has been developed and tested for a 150 x 150 km^2 area in central Spain (La Mancha). This dual-frequency approach has subsequently been used to map soil moisture degradation over the nine-year lifetime of Nimbus-SMMR (1978-1987). This research was carried out within the framework of the European EFEDA/RESMEDES project on desertification in the Mediterranean (see Appendix 1).

6.12.2 The Dual-Frequency Approach

The dual-frequency approach makes use of horizontally polarised (H) 6.6 GHz and both H- and V-polarised 37 GHz brightness temperatures. Microwave radiation penetrates vegetation of which the transmissivity increases with decreasing frequency. Therefore, 6.6 GHz, which is the lowest frequency of SMMR, is the best qualified for surface moisture monitoring under (sparsely) vegetated conditions. At 37 GHz, which is also observed by SMMR, microwave radiation is strongly absorbed by vegetation. Therefore, the 37 GHz brightness temperature is closely related to wet biomass and can be used to estimate the transmissivity, $\tau_{c,6.6}$, of the vegetation canopy at 6.6 GHz.

The radiative transfer model used to describe the brightness temperature of the land surface at 6.6 GHz is an approximative version of the $\tilde{\omega}$ - τ_c model (Mo et al. 1982) described in Chapter 3, Eq. 3.44. To simplify the relationship it is assumed (i) that the thermodynamic temperature of the canopy equals that of the surface and T_s

$= T_c = T_{sc}$ is the *effective thermodynamic* temperature of the surface - canopy system, (ii) that consequently the fraction of the radiation emitted by the surface, reflected back to the surface by the canopy and then reflected by the surface and transmitted through the canopy is neglected $[\tilde{\omega}(1 - \tau_c)(1 - \varepsilon_s)\tau_c \ll 1]$, and (iii) that one deals with a uniform transparent canopy. Under these premisses Eq. 3.42 reduces to:

$$\frac{T_{eff}}{T_{sc}} = \varepsilon_s \tau_c + (1-\tau_c)(1-\tilde{\omega})\left[1+\tau_c(1-\varepsilon_s)\right] \qquad (6.12.1)$$

where T_{eff} is the measured *effective radiative* temperature.

The single scattering albedo *($\tilde{\omega}$)* varies with vegetation type, but knowledge about the exact values is limited and almost absent for natural vegetation. It is therefore further assumed that $\tilde{\omega}$ for Castilla - La Mancha equals the value determined for the Botswana savannah environment (Van de Griend and Owe 1993a) which is, with $\tilde{\omega} = 0.09$, within the range for agricultural crops ($0.06 < \tilde{\omega} < 0.12$). The vegetation transmissivity was modelled in terms of the Microwave Polarization Difference Index (*MPDI*) for 37 GHz, defined by Becker and Choudhury (1988):

$$MPDI_{37} = 20 * \frac{T_{V(37)} - T_{H(37)}}{T_{V(37)} + T_{H(37)}} \qquad (6.12.2)$$

$T_{V(37)}$ and $T_{H(37)}$ are the vertical and horizontal brightness temperatures at 37 GHz frequency. The multiplication with 20 is for scaling purposes. It is assumed that the relationship between τ_c and *MPDI* that was found for Botswana (Van de Griend and Owe 1994b) would also give a reasonable first-order estimate for Spain. The resulting relationship is given by:

$$\tau_{c,6.6} = 0.477 + 0.289 * MPDI \qquad (6.12.3)$$

The surface emissivity at 37 GHz and V-polarisation increases with observation angle and approaches unity at 52° (which is the look-angle of SMMR) for both vegetation and bare soil, except if a very thin top-layer of a few millimetres is wet over a substantial part of the foot print. Therefore, the vertical brightness temperature at 37 GHz is closely related to the thermodynamic temperature of the emitting surface (Owe and Van de Griend 2000, Van de Griend 2001).

For night-time, the relationship between T_s and $T_{V(37)}$ was determined by comparing observed daily minimum air temperature with observed surface temperature during an intensive field campaign and by transferring this relationship to the SMMR-period of 1978-1987. The relationship is given by:

$$T_{S(0UTC)} \cong T_{eff(6.6)} = 1.293\, T_{V(37)} - 56.6\,[K] \qquad (6.12.4)$$

During daytime, the thermodynamic surface temperature may be significantly higher than the effective thermodynamic temperature of the emitting body at 6.6 GHz. From a sensitivity study carried out for the Iberian Peninsula, using both daytime and night-time observations (Van de Griend 2001), the following equation was determined for daytime observations:

$$T_{S(12UTC)} \cong T_{eff(6.6)} = 1.352 * T_{V(37)} - 63.4 \quad [K] \qquad (6.12.5)$$

6.12.3 Available Data and Validation

The dual-frequency approach was validated for a 150 x 150 km^2 area in central Spain (La Mancha) by comparing satellite-derived surface moisture with time series of areal rainfall on a daily basis for the whole lifetime of Nimbus/SMMR, i.e.1978-1987. The peculiar scanning pattern of SMMR requires a special sampling routine, which resulted in one area representative brightness temperature every 3 to 4 days. Analyses were done independently for both daytime and night-time observations. Fig. 6.12.1 shows the time series of satellite derived surface moisture (daytime and night-time) together with the rainfall for the year 1979. Calculations were done for all years 1979 till 1987. The satellite-derived soil moisture estimates follow the precipitation patterns systematically in the sense that each rainfall occurrence results in an increase of soil moisture. The fact that soil moisture data are absent from central Spain prevents actual validation. However, consistencies in *(a)* moisture range (5% < 9 < 20%), *(b)* close correspondence between areal rainfall and satellite-derived time series of surface moisture, and *(c)* between night-time and daytime calculations, give sufficient confidence with respect to the general performance of the method, at least in a relative sense. This makes the model especially useful for long-term change detection.

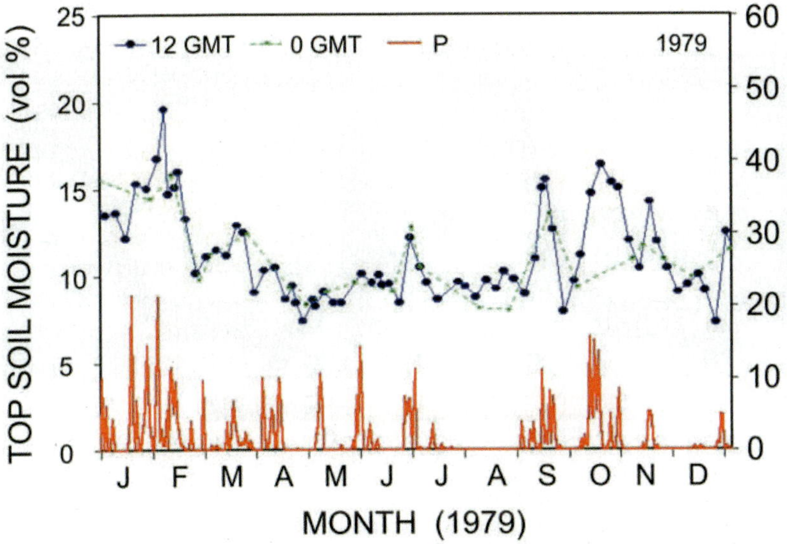

Fig. 6.12.1 Time series of rainfall (red line) and satellite derived daytime (blue) and night-time (green) surface moisture for La Mancha, Central-Spain, 1979

6.12.4 Degradation/Aridification Mapping over the Iberian Peninsula

The dual-frequency approach described above has been applied to the whole Iberian Peninsula for the period 1978-1987. For this purpose, the whole peninsula was split up in 32 x 32 pixels of 0.25 degree resolution. Time series of satellite derived soil moisture were produced for each pixel using the dual-frequency approach for both daytime and night-time. The time series were reduced to monthly-mean soil moisture values for each pixel. A trend analysis was used to determine temporal changes.

The spatial distribution of moisture change shows that only one area suffers serious long-term degradation. This is the Ebro-valley in northeastern Spain and its surroundings. The degradation is visible in both night- and daytime maps (Fig. 6.12.2).

In order to study the degradation of soil moisture in relation to vegetation biomass, the vegetation indices NDVI and MPDI were also included in the analyses. A special mapping routine, based on K-means cluster analysis was developed to identify the spatial distribution of the major characteristics of long-term time series, such as *amplitude, phase-shift* and *trend.* These parameters were determined by Fourier analysis on a pixel by pixel basis from monthly mean NDVI, MPDI and satellite-derived surface moisture values. This resulted in 6 significant vegetation classes (Fig. 6.12.3) of which two demonstrated significant moisture degradation, i.e., the Ebro-valley (class 1) and its direct surroundings (class 2).

Fig. 6.12.4 shows an example of long-term courses of NDVI, MPDI and soil moisture for a pixel in the center of class 1. This cell is more or less representative for a larger zone in North-Eastern Spain, which shows a significant decrease in soil moisture over the period 1987-1978. Fig. 6.12.5 shows the mean courses for the

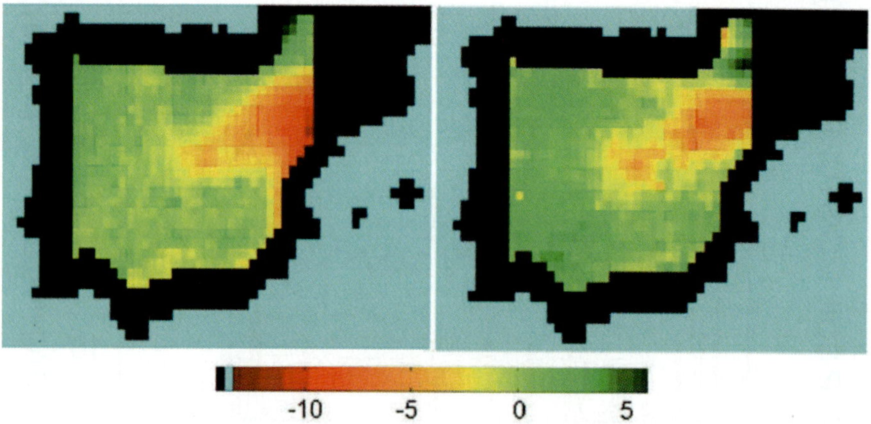

Fig. 6.12.2. Maps of the Iberian Peninsula showing the long-term trend in soil moisture change (vol.%) over a 9 year period, based on bi-weekly satellite observations over the whole life-time of Nimbus/SMMR (1978-1987). The maps show a significant degradation of surface moisture (red) in north-eastern Spain, i.e. the driest and most desertification threatened area of Spain, derived from independent daytime (left figure) and night-time (right figure) observations

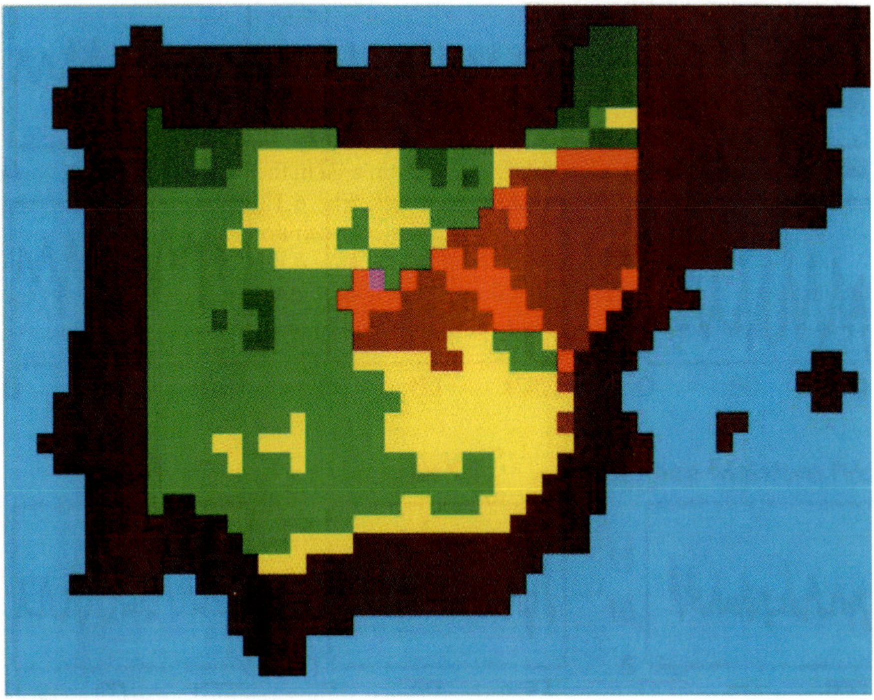

Legend:

Class	Color	Vegetation cover	Sensitive to Aridification
1		Sparse	Yes
2		Intermediate	Yes
3		Dense	Yes
4		Sparse	No
5		Intermediate	No
6		Dense	No

Fig. 6.12.3 Map showing the spatial distribution of 6 regions (classes) derived form K-means cluster analysis

individual classes.

An analysis of rainfall and climate data of the Ebro-region over the period 1978-1987 did not reveal any significant change in climatic conditions. It is therefore not completely clear what the consistent and gradual change in surface properties really represents. There is no doubt that there is a physical phenomenon responsible for the long-term changes observed in classes 1 and 2. A possible explanation might be a change in vegetation biomass, but both NDVI and MPDI are invariable during the whole period (Fig. 6.12.5). Another explanation might be found in the long-term irrigation practice in large parts of the Ebro-valley. The central part of the Ebro-valley strongly saline as a result of bad drainage conditions. Saline soils have a deviating characteristic of the dielectric constant resulting in a lower soil emissivity at 6.6 GHz for the same moisture content. Continuous irrigation is expected to leach saline soils thus leading to a decrease of the dielctric constant and an increase in emissivity for the same moisture content. In other words, continuous leaching would lead to higher emissivities which lead to dryer soils in the dual-frequency model. For this reason, we sampled 27 soils in a 60 x 60 km area in the central part of the Ebro-valley and determined the dielectric properties. Fig. 6.12.6 show the results. The cluster consists of a family of curves varying between non-saline and saline soils. Although it is difficult to come to a final conclusion, it might be well possible that the observed gradual change in soil mositure is in reality a gradual change in the dielectric constant, not as a result of soil moisture change, but as a result of de-salinization.

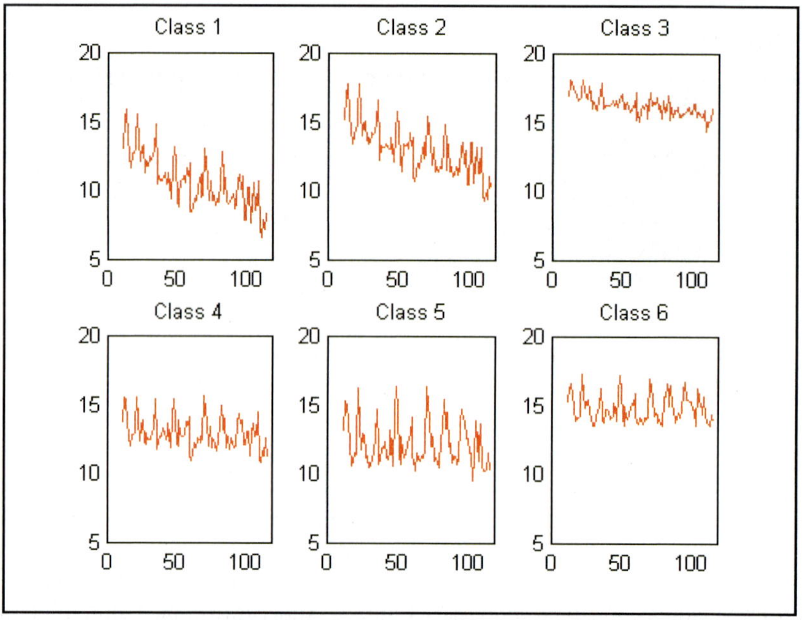

Fig. 6.12.4. The course of nighttime (0 GMT) surface moisture (vertical scale in Vol. %) over the period 1978-1987 (horizontal scale in months) for 6 different vegetation classes in Spain.

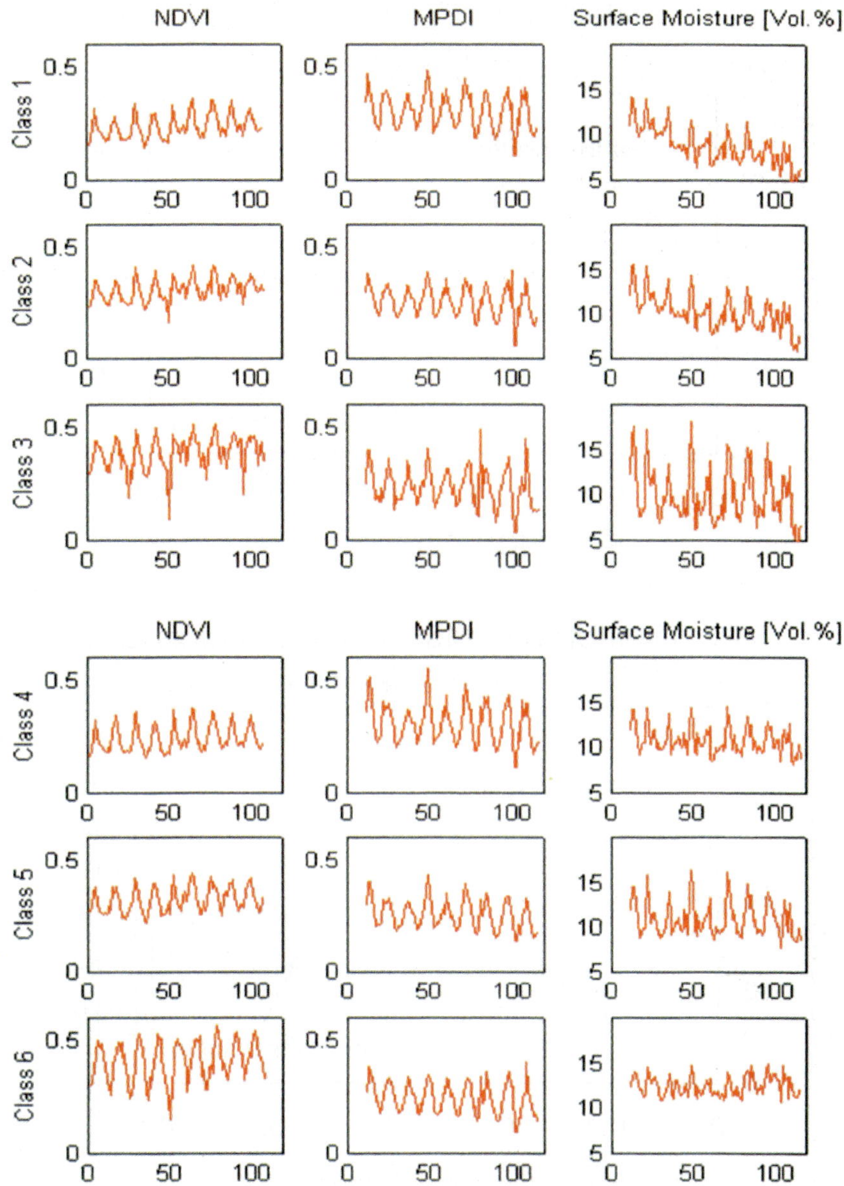

Fig 6.12.5 Long-term behavior of surface moisture and vegetation indices (day time) for each degradation class.

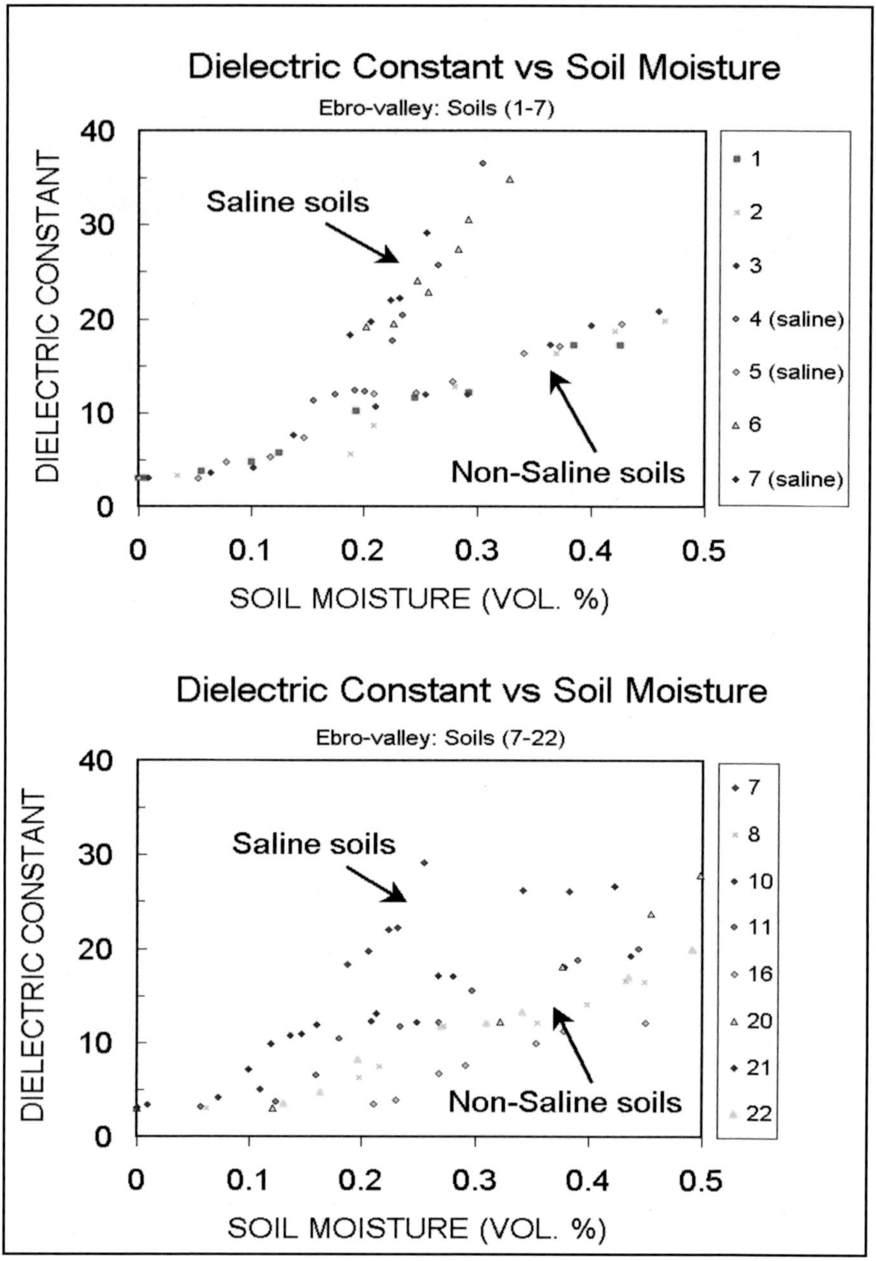

Fig. 6.12.6. Dielectric constant (real part) as a function of soil moisture for soils (arbitrarily numbered) in the central Ebro-valley showing the difference between highly saline soils and non-saline soils (upper figure) and soils in between both extremes (lower figure).

6.13 Integration of Conventional and Remote Sensing Data to Model Transpiration of Forest Mediterranean Ecosystems

6.13.1 Background

During the last decades there has been an increasing interest for modelling parameters and processes of different ecosystems. This is a new approach to ecological studies which can be particularly useful to obtain information on wide spatial and temporal scales (Waring and Running 1998). The importance of environmental modelling is also linked to the necessity of being able to detect and study the effects of large-scale perturbations such as global climate change and air pollution on forest and other terrestrial ecosystems. The estimated parameters, and in particular those related to the main vegetation processes such as transpiration, photosynthesis and productivity, can be a valid support to all decision-makers dealing with problems of managing environmental resources.

Even though modern ecosystem simulation models have reached high efficiency and accuracy, their applicability to operational cases is often limited by the fundamental drawback of requiring input parameters which are difficult to collect for large vegetated areas. This is especially the case for parameters describing forest composition and structure, such as tree species, density, Leaf Area Index, which are generally highly variable in space and time and difficult to measure with conventional methods (Lacaze et al. 1996). The same parameters are however those which are more directly linked to the spectral properties of vegetation, so that remote sensing techniques have been recently proposed as a promising source of information for this ecosystem simulation model (Running et al. 1989, Lucas et al. 2000).

The current work builds on these considerations and investigates the possibility to apply in Mediterranean environments a forest ecosystem simulation model (FOREST-BGC) driven by conventional and remote sensing data. FOREST-BGC was chosen for its wide applicability, for its suitability to accept data from different sources, and specifically from remote sensing images (Running and Coughlan 1988). The application of FOREST-BGC in Mediterranean areas is however a challenge, since the model was originally proposed and developed for simulating the main processes of coniferous forests under a range of climates (Running and Coughlan, 1988). Some modifications were therefore needed to adapt the model to a different climatic situation and forest ecosystems mainly composed of deciduous species. Additionally, in order to face the problems related to the topographic and land use variability of these environments, a new methodology had to be developed for deriving suitable inputs to the model by the integration of remotely sensed and ancillary data.

The current work was carried out in a hilly area around the Radicondoli village in Tuscany (Central Italy) mostly covered by deciduous woods. The investigation was focused on two stands covered by two oak species (*Quercus pubescens* Willd. and *Quercus cerris* L.) representative for most of local tree species. The parameters that are necessary to initialise and drive the model were collected from forestry literature, the forest management plan of the area, and field measurements. In particular, daily meteorological data (minimum and maximum temperature, rainfall and radiation) were extrapolated from an adjacent station, while monthly LAI profiles were directly measured in the two stands. In addition to this, LAI estimates of the two stands were obtained from NOAA-AVHRR NDVI data using an integration procedure previously developed (Lacaze et al. 1996, Maselli 2001). All these data served to drive FOREST-BGC for the growing season of two years (1997 - 1998) for which ground transpiration measurements were available, obtained by a sap flow method (Cermak et al. 1998). More particularly, the transpiration values from the first season were used for calibrating the model, while those of the second year were used for its testing.

The present work is organized as follows. First, a brief introduction to the functionality of FOREST-BGC is provided. The next section deals with a description of the study area and data. The methods used to retrieve the input data for the model are then presented, followed by the application of the model to two forest stands. Next, the calibration and validation of the model by comparison to daily measured transpiration values are described. Discussion and conclusions finally follow on the potential application of the model to monitor water and carbon fluxes through vegetation over wide Mediterranean areas.

6.13.2 The Model FOREST - BGC

The model FOREST-BGC was developed at the University of Montana (Running and Coughlan 1988, Running and Gower 1991, Running and Hunt 1993) where it had been originally applied to coniferous forest stands. It works simulating water, carbon, and nitrogen fluxes within homogeneous forests. The model needs to be initialised and driven by data that can be collected in the field or taken from existing literature. The basic input data are:

1. initial condition of the station at which the model is applied (i.e. its geographical position and soil characteristics),
2. ecophysiological parameters describing the species for which the estimation is required,
3. daily meteorological data (maximum and minimum air temperature, solar radiation, precipitation and dew point temperature).

On the basis of these data, the model works on two different time steps (Fig. 6.13.1): a daily one that is related to the possibility of obtaining estimates of water balance and production, and a yearly one that considers also processes of carbon and nitrogen allocation and decomposition processes.

One of the most important features of the model is related to the possibility of using remotely sensed information as input data, which makes the model suitable for regional scale monitoring applications.

In particular the original model has been recently modified to improve its performance when applied to deciduous forests. While in fact the original version, suitable for evergreen forests, uses a constant value of LAI for the whole year, the new version can receive daily values of LAI derived from various sources such as field measurements and satellite data (Maselli et al. 1999). This additional information is expressed as leaf carbon content derived by the Leaf Area Index and Surface Leaf Area (one of the ecophysiological parameters required as input by the

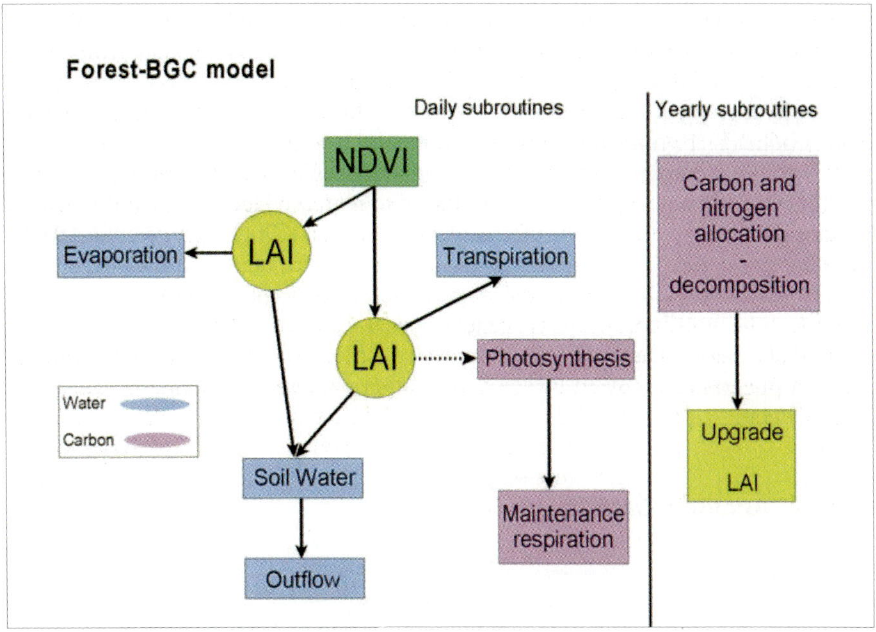

Fig. 6.13.1. Working scheme of the Forest - BGC model

model). The former parameter can be derived from suitably processed NDVI data.

The modification is based on the idea that LAI is the most important parameter describing the structural characters of the canopy (mainly leaf density and biomass) and, indirectly, of all processes connected with it: evaporation, transpiration, photosynthesis, and so on (Waring and Running 1998, Nemani and Running 1989). Additionally, LAI, together with climate data, has been indicated as one of the most important variables forcing and driving this model of forest ecosystem processes.

The outputs estimated by the model are about the most important ecophysiological parameters characterising the carbon, nitrogen and water fluxes within terrestrial ecosystems. In particular, the outputs of the model are useful to:

- quantify water cycles,
- establish the amount of C storage by plants,
- predict the behaviour of different ecosystems in case of changes of CO_2 concentration in the air,
- furnish important parameters to manage forest ecosystems.

6.13.3 Study Area and Data

The area around Radicondoli was chosen as a primary test site for the EU Projects RESMEDES and RESYSMED as representative of Mediterranean environment. It is situated in the centre of Tuscany, Central Italy, at 43.18 - 43.29° N latitude, 10.95 - 10.57° E longitude (Fig. 6.13.2). The terrain of the area is mainly hilly, with elevations ranging from 300 to 900 m. The climate is Mediterranean sub-humid (Rapetti and Vittorini 1995), with mean annual rainfall around 800 mm, temperature around 15 °C, mild winters and long dry summers.

The land cover is characterised by the alternation of agricultural fields, pastures and woodlands. Two main forests are present, which cover about 70 % of the area. Both forests are dominated by different deciduous species, with the presence of some conifers (mainly pines) artificially introduced to increase land productivity. Among the deciduous species two Mediterranean oak species (*Quercus pubescens* Willd. and *Quercus cerris* L.) are by far the most common, but other broadleaved species can also be found such as *Ostrya carpinifolia* Scop., *Acer campestre* L. and *Fraxinus ornus* L.. The understorey layer is generally characterised by the presence of young trees of the main canopy species (especially natural regeneration) and shrubs of different species (e.g. *Cornus sanguinea* L., *Rubus caesius* L., *Juniperus communis* L. and others).

6.13.4 Ancillary Data

Topography

Topographic maps (1:25,000) produced by the Italian Military Geographic Institute (I.G.M.I.) were available for the whole area. The same Institute also provided a digital elevation model (DEM) in the form of raster images with a pixel size of 20x20 m, derived from contour lines every 25 m. Land cover information was derived from the Land Cover classification produced during the EU Project CORINE

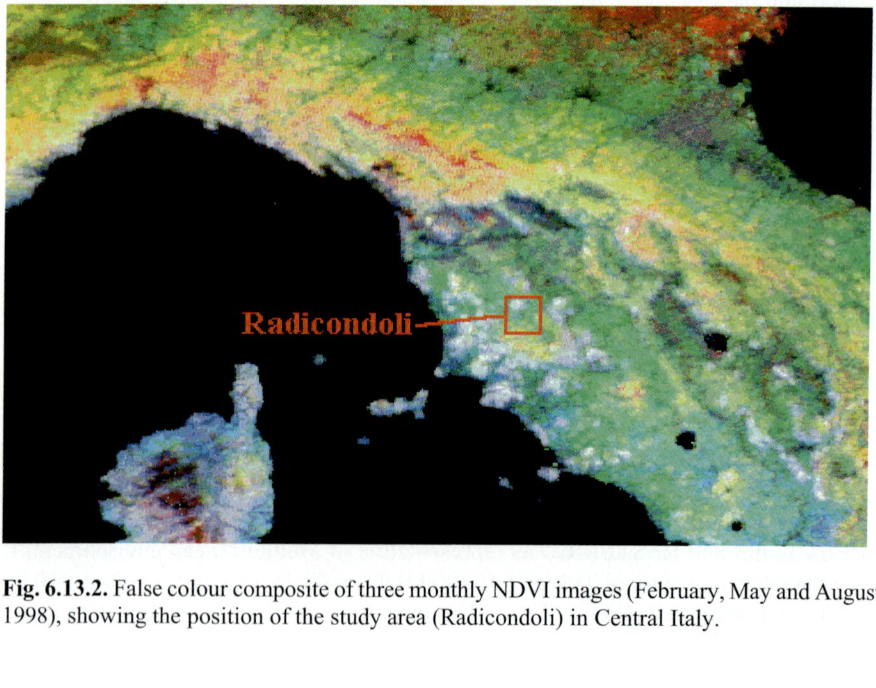

Fig. 6.13.2. False colour composite of three monthly NDVI images (February, May and August 1998), showing the position of the study area (Radicondoli) in Central Italy.

at a nominal scale of 1:100,000 (Annoni and Perdigao 1997). The CORINE nomenclature is hierarchical and comprises 3 levels common to all countries, with 5 categories for the first level, 15 for the second, and 44 for the third. The classification provided by the Tuscany Regional Service for Cartography in the form of a vector file with a nominal scale of 1:100,000 was used in the current study. Additional information for the study forests was obtained by digitising 1:10,000 scale maps produced by DREAM (1994) for the Forest Plan of Management of the area. This document contained a brief description of each forest stand in the area, including tree specific composition, density, volume, mean diameter and height, terrain and soil conditions.

Meteorological Data

Since no meteorological station has been continuously active within the study area in the last years, the main source of meteorological data was the station of Larderello (400 m a.s.l.). This station, belonging to the network of the Italian National Hydrological Institute, has been collecting daily values of maximum and minimum temperature, precipitation and solar radiation since 1951. In addition there are shorter data series collected by a portable station installed at the beginning of RESMEDES (spring 1996) within the Radicondoli forest area, at Collitalli (485 m a.s.l.). This station measures a wider series of daily parameters (soil temperature, wind speed and direction, air temperature, air humidity, precipitation and solar radiation), but its activity has not been continuous, so that only incomplete data series are presently available.

Surface Data: Leaf Area Index

Leaf Area Index (LAI) data were collected monthly during the growing seasons from 1996 to 2001 in fourteen forest stands selected all over the wooded area. The criteria used to select these stands were related to their representativity for the forests of the area and their accessibility. The LAI measurements were carried out using a LAI-2000 Plant Canopy Analyser (Li-Cor, NE, USA). Each stand was characterized by taking 6 optical measurements along linear transects of about 50 m. A calibration of the LAI values obtained by LAI-2000 was preliminarily performed for the main tree species by collecting the leaves and measuring their surface area with a planimeter. A correction factor was thus obtained for each tree species, which was then applied to all LAI-2000 estimates.

Surface Data: Transpiration

Transpiration (actually sap flow) was measured during the growing seasons 1997 and 1998 in two selected stands, one with *Quercus pubescens* and the other with *Quercus cerris*, as the dominant species. 12 sample trees per stand were chosen, calculated as representative on the basis of quantils of total (Cermak and Kucera

1990) for all specific diameter classes. Sap flow was measured at breast height (1.3 m) from two opposite sides of stems (or more in large trees). Two different methods to determine the sap flow rate were used according to the diameter of the sampled trees: for large ones (diameter at breast height, DBH, more than 15 cm) the stem segment heat balance method (THB, Cermak et al. 1973, 1982, Kucera et al. 1976) and for smaller trees (DBH < 15 cm) the heat field deformation method (HFD, Nadezhdina and Cermak 1998, Nadezhdina et al. 1998). Data were measured every minute and stored in 15 minute intervals using the EMS-measuring systems (for THB) and UNILOG dataloggers (equipment provided by Brno, Czech Rep.). Integration of data from individual measuring points to the whole tree was done on the basis of measurements of the radial pattern of flow in all studied stem sections (Nadezhdina et al. 2001), so that the conducting system in each stem (cross-section area of sapwood) was characterized by 24 (or more) points. The sap flow values integrated for the whole tree in all samples were up-scaled for the stand according to Cermak and Kucera (1990) and Cermak et al. (1998). Tree sap flow data were related to their basal area in each particular day and from the obtained curve transpiration of mean trees of DBH classes were calculated. These were multiplied by numbers of trees in classes per 1 ha and summarized.

Satellite Images

The satellite data used in the investigation were AVHRR images transformed to NDVI of the MEDOKADS data set (see Chapter 2). The original data were ten-day NDVI Maximum Value Composite (MVC) images of 6 years (1993-99), mapped in a geographic (Lat/Long) reference system with a 0.01° pixel size. The standard procedure for the production of these data comprised the georeferencing of the original images by a nearest neighbour algorithm, the radiometric calibration of the first two bands to derive apparent reflectance following Koslowsky (1999), and the computation of NDVI values to finally obtain the Maximum Value Composites (MVC) on a ten-day basis (Holben, 1986). The final products were therefore 36 ten-day NDVI MVC images for each of the 6 available years, which were used for the current analysis.

6.13.5 Data Processing

Evaluation Strategy

As previously explained, the performance of the different versions of Forest-BGC was evaluated against the sap flow transpiration measurements, which were available for the two stands in 1997 and 1998. These two stands differed in the composition of their species (as mentioned above, the first is covered by *Quercus cerris* and the second by *Quercus pubescens*) so that the tests performed were representative for two types of Mediterranean forest types. The model was applied in both study stands using inputs from different sources, with particular attention to the effect of LAI values on transpiration estimates. The measured transpiration data from first year

(1997) were used for model calibration, and those from the second year (1998) for model validation. These tests required the preliminarily processing of all ancillary and satellite data, which are described in the following sections.

Simulation of Meteorological Data for the Forest Test Sites

The first step concerned the extrapolation of the meteorological data needed for the model from the records of the available station (Larderello). This station was actually quite far from the study stands and at a different altitude. Thus, the temperature and rainfall values at these stands were extrapolated by the application of the model MT-Clim (version 4.3, Thornton and Running 1999). This model was developed to provide meteorological data for a "site station" that is situated at a different location with respect to the "base station", for which all meteorological parameters are known. The model was applied considering the different characteristics of the two study stands, and particularly their different topographical conditions (altitude, slope and aspect) and their annual rainfall averages. The application of the model was preceded by a test aimed at evaluating the expected accuracy of the data extrapolation. The test consisted in the simulation of the meteorological data at the Collitalli portable station from those of Larderello and in the subsequent comparison to the actual available data using RMSE and determination coefficient (r^2) as accuracy statistics.

Derivation of LAI Profiles From Different Sources

As previously mentioned, a major objective of the current research was the evaluation of the effect of different LAI inputs for the application of FOREST-BGC in Mediterranean areas. Tests were therefore performed to assess the performance of the original and modified model versions when using fixed, measured or estimated LAI inputs. First, the correction factor found with the planimeter was applied to all LAI values measured by the LAI-2000 instrument. The corrected values were used as reference for all model simulations. Since the model needs daily LAI values, the monthly measurements were then interpolated by a linear approximation. The derivation of LAI values from NDVI data was performed by a multi-step procedure based on previous methodological works. First, the ten-day MVC images were further composed over monthly periods, in order to reduce residual atmospheric disturbances (Holben 1986). Next, forest NDVI profiles were extracted from the images and then converted into LAI values after a calibration of the relevant relationship (Maselli 2001, Lacaze et al. 1996).

The extraction of forest NDVI values from the monthly images was carried out using the methodology proposed by Maselli (2001). This method can estimate spatially variable pure class spectral signatures (or NDVI values) on the basis of the knowledge of the cover fraction of all classes within each study pixel. In the current case the method was applied using the class cover fraction derived from the CORINE map, obtaining pure class forest NDVI monthly images from which profiles were extracted in correspondence of the two study stands. The conversion of these forest NDVI into LAI profiles was performed according to a general Beer's equation already applied for Mediterranean oaks by Lacaze et al. (1996), which is:

$$LAI = -\frac{1}{K} * \ln\left(\frac{NDVIcan - NDVI}{NDVIcan - NDVIback}\right) \tag{6.13.1}$$

The equation uses three coefficients, *K*, *NDVIcan* and *NDVIback*. The last two were obtained directly from the NDVI images, considering the minimum and maximum NDVI values of the area for the whole data set. The slope of the relationship Eq. 6.13.1, $1/K$, depends on the density of each stand. The NDVI is closely related to forest canopy LAI only if the closure is complete. In different cases there are problems due to the understorey layer, which influences the relationship between NDVI and forest LAI (Spanner et al. 1990). While in fact NDVI depends on the spectral response of all trees, shrubs and grasses within the stand, forest LAI as measured by the Plant Canopy Analyser corresponds only to tree canopies, which is also the layer simulated by Forest-BGC. The conversion of NDVI into tree canopy LAI can therefore be performed only after a specific calibration of the above K coefficient for each study stand. This was obtained by regressing the logarithm right term of Eq. 6.13.1 over the relevant LAI values for the available data of 1996, 1997 and 1998 of both study stands. In this way, a different K slope was obtained for each stand, which was then used in Eq. 6.13.1 for transforming the relevant NDVI data into LAI values.

Calibration and Validation of FOREST-BGC

The application of the model was preceded by a calibration and setting-up phase aiming at identifying the optimal configuration for the simulation of the two oak ecosystems. For this objective, the parameters describing the study forest stands were first retrieved by the Forest Management Plan (DREAM 1994) and by the digital elevation model of the area. Next, eco-physiological parameters describing the characters of the species were derived from existing literature, with particular reference to the publications describing the use of FOREST-BGC by Running and Coughlan (1988), Running and Gower (1991) and Waring and Running (1998). From this bibliographic investigation it was found that the modification of one single parameter, the "maximum canopy average leaf conductance", could be sufficient to adapt the model for the ecosystems considered here, if linked to time series of the LAI which show its variations during the simulation time. This parameter is obviously closely related to the transpiration behaviour of the tree species and can therefore control their xeric nature.

The identification of the optimal leaf conductance for the two species was carried out by using the transpiration measurements of the 1997 growing season. In practice, Forest-BGC was run using the above mentioned ancillary data with the original configuration (Running and Coughlan 1988) but with a range of maximum canopy average leaf conductance as modified according to Jones (1992). Since the effect of using different LAI inputs was also to assess, the calibration was repeated using the three versions with different LAI inputs, and in particular:

1. the original version that considers fixed LAI values during the growing season,

2. the modified version that uses variable daily LAI values derived by the linear interpolation of the monthly measurements,
3. the modified version using daily LAI values derived from NDVI profiles by the procedure previously described.

The simulated transpiration were then compared to the available measured values of 1997 in order to find error minima which could be indicative of the optimal model configuration. This configuration was then used for the model simulation of the two stands in 1998, with the transpiration outputs again compared to the measured values for final validation.

6.13.6 Results

Simulated Meteorological Data

No operational meteorological measurements were available directly at the experimental site. This made it necessary to use model-simulated data. To test their accuracy, they were applied to the nearest meteorological station Collitalli. The results of the comparison between daily meteorological data measured and simulated by MT-Clim for this station are summarised in Table 6.13.1.

Table 6.13.1. Results obtained by the application of MT-Clim to the Collitalli station using Larderello as base station. All statistics are referred to daily values

Meteorological Quantity	1996		1997		1998	
	r^2	RMSE	r^2	RMSE	r^2	RMSE
Max T	0.88	2.47°C	0.49	3.53°C	0.83	4.34°C
Min T	0.91	1.87°C	0.61	1.71°C	0.87	1.84°C
Precipitation	0.15	7.41 mm	0.35	6.57 mm	0.14	6.21 mm

Minimum and maximum temperatures were extrapolated with much higher accuracy than rainfall. This can be explained easily bearing in mind that the former parameters are distributed rather homogeneously on the land surface, mainly depending on the altitudinal gradient which is considered in MT-Clim. On the contrary, lower accuracy was obtained for rainfall, which obviously shows much higher spatial and temporal variability. This last parameter, however, was approximated much better when considering longer time periods (one week or month) which are more significant for the model performance. Thus, it was concluded that the extrapolation accuracy was sufficient for the scope of this investigation, and the procedure was used for all subsequent simulations.

Measured and Estimated Daily LAI profiles

The LAI profiles measured and derived from NDVI data by the procedure described are shown in Fig, 6.13.3. As can be noted, the stand with *Quercus cerris* has higher leaf coverage than that with *Quercus pubescens,* mainly dependent on its higher tree density. In this situation, the calibration of Eq. 6.13.1 for each stand was clearly necessary to approximate the real LAI values from NDVI data, due to the influence of the understorey on the NDVI/LAI relationship. In effect, the use of modified K coefficients for the two stands allowed a rather accurate estimation of their real values, even if with an anticipation of the LAI peak from August to June. The reasons and implications of this last fact for model working will be discussed in the conclusions.

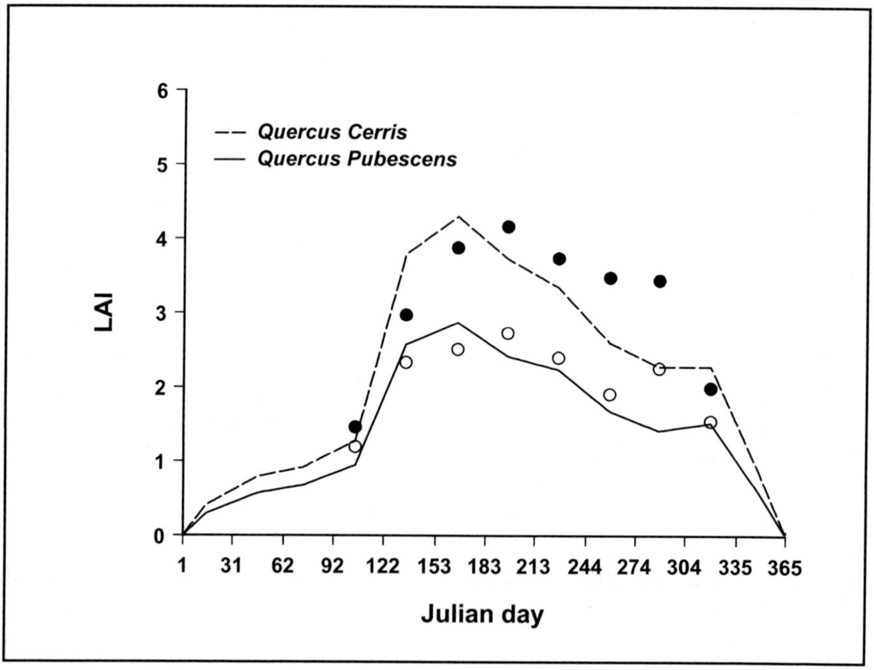

Fig. 6.13.3. Averaged LAI values of the three study years (1996-98) measured (points) and computed by equation 6.13.1 (lines) in the *Quercus cerris* and *Quercus pubescens* stands

Model Calibration

The results of the model calibration are presented in Fig. 6.14.4 for the two stands as root mean square errors (RMSE) of the estimated versus the measured transpiration

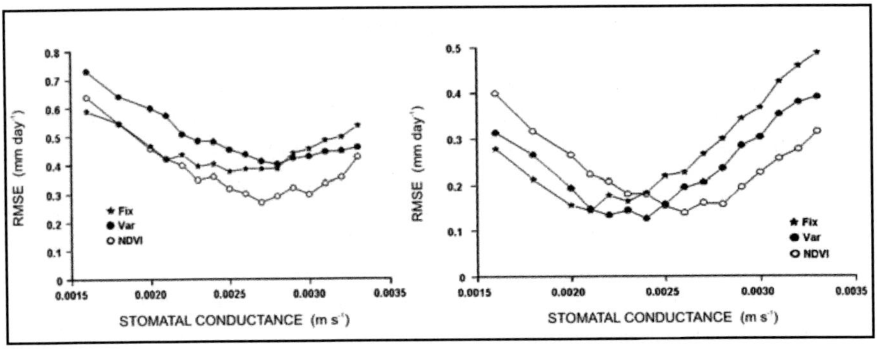

Fig. 6.13.4. RMSEs obtained using three different versions of the model presented in terms of transpiration. The triangles indicates errors with constant LAI during the growing season, the squares those with measured LAI and the rhombs those with LAI derived from NDVI. Left: *Quercus cerris*, right: *Quercus pubescens*.

data available for 1997. The left graph (*Quercus cerris* stand) shows that the errors coming from the use of the model with fixed and variable measured LAI were generally higher than those coming from the model with variable LAI derived from NDVI data. All errors had a rather clear minimum, which corresponded to lower stomatal conductance for the first two simulations with respect to the last one. In practice, the best model configuration was that with LAI derived from NDVI data and stomatal conductance equal to 0.0027 m s^{-1}.

Similar results were obtained for the *Quercus pubescens* stand, where an almost identical pattern of error minima was found regarding the three LAI inputs (i.e., minimum corresponding to higher stomatal conductance for variable LAI from NDVI). In this case, however, all LAI inputs produced almost the same error minima, so that the best LAI inputs cannot be clearly determined. As a possible explanation for this fact it can be recalled that the daily transpiration measurements for this stand were actually very few, which limited the accuracy of the error analysis with respect to the variations of stomatal conductance. In consideration of this fact, the same method as before for providing LAI values to the model was assumed to be the best, and the optimum configuration was again identified as that with LAI derived from NDVI data and stomatal conductance equal to 0.0026 m s^{-1}.

Model Validation

The simulated daily transpirations obtained for 1998 from the model runs with the best configuration (variable LAI profiles derived from NDVI and optimum stomatal conductance; i.e., 0.0027 m s^{-1} for *Quercus cerris* and 0.0026 m s^{-1} for *Quercus pubescens*) are shown in Fig. 6.13.5 together with the sap flow measurements.

As can be seen, the model was very efficient in simulating daily evapo-transpiration for these stands. Very accurate estimates have been obtained for the whole study period (for *Q. cerris* $r^2 = 0.861$, RMSE = 0.457 mm/day and for *Q. pubescens* $r^2 = 0.875$, RMSE = 0.264 mm/day).

The agreement between measured and simulated values was particularly good for *Q. cerris*, which showed a wider range of transpiration values which can be reasonably attributed to its higher LAI and stomatal conductance.

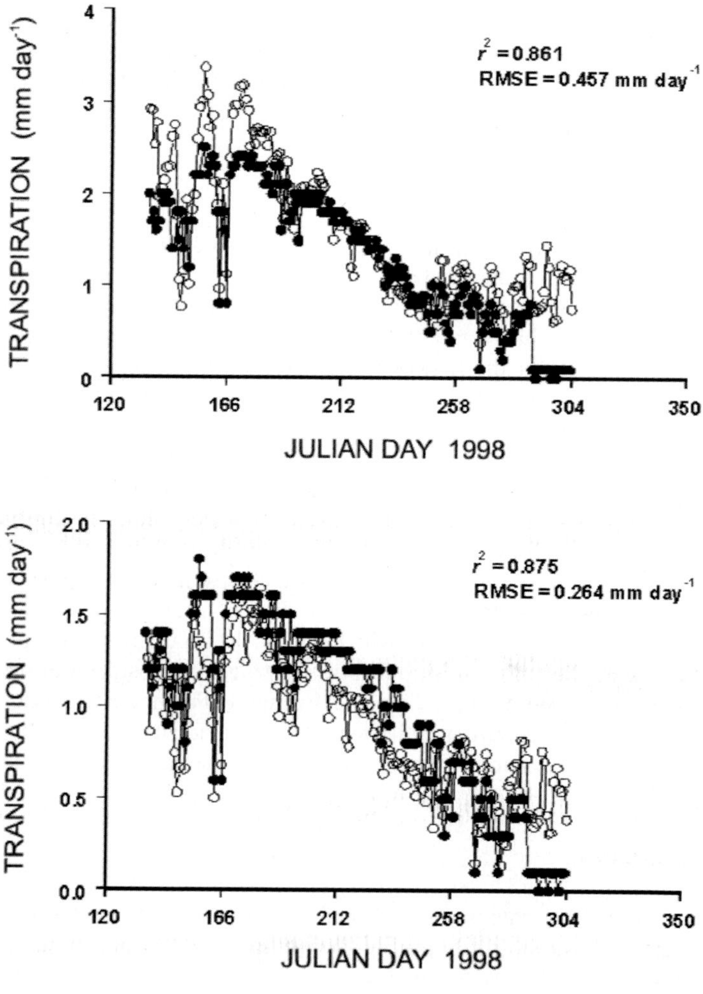

Fig. 6.13.5. Measured (open circle) and estimated (full circle) daily values of transpiration obtained for the growing season 1998 for the two study stands. Top: *Quercus cerris*, bottom: *Quercus pubescens*.

6.13.7 Study Area San Rossore

A second study area was San Rossore located along the sea-coast in Tuscany, central Italy (43.6-43.8° N, 10.4-10.6° E, Fig. 6.13.6). It is a flat area characterised by mainly sandy soils. The climate is Mediterranean with warm and dry summers and relatively warm winters; the average annual temperature is 15°C and the average rainfall is about 900 mm. The ecosystem is characterised by the presence of a Mediterranean pine forest composed by *Pinus pinaster* Ait and *Pinus pinea* L.. The whole forest is of artificial origin and it is often disturbed by the excessive density of wild animals and the presence of polluted air carried on the canopies by the sea wind. The understorey is dominated by species typical of the Mediterranean maquis such as *Phyllirea spp.* and *Erica arborea.*

The current simulation concerns a pine stand within which water and carbon fluxes are continuously measured since the beginning of 1999. The stand is placed near the middle of the study area and is mostly covered by *Pinus pinaster*, which is therefore the species considered for the model simulation.

The application required the preliminary processing of ground and remotely sensed data. First, meteorological data (daily minimum and maximum temperatures, main parameters characterising the ecophysiological behaviour of the study species and precipitation) were taken from a station close to the study area (Pisa). Next, the (Pinus pinaster Ait.) were derived from the existing literature (Running and Coughlan 1988, Waring and Running 1998, Kramer and Kozlowski 1979), while information on local stand conditions were retrieved from the local forest management plan. Leaf Area Index measurements were collected during the growing

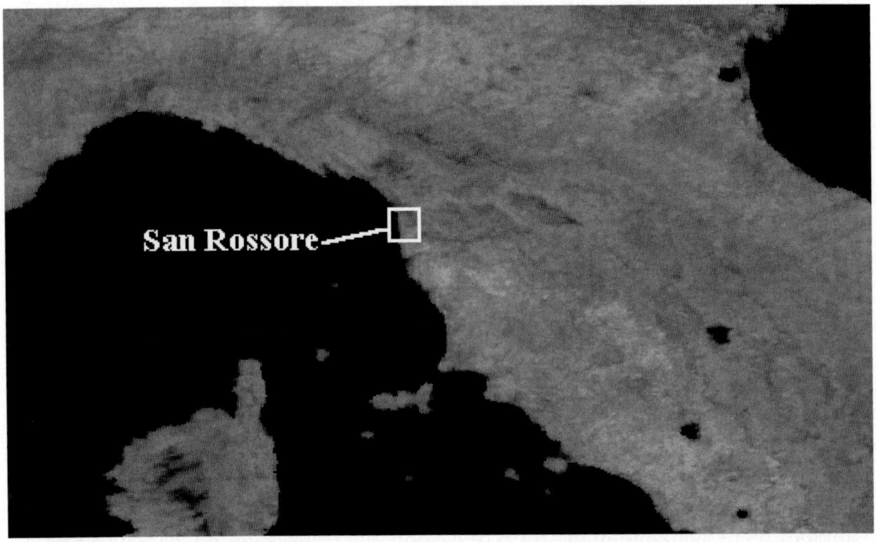

Fig. 6.13.6. NOAA-AVHRR NDVI image of May 1999 indicating the position of the study area.

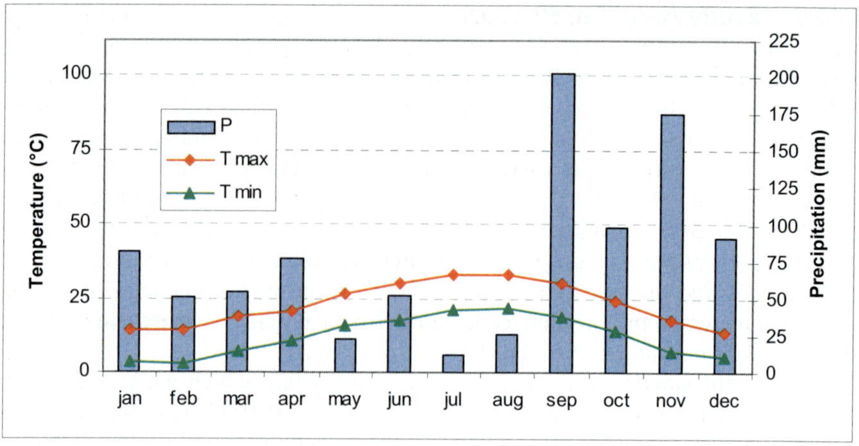

Fig. 6.13.7. Minimum and maximum temperatures and monthly precipitation for 1999 collected at the station of Pisa

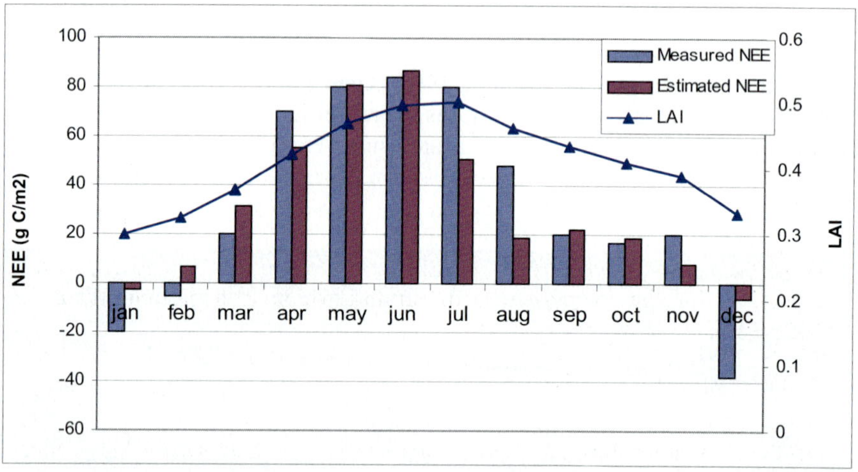

Fig. 6.13.8. Daily LAI profile derived by 10-day NDVI values for 1999 and comparison between measured and estimated NEE (r^2=0.827 RMSE=17.5 gC/m^2)

season 2000 by the use of LAI 2000 Plant Canopy Analyser by Li-Cor. These data were integrated with the NOAA-AVHRR images by the procedure described in Chiesi et al. (2001), in order to obtain monthly LAI profiles of several years which could be used as input to the model. The availability of these data allowed the simulation of local forest fluxes of 1999, for which monthly photosynthesis estimates were available in the form of carbon Net Ecosystem Exchange (NEE).

A summary of the used meteorological data is shown in Fig. 6.13.7, referring to 1999. The climatic pattern is typical for a Mediterranean zone, with a summer dry period with almost no rainfall (minimum values in July of 11.8 mm) and high temperatures (maximum values above 30°C during July, August and September). The LAI profile obtained by integrating the ground measurements and the monthly NDVI values of 1999 is shown in Fig. 6.13.8. In general, these LAI values are rather low (a maximum value of 2.8 is reached in June) and present limited inter-seasonal variations,. This is in accordance with the ecological character of the local evergreen forest, whose development is generally limited by the scarce water availability during the warm arid season.

The measured and estimated NEE values obtained for 1999 are shown in Fig. 6.13.8. As can be seen, both these values mostly follow the LAI profile of the forest, in accordance with the high importance of this in driving the main forest fluxes (Waring and Running 1998). Also, the agreement between measurement and simulated NEE values is quite good, with $r^2 = 0.827$ and RMSE = 17.5 g C m^{-2}. In particular, the simulation is quite accurate during the growing season from February to June, while there are problems for summer and winter periods. This behaviour can be probably explained considering the only approximate calibration of the model to fit the local climatic conditions during the dry period, which are obviously quite different from those for which Forest-BGC was created.

6.13.8 Conclusions

From the results obtained in the current study the following conclusions can be drawn:

1. MT-Clim can estimate all meteorological parameters that are necessary to initialise FOREST-BGC. The only necessary operation to use this model is the collection of general environmental information about the sites for which the simulation is required. More accurate procedures could however be useful, especially for the extrapolation of rainfall which is a particularly crucial parameter in Mediterranean areas.
2. The extraction of NDVI profiles by NOAA-AVHRR according to the methodology exposed gave good results; the following transformation of NDVI values into LAI can be easily obtained by calibrating the parameters of a general NDVI/LAI relationship. This transformation obviously requires the knowledge of the most important characters of the study forest stands, in particular those related to canopy density and LAI. This implies the necessity for ground measurements which are clearly a limitation to the application of the model to larger areas. The substitution of the manual calibration with semi-automatic procedures based on the use of higher resolution satellite data is presently under study.
3. In general, it was shown that the model FOREST-BGC can simulate Mediterranean forest ecosystems if appropriate eco-physiological parameters and LAI profiles are used. This last parameter, in particular, is determinant in controlling the main bio-geo-chemical processes of Mediterranean vegetation, since it forces the model to work with appropriate transpiring and photosynthesising leaf surfaces also during the arid season, which would be not the case when using the original model configuration which was conceived for temperate forests.

4. The necessary LAI profiles can be derived both from ground measurements and from NDVI data after suitable calibration. A remark is worth on the latter possibility, since it seems to improve the sensitivity of the model to the local Mediterranean features. It must be recalled, in fact, that LAI derived from NDVI is actually more related to plant photosynthetic activity than only to leaf area, in this sense being actually a "green" LAI (Spanner et al., 1990). This causes that, in Mediterranean areas, the green LAI shows a typical peak in late spring, followed by a decrease due to summer aridity, which is not visible in the measured LAI values. In practice, this behaviour can enhance the coupling of the LAI profiles with the actual transpiration and photosynthesis processes, which tends to improve the whole model functions during summer aridity. This is also confirmed by the higher leaf conductance found as optimal when using LAI from NDVI with respect to the other cases. Indeed, the greater adaptability of the model in the former configuration allows generally higher transpiration values during the summer season, which are better in accordance with ground measurements.
5. The possibility of using LAI data derived from remotely sensed images is particularly interesting from an operational point of view. This in fact indicates that proper simulations could be obtained by running the model within a GIS system which integrates conventional and remote sensing data. Thus, the current findings seem to open the possibility of implementing an automatic version of the model, running on a regional scale, which is the final direction of our research efforts.

6.14 Use of GAC NDVI Data for Cropland Identification and Yield Forecasting in Mediterranean African Countries

6.14.1 Introduction

The necessity for timely and accurate crop monitoring systems working at a regional scale is particularly felt in arid and semiarid countries, where temporal and geographical rainfall variability leads to high inter-year fluctuations in primary production and increases the risk of severe famines (Hutchinson 1991). These environmental situations, along with the wide extent of the areas to monitor and the generally poor availability of efficient communications, enhance the utility of remote sensing techniques, which can view large land surfaces synoptically and with high temporal frequency (Prince 1990).

Among the available remote sensing systems, the Advanced Very High Resolution Radiometer (AVHRR) mounted aboard the National and Oceanic Atmospheric Administration (NOAA) satellites is the only sensor which has acquired data useful for large scale crop monitoring during the last two decades (Prince 1991). The NOAA-AVHRR Normalized Difference Vegetation Index (NDVI) is in fact linked to several vegetation properties (percentage cover, leaf area index, active green biomass) and is an indirect indicator of primary productivity through its quasi-linear relation with the percentage of absorbed photosynthetically active radiation (%APAR) (Prince 1990, Los 1998). Such a relationship makes possible the early estimation of crop yield, since this parameter is mainly determined by the photosynthetic activity of agricultural plants in certain periods prior to harvest (Baret and Olioso 1989, Rossini and Benedetti 1993). Previous studies built on this property found useful statistical relationships between NDVI values at the peak of the growing season and final crop yield once the perturbing effects of the geographical variability in environmental features (natural vegetation and soil types and conditions, topography, etc.) had been removed (Maselli et al. 1993, Hayes and Decker 1996). The same and other investigations showed that yield forecasting can be obtained by the use of NDVI data of specific periods which depend on the eco-climatic conditions of the areas and the types of crop grown (Hayes and Decker 1996, Lewis et al. 1998, Maselli et al. 2000). Most recent research also indicated that a decisive improvement in crop yield forecasting capability is linked to the selective consideration of NDVI values from cropped areas, since other vegetation types, having different seasonal developments, may introduce noise in the relationships NDVI/yield (Genovese et al. 1999, Maselli et al. 2000). This should be particularly the case in Mediterranean countries, where areas cultivated with cereal crops are generally intermingled with other vegetation types, such as forests and maquis, which have different phenology and response to meteorological factors (Lacaze et al. 1996).

In the current work, carried out within the EU (European Union) project CAMELEO (Changes in Mediterranean Semi-arid Ecosystems on the Long Term through Earth Observation), the use of multitemporal NDVI data was tested for the estimation and forecasting of cereal crop yield in Mediterranean African countries. In particular, the first efforts were devoted to finding which was the most suitable period for yield forecasting by NDVI data in each country. The research was then directed to the identification of cropped areas which could be selectively considered for improving the forecasting capability. This was attempted first by focusing on different land cover types derived from an existing supervised classification. Next, a new statistical method based on correlation analysis was applied to identify single pixels specifically containing agricultural and non agricultural vegetation. The results obtained by the two methods were evaluated in terms of their correlation with final crop yield. Also, the areas identified as crop and forest land were compared to existing cartography. The utility of the optimal methodology found was finally evaluated for operational crop yield forecasting.

6.14.2 Study Area

Environmental Features

The study area corresponds to the four North African countries involved in the CAMELEO project, which are Morocco, Algeria, Tunisia and Egypt (Fig. 6.14.1). With the exclusion of the completely desert zones without seasonal changes, nearly the whole area is part of the arid and hyper arid ecosystems of the Mediterranean bioclimatic region, according to the classification of Emberger (Emberger 1955). In all four countries the Mediterranean climate is typical for a large East-West belt bounded by the Mediterranean Sea on the North and by the 125 mm (annual) isohyete on the South. North-South extension of this belt varies between 30 and 300 km. The Mediterranean area includes the Atlas mountains in Morocco and Tunisia and the coast massifs in Algeria, which form an effective barrier against the increase of width of the Sahara desert with its extreme aridity and high temperatures. Where mountains are absent, the Sahara desert almost reaches the Mediterranean coast. Annual rainfall ranges from 125 mm near the Sahara to 1000 mm in some upper

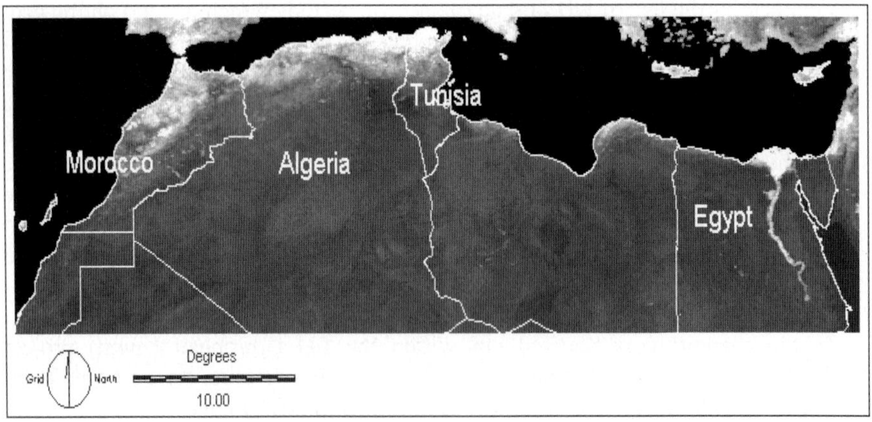

Fig. 6.14.1. Example of NDVI MVC image (April 1992) showing the four North African countries involved in CAMELEO (Morocco, Algeria, Tunisia and Egypt).

zones of the Atlas mountains and the High Tell mountains in Tunisia. Maximal annual rainfall is concentrated in the coldest months of the year (from November-December to March-April), and is the main limiting factor for plant production in the whole area with the exception of the Nile valley, where water is provided by the river floods.

Agricultural Features

In general, north African agriculture has been intensifying for most of the present century. Over the past three decades, this process has accelerated significantly (Swearingen 1998). Intensification has occurred in two key directions, horizontally and, above all, vertically. Horizontally, North African farmers have progressively expanded cultivation to lower rainfall areas. In addition, they have expanded livestock raising to more marginal areas that previously were little used. Vertically, farmers have intensified agriculture through irrigation development, reduction of fallow, and intensification of stock raising. In the near term, the intensification processes have fuelled economic development and enabled these countries to feed their growing populations. However, North African agriculture is now approaching its environmental limits which are due to available land and water resources.
As regards the crops grown, rainfed spring cereals are by far dominant in the three western study countries (Morocco, Algeria and Tunisia), with wheat representing 50-70% of total agricultural production. In these areas wheat is sowed in November-December and is harvested from April to June. The sowing period is generally longer in Algeria and Tunisia than in Morocco, where the whole growing season is anticipated (MARS 1997). A particular case is represented by Egypt, where the peculiar cultivation systems used are fed by the regular Nile's floods and relevant irrigation systems. Two growing seasons are therefore present, one with the peak in winter and the other with the peak in summer. During the former, wheat is mainly grown, while in the latter there is a prevalence of maize and rice, which together represent more than 50% of total agricultural production (FAO 1999).

6.14.3 Data

Cartographic Data

Digitized administrative boundaries of all North African countries were obtained from the Environmental Systems Research Institute (ESRI) ArcWorld™ 1:3M Continental Coverage CD ROM. These are very accurate national, regional and provincial boundaries reported as vector files in geographic projection.

A one-km digital land cover map produced by the U.S. Geological Survey (USGS), the University of Nebraska-Lincoln (UNL), and the European Commission's Joint Research Centre (JRC) was acquired by Internet facilities (Brown et al., 1993, Eidenshink and Faundeen, 1994). This is one portion of the global land cover characteristics data base, that was developed on a continent-by-continent basis. All continents in the global data base share the same map projections

(Interrupted Goode Homolosine), have one-km nominal spatial resolution, and are based on one-km AVHRR data spanning April 1992 through March 1993. Each continental data base has unique elements that are based on the salient geographic aspects of the specific continent. The legend for the four study countries includes eleven land cover classes, which from a preliminary analysis were found to be rather mixed (Maselli et al. 1998a).

Crop Yield Data

Agricultural statistics of the main crops grown in the study countries were taken from the Food and Agriculture Organization (FAO) statistical database through Internet facilities. In particular, time series of annual harvested area, total production and productivity for wheat, maize and total cereals over the period 1982-1994 were taken from the database. Unfortunately, these data were available only on a country level, i.e. total values for each country and each year.

Out of these data, total cereal yield was considered for further processing as the most interesting agricultural parameter to estimate at the national level. According to what was stated above about the intensification of agricultural practices, cereal yield showed a general increasing trend in all study countries during the thirteen years considered. A high inter annual variability however existed in Morocco, Algeria and Tunisia, presumably due to meteorological (rainfall) fluctuations. Again Egypt represented an exception, since mean cereal yield was far higher than in the other countries (5 versus 1 T/ha) and had lower inter annual variability thanks to the more regular water availability provided by Nile's floods.

Satellite Data

Following previous investigations, NOAA-AVHRR NDVI Maximum Value Composites (MVC) images were considered as the most appropriate for the estimation of crop yield (Groten 1993, Maselli et al. 1993, Hayes and Decker 1996). In particular, attention was directed to the collection of imagery covering the whole belt of North African countries (Morocco, Algeria, Tunisia, Egypt) with the highest possible spatial and temporal resolutions and for a relatively long time period (at least 10-12 years).

Only two data sets with these requisites were found, both derived from Global Area Coverage (GAC) images. The first were ten-day NDVI MVC images for the period 1981-1999 produced by Global Inventory Monitoring and Modelling System (GIMMS) (Los et al., 1994), and the second were ten-day NDVI MVC images for the period 1981-1994 produced by National Aeronautic and Space Administration (NASA) Mission to Planet Earth program (James and Kalluri 1994). Both of these data sets were freely distributed and were recovered by Internet facilities. After a preliminary evaluation, the latter set was chosen, even if shorter, for its higher geometric and radiometric consistency in the North African region (Maselli et al. 1998a).

The original NOAA AVHRR images used to generate this data set came from

Goddard Distributed Active Archive Center (DAAC) Pathfinder Land data set (PAL) and from NOAA's Satellite Active Archive (SAA). The final ten-day NDVI MVC imagery were produced through:

- Unpacking and storing of GAC orbital data; this includes also retrieval of ancillary data needed for subsequent processing.
- Navigation of each scan using an orbital model;
- Application of calibration (Rao 1993a and 1993b), and atmospheric corrections (Gordon et al. 1988);
- Calculation and appending of cloud flags.
- Resampling of data to 8 km $*$ 8 km resolution and conversion into Goode Homolosine projection (Goode 1925, Steinwand et al. 1995);
- Derivation of NDVI and computation of ten-day composites (Holben 1986). Only pixels within 42 degrees of nadir were used in the composites.

6.14.4 Data Processing and Results

Computer Facilities

The entire image processing chain was carried out using both commercial software packages (IDRISI and ENVI) and specific programs written in-house in Fortran and C codes. All commercial packages and some specific programs were used on PC Pentium platforms, while the other programs ran on a DIGITAL Microvax 3500 computer system.

Pre-processing

All georeferenced digital maps and NDVI images were first converted to IDRISI and ENVI formats. Then, if in a different projection, they were re-projected into the geographic reference system maintaining approximately the original resolutions. This operation could not be performed by the commercial packages for the Pathfinder data, which were originally in a projection unrecognised by IDRISI and ENVI (Interrupted Goode Homolosine), and a specific C program freely provided by NASA was used in this case. Since the ten days MVCs were still affected by cloud perturbations and other atmospheric effects, all of these images were further composed over one month periods, obtaining in this way 12 images for each of the 13 study years. These monthly MVCs were visually checked for both residual atmospheric disturbances and geometric distortions, and were found to be of acceptable quality under both points of view.

Correlation Analysis with Global NDVI Data

A first evaluation of the available NDVI data for crop yield estimation was made by computing the monthly NDVI MVC averages of each country and regressing them with the annual cereal yield values (Anderson 1984). Each regression analysis considered 13 points and was repeated for each study country for all months of the growing season. As previously explained, in the three western countries cereal crops are seeded in fall and harvested in spring of the following year, while in Egypt there are two growing seasons with the second ending in late summer. On the basis of this information and to maintain uniformity, all correlations were computed between crop yield of a season and monthly NDVI values from October of the previous year to September of the same year. In this way, it was possible to identify the most suitable month for final yield forecasting in each country as well as the efficiency of the process.

The results of all regression analyses are summarised in Fig. 6.14.2 in the form of determination coefficients (r^2). As can be seen, the maximum correlations between monthy NDVI values and final crop yield were found in spring-summer months and were rather high even though the NDVI values were computed over the whole countries. In practice, all maximum correlations were significant at the 99% confidence level, which, with 13 data points, corresponds to a determination coefficient of 0.47. The maximum correlations were found in different months for the four countries. The peak of correlation was in fact in March-April for Morocco, May for Algeria, May-June for Tunisia and July for Egypt, where also a secondary peak in winter was present. These correlation patterns can be explained on the basis of the above mentioned differences in the phenology of cereals, which is anticipated in the western zones and mainly linked to the summer growing season in Egypt.

Correlation Analysis with NDVI Data of Single Land Cover Classes

After the preliminary analysis at the national level, attention was focussed on the identification of areas where NDVI values were most correlated with crop yield. For this scope, the land cover classification produced by the USGS was used as indicator of homogeneous zones (Brown et al. 1993). The global land cover map was therefore windowed to an area corresponding to that of the GAC images. Fig. 6.14.3 shows a version of the classification degraded to the GAC pixel size (8x8 km). As previously mentioned, this reported 11 land cover classes for the study area, which were successively reduced to 8 for Egypt and 7 for the other countries by retaining only classes with more than 30 GAC pixels for each country and including a desert class. The legend of the modified USGS classification is summarized in Table

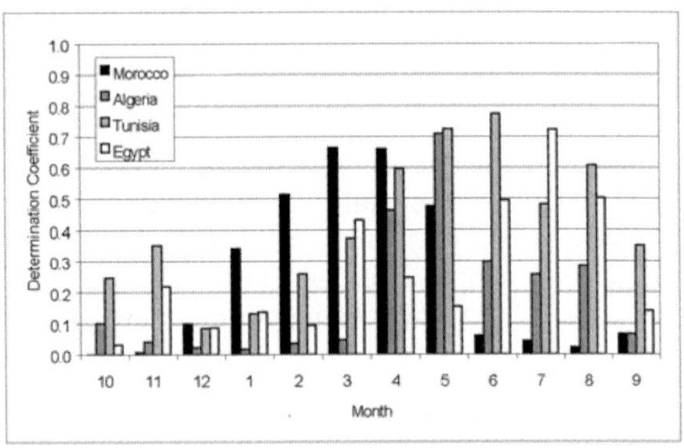

Fig. 6.14.2. Determination coefficients found between mean monthly NDVI values of each country and final crop yield in the 13 study years (1982-1994).

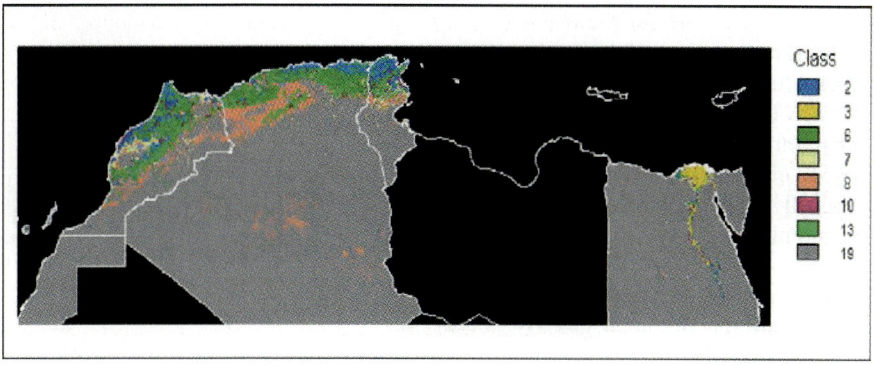

Fig. 6.14.3. USGS supervised classification windowed to the study area and degraded to the GAC pixel size (8x8 km)

6.14.1.

Since the USGS classification was available at 1 km resolution, a methodology was applied to extract the NDVI profiles of the pure classes at GAC resolution. This was a method recently proposed by our research group for the integration of data with different spatial and temporal resolutions (Maselli et al. 1998b). The method is based on an assumption of linearity in the composition of NDVI values from different cover classes which, even though strictly valid only for the original bands, was found to hold with a good approximation also for the already composed index (Kerdiles and Grondona 1995). First, a Boolean mask for each class of the 1 km USGS classification image was created. Next, the mask images of the eight classes were degraded to GAC resolution by simple averaging, obtaining abundance images which reported the proportion of each land cover class in each GAC pixel (Kerdiles and Grondona 1995). These abundance images were then regressed against the NDVI images by a multivariate linear procedure applied independently to each country. The extrapolation of the regression models found to complete class cover

Table 6.14.1 Legend of the USGS classes which are present in the study area

USGS Class	Description
2	Dryland cropland and pasture
3	Irrigated cropland and pasture
6	Cropland/Woodland mosaic
7	Grassland
8	Shrubland
10	Savanna
13	Evergreen broadleaf forest
19	Barren or sparsely vegetated land

(fraction equal to 1 for the class considered and to 0 for all others) allowed the computation of NDVI profiles analogous to end-member spectra for each USGS class and each country (Maselli et al. 1998b). Finally, the monthly NDVI values of the classes were regressed against the annual yield data in the same way as for the previous global analysis.

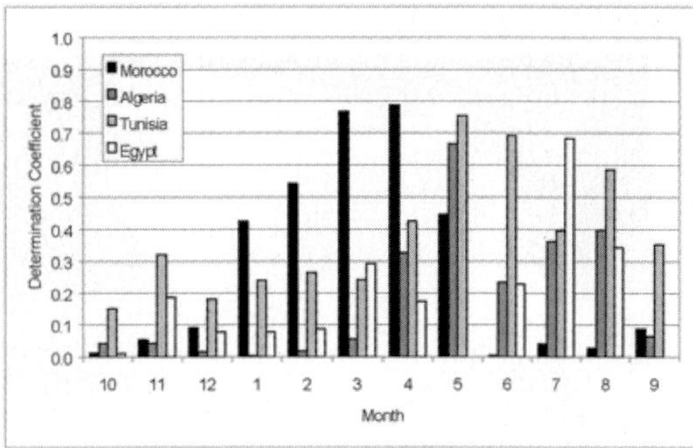

Fig. 6.14.4. Determination coefficients found between mean monthly NDVI values of the most correlated USGS land cover class of each country (class 6 for Morocco, Tunisia and Algeria; class 3 for Egypt) and final crop yield in the 13 study years (1982-1994)

The results of the NDVI/yield regressions were evaluated again by considering determination coefficients. To summarize the results obtained, the agricultural classes with highest correlations with cereal crop yield were selected, which corresponded to class 6 (cropland/woodland mosaic) for Morocco, Algeria and Tunisia and to class 3 (irrigated cropland and pasture) for Egypt. The determination coefficients of the NDVI/yield regressions for these classes are shown in Fig. 6.14.4. As can be seen, the seasonal correlation patterns were very similar to those obtained for the whole countries, and also the absolute values were almost the same for Algeria, Tunisia and Egypt, while a certain improvement was found for Morocco. This indicated that the use of the supervised classification brought only marginal improvements to the yield forecasting capability, probably due to the presence of mixed vegetation types within each land cover class (Maselli et al. 1998a).

Correlation Analysis with NDVI Data of Selected Pixels

On the basis of the previous considerations, a different approach was tested for the identification of cropland pixels and the selective extraction of their NDVI profiles. Since no reliable information was available on the spatial distribution of the cultivated areas at the GAC pixel resolution in all study countries, an image based approach was applied for the identification of cropland pixels. In practice, it was hypothesised that the pixels of each country most intensively cultivated had NDVI values in the best yield predicting period most correlated with the relevant final yields. This hypothesis relied on the assumption that the spatial distribution of cropland was approximately stable during the study years at least at a GAC pixel scale. The best monthly NDVI MVCs to predict crop yield were therefore selected for each country, which corresponded to images of April for Morocco, May for Algeria and Tunisia, and July for Egypt. A synthetic image was then generated for every country with each pixel expressing the correlation coefficient between these NDVI values in the 13 study years and the relevant values of final crop yield. This correlation image was considered as expressing information on the fractional cropland cover within each pixel. Since however it was noted that also areas with very low NDVI values could show high correlation with crop yield probably due to the presence of grasses with phenology similar to that of cereals, a vegetation mask was additionally applied. In practice, three images expressing vegetation quantity in the selected months (April, May and July) were computed by averaging the relevant NDVI MVCs over the 13 years. These mean vegetation images were stretched between 0.2 and 0.4 NDVI, which approximately correspond to the limits of desert and quite homogeneous vegetation cover (Maselli et al. 2000). Each resulting "mean vegetation" image, re-scaled between 0 and 1, was finally multiplied with two images, the first expressing the correlation with crop yield and the second the complementary to one of this correlation. A third image was obtained by subtracting the first two from 1. In this way three synthetic images were obtained for each country, all with values between 0 and 1, indicating respectively:

1. The fraction of vegetation correlated with crop yield (agricultural vegetation).
2. The fraction of vegetation not correlated with crop yield (non agricultural vegetation).
3. The fraction of barren land.

All these three images were subjected to a Gaussian filtering with 0.5 pixel standard deviation to reduce random noise. The three fraction images of each country were then re-composed to produce an unique image for each land use type (agricultural and non agricultural vegetation and barren land) covering the whole study area. The first two of these images are shown in Fig. 6.14.5 (the third is simply the complementary to 1 of these two).

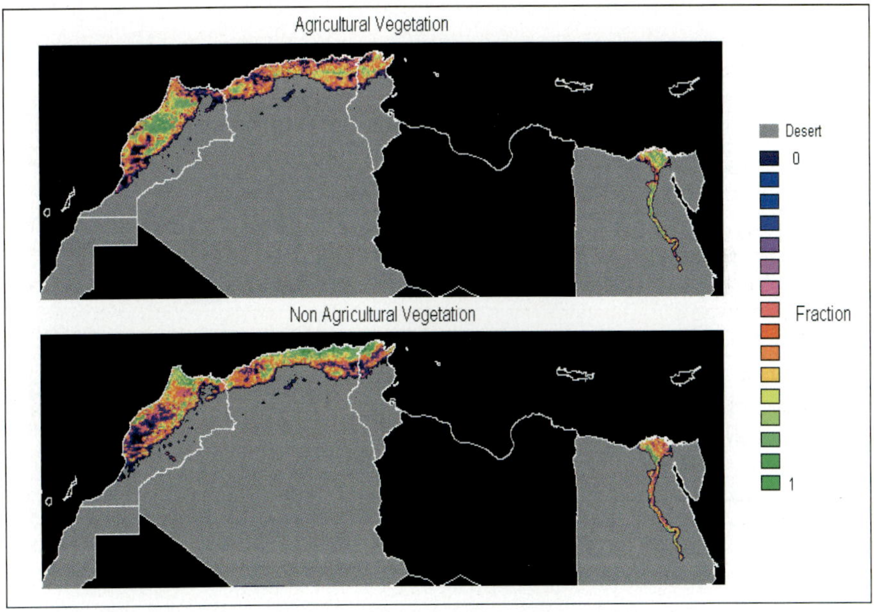

Fig. 6.14.5. Fraction images of agricultural and non agricultural vegetation obtained by the procedure described in the text

Based on the considerations exposed above, these images were considered similar to abundance images, which could be used to extract the NDVI profiles of the three "land use types" as done previously with the abundance images derived from the USGS classification. This again assumed a linearity in the composition of the NDVI values from each fraction of the image, which, even though not strictly demonstrated, was considered reasonable since the synthetic images were derived from linear correlation analysis with crop yield. The same methodology used before (multi-variate regression and NDVI extrapolation to full class fraction for each country) was therefore applied to compute NDVI profiles of the three land use types which again could be considered analogous to end-member spectra. The average annual profiles of the four countries over the study period are shown in Fig. 6.14.6. It is worth noting that in the three western countries these profiles were typical of:

1. Areas cultivated with cereal crops, with a clear NDVI maximum in the spring months and low NDVI values in summer.
2. Forest areas with a secondary low NDVI maximum in winter-spring months but with high NDVI during the whole year.
3. Desert areas, with NDVI always close to 0.

Egypt was again peculiar for the bimodal profiles of both agricultural and non agricultural vegetation. In this case the profile of the former land use type differed from that of the latter for its higher winter and summer NDVI peaks and lower spring and fall minima. These features, as expected, were indicative of a more intense vegetation activity of cropped areas during the two growing seasons. Globally, the methodology applied seemed to decompose the NDVI multitemporal data according to the main land use types of each country.

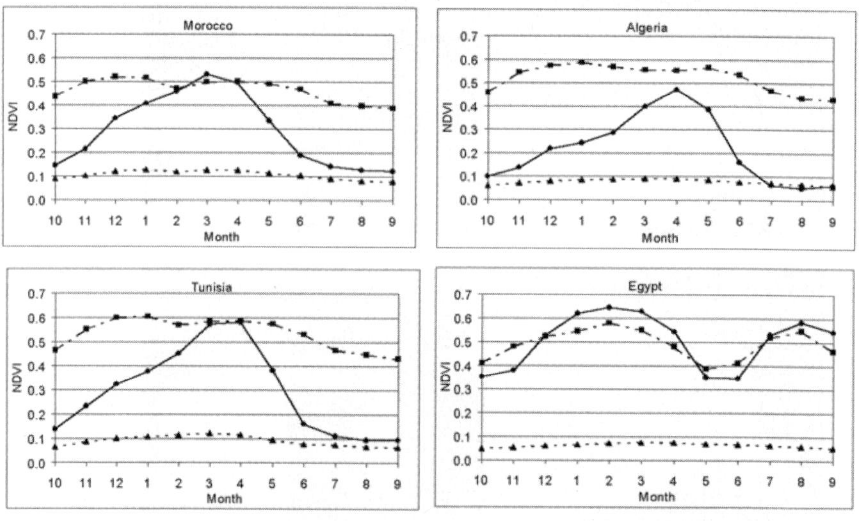

Fig: 6.14.6. Average NDVI profiles of the three land use classes (circle = agricultural vegetation, square = non agricultural vegetation, triangle = barren land) identified by the procedure described

Regression analysis was again applied between the estimated NDVI profiles and crop yield data of the 13 years. The determination coefficients found for the first land cover type in the four countries are shown in Fig. 6.14.7. As can be seen, a decisive improvement in correlation was obtained for Morocco, Algeria and Egypt. It is also worth noting that the other two land use types, and particularly non agricultural vegetation, showed very low and non significant determination coefficients with final crop yield (0-0.3). This was interpreted as a further indication that the methodology applied separated well the main land use components in the study area.

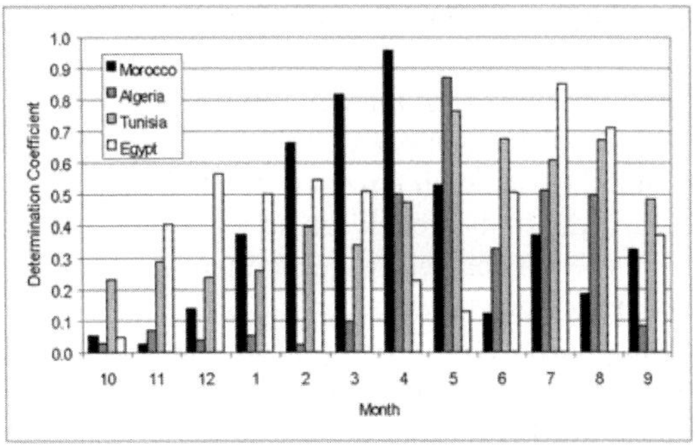

Fig. 6.14.7. Determination coefficients found between monthly NDVI values of agricultural vegetation of each country and final crop yield in the 13 study years (1982-1994)

From the comparison of the annual NDVI profiles of Fig. 6.14.6 with the relevant correlation trends of Fig. 6.14.7 another interesting observation was derived about the temporal relationships NDVI/yield. Where in fact wheat was the main spring crop (Morocco, Algeria and Tunisia), the NDVI peak preceded the period of maximum correlation with final yield of about one month. This is in accordance with previous investigations which found that the most influential period for wheat yield determination follows the flowering phase (which approximately coincides with the NDVI peak) of 20-30 days (Rossini and Benedetti 1993). On the contrary, this was not the case where maize and rice were the prevalent summer crops (Egypt). In this country the NDVI peak of August in fact followed the period of maximum correlation of July, presumably due to the different physiological behaviours of these crops.

Table 6.14.2. Fractions of the three land use types identified by the current methodology for each USGS land cover class

USGS Class	Agricultural Vegetation	Non agricultural Vegetation	Barren Land
2	0.33	0.43	0.24
3	0.57	0.40	0.03
6	0.44	0.40	0.16
7	0.20	0.15	0.65
8	0.04	0.03	0.93
10	0.29	0.31	0.40
13	0.27	0.31	0.42
19	0	0.00	1.00

Evaluation of Produced Maps

An evaluation of the three fraction images obtained in the previous section was first carried out by comparing them to the USGS land cover classification. In Table 6.14.2 the fractional covers of the three land use types are reported for each USGS class in the study area. It can be noted that, apart for USGS classes 8 (shrubland) and 19 (barren or sparsely vegetated land), which correspond to our non vegetated land use type, no complete correspondence exists. USGS class 7 (grassland) mostly corresponds again to barren land, while classes 2 (dryland cropland and pasture) 3 (irrigated cropland and pasture) 6 (cropland/woodland mosaic), 10 (savanna) and 13 (evergreen broadleaf forest) are differently split among our three land use types. In practice, the best correspondence with our agricultural vegetation type was found for the USGS class 3 (present almost exclusively in Egypt) and 6.

Evaluation of the Procedure for Operational Yield Forecasting

The identification of cropland areas is especially significant in the context of a operational yield forecasting system (Hutchinson 1991). This implies that, in addition to being correctly identified, cropland pixels should remain approximately stable in different years. Having NDVI images and ground data for only 13 years, this stability was checked by a leave-one-out approach. The same procedure for cropland identification (creation of correlation images, masking with mean vegetation images and Gaussian filtering) was therefore repeated 13 times using each time data from 12 years, with the exclusion of a rotating year. The three fraction images created for each time (with agricultural and non-agricultural vegetation and barren land) were then used to derive the three pure NDVI profiles for the remaining year by applying the multivariate regression method described previously to the relevant NDVI data. After 13 repetitions of this procedure, the NDVI values found for the best month of each country (April, May and July) were again regressed against the relevant crop yield. In practice this procedure simulated a situation in which the fraction images identified in a certain period are applied to crop yield forecasting in a following year.

The results for the four countries are shown in Fig. 6.14.8. Even if correlations lower than previously were obviously obtained, the prediction capability of the procedure remained high. Determination coefficients higher than 0.64 and root mean square errors (RMSE) lower than 18% of the mean yield values were in fact achieved for all countries using NDVI images taken 1-2 months earlier than crop harvesting.

It can also be noted that, similar to the previous analysis with data of all years, different NDVI/yield slopes were obtained for the four countries, which indicates an increasing efficiency of NDVI conversion to grain products all the way from Algeria to Tunisia, Morocco and, finally, to Egypt (Prince 1991). This is again in agreement with existing information, since Algeria is known to have the poorest agricultural technologies, while Egypt has the highest productivity thanks to its peculiar water provision systems (FAO 1999).

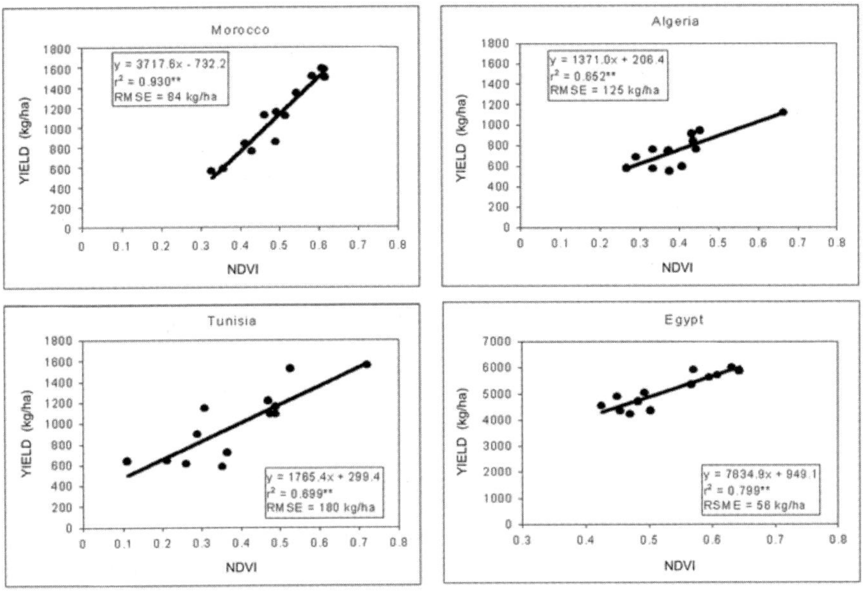

Fig. 6.14.8. Linear regressions between final crop yield and NDVI values obtained by the leave-one-out procedure for the four study countries (** = correlation significant at the 99% confidence level)

6.14.5 Conclusions

The importance of yield forecasting in semi-arid developing countries is clearly apparent and does not need to be further highlighted. Recent research demonstrated that NOAA-AVHRR NDVI MVCs are particularly suitable for this purpose, since they are a sensitive indicator of vegetation conditions during crop growth and can therefore be related to final productivity. In the current work we investigated on this subject in four North African countries involved in the EU project CAMELEO. For the investigation we had to rely on GAC NDVI data, which, even though not optimal for spatial and radiometric properties, were the most consistent set available over the whole study area for a relatively long period (13 years). Moreover, since only annual crop yield estimates for each country were available, the comparisons had to be made at national level, which prevented a higher spatial detail of the analysis.

In spite of these limitations, the first testing demonstrated that interesting relationships between final crop yield and NDVI values of suitable periods can be found even at a country level. In particular, different optimal linear relations were found for each country using NDVI data of spring (April-May) and summer (July) months.

Next, it was verified that the USGS land cover classification was not very effective for cropland identification, probably due to the definition of broad classes which contain mixed vegetation types. A specific method was therefore developed based on the assumption that cropped areas had NDVI values in the optimal yield forecasting period more correlated with final yield than non cropped areas. Correlation images were thus created and composed with mean NDVI images in order to estimate the distribution of agricultural and non agricultural vegetation. The consideration of these images as abundance estimates allowed the extraction of NDVI multitemporal profiles of three main land use types which were analogous to end-member spectra and particularly correlated and uncorrelated with final crop yield. Also, the areas identified as agricultural and non-agricultural vegetation were found to correspond to crop and forest land derived from independent sources. The application of a leave-one-out strategy finally confirmed the utility of the approach for operational crop yield forecasting.

The methodology proposed obviously relies on statistical considerations which may not be valid in other environmental situations. The whole approach is in fact based on the assumptions that the phenological development of the main cereal crops is approximately synchronous and constant within each country and that the distribution of these crops is stable in different years, at least at GAC pixel scale. These assumptions do not generally hold in temperate countries, where crops with different phenologies are usually intermingled, plant development is variable in time due to several factors and inter-year crop rotations are often applied. The existence of suitable conditions for the application of the method must therefore be ascertained in each single case considering both local environmental features and agricultural practices. Also, further modifications are possible in order to improve the performance of the procedure and to adapt it to different ecological situations. In general, however, the operational utility of the method can be notable in semi-arid areas where scarce reference data are available, since it only requires low resolution NDVI data and large scale ground yield assessments.

The current findings have also important implications for the evaluation of the GAC data set utilized. Even though these data are known to suffer from relevant technical limitations, they were proved to contain information which, if suitably extracted, can be very useful for land analysis not only at a regional level, but also at per-pixel and sub-pixel scales. In this sense, the availability of long image series represents a counterbalance for the decreased spatial resolution of the data. The use of multitemporal NDVI profiles in fact allows the extraction of sub-pixel information, at least in these semi-arid environments with few principal land use components.

A final remark is worth about the application of the current results within projects directed to the identification of possible degradation and/or desertification phenomena. Since vegetation variations can be generally ascribed almost exclusively to meteorological vagaries and human interventions, the identification of areas where these two factors are contemporaneously or separately active can be the basis for the analysis of relevant long term vegetation responses. As a first step, the identification of cropped and non-cropped areas could serve to concentrate desertification assessment on the latter, since the former are mainly controlled by human activities. Within a more advanced strategy, the NDVI signal correlated to crop yield could be removed on a per-pixel basis, and the remaining information could be related to meteorological driving factors (mainly rainfall) to detect possible variations in water use efficiency not related to agricultural intensification and therefore indicative of environmental changes (Davenport and Nicholson 1993).

6.15 Drought and Fire Impacts

6.15.1 Introduction

The necessity for efficient, economical and real time monitoring of earth resources is nowadays of global importance, specially in environments which are highly sensitive to climatic fluctuations and which are critically dependent on meteorological conditions. Recent studies have developed methodologies to remotely monitor the vegetation cover in semiarid regions and its response to climatic fluctuations (Tucker et al. 1985, Justice et al. 1986, Prince and Tucker 1986, Eklund 1998, Kalluri 1998, Washington-Allen et al. 1998). One common method involves the use of the NDVI. Numerous studies have suggested that there is a simple relationship with a high correlation between NDVI and rainfall (e.g., Maselli et al. 1993, Tamayo 1994, Rasmusen 1998b).

Fire impacts is one of the most important problem in the Mediterranean area. It exerts destructive influences on many ecosystems: organic matter is consumed, trees are removed, and soil is exposed to the erosive power of rain and wind. The effects of natural forest fires on land degradation are mainly due to the post-fire exposure of bare soil to rainfall. In the Mediterranean countries, the common occurrence of fires in summer results in the soil being unprotected against the usual heavy rains in autumn. Therefore, the longer the period without plant cover after the fire the higher the soil erosion. For the usual situations found in fire-prone systems in the Mediterranean, the critical post-fire period lasts some years. From the date of the fire onwards, the risk of soil erosion and runoff decrease exponentially up to three years after the fire, when the plant cover protection reaches values which are usually similar to the areas before the fire.

Control of wildfire requires an understanding of the relationships between weather and climate on the one hand and fire behaviour on the other. In this context, is very important to analyse fire management practices with information on these relationships and in the way remote sensing techniques can help fire managers.

The development and spreading from the ignition source depends largely on meteorological and climatological conditions. The general structure of vegetation in an area is largely determined by climatological factors such as rainfall, temperature and radiant energy input. The phenological development of plants in vegetation complexes depends considerably on the seasonal progression of climatic elements. Extended period of low humidity, low rainfall and high temperature produce conditions in which dead vegetation becomes highly flammable (Belda 1997).

The relationship between climatological factors and fire behaviour can be based on the actual occurrences of wildfires in an area or on the concurrence of meteorological and vegetation conditions necessary for the initiation of wildfires. The study of these two considerations will indicate those seasons of the year during which fire can be expected and for which fire management plans must be implemented. In this way, observations from the fire area or the closest standard weather observing station should be used. In different area, it is possible to use remote sensing from earth satellites (Meliá et al. 1996) to determine not only seasonal progression of fires but also the amount for area burned (García-Haro et al.

1996, García-Haro 1997, Viedma 1999).

Several climatological methods exist for the determination of fire areas and fire seasons. Reifsnyder (1962) suggested that correlation could be developed between fire occurrence statistics and climatic classifications. Hernández et al. (1991) determine synoptic characteristics patterns for day with great activity of forest fire in the west of Spain. Similar analysis was done in the Valencia Community by Belda et al. (1992). Belda (1997) applied this results to our study area between 1984 and 1994 and determined a relationship between annual rainfall and amount of burned area.

On the other hand, there are two principle factors which affect the regeneration of the area after the fire. Firstly, the size of the fire and the intensity of the damage affect the degree of biodiversity (Armesto and Picket 1985). Secondly, the impact of the fire on to the spatial structure of the vegetation communities which are susceptible to damage (the amount and connectivity of patches). Spectral and spatial responses of the vegetation after fires are suitable properties to analyse for characterising the process of regeneration. In this context, remote sensing and G.I.S. offer the optimum framework for monitoring and identifying the spatial patterns and changes in a forest after the fire (Chuvieco and Congalton 1988, Viedma 1999).

Melia et al. (1996) gave a theoretical justification and experimental demonstration (with assumptions and limitations) of the NDVI for natural vegetation monitoring and for detecting the changes after the forest fire. The work presented here is an attempt to apply this theory to natural vegetation in south-eastern of Spain. The goal was to test the relationship of climatic conditions and natural vegetation (the amount of green vegetation measured by NDVI at a regional scale) and to analyse this climatic influence on the natural post-fire regeneration of the burned area. For these purposes data from existing meteorological stations and Landsat Thematic Mapper (TM) images were used.

6.15.2 Climatological Characterization and Preliminary Analysis of Burned Area in A Study Pilot Area

The study pilot area, of about 20x20 km, is located in the north of the province of Alicante, on the Mediterranean coast of Spain, in the so-called Betics mountain chain, and is especially vulnerable to forest fires due to an extensive history of land use and to high levels of human pressure due to urban settlements. Fig. 6.15.1 shows the north of the province of Alicante and the south of the province of Valencia.

The complex topographic structure of the area develops microclimatic conditions that create different environmental conditions. In addition, the orographic complexity promotes agricultural activity in the vicinity of the shoreline, where a littoral plain is developed. The natural vegetation is a typical Mediterranean forest growing on limestone substrate. The Mediterranean forest is a heterogeneous formation composed mainly of Holm oaks (*Quercus rotundifolia*) considered to be the "climax vegetation", which have given way to coniferous species (*Pinus halepensis*, *Pinus pinaster*), and the secondary formations of shrublands, originating from the degradation of the climax vegetation, composed of: *Quercus coccifera*, *Ulex parviflorus*, *Rosmarinus officinalis*, *Thymus vulgaris*, etc. (Melia et al. 1996).

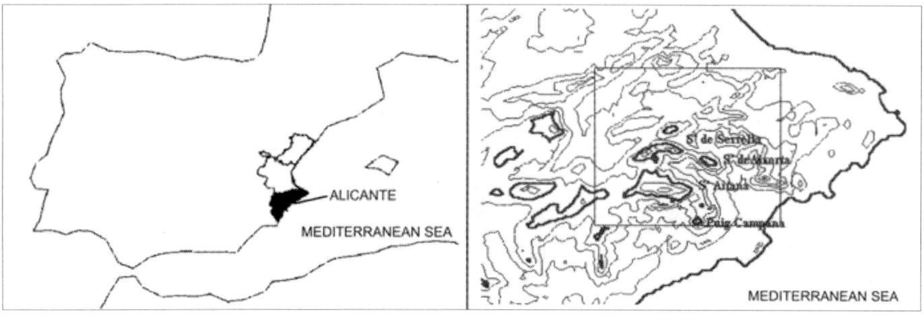

Fig. 6.15.1. Province of Alicante. Left: Position in Spain. Right: The map shows at the centre the area of the Landsat scene located between UTM X 730400 - X760400 and Y4274000 - Y4304000

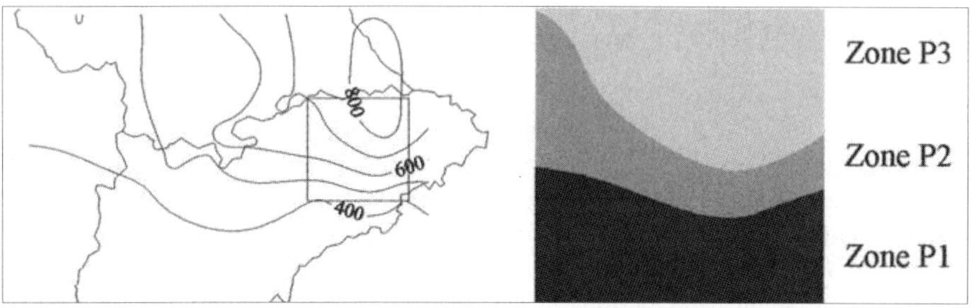

Fig. 6.15.2. Rainfall climatology of the province of Alicante. Left: Distribution of mean annual rainfall. Right: Climatic classification of the test area within the Landsat scene shown in Fig. 6.15.1 in three zones using annual Rainfall (R). Zone P1 with R < 600 mm, zone 2 with 600 mm < R > 700 mm, ande zone 3 with R > 700 mm.

The climatological study consists of the analysis of the mean, standard deviation and quartiles of the distribution of the amount of rainfall for annual, seasonal and monthly periods (Belda 1997). For temporal analysis of the natural vegetation and considering the study statistics and topographic conditions, it is distinguished between three different zones with different annual rainfall, R: zone P1 with R < 600 mm, zone P2 with 600 mm < R < 700 mm and zone P3 with R > 700 mm. The relief facilitates the so-called "screen effect", favouring rainfall in the slopes facing east which are affected by wet maritime flows (Belda 1997). In this area the maximum annual rainfall of the Comunidad Valenciana is recorded, reaching in the average 850 mm rainfall. In this way, the minimum rainfall recorded is connected to valleys and deep areas which channel the down wind flows. In general, these areas show the typical characteristics of the Mediterranean climate with strong gradients from North-South and East-West. The irregularity and intensity of the rainfall leads to an important erosion hazard on the soils.

A climatological detailed study, for the period 1950-1994, has been carried out in the Alicante area by using 25 meteorological stations (with dates of rainfall for the

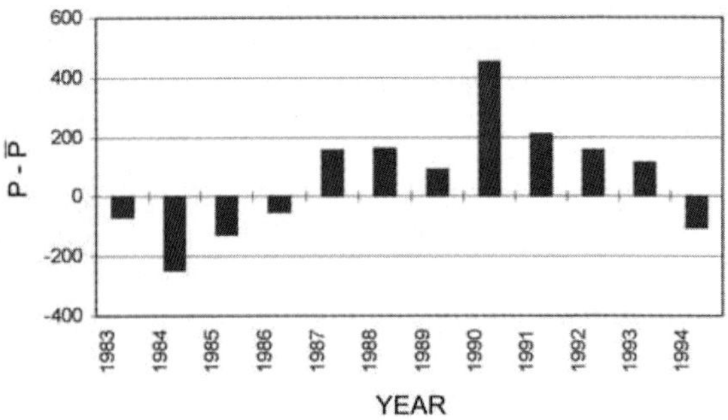

Fig. 6.15.3. Evolution of annual rainfall in the test area from 1983 to 1994. Shown is the mean of the difference between annual (P) and mean annual ($\overline{P}$) rainfall measured at 25 meteorological stations

period between 1950 and 1994) of the Spanish Meteorological Institute. Using this data, a long and homogeneous chronological series for rainfall is obtained. The grids were generated using the spatial resolution of the Landsat-5 scene as reference. Kriging was the method of interpolation used. Fig. 6.15.2 shows the map of the mean annual rainfall in the study area, and the climatic re-classification in the scene of Landsat image.

To evaluate forest damage in the study area, a forest fire database was used. This database was elaborated following the dates of the University of Alicante (Department of Ecology), allowing geographical dates for its treatment. The number of fires in the Province of Alicante, for 1973-1992, is 422 with 71.408 ha burned. In the study area, in the same period, the number of fires is 209 with 45.197 ha burned which represents 63% of the total surface burned. Monthly analysis shows that July and August are months with the highest values of burned surface.

The study of temporal analysis of rainfall, during the monitored period 1983-94, leads to the following statements. Autumn is the wettest period of the year, except the years 1983/84 and 1992/93. Wetter winters have been experienced in the years 1990/91 and 1992/93. More rainy springs have been registered in 1988/89 and 1989/90. Finally, only the summer of year 1991 can be considered as wet, characterising this year as irregular. To evaluate the annual variability of rainfall for the years 1983 to 1994, the deviation of the annual rainfall measured at 25 meteorological stations and the mean rainfall at these 25 stations were computed for every year (Fig. 6.15.3). Three periods can be distinguished: Firstly, the years between 1983 and 1986 can be characterised as a dry period, the year 1984 standing out as driest, secondly, the years between 1987 and 1993 are characterised as wet period, the year 1990 standing out as wettest and, thirdly, the year 1994 as a dry year.

6.15.3 Satellite Image Processing

Several Landsat-TM images are available from 1984 to 1994 to perform the multitemporal study. These images are: 7th April 1984, 29th June 1985, 9th April 1986, 3rd June 1987, 23rd July 1988, 7th May 1989, 7th June 1990, 14th June 1991, 28th March 1992, 19th June 1993, 21st May 1994. The data selection was performed taking into account the statistical information available, the seasonal incidence of fire (summers) and the absence of clouds.

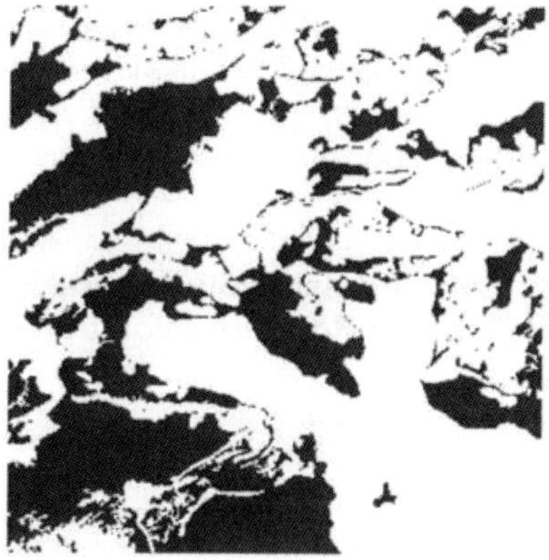

Fig. 6.15.4. Forest mask. Black areas indicate forests, white areas non-forested areas

The image processing consists of geometric and radiometric (atmospheric and topographic) corrections (Melia et al. 1996) to superimpose all the images and to make their multitemporal comparisons possible. Analysis of Thematic Mapper (TM) data related to forest canopy analyses is limited in mountain areas, where terrain-induced modulation of the solar radiation according to the pixel orientation is visually apparent in the scene. In order to take into account this topographic effect, a non-Lambertian reflectance model (Minnaert 1941) was employed, which parameterises the radiance, L, of a surface by means of the expression (Teillet 1986):

$$L(i,e) = a\cos^{k}(i)\cos^{k-1}(e) \qquad 6.15.1$$

where i is the sun incidence angle in relation to the normal of surface of a pixel, e is the sensor angle in relation to the normal of a pixel, a is a constant and k is a parameter, termed Minnaert constant, which indicates the extent to which the surface shows Lambertian behaviour, in which case $k = 1$. In order to compute the normal of the surface for each pixel it was necessary to use a Digital Terrain Model of 30 m resolution. Then, k was determined for each band empirically by linearising logarithmically equation (6.15.1), which simply is the slope of the linear regression. Finally, an image normalised with respect to the topographic effect was obtained by the application of a rotation of the regression to remove the dependence of radiance with the theoretical model throughout the scene.

Vegetation indices have been widely used and related to biophysical canopy parameters such as leaf area index (Baret and Guyot 1991, Asrar 1992, Rasmusen 1998a). Operational satellite sensors such as Landsat-TM and SPOT have multiple spectral bands which are used not only to classify pixels into distinct surface classes but also to estimate fractions of basic classes within each pixel (Garcia-Haro 1997) based on linear mixture model (compare section 6.8). This model has been shown to give a good performance, accounting for most of the spectral variability of the coarse resolution mixed pixels. (Melia et al. 1996). The Normalised Difference Vegetation Index, NDVI, used in this work was defined by the standard algebra of the bands 3 and 4, e.g.:

$$NDVI_{Landsat} = \frac{TM4 - TM3}{TM4 + TM3} \qquad (6.15.2)$$

To eliminate non forested areas, e.g. agricultural land, cities, etc, it is necessary to get a good classification of the area (Melia et al. 1996). A supervised classification was initially carried out to distinguish forest areas from the other area

(García-Haro 1997). This mask was finally corrected using aerial photography. Fig. 6.15.4 shows the forest mask for the Landsat scene defined in Fig. 6.15.1 and used in this study. The size of the forest surface is around 55.850 ha, and the non-forest surface (agricultural land, cities and roads) is about 34.150 ha.

6.15.4 Forest Evolution and its Relationship with Rainfall: Application to Post-fire Evolution

Site Selection

Using NDVI as an index to evaluate the amount of vegetation the NDVI evolution is analysed for natural vegetation in the study area, for 1984/94, following two steps. Firstly, NDVI values are correlated with natural vegetation for the whole study area of 20 × 20 km (Fig. 6.15.4) and also for the three separate climatic zones, P1 (dry), P2 (semi-wet) and P3 (wet) (Fig. 6.15.2). Secondly, a comparative study is made of the post-fire evolution in the three representative training areas, selected in each of the defined climatic zone with the same vegetal coverage. For comparison similar three non-burned pilot areas are defined in each climatic zone. Table 6.15.1 shows the six training areas selected with the most relevant characteristics. In order to investigate the relationship between the different parameters, a linear regression is made and this relation is analysed using the Pearson correlation coefficient (WMO 1975).

Table 1. Characteristics of training areas. The main vegetal community is dense shrub.

Zones	Training Areas	Mean Slope	Orientation	Characteristics
Zone P1	A	3.7	160°	not burned
	D	6.6	290°	burned 1984
Zone P2	B	9.5	192°	not burned
	E	9.6	305°	burned 1984
Zone P3	C	4.2	159°	not burned
	F	2.4	87°	burned 1985

Temporal Characteristics of NDVI and its Relationship with Rainfall

Fig. 6.15.5 shows in the upper part the NDVI evolution from 1984 to 1994 of the forest vegetation, for the whole area, indicating the variation of the mean and typical deviation. In the lower part of Fig. 6.15.5 the evolution of the mean NDVI is shown, for the different climatic zones. In the zones P1 and P2, the value of NDVI is between 0.21 (1984) and 0.29 (1990) with minimum relatives of 1986, 1988, 1991, 1992. For P3 zone, the value of NDVI is between 0.25 (1984) and 0.32 (1989), with relative minima in 1988 and 1993. The wetter zones have higher NDVI values. In the P1 and P2 zones evolution is similar with higher values in the P2 zone. The wet zone (P3) has higher NDVI values than the others zones, although it shows a more irregular behaviour. Fig. 6.15.6 shows the correlation between NDVI and rainfall for this period. A linear relationship is found between NDVI and rainfall. The correlation coefficients are presented in Eq. 6.15.1 for the whole area at the top followed by those of the zones P1, P2, and P3. It can be seen that the correlation is lower in the area with high rainfall than in the dry zone:

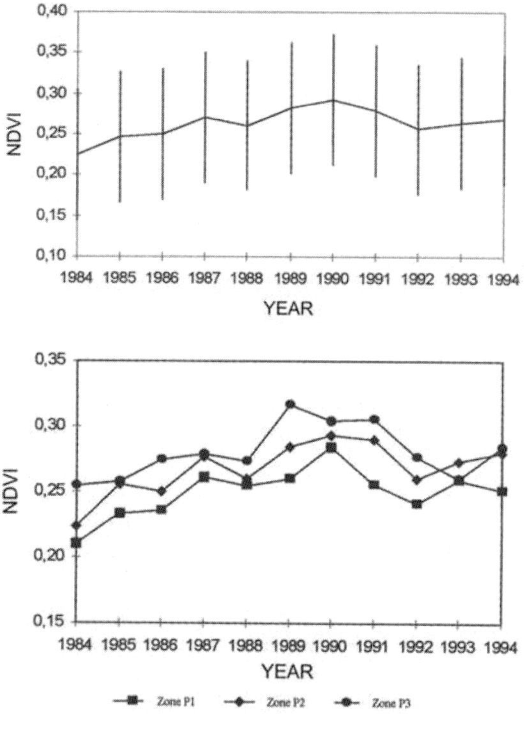

Fig. 6.15.5. NDVI evolution of forest vegetation for the period of 1984 to 1994. Top: Mean value and scatter for the whole area. Bottom: Mean values for the three climatic zones

$$
\begin{aligned}
NDVI_{tot} &= 0.210 + 8 \times 10^{-5} P \quad & r = 0.84 \\
NDVI_{P1} &= 0.195 + 1.1 \times 10^{-4} P \quad & r = 0.79 \\
NDVI_{P2} &= 0.218 + 7 \times 10^{-5} P \quad & r = 0.79 \\
NDVI_{P3} &= 0.239 + 5 \times 10^{-5} P \quad & r = 0.69
\end{aligned} \qquad (6.15.1)
$$

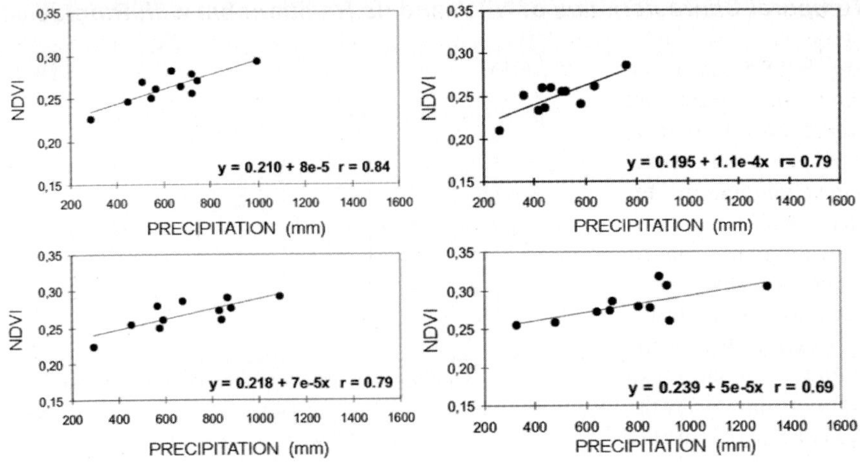

Fig. 6.15.6. Correlation between NDVI and rainfall for the period 1984 - 1994. Top left: Forest covers the whole scene. Top right: Forest in zone P1. Bottom left: Forest in zone P2. Bottom right: Forest in zone P3

Evaluation of the Training Areas

Fig. 6.15.7 shows the evolution of the NDVI in the not burned training areas during the whole period and for the different climatic zones. The NDVI evolution is regular and uniform. Its values are higher in wet zones than in dry ones, reaching 0.38 in the P3 zone. In the zones P1 and P2, the evolution of this index is parallel, with values higher in the wetter zone. One can see an increase from 1984 to 1990 in the P1 and P2 zones, and to 1988 in the P3 zone. In the B area, this trend continues for a number of years, which is a characteristic of this vegetal community (Melia et al. 1996). The C area, belonging to zone 3, shows higher and more irregular NDVI values.

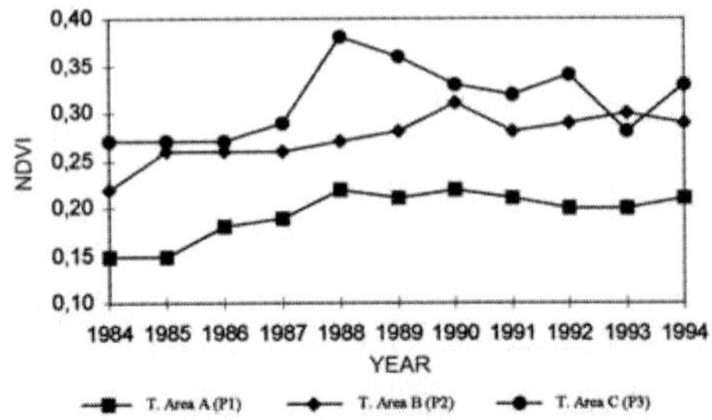

Fig. 6.15.7. Evolution of the NDVI for the period from 1984 to 1994 in the three unburned shrub-land training areas A (■), B (♦), and C (●)

In Fig. 6.15.8 the evolution of the NDVI is presented for the training areas that burned one time

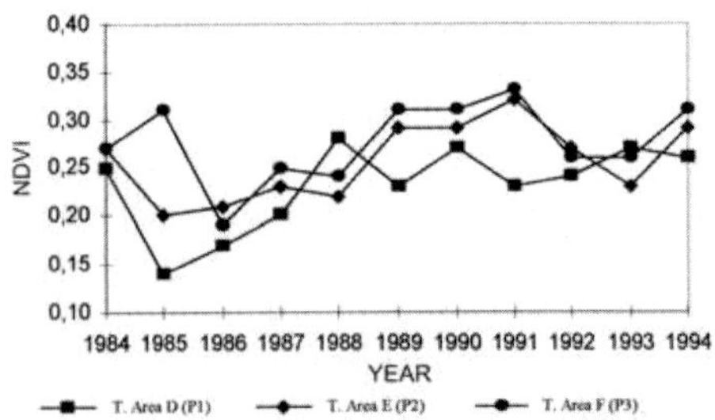

Fig. 6.15.8. Evolution of the NDVI for the period from 1984 to 1994 in the three shrub-land areas D (■), E (♦), and F (●) burned in 1984 (D and E) respectively 1985 (F)

during the period 1984-1994, again for the different climatic zones. Three different burned areas are considered, training areas D and E, burned in 1984, and training area F burned in 1985. Rapid post-fire growth is observed for the first years in the three zones, more uniform and smooth in dry zone. Higher NDVI values are reached in the zone with higher rainfall where regeneration is stronger but also more irregular. Finally, Fig. 6.15.9 shows the NDVI evolution in the training areas separated for different climatic zones.

Using the burned training areas in 1984 and 1985, an exponential function can be fitted to the NDVI evolution for the post-fire years. The general form of this function is:

$$NDVI(T) = A \ln T + B \tag{6.15.2}$$

Here A (NDVI/ln$T$ dimension) is the slope of this adjustment, and B (NDVI dimension) is the value for first year after fire. With this definition, the function is applied to the dense shrub-land considered in section 6.15.5. Similar growth is observed in the training areas, with similar slopes (≈ 0.05). The main difference lies in the parameter B, this value is lower in the dry zone (P1), and is similar for the wet (P3) and semi-wet (P2) zones. The growth after the first post-burning year is similar and reach higher values in the wet zones. The NDVI recovers the 90 % of the initial value (one year prior to the forest fire) between the third and the fourth year after the fire, 63 % of the initial value are reached prior to the second year. After five years the growth is in its saturation phase. Fig. 6.15.9 shows, for three different training areas, the curve of the exponential adjustment showing the function and correlation.

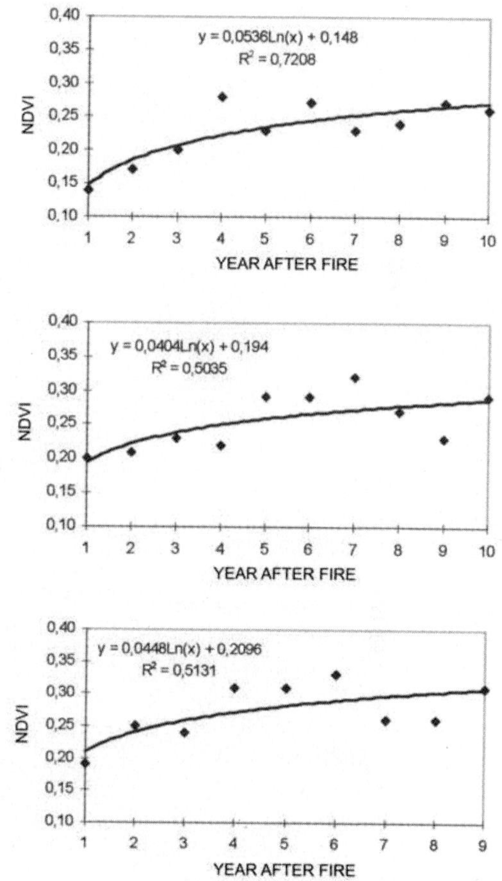

Fig. 6.15.9. Multiannual evolution of the NDVI in the training areas D (P1) burned in 1984 (top), E (P2) burned in 1984, and F (P3) burned in 1985 with the exponential adjustment as curve, function, and correlation coefficient

6.15.5 Multispectral Analysis of Burned Areas

In addition to the vegetation index which was used throughout the preceding sections, Landsat-TM data offer additional spectral bands to detect and to classify burned areas. Fig. 6.15.10 shows a scene of the Tuscan archipelago around the island of Elba and the mouth of the Cornia river near Piombino. Between 22 August and 23 September 1985, the dates of two cloudless Landsat scenes, fire broke out in Elba

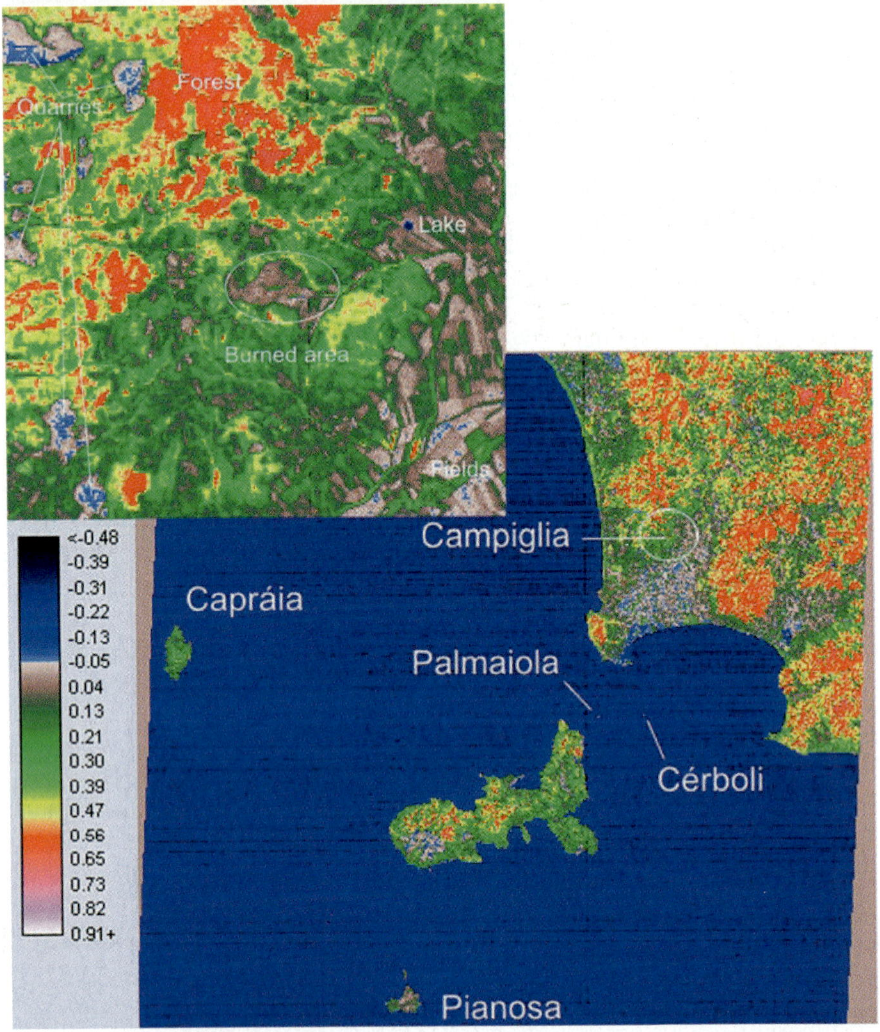

Fig. 6.15.10. NDVI scene of the Tuscan Archipelago of 23 September 1985. Red areas are forests, Brown and blue areas over land represent harvested fields, quarries or burned areas as in the south-west of the island of Elbe and near Campiglia M.ma of which an enlarged view is shown in the upper left section with the encircled burned area

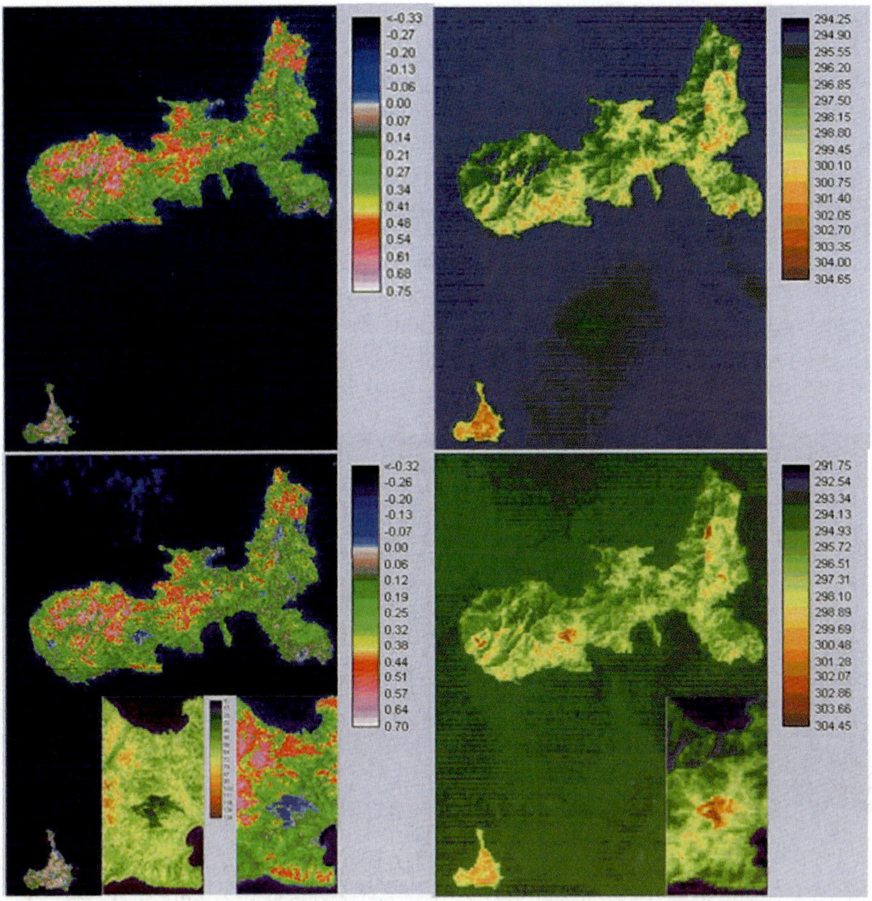

Fig. 6.15.11a. NDVI (left) and effective temperature (right) of the Islands of Elba and Pianosa derived from Landsat-TM data for the top of the atmosphere. The dates are 3 August 1984 (top) and 22 August 1985 (bottom). Inserts in the bottom images are enlargements of one of the burned areas. In the left scene at the left is an image of TM channel 4 with a separate colour scale, beside it the NDVI, and in the right scene the temperature.

and in an area near Campiglia Marittima at the mainland. Enlarged images of the two areas are shown in Figs. 6.15.11 and 6.15.12.

For Elba the scene of 22 August 1985 shows still an intact landscape as compared to the 1984 scene though some areas in the east seen have to suffered from earlier desertification processes. In the third scene from the top a new deserted area can clearly be seen in the south-west of the island reflected as dark area in the NDVI at the left and high temperatures in the thermal image at the right. The northern part of the island is cooler than the southern part because it lies in the shadow of mountains except for the scene of 3 August 1984 when the sun was higher. It can also be seen that by 1998 the vegetation seems to be even better developed that in 1984. The island of Pianosa which is seen in three scenes is a flat island with some agriculture. Only on 10 August the vegetation index is a little bit enhanced which causes a lower surface temperature.

Near Campiglia M.ma (Fig. 6.15.12) the fire broke out prior to 10 August 1985 as can be seen in the third scene from the top a little bit below the centre of the image as indicated in Fig. 6.15.10. The first image shows the situation one year prior to the fire showing an intact landscape which has its maximum vegetation index in late spring as the second scene from the top shows. By 1990 (see the lowest scene, deduced from a SPOT image) the area has largely recovered from the fire and is comparable to the top scene again. This confirms the finding above that after five years vegetation has almost recovered. In this case it is mostly low macchia, pine trees and olive groves with can be found. Differences between the top and bottom scenes can be attributed to a slightly changed land cover, and the pine species

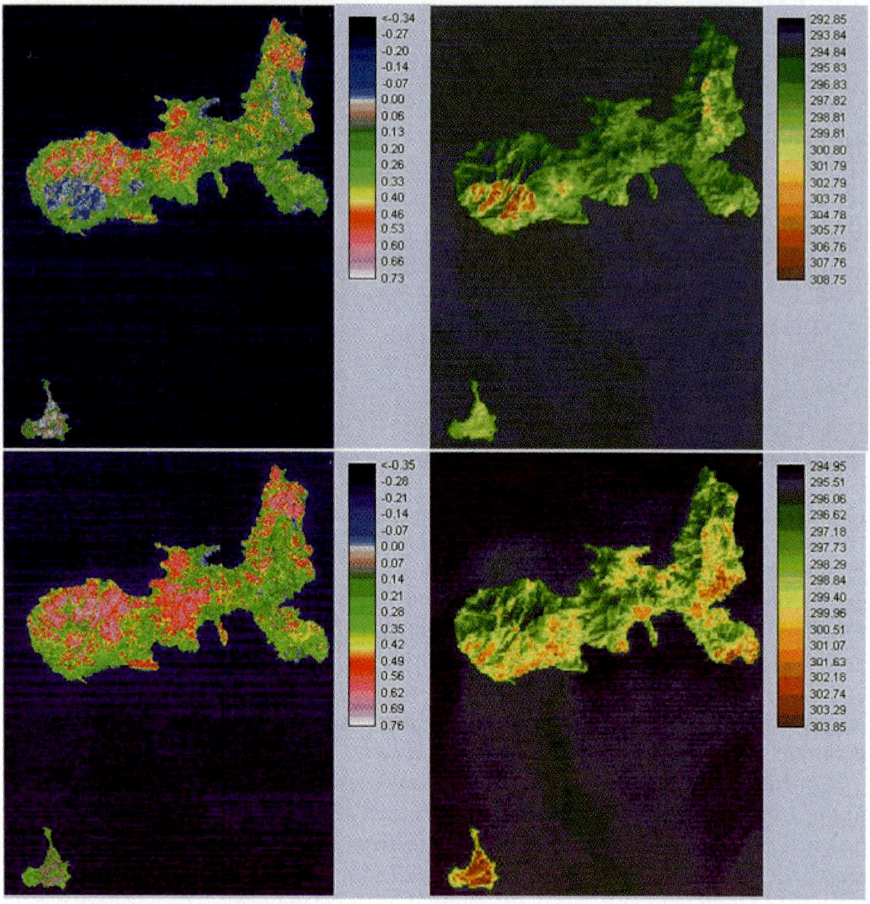

Fig. 6.15.11b. Same as Fig. 6.15.11a but for 23 September 1985 and 10 August 1998. Clearly to be seen in the 1985 image is the burned area in the south-west of the island. In 1998 the areas burned in 1985 can not anymore be identified

changed from *domestica* to *halipensis*. To ensure that the conclusions are not affected by the seasonal differences between the two scenes which are one month apart, two intact areas were selected to assess the natural changes which occurred between the two dates. The result was that for all channels except channel 6 the ratio between the September and August radiance values was in the average 0.81 with extreme values of 0.764 and 0.872. These differences can mostly be attributed to the change of the cosine of the solar zenith angle by a factor of 0.8216. It can therefore be assumed that changes due to the fires can be assessed by comparing the calibrated and corrected signals of the two dates.

The result of the spectral analysis is given in Fig. 6.15.13. The green line represents the spectral technical albedo for the top of the atmosphere measured on 22 August 1985 over the area which was burned between 22 August and 23 September 1985. It shows the typical slope of a vegetated surface. By 23 September 1985 the technical albedo of the same area had drastically changed in the near infrared (red line) although in the whole burned area green islands are imbedded which have been less affected by the fire. The differences become even more evident if one looks at smaller areas represented by the triangles which show extreme high values at the longest wavelengths for the burned area. It is remarkable that the signals in the visible do not change much. Green vegetation and burned vegetation both are very dark at short wavelengths. The increase of the signal towards the shortwave end is due to molecular and aerosol scattering in the atmosphere. At the right side of the diagram the temperatures are displayed. Here one can see that the burned area heats up well above the vegetated area.

On Elba the discrimination between burned and intact areas is relatively easy because most of the island is densely vegetated. This is different at the mainland test site with its agricultural land use. Depending on the soil, the harvested agricultural land may have a similar

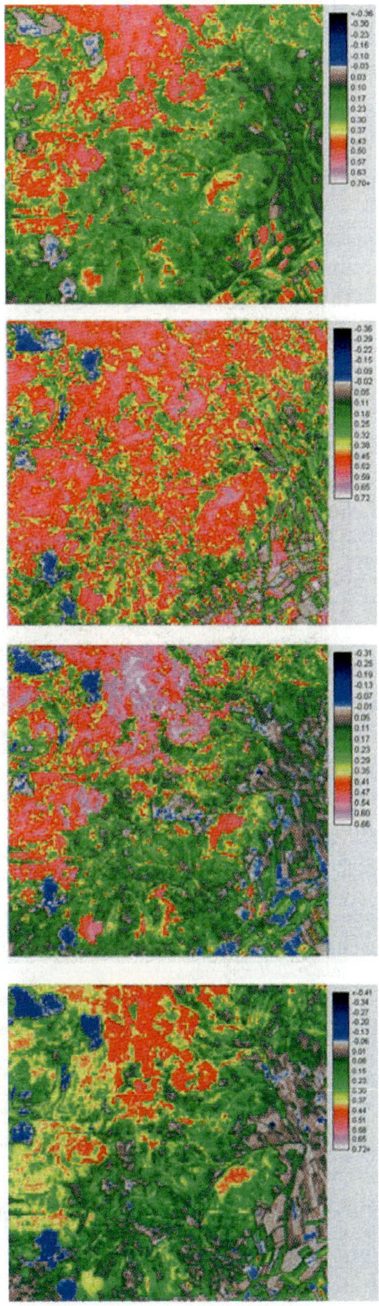

Fig. 6.15.12. NDVI of an area around Campiglia Marittima on 3 August 1984, 2 May 1985, 22 August 1985, and 5 August 1990 (from top to bottom)

vegetation index as the burned area. It is therefore of interest to look at differences between these land cover units. This is done in Fig. 6.15.14. where the radiances from different target measured at the top of the atmosphere are plotted. The green line stands for a forest which is compared with the average of two burned areas (red), the average of two harvested fields (brown) and a quarry (blue). It can be seen that the curves for the

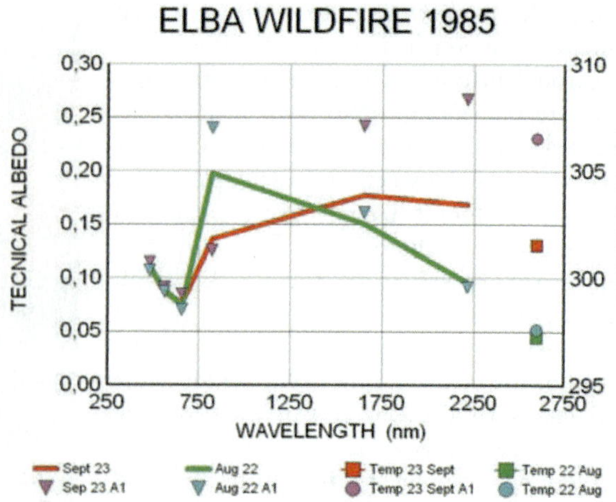

Fig. 6.15.13. Spectral technical albedo for the area in south-western Elba before (green line) and after (red line) the fire. Triangles give extreme values for the two cases. At the right side the inferred TOA temperatures are plotted for 22 August and 23 September for the same areas

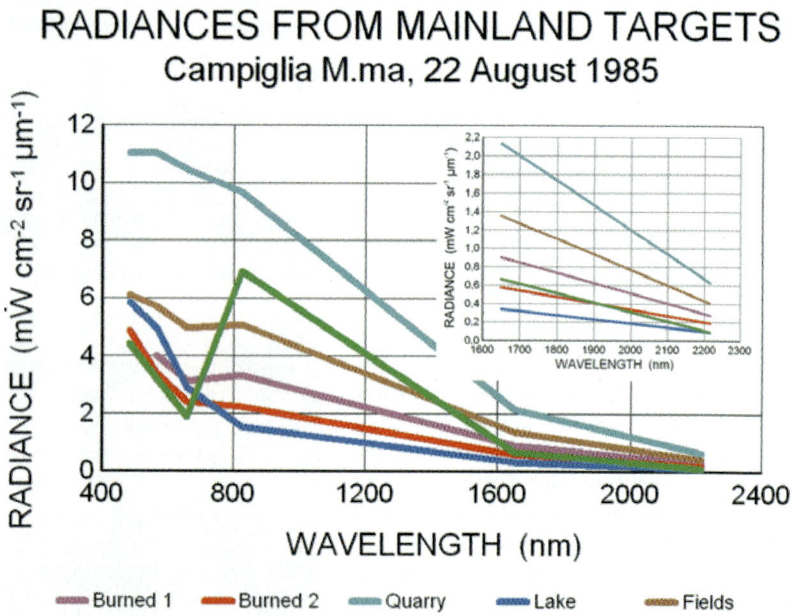

Fig. 6.15.14. Radiances from different targets near Campiglia M.ma on 22 August 1985 shortly after the fire. The insert shows the signals at the longer wavelengths with an enhanced scale. The green line is from a forested area

last three surfaces run more or less in parallel but the differences are large throughout the spectrum. The temperatures measured at the top of the atmosphere do not spread much. The burned area is 300.7 K warm, the harvested fields 298.6 - 299.9 K, the quarry 296.8 K, and the lake 297.5 K. Here not only the thermodynamic temperature plays a role but also the emissivity. The burned area is dark and probably has a high emissivity like the forest. The bright quarry with the shape of half a bowl openes towards the south may by at equal temperature with the burned area or even be warmer but the stones probably have a much smaller emissivity.

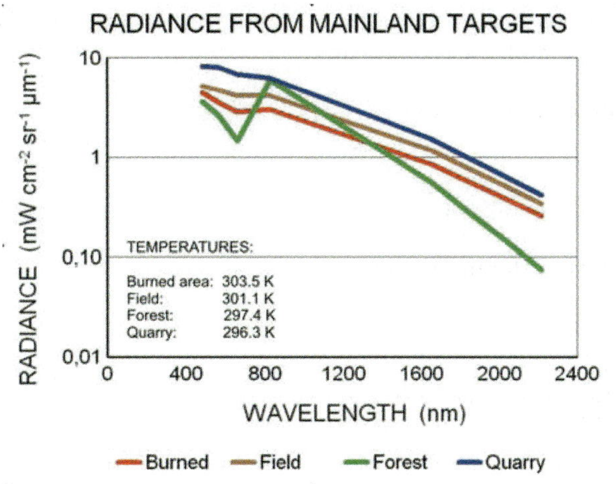

Fig. 6.15.15. Radiance from different targets near Campiglia M.ma. One month later than shown in Fig. 6.15.14

Fig. 6.15.15 shows similar targets one month later. The radiances of the reflected radiation now are lower than vin Fig. 6.15.14 because of the lower sun.

As can be seen, the differences between the radiances in channels 3 and 4 are small which results in NDVI values around zero. Thus it is not possible to distinguish between burned and harvested areas by the vegetation index alone. The spectral reflectance or technical albedo values and the temperature have in addition to be accounted for.

6.15.6 Discussion and Conclusions

Based on the analysis of the variation and typical deviation of the NDVI, it appears that in this area this index is a reasonable good estimate for the variation of the amount of vegetation for a long period. Spatial and temporal analysis indicate that the oscillation of the NDVI is low. In general, the values are higher in wet areas than in dry areas.

For the investigated type of vegetation, dense shrub-land, the NDVI shows reasonable uniformity with higher values in wet areas as compared to dry areas.

To identify burned areas it is often necessary to use the spectral reflectances and temperatures in addition to the NDVI.. Depending on the vegetation, the regeneration in burned areas follows an exponential curve which is stronger in wet zones and with high correlation coefficients. After five years, the curves become steady.

Temporarily, the relationship between NDVI and Rainfall is almost linear. This relation is worst in wet zones. Similar results have been found by Rasmusen (1998b) using AVHRR NOAA data.

6.16 Assimilation of Initial Soil Moisture Fields with Meteosat and NOAA Data

6.16.1 Introduction

Numerical Weather Prediction Models (NWP's) perform a short-term prediction of the weather on a global or regional scale. In operational applications these models are started typically a few times per day. A model run must be initialized using a reliable estimate of the actual state of the atmosphere, oceans and land surface.

From various simulation studies it was shown that the performance of NWP-models is partially dependent on the realism of the parameterization of sub-grid scale land-surface processes (for a review see Garratt 1993). Exchange of water vapour and sensible heat between land and atmosphere is to a large extent controlled by the availability of soil moisture, which has a strong feedback on vegetation development and evaporation. Soil moisture content is therefore among the most significant parameters for a reliable surface flux description (Shukla and Mintz 1982, Milly and Dunne 1994).

Most operational NWP's initialize the forecast cycle by means of a given soil moisture distribution over the forecast domain. This distribution is either obtained from a climatological database, or by treating the soil moisture content as a prognostic variable during the history of the forecast (for instance, the NWP of the European Centre for Medium-range Weather Forecasting, ECMWF; see Viterbo and Beljaars 1995). The climatological database is, in most cases, a monthly averaged global map of the actual soil moisture content relative to the field capacity. A well-known database is the one prepared by Mintz and Serafini (1989 and 1992). This database is derived from operational observations of precipitation and surface variables (radiation, temperature, specific humidity, wind speed). Surface evaporation is calculated using Thornwaite's (1944) formula.

Though simple to apply, a major disadvantage of a climatological soil moisture database is the inability to adapt to anomalous conditions. After exceptionally wet or dry spells, the actual soil moisture content will be relatively high (or low) as compared to the database used. This may result in significant over- (or under-) estimations of the surface evaporation, which has a strong feedback on the predictions of cloud cover, precipitation, and low-level temperature and humidity. For instance, Moene et al. (1995) showed that the predicted occurence of convective rainfall over the Iberian peninsula is closely related to the soil moisture initialization.

A second disadvantage of climatological soil moisture data is the unknown accuracy of the dataset. In particular for heterogeneous land surfaces the significance of an average soil moisture content in an area of the typical size of a NWP gridbox (say, 50 x 50 km) is doubtful. Extreme difficulties arise when representative soil moisture content is to be determined from limited field information.

On the other hand, models treating the soil moisture content as a prognostic parameter face a major difficulty in preventing drift of the predictions. At the timescale of several weeks to several months, complex feedback loops in the model may cause a systematic over- or underestimation of the actual land-surface evaporation. As an example, the ECMWF-NWP predicted a too strong soil moisture depletion in the summer of 1994 in the entire European area, which was associated with a systematic overestimation of low level temperature and radiation, and underestimation of precipitation and cloud cover (Viterbo 1994).

In order to prevent unrealistic drift, Bouttier et al. (1993a and 1993b) present a data-assimilation technique, which artificially 'updates' the model's soil moisture content by using observed model drift. This drift is expressed as differences between observed an predicted low-level temperature and atmospheric humidity. Observations are taken from operational synops stations. A series of statistical relationships are used, that express the sensitivity of (model errors of) low level temperature and humidity to the model's soil moisture content, following Mahfouf (1991). In particular, these statistical relationships depend on the land-surface parameterization scheme (in their case, the scheme of Noilhan and Planton 1989), which is a function of the description of the vegetation, the bare soil evaporation, and the ambient atmospheric conditions.

Also at ECMWF a similar 'nudging' is carried out, but here only the observed forecast error of the low level specific humidity is used (Viterbo 1994). The correction to soil moisture content is calculated as an empirical globally constant proportionality factor times the specific humidity bias.

A major disadvantage of the use of synop observations to assimilate soil moisture data is the inclusion of model errors of low-level quantities which are not related to soil moisture effects, for instance, result in mis-calculations of cloud cover. A second issue of concern is the limited global coverage of surface synop stations, in particular in remote areas. In some areas the reliability of the data-assimilation is severely hampered by the small number of surface observations. Moreover, this assimilation technique is a very new development, and is not yet well established in operational weather service environments.

Bastiaanssen et al. (1994) developed a methodology to estimate surface evaporation by using (cloudless) Meteosat and NOAA satellite images. The algorithm is semi-empirical, and makes an explicit use of the horizontal variability of surface temperature, surface albedo and vegetation index. The method is also suited for high resolution images (Bastiaanssen et al. 1996a). Since soil moisture availability strongly regulates surface evaporation, a strong coupling between the horizontal distribution of the evaporation as determined with the satellite algorithm, and the horizontal distribution of soil moisture may be expected. A case study carried out in La Mancha, Spain, during the 1991 EFEDA-campaign (Bolle et al. 1993) experimentally confirmed this hypothesis.

In this paper the possible applicability is explored of using evaporation maps derived from satellite data to assimilate soil moisture fields valuable for operational NWP purposes. For this, a simple 'poor mans' assimilation procedure is designed and tested for a single case, comprising the observed weather on the Iberian peninsula during a five day period in the summer of 1994. The study is mainly designed to detect whether the satellite data contains any useful information on the land-surface energy balance at all. A second aim of the study is to explore the theoretical and practical implications of the method, prior to operational implementation.

The simple soil moisture assimilation procedure is different from Mahfouf's (1991) and Viterbo's (1994) methods, in that soil moisture patterns are not attributed to regional differences between measured and predicted near surface quantities.

Rather, these differences are used as a validation tool to evaluate the performance of the satellite derived soil moisture fields. The Regional Atmospheric Climate MOdel (RACMO, Christensen and van Meijgaard 1992) is used for the present study.

The next section gives a brief outline of the satellite retrieval algorithm. Section 6.16.3 describes the simple assimilation procedure and the method to verify the results. A case study and its results are described in section 6.16.4, and in the final section 6.16.5 attention is payed to the uncertainties of the method, and the future developments necessary to optimize the assimilation procedure.

6.16.2 Assessment of the Land Surface Energy Balance Using Satellite Data

Bastiaanssen (1995, see also Bastiaanssen et al. 1998a and 1998b) gives a full description of the method to determine pixel by pixel evaporation rates using Meteosat and NOAA satelite imagery. The algorithm, labelled Surface Energy Balance Algorithm for Land (SEBAL) was briefly outlined in section 5.5.2.

SEBAL considers the surface energy balance equation introduced in section 1.3.5 in the notation

$$Q^* = G + H + \lambda E \qquad (6.16.1)$$

in which the symbols denote net radiation flux density (Q^*), soil heat flux density (G), sensible heat flux density (H), and latent heat flux density (λE). In the present case study, SEBAL is applied to solve the surface energy balance for a series of Meteosat pixels, by using Meteosat surface temperature and surface albedo, together with Normalized Difference Vegetation Index (NDVI) data obtained from the AVHRR radiometer onboard the NOAA platform.

Net radiation Q^* is solved by the following steps:

1. calculate the extra-terrestrial shortwave radiation as function of the pixelwise solar zenith angle at the moment of image acquisition,
2. calculate the downward shortwave radiation at the surface, K, as function of an assumed atmospheric transmittance,
3. calculate the net surface shortwave radiation by using the Meteosat albedo,
4. calculate the longwave downward radiation, $L\downarrow$, by an empirical function of atmospheric transmittance or, if available, actual values of $L\downarrow$ can also be given as input,
5. calculate the upward longwave radiation, $L\uparrow$, as function of Meteosat surface temperature and surface thermal emissivity dependent on NDVI,
6. parameterize the soil heat flux G is as a fraction of Q^*, where the fraction depends on:
   - the vegetation degree, which is parameterized using NOAA-NDVI
   - the surface soil moisture content, which is parameterized by using the Meteosat albedo information
   - the surface temperature, used to describe the phase lag between Q^* and G.

For the estimation of sensible and latent heat fluxes a series of steps is carried out. The first step is to assume that areas with very large evaporation rates will be recognizable as pixels with low surface albedo and surface temperature, associated with the presence of inland wetlands, storage reservoirs, shore lines, or dense vegetation stands. Areas with no evaporation will show the reverse signal, that is, high surface temperature and high albedo values. Threshold values of Meteosat albedo and surface temperature are used to define pixels with maximum evaporation equal to the available energy Q^* - G (albedo ρ < 0.1 and surface temperature T_s < 24 °C). Simultaneously, pixels with ρ > 0.23 are assumed to be associated with areas without any evaporation at all, and thus for which $H = Q^* - G$.

A scatter plot of actual Meteosat-measurements of albedo and surface temperature for a heterogeneous area is shown in Fig. 6.16.1. In this figure, data for the entire Iberian peninsula observed at noon on 1 July 1994 are displayed. Symbols shown denote average surface temperature per class of albedo values. Also plotted is a best fit regression polynomial, showing an optimum separating an ascending and a decending branch at $\rho = 0.22$. The ascending branch is controlled by evaporation, leading to higher albedo and surface temperature for smaller evaporation rates. The descending branch represents pixels where evaporation is assumed to be negligible, and is controlled by net radiation. It shows a further decrease of T_s as the albedo increases. Increasing the surface albedo of non-evaporating land surfaces leads to a reduction of net radiation, and thus of sensible heat flux. The slope of $\partial\rho/\partial T_s$ may be used to define a characteristic aerodynamic exchange resistance for these dry pixels, r_a^{dry}. Rewriting Eq. 6.16.1 and differentiating with respect to T_s gives:

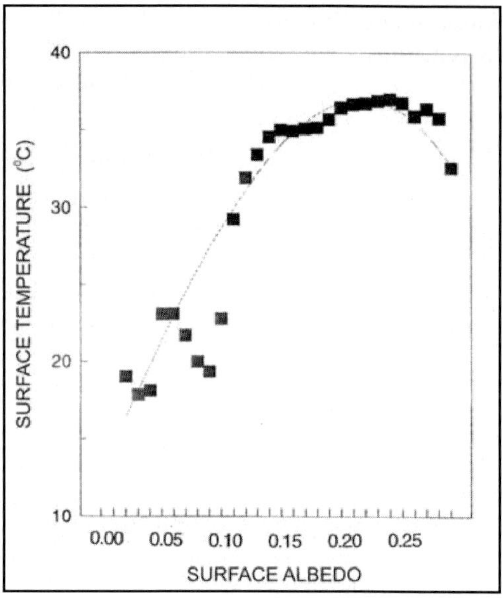

Fig. 6.16.1. Example of a scatterplot between satellite derived surface temperature and surface albedo. Shown are Meteosat data covering the entire Iberian peninsula on 1 July 1994

$$\frac{\partial H}{\partial T_s} = -K\frac{\partial \rho}{\partial T_s} - \frac{\partial L^{\uparrow}}{\partial T_s} - \frac{\partial G}{\partial T_s} = \rho_{air} c_p \left(\frac{1}{r_a^{dry}} + \frac{\partial}{\partial T_s} \frac{T_s}{r_a^{dry}} \right) \qquad (6.16.2)$$

where ρ_{air} is the density of dry air, and c_p the specific heat of dry air. The second equality applies to the definition of a reference height where the (potential) air temperature may be assumed identical for all dry pixel areas. Since $\partial\rho/\partial T_s$, $\partial L/\partial T_s$, and $\partial G/\partial T_s$ can be solved from the satellite measurement and the semi-empirical steps, r_a^{dry} can be found from Eq. 6.16.2.

The aerodynamic resistance r_a^{dry} thus obtained can be used to express a characteristic temperature gradient ΔT^{dry} for non-evaporating land surfaces in the image, defined by

$$Q^* - G = H = \rho c_p \frac{\Delta T^{dry}}{r_a^{dry}} \tag{6.16.3}$$

The next crucial step in SEBAL is to assume that for the entire image the relationship between the surface temperature, T_s, and the temperature gradient, ΔT, is quasi-linear (see Fig. 6.16.2). Then, two extreme regimes must be 'anchored' in the image: the dry area where $\lambda E = 0$, as applied to develop Eq. 2, and a wet area where it is assumed that $\lambda E >> H$, and so $\Delta T = 0$. Once these extremes are found from the regional scale satellite image, each pixel in the image can be associated with a value of ΔT corresponding to the observed value of T_s [1].

By use of the local terrain roughness on the basis of NDVI, together with a set of flux-profile relationships for both temperature and momentum transport, H is found for each individual pixel from ΔT. λE is the resulting term in the energy balance equation (Eq. 6.16.1).

Until now, SEBAL uses an empirical relationship to specify the roughness length for momentum, z_{0m}, as a function of NDVI obtained from NOAA images. The roughness length for heat, z_{0h}, is prescribed by adopting a fixed ratio of z_{0m}/z_{0h}. For the dry pixels, the wind speed at reference height, U_a, is not required, since r_a^{dry} follows directly from Eq. 6.16.2. On the other hand, it is implicitly assumed that U_a is similar for all pixels. A merit of this procedure is that z_{0h}, T_s and the air temperature are coupled correctly, and that realistic sensible heat fluxes are ensured.

The method SEBAL incorporates a series of crucial assumptions, which are all extensively discussed by Bastiaanssen (1995). In the context of using SEBAL for deriving surface evaporation maps for soil moisture initialization purposes, a few assumptions are important and worth repeating explicitly:

- The evaporation map is an instantaneous observation. Explicit use is made of the spatial variability within a single image, and no use is made of any temporal variability. The required scaling and anchoring of ΔT^{dry} is entirely based on one specific Meteosat image acquired at a single moment. This implies that the images must contain both pixels which represent areas with an evaporation at nearly

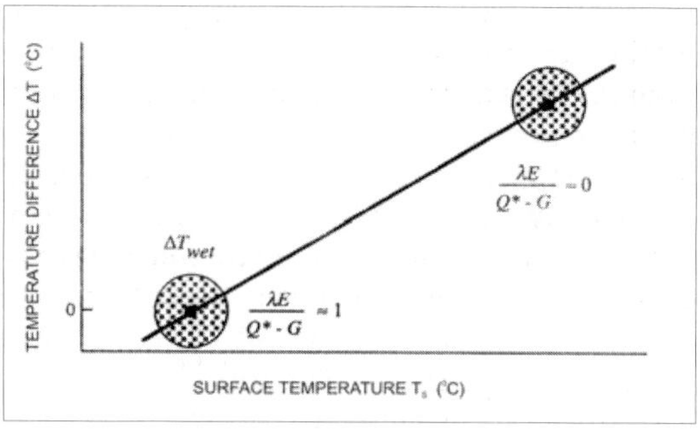

Fig. 6.16.2. Representation of the quasi-linear relationship between surface temperature T_s and the temperature gradient ΔT (from Bastiaanssen 1995)

[1]Note that by this procedure T_s is *used for interpolation purposes only*, and not used to calculate ΔT directly from a given air temperature. Hence *in situ air temperature measurements are not necessary*.

potential scale (H . 0) and pixels representing areas with no evaporation at all (λE . 0). Also, the radiation branch must be detectable to a minimum extent, implying that a range of dry areas with varying albedo must be captured in the image.

- The method only operates for cloud-free pixels in the image.
- There are several crucial surface parameters that must be explicitly provided: ratio of roughness lengths for momentum and heat, atmospheric shortwave transmittance, longwave downward radiation, and an estimate of surface temperature at a number of positions for the atmospheric correction of the Meteosat thermal infrared channel.

In the following section a simple assimilation procedure is addressed which was used to explore the applicability of SEBAL for deriving soil moisture initializations.

6.16.3 Outline of A "Poor Man's" Assimilation Procedure

Basic Assimilation Steps

Assuming that the evaporative fraction Λ, defined as $\lambda E / (H + \lambda E)$, derived by SEBAL is realistic, we can adjust Λ as calculated by the NWP-model by varying the initial soil moisture content, ω in the NWP. Λ is a convenient comparison parameter, since it is a surface wetness indicator which is normalized by the available energy, $Q^* - G$. It shows a fairly constant value during daytime, between approximately 10:00 and 14:00 UTC, and its temporal stability is better than λE or the resistance to evaporation. Insensitivity to temporal variability is necessary when SEBAL and the NWP are not synchronized. The purpose of the assimilation method is to define for each NWP grid point a value of ω which minimizes $| \Lambda_{NWP} - \Lambda_{SEBAL} |$, where the latter quantity is a simple averaged evaporation fraction of all valid Meteosat pixels located within the NWP gridbox (see below).

The NWP model used here, RACMO, is a limited area Numerical Weather Prediction Model (Christensen and van Meijgaard 1992). The physical package of RACMO is basically inspired on the climate model (ECHAM) of the Max Planck Institute for Meteorology, and as such provides a sophisticated tool for research purposes.

As a surface boundary condition over land, RACMO includes a single layer soil hydrology scheme (Dümenil and Todini 1992). Therefore, only one parameter (the moisture content in the soil moisture volume) needs to be adjusted to minimize $\Delta_{\mathrm{SEBAL}} - \Delta_{\mathrm{RACMO}}$. For the present study a crude estimate of ω was found by assuming a linear relationship between $\Delta\omega$ and $\Delta\Lambda$, where $\Delta\Lambda$ is the difference between Λ_{RACMO} (as calculated with a climatological soil moisture content, ω_c) and Λ_{SEBAL}. $\Delta\omega$ is a correction to ω_c required to minimize $\Delta\Lambda$.

For the present study three runs of RACMO were carried out: one run with a 'standard' initial soil moisture field (a climatology based on Mintz and Serafini (1992)), one 'wet' run with soil moisture content in the entire simulation domain initialized at 75% of saturation, and a 'dry' run, with soil moisture initialized at 35% of the saturation value. For each NWP grid box the optimal initial soil moisture was defined as (see Fig. 6.16.3):

$$\omega - \omega_c + \left(\Lambda_{SEBAL} - \Lambda_c\right)\frac{\omega_{sat} - \omega_c}{\Lambda_{sat} - \Lambda_c}, \quad as\ \Lambda_c > \Lambda_{SEBAL} \tag{6.16.3}$$

where the subscript *c* identifies the values obtained using the climatological soil moisture initialization, and *sat* and *dry* the saturated and dry soil moisture fields, respectively. This method in fact linearizes $\partial\Lambda/\partial\omega$ in two separate ranges: one for the area where $\omega > \omega_c$, and one for $\omega < \omega_c$. Any non-linear sensitivity $\partial\Lambda/\partial\omega$ within any of these two regimes is ignored.

Both SEBAL and NWP evaporative fractions were derived for 12:00 GMT, when surface fluxes are strong. The NWP fields were generated from a 12 hours forecast started at 0:00 GMT, and with soil moisture fields initialized as indicated above.

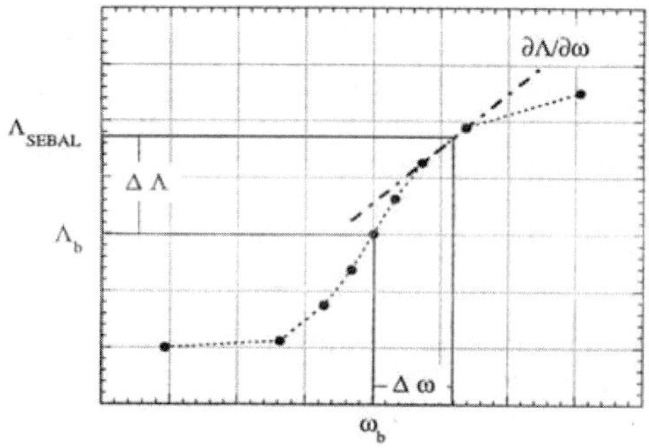

Fig. 6.16.3. Graphical representation of the "poor man's assimilation method. Shown are the linear relationships between ω and Λ as predicted by the Numerical Weather Prediction(NWP) model, and the simple regression method for obtaining an optimal soil moisture content, given Λ as derived from SEBAL

Horizontal Averaging

SEBAL calculates evaporative fractions on Meteosat pixel scale, which is at a much finer scale than the RACMO grid with a typical size of 50 x 50 km. For purpose of simplicity, the SEBAL fluxes calculated at all pixels within a single RACMO grid box were simply averaged. Then Λ_{SEBAL} is calculated from these average fluxes:

$$\overline{\Lambda_j} = \frac{\frac{1}{n}\sum_{i=1}^{n} \lambda E_i}{\frac{1}{n}\sum_{i=1}^{n}\left(\lambda E_i + H_i\right)} \tag{6.16.4}$$

where the summation is over all Meteosat pixels *i* within NWP gridbox *j*. Cloudy pixels were excluded from the averaging. Typically, for the case study included here approximately 70 SEBAL pixels were averaged to a single RACMO grid box value.

Verification of Results

In contrast to the methods described by Bouttier et al. (1993a and b) and Viterbo and Courtier (1995), low-level synops observations of temperature and specific humidity are not used as source of information to obtain a new initial soil moisture field, in spite of the relatively strong coupling between soil moisture and these quantities. Therefore, the routinely available data can be used as independent verification material to evaluate the performance of the assimilation procedure.

In the present study we carried out a series of 42-hour runs with RACMO, with soil moisture initialized as described above. In the current setup operational ECMWF analyses are used as boundary data. All runs are started at 0:00 UTC. The results of the simulation period between 24 hrs and 42 hrs are evaluated. In this time interval model spin-up effects are limited, and it is still short enough to avoid too large discrepancies from realtime large scale weather phenomena.

We express model results as an average bias and root-mean-square error of an ensemble of synop observations of temperature and specific humidity at 2m above surface in a selected area. The 2 m quantities in RACMO are calculated from model variables by an interpolation between the surface and the lowest model level following Monin-Obukhov theory (Geleyn 1988). The horizontal interpolation to synop station coordinates is carried out using a simple bi-linear interpolation (see next section for more details).

6.16.4 Selected Case Study and Results

Selected Case Study

For the present case study a five day period in 1994 (1 - 5 July) for the Iberian peninsula was selected, which was a relatively dry summer. The Iberian weather was characterized as cloud-free and warm (at 2 m height temperature of 35 °C were measured in central Spain), with weak winds from westerly directions. The selection of the case study was based on the following arguments:

- The SEBAL algorithm requires a series of extreme surface conditions, in terms of the surface evaporation. The method is unsuitable when extremely dry or wet pixels are not present in the captured images.
- The impact of soil moisture on atmospheric quantities is particularly present under conditions when the surface receives enough energy to evaporate or heat the air. Winter-time conditions are therefore less relevant for the issue of soil moisture initialization.
- An expected merit of assimilation of soil moisture using available data is the improvement of the regional distribution of soil moisture, and its impact on predicted atmospheric quantities. The case study is preferrably situated in an area with significant horizontal gradients. The Iberian peninsula shows a strong gradient in soil moisture climatology: wet in the North-west, rather dry in the central and South-eastern part of Spain (see Fig. 6.16.4).
- The case study area should be sufficiently large to allow for a statistically representative model performance evaluation.

- Since assimilated soil moisture is expected to improve the prediction of low-level quantities particularly in anomalous spells, the desired period should preferrably show an evaporation characteristic which is expected to be not too close to the (local) climatology.
- The area should be free of clouds during a significant part of the selected period.
- For reasons of practical origin, the case study could only be carried out for a limited area and period of time.

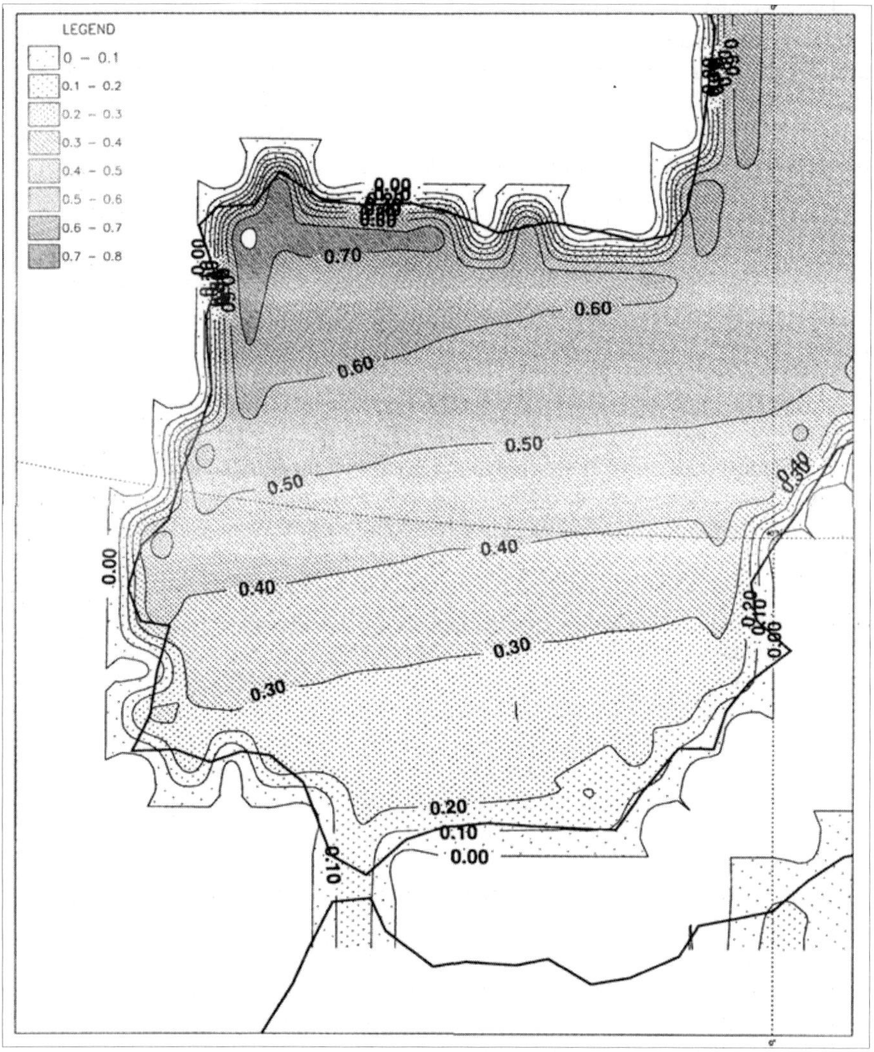

Fig. 6.16.4. Climatological soil moisture field for the Iberian peninsula for July, as used by RACMO operationally. Shown is the fraction of soil moisture saturation, defined as ω/ω_{sat} with ω_{sat} being the maximum soil moisture contend or field capacity (after Mintz and Serafini 1989)

During the selected period ground flux measurements, collected in the context of the EFEDA-II campaign (Bolle et al. 1993) were available. Ogink-Hendriks et al. (1995) showed field data of Λ collected in the vicinity of the city of Tomelloso, La Mancha. A brief validation of SEBAL-results of Λ using these data is shown below.

The data used to validate the runs with the new soil moisture field are taken from operational synop stations, extracted from the ECMWF MARS archive. In MARS, synop data are available at 6 hour time intervals. Stations at coordinates for which the fraction of the model grid box covered by land did not exceed 0.7 were excluded. Also stations for which the difference between the height of the station and the average height of the grid box interpolated to the station coordinate exceeded 300 m were excluded from the analysis (Navascues, 1996). These criteria are rather arbitrary, but exclude large biases owing to a poor representativity of the synops information. The filtering retained approximately 45 stations for the final intercomparison. The number of comparison data points varies per time slot and per parameter, since the data submission is not always identical.

Verification of Sebal-results with Ground Based Flux Measurements

In the context of the EFEDA-II campaign (Bolle et al., 1993), eddy correlation measurements of sensible and latent heat flux were conducted in an extensive vineyard near Tomelloso during June and July 1994 (see also Van den Hurk 1996). Three SEBAL Λ-maps were constructed around this period: for 1, 5 and 7 July 1994, around solar noon. Calculated values of Λ are shown in Fig. 6.16.5, together with SEBAL estimates at selected times. In spite of the difficulties associated with the different spatial scales of the methods, a reasonable agreement is shown. The weekly averaged value of Λ days 1

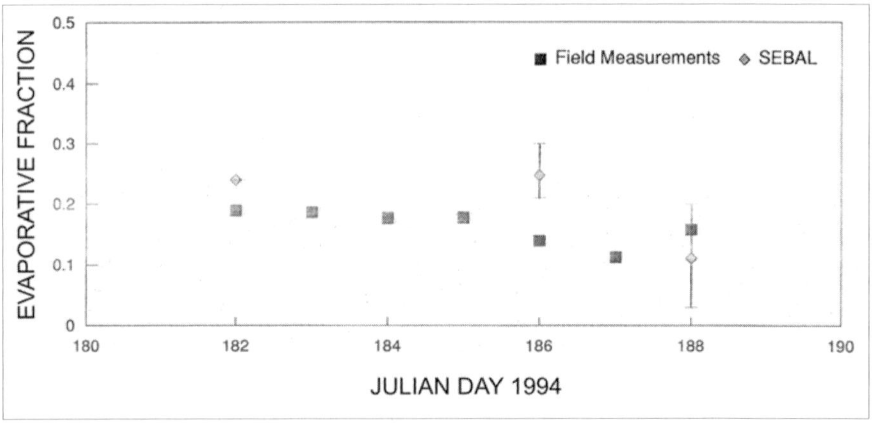

Fig. 6.16.5. Evaporative fraction Λ measured at a vineyard site in La Mancha during 1 - 7 July 1994. Shown are eddy correlation averages between 10:00 and 14:00 UTC (squares) and instantaneous SEBAL estimates at noon for three days (diamonds). Error bars represent standard deviation values for nine neighbouring Meteosat pixels, centred at the field measurement location

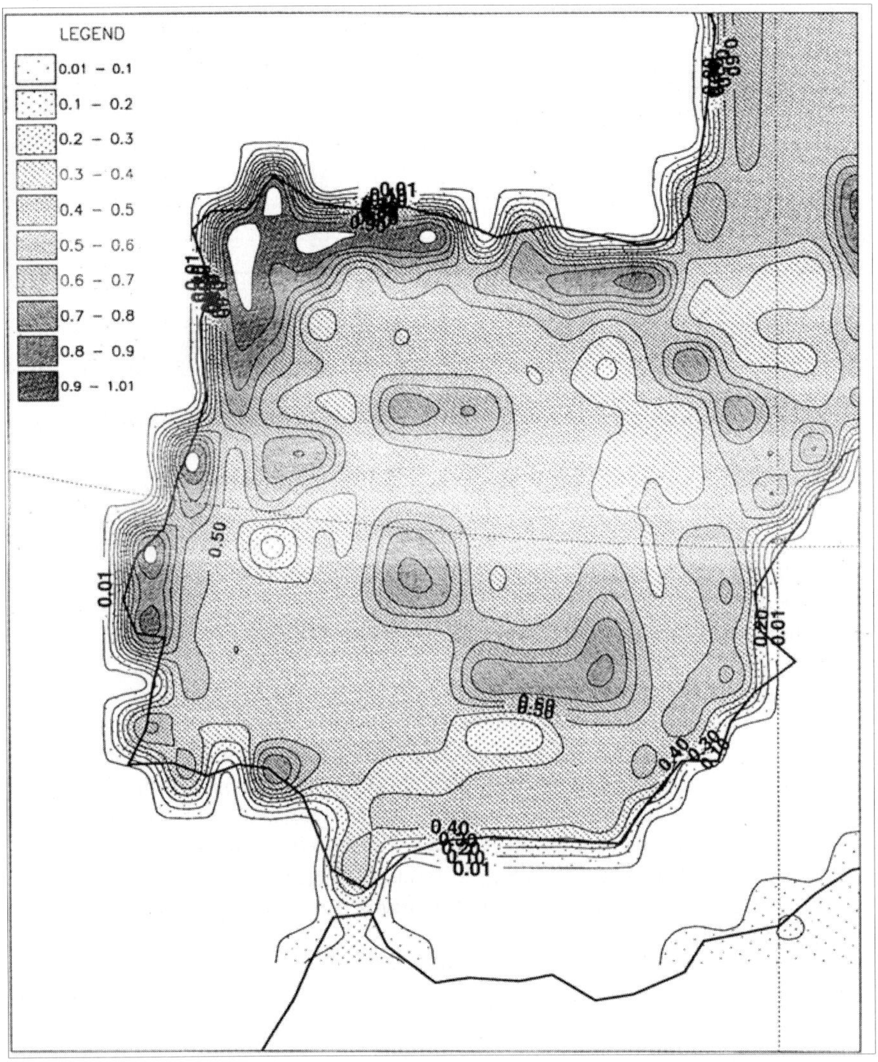

Fig. 6.16.6. New soil moisture field obtained using the simple assimilation procedure (see text). Shown are relative soil moisture contents, similar to Fig. 6.16.4

appeared to be 0.2 for SEBAL, and 0.16 for the field measurements.

Construction of A New Soil Moisture Field

Soil moisture corrections were calculated as indicated by Eq. 6.16.4 for only two days in July (1 and 5 July 1994). Grid values for which no soil moisture update was available (either because of excessive cloud cover in both SEBAL maps, or because of lying out

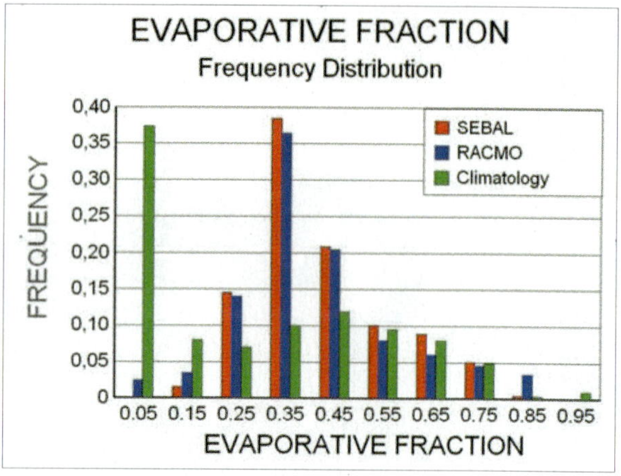

Fig. 6.16.7. Frequency distribution of the evaporative fraction valid for 1 July 1994, 12:00 UTC. Shown are distributions of SEBAL data (red bars), and the results from RACMO after 12 hours of simulation, starting from a climatological soil moisture field (blue bars) and from a soil moisture field derived from regression between multiple RACMO runs (green bars)

of the SEBAL domain) were kept at the July climatological value. Fig. 6.16.6 shows the new initial soil moisture map for the Iberian peninsula, to be compared with the climatological field (Fig. 6.16.4). Shown is the average of the two maps derived for the and 5 July, 12:00 GMT.

It is clearly seen from an intercomparison between Figs. 6.16.4 and 6.16.6 that the new soil moisture fields gives higher soil moisture values in the centre and most of the South of Spain. In spite of the fact that 1994 was a relatively dry summer for the European continent (Viterbo 1994), the SEBAL algorithm gave rise to suspect the local climatology to be too dry for that particular period. On the other hand, a clear reduction of soil moisture is enforced in the area East of Madrid, as well as near the Portuguese North-East border. A pronounced redistribution of soil moisture results from this simple assimilation method. The initial soil moisture increased by approximately 20% on average. Another clear aspect from Fig. 6.16.6 is the increased amount of details in the soil moisture fields. Rather localized features are visible, in contrast to the climatological field.

Fig. 6.16.7 shows the frequency distribution of the relative soil moisture content for all RACMO gridboxes on the Iberian peninsula (approximately 200), both for the climatological dataset and an average of the two new soil moisture fields. Also here a pronounced shift towards wetter soil reservoirs is shown. The number of grid boxes with a dry soil (15 - 25 %), as present in the climatology, is reduced. Also the algorithm causes a marked concentration of relative saturation values around 50 %, and a complete saturation of the moisture reservoir at some locations. This new soil moisture map was used as initialization for 5 new 42 hrs RACMO runs, starting at midnight on 1 - 5 July 1994. Since the two soil moisture fields derived for 1 and 5 July were nearly similar, a simple linear interpolation between these two time frames was applied to obtain an initial soil moisture field for each new RACMO run in this limited period.

Results of Simulations with the New Initial Soil Moisture Field

The shift towards wetter soil moisture values is a direct result of differences between Λ

values of SEBAL and RACMO. Fig. 6.16.8 shows the frequency distribution of the average evaporation fraction at 12:00 UTC, calculated from two SEBAL images (1 and 5 July) and 2 RACMO 12-hours forecasts verifying at the same time for the same grid points as for Fig. 6.16.7. The difference between SEBAL and RACMO using the climatological soil moisture initialization is very pronounced for low values of Λ. Where SEBAL does not generate Λ-values below 0.2, RACMO predicts much dryer locations in terms of evaporation fraction.

The difference is reduced on introducing the new soil moisture field, but still a clear difference from SEBAL is evident. Λ-values calculated by RACMO peak at slightly higher values than SEBAL, and also RACMO still generates some areas with very low evaporation fractions. However, the locations where low values of Λ occurred changed considerably by adopting the new soil moisture field. The original climatological soil moisture field resulted in $\Lambda < 0.1$ for the entire South of Spain, whereas the new runs of RACMO gave rise to these low values of Λ East and West of Madrid (compare Figs. 6.16.4 and 6.16.6). The difference between the SEBAL and RACMO Λ-frequency distributions in Fig. 6.16.8, together with the change of locations with extreme Λ-values indicate that the algorithm 'overreacts' on both the dry and wet side of the distribution. The sensitivity $\Delta\Lambda/\Delta\omega$, evaluated at the pursued value ΛSEBAL, is somewhat underestimated by assuming a linear dependence between $\Delta\Lambda$ and $\Delta\omega$ in a wide range of ω. Consequences of this will be discussed in more detail below.

Fig. 6.16.9 shows the bias and rms (corrected for bias) of the model performance in terms of 2 m temperature (upper panel) and 2 m specific humidity (lower panel). Both the climatological run and the run using the new soil moisture field are shown. The time shown is the verification time, achieved after 24-42 hours of model simulation.

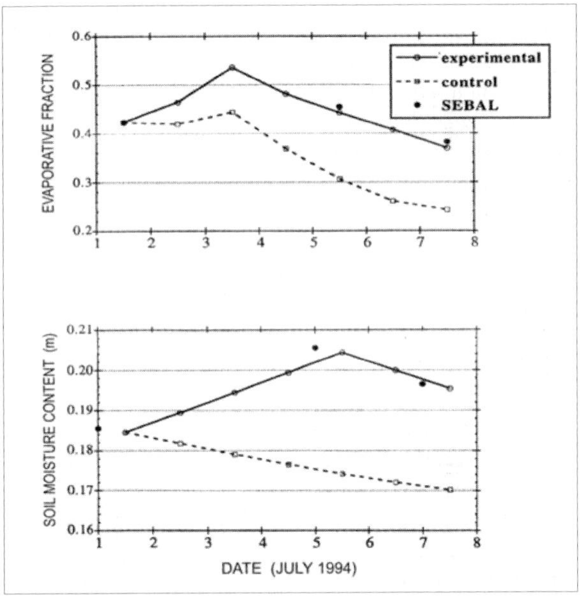

Fig. 6.16.8. Upper panel: Average evaporative fraction at noon for the Iberian gridpoints, calculated by SEBAL (symbols) and RACMO after 12 hours of simulation. Solid lines represent the experimental runs with daily adjustment of soil moisture, while dashed lines represent the control runs without soil moisture adjustment. Lower panel: Average soil moisture content for the same cases. SEBAL results are shown at initialization time, while RACMO results are shown for after 12 hours of simulation

Concentrating on the bias of the 2 m temperature, a clear diurnal variation is evident. For daytime conditions the new soil moisture initialization systematically decreases the (positive) bias by up to 0.5 °C. An increase of the (negative) bias by the same amount takes place during nighttime. The new soil moisture initialization cools the lowest layer of the atmosphere by 0.5 °C, owing to an increase of the average soil moisture content. During daytime the interaction between the surface and the lowest atmospheric levels is strong, since surface heat fluxes are generally large. During nighttime a much weaker coupling between soil moisture and low level atmospheric quantities exists, and the change of the model predictions of the 2 m temperature must be a result of the change of the entire boundary layer during daytime. Apart from the selection of validation stations, as outlined above, no attempt was carried out to correct for the difference between the station height and the average model grid box height. The strong vertical temperature gradients during nighttime will result in a considerable sensitivity of the model performance to the height difference.

From Fig. 6.16.9 it can also be seen that the root-mean-square (rms) error of the 2m temperature is slightly reduced by adopting the new soil moisture field. The reduction of rms implies that the change of the geographical distribution of soil moisture has a beneficial impact on the geographical

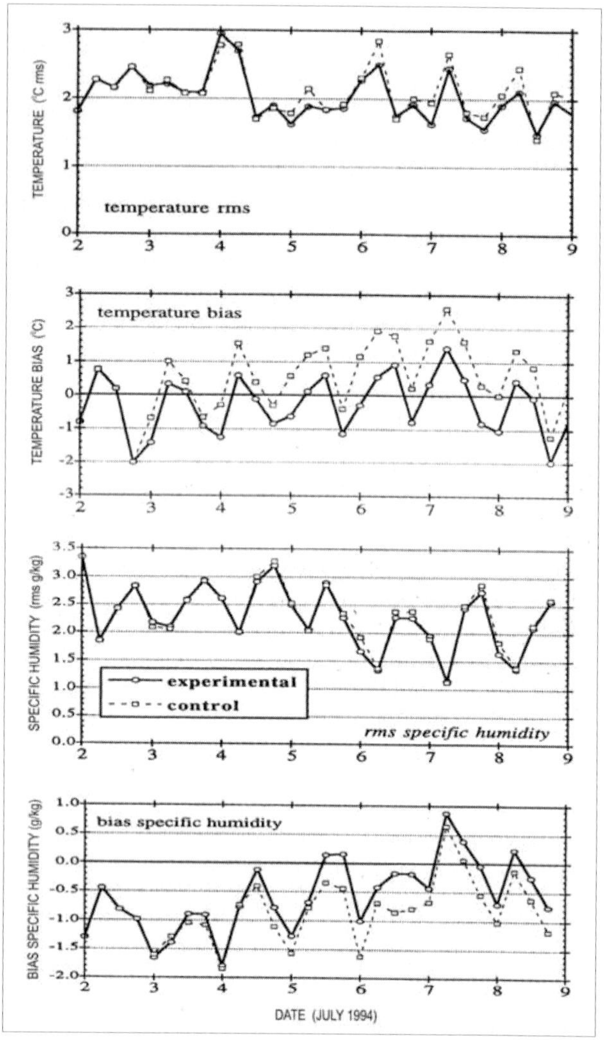

Fig. 6.16.9. Bias and root-mean-square error of 2 m temperature (upper panels) and specific humidity (lower panels). Shown are the model performances using the control run without soil moisture adjustment (dashed lines), and the experimental runs with daily adjustment (solid lines). The time indicated is the time of evaluation. Shown are simulations at t + 24 hrs (00:00 UTC) untilt + 42 hrs (18:00 UTC)

distribution of calculated surface fluxes.

Also the (negative) bias of the specific humidity is reduced systematically, by approximately 0.5 g/kg around noon, and up to 1 g/kg at 18:00 UTC. The night-time bias is less affected by the new soil moisture field, which is to be expected from the weak coupling between the surface and the atmospheric humidity when surface evaporation is low. For the last two simulation days the original bias is entirely removed, and for the other days considerably reduced. The rms-error of 2 m specific humidity is only slightly reduced (apart from the night of 5 - 6 July).

6.16.5 Conclusions and Guidelines for Further Development

The daytime reduction of the bias of both temperature and specific humidity is encouraging, regarding the rather high degree of simplicity of the method applied here. The bias reduction implies clearly that some significant information is present in the SEBAL evaporative fraction fields, that is compatible with information on forecast errors provided by synop observations. It is of particular importance to note that the sign of the SEBAL signal is compatible to what is found from the synop observations: an average underestimation of Λ is detected by both SEBAL (yielding higher Λ) and the synops data (confirming on an underestimation of Λ by showing that model gives too low specific humidity and too high temperature during daytime).

Also it is clear that the SEBAL signal appears to be strong enough to beat the climatology for the present case study, in which it was on beforehand known that the land surface was relatively dry. It detects an average underestimation of soil moisture of 20%. The climatological value of soil moisture for the selected area is approximately 45% of field capacity (0.35 m water), and in some areas as low as 20% of field capacity. A change of 20% is a relatively modest modification to the climatological soil moisture field, but has a clearly detectable impact on the model performance.

The first results from the feasibility study shown above are rather encouraging in the context of using Meteosat and NOAA data for assimilation of soil moisture fields. However, both significant improvements and some drawbacks are to be explored further. We will list the set of improvements that should be included, and pay attention to the drawbacks of the methodology.

First of all, the assumption of a linear relationship between $\Delta\omega$ and $\Delta\Lambda$ is both unrealistic and unpractical in the form as is applied here. The lack of realism is evident from the overreactive response of Λ to initialization with the new soil moisture field. Sensitivity analyses with RACMO reveal that the slope of $\partial\Lambda/\partial\omega$ as realized with the three runs (see Fig. 6.16.3) is underestimated in the range $0.35 < \omega/\omega_{max} < 0.75$ (sensitivity results not shown). Therefore, differences in Λ will result in too large corrections to ω in many cases, which lead to too large adjustments of calculated surface fluxes.

Also, it is too demanding in terms of computer resources to perform three complete (12 hours) simulation runs, only to generate an initial soil moisture field in an operational environment. It is more practical to incorporate the soil moisture initialization in an operational data assimilation system, where a short-term simulation and all available data are blended into an initial model state. With respect to the soil moisture assimilation, this implies that it is necessary to develop an explicit sensitivity function $\partial\omega/\partial\Lambda$ in which as much information as possible is to be inserted to produce an optimized soil moisture field. Apart from the Meteosat and NOAA information also available synops observations should be used, in particular 2m specific humidity. Bouttier et al. (1993a) suggest that inclusion of radiation data in their assimilation system would improve the performance significantly. Use of satellite information as demonstrated here allows for this inclusion.

A second major issue for improvement of the methodology is to improve the compatibility of the satellite retrieval system and the NWP environment. For the purpose of this feasibility study, SEBAL and RACMO were operated fully independent from each other. This implies that differences are present in the distribution of for instance surface albedo, roughness, soil heat flow and even vegetation cover. In a future study methodological incompatibilities must be minimized as much as possible by sharing datasets of this nature where possible.

Related to this issue, a possible route to improve the quality of SEBAL generated Λ-fields is to force the algorithm with analyzed wind fields at reference height, rather than estimating it from r_a^{dry}, as was explained at Eq. 6.16.2. Preliminary attempts have shown only a modest sensitivity of Λ to this adjustment, justifying the crude approximation carried out in this study. However, a more systematic investigation is to be carried out, in particular for weather situations with strong horizontal wind gradients.

Probably of more significance is to replace the parameterization of incoming longwave radiation (necessary to determine surface net radiation) by analyzed fields. It is not clear how the parameterization of $L\downarrow$ adopted here has affected the values of Λ calculated by SEBAL.

Third, the surface scheme in RACMO is provided with only a single soil layer. However, as is shown by Viterbo (1996), moisture in the deep soil layers has a relatively long variation time scale, as opposed to near surface soil moisture which can show considerable diurnal variations. It is expected that an improvement of the assimilation scheme can be obtained when the soil moisture increments are divided over surface and deeper layer by making use of the available vegetation cover degree, as specified in the NWP explicitly. Under conditions of large portions of bare soils, a mismatch in Λ between SEBAL and RACMO will lead to a correction of the moisture content in the top soil layer, i.e. bare soil evaporation dominates over canopy evaporation. For large canopy covers more weight will be paid to the correction of deeper soil moisture contents, controlled by canopy evaporation. This mechanism allows for a distinction between long-term drift (induced by errors in the deep reservoir) and short-term errors, which are more likely to be caused by errors of, for instance, predicted precipitation distributions.

This distinction between deep and surface soil moisture reservoir is important in the context of preventing NWP-models from drifting. The drift of the land-surface moisture content is in fact a result of an accumulation of model errors from the past into the soil reservoir, amplified by model feedback mechanisms. This type of drift has a relatively long time scale, of the order of up to 4 months (Betts et al., 1996). This time scale consideration has as practical implication that correction of the soil moisture content using available field information does not have to be carried out daily, but will probably be sufficient when employed a few times per month. This relaxes the fact that the use of SEBAL requires the occurence of cloud-free conditions in a given area. For the period in between two soil moisture corrections model drift must be confined as much as possible, which imposes the need for a realistic land-surface scheme. Also for this reason a multilayer soil model is to be preferred above a single layer semi-empirical scheme.

However, a disadvantage of the adoption of a multilayer soil model is the necessity to include the description of vertical soil moisture transport. For this, hydraulic transport parameters must be prescribed, which are of great importance for describing the hydrological connection between vertical layers. The net benefit of including multiple soil layers should therefore be addressed by means of a series of antecedent experiments.

A fourth issue that must be further explored is the averaging procedure that must be applied for matching SEBAL and RACMO coordinates. Probably averaging fluxes from pixel scale to NWP gridbox resolution is the best choice when one is interested in matching fluxes between the surface and the atmosphere. It is yet still unclear how the various non-linearities in the parameterization of fluxes as function of atmospheric and surface quantities must be incorporated in the averaging procedure.

Next, the case study reported above focussed on a semi-arid area on the Iberian peninsula. As commented before, SEBAL is expected to work relatively well in these areas, since the assumption of the presence of both very dry and very wet pixels is most likely met during most of summer season here. However, NWP models cover a much wider range of climate zones, and it is of importance that the assimilation scheme shows a proper behaviour in the entire simulation area. It must be verified that the assimilation procedure derived from the SEBAL algorithm can be applied in the whole of Europe.

Special attention must be paid to:

- the performance of SEBAL in other seasons than full summer. The strongest changes of soil moisture in Europe occur generally in Spring, as the growing season starts and evaporation becomes stronger. Model drifts often originate in this period (Viterbo, 1996), and a proper control of these drifts is of major practical importance,
- the role of the soil moisture parameterization in model drift in temperate climate regions,
- situations of long periods of cloud cover in certain areas, which might play a role in some of the North- and West-European areas in early and late summer,
- the treatment of NWP grid boxes which are only partially covered with land surface.

6.17 Methodology for Validation of Remote Sensing Data Products: The Valencia Anchor Station

6.17.1 Introduction

Calibration and validation is the activity that endeavours to ensure that remote sensing products are highly consistent and reproducible. This is an evolving scientific activity that is becoming increasingly important as more long-term studies on global change are undertaken, and new satellite missions are launched. In many geophysical fields, the region of the earth viewed by a satellite sensor rarely matches exactly that viewed by a surface radiometer. Recourse must be made to statistical comparisons in order to reduce sampling noise induced by the different space and time sampling characteristics of satellite and surface data. The special character of remote sensing measurements to correspond to area integrated values obliges independent *in situ* measurements to be representative of zones of a minimum number of pixels of the sensor under consideration. The large pixel size of some satellite missions such as GERB, CERES, SMOS, AMSR, introduces a number of scientific issues that make it relevant and necessary to develop a specific methodology and carry out specific measurements over large extended areas. These regions should be well controlled from the viewpoint of other complementary measurements, also at a large scale.

Calibration is the process of quantitatively defining the system responses to known, controlled signal inputs. *Validation* refers to the process of assessing, by independent means, the quality of the data products derived from the system outputs. These definitions are generally accepted and most often used in the remote sensing context to refer specifically and respectively to sensor radiometric calibration and geophysical parameter validation (see e.g., http://www.estec.esa.nl/vrwww/cal-val_glossary.html).

Agencies usually undertake *calibration* of their respective mission satellite systems during the sensor commissioning phase; however to extend this beyond that period is potentially very difficult. Therefore, well-instrumented benchmark test sites and data sets for calibration and validation should be supported, in particular for terrestrial applications, to provide information to supplement or substitute on-board calibration. One prerequisite is to ensure continuity, reliability and access to the data with minimal delay. The ideal requirements for these test sites include that they are large, flat homogenous areas, with nearly Lambertian reflectance including bright and low reflectance targets, and with high percentage of clear skies and dry conditions. Beside having a year-round availability, the test site should somehow have a "controlled" human disturbance as well. A history of long data records is also desirable.

The primary objective of *validation* is to test evaluation algorithms, to assess the quality, and as far as possible to quantify the accuracy of remote sensing data products. Ideally, validation activities seek to compare data products with more accurate independent measurements of the same quantity over a statistically significant number of samples and a wide variety of situations. The problem is that the space and time scales of *in-situ* and satellite data are rarely directly comparable. Typically, there are insufficient *in-situ* measurements to cover a satellite field of view, whether the field of view is some metres (as in the case of high resolution imagers), or several km (broadband radiometers, L-band radiometers, etc) wide. Thus, even a perfect remote sensing measurement will be expected to differ from an *in-situ* verification measurement because of the inability to match the observations in time or space. *In-situ* measurements are invariably taken at small space and time scales, while a satellite overpass is a few second snapshot of a large area. For broadband radiometers, a common example is the comparison of a surface flux radiometer to a satellite based estimate for the radiometer field of view. The matching error can be reduced (but not eliminated) by increasing the number of surface observations within the field of view (level 2 data) or grid box (level 3 data). This error can also be minimized by using very large ensembles of matched data.

Often satellite data are used in conjunction with models which describe certain processes. Satellite data are in these cases usually used to control one of the variables produced by the model to ensure that the model works correctly or they are needed to initiate the model run. As an example, mesoscale models can in principle be operated with little instantaneous input from ground observations if topography and synoptic scale flow are known. Whether they produce the correct interactions with the surface (momentum, energy and mass exchange) must be checked at least occasionally by direct measurements. For hydrological models the same is true. In their case it has to be validated whether they produce the correct soil moisture and run-off. Satellite data are evaluated with complicated algorithms that link the measured radiances to surface properties. These algorithms have to be tested against direct measurements. Climate models not only need carbon dioxide and other trace gas fluxes in addition to meteorological data, but they also need regularly updated aerosol information. These can to some degree be inferred from satellite data but need strong control by ground based measurements. The state of vegetation and the surface albedo can to some degree be estimated from satellite observations. The conclusions from such kind of measurements may, however, be misleading if no ground truth is available.

Field campaigns are best defined for process studies and hypothesis formation. They provide the most complete case study data set, but with very limited statistical significance. Dedicated sites, *Baseline Surface Radiation Network* (BSRN) surface sites, and other long-term sites are considered the best strategy for validation of cloud and radiation data. To apply this successful strategy to observations of the Earth surface, the concept of "Anchor Stations" (ASs) was developed. The term "anchor" means that the information derived from measurements made in space are firmly tied to the conditions measured at the surface.

6.17.2 Definition of Anchor Stations

Anchor Stations are instrumented sites to measure in representative areas quantities which are needed to run, calibrate, or validate models or algorithms e.g. such as those needed to evaluate quantitatively satellite data and to convert them into geophysical information. The stations must be instrumented with equipment for long time measurements of basic quantities (meteorological and hydrological background data) and in addition for quantities not readily assessed at operational meteorological and hydrological stations but needed for the interpretation of the signals recorded in space (see Box 6.17.1). Anchor Stations furthermore must offer the infrastructure to undertake

Box 6.17.1. Measurements desirable at Anchor Stations after Becker et al. (1988)

A. Continuously
- Atmospheric transmittance (spectral)
- Net radiation flux; preferable all four radiation fluxes plus in addition independently the net flux
- Bowen ratio (including the measurements of air temperature and humidity at two levels)
- Soil heat fluxes and temperatures at least at four levels
- Precipitation
- Soil moisture at three levels: near surface, intermediate, root layer
- Effective radiation temperature of the surface
- Ground water (piezometric) level

B. Occasionally (especially under extreme situations)
- Sensible and latent heat fluxes separately by advanced methods (e.g. eddy correlation, scintillometers, etc)
- Vegetation parameters such as:
  - leaf area index
  - fractional coverage
  - biomass area density
  - stem flow
  - stomata resistance
  - vegetation height
  - vegetation structure, leaf orientation, leaf size
- Land use at and around the station

C. From time to time at longer intervals
- Soil quality
- Radiometric characteristics (reflexion spectra, reflectance function)

D. At the central data bank the following derived quantities have to be computed from the measured values:
- Wavelength dependent optical depth and vertical distribution of water vapour mass as needed for the correction of satellite data. The goal must be to derive a climatology of aerosol optical properties for the region. If reliable algorithms become available to compute atmospheric corrections from measurements made in space, this network serves the validation of these algorithms and the quality control of atmospheric corrections. This task can be supported by measurements made in the areas of the anchor stations by aircraft in order to obtain a better area coverage
- Daily and monthly values of all flux components (radiation, sensible heat, latent heat, soil heat flux respectively heat storage in the soil)
- Soil moisture as the difference between precipitation, latent heat flux, and storage
- Ratio of directional reflectance to albedo (according to models parameterized with respect to the measured quantities that are still to be developed)
- Surface conductance

E. Provision has to be made for infrastructure, station maintenance, accomodation for guest scientists, calibration facilities (ewtl. exchangeable among anchor stations) more refined measurements for shorter intensive observation periods in the surrounding of the station. The quantities that in addition to background quantities (temperature, moisture, pressure) should be measured are fluxes (surface albedo, the four components of net radiation, soil heat flux, sensible and latent heat flux, carbon fluxes, water fluxes in the soil), Bowen ratio, spectral atmospheric optical depth, atmospheric water vapour, aerosol parameters, ground water level, soil moisture, vegetation parameters (type, height or z_0, resistances, biomass, NPP, LAI). Not all *Anchor Stations* need to measure all parameters permanently (Becker et al., 1988).

Because of the need for correcting the satellite data for the effect of aerosols and the need for aerosol data in some applications, Anchor Stations should also be equipped with instrumentation for spectral optical depth measurements or other kind of equipment that provides information about aerosol concentrations and types.

Anchor Stations must fulfil certain requirements with respect to geographical position, management, measuring capacity, maintenance expertise, and infrastructure. They must build up a scientific network that includes regular inter-calibration of the applied instruments, a common data bank, and exchange of personal to spread experience.

With respect to its geographical position, topography, soils and vegetation, an Anchor Station must be representative for the environment under consideration and adequate surroundings must be available for limited field experiments as they may become necessary to study processes. This could be, as an example, the investigation of the effect of heavy rains on soil, vegetation, and partitioning between infiltration and run off, or the introduction of new measuring techniques (the most elaborate task would be to install vertical profilers at these sites). An adequate homogeneous terrain must be accessible for sampling collateral data needed for re-calibration of satellite sensors respectively to check the quality of algorithms used to infer from the measurements made in space information about the surface. Ideally this should be an area of 10 km^2, but the re-calibration of high resolution satellite data from smaller areas, such as of the order of 0.1 - 0.3 km^2, is also possible and upscaling procedures may be used for the medium resolution satellite pixel.

The hydro-geology and soil composition underneath the site should be known (see Appendix A.1 and A.3). A radiosounding station should not be too far from the station which should have access to the data.

There must be adequate infrastructure available to maintain the station and to host guest scientists. One of the sites of a network should keep a mobile set of instruments and the necessary personal to perform instrument inter-comparisons and inter-calibrations at the other sites.

During the EFEDA and following-on field experiments several stations have been operated which partly met the requirements for Anchor Stations insofar as the standard meteorological and hydro-meteorological variables and shortwave radiation fluxes were continuously measured. In addition, in the surrounding of the stations, many of the additionally required variables were measured during the intensive field measurement periods. These stations were situated in Tomelloso, Barrax, El Bonillo (Spain), and Radicondoli (Italy). With the advent of new satellite measuring systems the necessity for ground truth increased and two new stations have been build in Spain. The equipment and activities at one of these stations is described in the following sections.

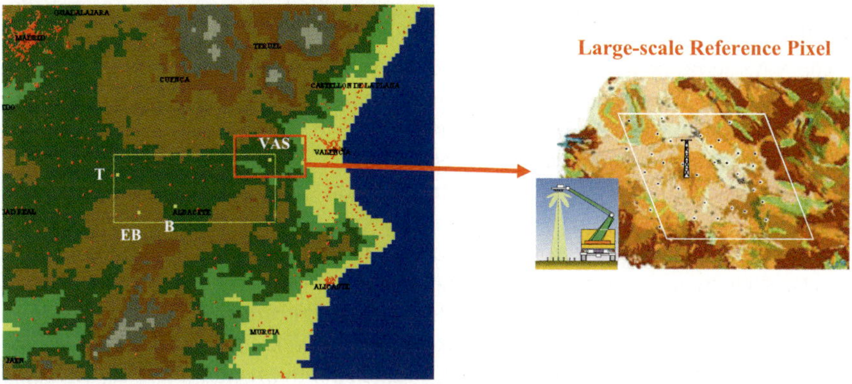

Fig. 6.17.1. Position of the Anchor Stations in Central Spain and the Valencia Region. VAS: Valencia Anchor Station, T: Tomelloso, B: Barrax, and EB: El Bonillo.

6.17.3 Spanish Anchor Stations

The University of Valencia (Spain) has recently set up an automatic meteorological station located towards the West of the Valencia region, in the natural region of Utiel-Requena Plateau, about 80 km from the city of Valencia. The main objective of this "Valencia Anchor Station (VAS)" is to define and characterise a large, reasonably homogeneous and flat area, mainly dedicated to vineyards, as reference for calibration and validation (Cal/Val) activities in low-resolution large-scale pixel size satellite sensors, as those corresponding to the missions mentioned above. The area is well documented and has previously been used in other projects. It is desirable that the Valencia Anchor Station area and the three stations in the Central Spain area (Tomelloso and near Barrax) operated by the University of Castilla-La Mancha, could define a still larger and reasonably homogeneous region of about 300 x 200 km^2, in order to be able to count on a minimum number of large size pixels of the order of 50 x 50 km^2 (Fig. 6.17.1). Recently another station was implemented in the province of Alicante (see section 6.17.7) which will extend the test area to the south.

The tasks currently carried out at the zone are related to the Earth Observation missions mentioned above. Their specific scientific objectives are described in the following section (see Lopez-Baeza et al., 2001, 2002 and 2003).

6.17.4 Scientific Objectives of the Valencia Anchor Station

Definition of A Large Scale Validation Area for Low Spatial Resolution Missions

Requirements for a test site are: information about the area, basic documentation, availability of retrospective measurements, maintenance and attention of the site. These conditions are accomplished for the University of Valencia site, that is, the Valencia Anchor Station (VAS), and for the University of Castilla-La Mancha sites, namely the Tomelloso, El Bonillo and Barrax Anchor Stations.

The VAS territory was scientifically documented from the viewpoint of soil as a natural resource. Physiographic maps together with maps of soils, land capability, potential erosion and actual erosion and provided by the former *Regional Ministry for Public Works, Land Planning and Transports* of the Regional Government of Valencia (Antolin, 1998) and are available in the environment of a Geographical Information System.

The Castilla-La Mancha region has thoroughly been studied in previous international projects, such as EFEDA and its follow-on projects (see Appendix 1) as well as RISMOP (*Radiometric Impact of Surface Moisture on Precipitation*) project. On top of that, the area has currently been used as the *vicarious validation area* for Meteosat. The University of Castilla-La Mancha holds additional information such as on soils, vegetation and land use classification.

The overall proposed area (Fig. 6.17.1) may be defined from the available information, taking into account the characteristics of soils, climate, physiography, etc. It does then result a large extended area, more than 300 km wide, reasonably homogeneous taking into account the low spatial resolution of the remote sensing instruments under consideration. The first principal task is to homogenize the large quantity of information available and complete intermediate zones by means of interpolation methods based on spatial statistical techniques (kriging and co-kriging) and by using remote sensing techniques. It may be stated that there hardly exists in Europe an area of similar characteristics, especially as large and homogeneous, and located in a climatic area of so much scientific interest from many different points of view.

Definition and Characterisation of A Large Scale Reference Pixel

Complementarily to the previous objective, specific tasks are currently carried out to facilitate Cal/Val activities of the large pixel size missions mentioned earlier. This is a objective that both the *GERB International Science Team (GIST)* and the *SMOS Science Team* have positively valued by accepting the VAS site as one of the basic areas to carry out these kind of activities. This objective includes several tasks starting from the classification of different homogeneous environmental units in the large pixel area and keep it updated in time. The area is about 50 km wide around the actual AS, approximately of the size of a GERB or a SMOS pixel where a number of distributed measurements of soil moisture content are being developed, together with other measurements of soil temperature, surface temperature, reflectance, albedo, net radiation and other conventional meteorological parameters.

Heterogeneity within each unit may be characterised by defining transects and using mobile instrumentation. Within the *Remote Sensing Research Unit* of the University of Valencia both the "Climatology from Satellites" - Group and "Field Radiometry" - Group are engaged in this work. Collaboration with the *Microwave Radiometry Group* from the Polytechnic University of Catalonia (Spain) has originated in a number of periodic field campaigns to characterise the area also at L-band frequencies (Vall-llosera

Fig. 6.17.2. LANDSAT-5 TM RGB image (5 July 2003) of the Utiel-Requena Plateau region (Valencia, Spain)

et al., 2004).

This area is currently characterised in a progressive way, starting with area of about 5 km x 5 km around the station and then scaling up to 10 km, 20 km, and so on until covering the whole zone. This work is carried out under different meteorological and climatological conditions during the year and account for spatial variability such as topographic effects.

Scaling Issues: Aggregation and Disaggregation, Time Interpolation and Spatial Averaging

To compare measurements resulting from the various sources at different scales, it is necessary to study in detail the effect of scale changes on the description of the involved processes. To obtain area averages at different scales, criteria have to be establish for aggregation and disaggregation of quantities measured and to validate large scale pixels. Data from point measurements, aircraft observations, and from different satellite platforms acquired at different spatial resolution have to be merged for this purpose.

Similarly, time interpolation to handle the time dependent diurnal cycles or getting monthly averages of different quantities may need the application of diurnal cycle models (see section 4.4.4), except for geostationary satellites. This is necessary for quantities such as surface temperature, solar zenith angle albedo dependence, and cloud properties to name only a few.

6.17.5 Specifications of the Valencia Anchor Station

The VAS is located at 39º34'15''N, 1º17'18''W, 813 m altitude (www.uv.es/anchors), at the Utiel-Requena Plateau, a natural region in the province of Valencia (Spain) (Fig. 6.17.2). This is an extensive undulated plain formed by quaternary sediments surrounded by mountainous regions at the northern and eastern sides and the *Gabriel River Basin* at the western and southern sides. The land shows erosion and depositional foot slopes. The topography is characterised by flat regions (slope <2%) and undulated areas (8-15%).

The major soils of the region - *Calcareous Cambisol* and *Haplic Calcisol* (see Appendix A.3) - are deep soils with carbonate accumulations and equilibrated textures, and amounts of low organic matter.

In spite of its relatively flat topography, the small altitude variations of the region clearly influence climate. This oscillates between semiarid in the areas of the towns of Utiel and Caudete de las Fuentes and dry to sub-humid towards Villagordo del Cabriel. The altitudinal differentiation between both climate types corresponds to levels of 800 – 850 m. Annual mean temperatures oscillate between 12 ºC of Villagordo del Cabriel and 14.2 ºC of Caudete de las Fuentes. Annual precipitation varies between 396 mm in Utiel and 451 mm of Caudete de las Fuentes and Villagordo del Cabriel. The duration of

Fig. 6.17.3a. General view of the Valencia Anchor Station

frost free periods is similar for the three town areas, from May to November. The maximum of the precipitation occurs in spring and autumn. The spring maximum is generally in May, whereas the autumn maximum is variable, in October for the areas of Caudete de las Fuentes and Utiel, and November for that of Villagordo del Cabriel.

The dominant land use in the plateau is dedicated to vineyards. However, the area surrounding the AS contains representative ecosystems of the Mediterranean areas, typical of semi-arid regions, including *Aleppo* pine (*Halepensis*) forests and Mediterranean shrubs.

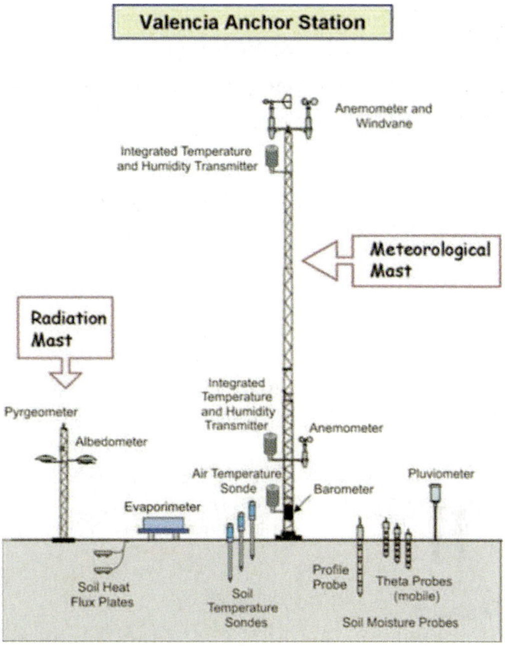

Fig. 6.17.3b. Scheme of the instrumentation of the Valencia Anchor Station

In principle, the VAS as such is a robust meteorological station where measurements are made at different levels both in the atmosphere and in the soil in order to be able to derive surface energy fluxes. Table 6.17.1 shows the parameters that are operationally measured at different levels together with the characteristics of those

Fig. 6.17.4. Surroundings of the Valencia Anchor Station (February 2004)

Table 6.17.1. Instrumentation of the Valencia Anchor Station

Quantity	Instrument	Resolution	Range	Model	Company
Wind Speed at 2 and 15 m	Anemometer	0.1 m^{-1}	0 - 50 m s^{-1}	VV-200	Geónica
Wind Direction at 15 m	Wind Vane	1º	0 – 360º	DV-200	Geónica
Air Temperature at 0.5 m	Temperature Sonde	0.1ºC	-50 to +50 ºC	STA-212-PVC	Geónica
Air Temperature and Humidity at 2 and 15 m	Integrated T and H Transmitter	t: 0.1ºC H: 5 %	-50 - +50 ºC 0 - 100 %	44212/50-U	Vaisala
Atmospheric Pressure	Barometer	1 mb	600 – 1100 mb	SPA-900	Druck Limited
Downwelling, Reflected and Net SW Radiation	Albedometer	1 W m^{-2}	0 – 1500 Wm^{-2}	CM 14	Kipp & Zonen
Downwelling and Net LW Radiation	Pyrgeometer	1 W m^{-2}	0 – 1500 Wm^{-2}	CG 1/2	Kipp & Zonen
Soil Temperature	Temperature Sonde	0.1 ºC	-50 to +50 ºC	STS-212-PVC	Geónica
Soil Heat Flux	Soil Heat Plate	1 W m^{-2}	0 – 1500 Wm^{-2}	RIMCO HP3/CN3	RIMCO Middleton Synchotac
Profile Soil Moisture Content	Delta-T Profile Probe (40 cm)	±0.5 m^3m^{-3}	0-10 m^3 m^{-3}	PR 1/4 – 1/6	Delta-T
Evaporation	Evaporimeter	0.76 mm	0 – 254 mm	255-100	Nova Lynx Corp.
Precipitation	Pluviometer	0.1 mm	0 – 26 cm	52202/52203	Young

Table 6.17.2. Mobile instrumentation of the Valencia Anchor Station (collaborating groups providing equipment are shown in *italics*)

Quantity	Instrument
Net Radiation	2 Middleton pyrradiometers
Global Irradiance	2 Kipp and Zonen pyranometers
Surface Albedo	2 Kipp and Zonen albedometers
Soil Heat Flux	4 Middleton heat flux plates
Soil Temperature Profile	Campbell Pt-100 Temperature probes
Surface Radiometric Temperature	Raynger MX4 Howholslt
Atmospheric Transmissivity	EKO Sun photometer MS-110
Soil Dielectric Constant	Tektronic TDR 1502C
Soil Moisture Content	25 Theta Probe type ML2X
Surface Spectral Reflectance (*Field Radiometry Group*)	GER-SIRIS and GER-3700 Spectroradiometers
Band Reflectance (*Field Radiometry Group*)	EXOTECH Radiometer
LAI (*Field Radiometry Group*)	LICOR-2000 Plant Canopy Analyser
Chlorophyll (*Field Radiometry Group*)	Minolta SPAD-502
Atmospheric Profile of Dry and Wet Temperatures, Wind Speed and bearing, and Pressure (*Spanish Institute of Meteorology*)	Radiosounding Ascents
Integrated Water Vapour Content (*Institute for Space Studies of Catalonia, IEEC*)	GPS Equipment
3-D Atmospheric Fields	MM5 (*IEEC*), HIRLAM (*INM*)
Atmospheric Chemistry (*Environmental Quality Department. Regional Government*)	Mobile Fully Equipped Air Quality Station
L-Band Radiometric Temperatures (*Polytechnic University of Catalonia*)	LAURA L-Band Radiometer

measurements (instrument, resolution, range, model and maker). Brand names are given just for information and do not mean any special endorsement from the authors. The station is composed of two masts (Fig. 6.17.3). One holds conventional meteorological instruments at the atmospheric levels of 2 and 15 m, and –10, -20 and –40 cm in the soil. Air temperature is also measured at 0.5 m. The second mast is dedicated to the measurement of the four radiation components at 2 m above the ground.

Table 6.17.2 shows the mobile instruments that can be used to measure specific parameters (surface reflectance and brightness temperature, vegetation parameters, soil characteristics, soil moisture content, soil dielectric characteristics, etc) and some meteorological quantities in the whole Anchor Station area by means of transects performed with vehicles (Fig. 6.17.5). Some of these instruments are currently used in collaboration with other research groups. There are two mobile meteorological stations that measure air temperature and humidity, global irradiance, surface albedo and net radiation at 2 m above the ground and soil temperature profile and soil heat flux at different levels within the soil (Fig 6.17.6). Atmospheric transmissivity is also measured during intensive observation periods. There are other instruments planned to be acquired

Fig. 6.17.5. Truck with a 15 m broom equipped with instruments measuring surface reflectance and microwave L-band brightness temperature for soil moisture determination

Fig. 6.17.6. Mobile equipment of the Valencia Anchor Station. Mobile radiation and meteorological stations respectively over vineyards and matorral areas

in the short future to improve atmospheric and aerosol and cloud measurements, as well as a singular and unique instrument to determine atmospheric radiation divergence.

6.17.6 Simulation of Top of the Atmosphere Ceres Radiances

The methodology that is currently been developed to obtain ground independent values of some geophysical parameters at GERB scale basically uses data from the VAS together with measurements from GPS to calculate precipitable water content, from a sun-photometer to calculate atmospheric transmissivity and aerosol optical depth, and from radiosounding ascents, specifically obtained over the Station and simultaneously to the satellite overpassing times. The two mobile meteorological stations help to account for non homogeneities by measuring some parameters in other regions of different characteristics to those of the Anchor Station. 3-D high resolution simulations (1 x 1 km^2) from the MM5 meteorological model are used as an interpolation tool to obtain atmospheric fields over the whole region. So far, this methodology has been used to reproduce top of the atmosphere (TOA) CERES longwave and shortwave radiances as a very convenient intermediate step between ground measurements and GERB observations.

The main objective of the CERES/GERB intercomparison and validation campaigns

Fig. 6.17.7. Simulation of a CERES footprint with 15° viewing zenith angle over the Valencia Anchor Station (see Fig. 2). CERES footprint at a sub-satellite point is about 20 x 20 km^2 large and GERB's would be about 60 x 60 km^2 at this latitude

is an attempt to establish the consistency between their broadband measurements. In order to make a valid comparison of radiance measurements by two different instruments, it is important to match their viewing geometries at a given time instance. In addition to this spatial and temporal alignment, selecting homogenous energy sources to be measured is required for valid comparison.

The application example shown here refers to the simulation of CERES TOA longwave and shortwave radiances acquired over the VAS during the GERB Ground Validation Campaign carried out in February 2004. In order to be able to reproduce CERES observations, it is necessary to have a good characterization of the surface and of the atmosphere. Fig. 6.17.7 gives an impression how the CERES footprint fits into the Utiel - Requena area.

Despite the fact that the earth viewing geometries of low earth orbit and geostationary orbit satellites are quite different, it is possible to match them utilizing a unique feature of CERES scanners, so called *Programmable Azimuth Plane Scan* (PAPS) mode (Fig. 6.17.8). This is accomplished by rotating the CERES scanning plane to align it with the GERB relative azimuth as the satellite flies over the portion of the Earth visible by the GERB instrument. There are no ideal energy sources to be measured by both instruments. Therefore, collecting a large amount of data for different scene types and averaging alleviates the problem of non-uniform radiation sources. VAS has offered a place for cross-referencing measurements from different instruments, as its surface is reasonably homogeneous.

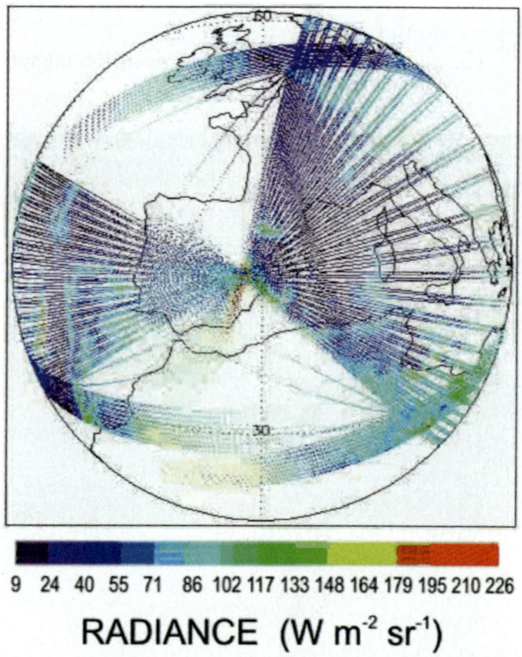

Fig. 6.17.8. CERES *Programmable Azimuth Plane Scanning* (PAPS) mode applied over the VAS. The figure shows the CERES acquisitions over two consecutive NASA Aqua satellite orbits for a satellite altitude of 707 km, tilt angle 0.0 degree, and rotation angle 0.0 degree (courtesy Dr. Z.P. Szewczyk. *CERES Science Team*)

Literature is full of references showing that TOA radiances are sensitive to the anisotropy of surface reflectance and of its diurnal variations. Thus, it is crucial to know surface spectral albedo and bidirectional reflectance. The details of this study, the data used, the Bidirectional Reflectance Distribution Function (BRDF) model employed (Ahmad and Deering 1992) are given in Velazquez et al (2004), and the simulations are explained in Lopez-Baeza et al. (2005). A full account of the treatment of homogeneous cases so far analyzed consisting of clear sky situations and measurements during winter when most of the surface Plateau is bare, is given in the academic report of Velazquez (2004).

Figs. 6.17.9 and 10 show the comparison between the CERES PAPS measurements of longwave (blue dots) and shortwave (red dots) TOA radiances and the simulated radiances (black small dots) obtained from the STREAMER radiative transfer code (Key and Schweiger 1998) fed with real ground and atmospheric measurements obtained in the February Campaign (Lopez-Baeza et al. 2005).

The observation configuration is fairly good in the first case (Fig. Fig. 6.17.9) because there is a large variety of viewing zenith angles (VZA) and azimuth angles. The simulation results (black small dots) adjust quite well to the CERES shortwave (red dots) measurements, both for the forward and backward observations. In the case of the longwave radiation (blue dots), it seems to be that the surface temperature calculated for

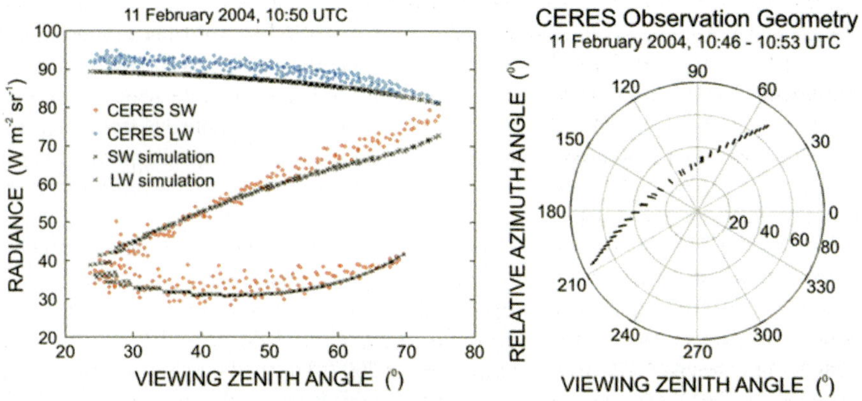

Fig. 6.17.9. Simulation of longwave and shortwave TOA CERES (Terra) radiances (left) and PAPS observation geometry (right) for 11 February 2004 over the Valencia Anchor Station (Velazquez, 2004)

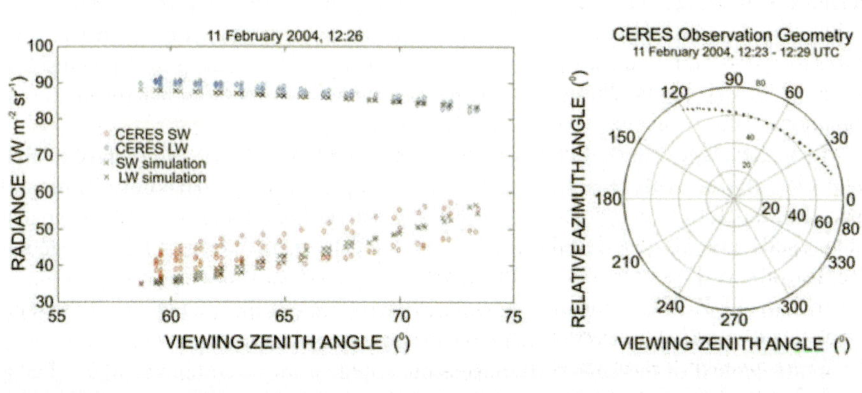

Fig. 6.17.10. Simulation of longwave and shortwave TOA CERES (*Aqua*) radiances (left) and PAPS observation geometry (right) for 11 February 2004 over the VAS. Note the different scaling of the VZA axis with respect to Fig. 6.17.9 (Velazquez, 2004)

the whole area from the station's data is lower than that observed by the CERES instrument because the simulated radiances are somewhat lower than the measured ones. Besides, the CERES longwave measurements seem to decrease more quickly than the simulated ones as the VZA increases, that is, the experimental data seem to have larger limb darkening effect (reduction of the emitted radiance as the VZA increases) than the simulation.

In the second case, (Fig. 6.17.10), the VZAs are very high and almost all the measurements are in the forward direction due to the azimuth geometry. This makes the shortwave radiances (purple dots) smaller and more similar with respect to each other. As for the longwave radiances (blue dots), it is observed that the simulations fits the data quite well, but the limb darkening effect is still present

6.17.7 Conclusions and Future Activities

The Valencia Anchor Station, installed by the University of Valencia in the natural region of the Utiel-Requena Plateau, near Valencia (Spain), was installed as a calibration/validation (Cal/Val) test area principally dedicated to large-scale low-spatial resolution remote sensing missions. The site is reasonably homogeneous from different viewpoints such as topography, land use, and climatic conditions and is proving to be a good and convenient instrument for validation of low resolution Earth Observation remote sensing instruments.

A specific methodology was developed to characterise an area of about 50 x 50 km^2 which is about the size of a pixel of GERB (Geostationary Earth Radiation Budget instrument), the broadband instrument onboard of the Meteosat Second Generation (MSG) satellite, or that of the forthcoming ESA Soil Moisture and Ocean Salinity (SMOS) Mission planned to be launched early in 2007. The methodology to obtain large scale independent surface and 3-D atmospheric fields is based on the use of MM5 high resolution simulations as a robust interpolation tool (Lopez-Baeza and the SCALES Project Team, 2003 and 2004).

The station is equipped with conventional standard meteorological instruments at two levels in the atmosphere, temperature and heat flux measuring instruments at different levels in the soil, specialised measurements of all radiation components and instruments measuring the soil moisture content. The equipment also includes instruments and facilities to carry out mobile and transect measurements to account for heterogeneities in the area.

Tasks which are currently being carried out at the VAS are (i) definition of a large scale validation area for low-spatial resolution missions, (ii) definition and characterisation of a large scale reference pixels, (iii) temporal interpolation and spatial averaging, (iv) studies on scaling issues, and (v) development of criteria for aggregation and disaggregation.

A number of field campaigns have been carried out in the area showing the high quality of the surface and atmospheric measurements obtained from the ground.

In the context of the ESA SMOS Mission, a large scale aircraft campaign is being organised to validate SMOS soil moisture retrieval algorithms. The part of the VAS CoSMOS (*Campaign for the Operation of SMOS*) campaign will take place during spring 2005, between April and June, consisting of a number of aircraft flights with an

L-band radiometer and a large deployment of soil moisture measurements, both automatic and manual, simultaneously performed in the ground, in an area of 10 x 10 km^2 surrounding the *Anchor Station* (CoSMOS Experimenters Team, 2005).

There will be further field campaigns to be organised in the area to consider more general atmospheric and surface conditions and to validate surface energy fluxes to be derived from GERB TOA radiances. Besides GERB and SMOS, it is expected that VAS could also contribute in the framework of other satellite missions such as SEVIRI (*Spinning Enhanced Visible and Infrared Imager*), ENVISAT (AATSR, *Advanced Along Track Scanning Radiometer*), CERES (*Clouds and the Earth's Radiant Energy System*) and AMSR (*Advanced Microwave Scanning Radiometer*).

In December 2004, an additional Anchor Station has been set up in the province of Alicante (Spain) by the Regional Ministry for Planning and Housing (Directorate General for Environmental Quality). The *Alacant* Anchor Station (Fig.11) is a twin station of the VAS and is situated 180 km south of the city of Valencia. Both Stations will work coupled to define a *Water Cycle Observatory* to monitor and compare meteorological parameters and fluxes of radiation, energy and water between both different climate areas. In fact, the Alacant Anchor Station area has been chosen in the objectively most degraded and hydrologically least sustainable crop area of the Valencia Community Region (see Project Lopez-Baeza, 2004) by using wavelet transform techniques on remote sensing NDVI data (Gimeno-Ferrer 2003).

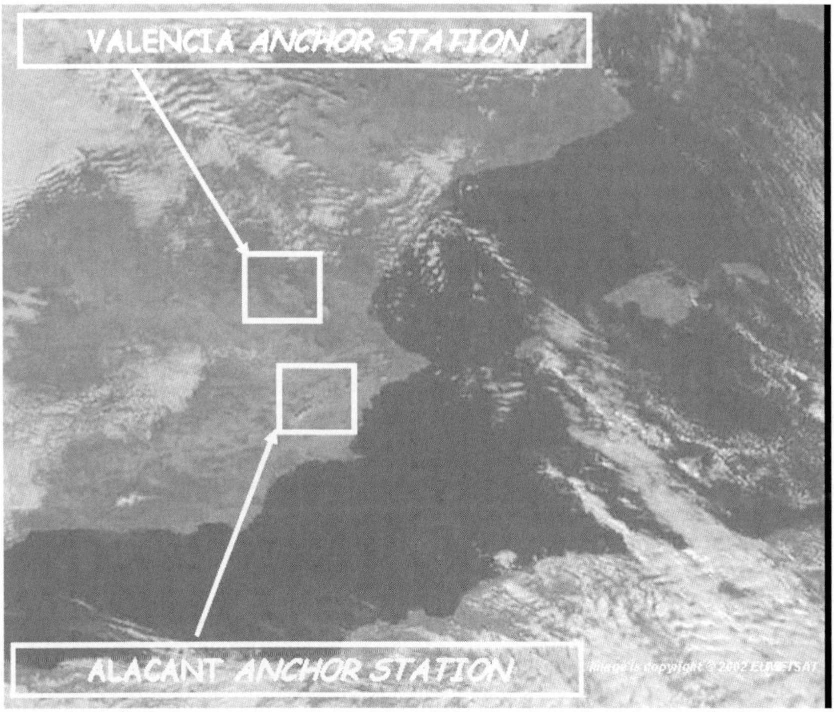

Fig. 6.17.11. Relative position of the Valencia and the Alacant Anchor Stations

6.18 Assessment of Land-surface Changes in Space and Time - General Conclusions

One of the goals of the work described in this book was to assess changes of land-surface processes in the course of time. The work focussed specifically on those processes that are related to "Global Change", such as the energy and water transfer between surface and atmosphere. Changes of these processes can be called trends, if the change continues in the average in the same direction over many years. Trends of land-surface processes develop gradually and are superimposed by the natural inter-annual variability. They can only be extracted from long term homogenous time-series after correction for higher frequency fluctuations such as caused e.g. by precipitation events, seasonality, or volcanic eruptions . For such analyses, data must be free of sensor drifts and degradations. Absolute calibration is necessary to avoid 'jumps' in data series in case satellites are replaced or, even worse, if new instruments are launched with different spectral characteristics. Also, for the same piece of land, a different solar angle will lead to different illumination characteristics and different shadowing properties at the surface, giving non-identical values for unaltered landscapes. In addition, the differing path lengths of solar irradiance through the atmosphere may also affect atmospheric corrections. To overcome the cloud problem, the analysis also has to be based preferably upon daily observations.

As for Earth observing satellites with high spatial resolution instruments there is another problem to be considered which is the irregular and poor temporal coverage, the low probability of cloudless scenes, and the high cost of these data. On the other hand, the spatial resolution of meteorological satellites is not sufficient to observe isolated clouds and small features such as found in the often highly structured areas of southern Europe. If one is interested in this scale, Landsat-TM or SPOT measurements have to be 'nested' into data-sets of meteorological satellites as exemplified in Section 6.6.

For a specific area, it is difficult to find enough cloudless high-resolution satellite images to establish a climatology for a long period of several years. Annual mean values are therefore difficult to establish with high-resolution data. Only in a few years there may be enough observations to study the seasonal variability in this way. It is equally difficult to find a data series, for a specific week or even a specific month, to demonstrate variability within a series covering several years.

One strategy to quantify changes that occur at smaller time scales, exclusively with high resolution satellite data, therefore is to first determine variability introduced by the seasons for one or more years and subsequently to select a time window within the year which (a) may be representative to assess a trend and (b) for which observations are available for a sufficient number of years to determine these changes. Knowledge about the seasonal variability is important because high-resolution data will not be available year by year for exactly the same day or week. Moreover, the development of vegetation is not exactly synchronized with the date but depends on the preceding weather record. So one has to know the course of the seasonal change to judge the representativeness of images taken within a time interval of say, plus/minus one month, around a certain day of the year.

Atmospheric effects may introduce additional noise to the data. Although it seems now possible to correct reasonably well for the gaseous constituents of the atmosphere, problems still remain with aerosols, contrails and thin cirrus. The aerosol problem is especially severe in the Mediterranean area because of (a) industrial pollution, (b) desert dust from the Sahara, and (c) dust from arid lands. Only a few stations report data on the optical depth and atmospheric aerosol parameters. The contrail problem escalates over Europe especially in the vicinity of airports, which could severely hamper remote sensing from space.

One has also to keep in mind that remote sensing is only one view at the Earth system. *In situ* measurements and models create differently natured data sets that mirror different aspects of reality. In order to obtain a consistent picture, the information inferred from the different measuring and modelling systems finally must be amalgamated. This requires a comparison and, if necessary, an assimilation of the data sets. From the viewpoint of remote sensing the comparison with in situ measurements often is called *validation* though it is an open question, which of the three views at the system is more valid than the others. The comparison of information deduced from remote sensing and measurements on the ground is not a straightforward task in inhomogeneous landscapes such as the Mediterranean. This is especially true for the secondary level of information with its complex data evaluation schemes. For these cases a specific strategy has been developed. For example, spotty measurements at the ground are compared with high resolution satellite data and these are scaled up to the lower resolution images of meteorological satellites. It is noteworthy to recall that one AVHRR pixel contains 1600 SPOT pixels. Alone a one-to-one comparison of satellite data, even with the same pixel size, is not possible because of the inability to superimpose satellite images of different systems to better than a fraction (typically one third) of a pixel. Even more complex is the comparison of the regular pixel pattern of satellite images with an irregular arrangement of ground stations. Consequently these comparisons have to be made with the largest possible number of pixels which requires relative homogeneous surfaces.

Although the technical specifications of each instrument are tested prior to launch, deteriorations may occur during the launch phase and during its operation. The selection of calibration sites to detect long-term instrument degradation is not easy and might require the installation of a number of so-called ‘anchor stations’, to measure appropriate collateral data.

The application of a strict analysis procedure starting with raw data and developing step by step validated and normalized data series led to a unique data base for the Mediterranean area for the end of the 20th century. The use of these data in numerous field of application is demonstrated in this Chapter 6. In Appendix 8 it is communicated how this data set is continued into the 21st century using the latest information about the performance of the new generation of AVHRR instruments.

Presently, the possibility of an increase of the vegetation cover in the northern hemisphere and a prolongation of the vegetation period is under discussion in a number of publications. For the Mediterranean region it turns out that any conclusions in this direction and in such generality have to be taken with caution because this process is counteracted by aridification during summertime. While finalizing this book the media report about a drought in Portugal and Spain which may develop towards a disastrous water shortage in summer. This is reflected in the NDVI by reduced greenness. In Fig. 6.18.1 the deviation of the NDVI from its long term mean value is shown for the month of May. This map shows the extension of the drought in the south by red colour and the strong contrast between the north and the south. By comparison with the map of the year of 1995, which is shown as well, it can be seen that in 1995 the affected area was much larger and the drought must have been more severe than the one in spring 2005. One reason for the intensity of the NDVI reduction in 1995 may be that also 1994 was already a dry and hot year. Thus the ground water deficit in 1995 must have been larger than 2005.

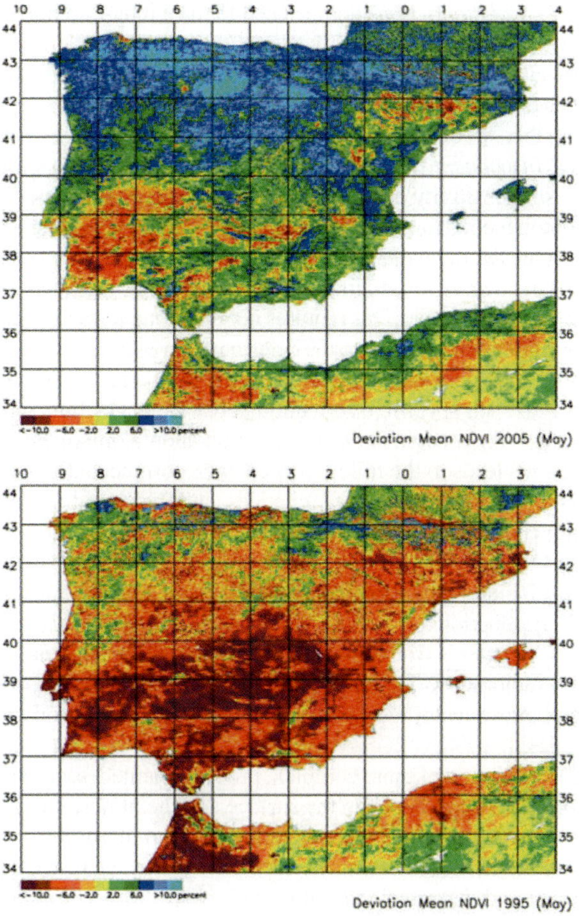

Fig. 6.18.1. Deviation of the NDVI in May from 16 years mean value for the Iberian peninsula in 2005 (top) and 1995 (bottom)

In summary, assessment of land-surface changes requires to pay attention to the following items towards the generation of high quality data sets:
(i) *Observation period* – define the minimum length of time period required to separate short-term fluctuations from long-term trends.
(ii) *Observation frequency* – determine the minimum frequency of observation required to follow (a) quick changes of the land-surface state variables due to varying meteorological conditions (i.e. temperature, clouds, dust, air mass changes, soil moisture changes after rain) which may require several times a day, and (b) to follow the real changes on the surface, like state of vegetation, albedo, and soil condition changes due to long lasting droughts which may require only one measurement a day.
(iii) *Spatial resolution* – satisfy the need for data-types that allow to separate regional from large scale changes, viz. SPOT, Landsat and NOAA-AVHRR for local and regional studies and Meteosat for studying large scale phenomena. Also, the corroborative data should be able to match the data inferred from each of the satellite instruments with their respective resolution and spectral range.

(iv) *Validation* – consider the need for data inter-comparison and arrange the measuring strategy accordingly.
(v) *Homogenization and continuity* – production of homogeneous data-series from successive satellites requires intercalibration and intercomparison of data-series and, if possible, an overlapping period of measurements.
(vi) *Instrument degradation and drift* – continuos updating of the information about sensor sensitivity and satellite orbit stability is a necessary prerequisite for generating homogeneous long time data sets.

One major challenge for the use of observations from space is to answer the question to what degree vegetation cover is influenced by global warming and what its impact is on the water cycle. This requires not only long term investigations of the annual average vegetation index but also the determination of the starting time, peak, and duration of the vegetation period. Fig. 6.9.7 points into the direction how satellite data can be processed to gain such specific information about the vegetation and several sections of the book address the assessment of information about components of the water cycle.

This leads to the following general recommendations for steps necessary to be taken for studying land-surface changes and for building an early warning system for irreversible land-surface degradation:
(1) Further improvement of the standardization and quality of satellite data products. This includes sensor calibration as well as atmospheric and geometrical corrections.
(2) Production of continuous primary data sets with high precision and stability to separate trends from noise, such as obtained for spectral reflectance, vegetation indices and radiometric TOA temperature which allow to follow, by means of indicators, the temporal development of land-surface properties from, say, weekly to decadal time scales.
(3) Derivation of more complex, process oriented, secondary data sets such as needed to compute fluxes between the surface and the atmosphere.
(4) Continuous quality check of information inferred from satellite data by comparison with in situ measurements in test areas, such as the so-called ‘anchor stations’ mentioned above and described in section 6.17.
(5) Generation of long-term data sets for the whole Mediterranean area.
(6) Proper archiving of original and collateral data to allow re-analysis with newly developed algorithms. This poses a heavy load on the archiving system and requires adequate funding.

Appendices

Appendix 1 The ECHIVAL Field Experiment in Desertification Threatened Areas (EFEDA)

A.1.1 The EFEDA Research Concept

Many of the algorithms discussed in this book were first applied and their performances were validated during the EFEDA field experiment in Castilla - La Mancha (see Figs. 4.9 and 5.43), Spain, in June 1991, and a series of smaller and more specialized experiments that took place again in Castilla La Mancha in 1994, in the Messara Valley of Crete (Vardavas 1996), in September 1994 and April/May 1995, at two locations in Tuscany, the Cornia river valley near Campiglia M.ma (Fig. 5.10), the area around Radicondoli (section 6.6) and Piano di Rosia (Fig. 5.33/5.34, Fig. 5.50) south of Florence, near Matera (Fig. 1.31), in the agricultural test area near Rutigliano (Bari), and in Israel (section 6.11). EFEDA had also a number of precursors like the experiments in La Crau, France, Ibecetene (Republic of Niger), and several measurements around Campigliua M.ma between 1983 and 1999. EFEDA was initiated by the former CEC in response to the international activity on land-surface processes of the International Satellite Land-Surface Climatology Project (ISLSCP) and the Hydrologic Atmospheric Pilot Experiment (HAPEX) initiative of the World Climate Research Programme. The experiments were organised according to the strategy developed by ISLSCP (Becker, Bolle, Rowntree 1988) which subsequently have been adopted by IGBP-BAHC (1993).

EFEDA was designed as a multidisciplinary activity to study the full range of water transfer processes from sub-surface transports up into the free troposphere including observations from space. Alone the exploration of the capabilities offered by remote sensing from satellites already requires a wide range of subsidiary information about the observed targets. Consequently, scientists representing the following disciplines were involved: geology, soil science, hydrology, geography, botany, ecology, meteorology (with major emphasis on atmospheric boundary layer research), aeronomy, modelling, and remote sensing. The experiment in Castilla - La Mancha included both a NASA and a German instrumented research aircraft and instrumented German motorgliders as well as several radiosonde stations.

To be able to compare the results with the output of wide meshed climate models and to contribute to an improvement of these models, the main experiment had to span over an area compatible with the grid width of climate models which is a scale of approximately 10^4 km^2. The extended plain of Castilla - La Mancha was an ideal area to conduct such an experiment with a minimum of orographic disturbance while some special investigations including hilly terrain were conducted in the other areas. Research concentrated at water transfer processes in the near to surface layers of the earth - atmosphere system: Sub-surface water exchange (geo-hydrology), processes in the upper, mostly unsaturated, layer (typically 30 cm) of the soil, topographic effects, exchange processes in the vegetation canopy, transfer in the atmospheric (or planetary) boundary layer, transport into the free troposphere, and, finally, observation of the accessible part of the system by remote sensing from aircraft and satellites. Each of these “layers” contribute to the geo-information system into which data can be mapped, such as the position of aquifers, types and structure of soils, surface characteristics, topography, land-cover and land-use, water vapour and aerosol distribution in the lower levels of the atmosphere, and radiation fields at the top of the atmosphere.

A.1.2 The Castilla - La Mancha Experiment

Research Components

Overview. Within the large plain of Castilla - La Mancha three representative experimental sites as indicated in Figs. 4.9 and 5.43 were selected and densely instrumented. One was the dry farming area (mainly vineyards) near Tomelloso, in another area (near Barrax) a mixture exists of dry and irrigated farming, and the third area (near Rada de Haro) was slightly hilly and mainly matorral with some dry farming. The space in-between these three sites was bridged by remote sensing measurements from aircraft and satellites as well as by singular measuring stations, some movable equipment and visual inspections.

Hydro-geology. The lower boundary for modelling the soil - atmosphere water transfer processes is determined by the aquifers which feed the overlying soil by capillary suction. Modelling of these hydro-geological processes requires knowledge about water conducting soils, the position of the aquifers and estimates about the volumes of water reservoirs. The aquifer system underneath the plain between Tomelloso and Albacete shown in the map Fig. A.1.1a is well explored. The legend of this map, shown in Fig. A.1.1b, inter alia lists the soils that have been identified. Vertical cross section through the experimental sites are shown in Fig. A.1.2. Large scale hydrological investigations have also been performed in the Messara valley (Fig. 1.17).

Soil Types. The surface of the soils often can be observed directly from space because the vegetation cover is thin (section 6.5). The soil structure was investigated at many sites of the experimental area. A description of the soil classes is given in Appendix 3. Fig. A.1.3a shows the soil map of the EFEDA experimental area. The legend is given in Fig. A.1.3.c which shows the detail of the classification that partly can not be distinguished in the overview map. Therefore, Fig. A.1.3b shows the enlarged section of the area around Rada de Haro. Not only the soil type was mapped for the whole region but also the degree of degradation (see Appendix 3).

Topography. Though topographic effects were minimized by the selection of the mostly plain area for the experiment, topographical maps are the backbone for the interpretation of remote sensing data in many areas, such as the more structured Rada de Haro region. One of the tasks within the EFEDA project therefore was the creation of high-fidelity Digital Elevation Models (DEM) with relief-derived attributes for the three experimental sites of the EFEDA project as described in the section *Digital Elevation Models*.

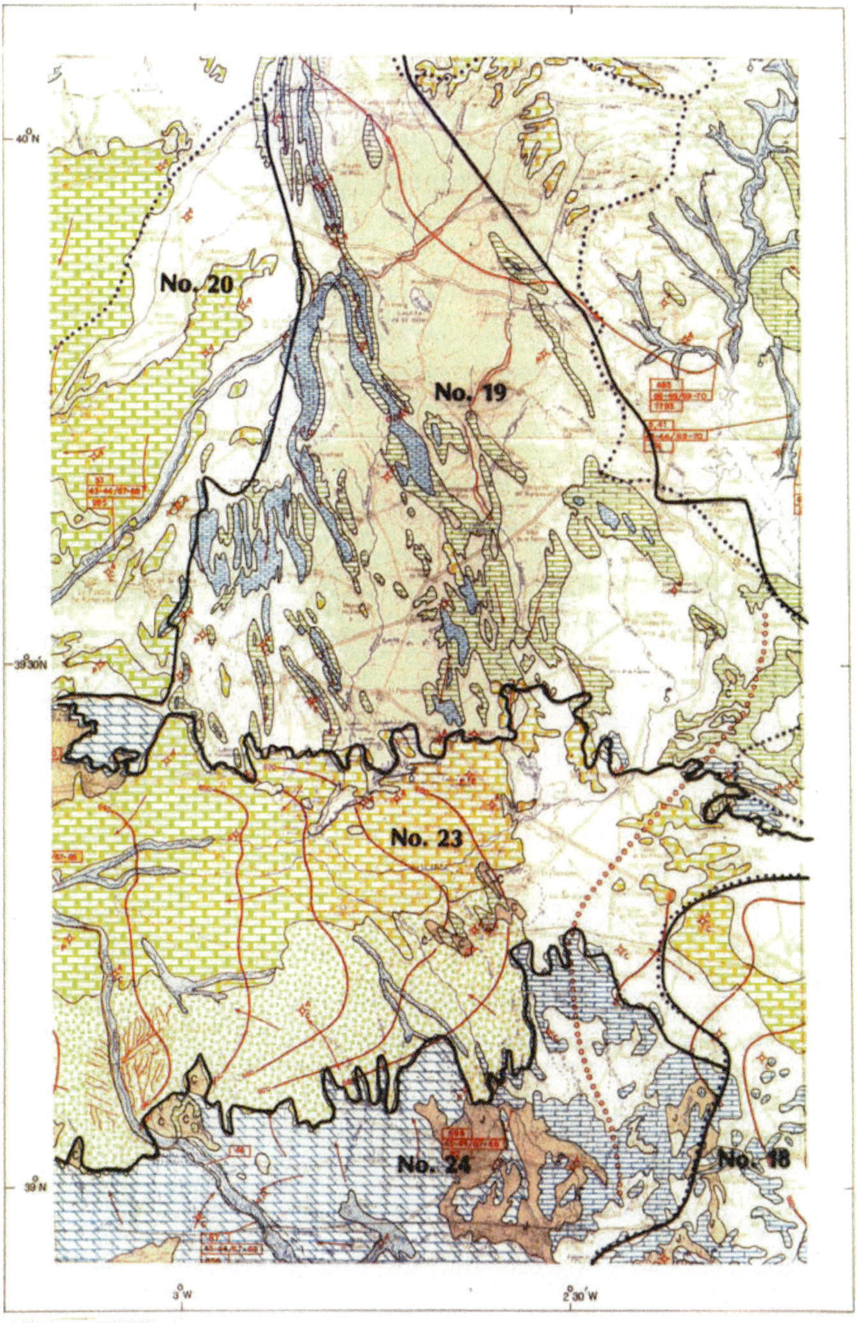

Fig. A.1.1a. Soil types and aquifer systems underneath the plain of Castilla - La Mancha

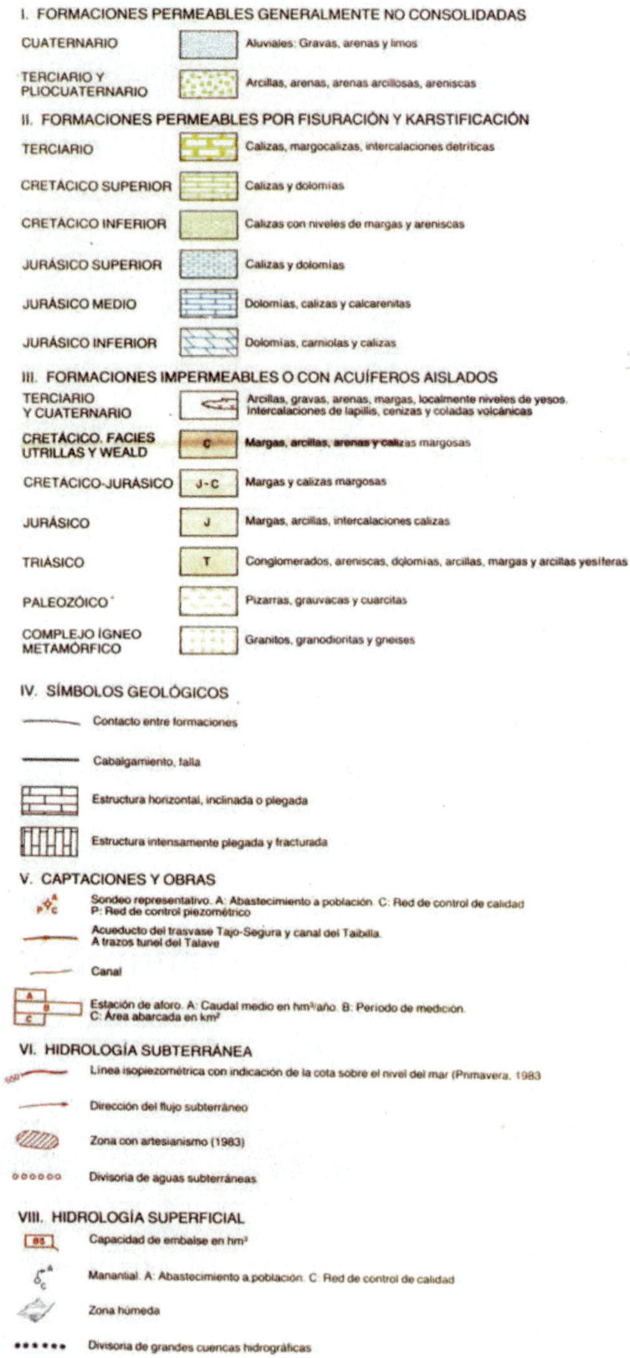

Fig. A.1.1b Legend of the map shown in Fig. A.1.1a

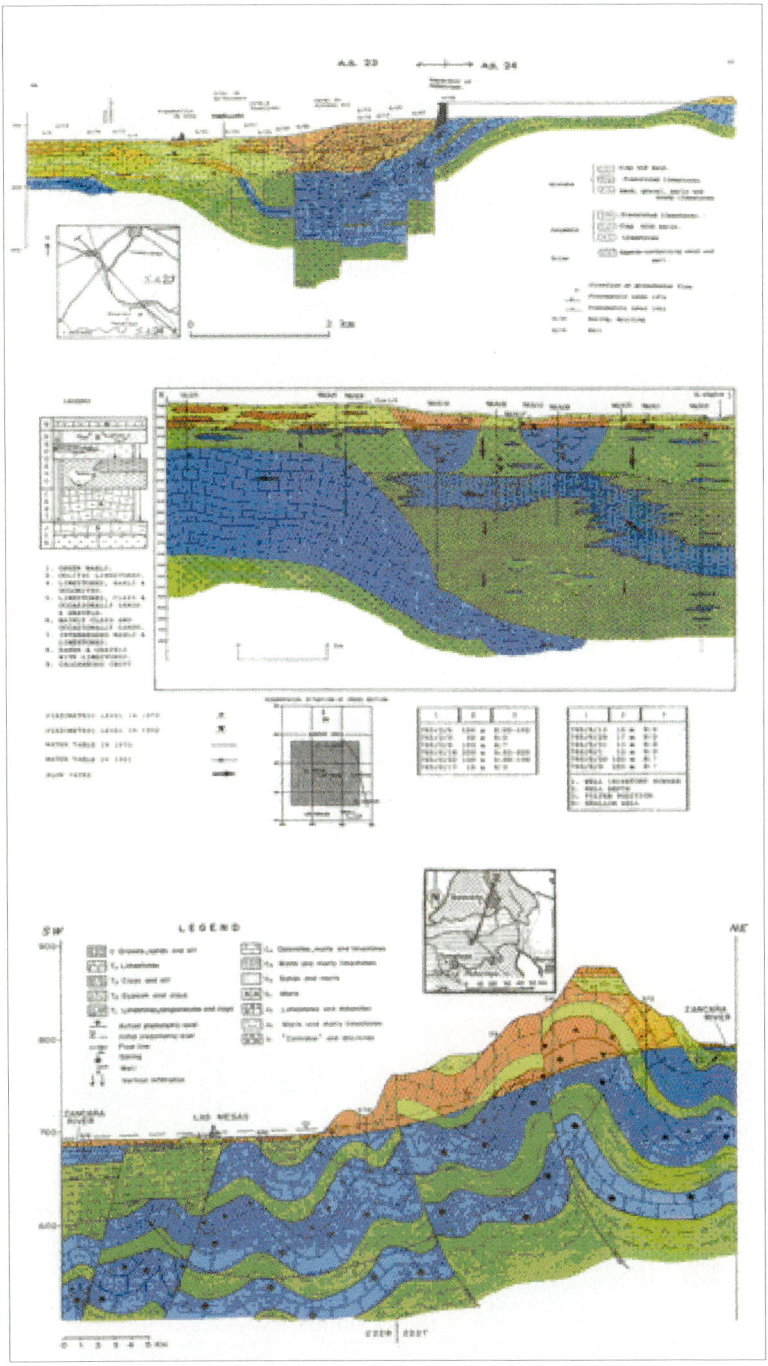

Fig. A.1.2. Hydro-geological cross section through the areas of Tomelloso, Barrax and Belmonte - Rada de Haro (Llamas and Cruces 1993)

Fig. A.1.3a. Soil map of the EFEDA experimental area after Díaz et al. (19xx)

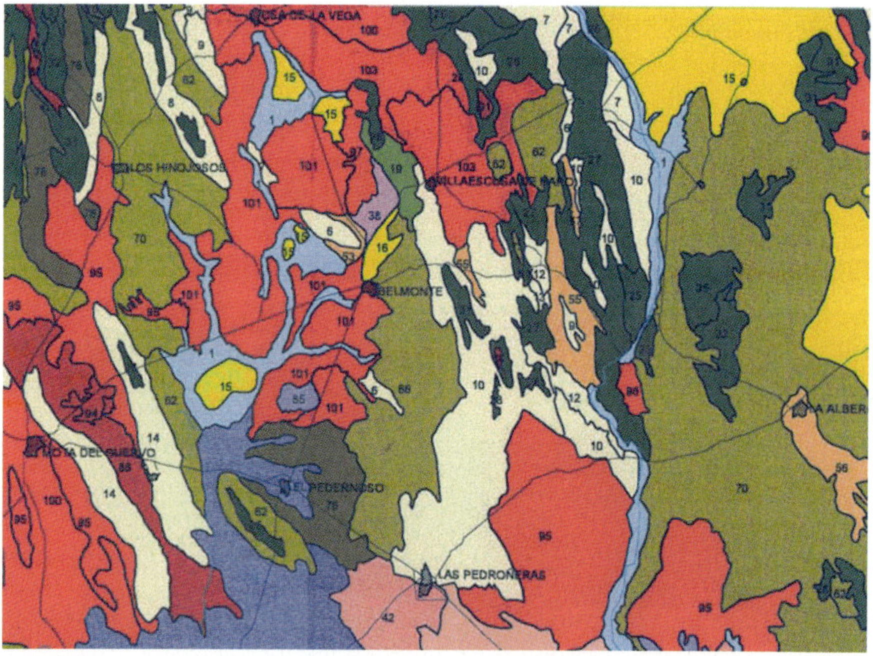

Fig. A.1.3b. Soil map of the area around Rada de Haro. Enlarged detail of Fig. A.1.3a around the Belmonte/Rada de Haro experimerntal site (compare Fig. A.1.7)

Symbol	Dominant Soil	Associated Soils
1	Calcaric Fluvisols	
2	Calcaric Fluvisols	Calcaric Arenosols
3	Calcaric Fluvisols	Calcic Fluvisols
4	Calcaric Fluvisols	Lithic Leptosols Calcaric Regosols
5	Salic Fluvisols	
6	Calcaric Regosols	
7	Calcaric Regosols	Gypsic Regosols
8	Calcaric Regosols	Eutric Leptosols
9	Calcaric Regosols	Lithic Leptosols
10	Calcaric Regosols	Calcaric Cambisols
11	Calcaric Regosols	Chromic Luvisols
12	Calcaric Regosols	Eutric Leptosols Lithic Leptosols
13	Calcaric Regosols	Lithic Leptosols Eutric Leptosols
14	Calcaric Regosols	Haplic Calcisols Lithic Leptosols
15	Gypsic Regosol	
16	Gypsic Regosols	Calcic Lluvisols
17	Gypsic Regosols	Luvic Gypsisols
18	Eutric Leptosols	
19	Eutric Leptosols	Calcaric Regosols
20	Eutric Leptosols	Lithic Leptosols
21	Eutric Leptosols	Haplic Cacisols
22	Eutric Leptosols	Chromic Luvisols
23	Eutric Leptosols	Lithic Leptosols Calcaric Regosols
24	Eutric Leptosols	Calcic Luvisols Chromic Luvisols
25	Rendziz Leptosols	Lithic Leptosols Calcaric Phaeozems
26	Lithic Leptosols	
27	Lithic Leptosols	Calcaric Regosols
28	Lithic Leptosols	Eutric Leptosols
29	Lithic Leptosols	Calcaric Arenosols
30	Lithic Leptosols	Haplic Calcisols
31	Lithic Leptosols	Chromic Luvisols
32	Lithic Leptosols	Eutric Leptosols Calcaric Regosols
33	Lithic Leptosols	Eutric Leptosols Rendzic Leptosols
34	Lithic Leptosols	Chromic Luvisols Eutric Leptosols
35	Lithic Leptosols	Calcaric Phaeozems Calcic Kastanozems
36	Haplic Arenosols	
37	Haplic Arenosols	Calcaric Regosols
38	Cambic Arenosols	
39	Luvic Arenosols	
40	Luvic Arenosols	Lithic Leptosols
41	Luvic Arenosols	Calcaric Arenosols
42	Luvic Arenosols	Chromic Luvisols
43	Luvic Arenosols	Haplic Arenosols
44	Calcaric Arenosols	
45	Calcaric Arenosols	Calcaric Fluvisols
46	Calcaric Arenosols	Cambic Arenosols
47	Calcaric Arenosols	Luvic Arenosols
48	Calcaric Arenosols	Luvic Calcisols
49	Calcaric Arenosols	Gypsic Solonchaks
50	Calcaric Arenosols	Calcic Luvisols
51	Calcaric Arenosols	Luvic Arenosols Eutric Leptosols Lithic Leptosols
52	Eutric Cambisols	Cambic Arenosols
53	Calcaric Cambisols	
54	Calcaric Cambisols	Calcaric Fluvisols
55	Calcaric Cambisols	Calcaric Regosols
56	Calcaric Cambisols	Lithic Leptosols
57	Calcaric Cambisols	Haplic Calcisols
58	Chromic Cambisols	
59	Chromic Cambisols	Luvic Arenosols
60	Chromic Cambisols	Chromic Luvisols Calcic Luvisols
61	Chromic Cambisols	Calcic Luvisols
62	Haplic Calcisols	
63	Haplic Calcisols	Calcaric Regosols
64	Haplic Calcisols	Eutric Leptosols
65	Haplic Calcisols	Lithic Leptosols
66	Haplic Calcisols	Calcaric Cambisols
67	Haplic Calcisols	Chromic Cambisols
68	Haplic Calcisols	Petric Calcisols
69	Haplic Calcisols	Chromic Luvisols
70	Haplic Calcisols	Calcic Luvisols
71	Haplic Calcisols	Lithic Leptosols Calcaric Regosols
72	Haplic Calcisols	Calcaric Arenosols Luvic Arenosols
73	Haplic Calcisols	Luvic Calcisols Calcaric Regosols
74	Haplic Calcisols	Calcic Luvisols Calcaric Regosols
75	Luvic Calcisols	
76	Petric Calcisols	
77	Petric Calcisols	Eutric Leptosols
78	Petric Calcisols	Lithic Leptosols
79	Petric Calcisols	Haplic Calcisols
80	Petric Calcisols	Luvic Calcisols
81	Petric Calcisols	Haplic Calcisols Luvic Calcisols
82	Petric Calcisols	Eutric Cambisols Haplic Arenosols Calcaric Arenosols
83	Calcic Solonchaks	
84	Calcic Solonchaks	Gleyic Solonchaks
85	Gypsic Solonchaks	
86	Gypsic Solonchaks	Calcaric Arenosols
87	Haplic Luvisols	
88	Chromic Luvisols	
89	Chromic Luvisols	Eutric Regosols
90	Chromic Luvisols	Lithic Leptosols
91	Chromic Luvisols	Calcaric Cambisols
92	Chromic Luvisols	Chromic Cambisols
93	Chromic Luvisols	Haplic Calcisols
94	Chromic Luvisols	Luvic Phaeozems
95	Calcic Luvisols	
96	Calcic Luvisols	Calcaric Regosols
97	Calcic Luvisols	Gypsic Regosols
98	Calcic Luvisols	Calcaric Arenosols
99	Calcic Luvisols	Chromic Cambisols
100	Calcic Luvisols	Haplic Calcisols
101	Calcic Luvisols	Calcic Gypsisols
102	Calcic Luvisols	Haplic Calcisols Petric Calcisols
103	Calcic Luvisols	Calcic Gypsisols Haplic Calcisols

Fig. A.1.3c. Legend for Figs. A.1.3a and b

Land-cover/land-use. Superimposed to the topography is the land-cover/land-use. Knowledge about the vegetation is one prerequisite for modelling energy transfer from the soil through the canopy into the atmospheric boundary layer (compare Fig. 1.26). One year prior to the EFEDA experiment the land use was mapped for the three experimental sites. Here one example is shown for the agricultural site near Barrax (Fig. A.1.4). For the Tomelloso area a map was constructed in connection with the land-use change study (section 6.10).

Transfer of heat and water vapour from the soil through the Vegetation into the atmosphere. Detailed experimental data of the surface energy budget and water transfer within the soils and through the plants into the atmosphere lead to an improvement and adaptation for the semi-arid conditions of Soil-Vegetation-Atmosphere-Transfer (SVAT) models. An extensive inter-comparison the performance of all SVAT models applied in EFEDA was conducted (Linder et al. 1995). Here one example is shown (Fig. A.1.5) to illustrate the quality by which quantities relevant for the exchange of energy between the surface and the atmosphere can be simulated by a SVAT model. For this illustration the modified BATS model (Dickinson et al. 1986) of Blümel (1992) was used.

Atmospheric Boundary Layer (ABL). The structure of the ABL is essential for the further transfer of energy and water from the canopy into the "free" troposphere (which means the layer which is not anymore affected by the roughness of the surface). The structure of the atmospheric boundary was studied with the dense radiosonde network that was for a certain time period established in the area, with sodar, and flux measurements from aircraft and motor gliders (Michels BI 1992, Entstrasser et al. 1994, Michels BI, Jochum AM 1995) that provided detailed information on fluxes and water vapour as well as aerosol distributions (Figs. 2.15 and 2.17). One example of radiosonde results is given in Fig. A.1.6 in which the results of a number of radiosonde ascents are combined to a temporal cross section.

Troposphere. Although the mission of EFEDA ended with the energy and water transfer through the ABL, the troposphere constitutes the upper boundary condition of the research as the hydro-geology constitutes its lower boundary. Tropospheric processes were important in respect of weather events (cloudiness, precipitation) and for the correction of remote sensing data (aerosol and water vapour). The latter affect the atmospheric optical depth (see Chapter 3). The aerosol optical depth was derived from transmission measurements of solar radiation. Figs. A.6.3.1-8 give a few examples. By the dense network of radiosonde stations established for some time in the EFEDA experimental area the vertical temperature and humidity profiles were determined also within the troposphere from which the atmospheric water content was computed which is especially needed for corrections of signals measured by satellite instruments in the thermal infrared spectral region.

The availability of these aerological data allowed to initiate and to test mesoscale models which simulated the atmospheric circulation system in the area and the effect of the land-sea circulation that develops at the Mediterranean coast (compare e.g. Bolle, 1998). The land-sea circulation just reached the eastern part of the EFEDA experimental area at some days.

Observations from Space. The evaluation of satellite data gained immensely from the possibility to validate the inference of primary quantities such as reflectance, temperature and vegetation indices by comparisons with ground measurements

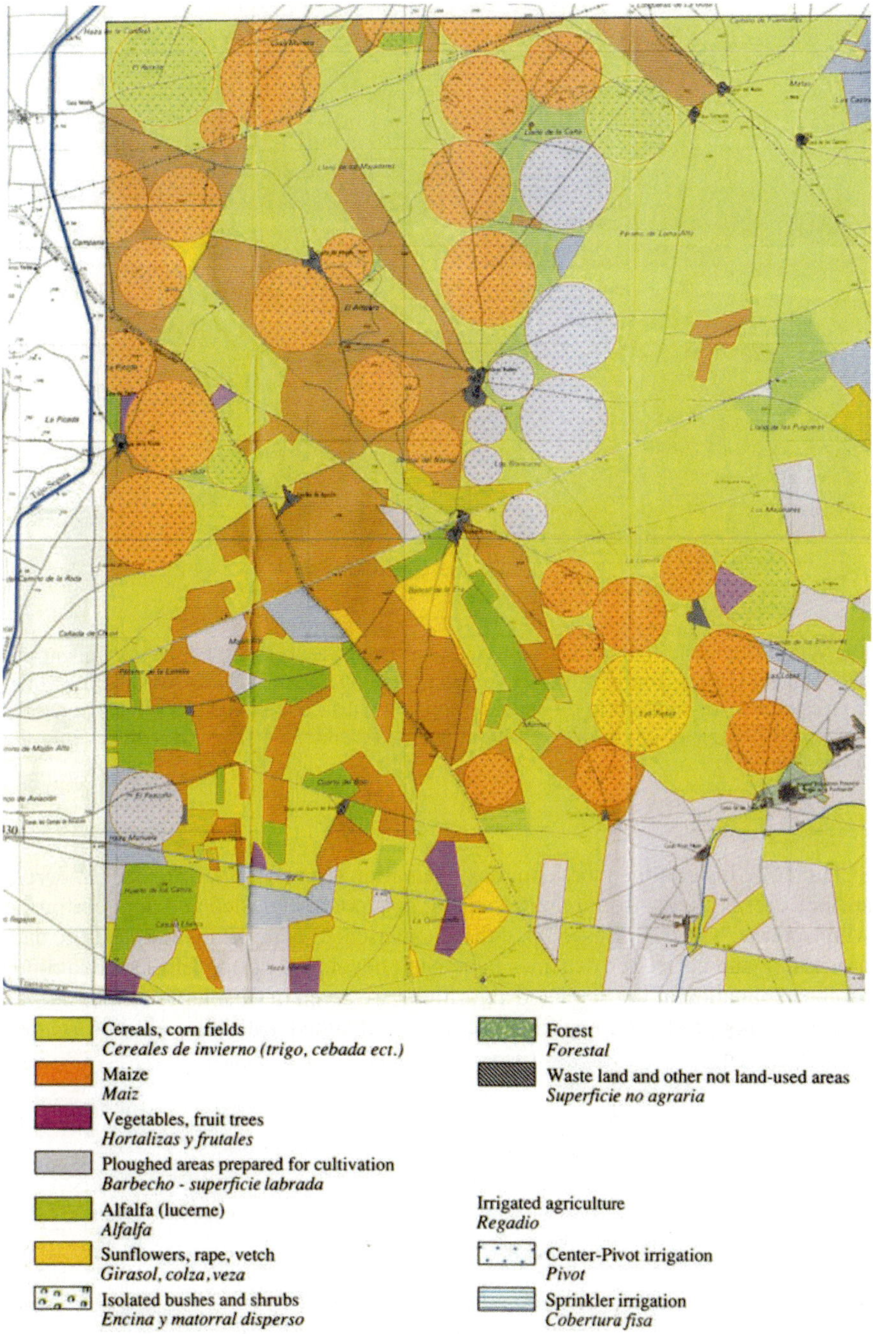

Fig. A.1.4. Land-use map of the area around the agricultural research station "Finca Las Tiesas" near Barrax (compare Figs. 4.32-34).

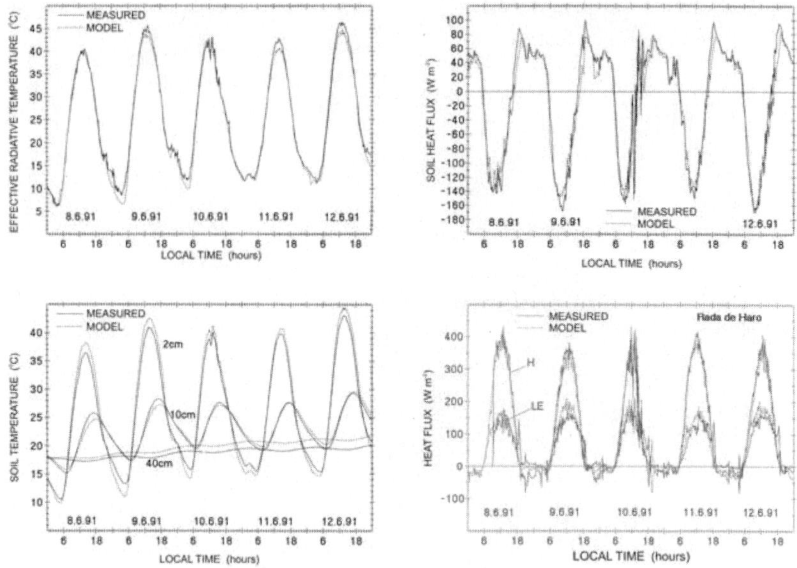

Fig. A.1.5. Comparison of modelled (dotted line) and measured (full line) SVAT quantities after Blümel (1992). Top left - Effective radiative temperature such as would be measured from satellites; top right - heat flux into the soil at its surface; bottom left - soil temperatures at 2 cm, 10 cm, and 40 cm depth; bottom right - sensible and latent heat fluxes into the atmosphere

(Chapter 4). In addition, more complex evaluation schemes such as needed to derive surface energy fluxes from remote sensing data could be tested and applied for the whole area (Chapter 5). Based upon these experiences completely new satellite data processing and correction schemes have been developed during EFEDA Examples for the application of the validated algorithms are given in Chapter 6. Emphasis was laid upon merging and comparing the results obtained with different observation systems.

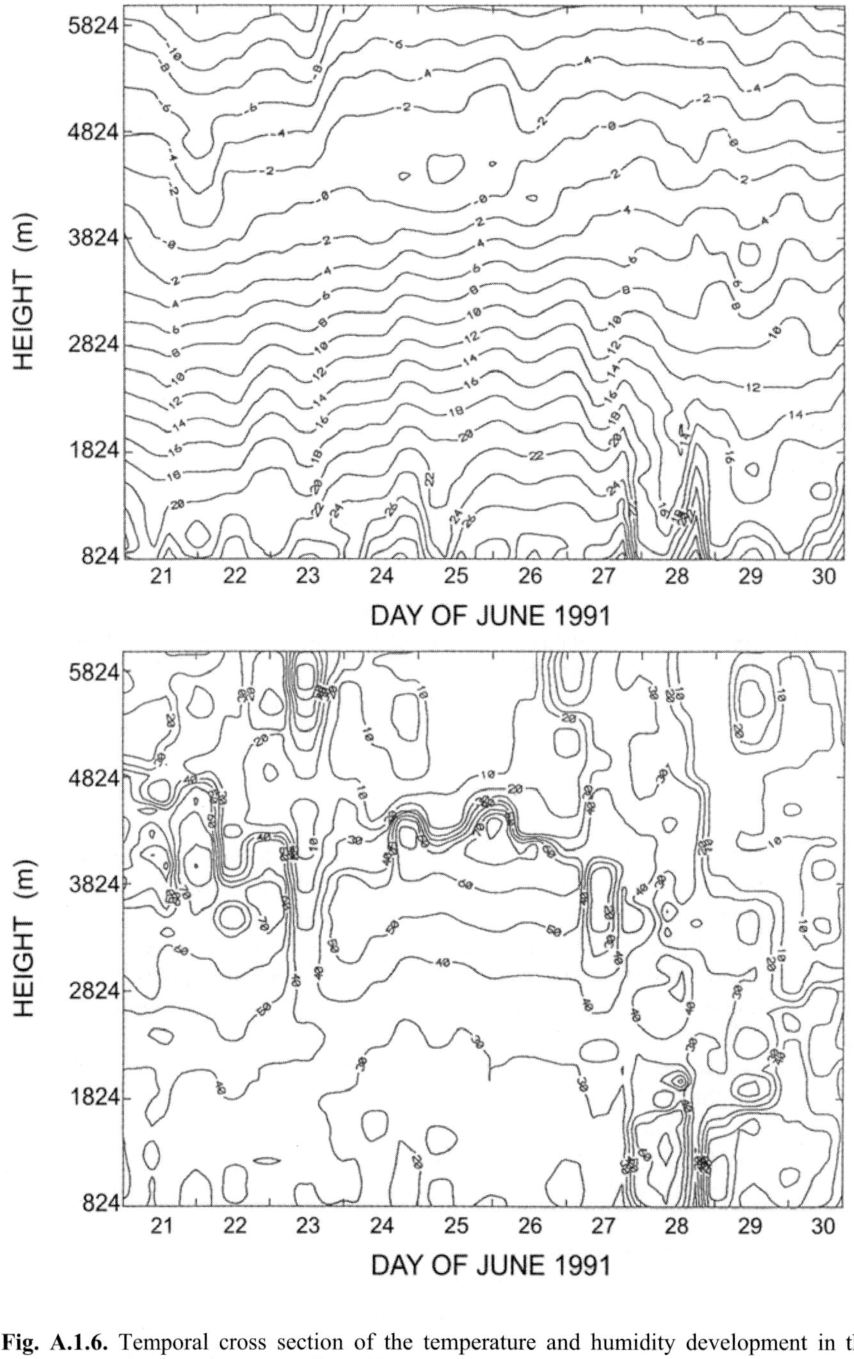

Fig. A.1.6. Temporal cross section of the temperature and humidity development in the atmosphere above Rada de Haro from 21 to 30 June 1991

A.1.3 Digital Elevation Models

Geographical Information Systems and Digital Elevation Models. To sample the individual research results and to combine them with each other, a Geographical Information System (GIS) oriented to a multidisciplinary study of the area under research. is extremely helpful. The basis of such a system constitutes the topographical database. While the term Digital Terrain Model (DTM) is applied to attributes of the landscape including others than the altitude, the term Digital Elevation Model (DEM) is preferred for models containing only elevation data. Following the work of Bitelli et al.(1993) the development of the topographical database for the EFEDA area is shortly described here.

Data structure. The DTM is created by means the complete transformation of input data, independently from the method that has been used for sampling, into a data structure able to adequately describe the land surface. Approaches for defining such a structure, or data model, are mainly based on the sub-dividing of space into regions, namely:
- *Regular sub-division:* total covering of the domain using a regular mesh (the cells used are regular polygons, generally square but in some cases also triangular or hexagonal) and
- *Irregular sub-division:* guided by the spatial position of the input data.

The more largely used data models are: the Grid structure, the TIN (Triangulated Irregular Network) model, and the Hybrid structure, a combination of both. In the *Grid model* data is organised in a regular structure of cells with a given fixed dimension, or sometimes with a variable dimension. Each cell is characterised by a height value (or by any other land attribute) which is associated with a node of the grid (for example the lower left corner of the cell) or with the centre of the cell. The *TIN* constitutes an irregular structure, formed by a set of triangles in a three dimensional space. Every measured data point is used as vertices of the triangles used for modelling the surface. The *Hybrid model* is obtained by integrating local TINs within a grid structure.

From DTMs the following products are obtained. *Elevation maps* represent the height stored in a grid structure (heights matrix). The graphical display is obtained by colour coding of the elevation values. *Contour maps* are obtained by organizing data either in grid or TIN structures. *Slope maps* are calculated using the first derivatives in X and Y directions over a distance of two or more adjacent cells. *Aspect maps* are derived from the gradients calculated in the X and Y directions. They provide the direction respect to north of the maximum slope of every cell in a grid-based DTM. *3D view* can be derived either from grid or TIN based DTMs. The following parameters are needed: Point of view (distance to the object, angle of elevatiions, etc.), the vertical exaggeration, and the rotating angle of the object. The projection could be of a perspective or assonometric type. A very interesting characteristics of the 3D views is the possibility to drape a raster image over the 3D projection of elevation. A *shaded relief map* can be derived from slope and aspect information which reproduces the light and shadow situation on a 3D model. Examples for these products will be given below in the context of the Belmonte/Rada de Haro experimental site.

Acquisition of elevation data through digitisation of existing maps constitutes one of the most largely used methods. However, the resulting DTM is a third generation product, being derived from a map which was already obtained from a photogrammetric survey. In this way there are two critical factors to consider: the accuracy of the data sources (map) and the digitising process, which will inevitably cause a precision reduction, depending to the characteristics of the instrument used, the way in which the digitising is accomplished and the ability of the operator. Contour lines digitization can be accomplished manually by means of digitising tables, or semi-automatically by line-follower or raster scanner technology.

Interpolation procedures. For EFEDA, it was decided to confine the analysis to three techniques all belonging to the family of the so ca1led "contour- specific" procedures, namely, those designed to work from a contour map input. The three procedures selected for this analysis are the as fellows:

Procedure A: Linear interpolation based on the Borgerfors transformation (Borgefors 1986, Gorte and Koolhoven 1990), as implemented in ILWIS 1.3 (Integrated Land and Water management Information System) GIS, developed at ITC, The Netherlands. The Borgefors distance transform is a two-pass, recursive, non linear local image operator. In the input image, feature pixels are coded as 0, while non-feature pixels have a value *M*, which is larger than the maximum (expected) distance. Then, in the "transformed image", all *M*-values are replaced by the distances values, while 0 values remain unchanged. During the first pass, the image is processed "forward", in normal scanning order. The distance at position *F*(*i,j*), row i, column j, is thus derived by adding constant values to the distances that are already stored in the three neighbouring pixels of the row above and in the left neighbour pixel of the current row. The minimum of these values is subsequently taken and in the case the value of the current pixel is smaller than such a minimum, this is not changed. During pass 2, the result of pass 1 is processed "backward", in reverse scanning order, considering the values of the 3 pixels of the row below and the pixel at the right. Then, the interpolation of a regular grid from the rasterized contours applies. The procedure involves three raster data sets, namely:

1) Input file: The dataset of the rasterized contours;

2) Work file: A raster dataset in which every pixel is characterised by a "four tuple": (Dlij, Vlij, D2ij, V2ij). These values are ca1culated in a similar process to the distance transform: a sequence of steps processing forward and backward. At the end of the process each value denotes:

D1ij - distance of pixel (i,j) to the nearest contour
V1ij - value of the nearest contour
D2ij - distance of pixel (i,j) to the second nearest contour
V2ij - value of the second nearest contour.

3) Output file: A raster dataset in which every element contains an interpolated value Hij (elevation), calculated from the final version of the work file as follows:

$$\mathrm{Hij} = \frac{(\mathrm{D1ij} * \mathrm{V2ij}) + (\mathrm{D2ij} * \mathrm{V1ij})}{\mathrm{D1ij} + \mathrm{D2ij}} \qquad \text{(A.1.1)}$$

This is a standard linear interpolation, with the peculiarity that it actually considers minimum distances, in *whatever direction* this may require, taking into account that every point lies between precisely two contour lines. In this way, the basically global interpolation problem is changed into a local sequential process, which makes the method relatively quick. Since the computation is sequential, only information about the current and preceding rows is needed and thus present in memory.

Procedure B. Linear interpolation based on morphology dependent criteria using steepest slope and moving average techniques.

This interpolator, named MDIP (Morphology Dependent Interpolation Procedure) was developed by Carlá and Carrara, tram *CNR,* Italy (Carlà and Carrara 1987, Carrara 1988). The procedure attempts to exploit the specific topologic and morphometric properties of contour lines. Each point is interpolated by means of the technique most suited to the morphologic characteristics of the neighboring area (this is the method by which a skilful reader should determine the height of that point on a contour map). The digital contour data file is partitioned and reorganised into sub-domains (named "meshes") of 4 x 4 em (l x 1 km a 1:25000 map), each corresponding to one grid unit of the UTM network: UTM coordinates are converted into x/y Cartesian coordinates, referenced to the origin of each mesh and expressed as multiples of 4/255 em. Consequently, data processing takes place by "plotting" sequentially each mesh with its eigth neighbours on the image-processing monitor, and converting, for each data subset elevation points into continuous lines by linear interpolation.

First, the area around each grid point (node) is explored along eight search directions, recording for each direction height and distance of the two intersections with contour lines, at different elevation, that are encountered first.

From the analysis of the heights and distances of the susbet of eight points (intersections) located nearest to the grid node and by comparing these data with the corresponding heights and distances of the second eight point subset, the grid node neighbourhood can be readily c l assified as slope, depression, and peak. The first term refers to any topographic configuration, which in general reflects a slope unit, where at least two points of the first sub set have different elevations, whose difference is necessarily equal to the contour interval of the map. The second term applies when the grid node is surrounded by the points of the first subset having the same elevation, which is lower than the heights of the second subset. This outcome indicates either a true geomorphologica l depression (such as a sinkhole) or, more frequently, inadequate information provided by the search along the eight directions. The latter is likely to occur when the gridpoint falls along the course of the narrow, meandering valley with shallow longitudinal profile. Being equivocal, this situation requires further inspection, to be carried out in a subsequent phase of the procedure. The third term applies where the eight heigths of the first subset are equal and greater than those of the second subset of points, that is the grid point lies on a hill top.

The great variety of morphological configurations foud in natur is thus reduced to the three basic classes “interpolation of a point on a slope”, interpolation of a point within a depression”, and “interpolation of a point on a peak”.

Procedure C. Linear interpolation from a set of rasterized contours, as implemented in INTERCON, a module of IDRISI 4.0 GIS, developed at Clark University, Worcester, U.S.A. (Eastman 1992), and is a modification of the CONSURF algorithm developed by D. Douglas at the University of Ottawa, Canada.

The set of rasterized contours is interpolated as follows. First profiles are calculated around the four edges of the map to produce a completely closed space (and so all regions of unknown height are bounded by known heights). The four corner points, which heights must be estimated by the user from the map, and any contour that intersect the edges are used to construct the bounding profiles; as it can be seen it is particularly important that contours intersect the edges of the map sheet exactly - then a set of horizontal profiles across each row are created using the edge profiles and any contour crossed by the horizontal scan. At each cell of unknown height the interpolated profile height is recorded along with the slope of the profile at that point. In a second phase profiles are then constructed in subsequent passes vertically along the columns and diagonally from left to fight, and fight to left across the image. This requires a total of six passes: in terms of time required one may say that this have been the more time-consuming of the three procedures considered. The final height recorded for each cell is that of the profile with the maximum slope.

Comparison of the DTMs generated by the three procedures. The "contour-specific" algorithms share some common features in terms of advantages and pitfalls. All interpolated elevation points which fall near the input contour lines invariably exhibit correct values; indeed, contours subsequently calculated from the DTM well coincide with the original ones; this does not occur when applying general interpolators. This advantage, inherent to the approach, incorporates also some drawbacks. The interpolation of each grid point takes place piecewise on subdomains strictly bounded by pairs of adjacent contour lines. Hence, grid points lying on the opposite side of each contour will be processed in a way which may be very different and totally dependent on the topology and morphology of the upslope and downslope contours, respectively. This may be correct only under the hypothesis that the original contours are error-free and constitute a somewhat physical boundary between different ground relief domains. In general both assumptions are not true, but, without other sources of information (i.e. a very dense set of spot heights), it is difficult or impossible to control or reject such constraints.

A second pitfall, somewhat related to the previous one, regards the fact that elevation values corresponding to the input contour labels exhibit a higher frequency in respect to the other ones. This phenomenon derives from the vector to raster conversion operation of the input data, which takes place before interpolation is performed. All grid cells which partly or totally coincide with input contour lines are coded with the corresponding contour label, while the other cells falling between contours will be assigned elevation values dependent on the type of algorithm selected. The number of cells having values intermediate between each pair of contours is then dependent on the grid spacing used, slope gradient and local morphology. As grid size or slope gradient increases, the number of elevation points with values different from contour labels decreases accordingly.

This undisirable effect can be partially reduced bur not eliminated by decreasing the grid size, applying a more appropriate interpolation algerithm and calculating elevation in decimeters, instead of metres.

A third limitation, which applies mainly to procedures B and C, concerns the search direction.

A fourth pitfall, concerning procedures A and C, refers to thge need to both rasterize contour lines and interpolate elevations with a grid spacing much smaller than required for the final DTM matrix.

Tests proved that a reliable DTM with a spatial resolution of 25 m requires a contour rasterization and interpolation with a resolution of at least 5 m.

Belmonte - Rada de Haro experimental site. For a sub-area of the Belmonte - Rada de Haro test site, 2.5 km x 2.5 km in size (Fig. A.1.7), contour lines were calculated from the DTMs obtained with interpolators A, B, and C, all with a grid spacing equal to 25 m. For a slightly smaller subarea, 2 x 2 km^2 the same DTM data are portrayed using a perspective view (Figs. A.1.8).

Visual inspection of the contour maps enables to evaluate the relative advantages and limitations of each procedure. The first interpolator (A) yields DTM values that well reflect, with a pleasant smoothing effect, the morphology of the area, as expressed by the input contour lines. However, the method fails in calculating elevations within closed contours, that is, in the peak areas which result as invariably flat areas. This is an intrinsic limitation of the approach, which to be overcome would require a second pass applying a different algorithm. In addition, in ridge zones or in narrow valleys interpolation between adjacent contours is far to be linear. Indeed, by examining the contours with values intermediate between those of the original input lines, it is apparent that the longitudinal profiles of ridges and valleys are concave and convex, respectively. Geomorphologically, this might be the case but not necessarily everywhere.

The second procedure (B) well interpolates elevations in any morphologic setting (slopes, valleys and peaks). In addition and most importantly, it enhances small relief variations such as first order channels and minor crests. This is not surprising since it was specifically designed for this task. However, locally spurious or unrealistic elevations occur locally, as witnessed by small dosed contours on valley bottoms. These artifacts are an inherent drawback of the procedure; when two adjacent elevation points are calculated by two different algorithms a jump in height may occur. To overcome this problem, a local filter would be necessary.

The third technique (C) failed in the main valley bottom, in some ridge zones and also in the peak areas. In general too many artifacts are generated. However, locally minor relief features are well enhanced.

As a result, it can be stated that procedure A and B provide the best results. The first might be more suitable where the modelling or simulation of elevation or its derivatives would require a fairly smooth, somewhat unrealistic representation of the ground surface. The second is preferable when the smallest variation in relief should be preserved for detailed geomorphologica1 analysis or other spatial applications.

The Belmonte - Rada de Haro experimental site exhibits a fairly irregular morphology with long valleys and long ridges. Although local relief is fairly low, local morphology is very complex. An impression of the area is given in Fig. 1.5. The DTM data obtained with procedures B and C show minor artifacts in the northern portion of the area, namely along the wide, flat-lying river bed of the Záncara main course. Therefore, this data subset was filtered with a simple moving average algorithm which successfully cancelled out every artifact without affecting elevations of surrounding areas.

Fig. A.1.9 shows the elevation map generated with procedure B from the DTM together with original contour lines and coloured in eleven steps between 730 and 885 m. The elevation map based upon a TIN model is shown in Fig. A.1.10 but in a perspective view. The frequency distribution of the elevation is given in Fig. A.1.11. The slope map is displayed in Fig. A.1.12 and its frequency distribution in Fig. A.1.13. The corresponding map of the aspect is displayed in Fig. A.1.14. The map of Fig. 4.1.10 with uniform rendering can be seen in Fig. A.1.15. The topographic map of Fig. A.1.9 draped with hill shading is displayed in a plane and a perspective view in Figs. A.1.16 and 17. The land use of the area combined with contour lines is

Fig. A.1.7. Map of the Rada de Haro site with original contour lines and the position of measuring devices (1-5), compare Fig. A.1.18

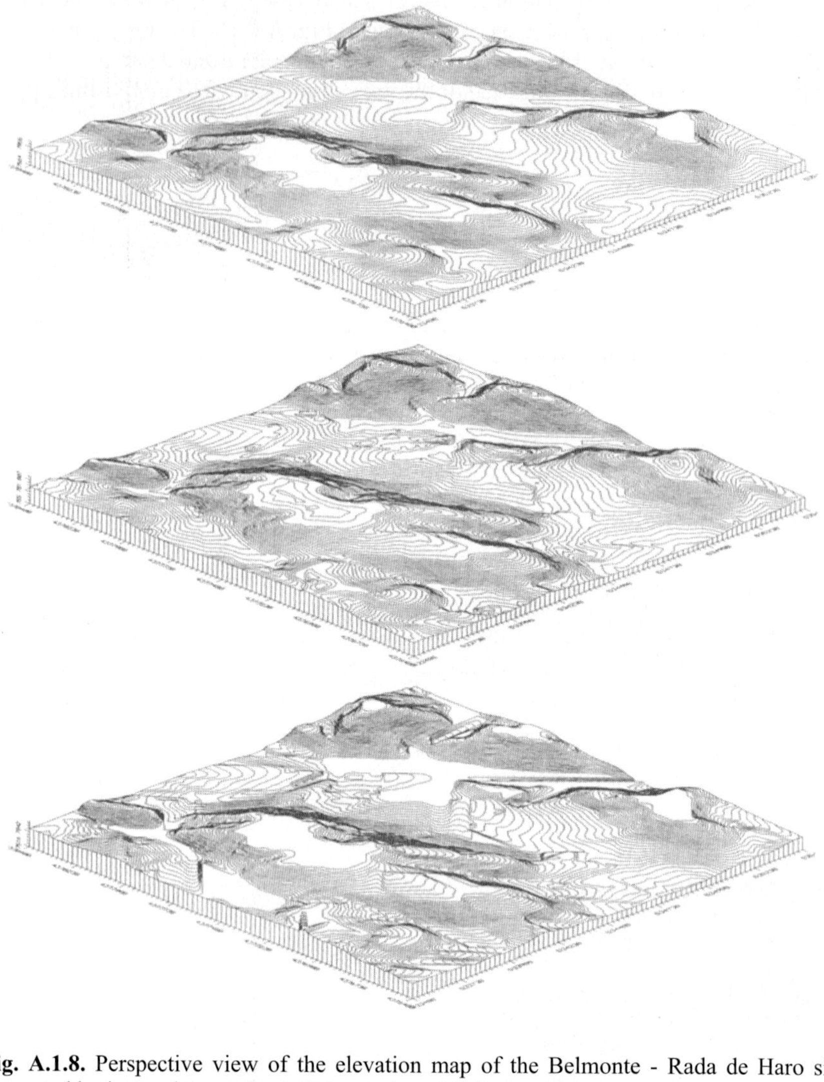

Fig. A.1.8. Perspective view of the elevation map of the Belmonte - Rada de Haro site generated by interpolator A (top), B (centre), and C (bottom)

Fig. A.1.9. Elevation map of the Belmonte - Rada de Haro site. Left - Elevation map from DTM generated by procedure B with original contour lines. Right - same as on the left side but coloured in eleven steps from 730 - 885 m elevation

Fig. A.1.10. As Fig. A.1.9 but generated from TIN model and draped with elevation classes in a perspective view

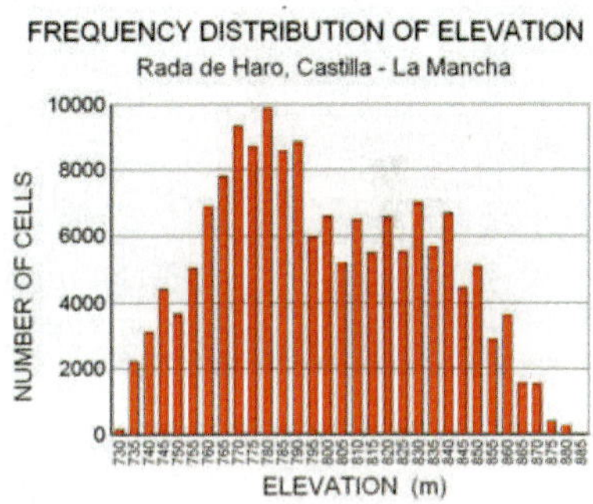

Fig. A.1.11. Frequency distribution of the elevation at the Belmonte - Rada de Haro site

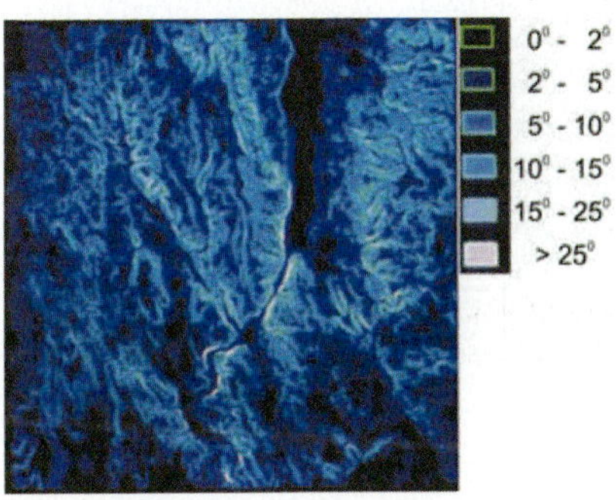

Fig. A.1.12. Slope map of the Belmonte - Rada de Haro site

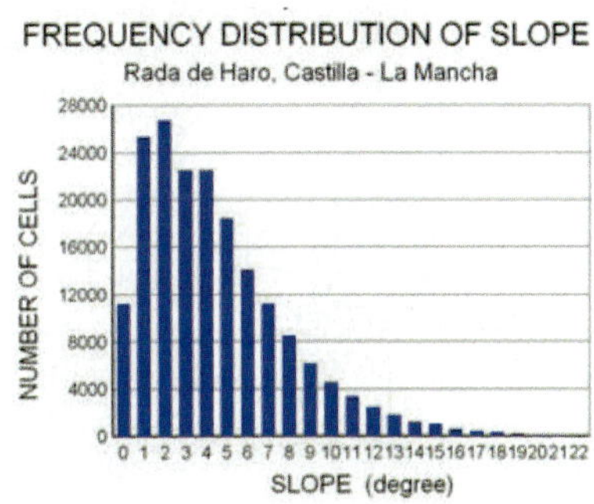

Fig. A.1.13. Frequency distribution of the slopes in the Belmonte - Rada de Haro area

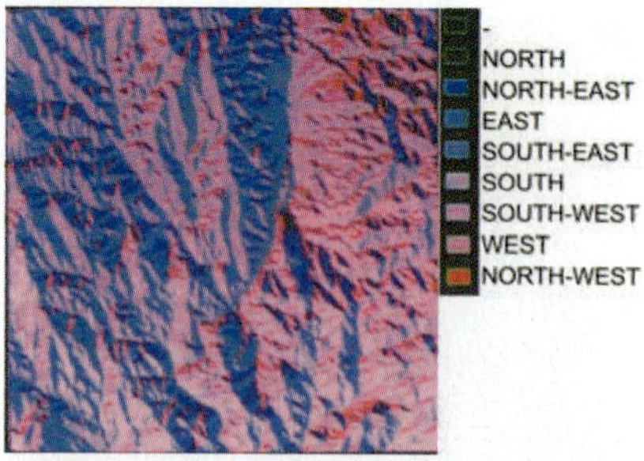

Fig. A.1.14. Aspect map of the Belmonte - Rada de Haro site

Fig. A.1.15. Elevation model as Fig. A.1.10 with uniform rendering of the area

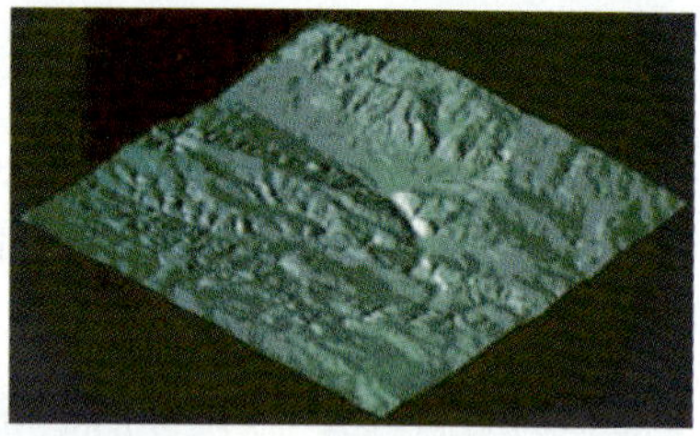

Fig. A.1.17. As Fig A.1.16 but perspective view

Fig. A.1.16. Topographic map of the Belmote - Rada de Haro site with hill shading

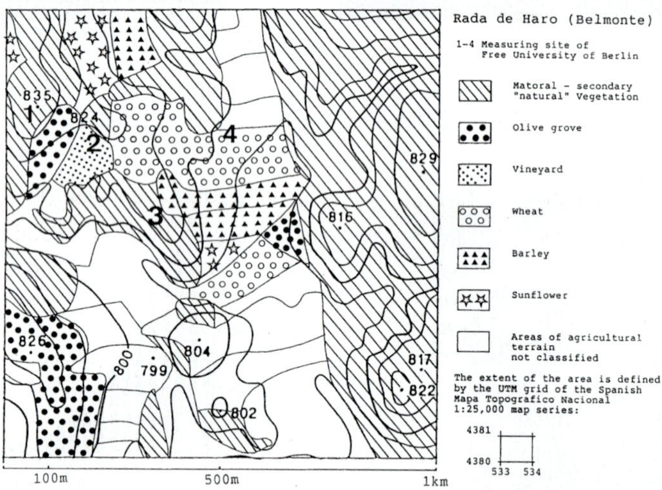

Fig. A.1.18a. Detailed topography and land use around the stations 1 - 4 in Fig. A.1.7

plotted in Fig. A.1.18a and the information about the soil types can be found in Fig. A.1.18b.

The instruments indicated in Fig. A.1.7 and A.1.18a have been positioned over matorral (1 and 3), vine (2), wheat (4) and sun flowers (5) which at the time of the experiment in June were very young.

Barrax and Tomelloso experimental sites. As compared to the Belmonte-Rada de Haro site the other two instrumented areas of the EFEDA experiment are situated in rather flat or gently ondulating landscapes. Only in the south - east part of the Tomelloso area morphology turns out to be more irregular and in the Barrax area some dolines ("hoyas") are present.

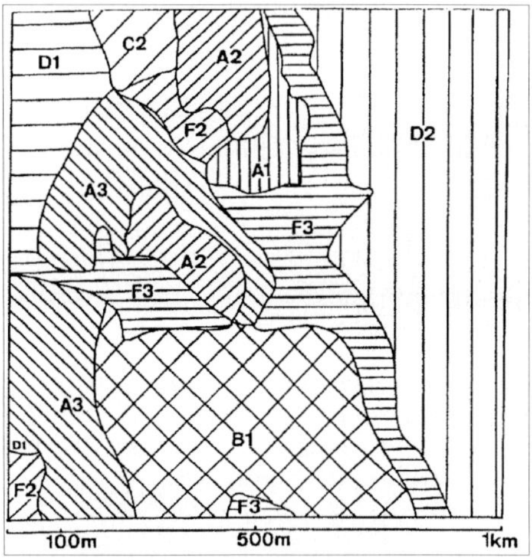

Fig. A.1.18b. Rada de Haro test area soil map. A soil based on Dolomite, B soil based on sandy limestone, C soil based on marl, D mixture under matorral and on agriculturally used areas, F soil based on colluvial and alluvial sediments

A.1.4 The Matera-Rutigliano Experimental Site

Puglia and the eastern part of the Basilicata exhibits a conspicuous low NDVI value feature (compare Figs. 6.9.5 and following). The area is known as karstic with outcropping limestone ("Murgia") and frequent occurrence of dolines (according to G. Pandiscia in RESRAPS 1994, section 5.4.1.4). The oldest part of this area is the area called "Le Murge" (see the map of Fig. A.1.19) with outcropping "Limestone of Altamura" and "Calcarenite of Gravina". The "Limestone of Altamura" consists of debris and organic limestone attributed to the Upper Cretaceous. The "Calcarenite of Gravina" found mainly along the borders of the "Murge" is a white-yellowish calcarenite of the Calabrian age. The limestone surface is slightly folded and in block forms. Two groups of faults cross the area from NW to SE and from SW to NE. These faults and others of smaller size underwent karst processes. Today, due to erosive phenomena, hollow incisions are visible which must have developed in periods of wetter climate. These incisions are called "gravine" (Gravina of Matera, Gravina of Lateraza, etc.). The superficial karst formations are represented by bowl-shaped dolines with red soil on their basis and diameters ranging between 50 and 100 metres, in exceptional cases also 400 - 500 metres. Due to cracking an karst processes the "Limestone of Altamura" is permeable. Infiltration and circulation of water vary strongly. After rain the surface dries quickly and latent heat flux drops accordingly. The "Calcarenite of Gravina" is only moderately permeable, owing to its porous structure.

In the southern part one finds rich clay deposits ("Sub-Apennine Clay"),

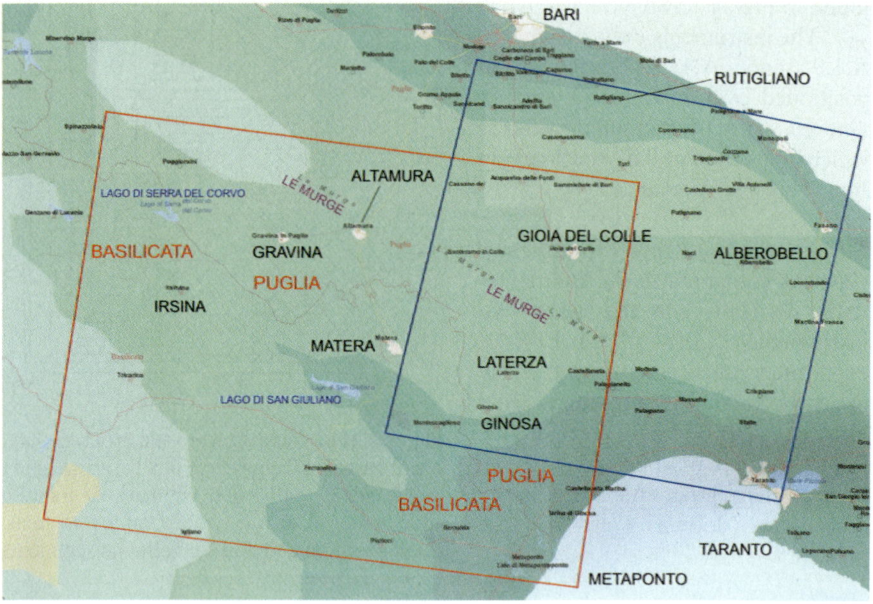

Fig. A.1.19. Map of southern Italy with the position of the areas of the NDVI scenes of Figs. A.1.21 and 22. The red rectangle shows the position of the scene of 03.06.1994 and the blue one the scene of 20.05.1994

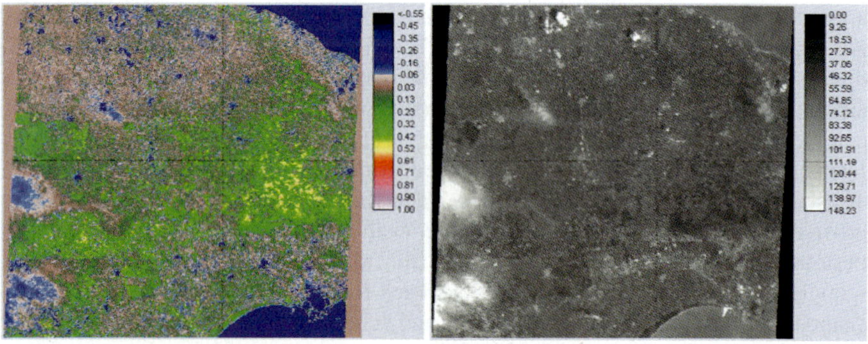

Fig. A.1.20. NDVI image derived from SPOT channel 2 and 3 data of 20.05.1994. The black and white scene (SPOT channel 1) shows the position of clouds. The area of Matera is partly covered by the cloud in the south-west corner of the figure. The small white pixels above that cloud and elsewhere are quarries

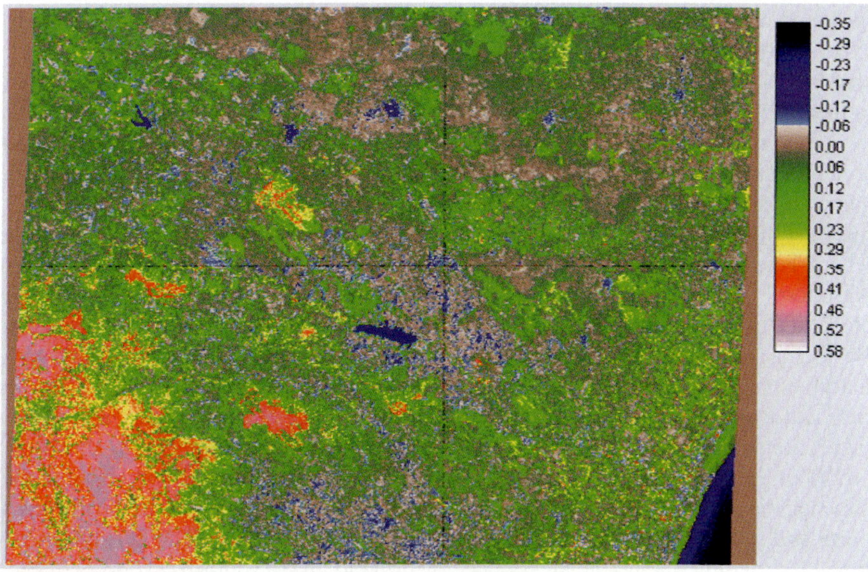

Fig. A.1.21. NDVI image derived from SPOT channel 2 and 3 data of 03.06.1994. Matera lies in the centre of the cross indicated by the thin black lines

accumulated layers of quartzy-calcareous sand ("Sand of Monte Marano") which is heterotropic to a calcarenite formation ("Calcarenite of Monte Castiglione"), and reddish polygenetic conglomerate ("Conglomerate of Irsina"). The landcape shows a hilly relief which slopes towards the sea. The here typical layers of clay, sand and conglomerate are the cause of various degrees of erosion (Fig. 1.31). This starts at surface level and drops deeply to the mid section where sand and conglomerate crop out. The underlaying less steep clay slopes are undergoing a process of gully and landsliding. The hilly reliefs are crossed by the valleys of the Bradano river, the Brasento river, and smaller water courses, with outcropping alluvial terraced deposites of sandy-pebbly orign.

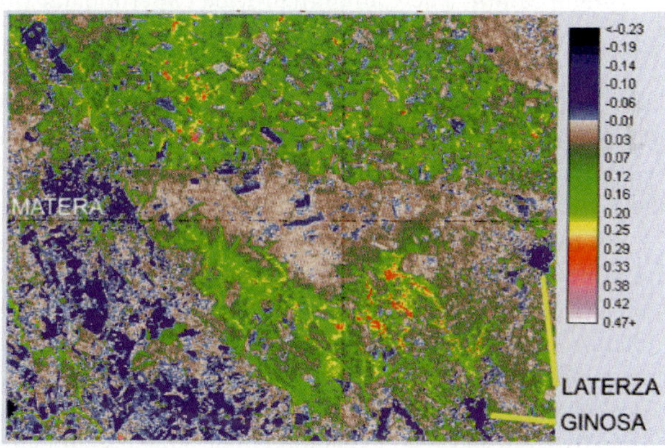

Fig. A.1.22. NDVI of the area near Matera showing the contrast between the hilly area in the centre and the lower lying agricultural areas. SPOT scene of 03. 06. 94

The mixture of "murge" and clay deposits splits the land into non-cropped areas (grass, wood, non-vegetated) covering 47% of the Basilicata region and an extensively cropped area (48%), mainly cultivated with winter wheat, and a small area (4.6%) of intensive tree-crops. These contrasting areas can clearly be identified in the NDVI scenes shown in Figs. A.1.20. and A.1.21.

Two short experimental studies have been conducted in the Matera - Rutigliano area to obtain first hand information about the relationship between the information gained by satellite imagery and the surface structure. Two positions were selected for this operation: a hill-side (480 m) murgia area (Murgia di Avino) south-east of Matera where Telespazio runs a receiving station (40.65°N, 16.7044°E), and at the Forest Agronomical Research Institute "Istituto Sperimentale Agronomico" (I.S.A.) located close to Rutigliano near Bari. The area near Matera was just at the border between Basilicata and Puglia. In this area exist large differences between the dry hillside with marginal vegetation (see Fig. A.6.5.13) and lower lying areas (Fig. A.6.5.11) with rich water occurrence and agriculture (mostly wheat). Measurements and visual inspections were also conducted in the surroundings of Matera. Fig. A.1.22 shows the NDVI of the area Matera - Laterza - Ginosa with the clear distinction between the elevated murge area and the agricultural areas in the deeper lying (370 m) fertile areas north and south of it. In the larger scale images Figs. A.4.20 and 21 it can be seen that the agricultural areas are surrounded by dryland. In the south-west corner of Fig. A.1.21 is an area with large NDVI (red colour) which indicates the woods one finds at the higher lying and wetter areas of western Basilicata. In comparing the NDVI values derived from SPOT data with those derived from AVHRR or Landsat-TM data one has to keep in mind that the filter functions differ. The scale of the SPOT NDVI is shifted to lower and even negative values. The bright spots seen in the black and white image of Fig. A.4.1.20 and corresponding dark blue areas in Fig. 4.1.22 are querries (Fig. A.6.5.9).

A.1.5 The Tuscan Experimental Sites

Campiglia M.ma. The Campiglia M.ma experimental site partly to be seen in Figs. A.1.24 and A.6.5.15 was used for several short periods between 1977 and 1999. In 1996 (22.04. - 03.05) the energy budget in a typical agricultural unit of the Cornia valley was investigated in parallel with the measurements made in the Radicondoli area (see section 6.6). A map of the area in Fig. A.1.23 shows the land-use assessed by Schwiebus (1997) together with Spot images showing the signals of channels 2 and 3 and of the NDVI computed from these values.

Sites in the hilly area north of this point have been used inter alia to investigate the effect of slopes on the energy budget and to study the spectral signatures of typical Mediterranean plants (see e.g. Figs. A.6.5.15-20). The components of the radiation budget measured at an eastward directed and a westward directed slope demonstrate the shading effect (Fig A.1.24). It can also be seen that there is a partial compensation of the albedo in the sense that for a hilly area the reflectance measured vertically from space and averaged over large enough an area that covers differently oriented hill-slopes represents approximately the average value of the landscape (Fig. 5.10).

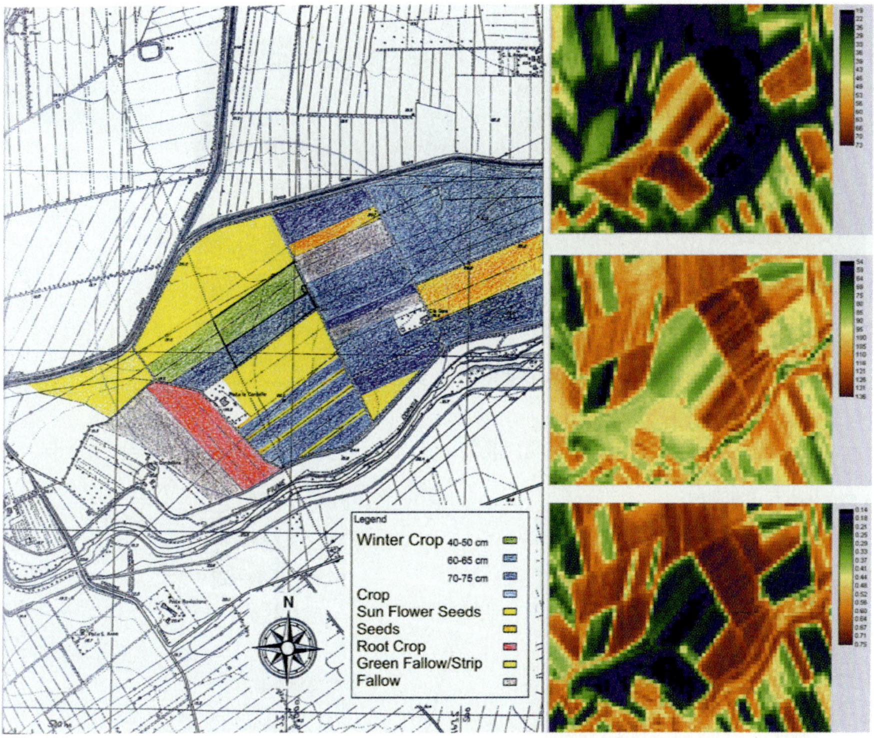

Fig. A.1.23. Experimental site near the Cornia river, Cafaggio, Campiglia M.ma, Tuscany in an agricultural area. To the right are the SPOT images of channel 2 (top) and 3 (centre) as well as the NDVI (bottom, scale from 0.14 - blue to 0.75 - dark red)

Radicondoli and Piano di Rosia. The Radicondoli and Piano di Rosia sites are inland sites (Figs. 5.32-34 and 5.50) at which radiation quantities as well as sensible and latent heat fluxes have been masured over forests, pasture and agricultures. A number of the results have already been reported in different chapters of this book. Specific highlights were the combination of AVHRR and Landsat-TM data as reported in section 6.6 to obtain annual trends and the combination of satellite data with ground based flux measurements using beside conventional evaporation measuring devices *inter alia* a scintillometer and sap flux equipment. Supplementing the measurements made mainly over agricultural land and macchia at other locations, here and at the coastal site of ***San Rossore*** (section 6.13.7), to a large extent, measurements have been made in and over forests.

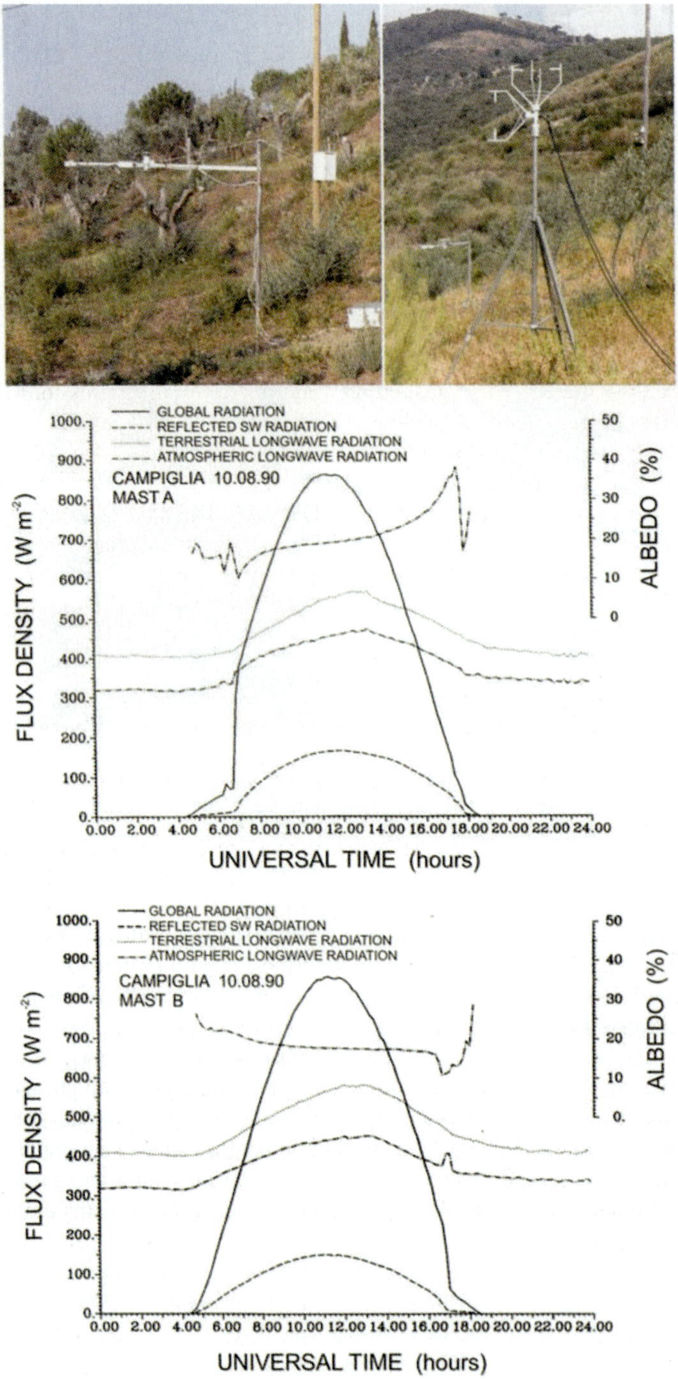

Fig. A.1.24. Components of the radiation budget measured at a westward (top) and an eastward (bottom) directed slope. The upper dashed-dotted curve is the albedo (scale to the right)

A.1.6 Reflectance and Albedo Measurements in Africa

Several investigations have been carried out in the Republic of Niger and in Sudan of which surface reflectance and albedo measurements are of interest for comparison with Mediterranean measurements in the context of desertification. Some of these measurements have already been quoted in previous chapters.

Fig. A.1.25 on the following page (for space reasons without caption) gives an overview of the conditions faced in the surrounding of Ibecetene in the northern part of the Republic of Niger southwest of Agadez. The central part of the figure is a composite of Landsat images and the small circle below the centre of the image is the position of Ibecetene. Agadez lies in the upperpart near the right margin. The area is characterized by a large outcrop of lateritic material as can be seen in the south westerly quadrant of the image. The surrounding diagrams give an impression of surface types with large differences in albedo and reflectance.

On the left side the albedo near the Ibecetene site covered with 50 % respectively 75 % of dry cram-cram (*Cencrus biflorus*), a short living grass, can be seen. Cram-cram is a common vegetation type in that area and can be seen in its dry state in Fig. A.1.26. Below that diagram in Fig. A.1.25 albedo measurements and their mean over a sand covered bush to be seen together with a reflectance measurements of the sandy surface and the albedo of a green bush. At the bottom two pairs of albedo measurements of lateritic soil can be seen which was found near Ibecetene.

One the right side results of albedo mesurements made on the way from Ibecetene through Abalak to Agadiz are presented in the first two graphs showing the large differences of soil albedo and in addition a measurement over a green bush. The lowest diagram shows the albedo of a small bright sand dune and dark laterite with was found in the neighbourhood of the dune near Kaita.

Fig. A.1.26 is a typical view of the area of Ibecetene. It shows a measurement with the spectral albedometer over a bush typical for that area and a soil which is covered by dry cram-cram. Results of albedo measurements made over thorny bushes in different states of dryness are shown in Fig. A.1.27. Some spectral measurements of the directional reflectance of laterite, reproduced in Fig. A.1.28, show that the reflectance of this very dark material is quite variable if looked at from different directions under various solar azimuth angles. In Fig. A.1.29 the measuring site near Kaita can be seen with the laterite in the foreground and in the background a bright dune the reflectance of which is discussed in Appendix 6. The picture is taken towards SW several degrees off the position of the sun. In the lower part of the figure a close-up of the lateritic material can be seen. The upper scene shows the border of the lateritic plateau with a view towards NW.

Similar measurements have been made in northern Africa, in Sudan. One set of data from this area is dealt with in Table A.1.1. Here from interference filter radiometer measurements at the ground the responses in the AVHRR channels 1 and 2 were simulated and then projected to the top of the atmosphere by applying an atmospheric model. These simulated data are then compared with actual AVHRR measurements over the area. At these places it has often been worked at the ground with interference filter radiometers which, in such remote areas, are more easily manageable than spectrometers.

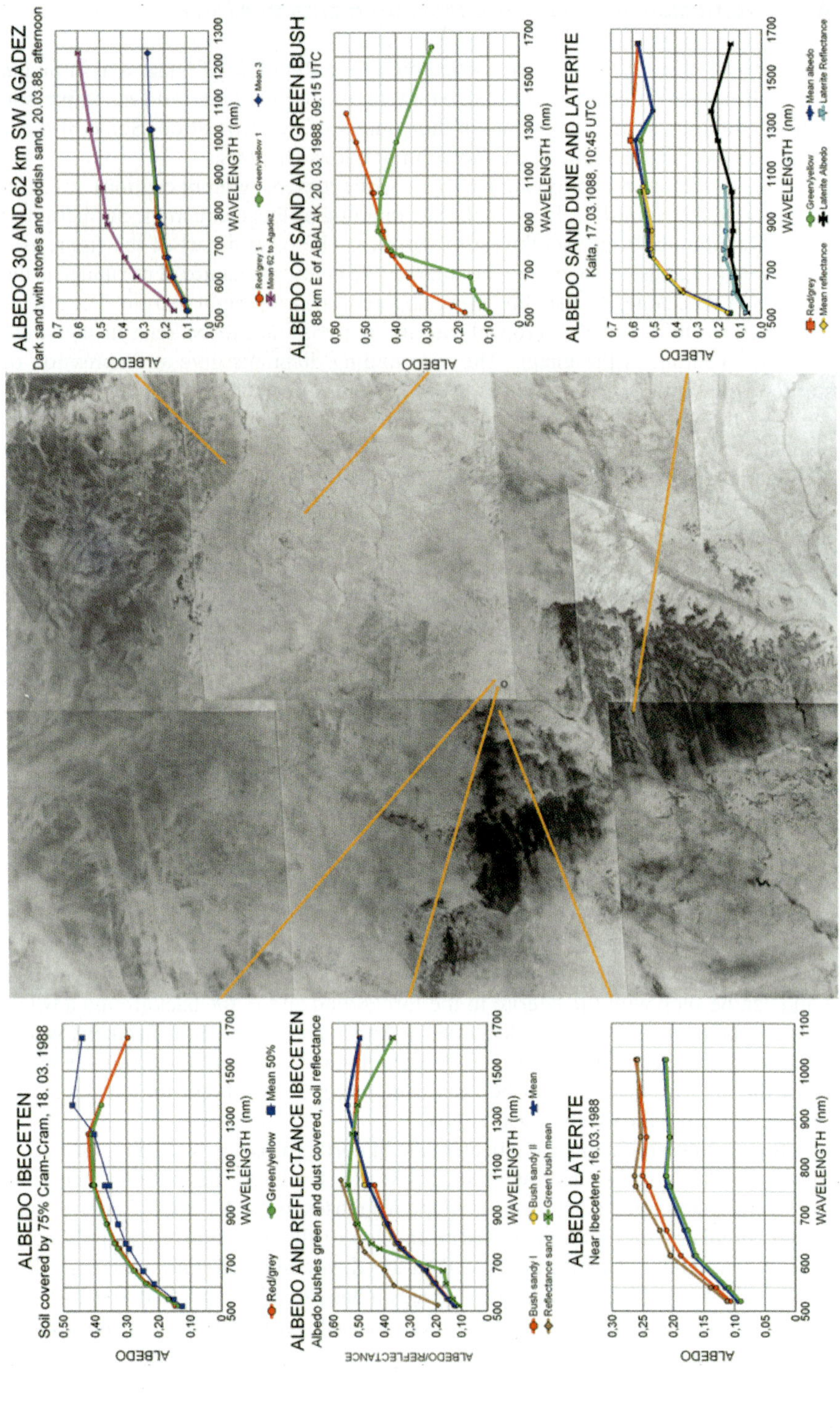
ALBEDO IBECETEN
Soil covered by 75% Cram-Cram, 18. 03. 1988
ALBEDO
WAVELENGTH (nm)
Red/grey
Green/yellow
Mean 50%
ALBEDO AND REFLECTANCE IBECETEN
Albedo bushes green and dust covered, soil reflectance
ALBEDO/REFLECTANCE
WAVELENGTH (nm)
Bush sandy I
Reflectance sand
Bush sandy II
Green bush mean
Mean
ALBEDO LATERITE
Near Ibecetene, 16.03.1988
ALBEDO
WAVELENGTH (nm)
ALBEDO 30 AND 62 km SW AGADEZ
Dark sand with stones and reddish sand, 20.03.88, afternoon
ALBEDO
WAVELENGTH (nm)
Red/grey 1
Mean 62 to Agadez
Green/yellow 1
Mean 3
ALBEDO OF SAND AND GREEN BUSH
88 km E of ABALAK. 20. 03. 1988, 09:15 UTC
ALBEDO
WAVELENGTH (nm)
ALBEDO SAND DUNE AND LATERITE
Kaita, 17.03.1088, 10:45 UTC
ALBEDO
WAVELENGTH (nm)
Red/grey
Mean reflectance
Green/yellow
Laterite Albedo
Mean albedo
Laterite Reflectance

Fig. A.1.26. Spectral albedometer measurements over a typical bush

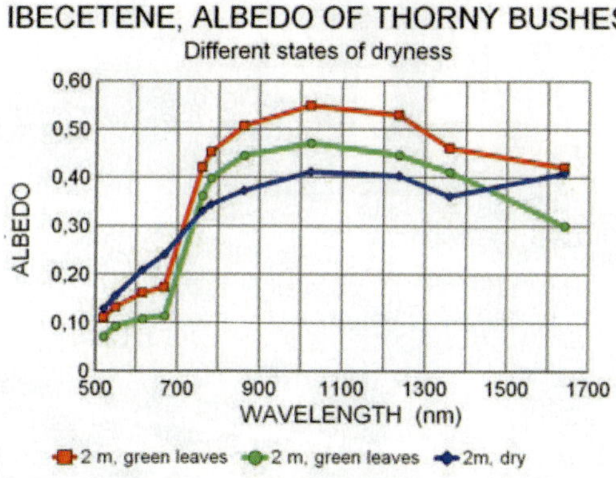

Fig. A.1.27. Spectral albedo of thorny bushes

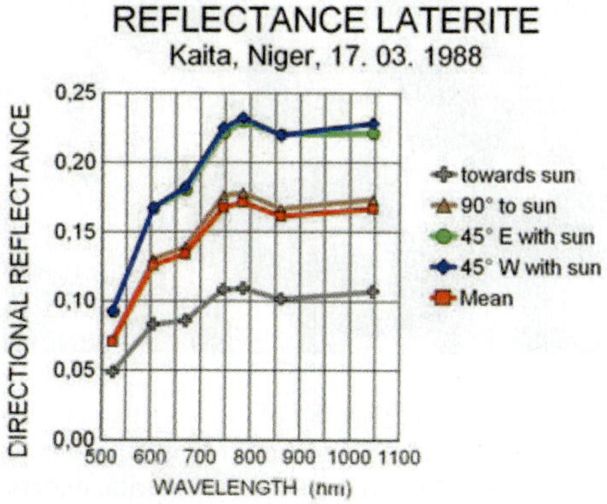

Fig. A.1.28. Spectral reflectance of laterite

Fig. A.1.29. Upper scene: Boarder of the lateritic plateau north of Kaita with a view towards NE. Centre: Measuring site near Kaita viewed towards SW. Bottom: Close-up of lateritic material in full sunshine

Table. A.1.2. Spectral albedometer measurements in Sudan, November 1989, and comparison after F. Fell (1991). The spectral values are mean values of four single measurements. Pseudo AVHRR channels are values interpolated from the ground measurements for the spectral bands of AVHRR at noon and for the actual overpass time of the NOAA satellite. At the time of the satellite overpass the solar zenith angle was around 74° and the aspect zenith angle was 9.2° - 12.8°. The discrepancy between the AVHRR channel 2 and the ground measurements may be due to incorrect water vapour data in the LOWTRAN correction code

	Wadi Abu Geidum	Djebel Fururi	Qoz El Harr	Buweiha/ Gezira
Geographical Coordinates	16.20°N 32.93°E	16.06°N 32.88°E	16.71°N 32.68°E	15.23°N 32.64°E
Size	≈ 25 km^2	≈ 15 km^2	≈ 85 km^2	>100 km^2
Date	16. 11. 89	19.11.89	24.11.89	28.11.89
Spectral channel				
521 nm	0.13 ± 0.01	0.15 ± 0.01	0.14 ± 0.01	0.12 ± 0.01
549 nm	0.16 ± 0.01	0.18 ± 0.02	0.19 ± 0.02	0.14 ± 0.01
615 nm	0.22 ± 0.02	0.31 ± 0.02	0.32 ± 0.02	0.19 ± 0.02
669 nm	0.24 ± 0.02	0.36 ± 0.02	0.37 ± 0.03	0.21 ± 0.02
760 nm	0.28 ± 0.02	0.42 ± 0.03	0.44 ± 0.03	0.23 ± 0.02
782 nm	0.28 ± 0.02	0.43 ± 0.03	0.45 ± 0.03	0.24 ± 0.02
862 nm	0.28 ± 0.02	0.43 ± 0.03	0.46 ± 0.03	0.25 ± 0.02
1024 nm	0.29 ± 0.02	0.46 ± 0.03	0.49 ± 0.03	0.26 ± 0.02
1238 nm	0.36 ± 0.01	0.54 ± 0.03	0.56 ± 0.04	0.29 ± 0.02
Pseudo noon AVHRR Ch. 1	0.22 ± 0.03	0.32 ± 0.04	0.33 ± 0.04	0.19 ± 0.02
Pseudo noon AVHRR Ch 2	0.28 ± 0.05	0.43 ± 0.07	0.45 ± 0.07	0.24 ± 0.04
Pseudo actual AVHRR Ch. 1	0.24 ± 0.03	0.34 ± 0.04	0.38 ± 0.04	0.20 ± 0.02
Pseudo actual AVHRR Ch. 2	0.30 ± 0.05	0.45 ± 0.07	0.53 ± 0.08	0.25 ± 0.04
AVHRR Ch. 1	0.26 ± 0.05	0.35 ± 0.06	0.39 ± 0.07	0.19 ± 0.04
AVHRR Ch. 2	0.36 ± 0.07	0.50 ± 0.09	0.56 ± 0.10	0.26 ± 0.05

More results of spectral reflectance measurements, partly relevant for the re-calibration of satellite instruments over deserts, are presented in section A.6.6.

A.1.7 General Results

The EFEDA and related projects lead to

(a) a general improvement of experimental and model techniques to study complex land-surface processes,

(b) a thorough understanding of the importance of and interactions between the land-surface processes during a drying period and thus for aridification which is one precurser of desertification, and

(c) the conclusion that the sustainable development of the investigated area is in a very critical phase, if

- the water resources are not recharged (e.g. due to climatic change respectively variability) to a degree that they are able to support intensive irrigation,
- the available water cannot be spent more efficiently (e.g. by application of modern irrigation techniques and/or change to different species), and
- the cultivation of species with roots reaching moist soil layer in some depth does not receive continued economic support.

The intensive measuring phase of EFEDA was a pilot experiment during which only processes could be investigated with short characteristic timescales and under dry conditions. But the occurrence of such dry conditions depends on long term climate variability respectively climate change and the impact of dry spells in alternation with wet events on the condition of the land-surfaces including its vegetation *inter alia* is a function of the time constants of sub-surface hydrological processes. These are much longer than could readily be studied within EFEDA. To receive full return from the achieved insights into the investigated land-surface processes it therefore is necessary to lead the developed methodology into an operational mode that relies upon very few but well equipped ground stations (so called "anchor stations") and to extend especially the hydrological studies over an number of years and into hilly terrain.

One important task during the EFEDA pilot experiment was the re-calibration of AVHRR satellite data and the validation of information derived from these data (Koslowsky, 1997). The continuation of the evaluation of satellite data and some scattered ground truth operations throughout the Mediterranean area (Castilla-La Mancha, Tuscany, Messara Valley on Crete, and Israel) during the years 1994 - 1999 nevertheless allow to embed the measurements made 1991 and 1994 in Spain into a larger space and time scale. Ten years of satellite data time series make it possible to analyse the interannual variability of the Castilla-La Mancha area in the context of changes all over the Iberian Peninsula. Moreover for few years carefully calibrated AVHRR data are now available for the whole Mediterranean area which enables first studies of synergisms throughout the basin.

The draw back of such regional to large scale EFEDA-type experiments so far was that for each single parameter which enters the process studies the measuring equipment had to be implemented or the relevant data had to be assessed by labourious efforts from various offices. To embed such regional short-term investigations into the larger scales of time and space variability a long term data set of at least a reduced spatial coverage must be available. Fortunately in the case of the Castilla-La Mancha experiment a long term hydrological data set was available at the University Complutense in Madrid. Also the meteorological support was excellent, weather maps and general meteorological data were made available for the time of the experiment. But for comparison with the larger scale satellite data so far no surface data backup exist. For the correction of the satellite data a study was undertaken within the relevant EC projects to determine the water vapour field over the Iberian peninsula based upon a ECMWF re-analysis. Such data must in the future be available in nearly real time to correct the satellite data.

EFEDA-type experiments gained a lot from already existing data and served as a vehicle to bring these data together, relate them to each other and to aggregate them into a new level of understanding and monitoring of processes. This aggregation often can be accomplished by models. EFEDA-type experiments per se, mainly because of their short duration, do not solve environmental problems, sometimes they indicate where problems may arise, but their main benefit is that they develop and prove new approaches and methodologies how to *attribute* changes to the "Global Change" problem. The problem is not to provide experimenters with additional data but to gather and analyse those data which have been identified by EFEDA-type experiments in a quasi-operational manner for time scales which are recognized as essential to isolate trends from interannual variability.

In an continuation of the efforts started with EFEDA and especially if it is desired to transform this approach into a large scale GEWEX-type study it would be necessary to gather in addition to meteorological and hydrological routine measurements at a certain number of "anchor stations" placed into representative areas around the Mediterranean and to merge the data of the operational system with this scientific network. The emphasis of the "anchor stations" must lie on those quantities which are not measured by the operational network. In the past there have been often a misunderstanding in this respect as standard climatological equipment was offered as "anchor station". But it is now feasible to infer from satellite measurements in addition to the "primary" quantities like surface reflectance, surface temperature, and various vegetation indices also information such as on vegetation quantities, fluxes between surface and atmosphere, and soil top moisture. These "secondary level" data, which are in the first instance Land-surface Change Indicators need to be "clamped" to reality by continuous observations at anchor stations - and from time to time field experiments to remove uncertainties - to enfold the whole observing potential of measurements made from space.

Considering the use of satellites it is mandatory that

- atmospheric data (three dimensional temperature, pressure, water vapor and aerosol fields) become available in quasi real time which allow in combination with topographic models to correct the measurements made in space for atmospheric effects, and
- that anchor stations be established around the Mediterranean Sea to continuously control at representative points the validity of the information deduced from the satellite measurements.

To embed short term EFEDA-type experiments into long term developments it is essential to obtain in addition to operationally available data also those data which are not operationally archived and are needed especially to expand local measurements into area averages. It is not as important if a few hectare of unmanaged forest burns down in an area where the environmental conditions allow a fast regrowth as it is if a large agricultural area slowly gets into an irreversible permanent water shortage situation because of extended droughts during the summers and not enough winter rain to compensate for the water evaporating during summer and extracted for households, irrigation, industry and tourism.

Appendix 2 Meteorological Terminology

Fig. A.2.1 summarizes meteorological phenomena that are observed in the Mediterranean area. The various terms and others used in the previous chapters are explained in the following Glossary.

Adiabatic: Heating (cooling) of a vertically moving air parcel due to compression (expansion) without exchange of energy with the surroundings.

Atmospheric Boundary Layer (ABL): Atmospheric layer in which the surface still influences the wind speed and direction, the concentration of water vapour, air pollution, and energy transfers. It ranges from the surface to a height of a few decametres in a stable atmosphere and up to 2-3 km in an atmosphere with strong convection. .

Azores subtropical high pressure cell: Quasi-stationary high pressure system with centre near the Azores (→ westerlies).

Blocking: Dynamical situation when the wave pattern of the westerly air flow (→ westerlies) makes a wide excursion like an Ω or an Ω upside down (→ cut off low). In the first case the Ω surrounds a high pressure cell, in the second case the inverted Ω surrounds a low pressure cell.

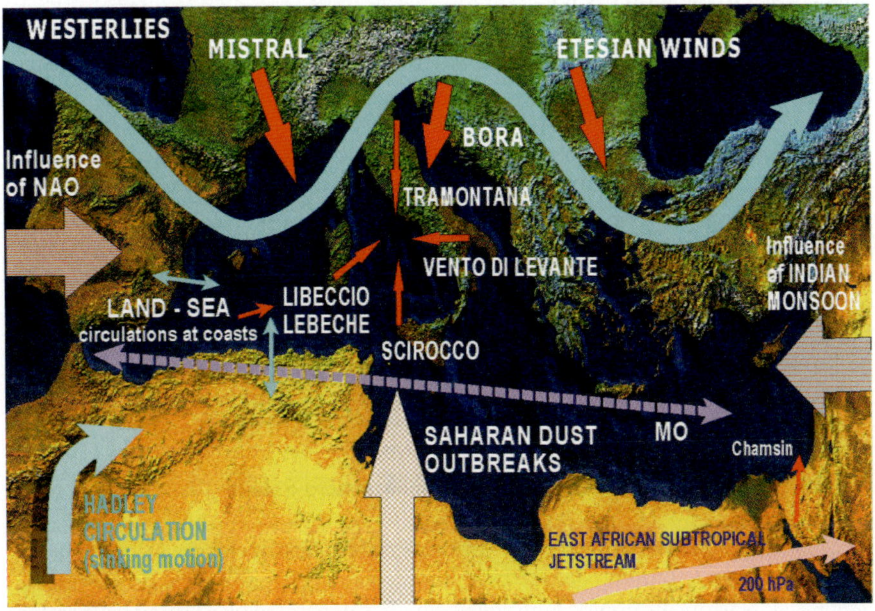

Fig. A.2.1. Meteorological phenomena experienced in the Mediterranean area

Bora: Air flow from mountains to sea level (→ katabatic winds).

Chamsin: Hot air flow from the Sahara into the eastern Mediterranean area loaded with desert dust.

Chinook (Föhn): Air flow lifted adiabatically (→) up to a mountain while losing its water vapour due to condensation and sloping down at the other side as dry air warming up by compression (→ adiabatic).

Cold air drop: Blocking (→) with a large southward excursion and cold air intrusion into the warmer air in the south.

Convection: Turbulent upwelling of air that is heated either from the hot underlying surface or by transformation of latent heat (energy used to evaporate water and set free at condensation) into sensible heat. If condensation occurs one speaks about wet convection otherwise of dry convection.

Cut-off low: Describes the situation if the blocking (→) leads to a complete cut-off of a cold air drop (→).

El Niño - Southern Oscillation (ENSO): El Niño is an event that every few years occurs and transports equatorial warm water to the South American coast from where it spreads in a large tongue westward over the Pacific Ocean. The Southern Oscillation describes the variability of the pressure gradient between Australia and South America. Both phenomena have been detected independently but are connected which leads to the term ENSO.

Etesian winds: Cold continental winds from northern directions in the eastern Mediterranean area.

Hadley cell: Circulation system driven by the strong convection (→) near the equator that forces upward wet warm air which cools adiabatically (→) due to expansion and loses its water vapour due to condensation. The latent heat (→) released during this process enables the air to raise further. The process stops at the tropopause (→) in 15 to 18 km height (sometimes with overshooting if convection is very strong) when the air becomes dry and extremely cold. The air is then forced to move towards the poles but because of the rotation of the Earth its poleward directed flow turns to the East and around 30° latitude the air descends. During the descent the air warms by adiabatic compression. Most of it flows directly back towards the equator in the ABL while a fraction is transported further northward by travelling eddies (→).

Harmattan: Dry wind from north-east over the eastern Sahara

Icelandic low: Quasi-stationary low pressure system near Iceland

Jetstream: Narrow air flow at high speed which is generated by a strong temperature gradient.

Katabatic wind: Cold dense air drains downslope into depressions and valleys (→ Bora)

Land-sea circulation: Wet cool air moves because of pressure differences at low levels from the cooler sea to the warmer land where it is warmed up, moves upward and returns at elevated levels to the sea (→ monsoon). Land-sea circulations are summer phenomena when the land warms up rapidly in the morning.

Latent heat: Energy transported by atmospheric water vapour in an amount that is needed to evaporate the water. This energy can be released as sensible heat (→) if the water vapour condenses into clouds due to adiabatic (→) cooling.

Libeccio (lebeche): Warm wind from the south-west in Italy (from west through the Strait of Gibraltar)

Mediterranean Oscillation (MO): Antiphase relationship between meteorological state quantities (e.g. pressure) in the western and eastern Mediterranean basin.

Mistral (maestrale): Northerly winds channelled through the Rhone valley and reaching Italy as cold winds from the north-west.

Monsoon: Intrusion of wet air from the sea to the land generated by pressure gradients that develop between warm land surfaces and the cooler sea. While large scale sea-land circulations are called monsoon, small scale circulation generated by the same principle are called land-sea circulation (→) systems or sea-breeze.

North Atlantic Oscillation: Variation of the pressure gradient between the Azores and Iceland which determines the direction of the westerlies.

Ridge: An elongated region of high barometric pressure that develops at the poleward side of the westerlies (→).

Rossby waves: Wave-like pattern of the zonal circulation at mid latitudes (→ westerlies) caused by the airflow across mountain ridges.

Sensible heat flux: Air flow which transports energy to be measured by its temperature along a temperature gradient.

Scirocco: Humid warm air flow from Africa reaching Italy from south to south-east.

Trade winds: Air flow in the ABL (→) from the sub-tropics towards the equator directed to south-west in the northern hemisphere.

Tramontana: cold and dry northern wind over Italy.

Travelling eddies: Mesoscale rotating and eastward progressing low pressure systems that determine weather at higher latitudes. They are generated in the west-wind zone

(→ Rossby waves). Due to its counter-clockwise rotation it transports in the northern hemisphere warm air from south to north and cold air from north to south.

Tropopause: Region up to which convective processes in the troposphere (→) reach. Above the tropopause the stratosphere starts. The height of the tropopause is approximately 18 km in the tropical region and much lower in polar latitudes.

Trough: An elongated region of low barometric pressure that develops at the equatorward side of the westerlies (→) and can lead to blocking (→) and a cut-off low (→).

Troposphere: Atmospheric layer between the ABL (→) and the tropopause (→).

Walker circulation: Longitudinal circulation in the tropics with high reaching convective cells over the continents (→ ENSO).

Westerlies: West wind belt between about 40° - 60° latitude generated by the temperature gradient between the subtropical warm air and the cold polar air. It meanders around the globe in a wavelike pattern (→ Rossby waves) primarily due to the effect of the mountain ridges the flow has to cross.

Appendix 3 Soils

A.3.1 Description of Soils

The soils of the EFEDA experimental area have been analysed in detail by Sanchez Diaz et al. (1992, 1994, 1996). 103 soil types which can be grouped into eleven major classes were identified and mapped.

In the description of soils the sequence of layers, called "*horizons*", plays an important role. These horizons can be described as follows:
H - dominated by (partly) undecomposed organic material, which is saturated with water for a prolonged period of the year
O - dominated by (partly) undecomposed organic material, which is not saturated with water for a prolonged period of the year
A - Topmost mineral layer, in which all or most of the original rock structure has been obliterated and humified organic matter has accumulated and mixed with the mineral fraction.
E - eluvial mineral horizon in which a major part of silicate clay, iron- and aluminium-oxides have been lost through percolation leaving a higher concentration of sand and silt particles in the upper layer than below.
B - subsurface mineral horizon with characteristic structure, texture and colour due to illuviation ("washing in") of masterial and/or weathering of the parent material.
C - contains unconsolidated parent-like material altered by pedogenetic processes.
R - consolidated bed-rock.

A.3.2 Soil Types

The soils found in the Castilla - La Mancha are briefly described here based on the FAO Soil Units key (FAO/Unesco Soil Map of the World, 1988).The accompanying photographs are taken from Sanchez Diaz (1992, 1994). In Fig. A.1.3. these soil classes are discerned by colors (the subdivisions within each class are distinguished by numbers). Spatial features down to 100 m are resolved.

Arenosols. Deep sandy soils developed from quaternary alluvial deposits or sand of eolian orign. Haplic Arenosols with only one ochric A horizon and not calcareous. Cambic Arenosols show colouring or a B horizon (weathered, > 8% clay, skeleton structures < 50%, CEC > 16 cmol(+) kg^{-1} clay or primary minerals > 10%).

Calcisols. Soils with carbonates. Luvic Calcisols with argic B horizon (enrichment of clay). Petric Calcisols with petrocalcic horizon (> 10 thick, +Mg, +Si, organic matter < 2%, CEC 20 cmol(+) kg^{-1}, variable in depth).

Cambisols. Young, slightly weathered fine textured soils. Cambic B horizon. Adequate supplies of nutrients, strong and stable structure. Eutric Cambisols have lost their carbonates, base saturation 50% or more. Calcaric Cambisols are calcareous in 20 - 50 cm depth. Chromic Cambisols have a strong brown to red B horizon.

Fig. A.3.1. Calcisols. From top left to bottom right: Haplic Calcisol from the Barrax area, irrigated Haplic Calcisol (degraded), Haplic Calcisol from the Rada de Haro area, Luvic Calcisol from the Tomelloso area, irrigated Petric Calcisol, and Petric Calcisol from the Tomelloso area

Fig. A.3.2. Calcaic Cambisols. To the left from dry land near Barrax, centre forest soil, and to the right from the Tomnelloso dry land (underneath cereals and vineyards)

Fig. A.3.3. Haplic Kastanozen (forest soil)

Fig. A.3.4. Leptosols. To the left Lithic Leptosol at the foot of a hill near Rada de Haro (compare Figs. 1.x and 1.y). To right Eutric Leptosol. These soils belong to the degraded type.

Fig. A.3.5. Luvisols. From top left to bottom right: Chromic Luvisol (degraded by argilization), dry land Calcaric Luvisol, Chromic Luvisol (forest soil), and Calcic Luvisol (recarbonated red soil, degrading by secondary recarbonation towards Calcisol, found at valley bottom)

Fig. A.3.6. Calcaric Phaenozems (degrading by mineralization)

Fig. A.3.7. Calcaric Regosol, a degraded soil underneath matorral vegetation

Fluvisols. Fine materials developed from recent alluvial deposits and found mostly in river valleys. They show no diagnostic horizon other than an ochric (organic carbon < 0.6 mass%) A horizon. Calcaric Fluvisols contain more than 15% calcium carbonate. Salic Fluvisols have salic properties with high electric conductivity (> 15 dS m^{-1} and > 1% salt). Eutric Fluvisols found in Italy, Spain, Portugal are structured into horizons A, C, 2A, 2C .....

Kastanozems. Humus rich soils with brown colour and carbonates or gypsum.

Leptosols, Lithic Leptosols. Shallow or stony soil limited in depth by hard rock or cemented layer within 10 cm of the surface. Eutric Leptosols with

Fig. A.3.8. Gypsic Solonchak

ochric A horizon and BS $\geq$ 50%. Rendzic Leptosols with mollic A horizon (dark humus, BS > 50%, C_{org} > 2%)

Luvisols. Argillic B horizon with high CEC and high BS signaling advanced stage of leaching. More than 24 cmol(+) kg^{-1} clay, 50% BS. Appearing as chromis, calcic and haplic.

Phaeozems. Humus rich soils with dark black colour.

Regosols. From tertiary hills and Keuper material. Unconsolidated material that is coarse structured or show fluvic (sedimentary) properties. No diagnostic horizon, exept ochric or umbric (accumulated humus, C_{org} > 2%, BS < 50%) horizon A. Low fertility, found in abandoned marginal dry farming regions. They appear as Calcaric Regosols (> 15% calcium carbonate equivalent), Gypsic Regosols ($\geq$ 5% gypsum), and Eutric Regosols (base saturation $\geq$ 50%).

Solonchaks. High salinity, no horizons. Differentiation between Solonchaks and Fluvisols is difficult to achieve. Calcis Solonchaks with a calcic horizon of soft powdery lime within 125 cm of the surface. Gypsic Solonchaks with a gypsic horizon. Gleyic Solonchanks with hydromorphical properties (saturated with water at least at some period of the year; evidence of reduction processes andsegregation of iron).

Vertisols. Over 35% clay. Soils with swell and shrink properties.

Xerosols (calcic or gypsic). Weak ocric horizon A. One or more of the following horizons: cambic, argilic, calcic, gypsic. Horizon A often lies in Horizon Ck (calcic) in which accumulation of calcic carbonate frequently is so strong that it brings about a continuous cementation. With argilic horizon: Luvic Xerosols which are remains from past wetter climatic conditions.

Other soil types less important for Spain but relevant for Africa (Niger) are:
Acrisols - Bt horizon with low CEC and low base saturation, eolian sand admixture
Alisols - Bt horizon with high CEC and low base saturation
Andosols - derived from volcanic material with a low bulk density
Anthrosols - strongly influenced by human activity
Chernozems - humus rich with dark black colour and carbonates or gypsum
Ferralsols - tropical fine textured, weathered, high content in sesquioxides
Gleysols - strong oxidation and reduction features due to water saturation
Greysems - Bt-horizon and uncoated silt and quartz grains
Gypsisols - with gypsum
Histosols - a large share of undecomposed organic matter
Lixisols - Bt horizon with low CEC and high base saturation
Nitisols - fine textured, weathered with shiny ped faces
Planosols - with seasonal surface water logging
Plinthosols - moted cleyey material that hardens upon exposure
Podsoluvisols - Bt-horizon and a distinct eluvial (E) horizon

Podsols - acid coarse textured souls having a strongly bleached horizon
Solonetz - strong sodium saturation

A.3.3 Soil Degradation

Soil degradation can take place due to biological, physical and chemical processes. Biological degradation (BD) is determined by the percentage of oxidable organic matter (OM) in the soil:

$$BD = \frac{1}{OM} \tag{A.3.1}$$

The physical degradation (PD) can be computed using the percentage of stable aggregates (SA), the water holding capacity (WHC),

$$WHC = pF_{0.33} - pF_{15} \tag{A.3.2}$$

where $pF_{0.33}$ is the matric potential 33 kPa and pF_{15} the matric potential at 1500 kPa, and the crust index (CI) which is obtained from the percentage of fine silt (Fst), the percentage of coarse silt (Cst), the percentage of clay (Cy), and the percentage of organic matter in the following way:

$$CI = \frac{1.5Fst + 0.75Cst}{Cy + 10 * OM} \tag{A.3.3}$$

The formula for the physical degradation then becomes

$$PD = \frac{CI}{SA + WHC} \tag{A.3.4}$$

The chemical degradation is given by

$$CD = \frac{\%salts + \dfrac{\%Na}{CEC}}{CEC} \tag{A.3.5}$$

where
CEC = the cation exchange capacity measured in cmol(+)/kg,
%salts = 336 ∗EC, EC being the electrical conductivity measured in dS/m,
%Na = percentage of sodium in the exchange complex.

The numbers which result from Eq. A.3.3 - A.3.5 for each pixel can be assigned to seven classes of degradation as listed in Table A.3.1. Then for each type of degradation, biological, chemical, and physical, a map has been constructed (not shown here) which shows the geographical distribution of the seven degradation classes. Based on these individual classes an integrative measure DI of degradation can be computed as the sum of the numbers computed for the biological, physical and chemical degradation. This, finally, leads to the integrative map of total soil degradation which is given in Fig. A.3.9.

The characteristic times for change processes have to be determined individually for soil types. A few examples are given here:

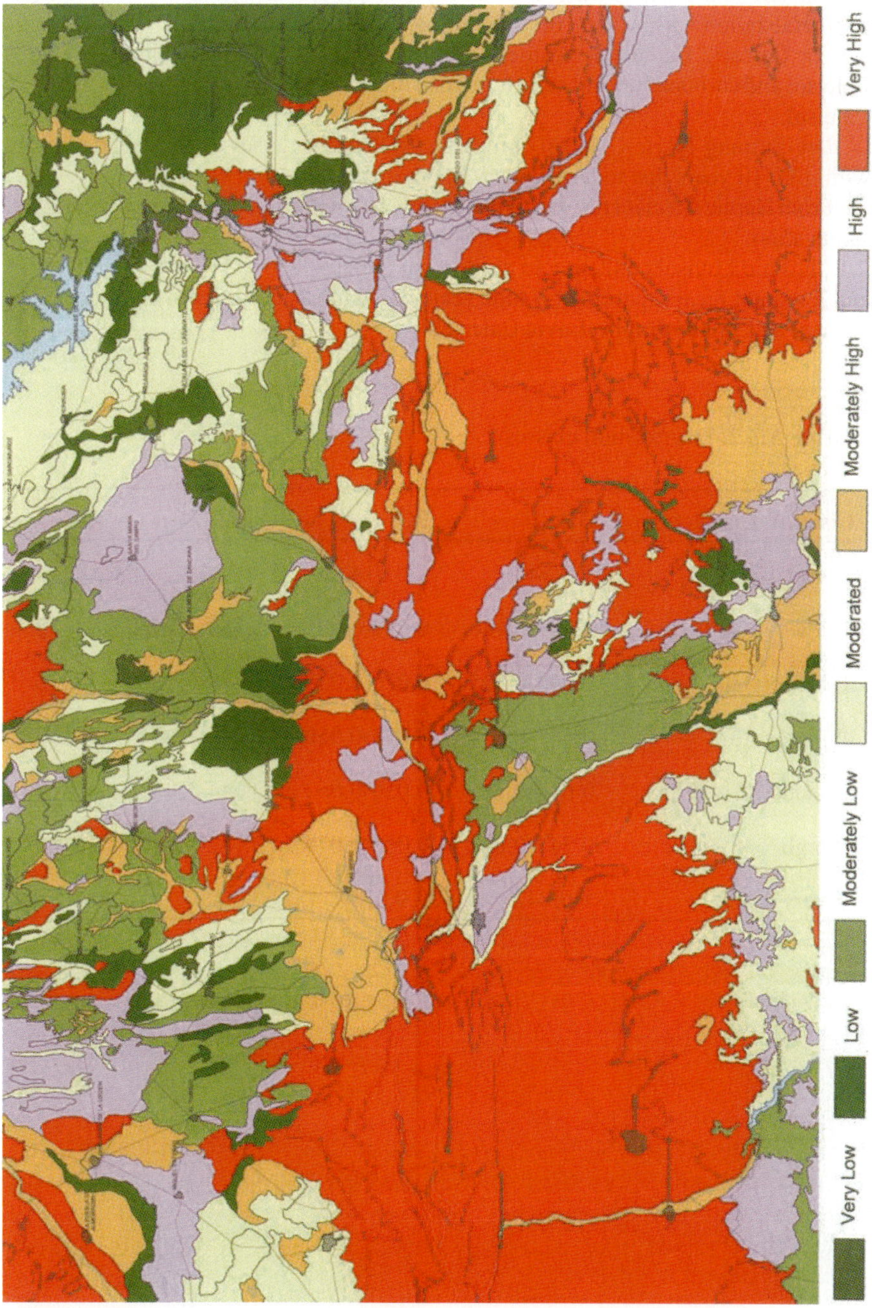

Fig.A,3.9. Soil degradation map of the EFEDA experimental area published by J Sanchez Diaz, R Boluda Hernandez, C Morelli Cot, and JC Colomer Marco

Calcaric Phaeozem undergoes a humification process within 100 - 1000 years and a mineralization within 1 to 100 years resulting in Calcaric Regosol.

For Calcaric Cambisol weathering takes 100 - 1000 years resulting in Haplic Calcisol and carbonation and cementation in more than 1000 years resulting in Petric Calcisol.

Dry farming Petric Calcisol converts into Petric Calcisol under irrigated crops within 1 - 10 years due to a secondary mineralization.

From Haplic Calcisol results Gypsic Solonchak within 10 to 100 years by primary salinization.

Chromic Luvisols undergo an argilization process within more than 1000 years and a second recarbonation at a time scale of 100 to 1000 years to become Calcic Luvisol and finally Haplic-Luvic Calcisol.

Table A.3.2. Degradation classes

Class of degradation	BD	PD	CD	DI
1: very low	< 0.347	< 0.014	< 1.350	< 5
2: low	0.347-0.446	0.014-0.026	1.350-1.778	38172
3: moderately low	0.446-0.545	0.026-0.038	1.778-2.206	38236
4: moderate	0.545-0.644	0.038-0.050	2.206-2.634	38299
5: moderately high	0.644-0.743	0.050-0.062	2.634-3.062	38362
6: high	0.743-0.842	0.062-0.074	3.062-3.490	38366
7: very high	> 0.842	> 0.074	> 3.490	> 15

Appendix 4 Characteristics of Earth Observation Satellites and Remote Sensing Instruments

Described Systems: Overview

A. Satellite Systems		B. Aircraft Systems
Aqua DMSP ENVISAT ERS Landsat Meteosat	Meteosat Second Generation Nimbus NOAA SPOT Terra	AIRSAR AVIRIS DIAL TMS

A. Satellites and Their Instruments

AQUA

Twin satellite to Terra. Observes land, ocean and atmosphere phenomena in the afternoon while Terra passes in the morning.. Instrumentation is similar to Terra. Launched on 4 May 2002

Altitude	680 km
Inclination	98.2°
Sidereal period	98" 24'
Instruments: Atmospheric Infrared Sounder (AIRS) Advanced Microwave Sounding Unit (AMSU) Humidity Sounder for Brazil (HSB) Clouds and the Earth's Radiant Energy System (CERES)	Moderate-Resolution Imaging Spectroradiometer (MODIS) Advanced Microwave Scanning Radiometer - EOS (AMSR-E)

DMSP (Defense Meteorological Satellite Program)

The first satellites operated under the name DMSP by the US Department of Defense have been launched in 1965. Since then over 35 DMSP satellites are started. The latest two are named DMSP-F12, DMSP-F13 and DMSP- F14

Altitude	820-860km
Orbit	sunsynchronous
Local Time	6 am and 10:30 am

Instruments:

Operational Linescan System (OLS)	Visible Telescope: 0.4 - 1.1 μm Infrared Telescope: 10.3 - 12.9 μm Resolution roughly 500 m
Microwave Imager (SSM/I)	7-channel, four frequency, linear polarized, passive microwave radiometer (19.35, 22.235, 37, and 85.5 Ghz, footprints 13x15 km to 43x69 km
Atmospheric Water Vapor Profiler (SSMT/2)	5-channel, total power microwave radiometer, three channels situated symetrically around the 183.31 Ghz water vapour line and two window channels. Spatial resolution approximately 48 km
Atmospheric Temperature Profiler (SSM/T)	7-channel microwave radiometer measuring atmospheric emission in the 50 to 60 Ghz oxygene band with a subtrack spatial resolution of 174 km
Precipitating Electron and Ion Spectrometer (SSJ/4)	Provides a complete energy spectrum of low energy particles between 30 eV and 30 keV
Ion Scintillation Monitor (SSIES)	Langmuir probe on 2.5 m boom
Magnetometer (SSM)	In combination with the SSI/ES (or SSI/ES-2) and the SSJ/4, the SSM provides heating and electron density profiles in the high-latitude ionosphere

Comparison of the microwave DMSP channels with AMSU-A, AMSU-B and MSU

Instrument	SSM/T	MSU	SSM/I	AMSU A	AMSU B	SSM/T2
Satellites	Current DMSP	Current NOAA	Current DMSP	NOAA 15, L,M	NOAA 15, L,M	Current DMSP
Channels	7	4	7	15	5	5
Frequency Range (Ghz)	50.5-59.4	50.3-57.9	19.35-85.5	23.8 - 89.0	89.0-183.3	90.0-183.3
NEΔT (K)	4	6	0.4-1.7	0.25-1.2	8	5
Best ground resolution	204 km	110 km	12.5-50 km	48 km	16 km	50-100 km

ENVISAT

Orbits per day	14 11/35
Repeat cycle	35 days
Period	100.59 min
LT descending node	10:00
Inclination	98.55 deg
Semi-major axis	7159.5 km
Mean altitude	799.8 km
Altitude range	780-820 km

ASAR (Advanced Synthetic Aperture Radar) operated in C-band (5.331 Ghz)

Instrument parameters	Imaging Mode	Alternating Polarisation	Wide Swath	Global Monitoring	Wave Mode
Swath width	up to 100 km	up to 100 km	>400 km	>400 km	5 km vignette
Operation time	up to 30 min per orbit			rest of orbit	
Data rate	≤100 Mbit/s			0.9 MBit/s	0.9 MBit/s
Power	1365 W	1395 W	1200W	713 W	647 W

MERIS (Medium Resolution Imaging Spectrometer)

Spectral range	390 - 1040 nm
Spectral resolution	2.5 nm max
Field of view	68.5°
Spatial resolution	300 m at nadir
Swath width	1150 km

Band No.	Band centre (nm)	Band width (nm)	Potential applications
1e+20	412.5	10	Yellow substance, turbidity
	442.5	10	Chlorophyll absorption maximum
	490	10	Chlorophyll, other pigments
	510	10	Turbidity, suspended sediments, red tides
	560	10	Chlorophyll reference, suspended sediments
	620	10	Suspended sediments
	665	10	Chlorophyll absorption
	681.25	7.5	Chlorophyll fluorescence
	705	10	Atmospheric correction, red edge
	753.75	7.5	Oxygen absorption reference
	760	2.5	Oxygen absorption, R-branch
	775	15	Aerosols, vegetation
	865	20	Aerosols correction over ocean
	890	10	Water vapour absorption reference
	900	10	Water vapour absorption, vegetation

MWR (Microwave Radiometer)

Operating frequencies	23.8 Ghz and 36.5 Ghz
Bandwidth	650 Mhz
Dynamic range	3-300 K
Absolute radiometric accuracy	< 3K
Radiometric sensitivity	< 0.5 K
Operation	Continuous
Data rate	16.7 kbps

MIPAS (Michelson Interferometer for Passive Atmospheric Sounding)
Fourier transform spectrometer operating in the mid infrared (4.15 - 14.6 micron) and sounding the atmosphere's limb emission.

SCIAMACHY (Scanning Imaging Absorption Spetrometer for Atmospheric Chartography

Channel	Spectral Range (nm)	Spectral resolution (nm)	Detector
1.2e+07	240-314	0.24	shelf silicon array (1024
	309-405	0.26	pixel)
	394-620	0.44	"
	604-805	0.48	"
	785-1050	0.54	"
	1000-1750	1.48	"
	1940-2040	0.22	InGaAs
	2265-2380	0.28	"
			"

Geometry	IFOV
Nadir	0.045 ±0.003° across track
	1.8±0.200° along track
Limb	0.045 ±0.003° across track
	1.8±0.200° along track

ERS 1 and 2 (European Remote Sensing Satellite 1 and 2)
Primarily built for ocean research but also used for land-surface research

Spacecraft	ERS-1	ERS-2
Launch date	17 July 1991	21 April 1995
Orbit	sun-synchronous	sun-synchronous
Mean Altitude	quasipolar	quasipolar
Inclination	785 km	785 km
Eccentricity	98.52°	98.49°
Repeat cycle	0.001 165	?
Mean local solar time	3, 35, 176 days	35 days
at ascending node	10h 30m	10h 30m

SAR (Synthetic Aperture Radar)

Frequency	5.3 Ghz
Bandwidth	15.5± 0.06 Mhz
Antenna size	10m x 1m
Incidence angle	23° nominal (35° in roll-tilt mode)
Spatial resolution	along track ≤ 30m; across track ≤26.3m
Radiometric resolution	≤ 2.5 dB
Maximum operation time	10 minutes per orbit
Swath width	102.5km (telemetered), 80.4km (full perform.)

ATSR (Along Track Scanning Radiometer)

Self calibrating conical scanning instrument designed for high accuracy temperature measurements due to dual view through the atmosphere. Enhanced version ATSR-2 since 21 April 1995

Spectral channels	1.6 μm, 3.7μm, 11μm, and 12 μm
Calibration	two onbord black bodies at different temperatures

Windscatterometer

Frequency	5.3 Ghz ± 52 kHz
Bandwidth	15.55 ± 0.01 Mhz
Peak power	4.8 kW
Polarization	Linear vertical
Antenna parameters	Fore MidAfter
aspect angle	45±5° 0±5° -45±5°
length	3.6m 2.5m 3.6m
dynamic range	42 dB 42 dB 42 dB
number of pulses for 50km	256 256 256
Swath location	200 - 700 km to side of orbital track
Swath width	400km within specifications 500 km telemetered
Incidence angle range	
forward directed beam	25° at node 1, 59° at node 19
mid beam	18° at node 1, 47° at node 19
backward directed beam	25° at node 1, 59° at node 19

Landsat

Altitude		696-741 km
Inclination		98.22°
Descending node		9:30 am local mean solar time
Spatial resolution		30 m (Channels 1-5,7), 120m (Channel 6)
image size		185 x 172 km
Swath width		185 km
Repetition rate		16 days
Period of revolution		99 min
Instrument	***Thematic Mapper (TM)***	***Multispectral Scanner Subsystem (MSS)***
Spectral band 1	0.450 - 0.515μm	0.5 -0.6 μm
Spectral band 2	0.525 - 0.605μm	0.6 -0.7 μm
Spectral band 3	0.630 - 0.690μm	0.7 -0.8 μm
Spectral band 4	0.750 - 0.900μm	0.8 -1.1 μm
Spectral band 5	1.550 - 1.75μm	
Spectral band 6	10.40 - 12.50μm	
Spectral band 7	2.080 - 2.35μm	
Ground IFOV	30m (bands 1-5,7) 120 m (band 6)	82 m
Due rate	85 Mbps	15 Mbps
Bit resolution	8	6
ETM P	Enhanced Thematic Mapper + additional on Landsat 7	
Spectral band	0.52 - 0.90μm	
Ground IFOV	15m	

Meteosat

Altitude		~ 36000 km geostationary orbit	
Inclination		0-1 °	
Repetition rate		0.5 hours	
Spectral bands (μm)	0.5 - 0.9 Visible	5.7 - 7.1 Infrared (water vapor)	10.5 - 12.5 thermal Infrared
Number of detectors	2		1
Detector type	Si	HgCdTe	HgCdTe
Field of view	0.065mrad	0.14mrad	0.14mrad
Resolution at nadir (km)	2.5	5	5
Radiometer performance	S/N > 200 at 25% albedo	NEΔT < 1.0K at 260K	NEΔT <0.4K at 290K

MSG (Meteosat Second Generation)

SEVIRI (Spinning Enhanced Visible and Infrared Imager)

Bands	Centre (μm)	Reso-lution	99% energy band (μm)	Dynamic range	Noise
HRV	0.7 (0.5-0.9)	1 km	similar to Meteosat	0-459 $W/m^2\mu m$	S/N > 4.3 for target of 1% of max dynamic range
VIS 0.6	0.635	3 km	0.56-0.71	0-533 $W/m^2\mu m$	S/N > 10.1 for target of 1% of max. dynamic range
VIS 0.8	0.81	3 km	0.74-0.88	0-357 $W/m^2\mu m$	S/N>7.28 for target of 1% of max dynamic range
IR 1.6	1.64	3 km	1.50-1.78	0-75 $W/m^2\mu m$	S/N>3 for target of 1% of max dynamic range
IR 3.9	3.92	3 km	3.48-4.36	0-335K	0.35K@300K
IR 8.7	8.70	3 km	8.30-9.10	0-300K	0.28K@300K
IR 10.8	10.8	3 km	9.80-11.80	0-335K	0.25K@300K
IR 12.0	12.0	3 km	11.0-13.0	0-335K	0.37K@300K
WV 6.2	6.25	3 km	5.35-7.15	0-300K	0.75K@250K
WV 7.3	7.35	3 km	6.85-7.85	0-300K	0.75K@250K
IR 9.7	9.66	3 km	9.38-9.94	0-310K	1.50K@255K
IR 13.4	13.40	3 km	12.4-14.4	0-300K	1.80K@270K

GERB (Geostationary Earth Radiation Budget Experiment)

Wavebands	With and quartz filter: 0.32 μm - 4.0μm Without quartz filter: 0.32μm - 30μm By subtraction: 4.0μm - 30μm	
	Shortwave	Longwave
Absolute accuracy (each pixel, excluding noise)	<2.4 $Wm^{-2}sr^{-1}$	<0.4 $Wm^{-2}sr^{-1}$
Noise (each pixel, 15minutes averaging)	< 0.8 $Wm^{-2}sr^{-1}$	< 0.15 $Wm^{-2}sr^{-1}$
Dynamic range	0-380 $Wm^{-2}sr^{-1}$	0-90 $Wm^{-2}sr^{-1}$
Signal/Noise (each pixel, 15 minutes averaging)	1250	400
Pixel size	44.6 x 39.3 km (NS x EW) at nadir	
Cycle time	Full Earth disc, both channels in 5 min	

Nimbus

Operation time	October 1978 - August 1994
Altitude	955 km
Orbit	Sunsynchronous
Orbital Period	104.16 min
Ascending Node	local noon

SMRR (Scanning Multichannel Microwave Radiometer)
(for retrospective research on land-surface processes primarily SMRR is of interest which was operational from October 1978 to August 1987)

Channel (H & V)	Frequency (GHz)	Wavelength (cm)	3 dB Beam Width (degrees)	NEΔT (K)	Resolution
1 & 2	6.63	4.54	42	4	148
3 & 4	10.69	2.8	26	5	91
5 & 6	18.0	1.66	16	7	55
7 & 8	21.0	1.36	14	7	46
9 & 10	37.0	0.81	8	11	27

NOAA Satellites and Instruments

NOAA Satellite Characteristics

Altitude	856/819 km*	Spatial Resolution (scan end)	6.5 x 2.3 km
Inclination	99.1/98.67°*		
Equator Crossing (ascending node)	15:26/19:33*	Swath Width	3016 km
		Bit resolution	10
Spatial Resolution (nadir)	1.09 km	Repetition rate	0.5 days

* Values for NOAA 14/NOAA 15 (July 1999). Since the orbital parameters are different for the different NOAA satellites and also change during the lifetime of the satellite, precise data can only be extracted from the data archives containing original data.

AVHRR Instrument Specifications

Spectral Channel	1	2	0.125	3B	4	5
Spectral range (μm)	0.58 -0.68	0.725 -1.1	1.58 - 1.64	3.55 - 3.93	10.3 - 11.3	11.5 - 12.5
Detector type	Si	Si	InGaAs	InSb	HgCdTe	
Resolution (km)	11	11	11	11	11	11
IFOV (mrad squared)	13	13	13	13	13	13
Noise equivalent temp. difference (NEΔT) 300 K	--	--	--	<0.12K	<0.12K	<0.12K
Sinal-to-noise ratio (mrad squ.) --> NOAA 15	>3:1 >9:1	>3:1 >9:1	>20:1	--	--	--

These values, taken from NOAA Technical Memorandum NESS95 and NOAA KLM Users Guide, are generalized. To obtain detailed and more precise information it is recommended to look to the NOAA Polar Orbiter Data Users Guide, edited by K.B.Kidwell. Here also the spectral response curves of the different AVHRR´s are given.

NOAA-TOVS Specifications

Earth Scan Angle	±49.5 deg	Longwave Detector	HgCdTe
Earth Swath Coverage	1127 km	Shortwave Detector	InSb
Field of View	1.40 deg all chan.	Visible Detector	Si
Resolution	18-20 km	Telescope Aperture	15.0 cm

Channel Number	Centre Wavelength		Half Power Bandwidth	NEΔN
	Wavenumber cm^{-1}	Wavelength μm	cm^{-1}	mW/m^2 sr cm^{-1}
1	*668.5/669	14.96/14.95	3/3	0.80/3.00
2	680/680	14.71/14.71	10/10	0.27/0.67
3	690/690	14.49/14.49	12/12	0.27/0.50
4	703/703	14.22/14.22	16/16	0.22/0.31
5	716/716	13.97/13.97	16/16	0.22/0.21
6	733/733	13.64/13.64	16/16	0.22/0.24
7	749/749	13.35/13.35	16/16	0.22/0.20
8	900/900	11.11/11.11	35/35	0.11/0.10
9	1030/1030	9.71/9.71	25/25	0.16/0.15
10	1225 / 802	8.16/12.47	60/16	0.16/0.15
11	1365/1365	7.32/7.32	40/40	0.22/0.20
12	1488/1533	6.72/6.52	80/55	0.11/0.20
13	2190/2188	4.56/4.57	23/23	0.002/0.006
14	2210/2210	4.52/4.52	23/23	0.002/0.003
15	2240/2235	4.46/4.47	23/23	0.002/0.004
16	2270/2245	4.41/4.45	23/23	0.002/0.004
17	2360/2420	4.24/4.13	23/28	0.002/0.002
18	2515/2515	3.98/3.98	35/35	0.002/0.002
19	2660/2660	3.76/3.76	100/100	0.001/0.001
20	14500/14500	0.69/0.69	1000/1000	0.1%/0.1% Albedo

* first value: HIRS/2 TIROS-N to NOAA 14, second value: HIRS/3 NOAA15 ff.

SSU Specifications

The SSU provides measurements in three channels using a pressure-modulated gas (CO_2) to accomplish selective bandpass filtration of sampled radiances. Temperatures in the stratosphere are derived from these radiances.

Detector	Tri-glycine Sulphate (TGS)
Angular Field of View	10°
Number of earth views per line	8
Total Scan Angle	±40° from nadir
Data Rate	480 Bps

Channel No.	Spectral range (cm^{-1})	Bandwidth (cm^{-1})	Resolution (km at nadir)	NEΔT (mW/m^2 sr cm^{-1})
1	669.99	2.0	147.3	0.25
2	669.93	1.0	147.3	0.5
3	669.36	0.4	147.3	1.25

MSU Specifications

The MSU detects and measures microwave energy from the troposphere up to 20 km in four frequency channels. The advantage of MSU is that data are not seriously affected by clouds.

Angular Resolution	7.5°
Number of steps	11
Cross Track Scan Angle	± 47.35°
Data Rate	320 Bps

Channel	Frequency (Ghz)	RF bandwidth (Mhz)	Resolution km at nadir	NEΔT (K)
1	50.30	220	105	0.3
2	53.74	220	105	0.3
3	54.96	220	105	0.3
4	57.95	220	105	0.3

These values, taken from NOAA Technical Memorandum NESS95 and NOAA KLM Users Guide, are generalized

ATOVS AMSU-A Specifications

Channel	Center Frequency (Mhz)	Bandwidth (Mhz)	NEΔT Accuracy (K)	Calibration (K)
1	23.800	251.02	0.211	2.0
2	31.400	161.20	0.265	2.0
3	50.300	161.14	0.219	1.5
4	52.800	380.52	0.143	1.5
5	53.596 ±115	168.20	0.148	1.5
6	54.400	380.54	0.154	1.5
7	54.940	380.56	0.132	1.5
8	55.500	310.34	0.141	1.5
$9 = f_0$	57.290344	310.42	0.236	1.5
10	f_0±217	76.58	0.250	1.5
11	f_0±322.2±48	35.11	0.280	1.5
12	f_0±322.2±22	15.29	0.399	1.5
13	f_0±322.2±10	7.93	0.539	1.5
14	f_0±322.2±4.5	2.94	0.914	1.5
15	89.000	1998.98	0.165	2.0

ATOVS AMSU-B Specifications

Channel	Frequency (Ghz)	Bandwidth (Mhz)	NEΔT (K)	Calibration Accuracy (K)
16	89.0 ± 0.9	3000	10	1.0±0.2
17	150.0 ± 0.9	2000	10	1.0±0.2
18	183.3±1.0	2 x 500	11	1.0±0.2
19	183.3±3.0	2 x 1000	10	1.0±0.2
20	183.3±7.0	2 x 2000	12	1.0±0.2

Besides the complete change of instrument (SSU and MSU replaced) the horizontal resolution of the AMSU instruments is now greatly improved to a nadir resolution of 48 km (AMSU-A) and 16 km (AMSU-B).

SPOT (Systeme Pour l'Observation de la Terre)

The instrumentation of the SPOT satellites offers some new capabilities of obtaining stereoscopic pairs of images

Satellite	SPOT 1-3	SPOT 4
Altitude	832 km	832 km
Inclination	99°	
Descendig node	10h30m am local time	10h30m local time
Spatial resolution	20/10m	20/10m, 1km
Swath width	60 km	60 km, 2200 km
Data rate	50Mbits/s	50 Mbit/s
Repetition rate	26 days	26 days

SPOT 1-3 HRV (High Resolution Visible)

Spectral bands	0.43 - 0.47 μm	0.43 - 0.47 μm
Multispectral mode	0.61 - 0.68 μm	0.61 - 0.68 μm
Panchromatic mode	0.79 - 0.89 μm	0.79 - 0.89 μm
IFOV	0.5 - 0.73 μm	1.58 - 1.75 μm
Bit resolution	4.13 deg	0.5 - 0.73 μm
Spatial resolution:	8	
multispectral		
panchromatic	20 m	
Number of pixels per line	10 m	
	3000/6000	

VEGETATION instrument for observation of vegetation with four cameras

Spectral bands	0.43 - 0.47 μm
	0.61 - 0.68 μm
	0.79 - 0.89 μm
	0.5 - 0.73 μm
	1.58 - 1.75 μm
	0.5 - 0.73 μm
Spatial resolution	1.15 km
Field of view (FOV)	101°
Swath width	2250 km

Terra

ASTER (Advanced Spaceborne Thermal Emission and Reflection Radiometer)

VNIR		SWIR		TIR	
Band	Band width (µm)	Band	Band width (µm)	Band	Band width (µm)
1	0.52-0.60 nadir looking	4	1.600-1.700	10	8.125-8.475
		5	2.145-2.185	11	8.475-8.825
2	0.63-0.69 nadir looking	6	2.185-2.225	12	8.925-9.275
		7	2.235-2.285	13	10.25-10.95
3a	0.76 - 0.86 nadir looking	8	2.295-2.365	14	10.95-11.65
		9	2.360-2.430		
3b	0.76 - 0.86 backward look				

	VNIR	SWIR	TIR
Ground resolution	15m	30m	90m
Data rates (Mbits/s)	62	23	4.2
Cross-track pointing (deg.)	± 24	± 8.55	± 8.55
Cross-track pointing (km)	± 318	± 116	± 116
Swath width	60	60	60
Detector type	Si	PtSi-Si	HgCdTe
Quantisation (bits)	8	8	12

MISR (Multi-angle Imaging Spectro-Radiometer)

Global coverage time	Every 9 days, with repeat coverage between 2 and 9 days depending on latitude
Crosstrack swath width	360 km common overlap of all 9 cameras
Instruments	Nine pushbroom cameras for foreward, nadir and backward viewing. Four cameras are designed with increasing viewing angle.
View angles	0, 26.1, 45.6, 60.0, and 70.5 degrees
Spectral coverage	blue (443 nm), green (555 nm), red (670 nm) and near-infrared (865 nm)
Detectors	Charge Coupled Devices (CCD`s), each camera with 4 independent line arrays (one per filter), 1504 active pixels per line
Radiometric accuracy	3% at maximum signal
Data rate	3.3 Mbps

MODIS (Moderate Resolution Imaging Spectrometer)

Spectral band	Bandwidth (µm)	Spatial resolution (m)	Primary Use
12	0.620-0.670	250	Land/Cloud
	0.841-0.876	250	Boundaries
3	0.459-0.479	500	Land/Cloud
4	0.545-0.565	500	Properties
5	1.230-1.250	500	
6	1.628-1.652	500	
7	2.105-2.155	500	
8	0.405-0.420	1000	Ocean Color/
9	0.438-0.448	1000	Phytoplankton/
10	0.483-0.493	1000	Biogeochemistry
11	0.526-0.536	1000	
12	0.546-0.556	1000	
13	0.662-0.672	1000	
14	0.673-0.683	1000	
15	0.743-0.753	1000	
16	0.863-0.877	1000	
17	0.890-0.920	1000	Atmospheric
18	0.931-0.941	1000	Water Vapor
19	0.915-0.965	1000	
20	3.660-3.840	1000	Surface/Cloud
21	3.929-3.989	1000	Temperature
22	3.929-3.989	1000	
23	4.020-4.080	1000	
24	4.433-4.498	1000	Atmospheric
25	4.482-4.549	1000	Temperature
26	1.360-1.390	1000	Cirrus Clouds
27	6.535-6.895	1000	Water Vapor
28	7.175-7.475	1000	
29	8.400-8.700	1000	
30	9.580-9.880	1000	Ozone
31	10.780-11.280	1000	Surface/Cloud
32	11.770-12.270	1000	Termperature

33	13.185-13.485	1000	Cloud Top
34	13.485-13.785	1000	Altitude
35	13.785-14.085	1000	
36	14.085-14.385	1000	

Scan rate	20.3 rpm, cross track
Swath dimensions	2230 km (cross track) by 10 km (along track at nadir)
Quantisation	12 bits
Data rate	6.2 Mbps (orbital average)

CERES (Clouds and the Earth's Radiant System) to study the role of clouds and of the energy cycles in global climate change

Spectral bands	Shortwave 0.3 - 5.0 μm	Longwave 8-12 μm	Total 0.3 to > 200 μm
Spatial resolution at nadir	20 km		
Data rate	20 kbps		

MOPITT (Measurement of Pollution in the Troposphere)

Spectral bands	2.3 μm (CH_4) 2.4 μm (CO) 4.7 μm (CO)
Swath width	640 km
Spatial resolution	22 km
Data rate	25 kbps (avg)

B. Aircraft Instruments

AIRSAR

Values in brackets () refer to the 40 MHz configuration of the stepped-frequency FM pulse (CHIRP)

Band	P-Band	L-Band	C-Band
Chirp Bandwidth (Mhz)	20(40)	20(40)	20(40)
Chirp Center Freq. (Mhz)	438.75 (427.5)	1248.75 (1237.5)	5298.75 (5287.5)
Antenna Polarization	H/V dual	H/V dual	H/V dual
Antenna gain (dBi)	14	18	24
Azimuth beamwidth (deg)	190	80	25
Elevation beamwidth (deg)	380	440	500
Antenna Size (m)	0.9 x 1.8	0.5 x 1.6	0.2 x 1.4
Data Rate (MB/s)	10	10	10
NEΔT (dB)	-45	-45	-35
Nominal Altitude (m)	8000	8000	8000
Nominal Velocity (Knots)	450	450	450
Slant Range Resolution (m)	10(5)	10(5)	10(5)
Azimuth Resolution (m)	1	1	1
Ground Range Swath (km)	10-15	10-15	10-15

AVIRIS

Instantaneous field of view (IFOV)		1 mrad	
Geometric resolution		20 m	
Field of View (FOV)		30° (11 km)	
Flight level		20 km	
Radiometric resolution		10 Bit	
Spectral range		0.41 - 2.45	
Number of spectral bands		224	
Spectrometer 1	0.41 - 0.70 μm	Spectrometer 3	1.25 - 1.86 μm
Spectrometer 2	0.68 - 1.27 μm	Spectrometer 4	1.84 - 2.45 μm

DIAL

Laser type	Nd:YAG pumped dye-laser
Pulse width	6 nsec
Pulse energy	10-15 mJ, max. 30 mJ
Bandwidth	$\leq$ 0.03 cm^{-1}
Spectral purity	typ. 70-75%
Repetition rate	10 Hz
on-/off-line frequency	4.5 Hz
Wavelength calibration (on-line)	Water vapor filled PAS-cell
Precision	0.01 cm^{-1}
Bandwidth control	Fizeau type spectrometer
Spectral resolution	0.007 cm^{-1}
Resolution of readout	0.002 cm^{-1}/pixel
Telescope	35 cm Cassegrain
Spectral filter	$\Delta\lambda$= 0.65 nm, 50% transmission
Detector	Photomultiplier Hamamatsu R928
Electronic bandwidth of amplifier	6 Mhz
Digitizer	12 Bit 20 Mhz

TMS

General characteristics	
Instantaneous field of view (IFOV)	1.25 mrad
Total field of view (FOV)	42.5°
Pixels/scan line	716
Geometric resolution	25m (65,000 ft aircraft altitude)
Swath width	15.4 km (65,000 ft aircraft altitude)

Spectral characteristics

Channel No.	Bandwidth (μm)	Equates to TM
Channel 1	0.42 - 0.45	non-TM band A
Channel 2	0.45 - 0.52	TM band 1
Channel 3	0.52 - 0.60	TM band 2
Channel 4	0.60 - 0.62	non-TM band B
Channel 5	0.63 - 0.69	TM band 3
Channel 6	0.69 - 0.75	non-TM band C
Channel 7	0.76 - 0.90	TM band 4
Channel 8	0.91 - 1.05	non-TM band D
Channel 9	1.55 - 1.75	TM band 5
Channel 10	2.08 - 2.35	TM band 7
Channel 11	8.50 - 14.0 (low gain)	TM band 6
Channel 12	8.50 - 14.0 (high gain)	TM band 6

Sources

ESA/CEOS (1995) Coordination for the next decade. 1995 CEOS yearbook
ESA (1992) ERS-1 User Handbook. ESA SP-1148, Noordwijk
Eumetsat (2002) Applications of Meteosat Second Generation, Darmstadt
Eumetsat (1999) Meteosat Second Generation Opportunities for Land Surface Research and Applications. EUM SP 01, Darnstadt
Goodrum G, Kidwell K, Winston W (1998) NOAA KLM User's Guide, U.S. Department of Commerce, NOAA, NESDIC, see also: September 2000 Revision (http://www2.ncdc.noaa.gov/docs)
Grody NC, Weng F, Ferraro RR (1998) Comparison between AMSU and DMSP Measurements. 9th Conf. On Satellite Meteorology and Oceanography, 25-29 May 1998. Paris
Kidder SQ, Vonder Haar TH (1995) Satellite Meteorology. Academic Press, San Diego
Kidwell K (1995) NOAA Polar Orbiter Data User's Guide, Suitland
Madrid CR, ed. (1978) The Nimbus 7 User's Guide, Greenbelt
NASA-GSFC/NOAA (1997) NOAA-K (http://poes.gsfc.nasa. gov/campaign/)
Schwalb A (1978) The TIROS-N/NOAA A-G Satellite Series. NOAA Technical Memorandum NESS 95, Suitland
Spotimage (1998) Vegetation. Understanding Global Change. ToulouseWMO (1994) Information on Meteorological and Other Environmental Satellites. WMO No. 411, 3rd edition, Geneva

Appendix 5 Useful Formulae and Data

A.5.1 The Spectrum of the Solar and Terrestrial Radiation

Fig. A.5.1 shows the nomenclature for the range of electromagnetic radiation that is of geophysical interest. The relationship between wavelength, wavenumber and frequency is presented in three sections. At the left the scale ranges from the X-ray region to the visible solar radiation, the middle part is dedicated to the infrared radiation and at the right one can see the spectrum of the microwave radiation with the approximate position of the bands used for different applications. As terrestrial radiation those electromagnetic waves (infrared and microwaves) are denoted which are generated in the Earth's atmosphere or at the Earth's surface.

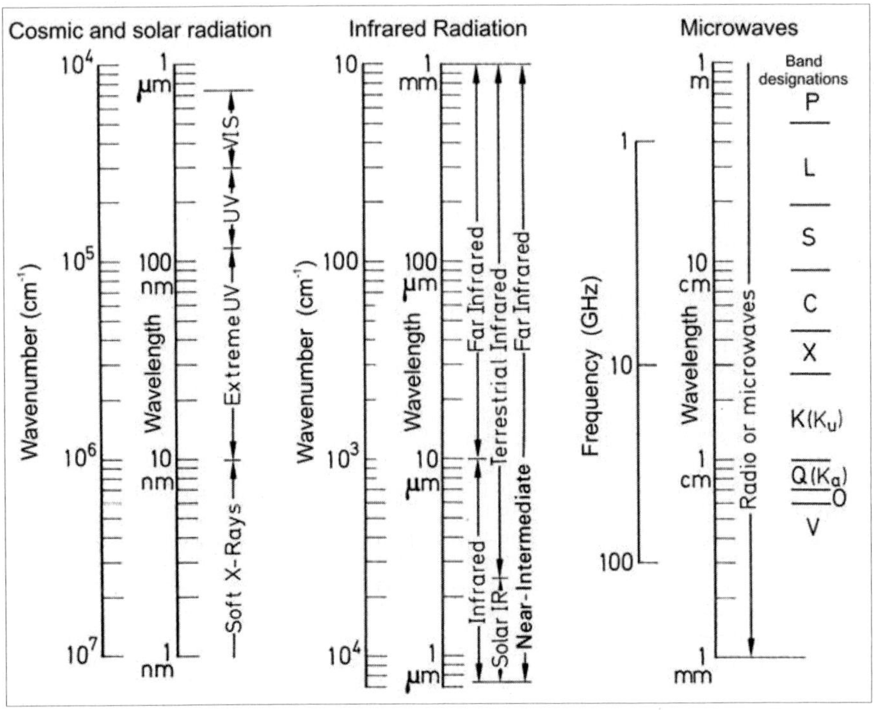

Fig. A.5.1. The range of the cosmic, solar and terrestrial radiation spectrum

A.5.2 Solar Radiation

The "*solar constant*", S_0, is the irradiance of the sun at a surface perpendicular to the direction of the solar beam measured in W m^{-2} outside the atmosphere at mean sun-earth distance. The solar constant is not an physical constant. It varies with the solar cycle within about 11 years by 0.15% - 0.18 %. Its difficult to measure average value seem now to settle at 1367.5 ± 1 Wm^{-2}. Its spectral distribution, $S_{0,\lambda}$, is presented in Table A.5.1. During the course of the year the distance sun-earth varies between 1.01674 and 0.9833 of the mean sun-earth distance of $149.6 * 10^6$ km. This gives rise to a variation of the solar irradiance within one year from + 3.32 % at the perihelion (2 January) to -3.38 % at aphelion (4 July) of the value at mean distance. The squared ratio of the mean distance (1 Astronomical Unit) to the actual distance at the day defined by the position of the radius vector θ can be expressed by Eq. A.5.2.1:

Table A.5.1. Spectral solar irradiance at the top of the atmosphere for a distance between sun and Earth of 1 Astronomical Unit after Thekaekara (1974) and Neckel und Labs (1984) . The values between 875 and 1250 nm have been corrected by an reduction of 2% for the absorption of solar lines according to the proposal of Labs und Neckel

λ nm	$S_0(\lambda)$ $Wm^{-2}\mu m^{-1}$	λ nm	$S_0(\lambda)$ $Wm^{-2}\mu m^{-1}$	λ nm	$S_0(\lambda)$ $Wm^{-2}\mu m^{-1}$	λ nm	$S_0(\lambda)$ $Wm^{-2}\mu m^{-1}$
405	1690	665	1557	925	876	1185	516
415	1727	675	1515	935	858	1195	508
425	1668	685	1475	945	840	1205	500
435	1676	695	1439	955	822	1215	491
445	1934	705	1399	965	804	1225	484
455	2038	715	1365	975	787	1235	476
465	2023	725	1342	985	753	1245	466
475	2016	735	1311	995	754	1350	374.3
485	1934	745	1278	1005	737	1450	318.6
495	1959	755	1256	1015	722	1550	272.8
505	1911	765	1220	1025	703	1650	227.9
515	1837	775	1200	1035	686	1750	186.3
525	1871	785	1182	1045	670	1850	152.9
535	1900	795	1147	1055	654	1950	126.9
545	1865	805	1123	1065	640	2050	106.4
555	1857	815	1104	1075	628	2150	89.5
565	1847	825	1078	1085	617	2250	75.9
575	1842	835	1051	1095	606	2350	64.8
585	1815	845	1021	1105	596	2450	55.8
595	1793	855	973	1115	585	2550	48.3
605	1759	865	964	1125	574	2650	42.9
615	1705	875	974	1135	564	2750	36.5
625	1690	885	955	1145	554	2850	32.0
635	1654	895	934	1155	545	2950	28.1
645	1609	905	914	1165	534		
655	1538	915	896	1175	525		

$$\left(\frac{\bar{d}}{d}\right)^2 = 1.000110 + 0.034221\cos\theta_0 + 0.001280\sin\theta_0 + 0.000719\cos 2\theta_0 + 0.000077\sin 2\theta_0 \qquad \text{(A.5.1)}$$

where θ_0 = 360 d_n/365 degree, d_n being the number of the day of the year (DOY) ranging from 0 at 1 January to 364 on 31 December for a normal year or 365 for a leap year.

The extraterrestrial solar irradiance at a horizontal plane tangential to the curvature of the Earth is given by

$$S_\lambda(\varphi,\delta,t) = S_{0,\lambda}\cos\zeta = S_{0.\lambda}\left(\frac{\bar{d}}{d}\right)^2 (\sin\varphi\sin\delta + \cos\varphi\cos\delta\cos\omega) \qquad \text{(A.5.2)}$$

where

$S_{0,\lambda}$ = Direct solar spectral irradiance at the top of the atmosphere in W m^{-2} μm^{-1}

$S_\lambda(\varphi,\delta,t)$ = Solar spectral flux density through a horizontal surface at the top of the atmosphere at the latitude φ, declination δ and time t in W m^{-2} μm^{-1}

$\bar{d}$ = mean distance Sun - Earth (1 A.U. = $149.5 \cdot 10^6$ km),

d = Sun - Earth distance at actual date,

δ = declination of the sun (accurate values for normal and leap years can be found in Linke's Meteorologischem Taschenbuch, A. Astronomische und Geodätische Tabellen, Tabellle 6; approximative formulae can be found in the internet),

ω = $15[t_{ref} + (\lambda' - \lambda_{ref})/15 + Z - 12]$,

t_{ref} = reference time, time in hours measured in zonal mean time,

λ' = local longitude,

λ_{ref} = longitude from where the reference time is counted,

Z = time equation.

ω, the hour angle, is the angle between the local south meridian and the meridian of the sun measured in the direction of the daily movement of the sun (positive in the afternoon, negative in the morning). The sun crosses the south meridian at 12:00 hours true local time (TLT). Because the of the variable speed of the Earth around the sun and the variable declination of the sun, the TLT differs from the Mean Local Time (MLT) with the exception of four dates during the year (Fig. A.5.2). The variable difference Z = TLT - LMT is called *Equation of Time*. LMT is related to the Zonal Mean Time (ZMT) which is the mean time to which a zone agrees (e.g. European Mean Time). The difference between LMT and ZMT is constant and equals the difference of the meridian λ' of the geographical location where the measurement takes place and the reference meridian λ_{ref} times 4' per hour (respectively 1h/15°). Eastern longitudes are counted positive, western ones negative. For the Mean European Time the reference meridian is 15°E. The ZMT is related to the Universal Time (UT) by the longitudinal difference between the ZMT reference meridian and 0° longitude (the Greenwich meridian) times 4'/h. UT is defined by the rotation of the

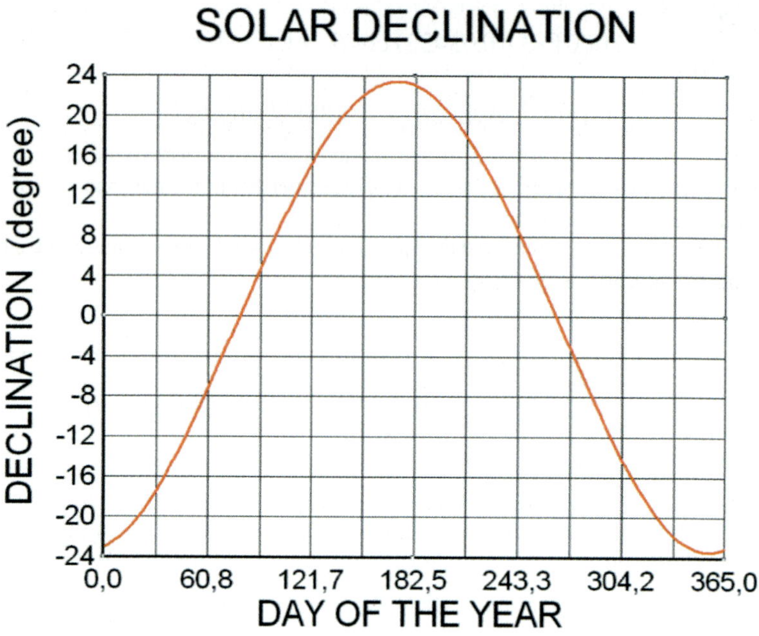

Fig. A.5.2. Solar declination

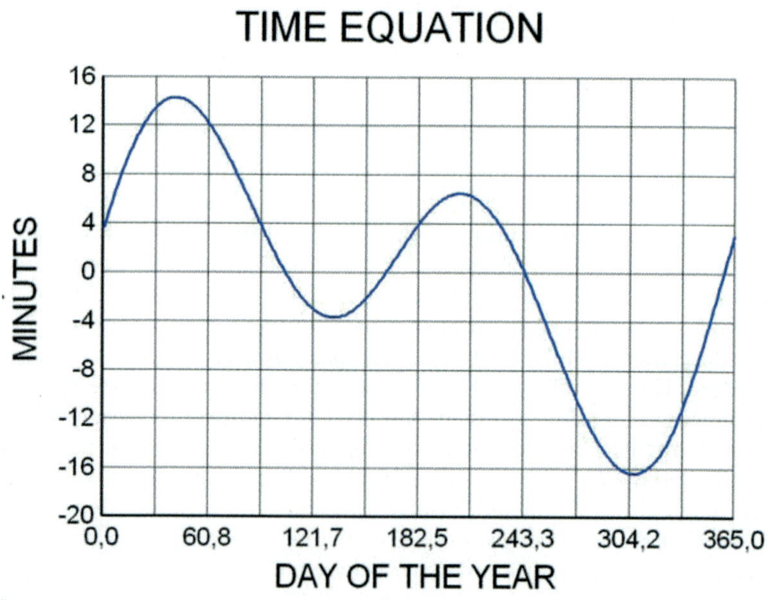

Fig. A.5.3. Time equation

Earth which is not absolutely constant. Therefore the Coordinated Universal Time (UTC) based upon an atomic clock was introduced as the general civil time standard.

To compute the solar zenith angle according to Eq. A.5.2 for a given location it is necessary to determine the declination and the hour angle for the specific day in line with the definitions given above. An example for such a computation is presented in Fig.A.5.4.

The extraterrestrial solar radiant exposure (the irradiance per exposure time) at a horizontal plane is obtained by integrating Eq. A.5.2 over the given time interval. The diurnal exposure can be computed by integrating Eq. A.5.2 from sunrise to sunset or from noon to sunset and multiplying by 2:

$$H_d = 2S_0\left(\frac{\bar{d}}{d}\right)\int_0^{\omega_s}\left(\sin\varphi\sin\delta + \cos\varphi\cos\delta\cos\omega\right)dt \qquad \text{(A.5.3)}$$

where $\cos\omega_s = -\,\mathrm{tg}\varphi\ \mathrm{tg}\delta$ is the cosine of the hour angle at sunset. With the transformation $dt = d\omega\,(dt/d\omega) = d\omega\,(86400\ \mathrm{s}/2\pi)$ it follows:

$$H_d = S_0\left(\frac{\bar{d}}{d}\right)^2\frac{86400}{\pi}\left(\sin\varphi\sin\delta\,\omega_s + \cos\varphi\cos\delta\sin\omega_s\right) \qquad \text{(A.5.4)}$$

and the dimension of H_d becomes $J\,m^{-2}\,day^{-1}$. $\left(\bar{d}/d\right)^2$ is given in Fig. A.5.5.

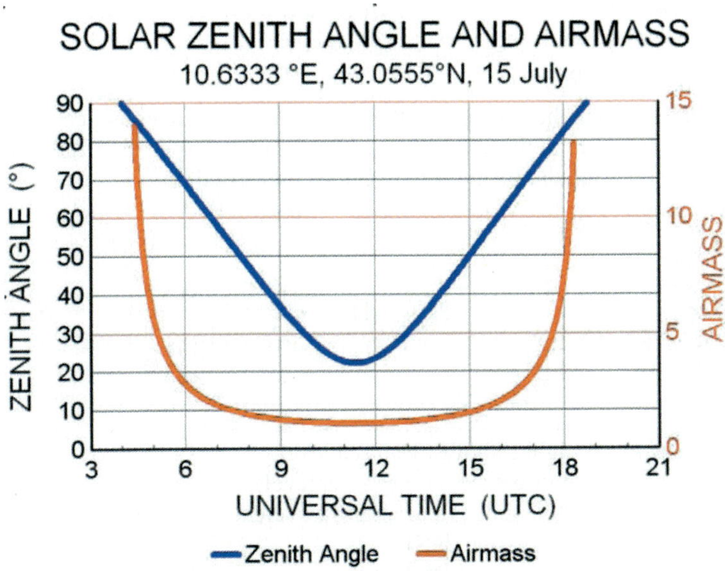

Fig. A.5.4. Solar zenith angle and air mass computed for the geographical position 10.6333 °E, 43.0555 °N and the Julian day 196 (15 July)

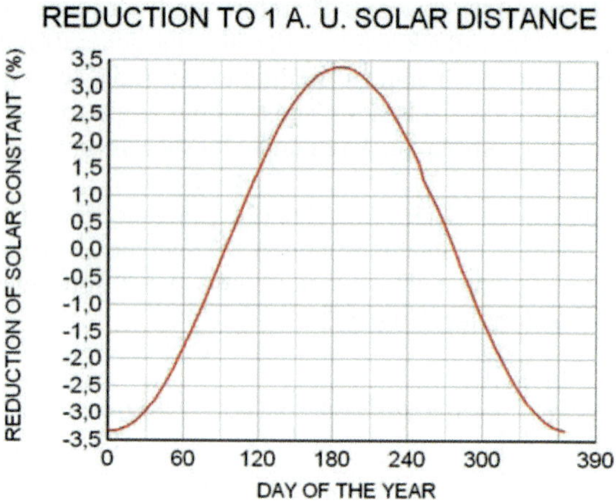

Fig. A.5.5. Variation of the solar irradiance during one year due to the ellipticity of the Earth orbit

A.5.3 Airmass

The relative airmass is the ratio of the air mass at a path the solar ray takes through the atmosphere if it is observed at the ground under an elevation angle $\gamma = 90 - \zeta$, where ζ is the solar zenith angle, to the vertical pass:

$$m(\gamma) = m_{abs}(\gamma) / m_{abs}(90°) \tag{A.5.5}$$

Due to the refraction in the atmosphere the relative air mass is a complicated function of the vertical structure of the atmosphere. According to Kasten (1989) the absolute air mass of an ISO Standard Atmosphere can be expressed as

$$m_{abs}(\gamma) = \rho_0 \int_0^{\infty} \left(\frac{\rho(z)}{\rho_0}\right) \left[1 - \left\{1 + 2(n_0 - 1)\left(1 - \frac{\rho(z)}{\rho_0}\right)\right\} \left\{\frac{\cos\gamma}{1 + \frac{z}{R}}\right\}^2\right]^{-\frac{1}{2}} dh \quad \text{(A..5.6)}$$

where
$\rho(z)$ = air density at height z above mean sea level,
ρ_0 = 1.22500 kg m^{-3} (air density at mean sea level),
n_0 = 2.76 10^{-4} (refractive index of air at the wavelength 700 nm and at $z = 0$),
R = 6356766 m (mean earth radius).

Kasten also provides an approximative formula to calculate the relative air mass:

$$f(\gamma) = \left[\sin\gamma + a(\gamma + b)^c\right]^{-1} \tag{A.5.7}$$

with a = 0.50572, b = 6.379°, c = – 1.6364.

For zenith angles < 70°, $m(\gamma)$ can be represented by $1/\sin\gamma$ with an accuracy of two decimal places or $m(\zeta) = 1/\cos\zeta$. The maximum value of the relative airmass for $\gamma = 0°$ is 38.0868 and the absolute air mass for $\gamma = 90°$ is 10356 kg m^{-2}.

A.5.4 Optical Depth of the Atmosphere

The optical depth of the atmosphere is the product of the atmospheric extinction coefficient and the vertical depth the atmosphere. It equals the natural logarithm of the vertical transmission through the atmosphere. The extinction of a non-absorbing atmosphere is caused by scattering. In the absence of aerosols there is always the molecular scattering which determines the lower limit of the optical dept. The molecular scattering coefficient per unit length is given by the Rayleigh formula:

$$\sigma'_R(\lambda) = \frac{32\pi^3[n(\lambda)-1]^2 f}{3N\lambda^4} \tag{A.5.8}$$

$N = 2.6867\cdot10^{25}$ m^{-3} is the Loschmidt number which gives the number of molecules per cubic metre,

n = refractive index of air

f = 1.054 is a correction factor for the non-sphericality of the molecules (Cabannes factor)

To obtain the extinction of the whole atmosphere the extinction coefficient has to be multiplied with the scale height of the atmosphere. This is the height of the atmosphere reduced to standard temperature and pressure (288 K, 1013.25 hPa, STP):

$$\sigma'_R(\lambda) = \frac{32\pi^3[n(\lambda)-1]^2 H\left(\frac{p}{p_0}\right) f}{3N\lambda^4} \tag{A.5.9}$$

The factor p/p_0 is introduced to reduce the scale height according to the surface pressure which is important in mountainous terrain. $(n\text{-}1)10^6 \cong 278.23$ at $\lambda = 0.5$ µm.

The scale height is defined by

$$H = \frac{RT_0}{gM_a} \approx 8435\,m \tag{A.5.10}$$

where:

R = 8.314 J mol^{-1} K^{-1} molecular gas constant,

T_0 = 288.15 K,

M_a = 0.028964 kg mol^{-1} molar mass of air up to about 90 km,

g = 9.8062 m s^{-2} acceleration due to gravity at 45° latitude.

In addition to the molecular scattering the absorption by gas molecules and the scattering and absorption due to aerosols have to be accounted for in determining the optical depth of the atmosphere. The absorption coefficients of ozone are given in Table A.5.2 together with the transmissivity in mid-latitude model atmospheres. The transmittance of water vapour can be computed with the LOWTRAN and MODTRAN radiative transfer codes if the atmospheric water vapour is known e.g.

from radiosonde ascents. The optical aerosol depth can be measured with photometers.

Table A.5.2. Absorption coefficients and transmittance of ozone

Wavelength (nm)	Absorption coefficient (cm^{-1})	Transmittance in Mid latitude atmospheres	
		Summer	Winter
440	0.001	0.0004	0.0003
520	0.055	0.0213	0.0173
550	0.092	0.0356	0.0289
670	0.036	0.0139	0.0113
750	0.014	0.0054	0.0044

A.5.5 Determination of the Optical Depth From the Ground

To measure the transmission τ of the atmosphere and deduce from it the optical depth δ requires only a simple uncalibrated but temperature stabilized sun-photometer with a narrow field of view (about 2 degree) and a selection of spectral narrow-band (interference) filters. Under cloudless conditions and unchanging atmospheric conditions it is possible to extrapolate the relative intensity values taken at different air masses to the air mass zero to obtain the extra-atmospheric value. The value determined for air mass one divided by this extra-atmospherical value gives τ for the vertical path and its natural logarithm the optical depth δ. With a known extra-atmospherical value $A_\lambda(0)$, measurements can then also be taken under partly cloudy conditions and the optical depth can directly be inferred.

The general relationship between the transmittance $\tau_\lambda(m)$, the optical depth $\delta_\lambda(m)$, the relative air mass m and the relative measured signal $A_\lambda(m)$ is

$$\frac{A_\lambda(m)}{A_\lambda(0)} = \tau_\lambda(m) = e^{-\delta_\lambda m} \qquad \text{(A.5.11)}$$

where m in a wide range is $1/\cos \zeta = \mu^{-1}$. For any two readings at different air masses one obtains under stable atmospheric conditions:

$$\ln A_\lambda(m_i) - \ln A_\lambda(0) = -\delta_\lambda m_i \qquad \text{(A.5.12)}$$

from which follows:

$$\delta_\lambda = \frac{\ln A_\lambda(m_i) - \ln A_\lambda(m_j)}{m_j - m_i} \qquad \text{(A.5.13)}$$

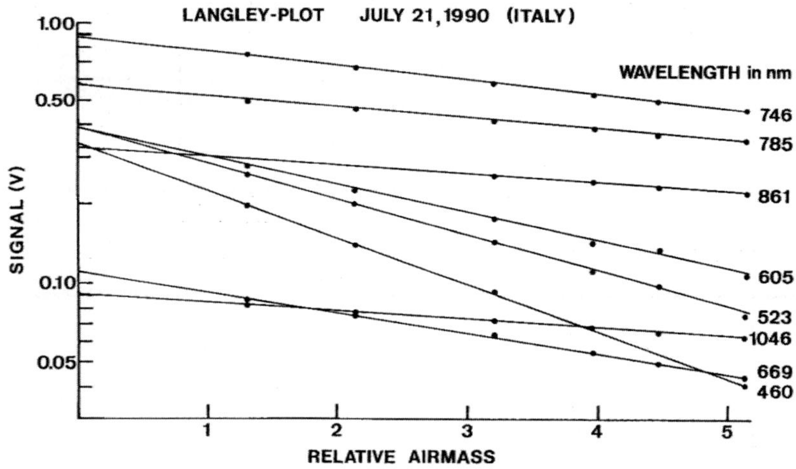

Fig. A.5.6. Example for a Langley plot

Thus in principle two measurements are sufficient to estimate δ. In practise one would make a series of measurements, plot it on a semi-logarithmic paper and draw a line through the points to smooth the scatter and to extrapolate linearly to the extra atmospheric value (Langley-plot, Fig. A.5.6). However, one has to be careful by applying this procedure and should not use an automatic smoothing procedure because the composition of the aerosol may change during the observation time and the resulting curve may in fact not be linear. In these cases only the morning hours of clear days should be used to determine the extrapolated extra-atmospheric value. If later, during the day or on following days, points depart from the linear line they can be connected with the extra atmospheric point and the slope of this line can be used to compute the instantaneous optical depth.The diurnal course of measured values almost always tends to be different in the morning and in the afternoon because the height of the atmospheric boundary layer and with it the aerosol content may increase or the aerosol type may change. Examples of optical depths are given in Appendix 6.

A.5.6 Relationship Between UTM and Longitude/Latitude Coordinates

The Universal-Transversal-Mercatorprojection (UTM) was introduced and described in section 2.5.3. To recall the subdivision into zones and bands the fraction of the world map of the UTM coordinates relevant for the Mediterranean is given in Fig. A.5.7.

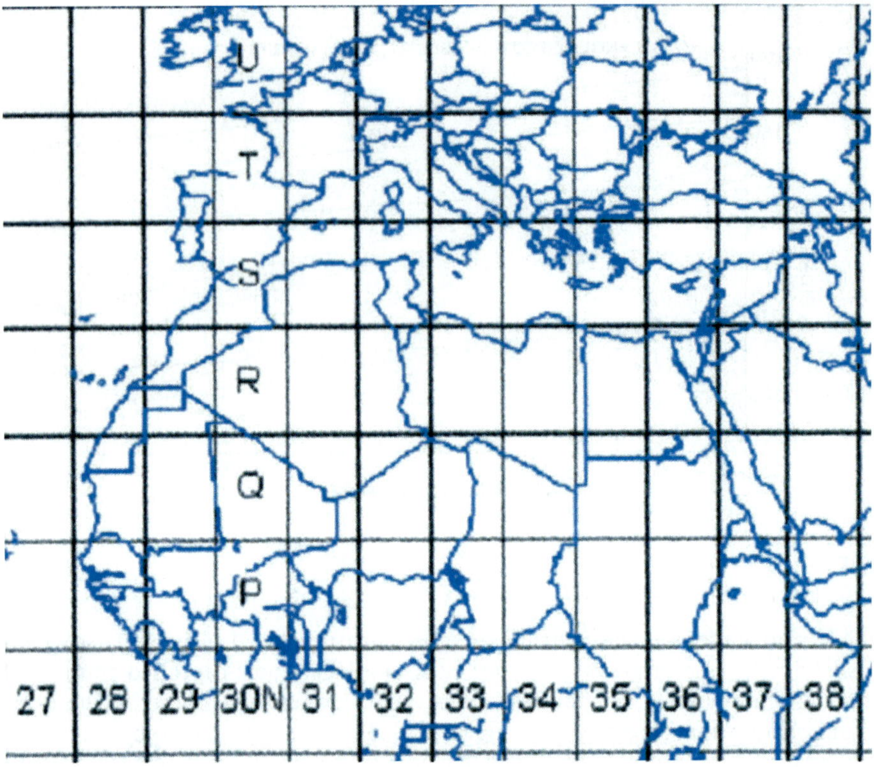

Fig. A.5.7. UTM zones 27 to 38 and bands N to U for the European sector

Each UTM zone-band box is subdivided into 100 km x 100 km quares. The lettering of the columns within the quares starts at 0° with the letter A. Since each zone at the equator is 6° or about 666 km wide, the first zone following the zero meridian ranges over at most eight columns. Within zone 31 these are lettered A to H and UTM zone 32 (6°E to 12°E) starts with column J and ranges to R. At 9°E column N starts. The same is applied for the rows in the square. Since 32 is an even number the lettering starts with F at the equator which is repeated at 4000 km. At 4500 km from the equator or 40.634623° the letter L starts, at 4800 km or 43.343597° the letter P (in fact it starts at 43°21.016187' = 43.350270, which means, one would have to count with 110.725955 km/degree), at 4900 km or 44.246589° starts letter Q, and at 5000 km or 45.149581° letter R. The position of the corner point NN/PN/NM/PM is at 10°14.037037', 43°20.8190082', that one of the corner point NP/PP/NN/PN at 10°12.940462', 42°26.543654'

As has ben mentioned earlier, the starting point of a row actually depends on the ellipsoid that is used. For Europe the system is based upon the Hayford (1924) ellipsoid. 1° latitude equals 110.569 km at the equator, 111.0 approximately at 38.2°N, and 111.7 at the pole. From the equator to the Mediterranean area one has to count in the average 110.743 km/degree.

As an example, at 43.24°N one is (43.24 x 111 km =) 4800 km distant from the equator. Here the row letter P starts (because 32 is an even zone, the numbering continues again with F at 4000 km from the equator, which equals 36.03°N). A position in Tuscany at 10°38'04"E, 43°03'23"N, that obviously is known to ±1" or ±30 m, has a position of 32 T PN 33100 68373 or 633100.25 m easting and 4768373.0 m northing in the UTM system. Conversion applets between longitude/latitude and UTM coordinates can be found in the internet. While the numerical UTM system uses only box codes (like 32 T), eastings (from the central meridian of each box) and northings (from the equator), the Military Grid Reference System (MGRS) uses the 100 km quare letter identification with additional digits. The first two digits locate the point in a 10 km x 10 km square within the 100 km x 100 km area. The following digits then locate the point within a 1 km square within the 10 km x 10 km area (two additional digits) and so on down to a 1 m square (10 digits).

A.5.7 Evaporation Equivalents

The evaporation rate generally is measured in mm water layer per day. The energy to evaporate 1 kg (10^6 mm^3) of water is temperature dependent and requires about 2.465 MJ at 15°C (Fig. A.5.x.). To evaporate 1 mm water per 1 m^2 hence requires 2.465 MJ m^{-2} or 2.465 MW s m^{-2}. To obtain the rate of energy necessary to evaporate this amount of water per day one has to divide this number by the number of seconds per day (86400) and it results that 1 mm evaporation per day is equivalent to 28.53 W m^{-2}.

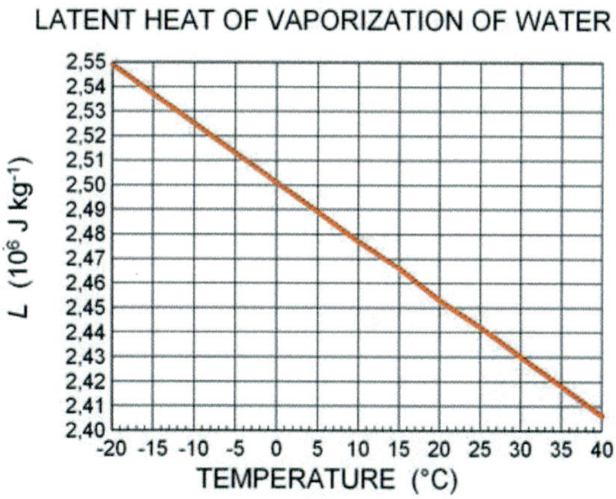

Fig. A.5.8. Dependence of the latent heat of vaporization of water on temperature

A.5.8 Relationship Between Dielectric Constant and Soil Moisture

Fig. A.5.9 gives the relationship between the relative dielectric constant and the volumetric soil moisture for the C-, L-, and P-bands which determines the potential to measure soil moisture.

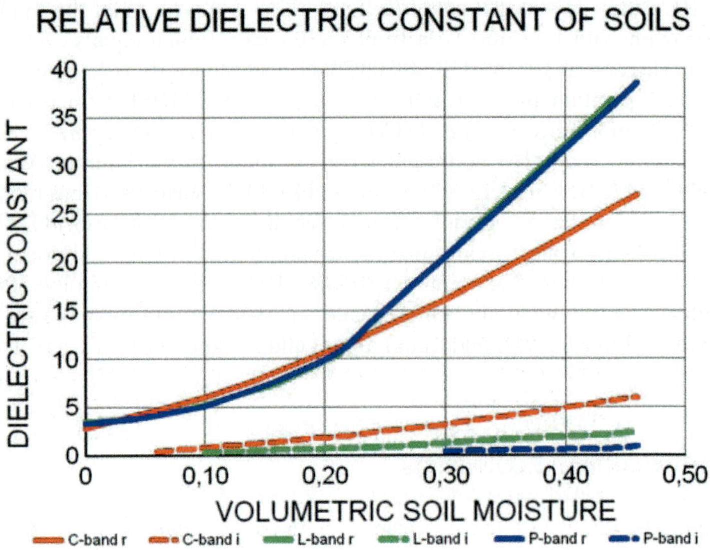

Fig. A.5.9. Dependence of the relative dielectric constant on volumetric soil moisture for the C-, L-, and P-bands

Appendix 6 Spectral Measurements

A.6.1 Introduction

This appendix contains supplemental information for several chapters in which it is dealt with different kinds of spectral measurements. It is subdivided into the following sections:

- Atmospheric spectra (A.6.2). In this section additional information is given on atmospheric infrared radiation, specifically of the thermal infrared sky spectrum. These measurements are in so far of historical interest because they date from a time as the contribution of the chlorofluorocarbons to the atmospheric radiation was still negligible. Today these spectra can be measured by application of improved interferometric technology with much higher spectral resolution of which one example is given in Chapter 3.
- Spectral atmospheric transmission values (A.6.3). These measurements are relative interference filter measurements of solar radiation from which the atmospheric optical depth can be derived as described in Appendix 5. Here some results are presented..
- Reflection spectra in the long-wave infrared region (A.6.4). These measurements are important for emissivity studies and a number of laboratory measurements are shown of the angular dependence of the reflectance of some surface materials.
- Reflection spectra of natural land surfaces (A.6.5). This sections contains reflection spectra measured with a spectrometer during several field campaigns. The spectra range from 400 to 2500 nm. Because these spectra are made in the field, the near atmospheric water vapour disturb the spectra in the region of its near infrared absorption bands which are therefore cut off.
- Spectral reflection values measured with interference filter devices, a radiometer and an albedometer (A.6.6). Measurements with radiometers require as those made with spectrometers a reference measurement to determine reflectance while the albedometer measures at the same time the radiation from below and above.

A.6.2 Atmospheric Infrared Spectra

The infrared emission and absorption of the atmosphere is a severe impediment for the correct assessment of the surface temperature by remote sensing from space. There are two strategies to overcome this problem: The split-window technique which is applicable if two channels are available in the thermal infrared and the computation of correction functions using spectral band models and instantaneous atmospheric data. The accurate simulation of the atmospheric spectrum in the thermal infrared region became possible after the determination of line positions, line intensities and line half-widths to which many scientists contributed (literature surveys can be found e.g. in Goody, 1964, or Bolle, 1982). All line data were compiled by McClatchey et al. (1973).

If the vertical distribution of the strongly variable atmospheric water vapour amount, which together with the less variable but permanently increasing carbon dioxide governs the spectrum, the vertical distribution of ozone, and the vertical temperature and pressure profiles are known, complete spectra can be simulated using the line data compilation. For other trace gases such as methane, nitrous oxide and carbon monoxide average values can be used. Fig. A.6.2.1 shows the result of such a simulation. The spectrum has originally been computed by narrowing the spectral resolution to the width of the Voigt-profile (10^{-5} cm^{-1}) and was afterwards degraded

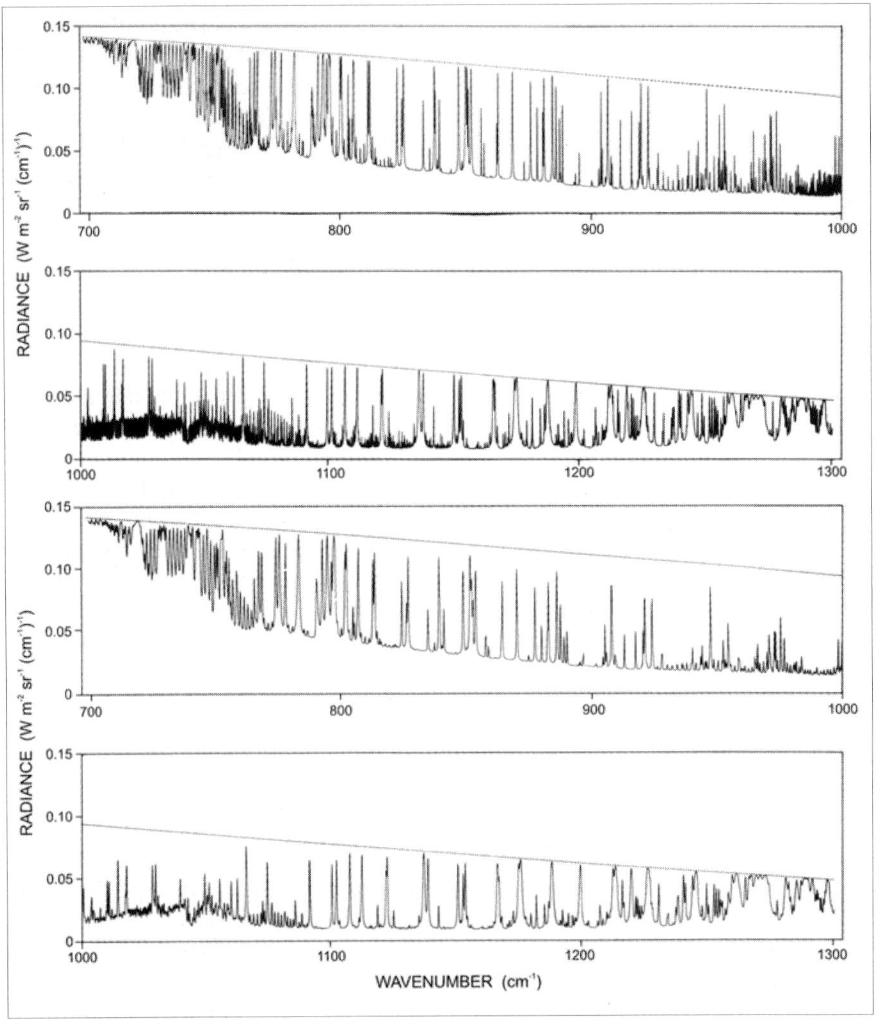

Fig. A.6.2.1. Zenith emission spectrum of the Atmosphere computed for Sant'Agata sui due Golfi, Italy, 455 m above sea level, 40°37'N, 14°11'E. Total amount of water vapour: 2.47 g cm^{-2}, total amount of carbon dioxide: 238.6 atm-cm, total amount of ozone: 0.295 atm-cm. The upper spectrum is degraded for a spectrometric slit width of 0.01 cm^{-1}, the lower one for a slit width of 0.2 cm^{-1}

to actually measurable spectral resolutions. Fig. A.6.2.2 shows the comparison of the computed spectrum with measured spectra. For this purpose the computed spectrum was degraded to slit widths of 1.3 and 2.6 cm^{-1} because in the wavenumber domain the spectral resolution of the measured spectrum decreases as the wavenumber increases. In the centre also the temperature and the assumed ozone profiles are shown, the actual water vapour mixing ratio and the temperature were measured with a local radiosonde.

At the wavelengths of 14 to 16 μm the band of carbon dioxide completely blocks the exchange of radiative energy between surface and space. At still longer wavelength, water vapour is the main absorber. At sea level the water vapour emission and absorption is so strong that beyond the carbon dioxide band at 15 μm the instrument can "see" only into slightly higher and cooler regions of the at the atmosphere. This changes as one gets higher up. Fig. A.6.2.3 shows spectra taken Jungfraujoch observatory at 3500 m altitude where windows open between the water vapour lines.

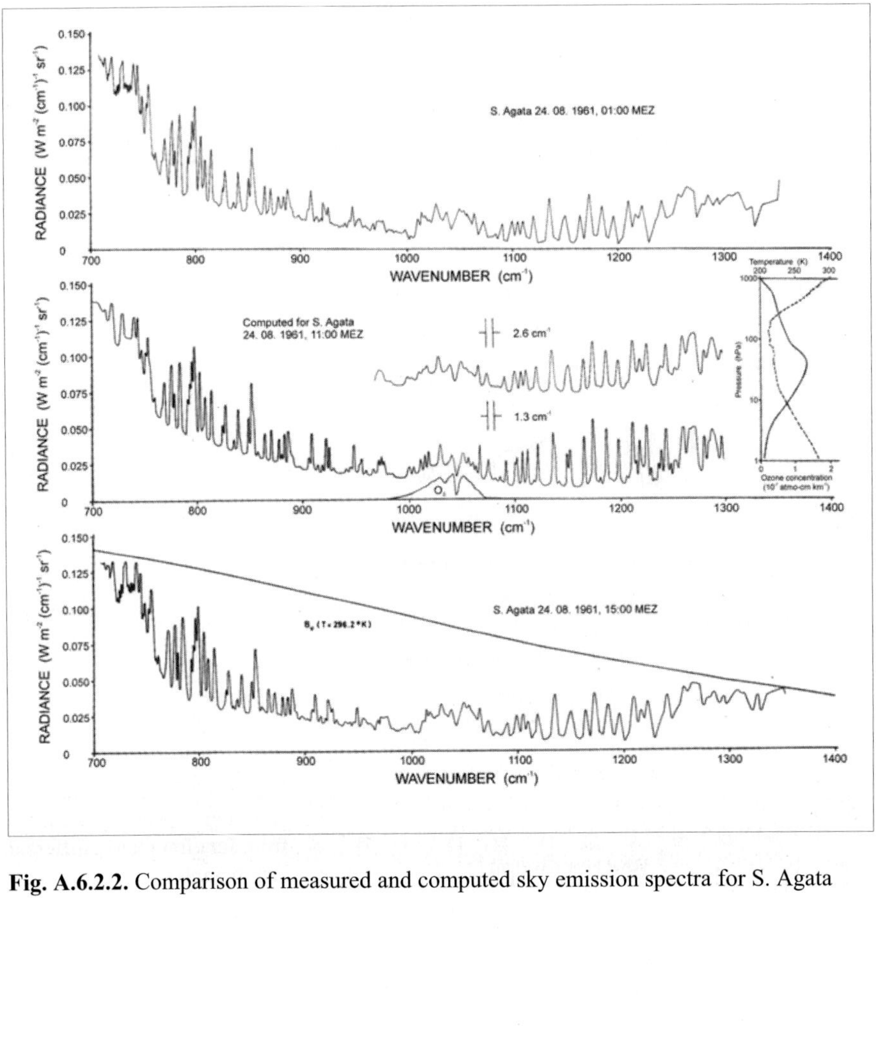

Fig. A.6.2.2. Comparison of measured and computed sky emission spectra for S. Agata

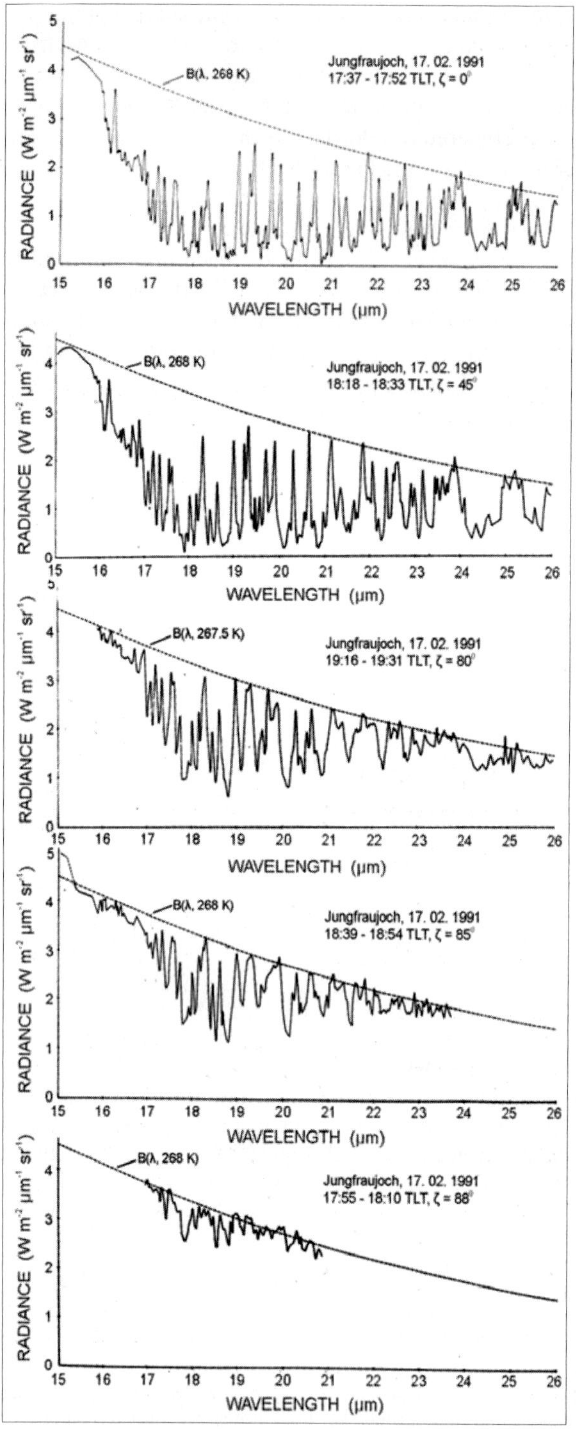

Fig. A.6.2.3. 15 - 26 μm sky emission spectra measured from Jungfraujoch at different zenith angles

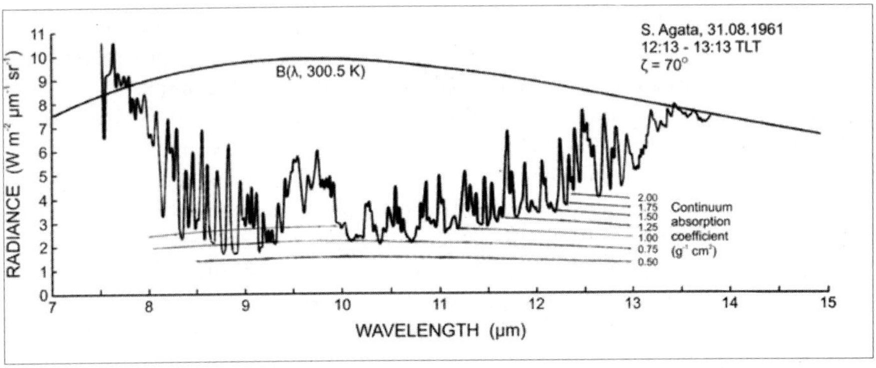

Fig. A.6.2.4. 7.5 - 14 μm sky emission spectrum measured at S. Agata with isolines of the emission due to the water vapour continuum

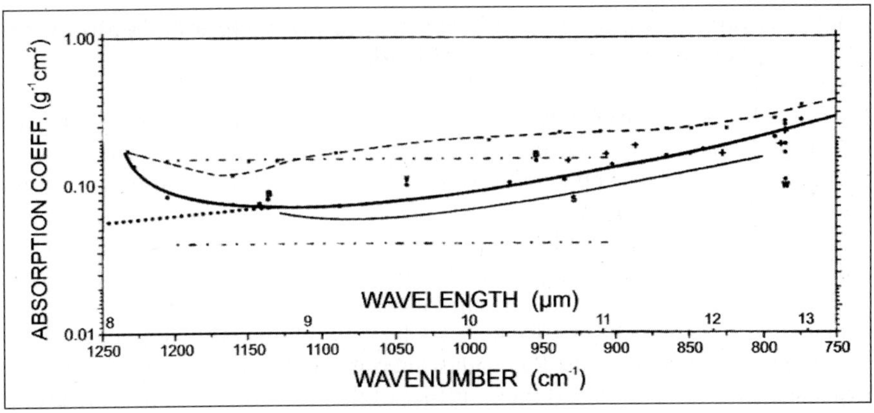

Fig. A.6.2.5. Quasi-continuum absorption coefficient derived from spectral measurements. The heavy line represents the measurements made in S. Agata, Italy. The dotted line is an extrapolation of the pure rotational continuum towards higher wavenumbers. The difference to the thick line is due to the contribution of the 6.3 μm rotation-vibration band. The dashed line connects measurements made in Beer Sheva, Israel. + - Jungfraujoch values, •B - values derived from emission measurements in Hamburg (Bolle, 1960), •V - values from Vigroux (1959) from emission measurements, thin line - mean values of Saidy (1960) from absorption measurements of solar radiation, dashed - dotted lines are extreme values reported by Walshaw (1960) obtained from emission measurements of the whole region 900 - 1200 cm^{-1}. Results of measurements in the 17 - 26 μm region which are not shown here can be approximated by the relationship $\log k = 0.1228\,\lambda - 2.2538$, where k is the absorption coefficient and 10 μm $< \lambda <$ 26 μm is the wavelength in μm

A problem remained for some time the determination of the so-called water vapour quasi-continuum which is superimposed to the line spectrum (see e.g. Grant 1990). This quasi-continuum results from the superposition of the wings of distant very strong water vapour lines and of bonds between water vapour molecules $(H_2O)_n$ which are hardly to be computed because the knowledge about the line shapes very far off the centre of a line and of the $(H_2O)_n$ emission features is weak.

From the high resolution spectra the continuum absorption coefficient can be estimated for a number of wavelength between the absorption lines. Fig. A.6.2.4 shows an S. Agata spectrum with iso-lines of the emission as calculated for different values of the continuum absorption coefficient. For these computations the vertical water vapour and temperature profiles measured by means of radiosondes made at the measuring site were used. In Fig. A.6.2.5 the continuum absorption coefficient as determined by different authors from emission and absorption spectra are compiled. The continuum absorption coefficient is reduced to standard pressure. It varies proportional to p/p_0. Its temperature dependence is small because the strength of very strong rotational lines around 50 μm which contribute strongly to the continuum are temperature independent.

A.6.3 Spectral Atmospheric Transmission and Optical Depth in the Wavelength Range of Solar Radiation

In Appendix 5 the method to estimate spectral atmospheric transmission was described. Because of the fundamental importance of these data for the correction of information derived from observations made from space, a few results obtained with this method are reported here.

Measurements were made with narrow field of view interference filter radiometers. One of it was equipped with two detectors, one for the visible and near infrared and the second one, a PbS cell, for the middle infrared. This extended the operational range to 3.7 μm. Table A.6.3.1 gives the result of one set of these measurements. In addition the Rayleigh scattering coefficient, the Rayleigh optical depth and the Rayleigh vertical transmittance of the standard atmosphere are given. The values are in so far of interest as at shorter wavelengths the optical depths were higher in the morning than at noon and at longer wavelengths (which can only be measured by the PbS detector) the values are lower in the morning than at noon. The change starts with the onset of the sea breeze by which obviously larger aerosol particles of a different kind were brought ashore on the expense of smaller rural ones at this near-coastal station. Land-sea winds and related aerosol-type changes occur everywhere in the Mediterranean basin and in principle it is necessary to determine the aerosol optical depth at the time of a satellite overpass. By comparing the measurements with the two detectors the table at the same time gives an idea what kind of accuracies can be obtained. Numerical examples are given in Table A .6.3.2.

A special case in the Mediterranean is the Po-Valley. Embedded between high mountain ranges it accumulates a large amount of aerosols from the large cities, Traffic and agriculture (ploughing, burning of debris). Fig. A.6.3.1 shows optical depths measured at a traverse across the Po-valley. The geographical positions of the measuring sites are given in Table A.6.3.3. The values are about twice as high as at the coastal station Cafaggio near Campiglia M.ma in Tuscany in July (Fig. 6.3.2) and by far higher that in the morning of a clear September day with a meteorological

Table A.6.3.1. Optical depths measured on 11.08.1990 near Campiglia Mma., Italy, with two independent detectors. For comparison the Rayleigh optical data are given. $H = R_m T/g$ is the atmospheric scale height (8431 m for T = 288.15 K), R_m is the specific gas constant.

Wavelength (nm)	Detector 1 (Si)		Detector 2 (PbS)		Rayleigh scattering coefficient	Rayleigh optical depth	Rayleigh transmissivity
	Morning hours	Noon	Morning hours	Noon	σ (10^{-8}cm^{-1})	$\delta = \sigma H$	exp $(-\delta)$
403	0.829	0.705	0.814	0.709	43.5	0.351	0.704
523	0.433	0.287	0.439	0.376	14.8	0.119	0.888
668	0.247	0.178	0.251		5.3	0.0427	0.958
747	0.185	0.115	0.187		1.87	0.0151	0.985
860	0.124	0.05	0.137		1.22	0.01	0.99
1032	0.09	0.131	0.091	0.135	0.96	0.008	0.992
2175	-	-	0.06	0.09	0.06	0	0.9995
3773	-	-	0.041	0.07			1

range of larger than 50 km which is shown in Fig. A.6.3.3. While in the morning the situation was rather stable, the optical depth increased during the afternoon. A few days earlier, on September 23, the situation was different with decreasing values during the morning hours (Fig. 6.3.4).

Another case can be seen in Figs. A.6.3.5 where measurements made at the top of a high altitude station on Crete (Skinarkas mountain, 1750 m). It shows the effect of uprising dust and pre-condensation states starting around noon with small optical depths in the morning and in the late afternoon. For comparison, the results of measurements made at another date in the Messara valley with a strong increase of turbidity during the day are given in Fig.A.6.3.6.

An extreme situation is reported in Fig. A.6.3.7. It shows a measurement made during a sandstorm at a station in the Republic of Niger (Ibecetene). The optical depth increased by nearly a factor of ten at the long- wave end of the measurements which indicates the presence of many large particles. The slope is completely different as compared to the Po-valley with as well high values at the shortwave end but a the strong decrease of the spectral curve towards longer wavelengths. In the Ibecetene case the aerosol optical depth is that of a pure aerosol atmosphere to which the Rayleigh optical depth has to be added to obtain the total optical depth.

In Fig. A.6.3.8 measurements made in Italy and Spain are compared. The values which are labelled as maximum and minimum (yellow and brown curves) refer to the measurements made in the Po-valley. It can be seen that the measurement made in Spain (Rada de Haro) nearly coincides with the measurements made in Cafaggio in the afternoon of 29 September. As had been shown in Chapter 1 by means of the

DIAL measurements, the boundary layer over Castilla - La Mancha was very heigh during summer and filled with aerosol.

Table A.6.3.2. Spectral optical depths measured at different locations and at wavelengths of minimal gaseous absorption. The measurements at the North Sea coast and NW of Berlin are average values of eight days.

Wave-length (nm)	Location and Date								
	Cam-piglia	Rada de Haro (1991)			Ibe-cetene	Hörnum, Sylt (North Sea)		Gülpe, NW of Berlin	
	23.07. 1990	03.06.	24.06.	25.06.	21.03. 1988	16.09. - 18.10. 1989		12.10. - 24.10. 1990	
						aver.	min.03.	aver.	max.17.
403						0.45	0.327	0.67	1.058
460	0.462	0.433	0.347	0.52	0.367				
523	0.337	0.372	0.258	0.291	0.294	0.23	0.108	0.41	0.683
605	0.271	0.299	0.209	0.243	0.264				
669	0.19	0.195	0.143	0.15	0.211	0.14	0.056	0.24	0.453
746	0.153	0.159	0.118	0.132	0.201	0.14	0.063	0.22	0.381
785	0.133	0.133	0.104	0.101	0.183				
861	0.104	0.0982	0.0783	0.0686	0.165	0.1	0.04	0.17	0.309
1032						0.1	0.018	0.14	0.26
1046	0.0721	0.0886	0.0707	0.132	0.168				

Table A.6.3.3. Position of measuring stations in Italy

Location	Latitude	Longitude	Time (MESZ)	Airmass
Villafranca	45°21'N	10°45.7'E	14:57-15:00	1.488-1.499
Roverbello (30m)	45°16.1'N	10°45.7'E	15:25-15:28	1.582-1.6
Montanara (30m)	45°07.9'N	10°43.3'E	15:57-15:59	1.748
Villa Pasquali, Sabionetta	45° 00.2'N	10°30.8'E	16:24-16:28	1.927-1.961
Torrile (30 m)	44°55.6'N	10°19'E	16:48-16:52	2.151-2.198
Mendesano (136m)	44°45.4'N	10°07.9'E	17:28-17:32	2.755-2.841
Cafaggio (117 m)	43°03.92'N	10°38.0'E	various dates & conditions	

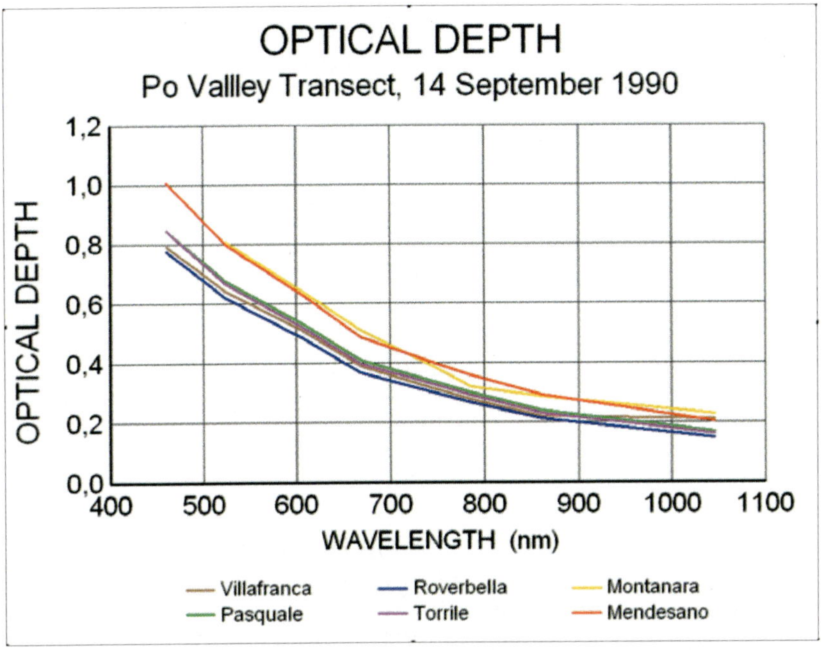

Fig. A.6.3.1. Optical depths measured on 14 September 1990 across the Po-valley. The geographical positions of the measuring sites are given in Table A.6

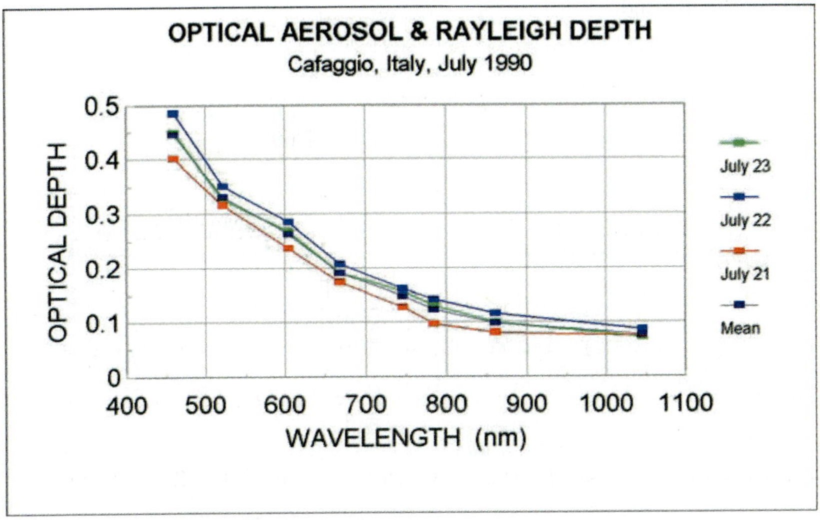

Fig. A.6.3.2. Wavelength dependence of the total optical depth measured at the station Cafaggio at three consecutive days in July 1990

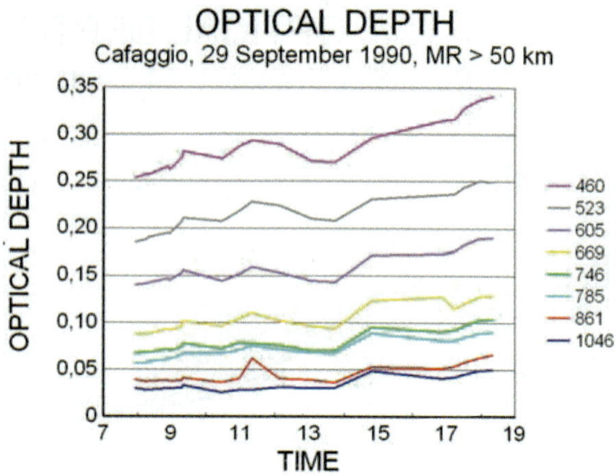

Fig. A.6.3.3. Development of the optical depth during 29 September 1990 at the station Cafaggio

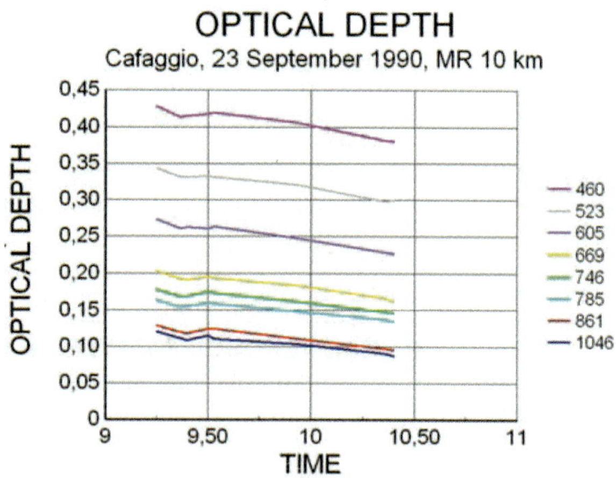

Fig. A.6.3.4. Optical depth at the station Cafaggio in the morning of 23 September 1990 for eight wavelength between 460 nm and 1046 nm

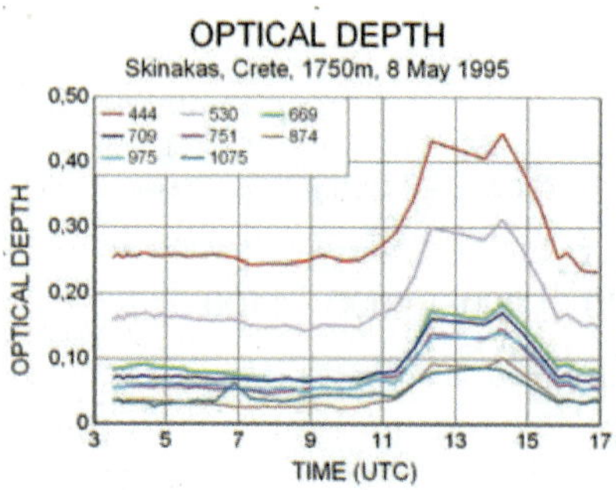

Fig. A.6.3.5. Spectral optical depth measured at the Skinakas (1750 m), Crete

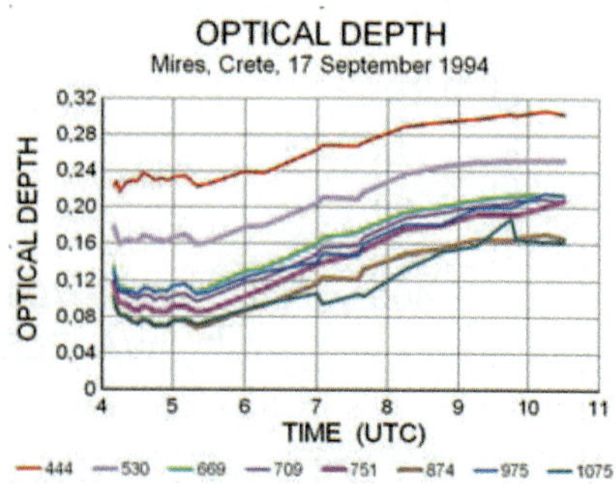

Fig. A.6.3.6. Spectral optical depth measured in the Messara valley

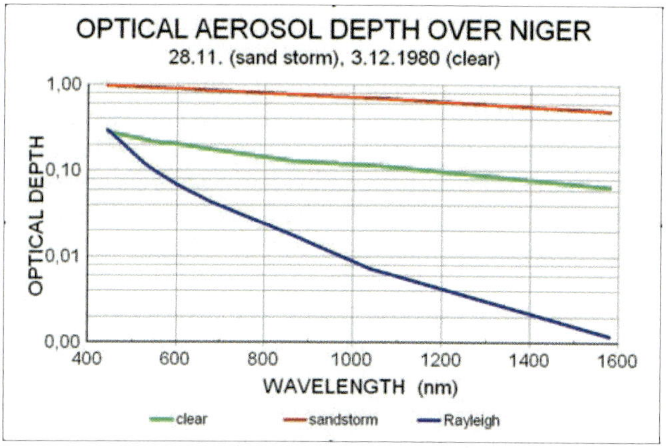

Fig. A.6.3.7. Aerosol optical depth measured in Ibecetene, Niger, at a clear day (green line) and during a sandstorm (red line). The Rayleigh optical depth (blue line) has to be added to obtain the total optical depth

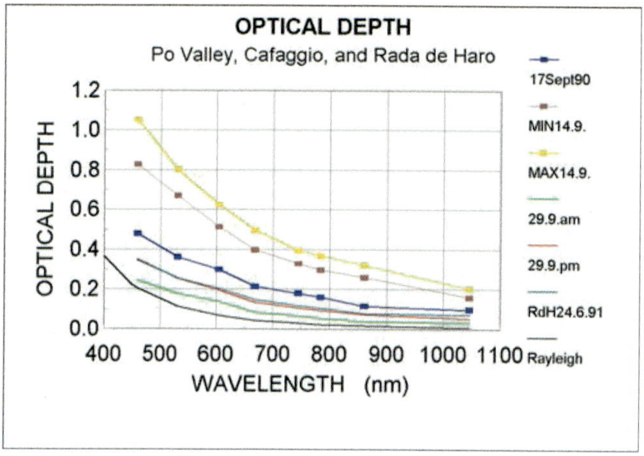

Fig. A.6.3.8. Comparison of different optical depth determinations

A.6.4 Angular Distribution of Spectral Longwave Infrared Surface Reflectance

The determination of thermodynamic surface temperatures depends on the knowledge of the surface emissivity in the spectral bands from 3.3 to 4.1 μm and from 8 to 13 μm The emissivity is not independent of the angle ob observation. Furthermore the measurement of the radiative temperature of the surface always includes a component of reflected solar and atmospheric radiation which has to be subtracted to obtain the true surface radiative temperature. There exist not many data about the angular dependence of the longwave infrared reflectance of natural surfaces. A few examples are reported in this section. The measurements have been made in the laboratory. Here the surface under investigation is illuminated by a parallel beam of infrared radiation. If M_0 denotes radiant exitance (W m^{-2})of the parallel beam, the irradiance E (W m^{-2}) at the sample depends on the angle of incidence, α, and can be defined as $E(\alpha) = M_0 \cos \alpha$. M_0 can be determined by measuring directly the exitance of the source at the location of the sample or $E(\alpha)$ is measured by replacing the investigated material by a mirror of known reflectance. The radiance L (W m^{-2} sr^{-1}) emerging from a natural surface illuminated by E depends on the angle of incidence, α, as well as on the angle of reflection, β, and is related to the irradiance by the directional incoming - directional outgoing reflectance factor $\rho_d(\alpha, \beta)$: $L(\alpha, \beta) = E(\alpha)\,\rho_d(\alpha, \beta)$. The dimension of $\rho_d(\alpha, \beta)$ is sr^{-1}. From a non-absorbing Lambertian surface the incident energy would isotropically be reflected into the hemisphere and the radiance emerging into any direction of the hemisphere would be $L = E(\alpha)/\pi$. If the reflectance $\rho(\alpha, \beta)$ is defined as the dimensionless ratio of the radiant energy re-emitted from a target divided by the energy that would be reflected by an ideal white Lambertian surface, the following relationship is obtained: $\rho(\alpha, \beta) = L(\alpha, \beta)/[E(\alpha)/\pi] = \pi\rho_d(\alpha, \beta)$. For an ideal white, isotropically re-emitting Lambertian surface one obtains $\rho(\alpha, \beta) = 1$. The following figures are expressed in this notation. Reflectance spectra with a specular component may in this presentation yield values exceeding the value of one.

Fig. A.6.4.1. shows the change of the specular reflectance of water with the specular angle. The Figs. A.6.4.9 add to this figure the degree of polarization of the reflected beam. Figs. A.6.4.2 illustrate the effect of reststrahlenbanden of quartz sand on the reflectance. It is remarkable that outside the strong reflection bands the angular dependence is weak. This effect is even more prominent if pounded brick material is investigated (Fig. A.6.4.8). Reststrahlenbanden are even visible in the reflection spectrum of humus with quartz and feldspar inclusions (Fig. A.6.4.6). Also the weathered surface of quartz minerals show the strong reflection around 10 µm (Fig. A.6.4.4). Limestone has two reflection maxima, at 2 and 10 µm (Fig. A.6.4.5). Its reflectance spectrum therefore differs from those seen before. In slate and claystone spectra again the 10 µm maximum is prominent (Figs. A.6.4.6 and A.6.4.7) . So far most reflection spectra originate from stony surfaces. Pounded material from red bricks behave similar but more marked than quartz sand: The reflectance remains over a large range of angles very isotropic and gets extremely small at 7 µm (Fig. A.6.4.8a). Only if the reflection is measured for incidence angles larger than 70° a reflection

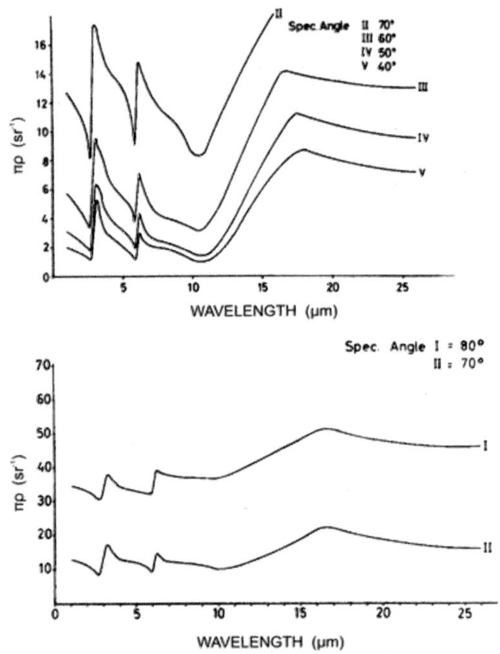

Fig. A.6.4.1. Specular reflectance of water, 1 - 26 µm wavelength. Top: Specular angle 40°, 50°, 60°, and 70°. Bottom: Specular angle 70° and 80°

band near 8 μm becomes visible and the reflection under the specular angle becomes very large and angle dependent (Fig. A.6.4.8b).

Depending on the angle of observation the reflected radiation generally is polarized. This becomes especially evident in the case of directional - directional reflection geometries. The degree of polarization is difficult to measure because of the lack of efficient polarisers for the longwave infrared part of the spectrum. Hanl (1969) measured the polarization for broad spectral regions filtered out by Grubb Parsons blocking filters of InSb combined with NaCl and LiF plates. The resulting spectral bands were 7.3 - 20 μm, 8 - 20 μm, and 12 - 20 μm with the effective central wavelengths of 9.5 μm, 10.6 μm, and 14 μm.

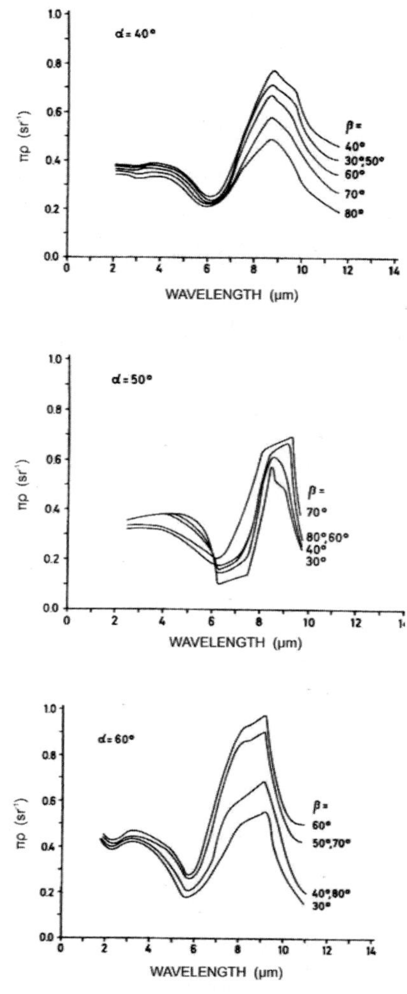

Fig. A.6.4.2a. Reflectance of quartz sand from the Negev desert (Israel). From top to bottom: Incidence angle $\alpha = 40°$, $50°$, and $60°$, reflection angle $\beta = 30°$, $40°$, $50°$, $60°$, $70°$, and $80°$ as indicated

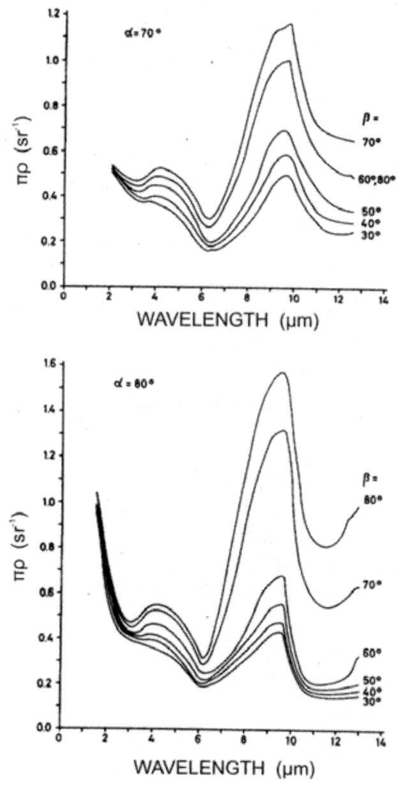

Fig. A.6.4.2b. Reflectance of quartz sand from the Negev desert (Israel). Top: Incidence angle $\alpha = 70°$. Bottom: Incidence angle $\alpha = 80°$. Reflection angle $\beta = 30°$, $40°$, $50°$, $60°$, $70°$, and $80°$ as indicated

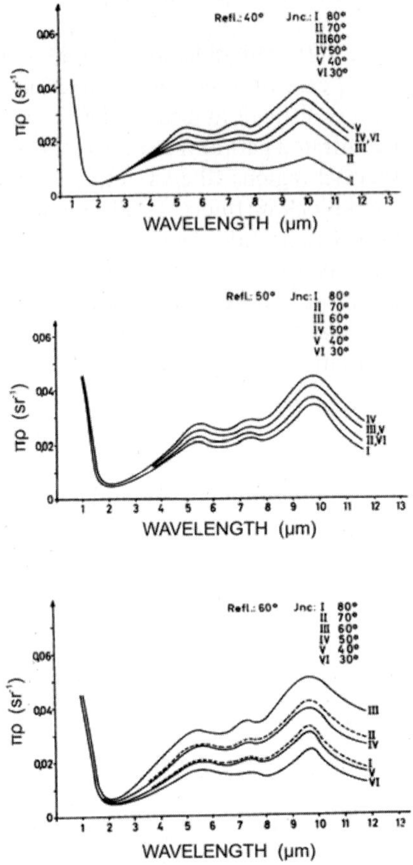

Fig. A.6.4.3. Reflectance spectra of humus (vegetable mould and fibres with small amounts of quartz and feldspar). Incidence angles α = 40°, 50°, and 60°. Reflection angles β = 30°, 40°, 50°, 60°, 70°, and 80°

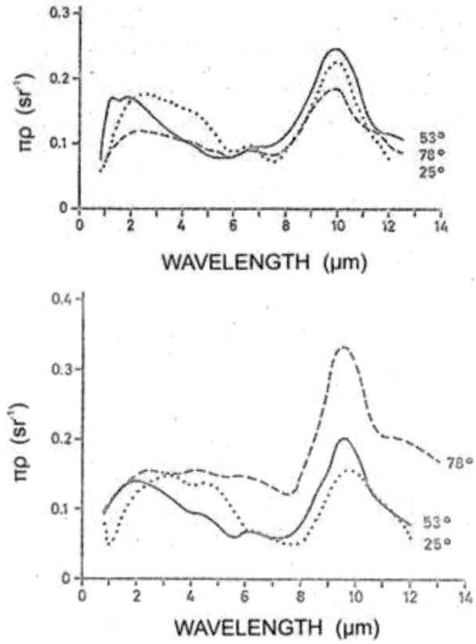

Fig. A.6.4.4. Reflection spectra of two samples of weathered agate-quartz from the Negev desert (Israel). Incidence angle 53°, reflection angles 25°, 53°, and 78°

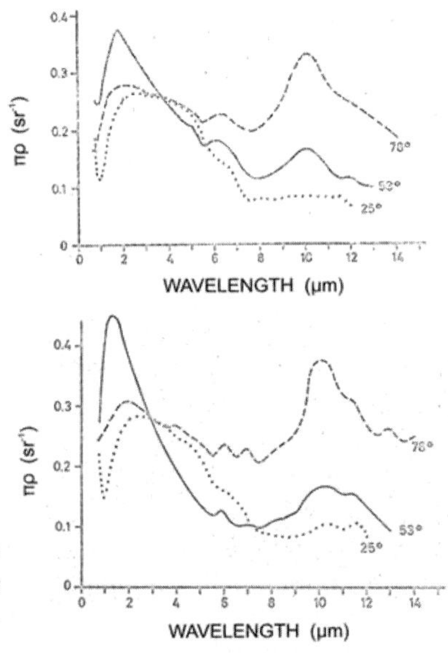

Fig. A.6.4.5. Reflectance spectra of two samples of marly limestone with silicate inclusions (chert, hornstein, lower figure). Reflection angles are 25°, 53°, and 78°

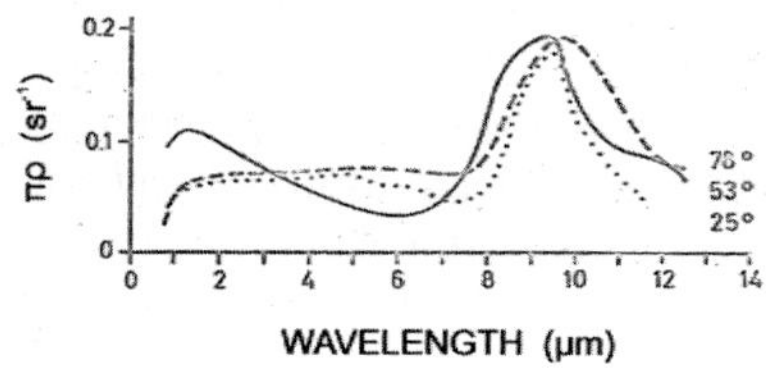

Fig. A.6.4.6. Reflectance spectra of slate with malachit and enriched with quartz for an incidence angle of 53° and reflection angles of 25°, 53°, and 78°

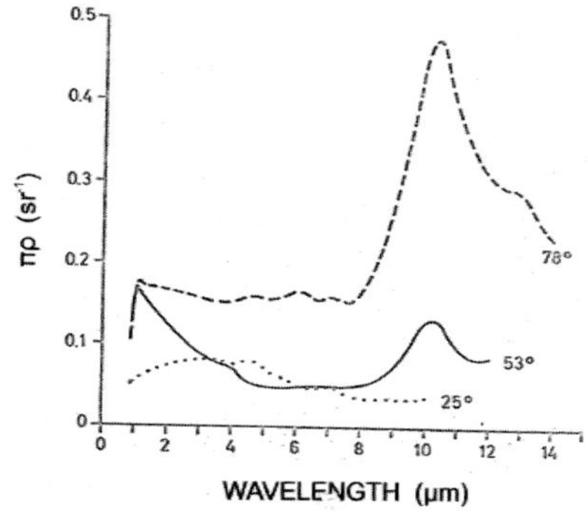

Fig. A.6.4.7. Reflectance spectra of silt- or claystone with cracks. Incidence angle 53°, reflection angles 25°, 53°, and 78°

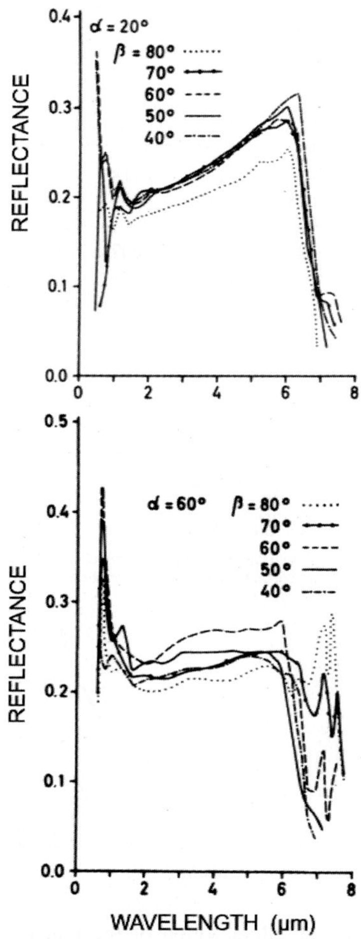

Fig. A.6.4.8a. Angular dependence of the reflectance from pounded bricks for incidence angles of 20° and 60°

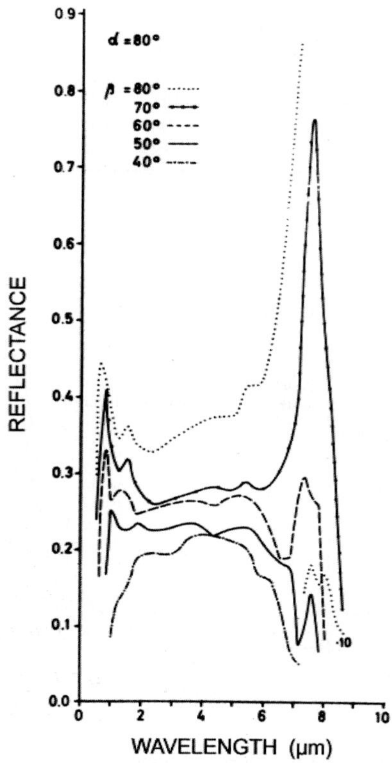

Fig. A.6.4.8b. As Fig. A.6.4.8a for 80°

To allow measurements under various incidence angles, the polarizer had to be very large. A construction of 30 polyethylene sheets each 30 μm thick and mounted under 55° was used. The degree of polarization is 0.98 throughout the spectral range. Measurements have been made on various materials of which results on water and bright beach sand are reported here. The results for the sand, the main component of which is quartz, are compiled in Table A.6.4.1. The results for water are shown together with theoretical values in the Figs. A.6.4.9 for selected wavelengths.

Table A.6.4.1. Degree of polarization of the radiation reflected by beach sand

Angle of reflexion	Angle of incidence								
	25°			50°			78°		
	Effective central wavelength in μm								
	9.5	11	14	9.5	11	14	9.5	11	14
25°	0.087	0.093	0.078	0.138	0.175	0.180	0.270	0.418	0.375
30°	0.101	0.110	0.138	0.171	0.246	0.239	0.283	0.509	0.412
40°	0.128	0.176	0.173	0.212	0.331	0.307	0.325	0.563	0.512
50°	0.178	0.231	0.193	0.259	0.424	0.373	0.351	0.602	0.526
60°	0.207	0.301	0.257	0.304	0.489	0.437	0.376	0.629	0.510
70°	0.256	0.386	0.379	0.337	0.561	0.480	0.363	0.569	0.473
78°	0.299	0.458	0.405	0.385	0.611	0.541	0.344	0.508	0.401

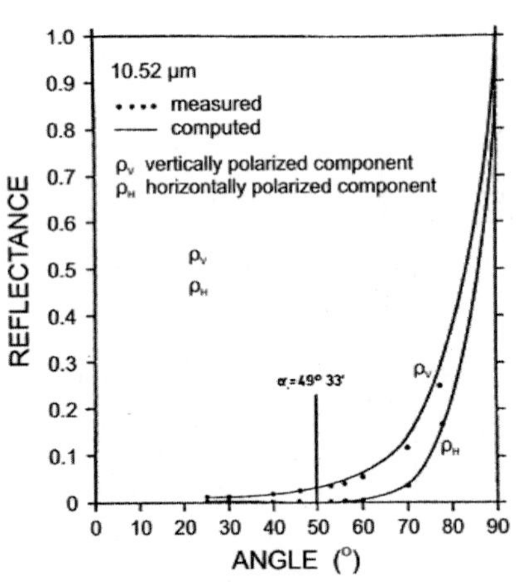

Fig. A.6.4.9a. Polarisation of the radiation reflected from a water surface around 10 μm in dependence of the specular angle

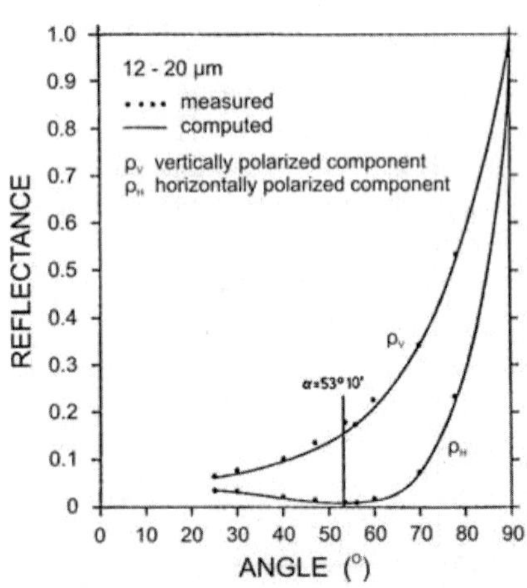

Fig. A.6.4.9b. Polarisation of the radiation reflected from a water surface for the broad band from 12 to 20 μm in dependence of the specular angle

A.6.5 Spectrometric Field Measurements of Surface Reflectance

Irradiance at Targets and the Role of Diffuse Sky Radiation and Shadows

The measurements reported here have been made with a SR-5000 spectrometer which differs from most spectrometers used in remote sensing applications in so far as it uses a telescope with a very narrow field of view and the detector can be focussed at the target. This allows to measure the radiance emerging from small targets but allows only little spatial integration.. The wavelengths are selected with a rotating circular interference filter. Typically the measurements are made alternatively against the target and a reference panel placed close to and, if possible, at the same inclination angle under which the target is seen. For flat soils the panel is accurately placed horizontally but in certain cases it is used at a different inclination angle. In the case of high trees if no tower is available a measure of the reflectance can be obtained if the measurement is made horizontally and if the

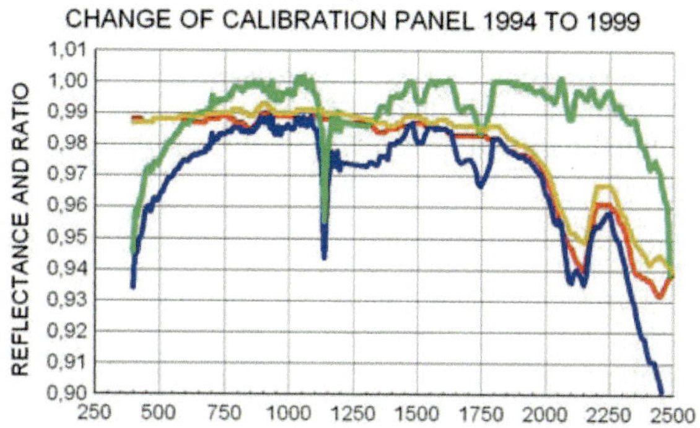

Fig. A.6.5.1. Degradation of the Spectralon reference panel. The yellow line is the manufacturer's calibration for a clean-kept panel, the red line those of a frequently used panel calibrated in 1994 . The blue line is the reflectance of the frequently used panel determined by comparison with the unused panel in 1999. The green line is the ratio of the reflectance of this panel in 1999 and 1994

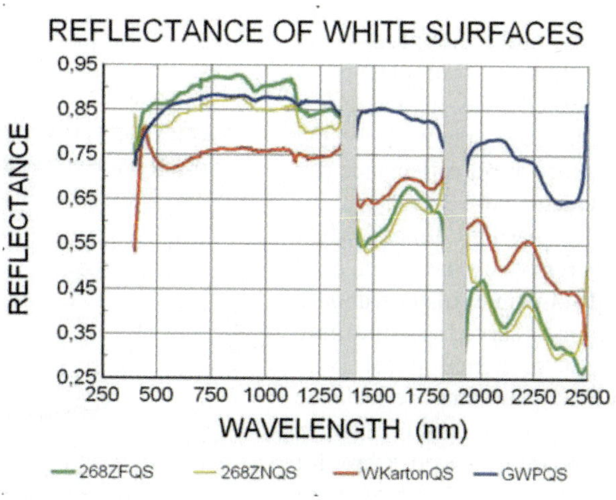

Fig. A.6.5.2. Reflectance of substitute reference panels. The blue line stands for a MgO covered panel, the red one for a Kodak cardboard box, the yellow one for thick drawing paper and the green one for drawing paper mounted on cardboard

reference panel is placed vertically in front of the tree and thus receives the same radiation from the sun and the surroundings as the target.

The best reference targets are calibrated Spectralon plates. Spectralon plates are not ideal Lambertian reflectors but come close to it. They are delivered with a spectral calibration. This calibration is done by illuminating them at an incidence of 8° from normal and measuring the hemispherically reflected radiation by a calibrated hemispherically integrating sphere. Since the radiation may only approximately be reflected isotropically, strictly speaking a directional - directional reflection measurement may not be absolutely correct. Fig. A.6.5.1 shows the factors by which the reference reflection measurement must be divided to obtain the irradiance at the panel or by which the target measurement has to be multiplied to become compatible with the reference measurement. While one of these panels was used in the field work, another one was kept clean and used only from time to time to re-calibrate the first one. This plate aged as can also be seen in Fig. A.6.5.1. If one starts with the manufacturer's calibration in 1994 and compares the permanently used plate with the stored away (and assumed unchanged) one in 1999 the decrease of the reflectance can be calculated.

In earlier times MgO covered panels have been used and sometimes also cardboard may have served as a substitute for a reference panel. Fig. A.6.5.2 contains reflectance values of a MgO plate, a cardboard of a Kodak box for photo paper and two types of drawing paper.

Depending on which reference panel has been used, these data allow to calculate

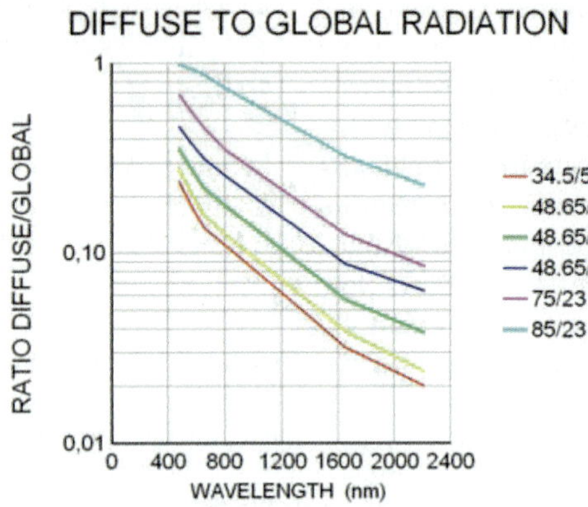

Fig. A.6.5.3. Calculated ratios of diffuse to global radiation. The first number of the calculated data are solar zenith angles ranging from 34.5 to 85 degree and the second number is the meteorological range in km.

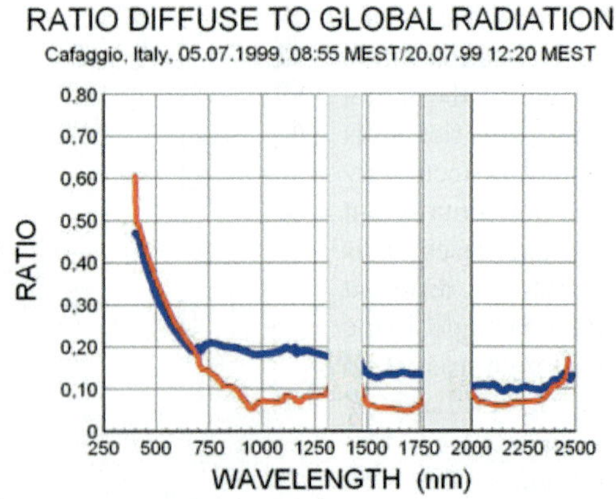

Fig. A.6.5.4. Two examples of a measured ratios of diffuse sky to global radiation.

the ratio of the measured reflected radiance to the radiance which would have emerged from a perfect white Lambertian panel. The obtained non-dimensional "reflectance" value is neither precisely a directional - directional reflectance nor a directional - hemispherical reflectance value because target and reference are illuminated by the directional solar radiation and the diffuse (and not isotropic) sky radiation. Under specific geometrical conditions (e.g. hilly terrain) target and reference panel even receive radiation reflected from the surface, The measurement gets close to the directional - directional reflectance in the infrared when the diffuse component becomes small.

Computed values of the ratio diffuse sky radiation to global radiation are shown in Fig. A.6.5.3 for different solar zenith angles and meteorological ranges. Experimentally the ratio of diffuse to directional radiation can be determined by shading the sun. In Fig. A.6.5.4 one example is given which was measured in a vegetated hilly terrain which is possibly responsible for the slightly enhanced values between 700 nm and 1500 nm in the blue curve which is the ratio of the diffuse radiation measured at a horizontal surface divided by the global radiation (05. 07. 1999, 08:55 MEST). The red line represents the ratio of the sky radiance at 70° zenith angle towards south at a meteorological range of 12 km divided by the global radiation measured at a horizontal surface (20. 09. 1999, 12:20 MEST).

To account for the change of the irradiance due to the change of solar height during the day it is advisable to take one reference measurement before and one after the target measurement and to average these two readings for the computation of the ratio.

Occasionally a remarkable fraction of the radiation received by a remote sensing device from a vegetated surface originates from areas shadowed by trees. The signal from a shadowed surface can be very complex if the tree canopy is not closed. Specifically cultivated olive trees have a highly variable LAI which is small after the spring cut and increases during the year. The ratio of a signal from a reference panel positioned underneath an olive tree of average LAI to the irradiance of the diffuse sky radiation measured outside the shadow is shown in Fig. A.6.5.5. as green line. The value of 0.5 for the shortwave end of the spectrum suggests that the signal results from the irradiance of the diffuse sky radiation from half of the

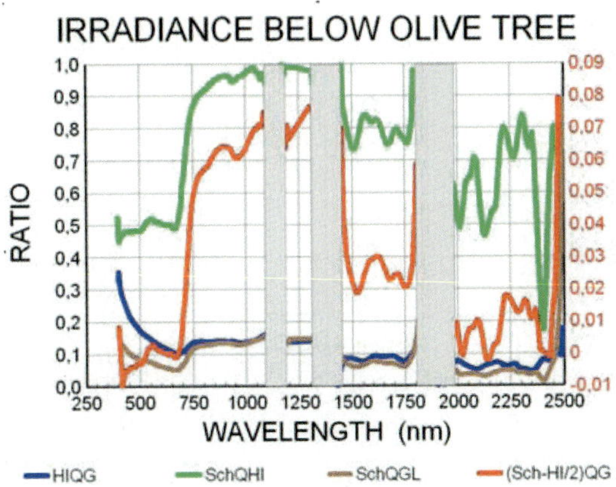

Fig. A.6.5.5. Measurements into the shadow of an olive tree. Green - ratio of the signal received from the shadowed area to diffuse radiation measured outside the shadow, brown - ratio shadow to global radiation, blue - ratio diffuse to global radiation, red - (shadow - diffuse)/global radiation (scale at the right)

hemisphere to which the area is open.

If of the diffuse radiation (blue line) 50% of the shadow signal is subtracted, the visible part of the spectrum indeed tends to zero. Beyond 700 nm there is a considerable contribution of near infrared radiation transmitted through the canopy (red curve, scale at the right).

Fig. A.6.5.6. Radiance from a shadowed area divided by radiance from a horizontal reference panel (blue and red curves, scale at the left) compared with the radiance from a wheat field (green colour, scale at the right) divided by the signal from the same horizontal reference panel. The distance to both targets is about 200 m. The upper picture shows the wheat field with the deep shadow under the trees at its upper boundary. The increase of radiance from the shadow at the shortwave end is due to scattered light from the air between instrument and target

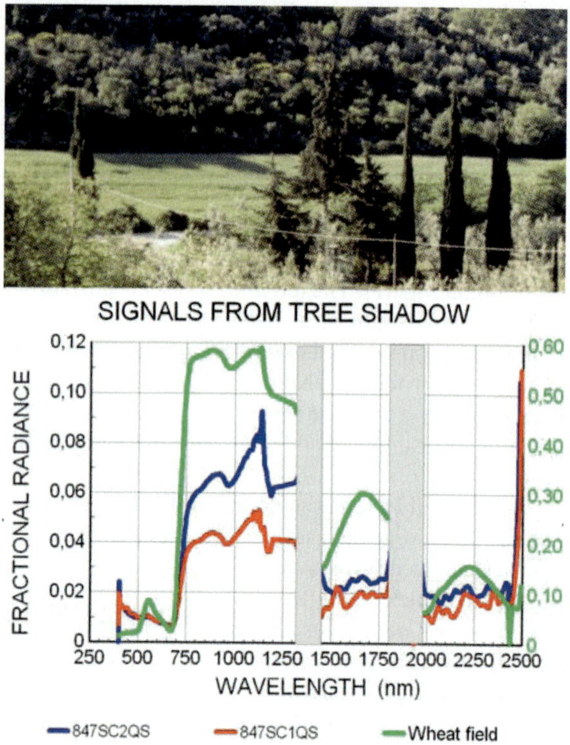

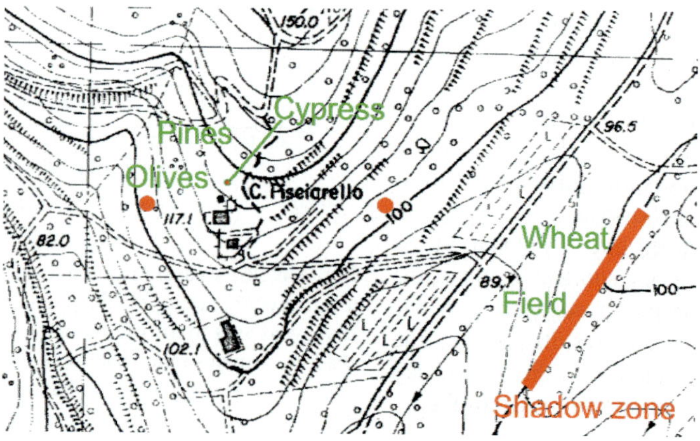

Fig. A.6.5.7. Measuring sites near Cafaggio, Italy

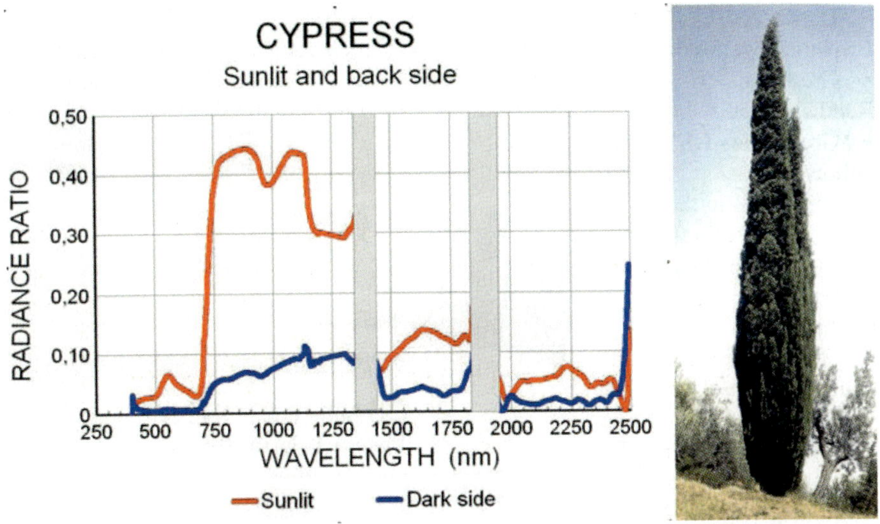

Fig. A.6.5.8. Radiance from sunlit (red) and shadowed (blue) side of a cypress divided by the radiance from a reference panel positioned vertically in front of the tree

A second case is shown in Fig. A.6.5.6. where the signal of a shadowed area (two measurements) is compared to the signal from a wheat field (green colour) in front of it. The measuring geometry can be seen in Fig. A.6.5.7. The picture at the top of Fig. A.6.5.6 was taken from the position of the spectrometer which was a little bit to the left of the right red spot. Fig. A.6.5.8. shows the reflection spectrum of the sunlit side (red colour) of a cypress (position indicated in Fig. A.6.5.7) as well as the spectrum measured at the back side of the cypress which is dark in the visible but again there is a considerable signal between 700 to 1600 nm. Both measurements are rationed to the illuminating total radiation which results from sun, sky and reflected radiation from the surroundings.

Examples of Rock and Soil Spectra

Spectra of soils and stones which have already been shown in preceding chapters. Some additional data are reported here. Conspicuous, mostly bright, features in remote sensing images are quarries. A unique opportunity to make measurements within a quarry arose near Matera. The quarry has a rectangular base with vertical walls. Fig. A.6.5.9 shows some of the spectra of rocks which were taken in a position as shown in the upper part of the figure. The white reference panel stands at the foot of the wall in the right half of the picture. The reflectance spans a wide range as can be seen in the lower part of the figure. Here the different colours have the following meaning:

- Red line - CST44: Bright whitish stone

- Dark blue line - CST43: Brownish - yellowish stone
- Grey line - CST42: Reddish stone
- Green line - CST41 and yellow line - CAVE41: Whitish spots in the wall
- Bright blue line - BO76: White stone dust at the bottom of the quarry (compare Fig. A.6.5.12).

Spectra of three other stones and slate with widely different reflectances found in The contrast of soils was found to be large in the Matera area. While at the elevated areas calcaric material is dominant, in the valley north of the road Matera - Laterza the darkest fertile soil was found with a reflectance similar to African laterite. This can be seen in Fig. A.6.3.11. The red and brown curve represent the dark soil (picture at the right, top). Contrasting to the low

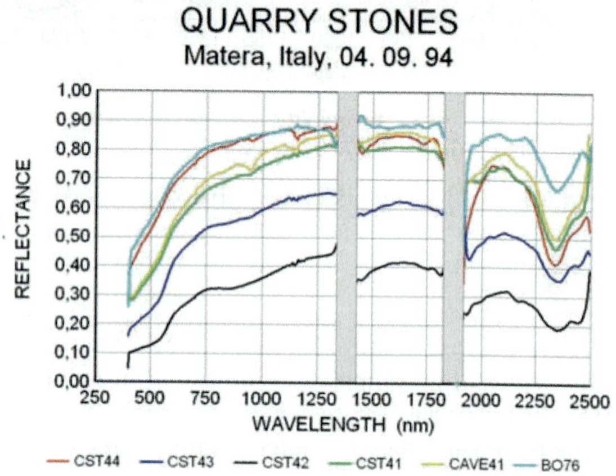

Fig. A.6.5.9. Reflection spectra of rocks in a vertical wall of a quarry near Matera, Italy

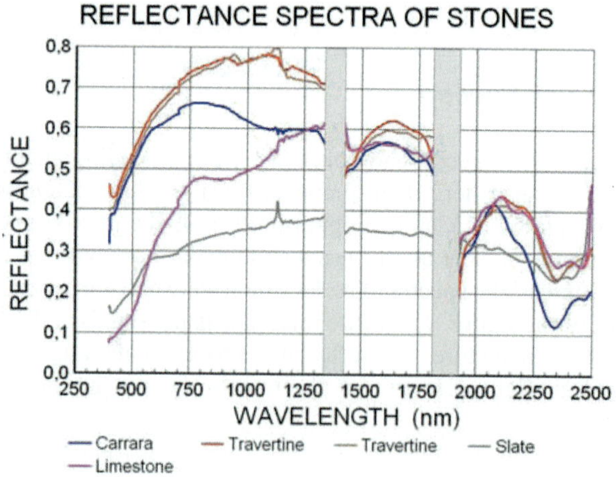

Fig. A.6.5.10. Reflectance spectra of stones. One probe each of Carrara marble, slate and limestone, two (rather similar) probes of travertine

values of less than 12.5% reflectance is the reflectance of the stony soil near the measuring site (blue line) and also of relatively bright yellowish- brown soil found elsewhere in the region (green line). Some more probes are shown in Fig A.6.3.12. Here the colours denote the following species of soils:

- B76-1: Soil from measuring site (red line and top of the picture at the right)
- B76-3: Whitish material from the quarry, horizontally deposited (blue line and bottom of the picture at the right)
- BO72: Red soil with small stones from the measuring site (green line, similar to the surrounding in picture at the right)
- BO74: Reddish-yellowish material from the quarry (brown line).

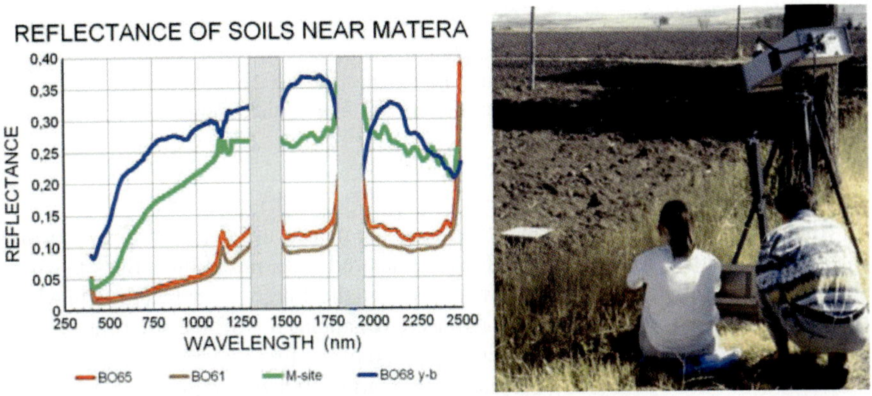

Fig. A.6.5.11. Reflectance spectra of soils found in the area of Matera, Italy.

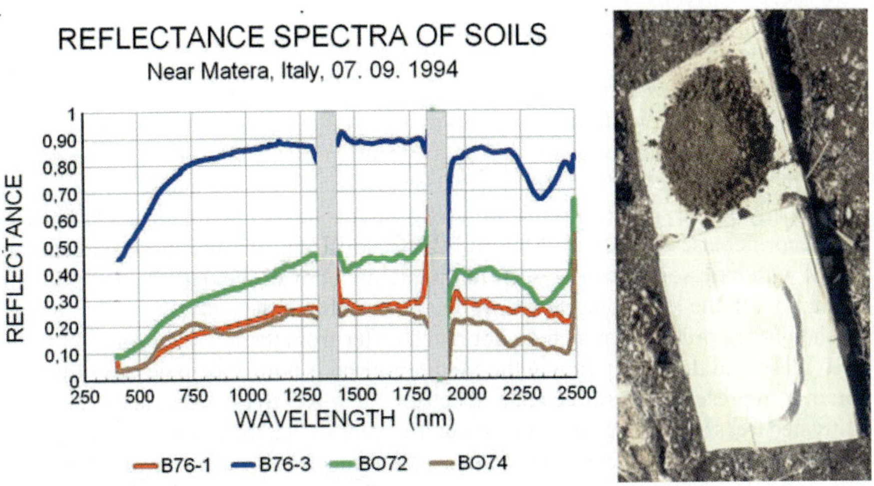

Fig. A.6.5.12. Soils from sites near Matera, Italy, 07. 09. 1994 (for identification see text)

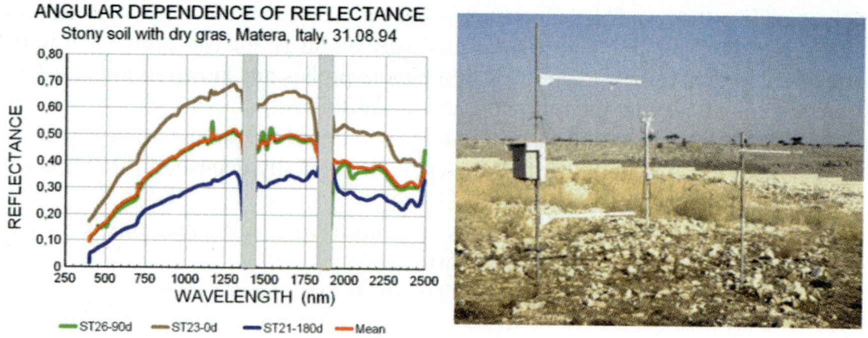

Fig. A.6.5.13. Directional reflectance of a stony soil with sparse dry grass (near Matera, Italy). The measurements were made with the sun (brown line), against the sun (blue line), and at 90° off the direction of the sun (green line). The red line is the average of the first two measurements

Italy can be seen in Fig. A.6.5.10.

Finally, in Fig A.6.3.13, the reflectance at the measuring site with stony calcaric material (compare also Fig. 1.31) is shown measured in the direction of the sun where the pebbles are fully illuminated (brown curve), in the direction against the sun where the spectrometer looks into shadows (blue curve) and at 90° off the solar beam. This last measurement at 90° coincides with the mean of the first two measurements. This demonstrates once again that a measurement made at 90° of the direction of the sun represents very well the albedo of the surface. This result was also found over grass.

Reflectance Spectra of Vegetation

Some spectra of vegetation have been shown in Chapter 1 which will be supplemented by a few additional examples. Fig. A.6.5.14 shows in the upper picture the scene of a matorral landscape and below some spectra. Matorral 1, 3- 5 represent different views at the low matorral vegetation of variable density and state. The more biologically active areas can be identified by the secondary maximum at 525 nm and the steep increase at 700 nm. Matorral 2 is a spectrum of very sparse vegetation mixed with bare soil as can be seen in the picture above. The dark field is without vegetation. In the lower diagram the reflection spectra of the bushes of different age and height are presented which do not much differ from the low matorral.

Widespread tree vegetation in the Mediterranean is holm oak, pine and olive. To obtain a representative reflectance spectrum of olive trees is not easy because cultivated trees in Tuscany are thinned out in winter so that the canopy becomes very transparent. A measurement from above would be much affected by the undergrowth. Fig. A.6.5.15 shows the reflectance spectra of regrown trees and the measuring geometry can be seen in the picture at the right. Two of the measured spectra taken at different trees and consecutive days are nearly identical, the third, covered with fruits, has obviously been caught in a more shadowed part. Pine trees have a denser canopy.The spectra of two species are shown in Fig. A.6.5.16. The red spectrum is from a nearby pine (see Fig. A.6.5.15, green shoots at the left), the other two are from a more distant stand which is about 250 m away.

Spectra of holm oak trees measured in Castilla-La Mancha are presented in Fig. A.6.5.17. The younger the tree the higher is the reflectance but still smaller than that of olive trees and pines. For comparison a spectrum of wheat measured near Matera in Italy is added in Fig. A.6.5.18. From these examples it can be seen that the shape of the reflectance spectra of these plants are rather similar except for the amplitude. This becomes also evident if the spectra of plants are normalized to 1.0 at 875 nm as shown for some species in Fig. A.6.5.19 (cherry, mastic, olive, rosemary).

The potential of the used spectrometer to measure the radiance emerging from small targets is illustrated in Fig. A.6.5.20. Here the reflected radiance of some tiny parts of succulents as compared to the radiance reflected from a horizontal reference panel is presented. The colours and numbers refer to the indicated parts of the pictured succulents. The amplitude of the radiances may not be completely correct because in this case the irradiance at the curved surfaces can only approximately be represented by the reference panel. It can be seen that most of the plants do not show the secondary peak at 550 nm. As D.M. Gates, the pioneer of leaf spectroscopy, pointed out, the spines protect the succulents from overheating in dry climates (1980) due to a reduction of the chlorophyll

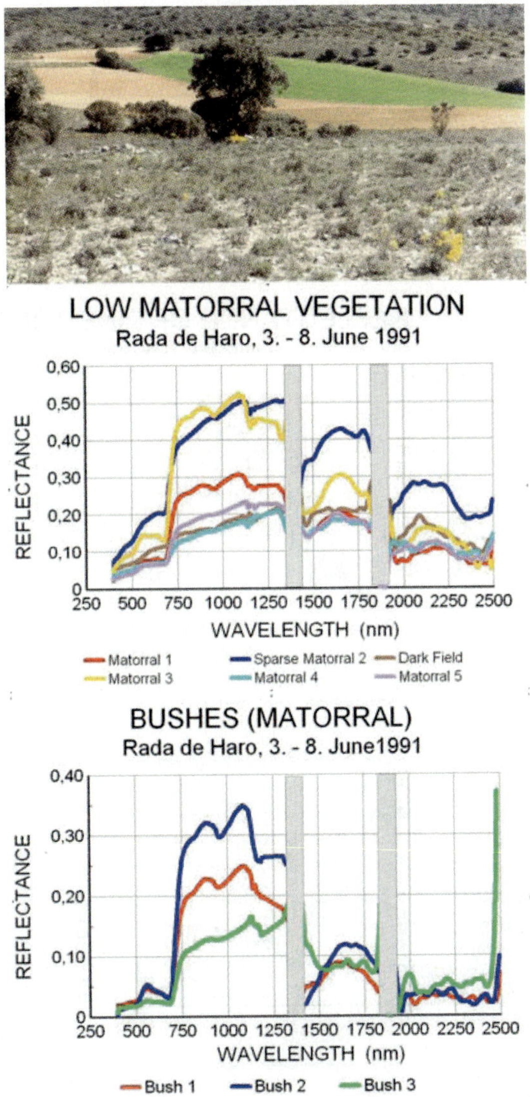

Fig. A.6.5.14. Reflectance of matorral vegetation

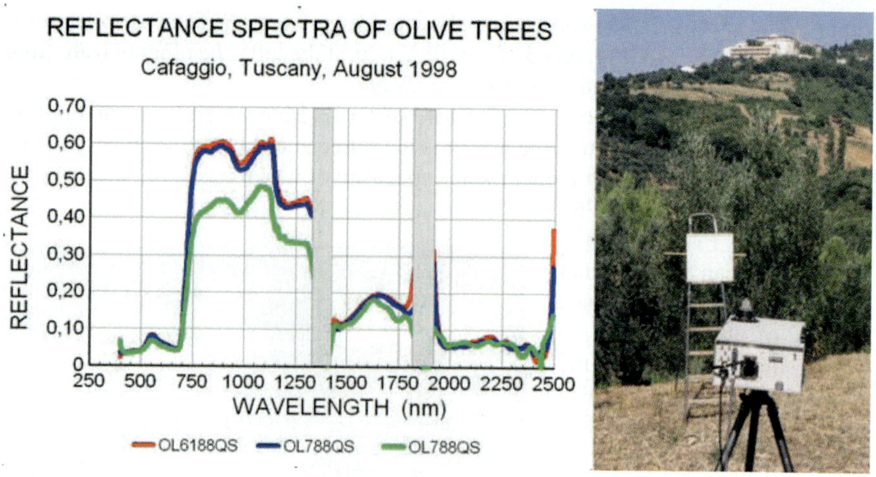

Fig. A.6.5.15. Reflectance spectra of three olive trees and, at the right, the experimental setup

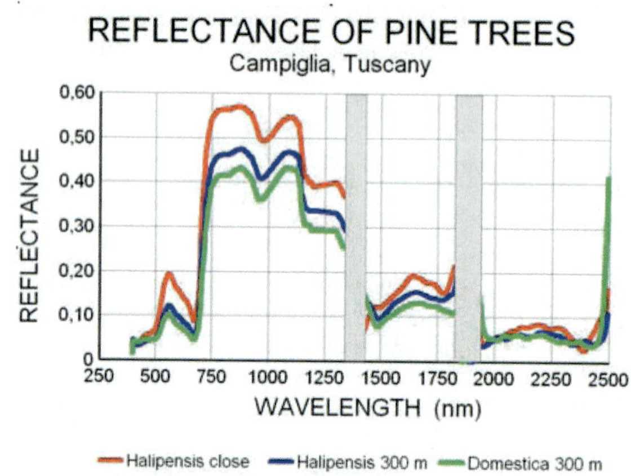

Fig. A.6.5.16. Reflectance spectra of pine trees (*Pinus halipensis and Pinus domestica*)

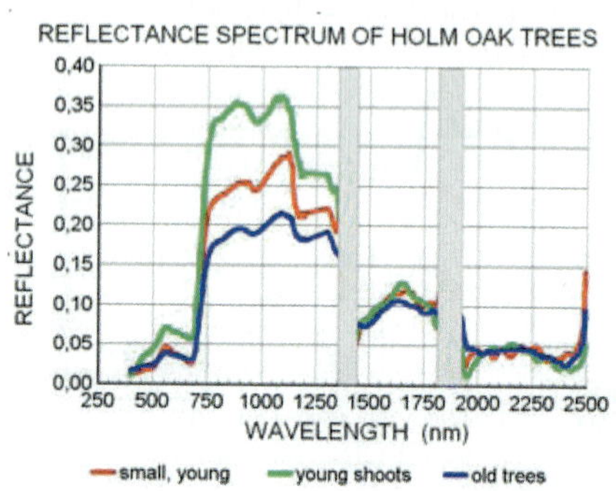

Fig. A.6.5.17. Reflectance spectra of holm oak trees of different age measured near Rada de Haro, Spain

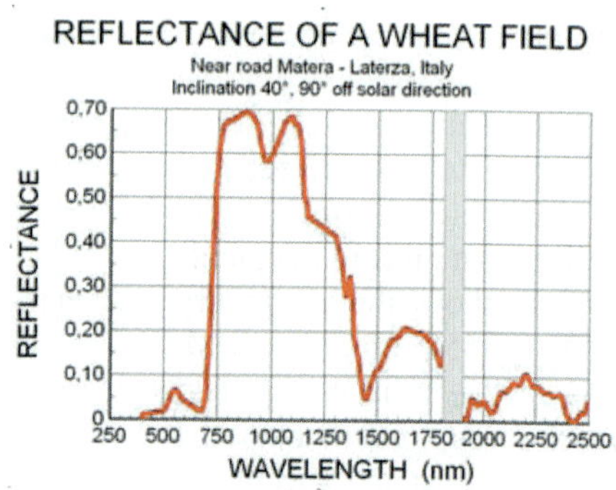

Fig. A.6.5.18. Reflectance spectrum of wheat at full green development

absorption. Only the small green *Opuntia* (#8, brown) shows this peak. The long plant (#9, yellow) is of the type *Espostoa procera*, the short ones of the type *Mammillaria schiedeana*.

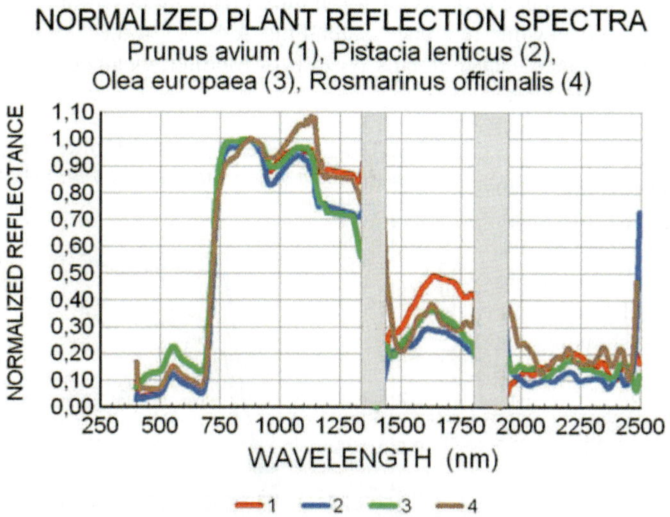

Fig. A.6.5.19. Reflectance spectra of different plants normalized to 100 % at 875 nm

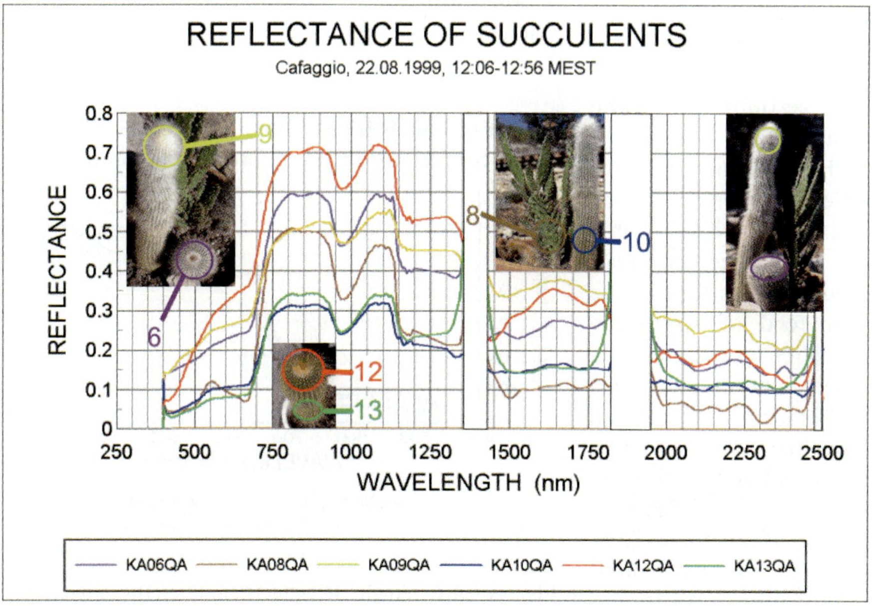

Fig. A.6.5.20. Reflection spectra of different parts of the succulents shown in the figure.

A.6.6 Spectral Surface Reflection and Albedo

Directional Reflectance Measurements

For directional reflectance measurements an interference filter radiometer was used in combination with a white reference panel. Figs. A.6.6.1 - A.6.6.7 show a series of radiometric reflectance measurements typical for Mediterranean surfaces from the Messara valley in Crete made at different observation angles. It follows one example from Italy featuring the difference between an oak and a pine forest in December and an example from Africa showing the difference between three types of sand.

In the figures, the symbols V represent vertical measurements, N, E, S and W the geographical directions of the measurements, z is the zenith angle under which the radiometer views the target, "vert" indicates a view near nadir, and az is the azimuth of the measurement.

Fig. A.6.6.1 presents reflectance spectra of vine and gras which shows that the difference is small though at this time of the year (end of April) the reflection maximum at 525 nm of vine is larger than that of grass. The red and blue lines are views into shadowed and sunlit parts of the vine. The reflectance spectra of olive trees (Fig. A.6.6.2) shows small differences between cultivated and not

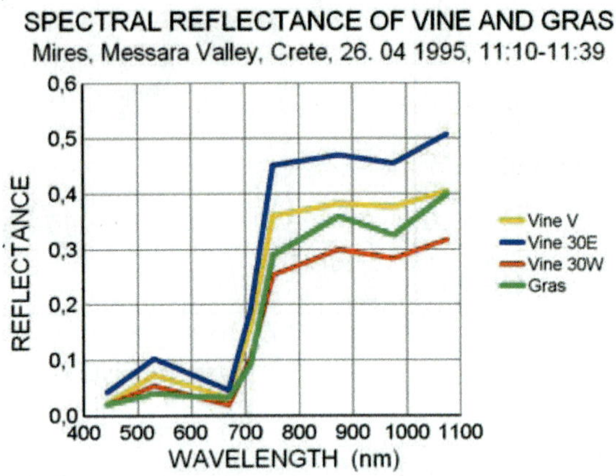

Fig. A.6.6.1. Angle dependent radiometric reflectance measurements near Mires in the Messara valley, Crete, of vine and gras

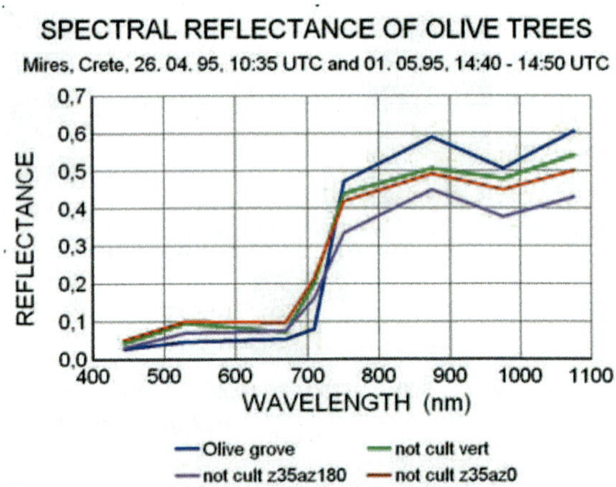

Fig. A.6.6.2. Angle dependent radiometric reflectance measurements at Mires in the Messara valley, Crete, of cultivated (blue) and not cultivated olive trees

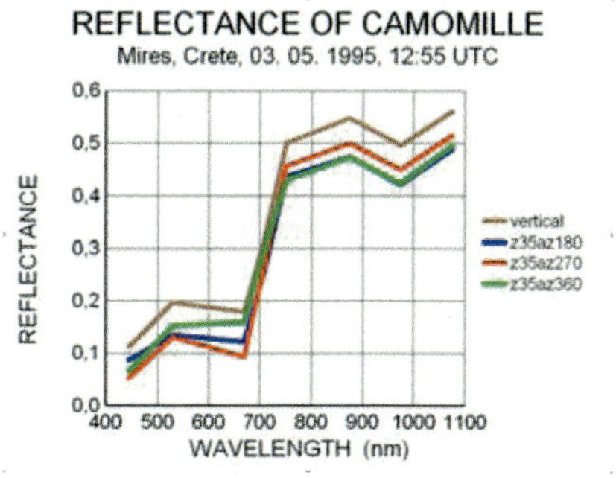

Fig. A.6.6.3. Angle dependent radiometric reflectance measurements at Mires in the Messara valley, Crete, of camomille

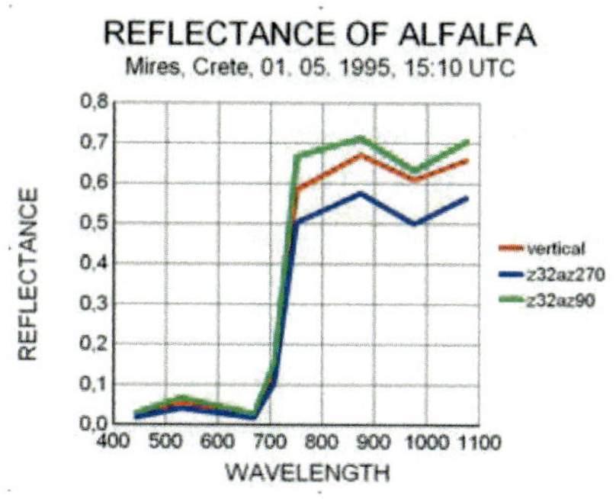

Fig. A.6.6.4. Angle dependent radiometric reflectance measurements at Mires in the Messara valley, Crete, of alfalfa

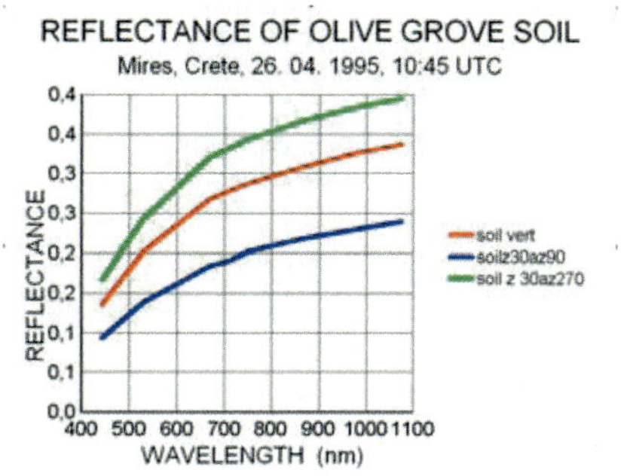

Fig. A.6.6.5. Angle dependent radiometric reflectance measurements near Mires in the Messara valley, Crete, of soil of an olive grove

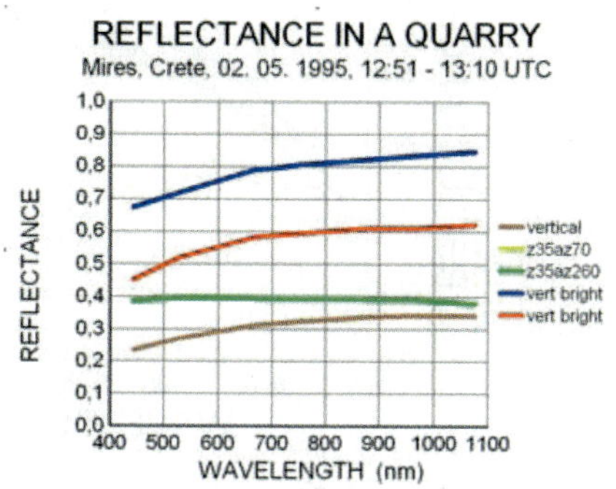

Fig. A.6.6.6. Angle dependent radiometric reflectance measurements in the Messara valley, Crete, in a quarry near Mires

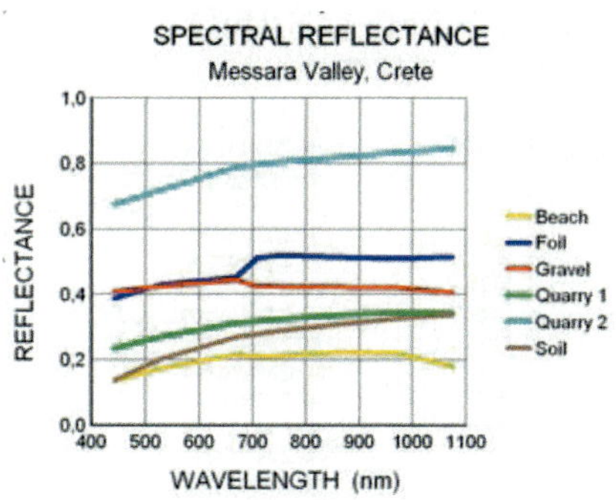

Fig. A.6.6.7. Directional reflectancefrom dark beach sand, gravel, two samples of a quarry (Fig. A.6.6.6), soil, and a foil covered field

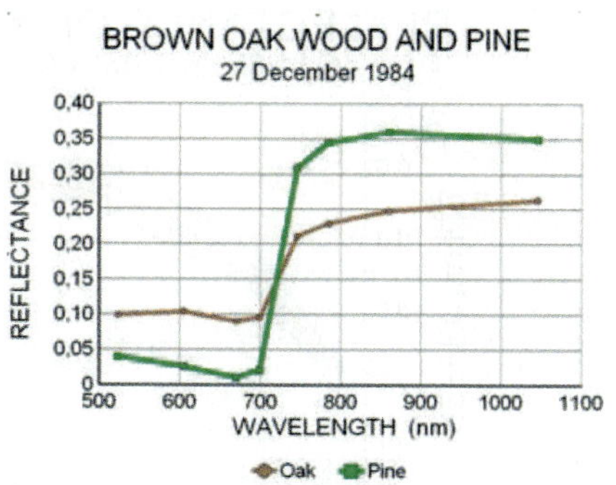

Fig. A.6.6.8. Directional reflectance from an oak and a pine forest in December near Cafaggio, Italy

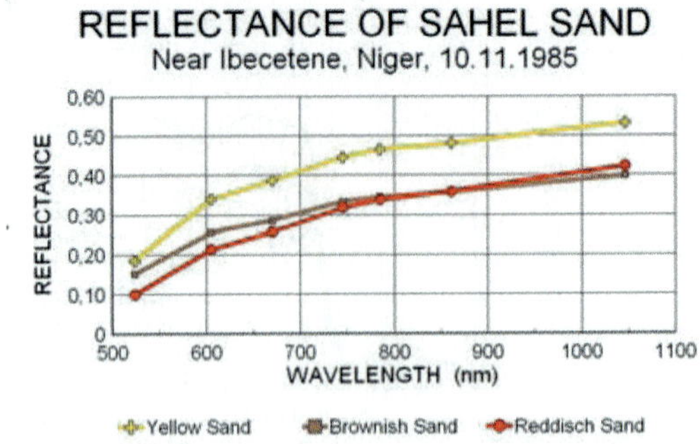

Fig. A.6.6.9. Directional reflectance in nearly zenith direction for three types of sand found near Ibecetene, Niger, with yellowish, brownish and reddish colour

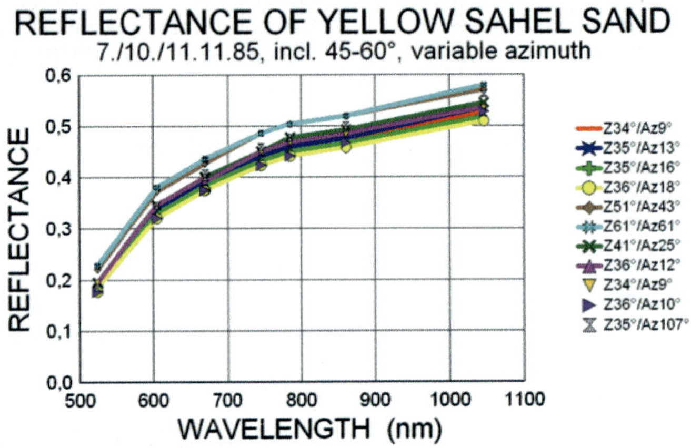

Fig. A.6.6.10. Directional spectral reflectance of yellow sand near Ibecetene measured on three days in November 1985 at 45° - 60° inclination of the instrument (Z = solar zenith angle, Az = azimuth difference between observer and solar direction)

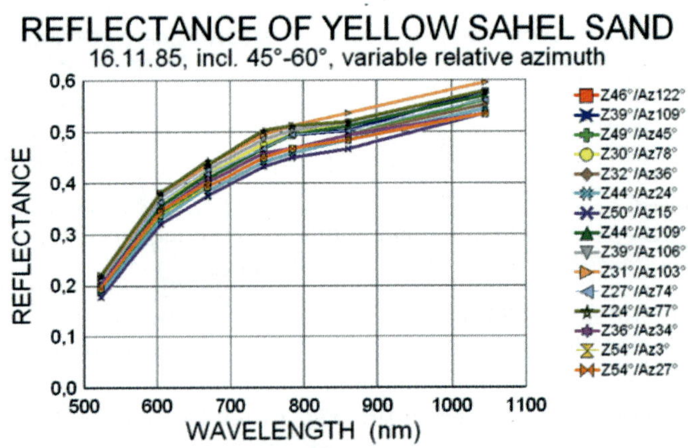

Fig. A.6.6.11. As Fig. A.6.6.10 for 16 November 1985

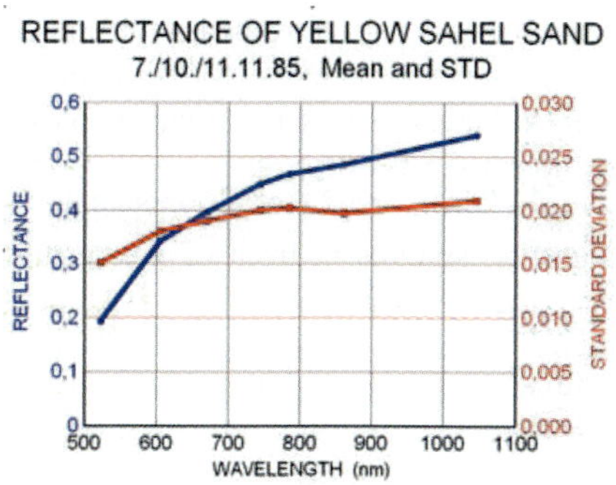

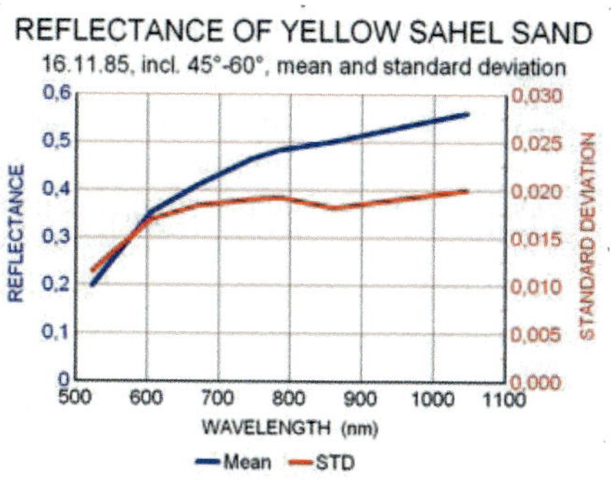

Fig. A.6.6.12a. Mean values and STD of the reflectances measured on 7., 10. and 11. November 1985

Fig. A.6.6.12b. Mean values and STD of the measurements of 16 November 1985

cultivated trees. In the following two figures (Figs. A.6.6.3 and 4) the directional reflectance of camomile and alfalfa can be seen.

It follows samples of soil material from underneath the olive trees (Fig. A.6.6.5) and of the bottom of a quarry seen under different aspect angles (Fig. A.6.6.6). In Fig. A.6.6.7) spectral reflectances of a dark beach sand, gravel, and a unique measurement of a foil covered field can be seen.

Fig. A.6.6.8 is a view into an oak wood and into a pine wood which shows the difference in the biological activity between evergreen an deciduous trees in winter. A set of reflectances of three different sands found near Ibecetene in Niger with brownish, reddish and yellowish colour is given in Fig. A .6.6.9.

Measurements made over deserts are often used to test the constancy of satellite instrument calibration. Because the illumination and observation geometries change during the course of the year, it is not granted that the "invariable" desert surfaces always show the same radiative properties. To test the reliability of the assumptiomn that these surfaces can be used for re-calibration purposes, measurements have been made under various illumination and observation angles over yellowish desert sand. Some results, shown in Figs. A.6.6.10 and 11, supplement the set of measurements already presented in Fig. 2.4. The mean values of these measurements and the standard deviations of each set are plotted in Fig. A.6.6.12. These data clearly show that using deserts as calibration control, one has to account for the angular dependent directional reflectance or to ensure that these control measurements are made under identical illumination and observation geometries. The standard deviation of about 2 % as presented here does not account for the effect of the undulating surfaces of sand dunes nor for the variable aerosol content in the atmosphere.

Albedometer Measurements

To validate reflectance determinations inferred from measurements made in space and to relate it to albedo, different instruments are used at the surface. Spectral and broadband simultaneously upward and downward looking instruments for albedo measurements and lightweight interference filter radiometers for directional reflectance measurements where the use of spectrometers such as described in section 6.17 and that one seen in some figures of Section A.6.5 is difficult.

Spectral albedometers equipped with interference filters are difficult to design for true cosine response. This would need integrating spheres in front of the sensor. Instead, diffusor plates are used to approximate the cosine response. The instrument described here was designed for to minimize the influence of its shadow on the measurement. It uses two pairs of detectors, one pair for wavelengths 500 - 1024 nm, the second pair for wavelengths 1024 - 1640 nm. In front of the detectors filter wheels turn automatically and inserted into the cover of the circular box are diffusor plates. The errors of the measurements made between 1300 and 1640 nm are somewhat larger than those made at the shorter wavelengths because these filters are inserted only in one of the filter wheels and the obtained signal is small. The instrument can be seen in Fig. A.1.26.

Spectral albedo is defined as

$$\rho(\lambda) = \frac{M\uparrow(\lambda)}{G(\lambda)} = \frac{M\uparrow(\lambda)}{S(\lambda)\cos\zeta + D(\lambda)} \tag{A.6.6.1}$$

where the following notation is used:
$\rho(\lambda)$ = spectral albedo,
$G(\lambda)$ = spectral flux density of global radiation,
$M\uparrow(\lambda)$ = from the surface upwelling spectral flux density,
$S(\lambda)\cos\zeta$ = spectral solar flux at the horizontal earth surface,
ζ = solar zenith angle
$D(\lambda)$ = spectral diffuse sky radiation.

If the wavelength dependence is momentary suppressed, the value measured by the interference filter albedometer is

$$\rho' = \frac{M'\uparrow}{G'\downarrow} = \frac{\kappa M\uparrow}{\gamma(\zeta)S\cos\zeta + \kappa D} \tag{A.6.6.2}$$

with $\gamma(\zeta)$ = gain factor for the direct solar radiation and κ = gain factor for the diffuse radiation. Here it is assumed that both the diffuse sky radiation and the upwelling radiation are isotropically distributed. This is certainly not the case for the diffuse sky radiation but since this component usually is much smaller than the global radiation it is, in this case, an agreeable assumption.

If now the ratio of diffuse sky radiation to global radiation is estimated either by a measurement (direct solar beam shadowed) or by radiative transfer calculations, a factor α can be defined as

$$\alpha_\zeta = D(\zeta)/G(\zeta) = \frac{G(\zeta) - S\cos\zeta}{G(\zeta)} \tag{A.6.6.3}$$

which allows to express also $S\cos\zeta$ in terms of G:

$$S\cos\zeta = G(\zeta)\left(1-\alpha_\zeta\right) \tag{A.6.6.4}$$

and, by using as well Eq. A.6.6.1, to modify Eq. A.6.6.2 in the following way:

$$\rho' = \frac{\kappa\rho G(\zeta)}{\gamma(\zeta)G(\zeta)(1-\alpha_\zeta)+\kappa G(\zeta)\alpha_\zeta} = \rho\frac{\kappa}{\gamma(\zeta)(1-\alpha_\zeta)+\kappa\alpha_\zeta} \quad \text{(A.6.6.5)}$$

For the spectral albedo finally follows:

$$\rho_i = \rho_i' \frac{\gamma_i(\zeta)[1-\alpha_i(\zeta)]+\kappa_{u,i}\alpha_i(\zeta)}{\kappa_{d,i}} = \rho_i' F_i(\zeta) \quad \text{(A.6.6.6)}$$

F_i' is the albedo response factor of the spectral channel i. The gain factor for the downward looking sensor, κ_d, may differ from the gain factor of the upward looking detector-filter combination, κ_u. As said before, for practical reasons these two factors are set equal. Furthermore, during a single measurement one detector is looking upward and another one is looking downward. The responses of the filter-detector combinations are not identical with the effect that $M_{i,1}' \uparrow / G_{i,2}' \neq M_{i,2}' \uparrow / G_{i,1}'(\lambda)$ if the indices 1 and 2 discern the two sensor systems.

To correct for the different sensitivities either a normalization factor has to be applied which can be determined by exposing both detectors to the same irradiance or pairs of measurements have to be made turning the albedometer by 180 degree. If this last procedure is chosen, one obtains two measurements:

$$\rho_{i,1} = \rho_{i,1}' F_{i,1} \quad and \quad \rho_{i,2} = \rho_{i,2}' F_{i,2} \quad \text{(A.6.6.7)}$$

and

$$\rho_i = \sqrt{\rho_{i,1} * \rho_{i,2}} = \sqrt{\rho_{i,1}' F_{i,1} * \rho_{i,2}' F_{i.2}} \quad or \quad = \frac{\rho_{i,1}' F_{i,1} + \rho_{i,2}' F_{i,2}}{2} \quad \text{(A.6.6.8)}$$

Because the differences in sensitivity are not very large, the geometrical and the arithmetical means are not very different.

Generally albedo measurements are made at high sun. Therefore some simplifications are possible. For the coefficient α the values in Table A.6.6.1 can be taken as first guess. The angle dependent response function $\gamma(\zeta)$ must be determined by measurements with a point source (Fig. 4.11). Then κ can be computed under the assumption of an isotropic distribution of the radiance by integration over the hemisphere according to

$$\kappa = \frac{1}{\pi} \int_{\varphi=0}^{2\pi} \int_{\zeta=0}^{\pi/2} \gamma(\zeta) \cos\zeta \sin\zeta \, d\zeta \, d\varphi \quad \text{(A.6.6.9)}$$

The results for the four sensor systems are 0.9024, 0.8781, 0.8705, and 0.8311. Without making too large an error, the mean value of 0.8705 may be taken for all sensors. The gain factor $\gamma(\zeta)$ for the direct solar radiation can directly be read off from the albedometer response diagram Fig. 4.11.

Examples of spectral albedo measurements have been given in previous chapters and Appendices such as Fig. A.1.25. A few more examples follow here to demonstrate some specific features.

In addition to the directional reflectance measurements shown in Figs. A.6.6.10

Table A.6.6.1. Ratio of diffuse sky radiation to global radiation for standard conditions

Wavelength	D/G				
	$\zeta = 0°$	$\zeta = 30°$	$\zeta \approx 45°$	$\zeta = 60°$	$\zeta = 80°$
520.6	0.155	0.17	0.2063	0.27	0.58
548.8	0.143	0.155	0.1881	0.24	0.525
614.9	0.105	0.12	0.1553	0.19	0.415
668.6	0.085	0.10	0.1356	0.155	0.34
759.9	0.07	0.075	0.1112	0..115	0.26
781.5	0.06	0.07	0.1066	0.107	0.25
862.2	0.05	0.055	0.0922	0.09	0.20
1024	0.03	0.032	0.0721	0.07	0.14
1237.7	0.025	0.026	0.0555	0.05	0.10
1359	0.022	0.023	0.0489	0.04	0.095
1639.5	0.019	0.020	0.0381	0.035	0.087

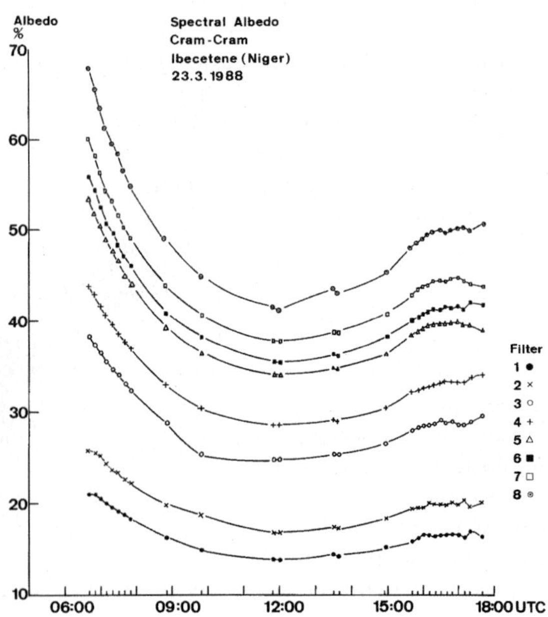

Fig. A.6.6.13. Change of the directional reflectance close to nadir with daytime (clear sky in the morning, increasing cloudiness after 15:00 hours) over a cram-cram covered soil for the wavelengths 523, 605, 669, 670, 746, 785, 861, and 1046 nm (from bottom to top)

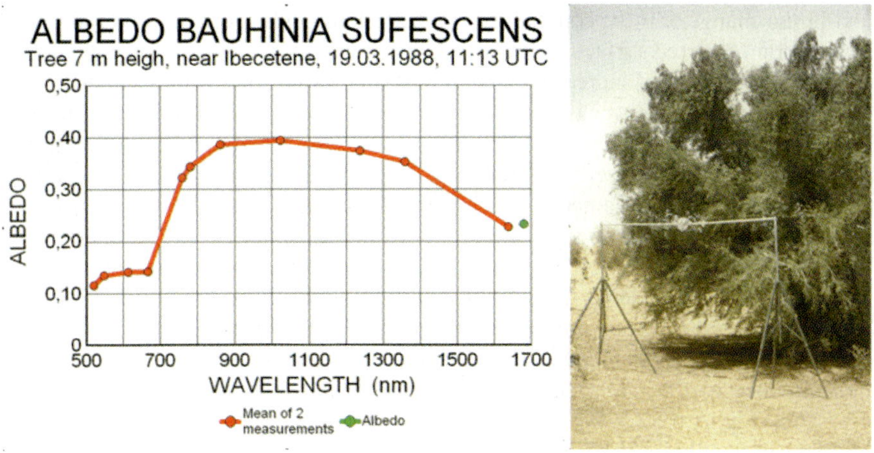

Fig. A.6.6.14. Sideways measured spectral albedo of *Bauhinia sufescens*, a tree near Ibecetene, Niger

Fig. A.6.6.15. Trees (Ibecetene, Niger) not suited for sideways albedo measurements and Sahel bushes at different stages of dryness the spectral albedo of which can be seen in Fig. A.6.6.16

and 11 the change of the spectral albedo with solar height is given in Fig A.6.6.13 for a cram-cram covered surface. While the morning was clear, towards evening clouds decreased the effect of increasing albedo with increasing solar zenith angles.

These are so far are measurements over homogenous terrain. Trees and bushes change this simplified view. The determination of the albedo of trees often needs elaborate stationary tower constructions. Only if the canopy is large enough, information can also be gained by placing the albedometer vertically at the side of the tree, measuring the albedo sideways as shown in Fig. A.6.6.14. In this diagram, as well as in some of the following figures, the data point shown near the right margin and indicated as "albedo" is the estimated broadband albedo of the object. If the tree is too transparent or does not fill the effective field of view of the radiometer (at least a circle with a radius that equals the distance between albedometer and target) or if the influence of the underlying surface is too large, such a measurement is not advisable (Fig. A.6.6.15, top).

In the case of holm oak trees (Fig. A.6.6.16) sometimes a dense canopy develops in the macchia which allows to position the albedometer on top of the plants. In this case two measurements, one made in April and the other made about two months later are shown.

More easy is the measurement of the albedo of bushes if their diameter is large enough. The lower three pictures of Fig. A.6.6.15 shows examples, again from Africa, of different stages of dryness. The measured spectral albedo of these bushes are ploted in Fig. A.6.6.17. It is clear that in these cases the underlying surface plays a role. This is especially the case if one looks at Spanish vine fields where the grapevines are grown in a chessboard pattern with 2 m distances between the individual plants. Spectra of such vine fields are reproduced in Figs. A.6.6.18 and 19. In this case the typical chlorophyll absorption at short wavelengths is superimposed by a high soil reflectance and the chlorophyll absorption is only noticeable if the measurement is made directly over a plant at its most intensive growth stage. Due to

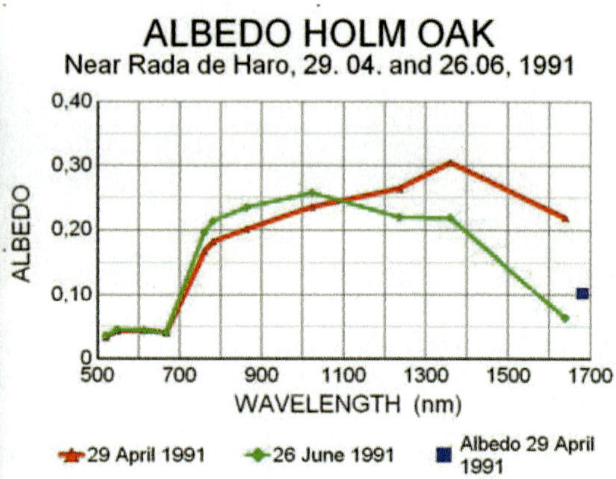

Fig. A.6.6.16. Spectral albedo of small and densely grown holm oak trees measured at the top of the canopy

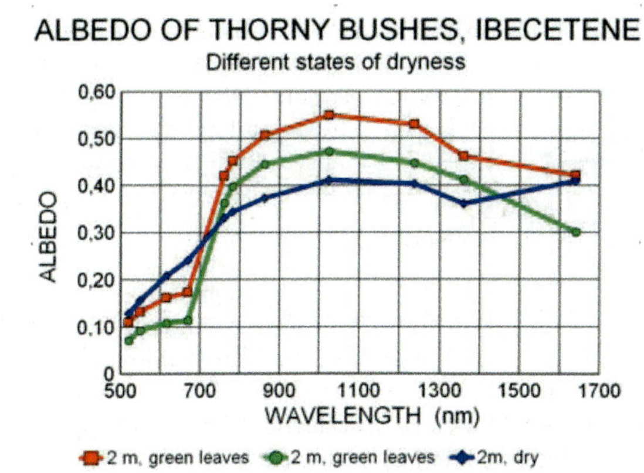

Fig. A.6.6.17. Spectral albedo of three bushes near Ibecetene, Niger, at different stages of dryness as can be seen in Fig. A.6.6.15. With increasing dryness the chlorophyll absorption decreases and the albde at longer wavelengths increases

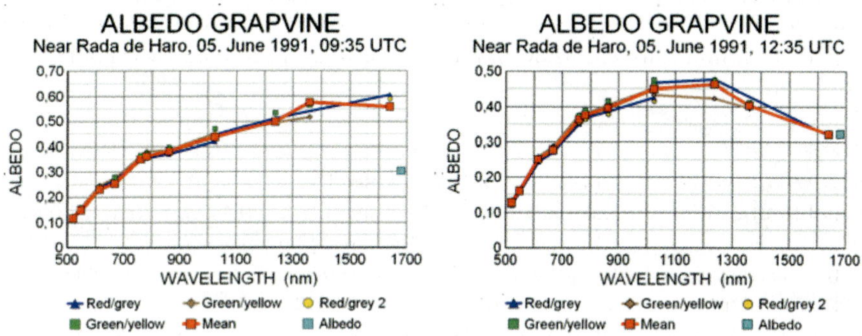

Fig. A.6.6.18. Spectral albedo measurements of four sets each discerned by colours made around 09:35 UTC (left) and 12:35 UTC (right) over a vine field near Rada de Haro, Spain, as well as mean values and estimated broadband albedo

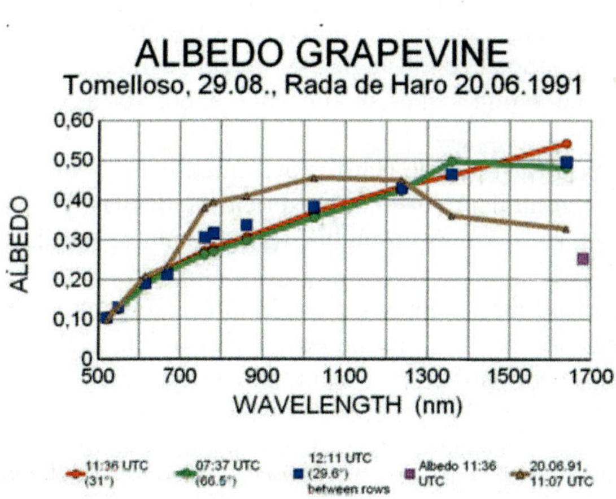

Fig. A.6.6.19. Albedo of grapevine measured at different times during 29. August 1991 near Tomelloso (solar zenith angles 29.6°, 31°, and 66.5°) as compared to a measurement on 20.06.1991 near Rada de Haro. In addition the estimated broadband albedo is noted for 29. August 1991

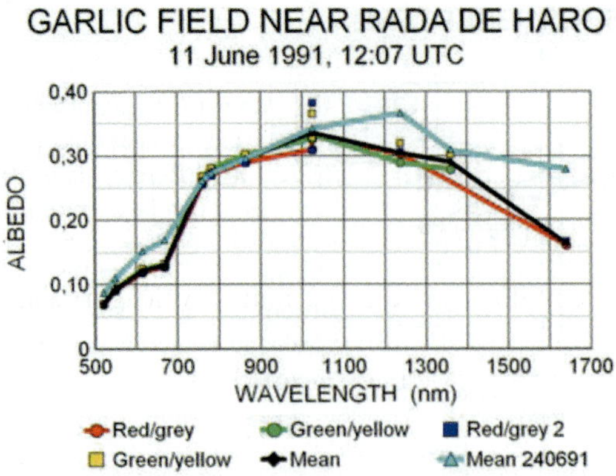

Fig. A.6.6.20. Different measurements of the spectral albedo of a garlic field and their mean of 11 June 1991. For comparison the albedo measured on 24 June is also shown. It can be seen that within the 12 days the chlorophyll absorption decreased while the longer wavelength albedo increased because of the stronger effect of the underlying bright soil

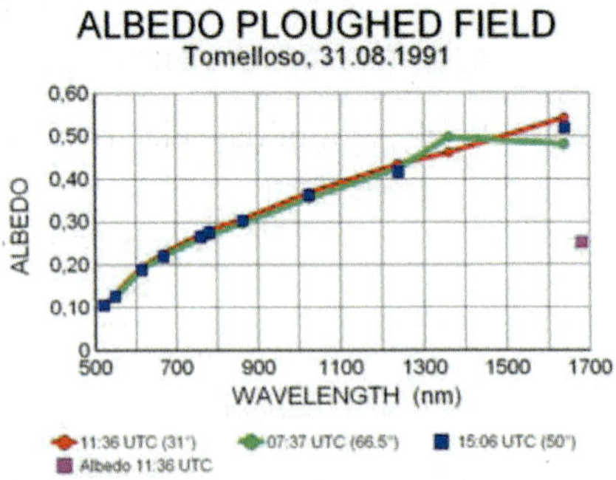

Fig. A.6.6.21. Spectral albedo of matorral near Rada de Haro measured for solar zenith angles of 31°, 50° and 68.5°. Also plotted is the estimated broad band albedo

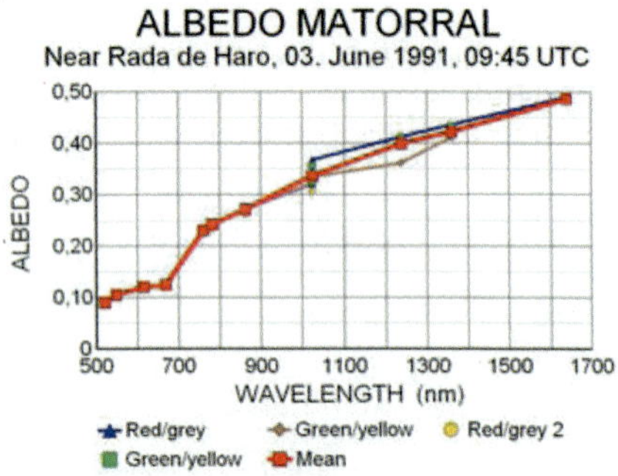

Fig. A.6.6.22. Spectral albedo of matorral. The figure shows four individual measurements and their mean value

the variable shadowing effect the albedo depends on the solar zenith angle. This is the case for all crops planted in rows such as garlic shown in Fig. A.6.6.20.

Contrasting to this dependency on solar zenith angle the values of a ploughed field do not show a dependence of solar zenith angles smaller than about 70° as can be seen in Fig. A.6.6.21.

Finally the albedo of matorral near Rada de Haro is presented in Fig. A.6.6.22 which shows a spectral dependence similar to the ploughed fields. It displays a small chlorophyll absorption and rises, at larger wavelengths, to high values, similar to those of bright soils.

In summary, these ground based measurements provide information about the necessity for bi-directional reflection corrections and possible errors in, at the one hand, calibration stability checks and, on the other hand, albedo estimates derived from directional observations from satellites. It could be shown, that, for certain vegetation covered soils, directional measurements made 90° off the solar direction can be looked at as representative for the albedo.

Appendix 7 Scintillometry

The scintillometer is a device to measure the fluctuations due to turbulence of a beam of light which is transmitted over a horizontal path. From the intensity fluctuations measured in a single spectral channel in the near infrared (e.g. 880 or 930 nm wavelength) a temperature structure parameter can be derived and in combination with a microwave scintillometer (e.g. 3.2 mm wavelength) also the humidity structure parameter is obtained. Similarity theory allows to scale these structure parameters as functions of atmospheric stability which are then used to determine sensible and latent heat fluxes.

The fluctuations are caused by heterogeneity of the refractive index along the path of propagation. The light intensity measured at the receiver is analysed in analogy to the theory of turbulence. The treatment of turbulence goes back to the work of Kolmogorov (1941) and Obukhov (1949) who defined a measure for the fluctuations of the velocities in an air volume. If **r** describes the position within this volume one can define a tensor which "measures" the changes of the velocity over the distances $\leq r$:

$$D_{ik}(\mathbf{r}) = \overline{\left[u_i(\mathbf{r}'+\mathbf{r}) - u_i(\mathbf{r}')\right]\left[u_k(\mathbf{r}'+\mathbf{r}) - u_k(\mathbf{r}')\right]} \tag{A.7.1}$$

In a developed turbulence it can be assumed that within a certain range the flux is, in a statistical way, locally homogeneous and isotropic. The characteristic lengths of the boundaries of this range are denominated as η, the *inner scale* (typically 1 - 10 mm). Below η occurs dissipation of kinetic energy and up to a scale length of L_0, the "external" length or *outer scale* (typically half of the height of the beam above the surface) the local homogeneity and isotropy can be assumed in a stochastic sense. Within the scale region of $\eta \ll \mathrm{r} \ll L_0$, the *inertial region*, energy dissipation can be neglected. The structural tensor depends only on $r = |\mathbf{r}|$ and the structural function can be written:

$$D(r) = \overline{\left[u(\mathbf{r}_2) - u(\mathbf{r}_1)\right]^2} \tag{A.7.2}$$

Within the inertial region the structural parameter D can only depend on r, the kinematic viscosity v, and the flux of kinetic energy, ε, through the different scales of eddies which is finally dissipated outside the inertial region. The dimension of D is m/s, the dimension of the kinematic viscosity m^2/s and that of the specific kinetic energy dissipation m^2/s^3. The only way to express a length scale and a velocity by ε and v is

$$\eta \sim v^{3/4}\,\varepsilon^{-1/4} \quad \mathit{and} \quad u_0 \sim v^{1/4}\,\varepsilon^{1/4} \tag{A.7.3}$$

which defines the lower boundary conditions of the inertial region. Now D can be expressed by u_0, η, and r respectively v, ε, and r:

$$D(r) \sim u_0^2\left[r/\eta\right]^{\alpha} \tag{A.7.4}$$

Within the inertial region, dissipation (v) is neglected and D should only depend on r and ε. This is the case if $\alpha = 2/3$. Now it can be written

$$D(r) = C^2 r^{2/3}, \qquad \eta << r << L_0 \tag{A.7.5}$$

where C^2 is called the structure constant or parameter which contains the ε-term and is given by

$$C_u^2 = \left[u(r_2) - u(r_1)\right]^2 r^{-2/3} \tag{A.7.6}$$

In analogy one can define a structure function for the fluctuations of the index of refraction. The refractive index depends on temperature, pressure and humidity. The temperature can be regarded as a conservative element of the air which is also passive because it does not influence the dynamics. For such a component the same structural function can be developed as for the air itself. For optical wavelength the effect of humidity can be neglected and one obtains

$$n = 1 + 77.6*10^{-6}(1 + 7.52*10^{-3}\lambda^{-2})p / T \tag{A.7.7}$$

If one considers horizontal plains the pressure fluctuations are much smaller than the temperature fluctuations. Consequently n becomes a function of T alone. Since T is a conservative quantity also n is a conservative quantity and its structure function can be expressed by

$$D_n(r) = \overline{\left[n(r'+r) - n(r)\right]^2} = C_n^2 r^{2/3} \tag{A.7.8}$$

where C_n^2 is the structure constant of the index of refraction.

Following Tatarskii (1961) the variance of the logarithm of the amplitude fluctuations ($\chi = \ln [A/A_0]$) can be expressed by

$$\sigma_\chi^2 = 0.124 \nu^{7/6} L^{11/6} \overline{C_n^2}, \qquad \eta << F << L_0, \sigma_\chi^2 < 0.3 \tag{A.7.9}$$

where ν is the optical wavenumber and L the path length. Equation (A.7.9) can be re-written in the form

$$\sigma_\chi^2 = \int_0^L C_n^2(x) W(x) dx \tag{A.7.10}$$

W is a weighting function which can be approximated by the relative position x/L within the beam:

$$W(x) \approx \left[\frac{x}{L}\left(1 - \frac{x}{L}\right)\right]^{5/6}, \qquad 0 < x < L. \tag{A.7.11}$$

The structure parameter C_n can be related to the structure parameter of the temperature. Kohsiek (1982) developed the following relationship:

$$C_n^2 \approx \frac{A_T^2}{T^2} C_T^2 \left(1 + \frac{0.03}{\beta}\right)^2 \tag{A.7.12}$$

with

$$A_T = -\left(\frac{0.776 \cdot 10^{-6}\, p}{T} + \frac{1.723q}{T}\right) \tag{A.7.13}$$

where *ß* is the Bowen ratio and *q* the specific humidity.

The Monin-Obukhov similarity theory allows to link the structure parameter of temperature with the sensible heat flux. The basis of the similarity theory (Monin and Yaglom 1971) applied to the flux-profile relationship is a dimensional analysis of the quantities which govern the process and which are combined to dimensionless and as such (approximately) invariant ratios. Starting point generally is the interaction between a flow of air over a surface which exerts a shear stress, τ_0, at the fluid. The ratio of the stress (force per unit area) and the density of the air happens to have the dimension of the square of a velocity which is called friction velocity, u_*. The shear causes a velocity gradient (d*u*/d*z*) in the fluid which depends on the distance, *z*, from the surface and the quantities u_*, d*u*/d*z*, and *z* combine to the dimensionless von Kármán constant *k*:

$$\frac{u_*}{z(d\bar{u}\,/\,dz)} = k \tag{A.7.14}$$

Considering the flow of heat, the friction velocity, the air density, the heat capacity, a temperature scale θ_* and the heat flux Φ_H can be combined to yield a dimensionless expression from which

$$\Phi_H = \rho c_p\, u_*\, \theta_* \tag{A.7.15}$$

results. Wyngaard et al. (1971) and De Bruin et al. (1995) found that the temperature structure parameter and the temperature scale are related as

$$\frac{C_T^2 (z-d)^{2/3}}{\theta_*^2} = 4.9\left(1 - 9\frac{z-d}{L_{MO}}\right)^{-2/3} \tag{A.7.16}$$

where *z* is the measuring height and *d* the displacement height (see Fig. 1.26). L_{MO} is the Monin-Obukhov length which again is related to u_*, θ_*, the temperature, the von Kármán constant, and the acceleration due to gravity:

$$L_{MO} = \frac{u_*^2\, T}{g\, k\, \theta_*}. \tag{A.7.17}$$

The Eq. A.7.14 - 17 can be solved iteratively to yield the sensible heat flux.

For very unstable atmospheric daytime conditions and a relatively high optical path the contribution of the friction velocity becomes small and the following formula can be applied:

$$\Phi_H = \rho c_p b(z-d)\left(\frac{g}{T}\right)^{1/2}\left(C_T^2\right)^{3/4}, \tag{A.7.18}$$

with

$$b = \sqrt{\left(\frac{1}{c_{T1}}\right)^{3/2} k\, c_{T2}} \cong 0.48, \quad c_{T1} = 4.9, \quad c_{T2} = 7, \quad 0.48 < b < 0.57. \tag{A.7.19}$$

k is the von Kármán constant, T the temperature, z the height above the surface, d the displacement height, and g the acceleration due to gravity.

In a similar way as derived here for the sensible heat flux also the latent heat flux can be obtained if two wavelengths are used of which one is sensible to water vapour. Further information may be obtained from De Bruin (2002), De Bruin et al. (1991, 1993), Kohsick (1982), Kohsick et al. 2002, Thiermann and Grassl (1992).

Appendix 8 AVHRR Time Series

A.8.1 Presentation of Large Scale Satellite Data Time Series

The sixteen years of Mediterranean AVHRR data which are now available provide already a considerable time series which can be looked at as a baseline data set for the end of the twentieth century. The data are stored in the MEDOKADS data bank with a resolution of about 1 km x 1 km. In principle it therefore is possible to follow temperature, reflectance, NDVI and albedo changes for any Mediterranean (and central European) area of a size larger than 1 km^2 and any time interval between 10 days (the minimum integration time to overcome the cloud problem) and 16 years between 1989 and 2004, the present length of the data bank. In practise the high spatial resolution will only be applied for small regions in which adjacent areas can undergo quite different changes that would be cancelled out if a larger integration area is chosen. In long time data series some details even become a little bit blurred because there always is a small jitter from image to image in the geographical pixel location.

Examples of these time series have already been shown in some of the preceding chapters to study regional and inter-annual changes of land-surface characteristics and surface energy budgets. This appendix focusses at the question to what degree longer term developments can be excerpted from the available 16 years data set. This leads into a new scale where the question of homogeneity and data normalization plays the important role. Mainly images of the whole Mediterranean area are presented which allow to visualize the relationships between the regions. In these images much detail of the original data gets lost for two reasons: Firstly, because of the strong degradation of the original images with one km spatial resolution into the printable format and, secondly, because of the colour code which was applied and in which each colour represents a certain band width of the presented data. Obviously also a sliding colour scale could be used but then the pictures would become very confusing and the colours could also not well be distinguished in print. Thus any image must only be regarded as a somewhat degraded or simplified view at the real data set.

On the other hand also by numbers the full information content can only be communicated for areas for which the amount of data remains manageable. In section 6.9 a procedure was introduced to quantify changes over a longer period of years for 1 degree x 1 degree windows which is probably a handy size for mesoscale studies. To recall the information presented in each box the code is repeated in the Fig. A.8.1.

Another possibility to present the results of time series analysis is the graphical presentation of mean values for well defined areas. Examples will be shown for time series and trends of annual mean values for larger geographical units such as the western, central and eastern Mediterranean land areas. These graphs are suitable for studies of the interaction of the Mediterranean with global climate variations. All regional details get lost.

It should also be recalled that AVHRR time series represent the conditions at the specific diurnal time window defined by the satellite overpass time and the AVHRR swath width which contains data of a range of solar zenith angles. Therefore the data represent a time window and neither a precisely defined time of the day nor a specific solar zenith angle which in addition varies for each scene during the year. Only because of the continuous repetition of the measurements an internally consistent data set is generated if the data are homogenized with respect to the idiosyncrasies of the individual instruments.

MEAN VALUE during prescribed time interval *% absolute* Example: 0.31 = 31% for 16 years (e.g. NDVI = 0.31)	CHANGE during prescribed time interval (16 years) *% relative* Example: 40.3% 0.403 of 0.31 = 0.1249 change in absolute units
TREND PER YEAR *% absolute/a* Example: 78.5 E-4 0.1249/16 = 0.00781	CHANGE RELATIVE TO WHOLE SCENE *% relative per time interval* Example: 35.3% Scene changes by 40.3 - 35.3 = 5% relative

Fig. A.8.1. Meaning of the four numbers used to present the information content within $1° * 1°$ boxes in numerical format

In Fig. A.8.2 the temperature development obtained by a linear trend analysis is presented. A discontinuity can be seen north of 41°N between 17°E and 20°E and south of 41°N between 7°E and 8°E. West of this border the data are selected from the second and third swath, east of it from the first and second swath of the three adjacent NOAA overpasses necessary to cover the whole Mediterranean basin. At this border the observation conditions change. In the case of temperatures the step of about one hour in observation time results in differences of the land temperatures. In the short wavelengths channels, differences due to the BRDF can result where measurements made from two consecutive orbits with opposite directions of observation are joint. The differences in observation time can be corrected if the diurnal course of the temperature is known (see section 4.4). Discontinuities due to different observations geometries in the shortwave range can be corrected if the BRDF is known (see section 4.2.2). In the development analysis this discontinuity may be enhanced by the loss of data between July 1990 and March 1991 east of the discontinuity border. In time series starting with 1992 the discontinuity nearly vanishes. The remaining discontinuity reflects the magnitude of uncertainty which is included in all parts of the maps already due to the cloud elimination problem. Since this effect is more randomly distributed it does not manifest itself as strikingly as the break due to the need to use three swathes to cover the whole region.

In the following sections first temperature time series will be dealt with. These are temperatures computed for the surface from the TOA data by applying the split window technique. Then follows the analysis of reflectances derived from TOA data the AVHRR channels 1 and 2. The quantities inferred from measurements made from space, reflectance, NDVI and albedo, are less dependent on the solar zenith angle than land surface temperature as long as the solar azimuth angle is not too large. Nevertheless the illumination and observation geometries for which no corrections are applied to these huge amounts of data cause a certain band width of measuring conditions.

A.8.2 Temperature Time Series

Seasonal data of the temperature development as seen at the time of the NOAA satellite overpass have already been shown in Figs. 6.9.1 and 6.9.2 for the years 1995 - 2000 and the annual mean temperature from 1989 to 2004 was presented in Fig. 1.1. Fig. A.8.2 adds the development throughout this period. It is followed in Figs. A.8.3 - A.8.6 by enhanced maps of the four Mediterranean regions Iberian peninsula, Italy, Greece, and Near East. It is astonishing how much detail remains visible even in such a long time period. This indicates that year for year a nearly identical temperature pattern appears.

For the long time scale discussed here and especially for temperature measurements it has, however, to be taken into consideration that the shift of the NOAA satellite orbits causes a problem. Due to this shift the time of observation changed from about 13:00 ± 1 hours local time to 16:00 ± 1 hours local time during the lifetime of NOAA-11. While nominal observation times are around the daily local temperature maximum, these shifted to late afternoon in the years of 1993 and 1994. This is similar for the NOAA-14 and -16 satellites.

The overview of Fig. A.8.2 gives the impression that the western part of the Mediterranean including parts of the sea is slightly cooling with the exception of parts of Morocco while in the eastern part warming is predominant over the continent. For the reasons mentioned in section A.8.1 there is a discontinuity between cooling and warming at a line between 15 and 20 degree east which indicates that the trend data from swath 1 and 3 do not completely fit together. While also the western SST seems partially to cool, the SST in the eastern part of the Mediterranean seems to remain constant within ± 0.1 °C per year (compare Figs.A.8.3 and A.8.6). *Nota bene*, the SST is not affected by the drift of the NOAA

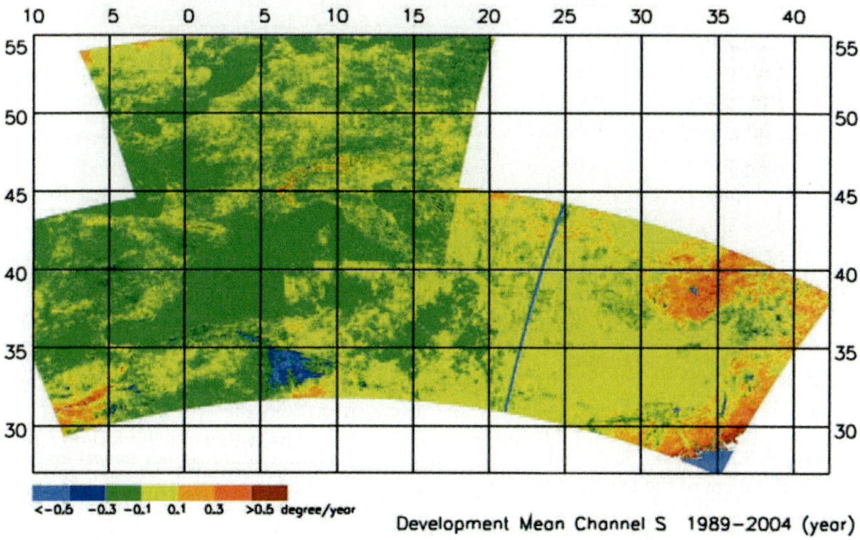

Fig. A.8.2. Development of the Mediterranean temperature from 1989 to 2004

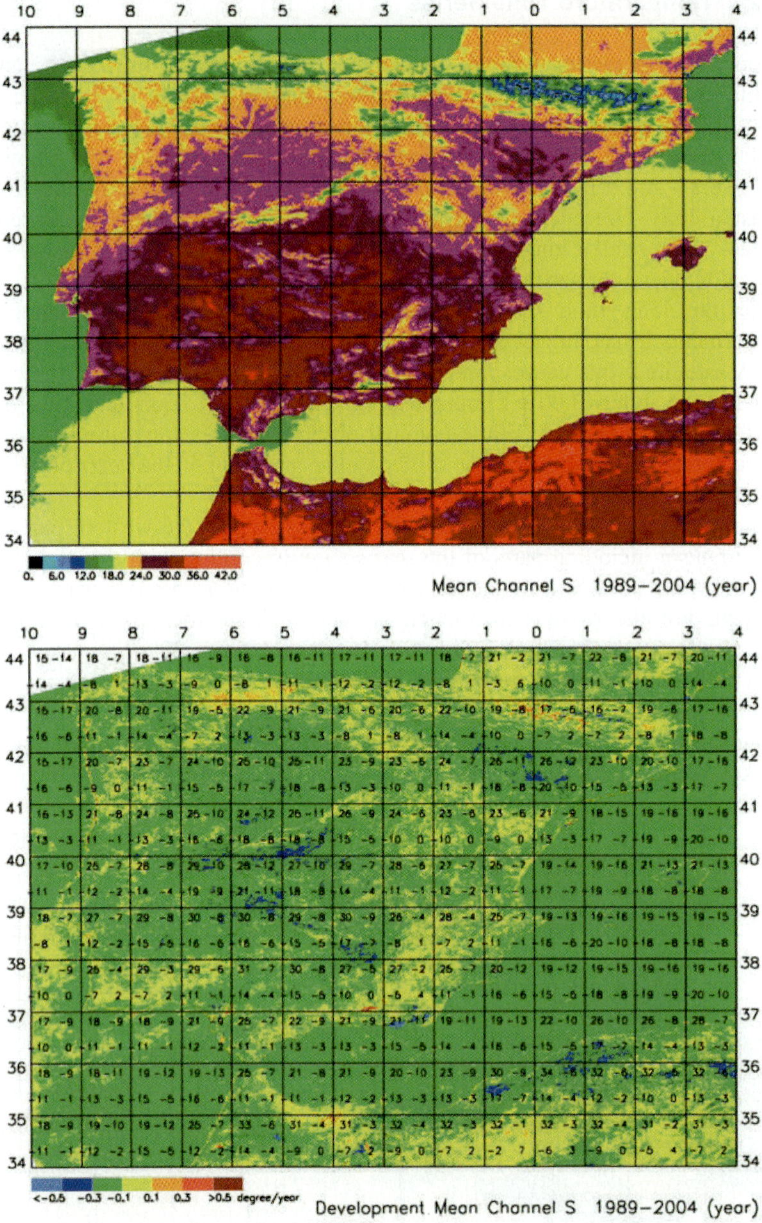

Fig. A.8.3. Iberian peninsula 16 years annual mean temperature (upper panel) and development of the temperature (lower panel) 1989 - 2004

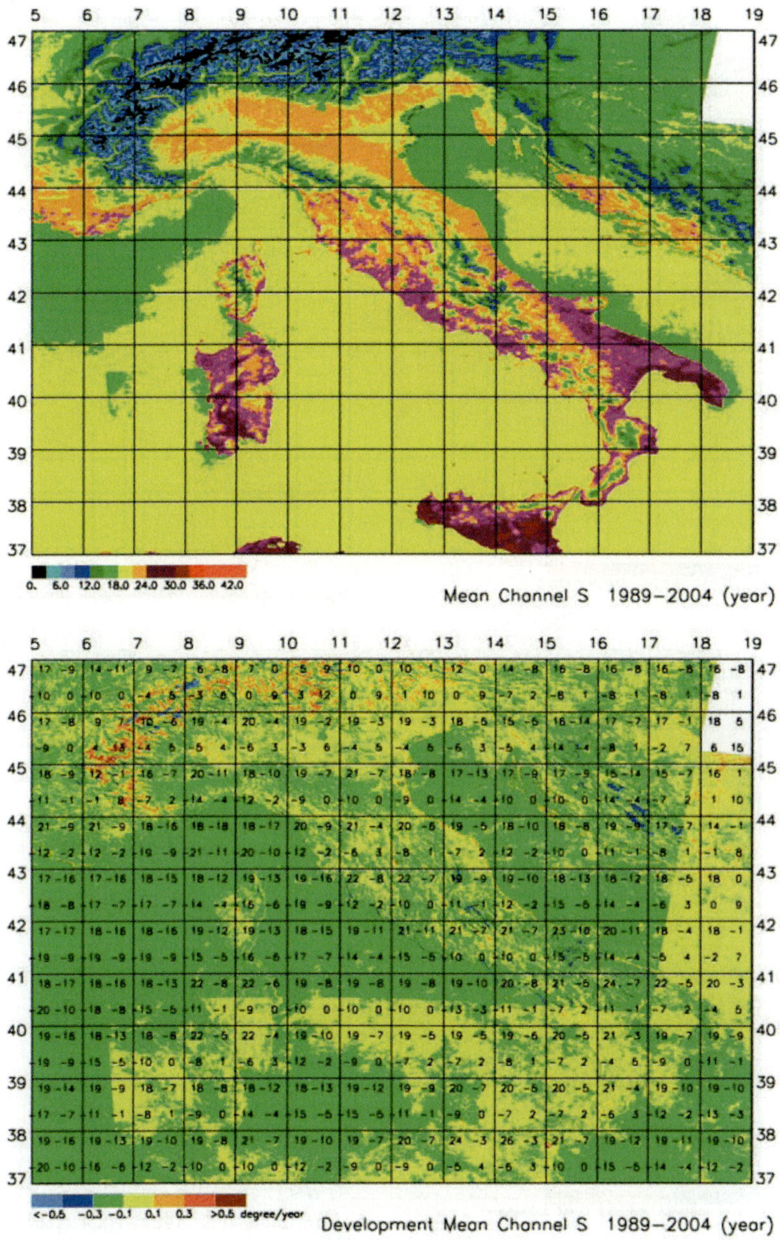

Fig. A.8.4. Central Mediterranean 16 years annual mean temperature (upper panel) and development of the temperature (lower panel) 1989 - 2004

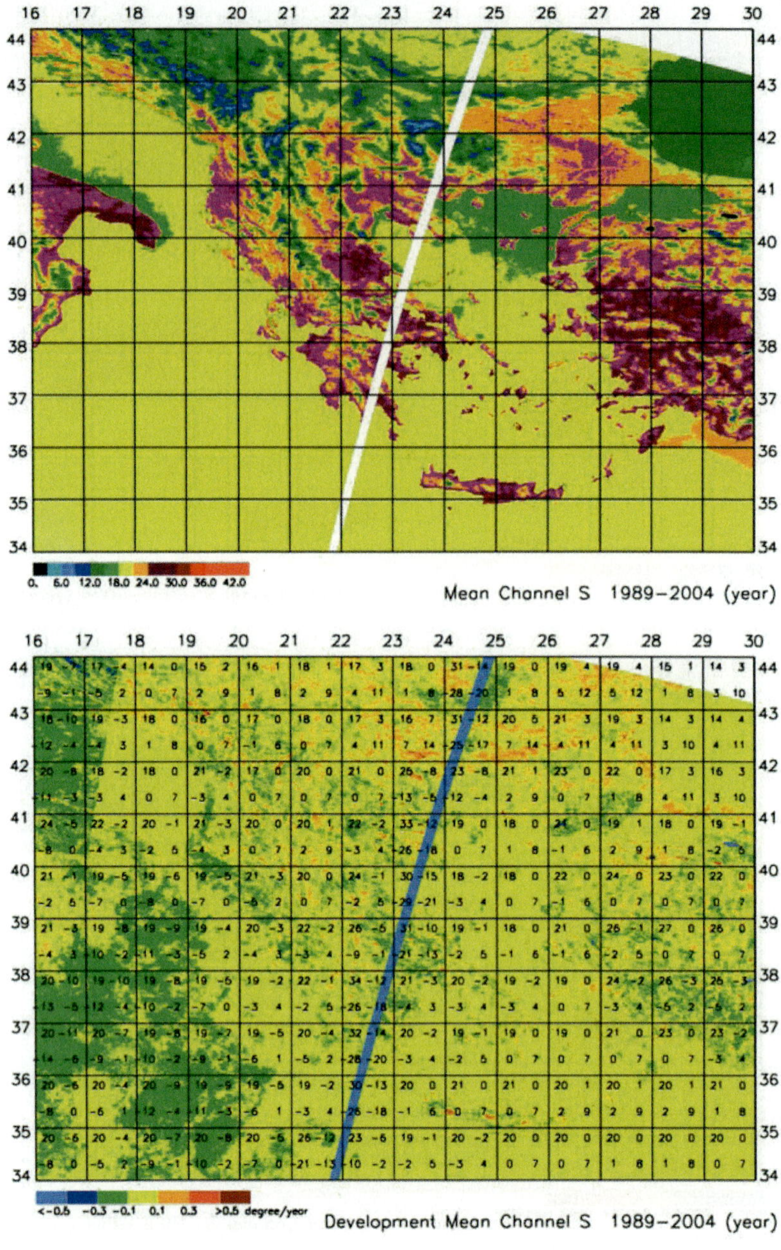

Fig. A.8.5. Greece and the Balkans 16 years annual mean temperature (upper panel) and development of the temperature (lower panel) 1989 - 2004

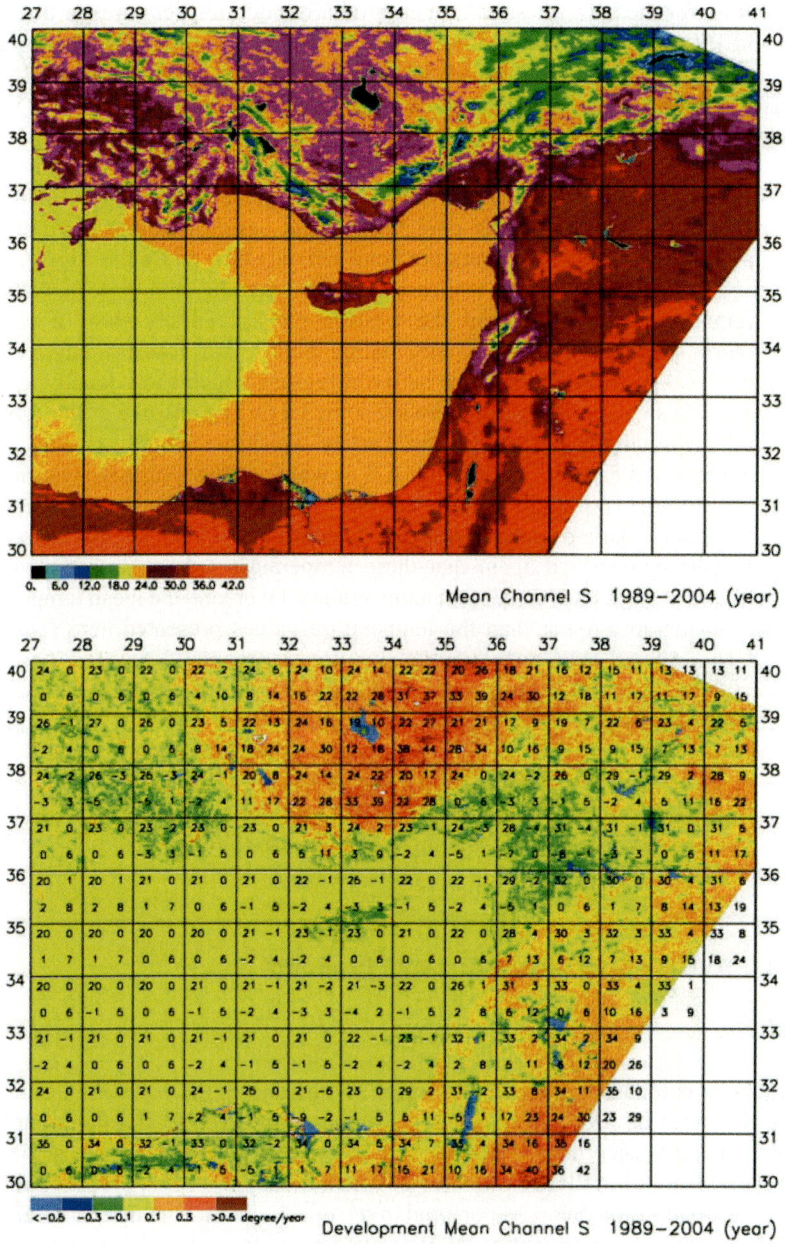

Fig. A.8.6. Eastern Mediterranean 16 years annual mean temperature (upper panel) and development of the temperature (lower panel) 1989 - 2004

orbits while for the LST the smaller measured afternoon temperatures in the years 1993 and 1994 may have affected the trend for the used swathes in different ways.

The enhanced regional figures (A.8.3-6) show many details more clearly. At the Iberian Peninsula there is a sharp partition between the temperature regime in the northern and southern part though the tendency towards cooler temperatures (Fig. A.8.3 lower panel) is general with the exception of the most south-easterly parts. Such differences may be due to the development of regional water resources. In Italy the Apennines is the outstanding cool feature and beside Sardinia is Puglia the hottest part. A clear positive temperature trend can only be seen for the Alps. Also in the mountainous areas of Italy a tendency towards warming is to be seen while the plains remain stable. Greece and the western part of Turkey show a marked temperature structure due to its mountains, plains and forests. The general impression is, that there is a balance between regions which show locally a small warming and a small cooling. But there are some larger regions like at the Adriatic coast and the southern part of western Turkey which are slightly cooling. This tendency repeats in eastern Turkey (Fig. A.8.6) which shows a massive warming in central Turkey and cooling in its more eastern part while in other land areas in the Near East again a marked warming appears. In contrary to Crete, Cyprus is cooling.

It must be emphasized again that these temperature data represent cloudless situations and close to daily maximum temperatures. Over land the mean temperature will be considerably lower than the temperature values presented here (compare Figs. 4.16 and 4.17). Consequently the developments discussed here are developments of maximum temperatures and not of annual mean temperatures. For the SST the difference will not be as essential and in this case it can be said that in the eastern part of the Mediterranean no changes are detected while in its western part a small cooling seems to occur. In Fig. A.8.4 the discontinuity discussed in section A.8.1 is clearly to be seen. But the numbers imprinted into the 1° x 1° boxes show only small differences and that here one is close to the -0.1 K/a border when yellow colour switches to green.

A.8.3 AVHRR Short-wave Channel Products

Goal

This section is focussed at the development in time during the years 1989 - 2004 of the reflectances in AVHRR channels one and two, the NDVI and the broadband albedo computed according to Eq. 6.9.1. A key question arising for the use of long term data series of AVHRR products is whether they are an appropriate tool to determine trends of the biomass potential in relation to global climate change and the related question of desertification. NDVI data derived from AVHRR measurements have widely been used throughout the chapters of this book for regional and interannual studies. The possibility of an increasing vegetation cover at higher latitudes (and of increasing desertification at lower ones) and its monitoring by the development of the NDVI is an important matter of discussion among scientists. The statement that an observed increase of the NDVI is caused by an increase of biomass due to global warming and/or increasing carbon dioxide amount in the atmosphere would be an important message. A prerequisite for such a conclusion are homogenous and normalized data series which are long enough to filter out the "noise" in the data. Different issues have to be considered in this context. First of all the instrumental degradation processes and the re-calibration which have been dealt with in Chapter 2. But there are also other issues such as orbital changes, changes of filter characteristics, changes of atmospheric turbidity, and changes of cloudiness and precipitation pattern which influence the data sampling. Prior to drawing conclusions one has therefore to ensure that the quality and normalization of the data satisfies the requirements.

Potential Error Sources

The Re-calibration Problem. The control of the stability of the calibration factors is done by observing targets which can be considered as invariable like certain desert areas. The stability of these areas can not be taken as granted because there are no anchor stations at which the real reflectance and albedo as well as turbidity of the atmosphere are monitored. As has been discussed in Appendix A.6 also the effect of the observation geometry on the reflectance has to be considered. Though the average of reflectances measured over 8 - 10 consecutive days may provide a representative value for the observation angle, the solar aspect angle changes in the course of the year. The FU-Berlin uses an area of the size of 30 km x 30 km at the position of 31.5°N, 1.8°E for determining the progressive instrumental degradation. This area still is well within the margin of the area from which signals can be received in Berlin. It seems to remain fairly stable but in its surrounding reflectance changes have been observed. This raises the question, whether or not the whole area may change. The need for anchor stations is obvious. The degradation curve of NOAA-14 which follows from these observations is shown as red line in Fig. 2.6. It can be seen that it is flattening during the last years both in channel one and channel two while the NOAASIS line (blue) remains straight. It can further be seen that the difference between both lines is larger for channel two than for channel one. This has the consequence that, if the NOAASIS calibration is applied, the signal of channel two will be more enhanced than that of channel one as compared to the application of the FU calibration. The result will be a larger NDVI in the case of the NOAASIS calibration. Seen from the viewpoint of the FU-Berlin calibration the NDVI computed with the NOAASIS calibration will run away to higher values.

How could the reliability of the calibration be proved without anchor station information? One possibility is to look at the reflectance of the sea surface. Does this remain constant as it should be the case? Re-calibration is made with targets of high albedo. If these high reflectances are miscalculated by 0.01, which is about 2% of the target reflectance, it means that the sea surface reflectance, which is of the order of 0.06, changes by 0.0012. Such a small change is hard to detect. Especially the coloured maps do not show small changes because of the bandwidth covered by each colour.

Another possibility is to look at areas with high NDVI values such as dense evergreen forests the NDVI of which should be constant as long as there are no dramatic environmental changes such as fires or long drought periods.

A third possibility is to use the data of other satellites with calibrated sensors such as Landsat or SPOT for comparisons. These and other topics related to the validation of long term data series will be discussed in the following paragraphs.

Compatibility of Filter Characteristics. The identity or at least the compatibility of the spectral characteristics of the instruments in time is one of the critical points. Since May 1998 a new series of NOAA satellites is in orbit, the NOAA-K, L, and M series which are numbered NOAA-15 (launch 13 May 1998), NOAA-16 (launch 21 September 2000) and NOAA-17 (launch 24 June 2002). These satellites carry an improved AVHRR instrument, the AVHRR-3. Though the general optical characteristics of channel one and two are unaltered, the flanks of the filters got steeper with the effect that they are more selective and less overlapping as was the case before. Trishchenko et al. (2002) investigated the effect of this change on the surface reflectance and the NDVI. If the NOAA-9 instrument is taken as reference, the signal obtained over vegetated surfaces with the AVHRR-3 instruments deviates in channel one towards smaller values and for channel two the opposite happens. That means, the difference between the NIR and VIS(red) signals becomes larger with the new instruments as compared to those used up to NOAA-14. Already large NDVIs in the old NOAA images (over dense forests) are enhanced in the new NOAA images. The difference is up to nearly 6% for NDVIs as large as 0.9 and tends to zero for small NDVIs. A reduction of images recorded with NOAA-K, L, M instruments to those of NOAA-9 to compose a coherent time series therefore becomes a troublesome task because each pixel has to be corrected individually. Due to the steepening of the flanks of the filter functions, the AVHRR-3 signals get more similar to the Landsat-TM channel three and four signals.

The effect which an adjustment of the AVHRR-3 data has for the long term NDVI data series can be seen in Fig. A.8.7. Here the annual mean NDVI data series are shown for central Europe, the whole Mediterranean area, and the North African deserts. Also shown is the occurrence of volcano eruptions and the operational times for NOAA-11, -14, and -16. The trend of the NDVI computed from NOAA-16 data since the year 2000 nearly vanishes.

Atmospheric Turbidity. The large scale long term AVHRR data sets are TOA data because up to date it is not possible to apply to these data series reliable atmospheric corrections. Too large is the variability of aerosols and water vapour and too little information is available to correct basically each single pixel for the aerosol effect. Nevertheless it has to be considered to what degree aerosols could distort the data. In Table 4.2. atmospheric correction functions have been presented which allow the conversion of reflectances measured for the top of the atmosphere into surface reflectances. For the AVHRR channels two quadratic approximation functions have been given, one for a relatively clear atmosphere (meteorological range 70 km) and one for a turbid atmosphere (meteorological range 21.5 km). These functions are plotted in Figs. A.8.8a and b. The general feature is the following: Over dark surfaces the radiation scattered back from the atmosphere is larger than the effects

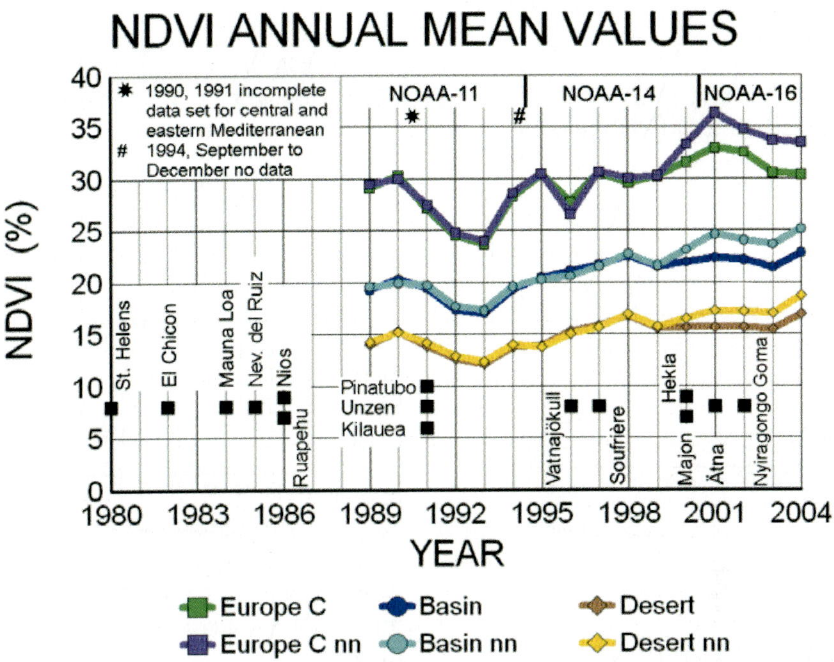

Fig. A.8.7. Mean annual area averaged NDVI for central Europe, the Mediterranean basin and desert areas in North Africa, volcanic eruptions and the operational times of the NOAA-11, -14 and -16 spacecraft. The NDVI is plotted as uncorrected and corrected for the filter function effect (normalized to NOAA-9). The colours stand for central Europe (green uncorrected, blue corrected), the Mediterranean basin (violet uncorrected, red corrected), and desert (yellow uncorrected, brown corrected). For the year 1994 the mean is computed for period January to the beginning of September only due to the failure of NOAA-11. For the years 1990/91 the full year mean is available only for the area west of about 20^{0}E. Incomplete data for the Balkans and the eastern Mediterranean were added in a second step (data from July 1990 to March 1991 are lost). Notwithstanding the spectral correction there seems to occur a jump in the NDVI data with the advent of NOAA-16 which is specifically noticeable in the central European data. Another strong feature mainly in the central European data is the dip 1992 - 1993 which can be attributed to the Pinatubo eruption

diminution of the signal which returns from the surface. Over bright surfaces the reduction of the surface signal is larger than the radiation scattered back from the atmosphere into space. Thus dark surfaces appear brighter when observed from space and bright surfaces darker. In each case exists a "neutral point" at which the atmosphere has no effect, the atmospheric backscatter just compensates the loss of radiation from the surface. For AVHRR Channel 1 the neutral or "cross-over" point lies at $\rho_{surface,\,c}(1) = 0.1648$ for the clear atmosphere and at $\rho_{surface,\,t}(1) = 0.1446$ for the turbid atmosphere. For AVHRR Channel 2 the points lie at $\rho_{surface,\,c}(2) = 0.0595$ and $\rho_{surface,\,t}(2) = 0.0709$.

If the aerosol content increases from a clear to a turbid atmosphere the following

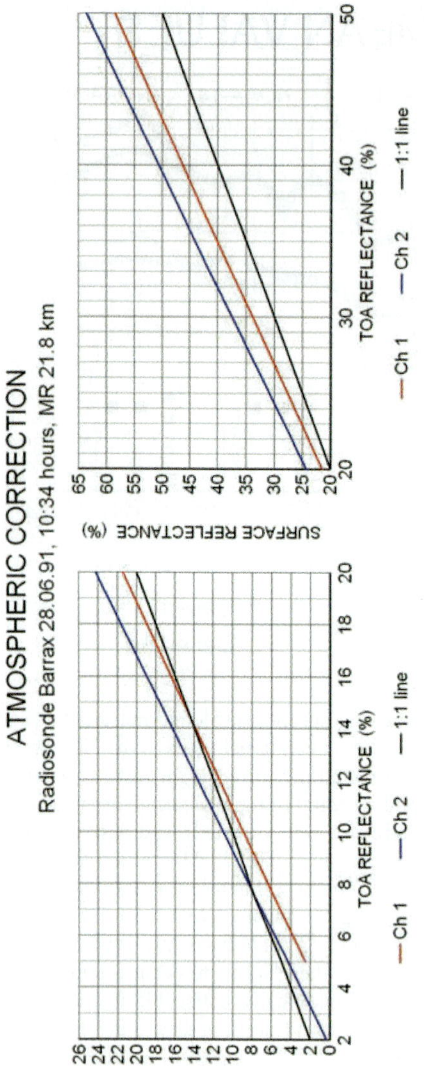

Fig. A.8.8a. Atmospheric correction functions for AVHRR channel 1 and 2 and a turbid atmosphere

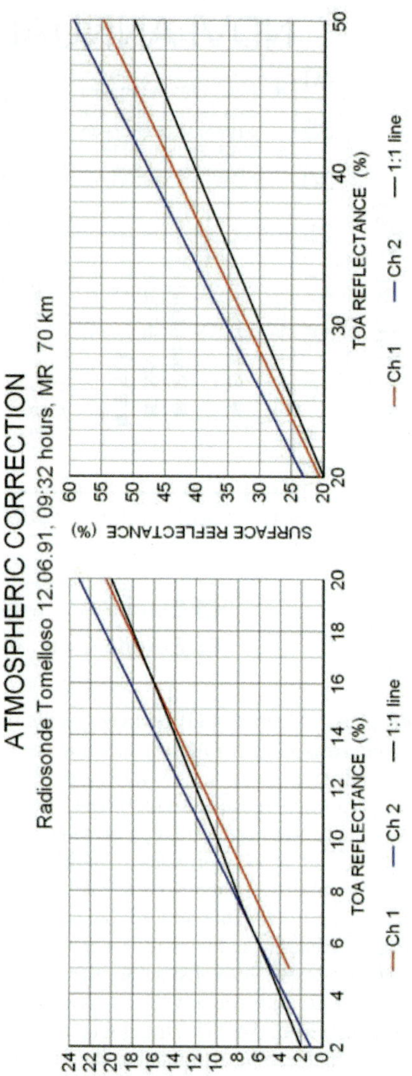

Fig. A.8.8b. Atmospheric correction functions for AVHRR channel 1 and 2 and a clear atmosphere

can be observed:

- Over dark surfaces with a reflectance lower than 0.14, such as lateritic soils or water, the TOA signal in channel 1 is larger than that of the surface and increases with increasing aerosol content. In channel 2 the neutral point lays lower so that for most surfaces (excluding water) the TOA signal will be nearly unchanges or lies slightly below the surface signal and decreases with increasing aerosol content. The NDVI will become smaller.

- Over bright surfaces such as desert soil the signal will decrease in both channels but to a slightly smaller degree in channel 1 than in channel 2 because its neutral point lies at a higher reflectance. Depending on the value of the reflectance the NDVI may remain nearly constant or slightly decrease.
- Over vegetation the surface is dark in channel 1 and bright in channel 2. Over dense vegetation the signal in channel 1 will become larger and the signal in channel 2 will significantly decrease. Consequently the NDVI measured at the top of the atmosphere will become smaller.

An increase of the aerosol content over Mediterranean areas will therefore have the following consequences for the TOA NDVI: Over desert areas the NDVI will remain fairly stable or slightly decrease depending on its albedo. Over evergreen forests the NDVI will decrease. Over agricultural areas the NDVI will decrease in spring but in summer, especially after the harvest when the surface becomes bare soil or stubble fields, it may remain nearly constant. Depending on the individual reflectances there can be a small decrease as well as a small increase of the NDVI if the aerosol amount increases. Examples of spectral reflectance values are given e.g. in Figs. 1.29 and 1.30 for several surfaces from which this effect can be derived in combination with the Figures A.8.8a and A.8.8b.

Over a stubble field in Castilla - La Mancha the reflectance was 0.332 at 630 nm which corresponds to AVHRR channel 1 and 0.509 at 910 nm which is representative for AVHRR channel 2. The surface NDVI then is of the order of 0.21. Extrapolated to the top of the atmosphere one obtains for the clear atmosphere 0.310 and 0.428 respectively 0.297 and 0.402 for the turbid atmosphere resulting in NDVIs of 0.160 respectively 0.150. Over a bright laguna a reflectance of 0.626 was measured at a wavelength of 630 nm and 0.689 at a wavelength of 910 nm. The surface NDVI is 0.048. Extrapolated to the top of the atmosphere one obtains 0.570 and 0.578 for the clear atmosphere or 0.535 and 0.541 for the turbid atmosphere. This results in TOA NDVIs of 0.007 and 0.0056 which signals an unmeasurable decrease with increasing turbidity for this extremely bright target. One consequence is that if the desert NDVIs remain constant as aerosol increases then there is a possibility that the NDVI of the whole scene increased for other reasons because in reality the desert NDVI tends to become smaller with increasing turbidity.

In case of a volcanic eruption there may be, for a short time and regionally limited, an increase of aerosol particles in the troposphere. If the eruption reaches the stratosphere, small particles may remain in the stratosphere for a couple of years. Though the consideration made so far deals with tropospheric aerosol distributions the conclusion will in principle remain the same though the stratospheric aerosol mainly affects the shorter wavelengths of AVHRR channel 1. It can therefore be expected that a strong volcanic eruption will enhance the signal of channel 1 over vegetation and reduce the signal in channel 2. The TOA NDVI will decrease.

One possible criterion for the quality of AVHRR data series mentioned before was the vegetation index over dense forests. In Appendix 4 spectra of pines have been shown. The reflectance of the more distant trees in AVHRR channel 1 is of the order of 0.08 and 0.43 in channel 2. The surface NDVI is 0.686. Extrapolated to the TOA one obtains the triples 0.092, 0.37, 0.60 for the clear atmosphere and 0.093, 0.349, 0.579 for the turbid atmosphere.

Jagdhuhn (1994) found for three forest test areas in Tuscany (Fig. 5.10) surface reflectances of in the average 0.066 for AVHRR channel 1, 0.323 for AVHRR channel 2, 0.050 for TM channel 3 and 0.332 for TM channel 4. This results in surface NDVI values of 0.667 for AVHRR and 0.738 for TM. The TOA reflectance would be for AVHRR 0.079 and 0.276, for TM 0.07 and 0.282 in the case of the clear atmosphere. For the turbid atmosphere one obtains the pairs 0.082, 0.261 and 0.07, 0.268. This results in TOA NDVIs of 0.554, 0.602, 0.517, and 0.589 which is close to the calculation starting from surface pine reflectances.

Measurements made by Kessler (1985) from a tower on top of a pine forest in Germany resulted in minimum albedo values of about 0.095 in June and nearly constant values around 0.105 from July to September. These values are supported by a measurement over one stand in Italy near Radicondoli as can be seen in Fig. A.8.9.

The diagram Fig. A.8.10 presents the relationships between the spectral reflectances and the broad band albedo if a NDVI of 0.7 is prescribed.

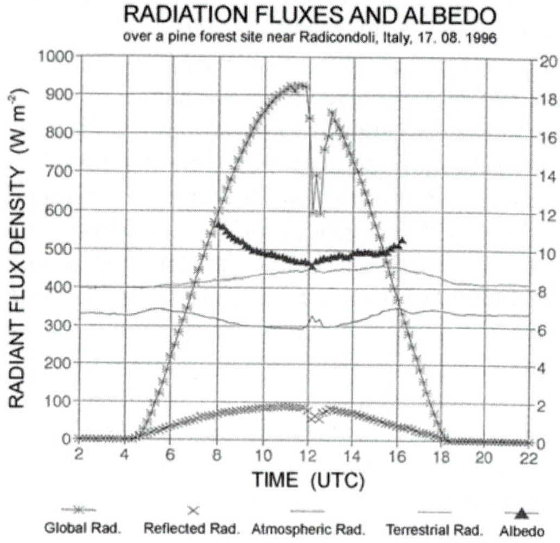

Fig. A.8.9. Radiantion fluxes and albedo measured over a pine stand near Radicondoli, Italy

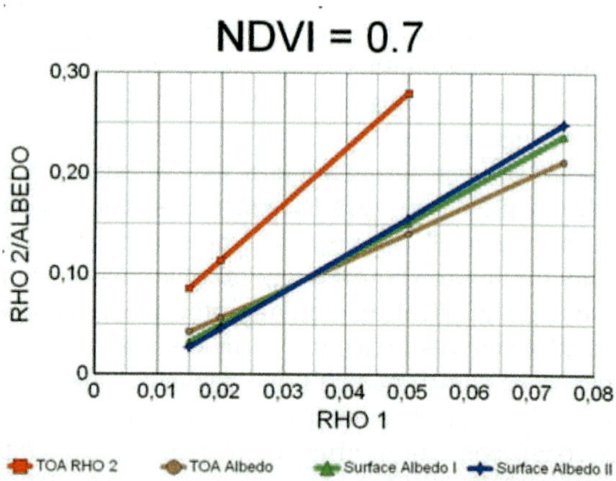

Fig. A.8.10. Diagram to determine pairs of AVHRR channel 1 and 2 reflectances which lead to an NDVI of 0.7 calculated by inverting Eq. 4.15. In addition surface albedo values are given which result from the pairs of reflectances if Eq. 6.9.1 is applied

It shows that for an albedo of 0.1 the effect of the atmosphere on the albedo is very small; the TOA albedo would for both types of atmospheres (clear and turbid) about the same. Eq. 6.9.1 allows to determine the pair of reflectances which would result in an NDVI of 0.7, the value which approximates Jagdhuhn's data, provided the formula applies also for surface data. The resulting combination is $\rho_1 = 0.035$ and $\rho_2 = 0.2$. These numbers are smaller than those obtained by Jagdhuhn but match rather precisely those of old holm oaks (Fig. A.6.5.17). If these reflectance values are extrapolated to the top of the atmosphere using Figs. A.8.8a and b one obtains the values 0.058 and 0.168 which yield the same broad band albedo but a NDVI of 0.487. It can therefore be concluded that AVHRR TOA NDVI values for extended forests larger than 0.5 are seldom and if the NDVI approaches the value of 0.7 for larger areas the data become unreliable.

Varying Meteorological Conditions and the Vegetation Period. The variability of reflectances and of the NDVI of individual months can be affected by the highly variable precipitation events and therefore generate irregular time series. Longer lasting cloud fields which prevent the measurement of reflectances during one or more decades may disturb the statistics. Precipitation may clean the atmosphere by washing out aerosols and change the reflectance of the surface. Observations show that precipitation maxima may shift for some years from spring to autumn and back thus affecting different stages of the vegetation period. Annual mean values reflect both the variable precipitation (primarily of spring) and the length of the vegetation period which in addition depends on temperature. High temperatures normally cause high evaporation rates leading to early senescence but during the hot summer of 2003 little effect can be seen. Obviously enough water had been stored from spring rains.

Several authors report an earlier start of the vegetation period as time goes by and take this as an argument for an increasing vegetation cover in a warming world. However, in the Mediterranean environment one has to consider that because of the lack of water the vegetation period also may end earlier. One has to consider annual values to investigate whether or not the vegetation cover increases and decadal values to estimate the real duration of the vegetation period. A newly upcoming test is the situation in the Iberian Peninsula versus the central Mediterranean area. In the winter to spring period of 2005 there have been heavy rains in Italy while there is a severe drought at the Iberian Peninsula. Furthermore it was cool in Italy causing a delay of the start of the vegetation period but this did not affect the summer situation. This should show up in the NDVI.

Mean Value and Trend Maps

The message resulting from the preceding discussion is that trends derived from AVHRR data may have deficiencies because of changes of the re-calibration target, wrong extrapolation of re-calibration factors, a change of the aerosol content over the Mediterranean, shift of observation times, changes of the aspect angle due to a shift of the satellite orbit, time series are too short for trend analysis or, for a sequence of years, phase shifts of meteorological events.

Because NDVI values computed from NOAA-16 measurements with the best known calibration factors still violate boundary conditions discussed in the preceding paragraphs, only filter function corrected data normalized to NOAA-9 conditions are presented here.

Though with this correction the MEDOKADS data set is carefully designed, it can not be excluded that residual errors remain in the data. Maselli (2005) compared MEDOKADS data of Italy with SPOT-VEGETATION (VGT) NDVIs from 1999 to 2004. The VGT data do not show a trend of the NDVI after 1999. The mean value for Italy from 1999 to 2004 is 0.533 ± 0.006. The minimum which shows up in the AVHRR values of 2000 is not present in the VGT data and must be neglected here because it seems to be a calibration error as will be discussed later. This comparison confirms that the NOAA-16 data yield too high NDVI values as compared to older AVHRR data. The ratio between AVHRR NDVI and VGT NDVI jumps from 0.679 in 1999 to in the average 0.772 for the pariod 2000 - 2004. This would require a reduction by a factor of 0.8727 for the NOAA-16 data. It could, however, also be argued that the AVHRR mean value of 1999, which was used here as starting point, is already affected by the decline which leads to the low value of the year 2000 because the values of the preceding years are slightly higher. This may change the 1999 ratio by 2.55 % to 0.6963 instead of 0.679 and the NOAA-16 correction factor from 0.8727 to 0.9020.

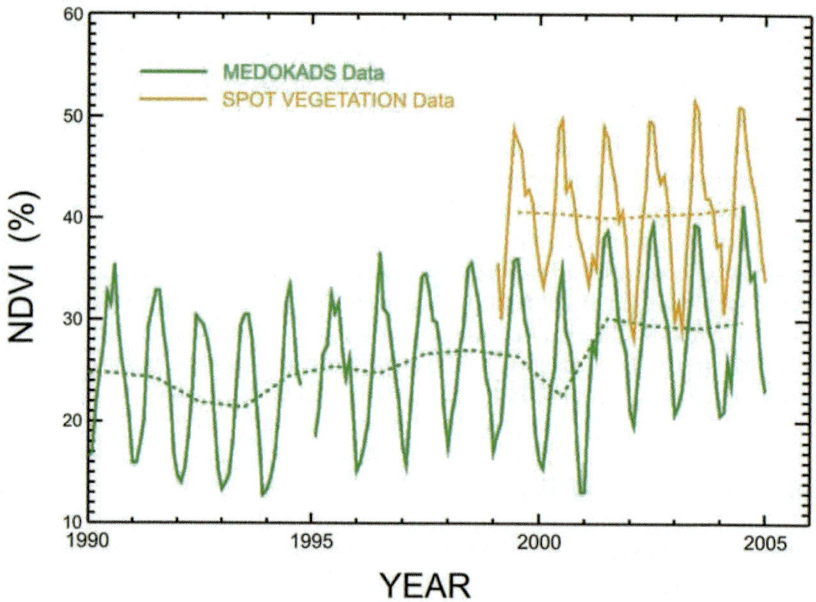

Fig. A.8.11. Monthly NDVI data derived from SPOT-VEGETATION and NOAA-AVHRR data for the whole Mediterranean basin

In a more recent study, Friedrich and Koslowsky (see Fig. A.8.11) showed that for the whole Mediterranean basin the VGT NDVI data remain stable within ± 0.2 % from the beginning of the VGT mission to 2005 (maybe a small trend of 0.06 %/a could be derived from the data). This work is based upon monthly data and allows to compare either data of 1999 (case a) or the AVHRR mean value of the stable period 1997-1999 (case b) with the VGT mean values of 1999-2005. The ratio AVHRR/VGT in these two cases is 0.6517 (a) respectively 0.6689 (b). The ratio between the 2001-2005 AVHRR and VGT data is 0.7217. To reduce this ratio to those obtained prior to the year 2000, it requires reduction factors of 0.9030 (a) respectively 0.9268 (b). The average would be 0.9149.

The applied filter function correction (normalization to NOAA-9) yields correction factors of 0.9385 for the Mediterranean and 0.9346 for central Europe. Here the average is 0.9366. Therefore, if the mean value is taken, it looks like that the filter function corrected NOAA-16 data need in the average an additional reduction of 2.32 %. For central Europe, the values would be 3.38 % (a) or 0.83 % (b). For the Mediterranean region, the values would be either 3.78 % (a) or 1.25 % (b) as compared to 7 % (a) or 3.89 (b) in the case of the comparison of the data for Italy. Thus it is very likely that the reduction factor for the Mediterranen region lies within 3.8 ± 2 % in addition to the normalization to NOAA-9 data.

With these uncertainties in mind, some results of trend analyses derived from the filter function corrected MEDOKADS data set for specific regions will now be discussed. These regions are defined by the land surfaces within the rectangular areas defined in Table A.8.1. Only data which are normalized to NOAA-9 will be used in the following figures. The effect of the normalization for the NOAA-16 data

Table A.8.1. Geographical coordinates of areas for which annual mean values and time series of AVHRR data products have been established

Code	Description	Coordinates
Med W	Western Mediterranean	10°W - 50°E / 30°N - 46°N
Med C	Central Mediterranean	5°E - 30°E / 30°N - 46°N
Med E	Eastern Mediterranean	25°E - 40°E / 30°N - 46°N
Med B	Mediterranean Basin	10°W - 42°E / 27°N - 46°N
SE Spain	South-east of Spain	5°W - 1°W / 35°N - 40°N
Desert	Desert in North Africa	10°W - 42°E / 27°N - 33°N
Europe C	Central Europe	5°E - 30°E / 45°N - 55°N
MEDOKADS	Whole area for which data are available	10°W - 42°E / 27°N - 55°N

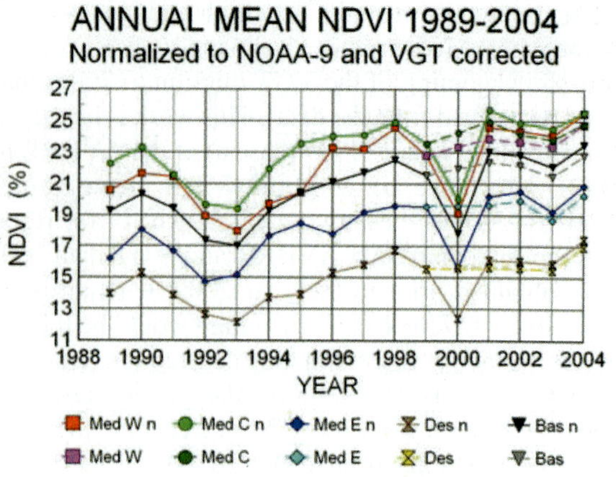

Fig. A.8.12. Mean annual NDVI time series for western, central and eastern Mediterranean as well as of the whole Mediterranean basin and north African deserts normalized to NOAA-9. Thin dashed lines from 1999 to 2004 are the VGT corrected values with an interpolated annual mean value of the year 2000.

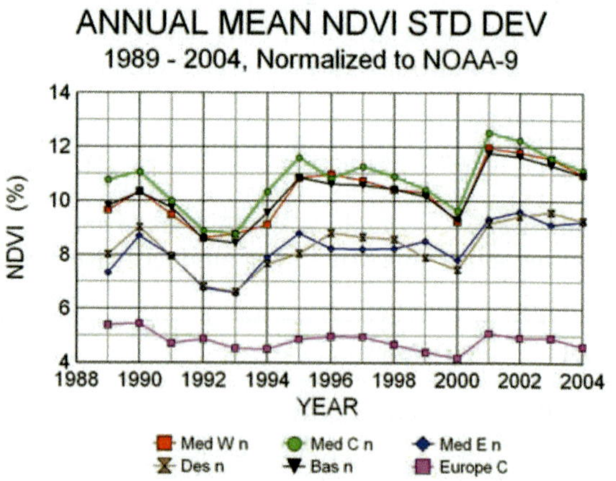

Fig. A.8.13. Standard deviation of the annual means of the normalized NDVI

can be seen in Figs. A.8.7 and A.8.16. Prior to the year 2000 the correction due to normalization is too small to be realized in the diagrams.

Fig. A.8.12 shows Mediterranean annual mean values of the NDVI normalized to NOAA-9 for the years 1989 - 2004. The letters W (red line), C (green line), E(blue line), Des (brown line) and Bas (black line) denote western, central, eastern Mediterranean, north African deserts and the whole basin. The letter "n" in the legend indicates "normalized" in case both normalized and not normalized data are shown. Dashed lines in Figs. A.8.12 and 14 mark data which are additionally decreased by 3.8 %, the estimated "VGT correction", for the years 1999-2004 with an interpolated value for the year 2000.

It can be seen that the VGT correction would bring the NOAA-16 values almost at level with the average of those from 1996 to 1998. In 2004 all values increase. Since this is also the case for deserts, it is likely to be an error, whereas the 2003 dip in all curves except for the deserts seems to be a real drought effect. The NDVIs for central and western Mediterranean are for many years close to each other but the drought of 1995 at the Iberian Peninsula and Morocco shows clearly up in the reduced NDVI value (red line). The standard deviation of the annual values is given for the three Mediteranean regions (west, central and east), desert, basin, and central Europe in Fig. A.8.13. They show the

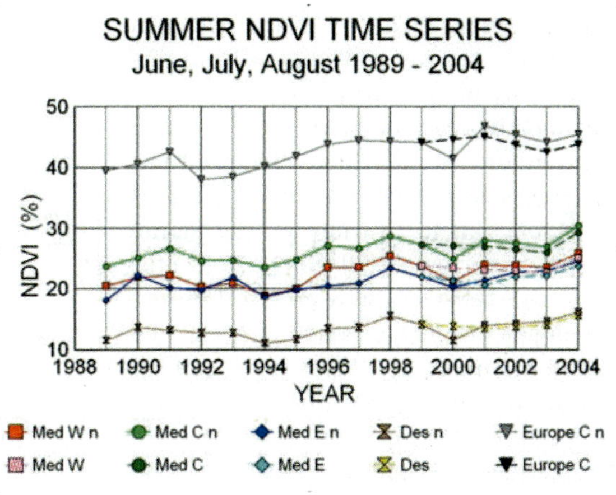

Fig. A.8.14. Summer NDVI values for Mediterranean regions and central Europe (dashed lines stand for spectrally not normalized data)

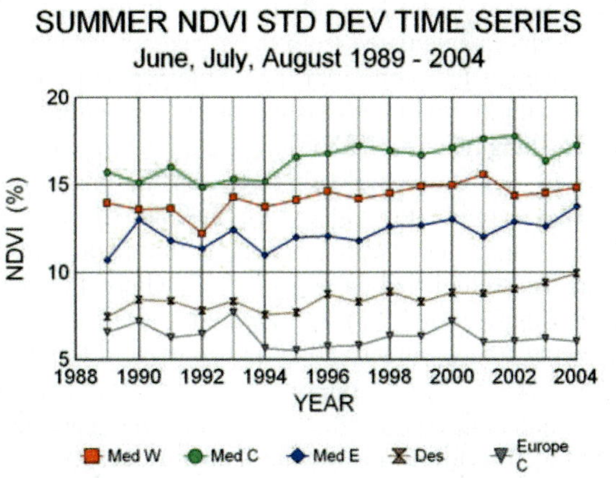

Fig. A.8.15. Standard deviation of the summer NDVI values for Mediterranean regions

large difference between the less variable central Europe and the highly variable northern Mediterranean areas. Desert and the eastern Mediterranean take intermediate positions.

A small increase of the NDVI can be noted between 1990 and 2001. Because this is also the case for the deserts, this probably is an artefact. The depression 1992-1993 is attributed to the Pinatubo eruption.

It is interesting to compare the full year time series with the NDVI time series of the summer months June, July and August (Fig. A.8.14). Except for central Europe (compare Fig. A.8.7) and central Mediterranean the values do not differ much. Table A.8.2 quantifies this for the mean values. Part of the differences may be caused due to the varying illumination geometry. Contrary to the annual mean values, the "Pinatubo effect" does not clearly show up in the summer values. This could be due to the smaller solar zenith angle. The standard deviation (Fig. A.8.15) of western and central Mediterranean as well as central European areas is higher in summer than in the annual mean and relatively constant during the course of the years. This shows that in summer the differences between dry or agricultural areas and forests become larger. For the eastern Mediterranean areas and north African deserts there is little change.

It is obvious from these diagrams that *all* regions follow more or less the same pattern. Even small peaks for specific years show up in almost all lines synchronously. Only for the western region one can see a deviation from this general trend for the 1994/1995 period indicated by a small depression. Furthermore, in all regions the trend has the same size, namely 1% per year. This is very remarkable in view of the stable annual mean values of SPOT VGT for Italy and because one expects different developments in the zones between the northern areas of Africa and central Europe, depending *inter alia* on the North Atlantic Oscillation and the Mediterranean Oscillation (see e.g. Bolle 2001).

Table A.8.2. NDVI mean values 1989-2004 in percent for annual and summer data

Region	Annual	JJA
Med W	22.102	22.388
Med C	23.164	26.205
Med E	18.196	19.829
Europe C	29.350	42.370
Bas	20.680	22.891
Des	14.887	13.377

The question remains to what degree the observed variability of the NDVI may depend on influences so far not taken into account. For that purpose it might be advisable to go back to the primary data measured in AVHRR channel 1 and 2. In Fig. A.8.16 the annual mean values of the reflectances calculated from the signals in channel 1 and 2 as well as the ratio of both signals are plotted for the whole period from 1989 to 2004, in the case of the ratio with and without (dashed lines) correction for the filter function effect. The astonishing result is that the single channel measurements show quite large sawtooth shaped variations but their ratio remains fairly constant. With the start of a new NOAA satellite the values are low and they increase towards the end of the measuring period. This leads to large fluctuations of the albedo but not of the NDVI. This is exemplified in Fig. A.8.17 where the mean annual albedo (calculated according to Eq. 6.9.1) is compared to the annual NDVI for Italy for the years of extreme albedo values and 2002. To make differences more clearly visible, not the annual mean albedo values but their deviation from the long time mean are plotted. The figure underscores that albedo and NDVI are not correlated. This effect must be attributed to the BRDF of the land surfaces. As has been shown earlier (sections 2.4.1 and 4.2.2), observations at increasing solar zenith angles go along with increasing reflectances and albedo. The influence of the filter function correction on the single channels is small but shows up clearly in the ratio. This effect makes it impossible to study broad band reflectance or albedo variations with data not corrected for the BRDF because the fluctuations are for both channels in the same direction and will add up to as well large changes of the albedo. Only one mean value map of the albedo is therefore communicated in Fig. A.8.18.

It should be mentioned here that using the annual mean values of channel 1 and 2 to compute ratios will result in smaller values than presented here. The mean ratios shown in Fig. A.8.16 are computed by summing up and averaging daily values. For example, the annual normalized of the ratio produced from daily data and then averaged from 1989 to 2004 yield a mean value of 1.6195. The ratio computed from the average over the same period of annual mean channel 1 and channel 2 values gives only a value of 1.4315.

The dip in the NDVI values in the year 2000 obviously is caused by a steep reduction of the channel 2 signal while the decrease of the channel 1 signal is smoother. This must be due to a calibration error and should not be considered as a excursion of the surface condition. This was verified by a comparison of AVHRR and SPOT-VEGETATION data for the whole Mediterranean basin which is shown in Fig. A.8.11.

A special case are the data for SE Spain. This area shows a ratio similar to the western Mediterranean area bur the values for the single channels are both very low. This would indicate a very dark surface which is unlikely for this dry area. This case therefore needs additional investigations in the context with the newly in this area

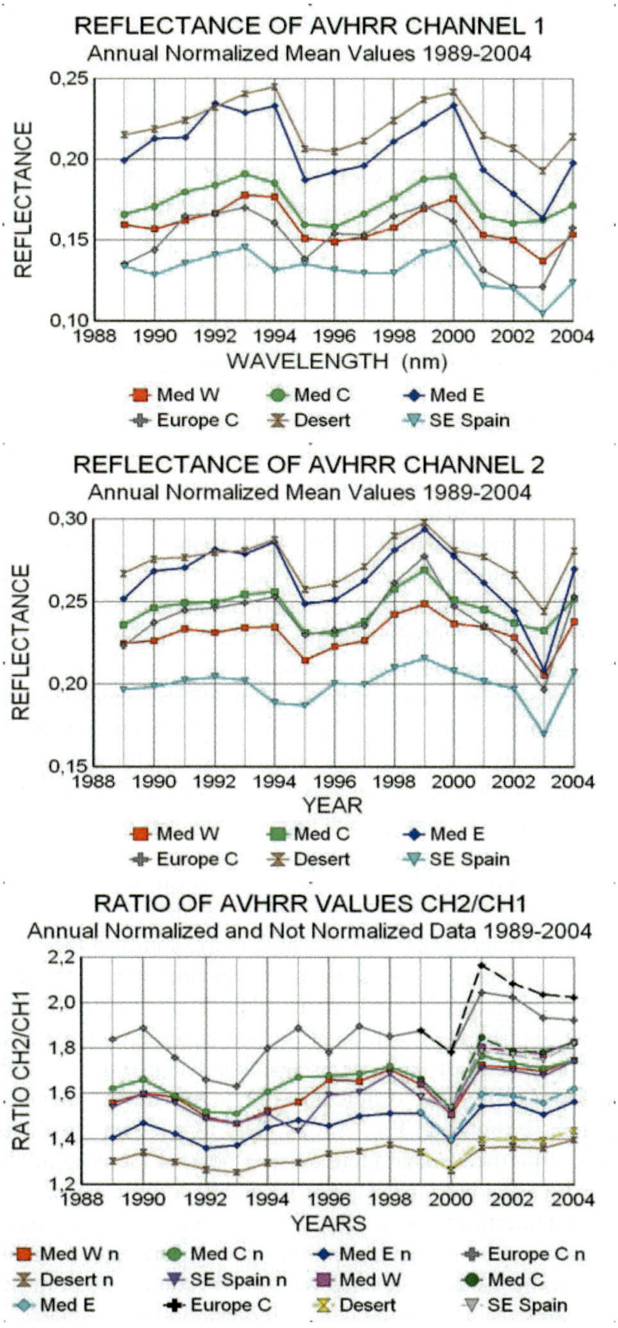

Fig. A.8.16. Time series of June, July and August mean values for the different Mediterranean regions without (full line) and with (broken line) filter function correction

establishe anchor station (section 6.17).

It could be argued that the increase of reflectance from 1996 to 1999 is caused by the volcanos Vatnajökul and Soufrière because small aerosol particles remain in the stratosphere for a couple of years and get more and more evenly distributed as time goes on. However, over high reflectance surfaces it would be expected that the signal is reduced by the aerosol. In view of the large fluctuations of the signal in the individual channels, it becomes difficult to analyse the effect even of the strong eruptions of Pinatubo, Unzen, and Kilauea in 1991. As compared to 1999, in 1992 the channel 2 value indeed is markedly reduced while the channel 1 value remains nearly at the same level. Under the assumption that the reflectance of the surface remains unchanged, increasing turbidity in fact reduces the reflectance over the mostly high albedo in channel 2 more than over the lower albedo which is typical for channel 1. This is observed in 1992.

Unfortunately the AERONET data set has large gaps and data back to 1997 are only available for one station in the Mediterranean area. Maximum monthly averages of the aerosol optical depth range around 0.6 at 500 nm and minimum monthly averages around 0.1. The corresponding values for annual averages are 0.35 and 0.17 or transmittances of 0.7 to 0.84. A couple of stations seem to indicate minimum optical depths at 500 nm in the years 2001 or 2002 as compared to high values in the years 1999, 2000, and 2003. This would correspond to the decline of the reflectance after the year 2000. Increasing turbidity reduces the TOA NDVI. The consequence for the NDVI would be that around 2001/2 enhanced TOA NDVI values should be expected which should drop again in 2003/4. Fig. 8.16 shows that indeed the values are higher in 2001 than in 2000 but they remain at the high level in 2003 and 2004. This clearly does not explain the very steep dip at the year 2000 which goes across all Mediterranean regions and which is just the last year of NOAA-14. Signals from the satellite can only be obtained as long as it is above the horizon. Due to its swath width the instrument can look further east and sees preferably illuminated bright areas with the sun further west (the question of directional reflectance differences is treated *inter alia* from the viewpoint of surface measurements in section A.6).

The mean annual NDVI which corresponds to the albedo shown in Fig. A.8.18 can be seen in Fig. A.8.19 and the development of these values in Fig. A.8.20 and A.8.21. Fig. A.8.20 gives the trend in absolute units ($100 * \Delta$NDVI). This trend may be biassed by the Pinatubo effect and uncertainties in the spectral calibration. If the average trend of the area as documented in Fig. A.8.20 is subtracted from each individual pixel the differences in regional developments can be seen more clearly which is shown in Fig. A.8.21. The map of the development of the NDVI (Fig. A.8.20) indicates that there exists a general increase of about 0.4% per year in the northern Mediterranean countries and no to little effect in North Africa and the Near East. Deserts and desert-like areas remained fairly stable and as such show a small negative relative deviation from the general trend in Fig. A.8.21. The positive deviations from the general trend observed in parts of the eastern Mediterranean are probably due to missing data as indicated in Fig. A.8.7.

If the filter function correction would not have been applied the increase would have been larger. The index of large areas in the northern Mediterranean would be shifted to the blue colour which means to 0.7 and even 0.9 %/a change. This would

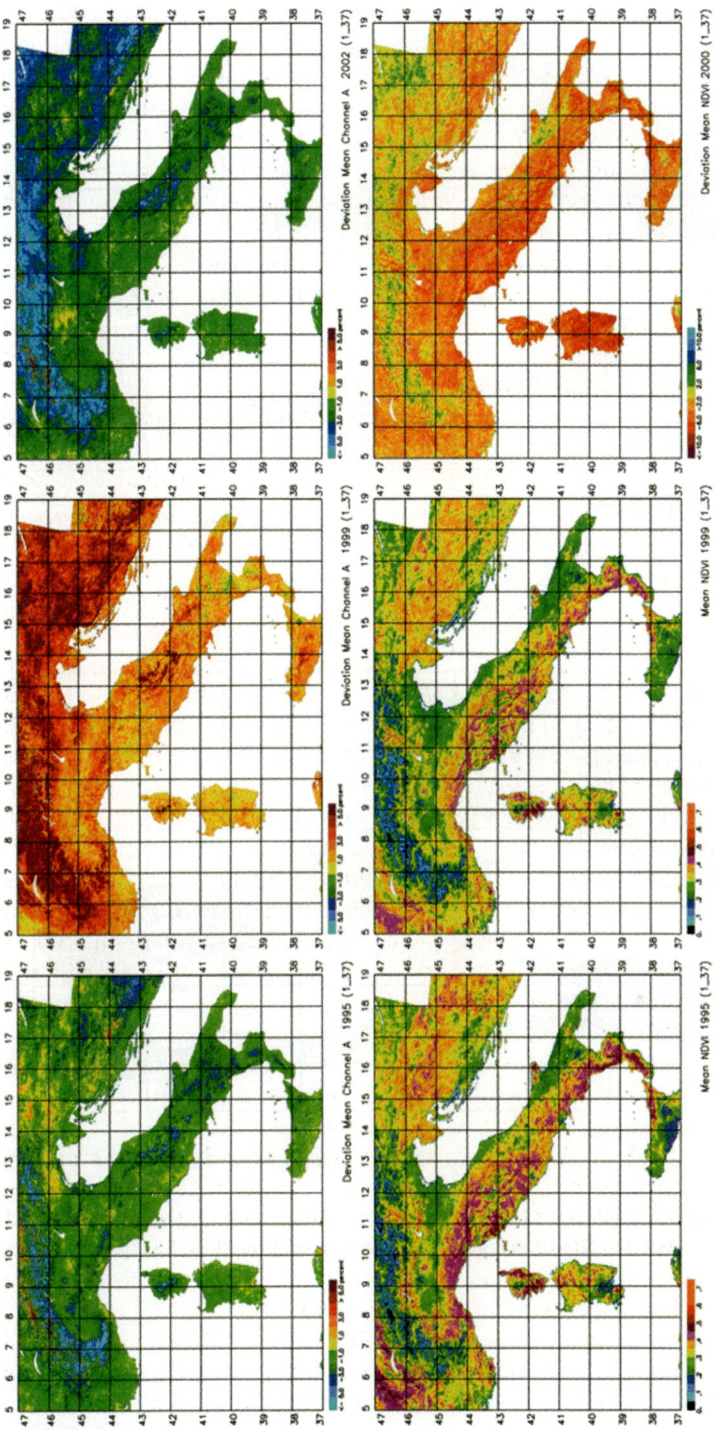

Fig. A.8.17. Comparison of the deviation from the mean of annual albedo (upper row) and NDVI for the years 1995, 1999, and 2002 for Italy showing that the large albedo fluctuations have no effect on the NDVI (the high NDVI values in 2002 may be caused by an insufficient correction of the filter width effect)

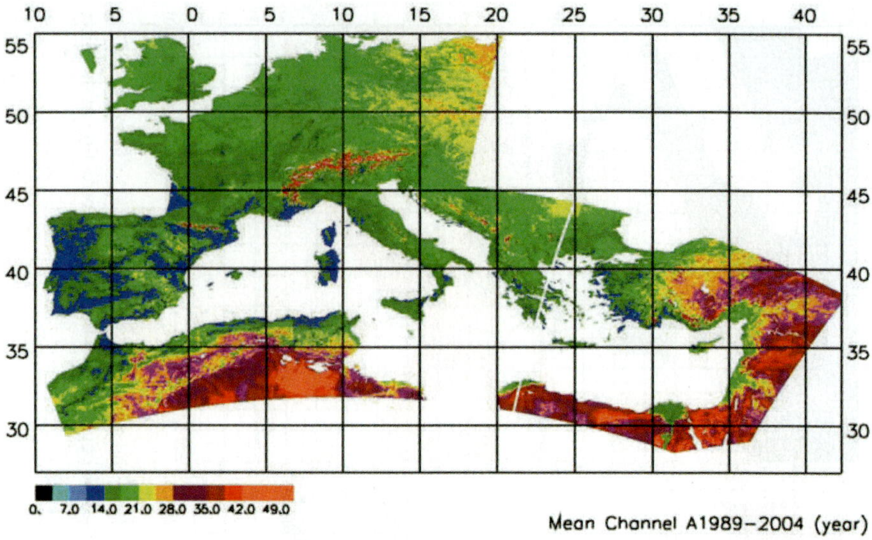

Fig. A.8.18. Albedo mean value for the years 1989 - 2004

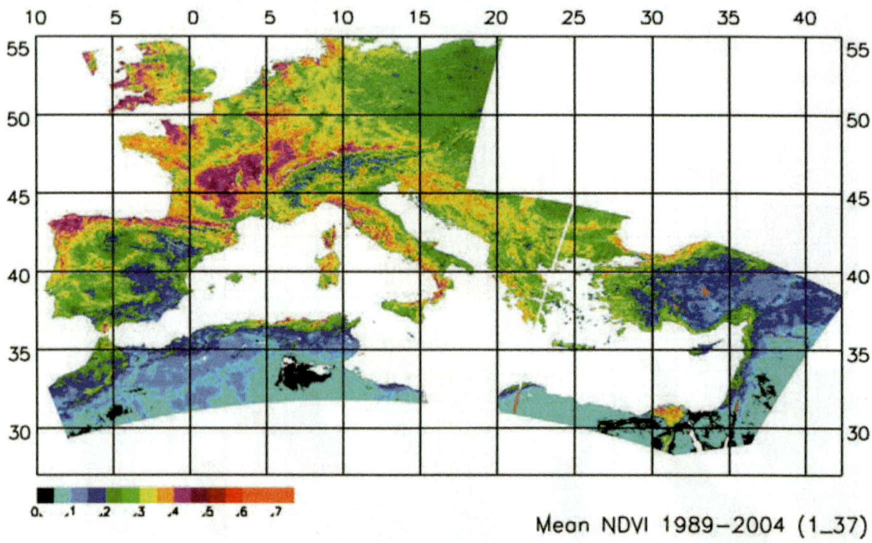

Fig. A.8.19. The mean annual NDVI for the whole Mediterranean averaged from 1989 to 2004 and spectrally normalized to NOAA 9 conditions (corrected for the filter function)

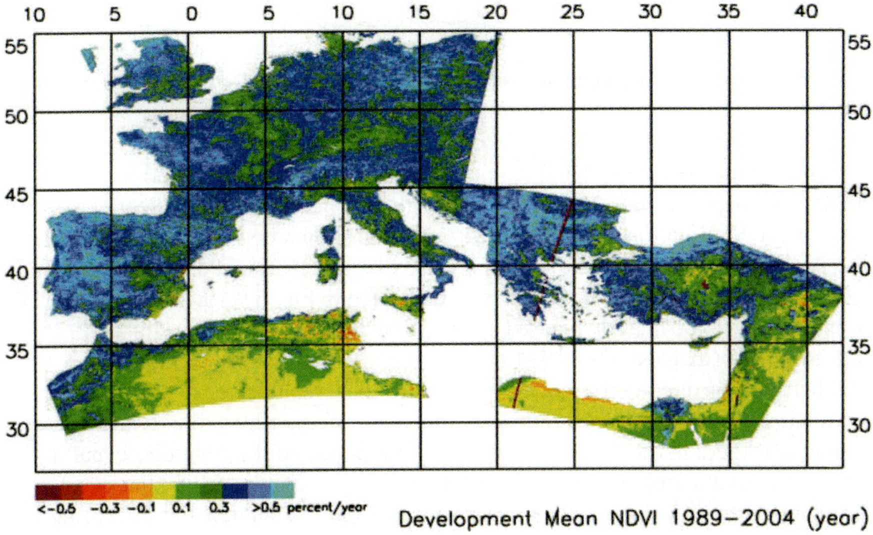

Fig. A.8.20. NDVI development from 1989 to 2004 for central Europe and the Mediterranean in absolute units (100 ∗ΔNDVI)

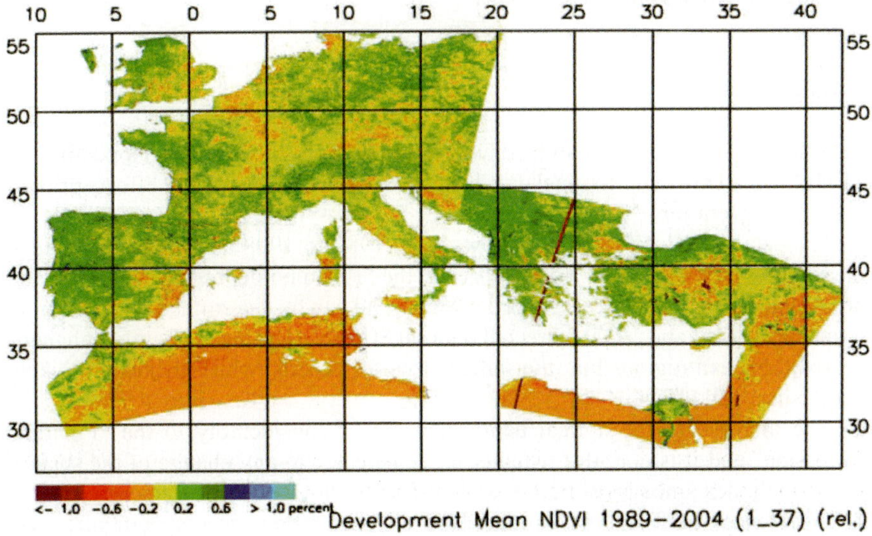

Fig. A.8.21. Mean departures of the local NDVI from the 1989 -2004 mean in relative units (per cent deviation of the local NDVI value change from the area mean value)

have been an unreasonable large change which would also cause the NDVI to rise to 0.7 over large forest areas after the year 2000.

Striking Deviations From Mean Values During the Period from 1989 to 2004

To demonstrate the regional and temporal variability within the period from 1989 to 2004 a few maps are added here to what has already been reported in Section 6.9 and in the preceding section in which the *difference* is shown between the annual NDVI distribution and the 16 years mean values. For the Iberian Peninsula the whole sequence from 1989 to 2004 was already presented in Fig. 6.9.10. The annual NDVI "trend" seems to be heavily biassed by the post Pinatubo eruption years 1992, 1993, and, to a lesser degree, also 1994 (Fig. A.8.22).It can not be decided at this time whether this is due to the aerosol effect or to a real decrease of the biomass production during these two years. Probably both effects play a role. It is interesting to note that there is no or only a very small indication of the Pinatubo event in the data of the summer months (Fig. A.8.12) which seems to indicate that the increase of the channel one signal due to small aerosols over dark vegetated areas in spring and autumn plays a major role.

After the recovery from the Pinatubo event a severe drought occurred in southern Spain and northern Morocco indicated by low NDVIs and high temperatures. A precurser of the drought could already be seen in 1994 (Fig. A.8.23). Strong regional effect can be identified of vegetation degradation in the period 2002 - 2004 around the unusually hot and dry year 2003 (Fig. A.8.24).

Conclusions

The following conclusions can be drawn from the discussions in this Appendix:
(1) MEDOKADS single channel data have to be corrected for BRDF effects prior to further interpretation.
(2) The data series of the NDVI shows from 1989 to 1998 an increasing trend if linear regression is applied. To this "trend" the "Pinatubo depression" of 1992 and 1993 contributes a great deal. Since 1998 no trend can be determined with certainty in the normalized and VGT corrected data set. The 2004 value is slightly enhanced in Mediterranean regions but the same is true for deserts, which makes it unlikely that this is a real effect.
(3) The increase of the optical depth contributed considerably to the "Pinatubo depression" and it is doubtful whether it is connected to any change of the surface vegetation index since measurements over deserts show the same tendency.
(4) Differences between the summer and all-year reflectances specifically over deserts indicate that the varying illumination geometry may influence the data. The seasonal variability of the signal over the desert re-calibration area (Fig. 2.3) may be caused by shadowing effects due to sand dune arrays with a preferred SW to NE direction.

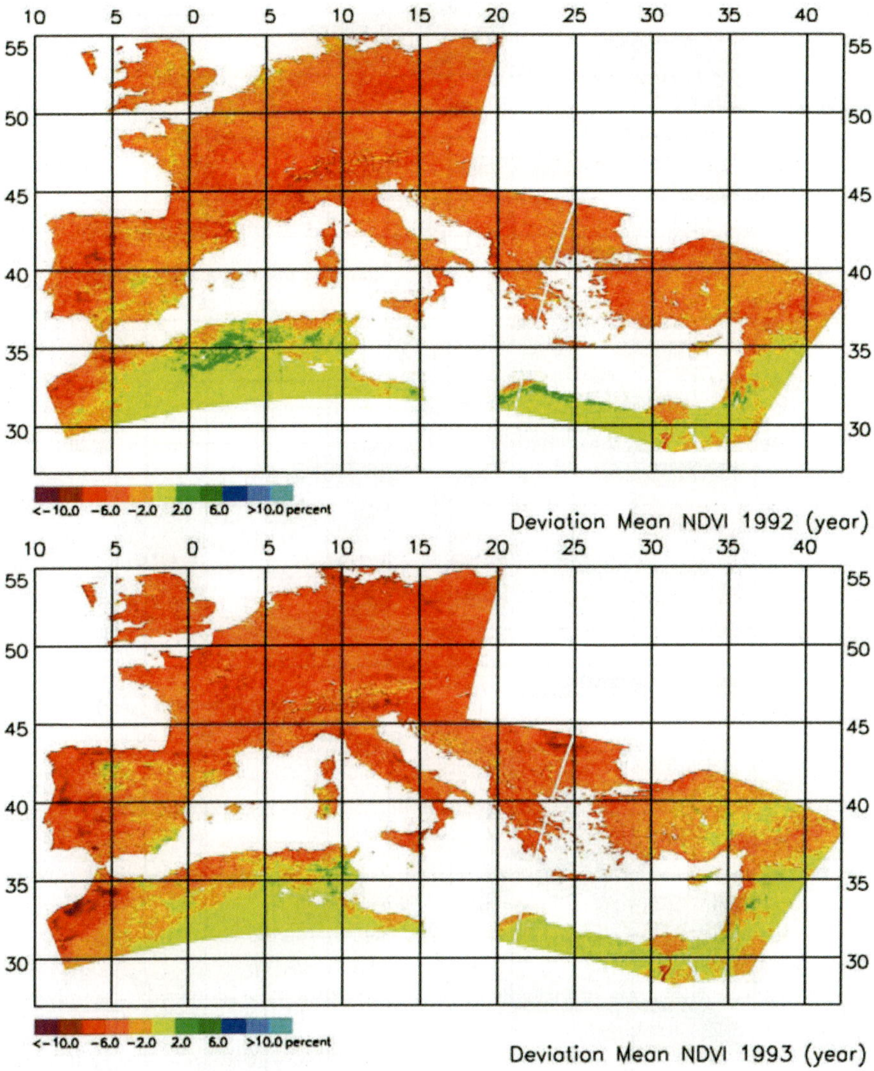

Fig. A.8.22. Deviation of the annual vegetation index from long-term mean in the years 1992 and 1993 after the Pinatubo eruption which biases the trend analysis

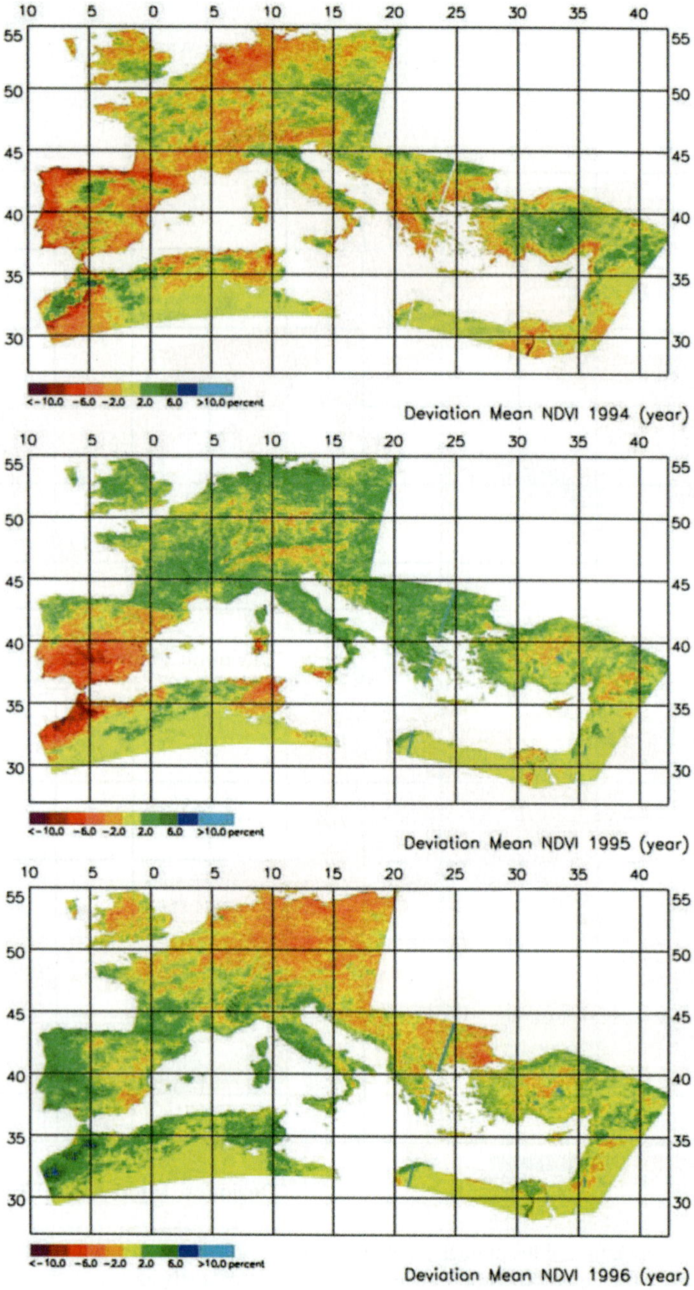

Fig. A.8.23. Deviation of annual NDVI mean values for 1994, 1995, and 1996 from the 16 years mean 1989 - 2004 showing a widespread reduction of the NDVI in 1994 and the drought in Spain and northern Morocco in 1995

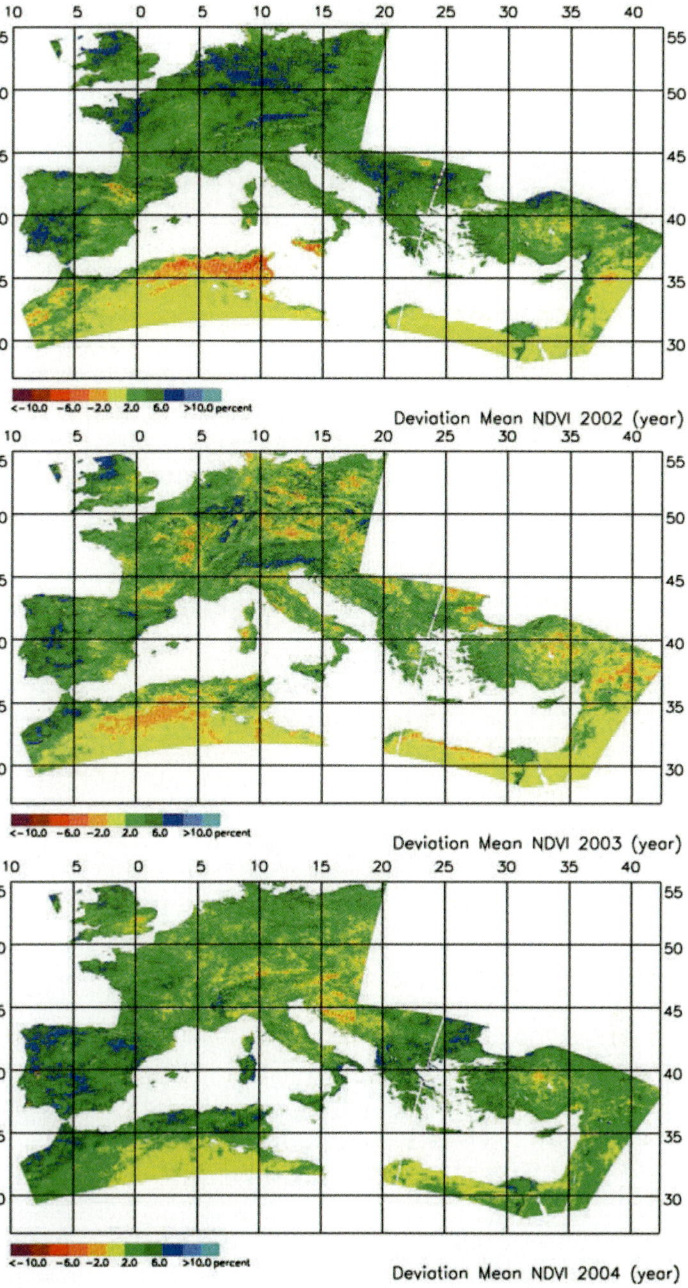

Fig. A.8.24. Deviation of the annual mean NDVI from the 16 years mean in the extremely hot and dry year 2003 (centre) as compared to 2002 and 2004

References

Abdullah A, Cracknell AP, Vaughan RA (1994) Atmospheric correction for thermal infrared data on a pixel basis - Comparison between AVHRR and MOS VTIR data. In: Vaughan RA (ed) Remote Sensing from Research to operational applications in the New Europe. Springer Hungaria, Budapest, pp. 377-384

Abel P (1990) Prelaunch calibration of the NOAA-11 AVHRR visible and near IR channels. Remote Sens. Environ. 31: 227-229.

Abel P, Guenther B, Galimore RN, Cooper JW (1993) Calibration results for NOAA-11 AVHRR channels 1 and 2 from congruent path aircraft observations. J. Atmos. Ocean. Tech. 10: 493-508.

Adler FR, Negri AJ (1988) A Satellite Infrared Technique to Estimate Tropical Convective and Stratiform Rainfall. J. Appl. Meteorology 27:30-51

Ahmad SP, Deering DW (1992) A simple analytical function for bidirectional reflectance. J. Geophys. Res. 97(D17): 18867-18886

Alexandris S, Allen PM, Black I, Blatsou C, Calamaras N, Giannopoulos P, Lemon M, Mimides T, Poulovassilis A, Psyhouyou N, Seaton RAF (1998) Agricultural production and water quality in the Argolid valley, Greece: A policy-relevant study in integrated method. In: Van der Leeuw SE (ed.) Understanding the Natural and Anthropogenic Causes of Soil Degradation and Desertification in the Mediterranean Basin. Volume 6: Synthesis, 281-326, Final Report on Contract EV5V-CT91-0021, ISBN 92-828-3226-0, European Communities, Luxembourg

Allan JA (1990) Sensors, platforms and applications; acquiring and managing remotely sensed data. In Steven MD and Clark JA (eds.) Applications of Remote Sensing in Agriculture. Butterworths, Cambridge, pp. 3-18

Anderson GL, Hanson JD (1992) Evaluating hand-held radiometer derived vegetation indices for estimating above ground biomass. Geocarto International (1) 1992: 71-78.

Anderson MC, Norman JM, Diak GR, Kustas WP, Mecikalski JR (1997) A two-source time-integrated model for estimating surface fluxes using thermal infrared remote sensing. Rem. Sens. Environment 60, 195-216

Anderson TW (1984) An Introduction to Multivariate Statistical Analysis, Second Edition. Wiley and Sons, New York, 102 pp.

Ångström A (1916) Über die Gegenstrahlung der Atmosphäre. Meteorol. Z. 33, 529-538

Annoni A, Perdigao V (1997) Technical and methodological guide for updating CORINE Land Cover Data Base. European Commission, EUR 17288EN, Space Application Institute of Joint Research Centre, Ispra, Italy.

Annual Bulletin on the Climate in WMO Region VI (1999) Deutscher Wetterdienst, Hamburg

Antolin C (1998) Soil as a Natural Resource in the Valencia Autonomous Community. Vol. I . Colección Territori 8, Regional Ministry for Urbanism, Planning and Public Works (C.O.P.U.T) - Generalitat Valenciana, Regional Government of the Valencia Autonomous Community, Valencia, 187 pp. Idem. Vol.II 4 pages. Foldable maps. 74 pages. of plates. + 1CD-ROM.

Arlt H, Bolle H-J (1968) Angle Dependent Reflectivity of Natural Surfaces, 1 - 12 Micron. Final Scientific Report, Investigation of the Infrared Emission Spectrum of the Atmosphere and Earth, Contract AF 61(052)-778, Air Force Cambridge Research Laboratories, OAR

Armesto JJ, Picket STA (1985) Experiments of disturbance in old-field plant communities: impact on species richness and abundance. Ecology 66(1): 230-240.

Ashcroft PM, Catt JA, Curran PJ, Munden J, Webster R (1990) The relation between reflected radiation and yield on the Broadbalk winter wheat experiment, Int. J. Remote Sens. 11(10): 1821-1836.

Asrar G (1992) Theory and applications of optical remote sensing. New York, John Wiley & Sons. 734

pp.
Bach H, Demircan A, Mauser W (1994) The use of AVIRIS data for the determination of agricultural plant development and water content. In: Wooding M, Lodge F, Attema E (eds.) MAC Europe '91 Final Results Workshop. European Space Agency, Paris, 254 pp.
Bannari A, Morin D, Bonn F, Huete AR (1995) A review of vegetation indices. Rem. Sens. Rev. 13: 95-120
Baret F, Guyot G, Major DJ (1989) Crop biomass evaluation using radiometric measurements. Photogrammetria (PRS), 43: 241-256
Baret F (1995) Use of spectral reflectance variation to retrieve canopy biophysical characteristics. In: Danson FM, Plummer SE (eds) Advances in Environmental Remote Sensing. John Wiley & Sons, Chichester, UK, ch. 3: 33-51.
Baret F, Olioso A (1989) Estimation à partir de mesures de réflectance spectrale du rayonnement photosynthétiquement actif absorbé par une colture de blé. *Agronomie,* 9: 885-895.
Baret F, Guyot G (1991) Potentials and limits of vegetation indices for LAI and APAR assessment, Remote Sens. Environ. 35: 161-173.
Baret F, Jacquemoud S, Guyot G, Leprieur C (1992) Modeled analysis of the biophysical nature of spectral sgifts and comparison with information content of broad bands. Remote Sens. Environment 41: 133-142
Barrett EC, D'Souza G, Power CH (1986) Bristol Techniques for the Use of Satellite Data in Raincloud and Rainfall Monitoring. J. Br. Interplanet. Soc. 39: 517-526
Barrett EC (1989) Satellite remote sensing of rainfall. In: Toselli F (ed.) Applications of Remote Sensing to Agroclimatology. Kluwer, Dordrecht, pp 305-326
Barth HK, Runge J (1992) Land use and vegetation of the pilot areas of the EFEDA-project La Mancha, Spain. Institute of Physical Geography University of Paderborn, Germany
Bastiaanssen WGM (1995) Regionalisation of surface flux densities and moisture indicators in composite terrain - A remote sensing approach under clear skies in Mediterranean climates. Ph.D. thesis, Agricultural University, Wageningen, The Netherlands, 273 pp. [Available from DLO-Staring Centre, PO Box 125, 6700 AE Wageningen, The Netherlands.]
Bastiaanssen WGM, Roebeling RA (1993) Analysis of land-surface exchange in two agricultural regions in Spain using Thematic Mapper Simulator data. In: Bolle H-J, Feddes RA, Kalma JD, Exchange Processes at the Land Surface for a Range of Space and Time Scales. IAHS Publication No. 212, Institute of Hydrology, Wallingford, Oxfordshire, UK
Bastiaanssen WGM, Hoekman DH, Roebeling RA (1994) A methodology for the assessment of surface resistance and soil water storage variability at mesoscale based on remote sensing measurements. IAHS Special Publication no. 2, IAHS Press, Institute of Hydrology, Wallingford, UK, pp 63
Bastiaanssen WGM, Menenti M, Dolman AJ, Feddes RA, Pelgrum H (1996a) Remote sensing parametrization of mesoscale land-surface evaporation. In: Raschke E (ed.) Radiation and Water in the Climate System: Remote Measurements. NATO ASI Series Vol. I 45, Springer-Verlag Berlin Heidelberg
Bastiaanssen WGM, Van der Wal T, Visser TNM (1996b) Diagnosis of regional evaporation by remote sensing to support irrigation performance assessments. Irrigation and Drainage Systems 10, no1: 1-23
Bastiaanssen WGM, Pelgrum H, Droogers P, de Bruin HAR, Menenti M (1997) Area-average estimates of evaporation, wetness indicators and top soil moisture during two golden days in EFEDA. Agricultural and Forest Meteorology 87: 119-137
Bastiaanssen WGM, Menenti M, Feddes RA, Holtslag AAM (1998a) The Surface Energy Balance Algorithm for Land (SEBAL): Part 1 formulation. J. Of Hydrology 212-213: 198-212
Bastiaanssen WGM, Pelgrum H, Wang J, Ma Y, Moreno J, Roerink GJ, van der Wal T (1998b) The Surface Energy Balance Algorithm for Land (SEBAL): Part 2 validation. J. of Hydrology 212-213: 213-229
Bastin GN, Pickup G, Pearce G (1995) Utility of AVHRR data for land degradation assessment: A case study. Int. Journ. Remote Sensing 16: 651-672
Bausch WC (1993) Soil background effects on reflectance based crop coefficients for corn. Remote Sens. Environ., 46: 213-222
Bazoffi P, Boscagli A, Brandi G, Busoni E, Calzolari C, Chiarucci A, Chiaverini I, Chisci G, Colica A, De Dominicis V, Del Sette M, Grasselli O, Lulli L, Maccherini S, Mazzanti R, Morandi M, Papini R, Pellegrini S, Rocchini A, Rodolfi G, Torri D, Vecchio B, eds.: Rodolfi G, Torri D (1997) Suppl. Geogr. Fis. Dinam. Quat. III, T2, pp. 151-166

Becker F, Nerry F, Ramanantsizehhena, Stoll MPH (1986) Mesure d'Emissivite Angulair par Reflexion dans l'Infrarouge Thermique - Implications pour la Teledetection. Int. J. Remote Sensing 7(2): 1751-1762

Becker F, Bolle H-J, Rowntree PR (1988) The International Satellite Land-Surface Climatology Project. ISLSCP-Report No. 10. ISLSCP Secretariat

Becker F, Choudhury BJ (1988) Relative sensitivity of Normalized Difference Vegetation Index (NDVI) and Microwave Polarization Difference Index (MPDI) for vegetation and desertification monitoring. Remote Sensing of Environment, Vol. 24: 297-311.

Becker F, Li Z-L (1990a) Temperature Independent Spectral Indices in Thermal Infrared Bands. Remote Sens. Environ. 32: 17-33

Becker F, Li Z-L (1990b) Towards a local split window method over land surface. International Journal of Remote Sensing, 11:369-394

Becker F, Li ZI (1993) Infrared remote Sensing of Surface Temperature and Surface Spectral Emissivities. In: Chédin A, Chahine MT, Scott NA (eds.) High Spectral resolution Infrared Remote Sensing for Earth's Weather and Climate Studies. Springer-Verlag, Heidelberg, pp 265-284

Becker F, Li Z-L (1995) Surface temperature and emissivity at different scales: definition, measurement and related problems. Remote Sensing Reviews, 12:225-253

Belda F (1997) Climatologia y teledeteccion en zonas forestales de la provincia de Alicante. Aplicacion a zonas incendiadas. Ph.D. Thesis. University of Valencia. Spain.

Belda F, Alvarez A, y Valcarcel M (1992) Situaciones sinópticas meteorológicas en los incendios forestales de la provincia de Valencia. I Congreso Iberoamericano sobre técnicas aplicadas a la Gestión de emergencias para la reducción de desastres naturales. Universidad Politécnica de Valencia.

Bell EE (1957) An atlas of reflectivities of some common types of materials. Inter. Engineering Report 659-6. Wrigth Air Development Center. Contract AF 33(616)-3312

Bemporad A (1904) Zur Theorie der Extinktion des Lichtes in der Erdatmosphäre. Mitt. Großherz. Sternwarte Heidelberg No. 4

Benedetti R, Rossini P (1993) On the use of NDVI profiles as a tool for agricultural statistics: The case study of wheat yield estimate and forecast in Emilia Romagna. Remote Sens. Environ. 45: 311-326.

Benedetti R, Rossini P, Taddei R (1994) Vegetation classification in the Middle Mediterranean area by satellite data. Int. J. Remote Sensing 15(3): 583-596

Berger M (1996) Flächenmittel der Energieflüsse in einem semi-ariden Testgebiet. Dissertation zur Erlangung des Grades Doktor der Naturwissenschaften, Freie Universität Berlin, pp. 145, Berlin, and (1997) Geowissenschaften, ShakerVerlag

Berger FH (1992) Die Bestimmung des Einflusses von hohen Wolken auf das Strahlungsfeld und das Klima durch Analyse von NOAA AVHRR-Daten. Dissertation, Freie Universität Berlin, Institut für Meteorologie, Wiss. Met. Abh., Neue Folge, Serie A 6,3.

Bertram F-W (1967) Die thermische Emission der Atmosphäre im Spektralbereich zwischen 3.1 und 4.2 µm. Pageoph 68: 196-213

Betts AK, Ball JH, Beljaars ACM, Miller MJ, Viterbo P (1996) The land surface-atmosphere interaction: a review based on observational and global modeling perspectives. J. Geoph. Res.101,no D3: 7209-7225.

Bill RG Jr, Cook AF, Allen LH Jr, Bartholic JF (1980) Predicting Fluxes of Latent and Sensible Heat of Lakes From Surface Water Temperatures. J. Geophys. Res. 85, C1: 507-512

Bitelli G, Carrara A, de Torres Curth MS, Folloni G (1993) Topographical database for EFEDA GIS. In: Bolle H-J, Streckenbach B (eds.) EFEDA Final Report Contract EPOC-CT 90-0030 and Annex to EFEDA Final Report Contract EPOC-CT 90-0030, Berlin, August 1993

Blümel K (1998) Estimation of sensible heat flux from surface temperature wave and one-time-of-day air temperature observation. Boundary-Layer Meteorology 86: 193-232

Blümel K (1992) Ableitung von Landoberflächendaten aus Satellitenmessungen, ihre Validierung und Nutzung in der Klimamodellierung - Teil II Modellierung der Wärmeflüsse am Erdboden mit Berücksichtigung von Vegetation. Abschlußbericht, BMFT, Föderungskennzeichen LOF 32/87 - 07 KF 114-6

Bneder (1994) Carte de Synthèse des grands ensembles agro-pedo-climatiques 1:1.000.000. Bureau national d'Etudes de développement rurale, Alger, Algerie.

Bolle H-J (1982) Radiation and energy transport in the earth atmosphere system. In: Hutzinger O (ed) Handbook of Environmental Chemistry, vol. 1/part B. Springer-Verlag, Berlin-Heidelberg.

Bolle H-J (1990) In: Dugdale et al. (Eds.) Remote Sensing of Energy and Mass Exchanges at the Surface of the Earth, Twinning project no. ST 2 J 0222, Final Report

Bolle H-J, Andre J-C, Arrue JL, Barth HK, Bessemoulin P, Brasa A, de Bruin HAR, Cruces J, Dugdale G, Engman ET, Evans DL, Fantechi R, Fiedler F, van de Griend AA, Imeson AC, Jochum A, Kabat P, Kratzsch T, Lagouarde JP, Langer I, Llamas R, Lopez-Baeza E, Melia Miralles J, Muniosguren LS, Nerry F, Noilhan J, Oliver HR, Roth R, Saatchi SS, Sanchez Diaz J, de Santa Olalla M, Shuttleworth WJ, Sogaard H, Stricker H, Thornes J, Vauclin M, Wickland D (1993) EFEDA: European field experiment in a desertification-threatened area. Annales Geophysicae, 11: 173-189.

Bolle H-J, Streckenbach B (eds.) (1993) EFEDA Final Report Contract EPOC-CT 90-0030, Berlin and Annex to EFEDA Final Report Contract EPOC-CT 90-0030, Berlin

Bolle H-J (1996) The role of remote sensing in understanding and controlling land degradation and desertification processes: The EFEDA research strategy. In: Hill J and Peter D (eds) The use of remote sensing für land degradation and desertification monitoring in the Mediterranean basin; state of the art and future research. European Commission, Directorate-General XIII, Luxembourg, EUR 16732,1996, pp 45-78.

Bolle H-J (1998) The EFEDA Project. In: Mairota P, Thornes JB and Geeson N (eds.) Atlas of the Mediterranean Environments in Europe. John Wiley & Sons, Chichester, ISBN 0-471-96092-6, pp.12-15.

Bolle H-J, ed (2003) Mediterranean Climate - Variability and Trends. Springer-Verlag, Berlin Heidelberg New York, ISBN3540438386, pp 372

Bolle H-J (2003) Climate and Climate History of the Mediterranean Basin. In: Brauch HG, Liotta PH, Marquina A, Rogers PF, El-Sayed Selim M (Eds) Security and Environment in the Mediterranean, Springer, Berlin, Heidelberg, 1134 pp

Bolz HM, Falkenberg G (1949) Neubestimmung der Konstanten der Ångströmschen Strahlungsformel. Z. Meteorologie 3: 97-100

Borgeaud M, Noll J (1994) Analysis of Theoretical Surface Scattering Models for Polarimetric Microwave Remote Sensing of Bare Soils. Int. J. Remote Sensing 15(14): 2931-2942

Borgerfors G (1986) Distance transformations in digital images. J. Appl. Meteorology 32: 1352-1364

Born M, Wolf E (1959) Principles of Optics. Pergamon Press, London, 803 pp.

Botkin DB, Running SW (1984) Role of vegetation in the biosphere. Machine Processing of Rem.13: 313-327

Bouttier F, Mahfouf J-F, Noilhan J (1993a) Sequential assimilation of soil moisture from atmospheric low-level parameters.Part I: Sensitivity and calibration studies. J. Applied Meteorol. 32: 1335-1351.

Bouttier F, Mahfouf J-F, Noilhan J (1993b) Sequential assimilation of soil moisture from atmospheric low-level parameters.Part II: Implementation in a mesoscale model. J. Applied Meteorol. 32: 1352-1364.

Brenner AJ, Cueto Romero M, García-Haro J, Gilabert MA, Incoll LD, Martínez-Fernández J, Porter E, Puignaire FI, Younis MT (1995) A comparison of direct and indirect methods for measuring leaf and surface areas of individual bushes. Plant, Cell and Environment, 18: 1332-1340.

Bach H, Demircan A, Mauser W (1994) The use of AVIRIS data for the determination of agricultural plant development and water content. In: Wooding M, Lodge F, Attema E (eds.) MAC Europe ′91 Final Results Workshop. European Space Agency, Paris, 254 pp.

Braud I, Noilhan J, Bessemoulin P, Mascart P, Haverkamp R., Vauclin M (1993) Bare-ground surface heat and water exchanges under dry conditions: Observations and Parameterization. Boundary Layer meteorology 66: 173-200

Brest CL, Goward SN (1987) Deriving surface albedo measurements from narrow band satellite data, Int. J. Remote Sensing 8 (3): 351-367.

Briggs S, Settle J, Gallagher F, Groom G, Wyatt F (1995) Spatial resolution of PRISM. Final Report, ESA/ESTEC contract 135/015, pp 48 + Annexes

Brisson N, Mary B, Ropoche D, Jeuffroy MH, Ruget F, Nicoullaud B, Gate P, Devienne F, Antonioletti R, Düürr C, Richard G, Beaudoin N, Recous S, Tayot X, Pléénet D, Cellier P, Machet JM, Meynard JM, Deléécolle R (1998) STICS: A generic model for the simulation of crops and their water and nitrogene balance. I. Theory and parameterization applied to wheat and corn. Agronomie 18: 311-346 (Part II, Validation, 2002, Agronomie 22: 69-92)

Brown J, Loveland TR, Merchant JW, Reed BC, Ohlen DO (1993) Using Multisource Data in Global Land-Cover Characterization: Concepts, Requirements, and Methods. Photogrammetric Engineering & Remote Sensing, 59: 77-987.

Bruckert D, Stoll MP, Kologo N, Nerry F (1995) A PC based Automatic AVHRR and Meteosat Receiving

Station Developed at "University Louis Pasteur-Strasbourg". An Example of Application. Proceedings of the 1995 Meteorological Satellite Data User's Conference, Winchester, United Kingdom, September 4-8, p. 594

Brunfeldt DR, Ulaby FT (1984b) Measured microwave emission and scattering in vegetation canopies. IEEE Transactions on Geoscience and Remote Sensing, GE-22(6): 520-524.

Brunt D (1932) Notes on radiation in the atmosphere: I. Quart. J.Roy. Met. Soc. 58: 389-420

Brutsaert W (1982, 1984 2nd ed.) Evaporation into the Atmosphere - Theory, History, and Applications. D. Reidel Publishing Company, Dordrecht/Boston/Lancaster, ISBN 90-277-1247-6, pp 299.

Bryson RA, Baerreis DA (1967) Possibilities of major climatic modification and their implications: Northwest India, a case for study. Bull. Am. Met. Soc. 48: 136-142

Budyko MI (1974) Climate and Life. Academic Press, N.Y., 508 pp.

Buettner KJK, Kern CD (1965) The Determination of Infrared Emissivities of Terrestrial Surfaces. J. of Geophysical Research 70, No. 6: 1329-1337

Buzzi A., Fantini M, Malguzzi P, Nerozzi F (1994) Validation of a limited area model in cases of Mediterranean cyclogenesis: surface fields and precipitation scores. Meteorol. Atmos. Phys. 53: 137-153

Buzzi A., Tartaglione N, Malguzzi P (1998) Numerical simulations of the 1994 Piedmont flood: Role of orography and moist processes. Mon. Wea. Rev., 126: 2369-2383.

Cabannes J (1921) Sur la diffusion de la lumière par les molécules des gaz transparents. Ann. Phys, 15: 5-150

Calvet J-C, Noilhan J-L, Roujean J-L, Bessemoulin P, Cabelguenne M, Olioso A, Wigneron J-P (1998) An interactive vegetation SVAT model tested against data from six contrasting sites. Agricultural and Forest Meteorology 92, pp. 73-95

Carlà R, Carrara A, Federici G (1987) Generazione di modelli digitali del terreno as alta precisione. Quad. Dip. Ing. Civile, Univ. Firenze, 0/87, Firenze

Carslaw HS, Jaeger JC (1959) Conduction of Heat in Solids. Oxford University Press, New York, 2nd Edition

Carrara A (1988) Drainage and divide networks derived from high-fidelity digital terrain models. In: Chung CF et al.(eds.) Quantitative Analysis of Mineral and Energy Resources. Deidel Publishing Company, pp 581-597

Carrara A, Catalano E, Sorriso-Valvo M, Reali C, Osso I (1978) Digital terrain analysis for land evaluation. Geol. Appl. Idrogeol. 13

Caselles V, J A Sobrino, F Becker (1988) Determination of the effective emissivity and temperature under vertical observation of a citrus orchard. Application to frost nowcasting, Int. J. Remote Sensing 9: 715-727

Castellari S, Pinardi N, Leaman K (1998) A model study of air-sea interactions in the Mediterranean Sea. J. of Marine Systems 18: 89-114

CEC (1993) CORINNE Land Cover technical guide. European Union, Directorate General Environment, Nuclear Safety and Civil Protection, Luxembourg, ISBN 92-826-2578-8

Chanzy A, Bruckler L, Perrier A (1995) Soil evaporation monitoring: A possible synergism of microwave and infrared remote sensing. J. of Hydrology 165: 235-259

Chandrasekhar S (1950) Radiative Transfer. Oxford University Press, Oxford, UK

Charney JG (1975) Dynamics of deserts and drought in the Sahel. Quart. J. R. Met. Soc. 101: 193-202

Charney JG, Stone PH, Quirk WJ (1975) Drought in the Sahara: A biogeophysical feedback mechanism. Science 187: 434-435

Chedin A, Scott N, Berroir A (1982) A single-channel double viewing method for SST determination from coincident Meteosat and TIROS-N measurements. Journal of Applied Meteorology, 21: 613-618.

Chen MF, Fung AK (1988) A numerical study of the regions of validity of the Kirchhoff and small perturbation rough surface scattering models. Radio Science 23(2): 163-170

Cermak J, Deml M, Penka M (1973) A new method of sap flow rate determination in trees. Biol.Plant. (Praha) 15: 171-178.

Cermak J, Ulehla J, Kucera J, Penka M (1982) Sap flow rate and transpiration dynamics in the full-grown oak (*Quercus robur* L.) in the floodplain forest exposed to seasonal floods as related to potential evapotranspiration and tree dimensions. Biol. Plant. (Praha) 24: 446-460.

Cermak J, Kucera J (1990) Scaling up transpiration data between trees, stands and watersheds. Silva Carelica 15: 101-120.

Cermak, J., Nadezhdina, N., Raschi, A., Tognetti, R., 1998. Transpiration of typical natural plant stands

in Tuscany. Part of the project EC "MEGARICH"- ENV4/CT97/0503.
Cihlar J, Manak D, Voisin N (1994) AVHRR bidirectional reflectance effects and compositing, Remote Sens. Environ. 48: 77-88
Chiesi M, Maselli F, Bindi M, Fibbi L, Bonora L, Raschi A, Tognetti R, Cermak J, Nadeshdina N (2002) Calibration and application of FOREST-BGC in a Mediterranean area by the use of conventional and remote sensing data. Ecological Modelling 154: 251-262
Choudhury B, Kerr YH, Njoku EG, Pampaloni P, eds. (1995) Passive microwave remote sensing of land-atmosphere interactions, VSP Int. Science Publishers, Utrecht, The Netherlands, pp. 509-530
Choudhury BJ, Schmugge TJ, Newton RW, Chang A (1979) Effect of surface roughness on the microwave emission from soils. Journal of Geophysical Research, 84: 5699-5706
Christensen JH, van Meijgaard E (1992) On the constructionof a regional atmospheric climate model. KNMI Scientific Report No.147, 22 pp.
Chuvieco, E. and Congalton, R.G. (1988) Mapping and inventory of forest fires from digital processing of TM data. Geocarto International 3: 41-53.
Clapp RB, Hornberger GM (1978) Empirical relations for some soil hydraulic properties. Water Resour. Res. 14: 601-604
Claussen M, Gayler V (1997) The greening of the Sahara during the mid-Holocene: results of an interactive atmosphere-biosphere model. Global Ecol. Biogeogr. 6: 369-377
Clothier BE, Clawson KL, Pinter Jr. PJ, Moran MS, Reginato RJ, Jackson RD (1986) Estimation of soil heat flux from net radiation during the growth of alfalfa. Agric. Forest Meteorology 37: 319-329
Coll C, Caselles V (1997) A split window algorithm for land surface temperature from advanced very high resolution radiometer data: Validation and algorithm comparison. J. Geophys. Res. 102: 16697-16713
Coll C, Casalles V, Sobrino JA, Valor E (1994) On the atmospheric dependence of the split-window equation for land surface temperature. International Journal of Remote Sensing, 15:105-122.
Collin JJ (1995) Some aspects of desertification seen from the ground water viewpoint. In Fantechi R, Peter D, Balabanis P, Rubio JL (eds.) Desertification in a European context: Physical and socio-economic aspects. European Commission EUR 15415: 111-121
Conese C, Maselli F (1994) Evaluation of Contextual, Per-pixel and Mixed Classification Procedures Applied to a Sub-Tropical Landscape. Remote Sensing Reviews 9: 175-186
CoSMOS Experimenters Team, 2005, CoSMOS-2005 Campaign Experimenters Handbook, edited by RSAC, UK
Cox, Peter M (2001) Nature 408 (6809) 184-187
Cramer W, Bondeau A, Woodward FI, Prentice IC, Betts RA, Brovkin V, Cox PM, Fisher V, Foley JA, Friend AD, Kucharik C, Lomas MR, Ramankutty N, Sitch S, Smith B, White A, Young-Molling C (2001) Global response of terrestrial ecosystem structure and function to CO_2 and climate change: results from six dynamic global vegetation models. Global Change Biol. 7: 357-373
Croke B, Cleridou N, Kolovos A, Vardavas I, Papamastorakis J (2000) Water resources in the desertification-threatened Messara Valley of Crete: Estimatiion of the annual water budget using a rainfall-runoff model. Environmental Modelling & Software 15: 387-402
Danson FM (1995) Developments in the remote sensing of forest canopy structure. In Danson FM, Plummer SE (eds) Advances in Environmental Remote Sensing, John Wiley & Sons, Chichester, UK, ch. 4: 53-69.
Daughtry CST, Gallo KP, Goward SN, Prince SD, Kustas WP (1992) Spectral estimates of absorbed radiation and phytomass production in corn and soybean canopies. Remote Sens. Environ. 39: 141-152.
Davenport ML, Nicholson S E (1993) On the relation between rainfall and the Normalized Vegetation Index for diverse vegetation types in East Africa. International Journal of Remote Sensing, 14: 2369-2389.
Demetriades-Shah TH, Steven MD, Clark JA (1990) High resolution derivative spectra in remote sensing, Remote Sens. Environ. 33: 55-64.
Dash P, Göttsche F-M, Olesen F-S, Fischer, H (2002), Land surface temperature and emissivity estimation from passive sensor data: theory and practice- current trends. International Journal of Remote Sensing, 23: 2563-2594.
De Bruin HAR (2002) Introduction: Renaissance of Scintillometry. Boundary-Layer Meteorology 105: 1-4
De Bruin HAR, Bink NJ, Kroon LJM (1991) Land Surface Evaporation. In: Schmugge and Andre (eds) Fluxes in the Surface Layer Under Advective Conditions. Springer N.Y.
De Bruin HAR, Kohsick W, van den Hurk BJJM (1993) A verification of some methods to determine the

fluxes of momentum, sensible heat, and water vaour using standard deviation and structure parameter of scalar meteorological quantities. Boundary-Layer Meteorology 63: 231-257

De Bruin HAR, van den Hurk BJJM, Kohsiek W (1995) The scintillation method tested over a dry vineyard area. Boundary Layer Meteorology 63: 231-257

De Dios del Pino Corredera J (1997) Análisis de homogeneidad y variabilidad de la serie de precipitación de Murcia. In: Marzol MV, Dorta P, Valladares, eds. Clima y aqua: La gestión de un recurso climático. La laguna, Madrid

Deering DW, Middleton EM, Irons JR, Blad BL, Walter-Shea EA, Hays CJ, Walthall C, Eck TF, Ahmad SP, Banerjee BP (1992a) Prairie grassland bidirectional reflectances measured by different instruments at the FIFE site, J. Geophys. Res. 97(D17):18887-18903.

Deering DW, Eck TF, Otterman J (1990) Bidirectional reflectances of selected desert surfaces and their three-parameter soil characterisation. Agricultural and Forest Meteorology 52: 71-93

Derrien M, Farki B, Harang L, LeGléau H, Noyalet A, Pochic D, Sairouni A (1993) Automatic cloud detection applied to NOAA-11/AVHRR imagery. Remote Sens. Environ. 46: 246-267.

Dickinson RE, Henderson-Sellers A, Kennedy PJ, Wilson MF (1986) Biosphere/Atmosphere Transfer Scheme (BATS) for the NCAR Community Climate Model. NCAR Technical Note TNXXX, National Center for Atmospheric Research, Boulder, CO, 80307

Dickinson RE (1984) Modeling evapotranspiration for three-dimensional global climate models. In: Climate processes and Climate Sensitivity. Geophysical monograph 29, Maurice Ewing Volume 5, American Geophysical Union, Washington, DC, 58-72

Dobson MC, Ulaby FT, Hallikainen MT, El-Rayes MA (1985) Microwave Dielectric Behavior of Wet Soil - Part II: Dielectric Mixing Models. IEEE Trans. on Geoscience and Remote Sensing, Vol. GE-23 (1): 35-46.

Doelling DR, Chakrapani V, Minnis P, Nguyen (2001), The Calibration of NOAA AVHRR Visible Radiances with VIRS, Proc. AMS 11th Conf. Atmos. Radiation, Madison, WI, Oct 15-18.

Dogniaux R (1940) Propositiones relatives à la normalisation des sources à considérer pourla determionation des caractéristiques photométriques des matériaux utilisés en vitrages (Uccle, 1965), see Moon P (1940) J. Franclin Inst 230: 583

D.R.E.A.M. (1994) Complesso forestale Le Carline-La Selva: piano di assestamento forestale 1994-2004. Published by Regione Toscana, Pistoia, Italy.

Dümenil L, Todini E (1992) A rainfall-runoff scheme for usein the Hamburg climate model. Adv. In: O'Kane JP (ed) Theor. Hydrology, European Geoph. Soc. Series on Hydrological Sciences, Vol. 1, Elsevier Science Publishers, Amsterdam, pp 129-157.

Duffie JA, Beckman WA (1980) Solar Engineering of Thermal Processes. John Wiley & Sons, New York

Dubois PC, van Zyl JJ, Engman T (1995) Measuring Soil Moisture with Imaging Radars. IEEE Trans. Geosci. & Remote Sensing 33: 915-926

Duggin M, Schoch L, Cunia T, Piwinski D (1984) Effects of random and systematic variations in unresolved cloud on recorded radiance and on target discriminability. Appl. Optics 23: 387

Duggin M, Piwinski D (1984) Recorded radiance indices for vegetation monitoring using NOAA AVHRR data; atmospheric and other effects in multitemporal data sets. Appl. Optics 23: 2620

Eastman IR (1992) IDRISI Version 4.0 Technical reference Manual. Clark University, Worcester

Ebert E (1987) A pattern recognition technique for distinguishing surface and cloud types in the polar regions. J. Clim. Appl. Meteor 26: 1412-1427.

Ehret G, Kiemle C, Renger C, Simmet G (1993) Airborne Remote Sensing of Tropospheric Water Vapor Using a Near Infrared DIAL system. Appl.Opt. 32: 4534-4551

EFEDA (1993) see Bolle H-J, Streckenbach B (eds) EFEDA Final Report and Annex

Eidenshink JC, Faundeen JL (1994) The 1 km AVHRR global land data set - first stages in implementation: International Journal of Remote Sensing, v. 15(17): 3,443-3,462.

Eklundh L (1998) Estimating relations between AVHRR NDVI and rainfall in East Africa at 10-day and monthly time scales. International Journal of Remote Sensing 19: 563-568.

Ellsaesser HW, MacCracken MC, Potter GL, Luther FM (1976) An additional model test of positive feedback from high desert albedo. Quart. J. R. Met. Soc. 102: 655-666

Emberger L (1955) Une classification biogéographique des climats. Natur. Monspel., série Botanique 7: 3-43.

Enne G, Peter D, Pottier D (2001) Desertification Convention - Data and information requirements for interdisciplinary research. European Commission, Directorate-General for Research, Luxembourg, EUR 19496

Entstrasser N, Strodl C, Kloster O, Rüba H, Willeke H, Löffler H, Jochum AM (1994) EFEDA 1994 Field Phase: DLR powered gliders - Data quicklook results. Institut für Physik der Atmosphäre, DLR, Oberpfaffenhofen

Engelsen O, Pinty B, Verstraete MM, Martonchik JV (1996) Parametric bidirectionel reflectance factor models: Evaluation, improvements and applications. Report EUR 16426EN, European Commission, Joint Research Centre, Space Applications Institute, Ispra, Italy, 120 pp

EOSAT (1986) Landsat Technical Notes No. 1

Epema GF (1990) Determination of Planetary Reflectance for Landsat-5 Thematic-Mapper Tapes Processed by Earthnet (Italy). ESA Journal 14: 101-108

ESA (1998) Emerging Scatterometer Applications. From Research to Applications. ESA SP-424

Escadafal R (1994) Soil spectral properties and their relationships with environmental parameters - Examples from arid regions. In: Hill J and Megier (eds.) Imaging Spectrometry - A Tool for Environmental Observations. Kluver Academic Publishers, Dordrecht, pp. 71-87

Falckenberg G (1928) Absorptionskonstanten einiger meteorologisch wichtiger Körper für infrarote Wellen. Meteorologische Zeitschrift 45:334-337

FAO-UNESCO (1988) Soil Map od the World: Revised Legend. World Soils Resources Report, FAO, Rome

FAO (1999) FAOSTAT 98 CD-ROM. FAO Statistical Databases. ISBN 9250042302, Rome, Italy.

Fell F (1991) Ein Test der aus Satellitendaten in den Spektralbereichen der AVHRR-Kanäle eins und zwei bestimmten Bodenalbedo durch vergleichende Bodenmessungen im Sudan, Diplomarbeit. Institut für Meteorologie, Freie Universität Berlin

Fell F (1993) personal communication

Femenia A., Fuster JM., Lopez-Baeza E., Fortea JC. (1995). Clasificacion de nubes en imagenes METEOSAT mediante redes neuronales. X Symposium Nacional de la Union Cientifica Internacional de Radio (Universidad de Valladolid). Valladolid, 27-29 September.

Femenia A., Lopez-Baeza E., Fortea J.C., Fuster JM. (1995). Utilizacion del paquete matematico MA TLAB para c lasificacion de nubes en imagenes METEOSAT mediante redes neuronales. I Congreso de Usuariosde MATLAB (UNED). Madrid, 22-23 May.

Ferrazzoli P, Guerriero L, Paloscia S, Pampaloni P (1992) Modelling polarization properties of emission from soil covered with vegetation. IEEE Trans. on Geosc. and Remote Sensing, Vol. 30 (1): 157-165.

Ferrier G, Wadge G, Moore JM, Liu JG, Machin R, Murphy R (1994) Using the AVIRIS imaging spectrometer to map alteration due to gold mineralization in southern Spain. In: Wooding M, Lodge F, Attema E (eds.) MAC Europe ′91 Final Results Workshop. European Space Agency, Paris, 254 pp.

Ferrier G, Wadge G, Moore JM, Liu JG, Machin R, Murphy R (1994) Using the AVIRIS imaging spectrometer to map alteration due to gold mineralization in sozthern Spain. In: Wooding M, Lodge F, Attema E (eds.) MAC Europe '91 Final Results Workshop. European Space Agency, Paris, pp. 254

Feußner K (1953) In: Linke, Met. Taschenbuch II, Tab. 126, p 541

Field C, Mooney HA (1986) The photosynthesis-nitrogen relationship in wild plants. In: Givnish TJ (ed.) On the Economy of Plant Form and Functions. Cambridge University Press, New York, pp 25-55

Fiedler F, Becker KH, Geiger H, Koppmann R (2000) Feldexperimente zur Untersuchung von Photosmog-episoden (z.B. TRACT, BERLIOZ, POPCORN). Promet 26, 132-150

Fisher PF, Pathirana S (1993) The Evaluation of Fuzzy Membership of Land Cover Classes in the Suburban Zone. Remote Sens. Environment 34: 121-132

Foody GM (1992) A Fuzzy Sets Approach to the Representation of Vegetation Continua from Remotely Sensed Data: An Example from Lowland Heath. Photogrammetric Engineering & Remote Sensing 58(2):221-225

Foody GM (1994) Sub-pixel land cover composition estimation using a linear mixture model and fuzzy membership functions. Interrnational Journal of Remote Sensing 15: 619-631

Fortea JC, Valiente JA, Lopez-Baeza E (1997) Surface albedo from Meteosat-4 data. Accurate atmospheric correction using ground data. VII Congreso Nacional de Teledetección, Santiago de Compostela, Spain, 24-28 Junio

Fortea JC,Lopez-Baeza E (1997) Estimation of the Shortwave Surface Net Radiation Over the Iberian Peninsula: Preliminary Study of the Cloud Cover Effect at METEOSAT Scale. Proceedings of the 1997 Met. Satellite Data Users' Conference. Session IV: Derivation of Geophysical Parameters, EUM P 21, ISSN 1011-3932. EUMETSAT, Darmstadt, ISBN 92-9110-021-8, pg. 301-308. Brussels (Belgium), 29 Sept.-3 Oct. 1997

Fourier J (1818) Note relative aux vibrations des surfaces élastiques et au mouvement des ondes. Bulletin

des Sciences par Société Philimatique 1:126-136
Fourier JB (1820) Théorie analytique de la chaleur. Paris
Fowle FE (1917) Water Vapor Transparency to Low-temperature Radiation. Smithsonian Miscellaneous Collections, Vol. 68(8), Publication 2484, City of Washington, Smithsonian Institution, 68 pp
Francey RJ, Garratt JR (1978) Eddy Flux Measurements over the Ocean and Related Transfer Coefficients. Boundary-Layer Meteorology 14: 153
François C, Ottlé C (1996) Atmospheric corrections in the thermal infrared: global and water vapor dependent split-window algorithms- applications to ATSR and AVHRR data. IEEE Transactions on Geoscience and Remote Sensing, 34: 457-470
Freeman A, Durden SL (1992) A three-component scattering modell to describe polarimetric SA data. Proc. SPIE Conference Radar Polarimetry, San Diego, CA, pp 213-224
Frulla LA, Milovich JA, Gagliardini DA (1995) Illumination and observation geometry for NOAA-AVHRR images. Int. J. Remote Sensing 16: 2233-2253
Fung AK, Zongquian LI, Chen KS (1992) Backscattering from randomly rough dielectric surface. IEEE Transactions on Geoscience and Remote Sensing 30:356-369
Gaertner MA, Fernández C, Castro M (1993) A two-dimensional simulation of the Iberian summer thermal low. Monthly Weather Reviev 121: 2740-2756.
Gao W, Coulter RL, Lesht BM, Qui J, Wesely ML (1998) Estimating clear-sky regional surface fluxes in the southern great plaines atmospheric radiation measurement site with ground measurements and satellite observations. J. Appl. Meteor. 37: 5-22
Garand L, Weinman A (1986) A Structural Stochastic Model for the Analysis and Synthesis of Cloud Images, J. Clim. Appl. Meteor. 25:1052-1068
García Novo F, Jiménez Cañete JM (1955) Analusía sites: Effects of long-term human activities on the natural resource. In: Dubost M (ed.) MEDIMONT: A multinational, multidisciplinary research programme on the rôle and place of the mountains in the desertification of the Mediterranean mountain regions. Second and Final Scientific Report, Contract EV5V-CT91-0045, pp. 39
García-Haro FJ, Gilabert A, Meliá J, Belda F (1996) Methods for monitoring burned areas using thematic mapper data (poster). Application of Remote Sensing in European Forest Monitoring. International Workshop, Vienna.
García-Haro, F.J. 1997. Modelizacion y estimacion de parámetros relacionados con la cubierta vegetal de teledeteccion. Ph.D. Thesis. University of Valencia. Spain.
Garratt JR (1993) Sensitivity of climate simulations to land-surfaceand atmospheric boundary-layer treatments: a review. J. Climate 6: 419-449
Gates DM (1980) Biophysical Ecology. Springer-Verlag, New York Heidelberg Berlin
Gates DM, Keegan HJ, Schleter JC, Weidner VR (1965) Spectral properties of plants. Appl. Optics 4: 11-20
Gates DM, Tantraporn W (1952) The reflectivity of deciduous trees and herbaceous plants in the infrared up to 25 microns. Science 115: 613-616.
Geerders PJF, Roozekrans JN (ed.) (1995) Workshop of European NOAA-AVHRR data providers, 1-4 May 1995, Centre for Earth Observation (CEO), JRC, Ispra, Italy.
Geleyn JF (1988) Interpolation of wind, temperature and humidityvalues from model levels to the height of measurements. Tellus 40A: 347-351.
Gemmell FM, Colls JJ (1992) The effects of sulphur dioxide on the spectral characteristics of leaves of *vicia faba L.* Int. J. Remote Sens. 13(14): 2547-2563
Genovese G, Vignolles C, Nègre T, Passera G (1999) The use of CORINE land cover to improve vegetation monitoring through NOAA-AVHRR/NDVI profiles. Proceedings of the International Symposium "Modelling Cropping Systems", Lleida (Spain), 21-23 June 1999.
Gerstl SA, Simmer C (1986) Radiation physics and modelling for non-Lambertian surfaces, Remote Sens. Environ. 20: 1-29.
Gesell G (1989) An Algorithm für Snow and!ce Detection Using AVHRR Data: An Extension to the APOLLO Software Package. International Journal of Remote Sensing, 10(4-5):897-905
Gibson JK, Kallberg P, Uppala S, Noumura A, Hernandez A, Serrano E (1997) ERA Description. ECMWF Re-Analysis Project Report Series, 1. ECMWF, Reading, UK, 77 pp.
Gilabert MA, Gandía S, Meliá J (1996) Analyses of spectral-biophysical relationships for a corn canopy. Remote Sensing of the Environment 55:11-20
Gilabert M-A, Meliá J (1993) Solar angle and sky light effects on ground reflectance measurements in a citrus canopy, Remote Sens. Environ. 45 (44?):281-293.

Gilabert MA, Conese C, Maselli F (1994) An atmospheric correction method for the automatic retrieval of surface reflectances from TM images. International Journal of Remote Sensing 15 (10): 2065-2086

Gimeno-Ferrer JF (2003) Estudio de la desertificación en la Península Ibérica mediante teledetección aplicando la transformada wavelet [Remote Sensing Study of Desertification in the Iberian Peninsula by Applying the Wavelet Transform].Diploma for Advanced Studies Research Report, University of Valencia, Sept. 2003

Gitelson A, Kogan F, Spivak L, Zakarin E, Lebed L (1996) Estimation of Seasonal Dynamics of Pasture and Crop Productivity Using *NOAA*/AVHRR data. 1996 International Geoscience and Remote Sensing Symposium (IGARSS'96), May 27-31, Lincoln, Nabraska, IEEE, Val. 1, pp. 209-211

Gitelson AA, Kogan F, Zakarin E, Spivak L, Lebed L (1998) Using AVHRR Data for Quantitative Estimation of Vegetation Conditions: Calibration and Validation. Advances in Space Research.

Gitelson A, Schmidt H (1999) Monitoring Vegetation Dynamics in Israeli Transition Zone with Advanced Very High Resolution Radiometer Data. Journal of Arid Land Studies, 9: 267-276.

Gloersen P, Campbell WJ, Cavalieri DJ, Comiso JC, Parkinson CL, Zwally HJ (1992) Arctic and Antarctic Sea Ice, 1978-1987: Satellite Passive-Microwave Observations and Analysis.National Aeronautics and Space Administration Scientific and Technical Information Program. Washington, D.C. NASA SP-511

Gobron N, Pinty B, Verstraete M, Govaerts Y (1999) The MERIS Global Vegetation Index (MGVI): description and preliminary application. Int. J. Remote Sensing, 20(9): 1917-1927

Gong P, Howarth PJ (1992 a) Land-use classification of SPOT HRV data using a cover-frequency method. Int. J. Of Remote sensing 13(8): 1459-1471

Gong P, Howarth PJ (1992b) Frequency-based contextual classification and gray-level vector reduction for land-use identification. Photogrammetric Engineering and Remote Sensing 58(4): 423-437

Goel NS, Qin W (1994) Influences of canopy achitecture on relationships between various vegetation indices and LAI and FPAR: A computer simulation. Remote Sens. Rev. 10, 309-347

Goode, J.P., 1925. The Homolosine projection: a new device for portraying the Earth's surface entire. Association of American Geographers, Annals 15: 119-125.

Goodman B, Martin DW, Menzel WP, Cutrim EC (1994) A Non-linear Algorithm for Estimating Three-hourly Rain Rates over Amazonia from GOES-VISSR Observations. Rem. Sens. Rev. 10: 169-177

Goodrum G, Kidwell K, Winston W, eds. (2000) NOAA KLM User's Guide with NOAA-N, N' Supplement, September 2000 Revision. U.S. Department of ommerce, NOAA, NESDIS

Goody RM (1964) Atmospheric Radiation. I Theoretical Basis. Clarendon Press, Oxford, 436 pp.

Gordon HR, Brown JW, Evans RH (1988) Exact Rayleigh Scattering Calculations for use with the Nimbus -7 coastal zone colour scanner. Applied Optics, 27: 2111-2122.

Gorte B, Koolhoven W (1990) Interpolation between isolines based on the Borgerfors distance transform. ITC Journal 3: 245-247, Amsterdam

Göttsche F-M, Olesen F-S (2001) Modelling of diurnal cycles of brightness temperature extracted from METEOSAT data. Remote Sensing of Environment, Vol. 76(3): 337-348.

Göttsche F-M, Olesen F-S (2002) Evolution of neural networks for radiative transfer calculations in the terrestrial infrared. Remote Sensing of Environment, Vol. 80(1): 157-164.

Govaerts YM, Verstraete MM, Pinty B, Gobron N (1999) Designing optimal spectral indices: a feasibility and proof of concept study. Int. J. Remote Sensing, 20, No. 9: 1853-1873

Goward SN, Turner S, Dye DG, Liang S (1994) The University of Maryland improved Global Vegetation Index product, Int. J. Remote Sensing 15(17): 3365-3395.

Goward SN, Huemmrich KF (1992) Vegetation canopy PAR absorptance and the Normalized Difference Vegetation Index: An assessment using the SAIL model. Remote Sens. Environ. 39: 119-140

Greenwald T J, Stephens GL, Vonder Haar TH, Jackson DL (1993) A physical retrieval of cloud liquid water over the global oceans using Special Sensor Microwave/Imager (SSM/I) observations. J. Geophysical Research 98: 18471-18488.

Groten SME (1993) NDVI-crop monitoring and early yield assessment of Burkina Faso. International Journal of Remote Sensing 8: 1,495-1,515.

Grove CI, Hook SI, Paylor ED (1992) Laboratory reflectance spectra of 160 minerals, 0.4 - 2.5 micrometers. JPL Publication 92-2, 280 pp.

Gu D, Gillespie AR, Kahle AB, Palluconi FD (2000) Autonomous Atmospheric Compensation (AAC) of high resolution hyperspectral thermal infrared remote-sensing imagery. IEEE Transactions on Geoscience and Remote Sensing, 38: 2557-2569

Gutman GG (1988) A simple method for estimating monthly mean albedo of land surfaces from AVHRR data. J. Appl. Meteor. 27: 973- 988

Gutman G (1989) On the relationship between monthly mean and maximum-value composite normalized vegetation indices. Int. J. Remote Sensing, 10: 1317-1325.
Gutman GG (1991) Vegetation indices from AVHRR: An update and future prospects, Remote Sens. Environ. 35: 121-136
Gutman GG, Ignatov AM, Olson S (1994) Towards better quality of AVHRR composite images over land: Reduction of cloud contamination. Remote Sens. Environ. 50: 134-148
Gutman GG, Ignatov AM (1995) Global land monitoring from AVHRR: potential and limitations, Int.J. Remote Sensing 16(13): 2301-2309.
Gutman GG (1999) On the use of long-term global data of land reflectances and vegetation indices derived from the advanced very high resolution radiometer. J. Geophys. Research. 104 D6: 6241-6256
Hayes MJ, Decker WL (1996) Using NOAA AVHRR data to estimate maize production in the United States Corn Belt. International Journal of Remote Sensing, 17: 3189-3200.
Hallikainen MT, Ulaby FT, Dobson MC, El-Rayes MA, Wu L (1985) Microwave dielectric behaviour of wet soil - Part I: Empirical models and experimental observations. IEEE Transactions on Geoscience and Remote Sensing 23: 25-34
Hanel RA, Conrath BJ, Kunde VG, Prabhakara C, Revah I, Salomonson VV, Wolford G (1971) The NIMBUS 4 Infrared Spectroscopy Experiment, IRIS-D. Goddard Space Flight Center, Greenbelt, Maryland, X-622-71-272 (see also: Hanel RA, Conrath BJ, 1970, Thermal emission spectra of the Earth and Atmosphere from the NIMBUS 4 Michelson Interferometer experiment. Nature 5267: 143-145)
Hanl G (1969) Polarisationsefekte bei Reflexion infraroter Strahlung an Oberflächen natürlicher Materialien. Diplomarbeit für meteorologie, Ludwig-Maximilians-Universität München
Vauclin M, Haverkamp R, Carrillo E, Gaudet JP, Laty R, Laurent JP, Thony JL, Vanderwaare JP (1992) Basic Description of Soil Data Sets and Some Preliminary Results of the EFEDA-Spain Field Campaign. First EEC EFEDA Report, Berlin, pp. 26-37
He C, Kittleson K, Bartholic J (1987) Evapotranspiration monitored from satellite as an indication of shift and impact of vegetation change. Proc. 21st Int. Symp. On Remote Sensing of Environment. Ann Arbor, MI, Environmental Research Institute of Michigan, pp 695-708
Heidinger A, Cao C, Sukkivan J (2002) Using MODIS to calibrate AVHRR reflectance channels. J. Geophys. Res. 107(23): 4702-4711
Henderson-Sellers A, Pitman AJ (1992) Land-surface schemes for future climate model: specification, aggregation and heterogeneity. J.Geophys. Res.97, D3: 2687-2696.
Hernández, E., Valero, F., Martín, F., Alvarez, J., Enríquez, S., y Gutiérrez, E., 1991. Situaciones sinópticas meteorológicas en los incendios forestales durante los años, 1981-87. Revista de Geofísica. 47: 45-58.
Herzberg G (1939) Molecular Spectra and Molecular Structure, I. Spectra of Diatomic Molecules. Prentice Hall Inc. (Reprint 1950 D. Van Nostrand Company, Inc., Princeton, N.J. USA), 658 pp.
Herzberg G (1945) Molecular Spectra and Molecular Structure, II. Infrared and Raman Spectra of Polyatomic Molecules. D. Van Nostrand Company, Princeton, New Jersey, U.S.A., 632 pp
Hill J (1991) Analysis of GER imaging spectrometer data acquired during the European Imaging Spectrometry Aircraft Campaign (EISAC) '89 - Quality assessment and first results. EARSeL Advances in Remote Sensing 1: 64-77
Hill J (ed) (1996) Integrated approaches to desertification mapping and monitoring in the Mediterranean basin. Final Report of the DEMON-1 Project, JRC, European Commission
Hill J, Mégier J (1991) The use of imaging spectrometry in Mediterranean land degradation and soil erosion hazard assessment. Proceedings of the 5th International Colloquium on Physical Measurements and Signatures in Remorte Sensing, Courchevel, France
Hill J, Smith MO, Mehl W, Lacaze B, Sommer S (1994) The use of high spectral resolution imaging systems for mapping land degradation in Mediterranean ecosystems. In: Wooding M, Lodge F, Attema E (eds.) MAC Europe ´91 Final Results Workshop. European Space Agency, Paris, 254 pp.
Hill J, Lacaze B, Riezebos H, de Jong S, Caselles V, Valor E, Mehl W, Sommer S, Tsiourlis G (1997) DeMon-2: Desertification in the European Mediterranean - The DeMon-2 Approach in Using Satellite Remote Sensing for Assessing and Monitoring Environmental Degradation. In: Spiteri A (ed.), Proc. 16th EARSeL Symp., Malta, 20-23 May 1996, pp. 305-310
Hobbs TJ (1995): The use of NOAA-AVHRR NDVI data to assess herbage production in the arid rangelands of Central Australia. International Journal of Remote Sensing, 16: 1289-1302.
Hoekman DH (1991) Speckle ensemble statistics of logarithmically scaled data. IEEE Trans. Geosci. And

Remote Sensing 29: 180-182
Holben BN (1986) Characteristics of maximum-value composite images from temporal AVHRR data. Int. J. Remote Sens. 7: 1417-1434
Holben BN, Kimes D, Fraser RS (1986) Directional reflectance response in AVHRR red and near-IR bands for three cover types and varying atmospheric conditions. Remote Sensing of Environment 19: 213-236
Holben BN, Kaufman YJ, Kendall JD (1990) NOAA-11 AVHRR visible and near-IR inflight calibration. Int. J. Remote Sensing 11(8): 1511-1519.
Hoogenboom G, Wilkens W, Thornton PK, Jones JW, Hunt LA (1999) Advances in the development and application of DSSAT. In: Proceedings International Symposium on Modelling Cropping Systems
Horler DNH, Dockray M, Barber J (1983) The red edge of plant reflectance. Int. J. Remote Sens. **4**(2): 273-288.
Horler DNH, Ahern FJ (1986) Forestry information content of Thematic Mapper data. International Journal of Remote Sensing 8: 1785-1796.
Houghton JT, Callander BA, Varney SK, eds. (1992) Climate Change 1992: The supplementary report to the IPCC scientific assessment. Cambridge University Press, 200 pp
Houghton JT, Meira Filho LG, Bruce J, Hoesung Lee, Callander BA, Haites E, Harris N, Maskell K (1994) Climate Change 1994, Radiative Forcing of Climate Change and An Evaluation of the IPCC IS92 Emission Scenarios. Cambridge University Press, Cambridge, UK
Houghton JT, DING Y, Griggs DJ, Noguer M, Van der Linden PJ, Xiaosu D (eds.) (2001) Climate Change 2001: The Scientific Basis. Cambridge University Press
Hsu KL, Gao X, Sorooshian S, Gupta HV (1997) Precipitation Estimates from Remotely Sensed Information Using Artificial Neural Networks. J. Appl. Meteorol. 36: 1176-1190
Huete AR (1986) Separation of soil-plant spectral mixtures by factor analysis. Remote Sensing of Environment 19: 237-251
Huete AR, Jackson RD, Post DF (1985) Soil spectral response of a plant canopy with different soil backgrounds. Remote Sens. Environm. 15: 155-165
Huete AR (1988) A Soil Adjusted Vegetation Index (SAVI). Remote Sens. Environ. 25: 295-309.
Hulme M, Conway D, Jones PD, Barrow EM, Jiang T, Turney C (1995a) A 1961-1990 climatology for Europe for climate change modelling and impact applications. International Journal of Climatology 15: 1333-1363
Hulme M, Conway D, Jones PD, Jiang T, Zhou X, Barrow EM, Turney C (1995b) A 1961-1990 Gridded Surface Climatology for Europe. Climate LINK Project, Climatic Research Unit, University of East Anglia, Norwich, UK
Hulme M, Conway D, Jones PD, Jiang T, Barrow EM, Turney C (1995c) Construction of a 1961-1990 European climatology for climate change modelling and impact applications. Int. J. Climatology 15: 237-258
Hucek R, Jacobowitz H (1955) Impact of scene dependence on AVHRR albedo models. J. Atmos. Oceanic Technol. 12: 697-711
Hutchinson CF (1991) Uses of satellite data for famine early warning in sub-Saharan Africa. International Journal of Remote Sensing 12: 1405-1421.
Hurley P, Boers R (1996) The effect of cloud cover on surface net radiation - Simple formulas for use in mesoscale models. Boundary Layer meteorology 79: 191-199
IAMAP-IRC (1986) s. WMO
Idso SB, Jackson RD, Reginato RJ (1976) Determining emittances for use in infrared thermometry: A simple technique for expanding the utility of existing methods. J. Appl. Meteorology 15: 16-20
Idso SB, Jackson RD, Reginato RJ (1976) Compensating for environmental variability in the thermal inertia approach to remote sensing od soil moisture. J. Appl. Meteorology 15: 811-817
IGBP (1990) The International Geosphere Biosphere Programme: A study of global change. The Initial Core Projects. Report nº 12. International Council of Scientific Unions, Stockholm
IGBP-BAHC (1993) Biospheric Aspects of the Hydrological Cycle. The Operational Plan. IGBP Report No. 27, Stockholm
IGN (Instituto Geografico Nacional) (1982). Land uses map of the Tomelloso area, scale 1:25000.
Iqbal M (1983) An Introduction to Solar Radiation. Academic Press Toronto Canada, 390 pp
Jackson TJ, O'Neill PE (1990) Attenuation of soil microwave emission by corn and soybeans at 1.4 and 5 GHz. IEEE Transactions on Geoscience and Remote Sensing, 28(50): 978-980.
Jacquemoud S, Baret F, Hanocq JF (1992) Modeling spectral and bidirectional soil reflectance. Remote

Sensing Environment 41: 123-132
Jacquemoud S (1993) Inversion of the Prospect & Sail canopy reflectance model from AVIRIS equivalent spectra: Theoretical study. Remote Sensing of Environment 44(2&3): 281-292
Jacquemoud S, Baret F (1990) Prospect: A Model of Leaf Optical Properties Spectra. Remote Sensing Environment 41: 123-132
Jagdhuhn S (1994) Bestimmung der Bodenalbedo aus Landsat-TM- und NOAA-AVHRR-Daten, Diploma Thesis, Institut für Meteorologie, Freie Universität Berlin
James ME, Kalluri SNV (1994) The pathfinder AVHRR land data set: An improved coarse resolution data set for terrestrial monitoring. *Int. J. Remote Sensing* 15(17): 3347-3363.
James P, Stohl A, Eckardt S, Spichtinger N, Forster C (2002) Trajectory modelling of water vapour transport during the severe flooding episodes in central Europe in August 2002. AFO2000 (German Atmospheric Research Programme) Status Seminar 2002, Schliersee, German Ministry for Education and Research (BMBF)
Jensen A, Lorenzen B, Spelling Østergaard H, Kloster Hvelplund E (1990) Radiometric estimation of biomass and nitrogen content of barley grown at different nitrogen levels. Int. J. Remote Sens. 11(10): 1809-1820
Jochum AM (1993) Evaporation and energy fluxes during EFEDA: Horizontal variability and area averaging. In: Bolle H-J, Feddes RA, Kalma JD (eds.) Exchange Processes at the Land Surface for a Range of Space and time Scales. IAHS Publ., Wallingford, UK, 212: 373-380
Jochum A, Strodl C, Moreno J, de Dios J (1998) Remote Sensing of Mediterranean Desertification and Environmental Stability (RESMEDES), Final Report Project ENV4-CT95-0094
Jochum AM, Kabat P, Hutjes R (2000) The role of remote sensing in land-surface experiments within BAHC and ISLSCP. Observing Land from Space: Science, Customers and Technology. Verstraete M, Menenti M, Peltoniemi J (Eds) Kluwer Academic Press, pp 91-103
Jones HG (1992) Plants and Microclimate. A quantitative approach to environmental plant physiology. 2nd edition, pp 145. Cambridge University Press, Cambridge
Junge C (1963) Air Chemistry and Radioactivity. Academic press, N.Y.
Justice, C.O., Holben, B.N., and Gwynw, M.D.. 1986. Monitoring East African vegetation using AVHRR data. International Journal of Remote Sensing, 7: 1453-1474.
Justice CO, Hiemaux PHY (1986) Monitoring the grasslands of the Sahel using NOAA AVHRR data: Niger1983. Int. J. Remote Sensing, 7: 1475-1497.
Kalluri SNV, Townshed JRG, Doraiswamy P (1998) A simple single layer model to estimate transpiration from vegetation using multi-spectral and meteorological data. International Journal of Remote Sensing 19: 1037-1054.
Kanemasu ET, Demetriades-Shah TH, Su H, Lang ARG (1990) Estimating grassland biomass using remotely sensed data. In: Steven MD, Clark JA (eds.) Applications of Remote Sensing in Agriculture
Kasten F, Young AT (1989) Revised optical air mass tables and approximation formula. Applied Optics 28(22): 4735-4738.
Kaufman YJ, Tanré D (1992) Atmospherically resistant vegetation index (ARVI) for EOS-MODIS. IEEE Trans. Geosci. Remote Sens. 30: 261:270.
Kaufman YJ (1987) The effect of subpixel clouds on remote sensing. Int. J. Remote Sensing 8: 839-857
Kaufman YJ, Holben BN (1993) Calibration of the AVHRR visible and near-IR bands by atmospheric scattering, ocean glint and desert reflection. Int. J. Remote Sensing 14: 21-52.
Kerr YH, Lagouarde JP, Imbemon J (1992) Accurate land surface temperature retrieval from AVHRR data with use of an improved split window algorithm. Remote Sens. Environ. 41: 197 -209
Kerr YH, Waldteufel P, Wigneron J-P, Font J, Berger M (2001) Soil moisture retrieval from space: The Soil Moisture and Ocean Salinity (SMOS) Mission. IEEE Trans. Geosc. & Remote Sensing, 39(8): 1729-1735
Kerdiles H, Grondona MO (1995) NOAA-AVHRR NDVI decomposition and subpixel classification using linear mixing in the Argentinean Pampa. International Journal of Remote Sensing, 16: 1,303-1,325.
Kessler A (1985) Über die kurzwellige Albedo eines Kiefernwaldes. Meteorol. Rundschau 38: 82-91
Key JR, Schweiger AJ (1998) Tools for Atmospheric Radiative transfer: Streamer and FluxNet. Comput. & Geosciences 24(5): 443-451
Kidwell KB (1995) Polar Orbiter Data Users Guide. U.S. Dept. of commerce, NOAA/National Environmental Satellite Data and Information Service, National Climatic Data Center, Satellite Data Services Division, Washington DC, pp 410
Kidwell KP, ed. (1990) Global Vegetation Index Users's Guide. Satellite Data Services Div.,

NOAA/NESDIS, Washington, D.C.

Kimes DS (1983) Dynamics of directional reflectance factor distributions of vegetation canopies. Appl. Opt. 22: 1364-1372.

Kimes DS (1984) Modelling the directional reflectance from complete homogeneous vegetation canopies with various leaf-orinetation distributions. J.Opt.Soc.Am. 1(7): 725-737

Kimes DS, Newcomb WW, Tucker CJ, Zoneveld IS, van Wijngaarden W, de Leeuw J, Epema GF (1985) Directional reflectance factor distributions for cover types of Northern Africa. Rem. Sens. of Environment 18: 1-19

Kimes DS, Newcomb WW, Nelson RF, Schutt JB (1986) Directional reflectance distributions of hardwood and pine forest canopy. IEEE Trans. Geosc. Remote Sens. 24: 281-239

Kimes DS, Sellers PJ, Diner DJ (1987) Extraction of spectral hemispherical reflectance (albedo) of surfaces from nadir and directional reflectance data. Int. J. Remote Sensing 8: 1727-1746

Kimes DS, Holben BN (1992) Extracting spectral albedo from NOAA-9 AVHRR multiple view data using an atmospheric correction procedure and an expert system. Int. J. Remote Sensing 13(2): 275-289

Kimes DS, Kerber AG, Sellers PJ (1993) Spatial averaging errors in creating hemispherical reflectance (albedo) maps from directional reflectance data. Remote Sens. Environm. 45: 85-94

Kirdiashev KP, Chikhlantsev AA, Shutko AM (1979) Microwave radiation of the earth's surface in the presence of vegetation cover. Translation of "Svch-izlucheniye Zemnoy Poverkhnosti Pri Nalichii Rastitel' Nogo Pokrova", Radiotekhnika i Elektronika, 24: 256-264; NASA-Washington, D.C. 20546, June 1979.

Kislovskii LD (1959) Opticheskie kharakteristiki vody I l'da v infrakrasnom I mikrovolnovom diapazone spektra (Optical characteristics of water and ice in infrared and microwave spectral frequencies. Optika I spektroskopiya 7: 201

Kneizys FX, Shettle EP, Gallery WO, Chetwynd JH, Abreu LW, Selby JEA, Clough SA, Fenn RW (1983) Atmospheric transmittance/radiance: computer code LOWTRAN 6. Env. Research Papers No. 846, AFGL-TR_83-0187, Opt. Phys. Division, Air Force Geophysics Laboratory, Hanscom AFB, Massachusetts, U.S.A., 200 pp

Kneizys FX, Shettle EP, Anderson GP, Abreu LW, Chetwynd JH, Selby JEA, Clough SA, Gallery WO (1989) Users guide to LOWTRAN 7, AFGL-TR-88-0177, Air Force Geophysics Laboratory, Hanscom,Massachusetts, U.S.A., 146 pp

Knipling EB (1970) Physical and physiological basis for the reflectance of visible and near-infrared radiation from vegetation. Remote Sens. Environ. 1: 155-159.

Knorr W, Heiman M (2001) Uncertainties in global terrestrial biosphere modeling. Part I: A comprehensive sensitivity analysis with a new photosynthesis and energy balance scheme. Global Biogeochemical Cycles 15: 207-225

Knorr FJ, Futrell JH (1979) Separation of mass spectra of mixtures by factor analysis. Analytical Chemistry 51: 1236-1241

Kondratiev KY, Niilisk HJ (1960) On the question of carbon dioxide heat radiation in the atmosphere. Geophysica Pura e Applicata 46: 216

Köpke P (1977) Bestimmung der atmosphärischen Trübung mittels geostationärer Satelliten. Wissenschaftliche Mitteilung Nr. 30, Meteorologisches Institut, Ludwig-Maximilians-Universität München

Köpke P (1989) Removal of atmospheric effects from AVHRR albedos. J. Appl. Meteor. 28(12): 1341-1348

Kogan FN (1997) Global drought watch from space. Bulletin of the American Meteorological Society 78: 621-636

Kogan FN (1990) Remote sensing of weather impacts on vegetation in non-homogeneous areas. Int. J. Remote Sensing 11: 1405-1419.

Kohsick W (1982) Measuring C_T^2, C_Q^2, and C_{TQ} in the unstable surface layer, and relation to the vertical fluxes of heat and moisture. Boundary-Layer Meteorology 24: 89-107

Kohsick W, Meijninger WML, Moene AF, Heusinkveld BG, Hartogenesis OK, Hillen WCAM, De Bruin HAR (2002) An Extra Large Aperture Sxintilllometger For Long Range Applications. Boundary-Layer Meteorology 105: 119-127

Kolmogorov AN (1941) Dissipation of energy in the locally isotropic turbulence. Comptes rendus (Doklady) de l'Académie des Sciences de l'U.R,S,S. 32: 16-18

Kolmogorov AN (1941) The local structure of turbulence in incompressible viscous fluid for very large

Reynold's numbers. Comptes rendus (Doklady) de l'Académie des Sciences de l'U.R,S,S. 30: 301-305
Körner C, Sarris D, Christodoulakis D (2005) Long-term increase in climatic dryness in the East-Mediterranean as evidenced for the island of Samos. Reg. Environ Change 5:27-36
Koslowsky D (1977) Digitale Bearbeitung der am Meteorologischen Institut der FU empfangenen Scanning-Radiometer(SR)-Daten der NOAA-Wettersatelliten,Teil IV: Geographische Entzerrung und Darstellung in stereographischer Landkartenprojektion, Wiss.Met.Abh., Neue Folge, Serie B, SO 86/77, Institut für Meteorologie, Freie Universität Berlin.
Koslowsky D (1989) On-line processing of NDVI maps. Proceedings of the 4th AVHRR Data Users' Meeting, Rothenburg, 5-8 September 1989, EUMETSAT, pp161-164
Koslowsky D (1996) Mehrjährige validierte und homogenisierte Reihen des Reflexionsgrades und des Vegetationsindexes von Landoberflächen aus täglichen AVHRR-Daten hoher Auflösung. Dissertation published as Meteorologische Abhandlungen N.F., Serie A, Monographien, Band 9, Heft 1, pp.237, Berlin
Koslowsky D (1997) Signal degradation of the AVHRR shortwave channels of NOAA 11 and NOAA 14 by daily monitoring of desert targets. Adv. Space Res. 19(1), 1355-1358.
Koslowsky D (1998) A cloud screening algorithm für daytime AVHRR data using dynamic thresholds. Adv.Space Res. 19(3): 533-536
Koslowsky D (1999) Remote Sensing and Radiometric Properties of the Surface: Assessment of Desertification from Space. Final Report, Contract No. EV5V-CT93-0284, Berlin 1997
Kossmann M, Corsmeier U, de Wekker SFJ, Fiedler F, Vögtlin R, Kalthoff N, Güsten H, Neininger B (1999) Observation of handover Processes between the Atmospheric Boundary Layer and the Free Troposphere over Mountainous Terrain. Contr. Atmos. Phys, 329-350
Kramer PJ, Kozlowski TT (1979) Physiology of woody plants. Academic Press. New York.
Kriebel KT (1974) Reflection properties of vegetated surfaces - tables of measured spectral biconical reflectance factors. Münchener Univ. Schriften, Fachbereich Physik, Met. Institut, Wiss. Mitt. no 29
Kriebel KT (1978) Measured spectral bidirectional reflection properties of four vegetated surfaces. Appl. Optics 17(2): 253-259
Kriebel KT, Saunders RW, Gesell G (1989) Optical Properties of Clouds Derived from Fully Cloudy AVHRR Pixels. Contributions to Atmospheric Physics, 62(3): 165-171
Kriebel KT, Gesell G, Kästner M, Mannstein H (1999) The cloud analysis tool APOLLO: Improvements an Validations, Report No 126, Institut für Physik der Atmosphäre, DLR, Oberpfaffenhofen.
Kruse and Lefkoff 1993 6.8
Kucera J, Cermak J, Penka M (1976) Improvement of the method of sap flow rate determination in adult trees based on heat balance with direct electric heating of xylem. Biol. Plant. 18: 105-110.
Kustas WP, Daughtry CST, Van Oevelen PJ (1993) Analytical treatment of the relationships between soil heat flux/net radiation ratio and vegetation indices. Remote Sens. Environ. 46: 319-330
Kustas WP, Moran MS, Humes KS, Stannard DI, Pinter Jr. PJ, Hipps LE, Swiatek E, Goodrich DC (1994) Surface energy balance estimates at local and regional scales using optical remote sensing from an aircraft platform and atmospheric data collected over semiarid rangelands. Water Res. Research 30(5): 1241-1259
Labed J (1990) Approche experimentale des problems lies al'emissivite et la temperature de surfaceainsi qu'a leur variabilite spatio-temporele. Dans le cadre de la teledetection spatiale dans l'infrarouge thermique. These Docteur de l'Un iversite de Strasbourg I
Lacaze B, Caselles V, Coll C, Hill J, Hoff C, de Jong S, Mehl W, Negendank JFW, Riezebos H, Rubio E, Sommer S, Teixeira Filho J, Valor E (1996) DeMon - Integrated Approaches to Desertification Mapping and Monitoring in the Mediterranean Basin. Final Report, EUR 16448 EN, ESCS-EC-EAEC, Brussels, Luxembourg
Lacaze B, Caselles V, Coll C, Hill J, Hoff C, De Jong S, Mehl W, Negendank JFW, Riesebos H, Rubio E, Sommer S, Teixeira Filho J, Valor E (1996) DeMon – Integrated approaches to desertification mapping and monitoring in the Mediterranean basin, Final Report of DeMon-1 Project, Joint Research Centre of European Commission, Ispra (VA), Italy.
Lambin EF, Strahler AH (1994) Change-vector analysis: a tool to detect and categorize land-cover change processes using temporal-resolution satellite data. Remote Sensing of Environment, 48: 231-244
Langer I (1997) Contribution to EFEDA Phase II Final Report Contract No. EV5V-CT93-0284 and successive updates
Langley SP (1881) Proc. Amer. Acad. Arts Sci. 8: 342
Larcher W (1995) Physiological Plant Ecology. Springer-Verlag, New York

Lee T, Kaufmann YJ (1986) The effect of surface non-Lambertianity of remote sensing on ground reflectance and Vegetation Index. IEEE J. Geosci. Remote Sensing GE-24: 699-708

Legates DR (1987) A climatology of global precipitation. Publ. in Climatology, 40(1), Neward, DE, 85 pp.

Legates DR, Willmott CJ (1990) Mean seasonal and spatial variability in gauge-corrected, global precipitation. Int. J. Climatology, 10: 111--127.

Leith H (1974) Phenology and seasonality modelling. Springer-Verlag, New York

Lemon M, Seaton R, Blatsou C, Calamaris N (1995) Agricultural Policy and Environmental Degradation: The case of the Argolid Valley, Greece. Medit 4: 26-34

Lenoble J (1985) Radiative Transfer in Scattering and Absorbing Atmospheres: Standard Computational Procedures. A. Deepak Publishing, Hampton, Virginia, 300 pp.

Leroy M, Roujeau JL (1994) Sun and view angle corrections on reflectances derived from NOAA/AVHRR data. IEEE Trans. Geosc. Re,ote Sens. 32: 684

Lewis JE, Rowland J, Nadeau A (1998) Estimating maize production in Kenya using NDVI: some statistical considerations. International Journal of Remote Sensing, 19: 2609-2617.

Li Y, Demetriades-Shah TH, Kanemasu ET, Shultis JK, Kirkham MB (1993) Use of second derivatives of canopy reflectance for monitoring prairie vegetation over different soil backgrounds. Remote Sens. Environ. 44: 81-87.

Li Z, Leighton HG (1993) Global climatologies of solar radiation budget at the surface and in the atmosphere from 5 years of ERBE data. J. Geophys. Res. 98(D3): 4919-4930

Li Z-L, Becker F (1993) Feasibility of land surface temperature and emissivuity determination from AVHRR data. Remote Sens. Environm. 43: 67-85

Lionello P, Malanotte-Rizzoli P, Boscolo R (eds.) (2005) Mediterranean Climate Variability. Elsevier, Amsterdam, The Netherlands

Liou KN (1980) An Introduction to Atmospheric Radiation. Academic Press, New York

Linke F (1942) Die Sonnenstrahlung und ihre Schwächung in der Attmosphäre. In: Handbuch der Geophysik, Bornträger, Berlin,8, pp 239-332

Linder W, Noilhan L, Berger M, Blümel K, Blyth E, Braud I, Dolman A, Fiedler F, Grunwald J, Harding R, v d Hurk B, Jaubert G, Mueller A, Ogink M (1995) Intercomparison of Surface Schemes Using EFEDA Flux Data. Technical Report, Meteo-France, CNRM/GMME/MC2, 42 Coriolis, 31057 Toulouse Cedex, 9 pp

Lambert JH (1760) Photometria sive de mensure de gratibus lumi-nis, colorum umbrae. Eberhard Klett

Llamas MR, Cruces J (1993) In: Bolle H-J, Streckenbach B, eds. (1993) EFEDA Final Report Contract EPOC-CT 90-0030, Berlin

Lloyd D (1990) A phenological classification of terrestrial vegetation cover using shortwave vegetation index imagery. Int. J. Remote Sensing 11: 2269-2279

Mainguet M (1991) Desertification - Natural Background and Human Mismanagement. Springer Verlag Berlin Heidelberg New York

Lönquist 0 (1950/51) Ark. Geof. 1: 79, 441 and (1954) Ark. Geof. 2: 246

Melia J, Lopez-Baeza E, Segarra D, Bodas A, Fortea JC, Lanjeri S (1999) Synthesis of Change Detection Parameters Into a Land-Surface Change Indicator for Long-Term Desertification Studies (RESYSMED). Final Report Contract No.: ENV4-CT97-0683. Firenze and München

Lopez-Baeza E (2001) MIDAS: Medidas Radiometricas en Microondas y Desarrollo de Algoritmos para la Mision SMOS [Microwave Radiometric Measurements and Algorithm Development for the SMOS Mission]. Spanish Ministry for Science and Technology, National Programme on Space Research (28-12-2001 to 27-12-2003)

López-Baeza E, Antolin C, Bodas A, Castell N, Ferrer F, Gimeno JF, Rausell J, Saleh K (2001) La Station de Reférence VALENCIA ANCHOR STATION de Donnés et de Produits de Télédétection à Grande Échelle. XIV Colloque Internacional de Climatologie. Seville 12-15 Septembre 2001

Lopez-Baeza E (2002) SCALES: Area de Calibracion/Validacion para Experimentos de Campo en Gran Escala con SEVIRI/GERB [SEVIRI & GERB Cal/Val Area for Large-scale Field Experiments]. Spanish Ministry for Science and Technology, National Programme on Space Research (01-10-2002 to 30-09-2005)

Lopez-Baeza E, (2003a) Definicion del Proceso de Datos de la Mision Espacial SMOS en la Estacion de Villafranca del Castillo (SMOS-GS-B). "Parte UVEG" (MIDAS-2) [Data Process Definition for the SMOS Space Mission at Villafranca del Castillo Station (SMOS-GS-B). "UVEG Part" (MIDAS-2)]. Spanish Ministry for Science and Technology, National Programme on Space Research (23-12-2003 to 22-12-2004 extended to 22-06-2005)

Lopez-Baeza E, The SCALES Project Team (2003b) SCALES: Area de Calibracion/Validacion para Experimentos de Campo en Gran Escala con SEVIRI/GERB, Informe de Primer Año, [SEVIRI & GERB Cal/Val Area for Large-scale Field Experiments, First Year Report]. Spanish Ministry for Science and Technology, National Programme on Space Research, Sept. 2003

Lopez-Baeza E, The SCALES Project Team (2003c) SCALES: Area de Calibracion/Validacion para Experimentos de Campo en Gran Escala con SEVIRI/GERB, Informe de Segundo Año, [SEVIRI & GERB Cal/Val Area for Large-scale Field Experiments, Second Year Report]. Spanish Ministry for Science and Technology, National Programme on Space Research, Sept. 2004

Lopez-Baeza E (2004) Tecnicas de Teledeteccion para la Observacion de Parametros Medioambientales en la Comunidad Valenciana en el Período 2004-2006 [Remote Sensing Techniques for the Observation of Environmental Parameters in the Valencian Community During the Period 2004-2006]. Regional Government of Valencia, Regional Ministry for Planning and Housing, Directorate General for Environmental Quality.

Lopez-Baeza E, Alonso S, Bodas A, Comeron A, Diaz-Pabon R, Domenech C, Gimeno-Ferrer JF, Jorge J, Labajo A, Pineda N, Pino D, Rius A, Rocadenbosch F, Saleh K, Sicard M, Tarruella R, Torrobella J, Velazquez A (2005) A high-quality dataset of land-surface and atmospheric measurements for the comparison/cross-calibration of data from large scale optical Earth Observation sensors in Space. The Valencia Anchor Station. Workshop on Inter-Comparison of Large Scale Optical and Infrared Sensors. ESA/ESTEC Noordwijk, The Netherlands 12 – 14 October 2004

Lord Rayleigh (1899) On the transmission of light through an atmosphere containing small particles in suspension, and on the origin of the blue sky. Phil. Mag. 47: 375-384. See also (1902) Scientific Papers. Cambridge, England, and (1918) On the scattering of light by a cloud of similar small particles of any shape and oriented at random. Phil. Mag. 35: 373-381

Lorentz HA (1880) Wiedem. Ann. 9: 681, see also Lorentz HA (1905) Over de absorptie- en emissiebanden van garformige Lichamen. I. au. II. Versl. Amsterdam Acad. 14: 518, 577; Lorentz HA (1916) The Theory of Electrons. Stechert & Co., New York

Lorenz D (1973) Die radiometrische Messung der Boden- und Wasseroberflächentemperatur und ihre Anwendung, insbesondere auf dem Gebiet der Meteorologie. Z. Geophysik 39: 627-701

Los SO (1993) Calibration adjustment of the NOAA AVHRR normalized difference vegetation index without recourse to component channels. Int. J. Rem. Sens. 14: 1907-1917.

Los SO (1998) Linkages Between Global Vegetation and Climate. An Analysis based on NOAA Advanced Very High Resolution Radiometer Data. PhD. Dissertation, Vrije Universiteit, Goddard Space Flight Centre, Greenbelt Maryland, 1998

Los S (1998) Estimation of the Ratio of Sensor Degradation Between NOAA AVHRR Channels 1 and 2 from Monthly NDVI Composites. IEEE Transactions on Geoscience and Remote Sensing 36(1):206-213.

Los SO (1998) Linkages between Global Vegetation and Climate, An Analysis based on NOAA Advanced Very High Resolution Radiometer Data. PhD Thesis, Faculty of Earth Sciences, June 1998, Vrije Universiteit Amsterdam, The Netherlands

Los SO, Justice CO, Tucker CJ (1994) A global 1 degree by 1 degree NDVI data set for climate studies derived from the GIMMS continental NDVI. Int. J. Rem. Sens., 15: 3493-3518.

Lovejoy S, Austin GL (1979) The sources of error in rain amount estimating schemes from GOES visible and IR satellite data. Mon. Wea. Rev. 107: 1048-1054

Lucas NJ, Curran PJ, Plummer SE, Danson FM (2000) Using the red edge inflection point ti drive an ecosystem simulation model for calculating the stem carbon production of an upland coniferous plantation. Int. J. Remote Sensing 21: 619-631

Mahfouf J-F (1991) Analysis of soil moisture from near-surface parameters: a feasibility study. J. Applied Meteorol. 30: 1534-1547.

Main IG (1990) Vibrations and Waves inPhysics. Cambridge University Press, Cambridge, UK

Malingreau JP (1986) Global vegetation dynamics: Satellite observations over Asia. Int. J. Remote Sensing 7: 1121-1146

Marceau DJ, Howarth PJ, Gratton DJ (1994a) Remote sensing and the measurement of geographical entities in a forest environment 1: The scale and spatial aggregation problem. Remote Sensing of Environment 49: 93-104

Marceau DJ, Howarth PJ, Gratton DJ (1994b) Remote sensing and the measurement of geographical entities in a forest environment 2: The optimal spatial resolution. Remote Sensing of Environment 49:

105-117

MAPA (1997). Anuario de Estadística Agraria. Ministerio de Agricultura, Pesca y Alimentación. Madrid (España).

MARS 1997. MARS Bulletin, 1996-1997 Season. Situation from the 1 September to 31 December 1996. Vol.5, 1. J.R.C. Ispra (Italy).

Mariano AJ and Brown OB (1992) Efficient objective analysis of dynamically heterogeneous and nonstationary fields via the parameter matrix. Deep-Sea Res. 39 (7/8): 1255-1271

Markham BL, Barker JL (1985) Spectral of the LANDSAT Thematic Mapper sensors. International Journal of Remote Sensing, 6(5): 697-716.

Markham BL, Halthore RN, Goetz SJ (1992) Surface reflectance retrieval from satellite and aircraft sensors: Results of sensor and algorithm comparisons during FIFE. J. Geophys. Res. 97(D17): 18785-18795

Maselli F (2001) Definition of spatially variable spectral end-members by locally calibrated multivariate regression analyses. Remote Sens. Environ. 75: 29-38

Maselli F (2005) private communication

Maselli F, Conese C, Petkov L, Resti R (1992) Inclusion of Prior Probabilities Derived from a Nonparametric Process into the Maximum Likelihood Classifier. Photogrammetric Engineering and Remote Sensing 58: 201-207

Maselli F, Conese C, Petkov L, Gilabert MA (1993) Environmental Monitoring and crop forecasting in the Sahel through the use of NOAA NDVI data. A case study: Niger 1986-89. International Journal of Remote Sensing, 14: 3471-3487

Maselli F, Conese C, Petkov L (1993) Use of Probability Entropy for the Estimation and Graphical Representation of the Accuracy of Maximum Likelihood Classifications. ISPRS Journal of Photogrammetry and Remote Sensing 49(2): 13-20

Maselli F, Rodolfi A, Conese C (1996) Multi-Scale Classification of Remotely Sensed Data by the Maximization of Fuzzy Membership Grades. Remote Sens. Reviews 13: 187-205

Maselli, F., Rembold, F., Rossini, P., 1998a. Preliminary evaluation of different NOAA-AVHRR data sets for the assessment of land surface changes in North African countries. In CAMELEO (Changes in Arid Mediterranean Ecosystems on the long term through Earth Observation) Annual Progress Report Year 1998, Edited by R. Escadafal, H. Bohbot, J. Mégier, Space Applications Institute, JRC, Ispra, Italy.

Maselli F, Gilabert MA, Conese C (1998) Integration of high and low resolution NDVI data for monitoring vegetation in Mediterranean environments. Remote Sensing of Environment 63: 208-218

Maselli F, Bonora L, Chirici G, Bindi M, Lori C (1999) Synthesis of change detection parameters into a land-surface change indicator for long term desertification studies in the Mediterranean area. Final report of Resysmed project. EU contract n. ENV4-CT97-0683.

Maselli F, Petkov L, Conese C, Battista P, Bonora L, Chirici G (1999) Synthesis of change detection parameters into a land-surface change indicator for long term desertification studies in the Mediterranean area.- Final report of the EU-progetto RESYSMED. Contract n. ENV4-CT97-0684

Maselli F, Romanelli S, Bottai L, Maracchi G (2000) Processing of GAC NDVI MVCs for yield forecasting in the Sahelian region. International Journal of Remote Sensing 21(18): 3509-3523

Maselli F, Rembold F (2001) Analysis of GAC NDVI data for cropland identification and yield forecasting in Mediterranean African countries. Photogrammetric Engineering and Remote Sensing 67(5): 593-602

Mätzler C, Standley A (2000) Relief effects for passive microwave remote sensing. Int. J. Remote Sensing 21(2):2403-2412. See also Mätzler C, Rosenkranz PW, Battaglia A, Wigneron JP (eds.) (2005) Thermal Microwave Radiation - Applications for Remote Sensing. IEE Electromagnetic Waves Series, London, UK

McClatchey RA, Fenn RW, Selby JEA, Volz FE, Garing JS (1971) Optical Properties of the Atmosphere. AFCRL-TR-71-0279, Environment Research Paper No. 354, L.G. Hanscom Field, Bedford, Massachusetts, U.S.A. 3rd Edition 1972, Environmental Research Paper No. 411, Air Force Cambridge Research Laboratories, U.S.A.

McCree KJ (1972) The action spectrum, absobance and quantum yield of photosynthesis in crop plants. Agric. Meteorology 9: 191-216

McMillin LM (1975) Estimation of sea surface temperature from two infrared window measurements with different absorption. Journal of Geophysical Research 80: 5113-5117.

Meliá J, Segarra D, Belda F, Caballer P, Fortea JC, García J, López E, Moreno J, Serrano JC, Viedma O

(1996) Evaluacion mediante teledetección de los procesos de reforestacion en zonas afectadas por incendios. "La restauracion de la cubierta vegetal en la Comunidad Valenciana" pp. 149-212. CEAM. Generalitat Valenciana

Mie G (1908) Beiträge zur Optik trüber Medien. Ann. Physik 25: 377-445

Menenti M (1984) Physical Aspects and Determination of Evaporation in Deserts Applying Remote Sensing Techniques. Thesis, Instituut voor Cultuurtechniek en Waterhuishouding, Wageningen, NL

Menenti M, Bastiaanssen WGM, van Eick D, Abd El Karim MH(1989) Linear relationships between surface reflectance and temperature and their application to map actual evaporation of groundwater. Adv. Space Res. 9:165-176.

Menenti M, Choudhury B (1993) Parameterization of land surface evaporation by means of location dependent potential evaporation and surface temperature range. Proc. IAHS Conference on Land Surface Processes. IAHS Publ. 212: 561-568

Menenti M, Ritchie JC (1994) Estimation of effective aerodynamic roughness of Walnut Gulch watershed using laser altimeter measurements. Water Resour. Res. 30: 1329-1337

Meyer D, Verstraete M, Pinty B (1995) The effect of surface anisotropy and viewing geometry on the estimation of NDVI from AVHRR. Rem. Sens. Rev. 12: 3-27.

Michaud JD, Shuttleworth WJ (Eds) (1998) Aggregate description of land-atmosphere interactions. Special issue of Journal of Hydrology 190(3-4): 176-412

Michels BI (1989) CHESS or How to calculate atmospheric correction from global radiation data in desert areas. SC-DLO Internal Note 19, Wageningen, The Netherlands: 51

Michels BI (1992) Fluxes of Heat and Water Vapour in a Convective Mixed Layer During EFEDA. Forschungsbericht DLR-FB 92-21, DLR Institut für Physik der Atmosphäre, Oberpfafenhofen

Michels BI, Jochum AM (1995) Heat and moisture flux profiles in a region with inhomogeneous surface evaporation. Journal of Hydrology 166: 383-407

Millán M, Estrela MJ, Caselles V (1995) Torrential precipitation on the Spanish east coast: the role of the Mediterranean sea surface temperature. Atmospheric Research 36: 1-16

Millán M (2000) Multi-annual drought episodes in the Mediterranean (Valencia region) from 1950 - 1996. A spatio-temporal analysis. International Journal of Climatology 20: 1599 - 1618

Miller JR, Wu J, Boyer MG, Belanger M, Hare EW (1991) Seasonal patterns in leaf reflectance red-edge characteristics. Int. J. Remote Sens. 12(7): 1509-1523.

Milly PCD, Dunne KA (1994) Sensitivity of the global watercycle to the water-holding capacity of land. J. Climate 7:506-526.

Minnaert N (1941) The reciprocity principle in lunar photometry. Astrophysical Journal 93: 403-410.

Mintz Y, Serafini YV (1989) Global climatology of soil moistureand water balance. LMD note 148, Laboratoire de Météorologique Dynamique du CNRS, Paris.

Mintz Y, Serafini YV (1992) A global climatology of soil moistureand water balance. Climate Dynamics 8: 13-27.

Mo T (1995) Calibration of the Advanced Microwave Sounding Unit-A for NOAA-K. NOAA Technical Report NESDIS 85

Mo T, Choudhury BJ, Schmugge TJ, Jackson TJ (1982) A model for microwave emission from vegetation-covered fields. Journal of Geophysical Research, Vol. 87: 11229-1237

Moene AF, de Bruin HAR, Holtslag AAM (1995) Validation ofthe surface parametrization of HIRLAM using surface-based measurementsand remote sensing data. KNMI scientific report WR 95-07, KNMI, De Bilt, 45 pp.

Mohan M, Siddiqui TA (1998) Applied modeling of surface fluxes under diferent stability regimes. J. Appl. Met. 37: 1055-1067

Möller KD, Rothschild WG (1971) Far-Infrared Spectroscopy. Wiley-Interscience, New York London Sydney Toronto.

Möller F (1943) Das Strahlungsdiagramm. Reichsamt für Wetterdienst

Monin AS, Yaglom AM (1971) Statistical fluid mechanics: Mechanics of turbulence, 2 vol I, MIT Press, London

Monin AS, Obukhov AM (1954) Basic Laws of Turbulence Mixing in the Ground Layer of the Atmosphere. Tr. Geofiz. Inst. Aka. Nauk, SSSR, 24: 163-187

Moreno JF (1994) Spectral/spatial integration effects on information extraction from multispectral data: Multiresolution approaches. In: Multispectral and Microwave Sensing of Forestry, Hydrology and Natural Resources, SPIE Vol. 2314

Moreno JF, Meliá J (1994) An optimum interpolation method applied to the resampling of NOAA AVHRR data. IEEE Transactions on Geoscience and Remote Sensing, 32 (1): 131-151.

Moreno JF, Green RO (1996) Surface and atmospheric parameters retrieval from AVIRIS data: The importance of non-linear effects. In: Summaries of the Sixth Annual JPL Airborne Geoscience Workshop, March 4-8, JPL Publ. 96-4, Vol 1

Myneni RB, Ross J, Eds. (1991) Photon-Vegetation Interactions. Springer Verlag. Berlin Heidelberg New York

Myneni RB, Asrar G (1994) Atmospheric effects and spectral vegetation indices. Remote Sens. Environ. 47: 390-402

Myneni RB, Williams DL (1994) On the relationship between FAPAR and NDVI. Remote Sens. Environ. 49: 200-211.

Nadezhdina N, Cermak J (1998a) The technique and instrumentation for measurements the sap flow rate in plants. Patent No. 286438 (PV-1587-98). (U.S. Patent and Trademark rec. No.69055, 1997)

Nadezhdina N, Cermak J, Nadezhdin N (1998b) Heat field deformation method for sap flow measurements. Proc. 4th. International Workshop on Measuring Sap flow in Intact Plants. Zidlochovice, Czech Republic, Oct.3-5, 1998. 72-92 pp. IUFR Publications, Publ. House Mendel. Univ. Brno

Nadezhdina, N, Cermak J, Ceulemans R (2001) Radial pattern of sap flow in woody stems related to positioning of sensors and scaling errors in dominant and understorey species. Tree Physiology (in press).

Navascues B (1996) Analysis of 2m temperature and specific humidity. Proceedings of HIRLAM 3 workshop on soil processes andsoil/surface data assimilation, Madrid, 27-29 November 1995, pp 38-44.

Neckel H, Labs D (1984) The Solar radiation between 3300 and 12500 Å. Sol Phys. 90: 205-258

Nemani RR, Running SW (1989) Testing a theoretical climate-soil-leaf area hydrologic equilibrium of forests using satellite data and ecosystem simulation. Agricoltural and Forest Meteorology, 44: 245-260.

Nerry F, Petitcolin F, Stoll MP (1998) Bi-directional Reflectivity in AVHRR Channel 3: Application to a Region in Northern Africa. Remote Sens. Environ. 66: 298-316

Nerry F, Labed J, Stoll M-P (1988) Emissivity signatures in the thermal IR band for remote sensing: calibration procedure and method of measurement. Applied Optics 27:758-764

NES 107 (1988) s. Planet

Nguyen L, Minnis P, Ayers JK, Doelling DR (2001) Intercalibration of Meteorological Satellite Imagers Using VIRS, ATS-2 and MODIS, Proc. AMS 11th Conf. Atmos. Radiation, Madison, WI, Oct 15-18.

Nicodemus FE, Richmond JC, Hsia JJ, Ginsberg IW, Limperis T (1977) Geometrical considerations and nomenclature for reflectance, Natl. Bur. Stand. Monogr. 160, U.S. Govt. Printing Office, Washington, D.C..

Njoku EG, Brown OB (1993) Sea surface temperature.. In: Gurney RJ, JL Foster, Parkinson CL (eds), Atlas of satellite observations related to global change. Cambridge University Press, Cambridge, pp. 237-250

Njoku E (1995) Surface temperature estimation over land using satellite microwave radiometry. In: Choudhury et al. (eds) Passive microwave remote sensing of Land-Atmosphere interactions. VSP Press, Utrecht, The Netherlands, pp 509-530.

Njoku E, Jackson, TJ, Lakshmi V, Chan T, Nghiem S (2003) Soil Moisture Retrieval from AMSR-E. IEEE Transactions on Geoscience and Remote Sensing. 41: 215-229

Noilhan J, Planton S (1989) A simple parametrization of landsurface processes for meteorological models. Mon.Weather Rev.117: 536-549

Obukhov AM (1946) Turbulence in an atmosphere with nonuniform temperature. Tr. Akad. Nauk, SSSR Inst. Teoret. Geofi. No. 1. Translation: (1971) Boundary Layer Meteorology 2: 7-29

Ogink-Hendriks MJ, Kabat P, Elbers JA, Bastiaanssen WGM, van der Elsen HGM (1995) Contribution to the EFEDA field campaignsin 1991 and 1994. Report 112, DLO-Winand Staring Centre, Wageningen, The Netherlands, 150 pp.

Oh Y, Sarabandi K, Ulaby FT (1992) An empirical model and an inversion technique for radar scattering from bare soil surfaces. IEEE Transactions on Geoscience and Remote Sensing 30: 370-381

Olesen F-S, Kind O, Reutter H (1995) Hight resolution time series of IR data from a combination of IR data of AVHRR and METEOSAT. Advances in Space Research, 16 (10): 141-146

Oliver HR, Sene KJ (1992) Energy and water balances of developing vines. Agricultural And Forest Meteorology 61: 167-185

Otterman J (1974) Baring high-albedo soils by overgrazing: a hypothesized desertification mechanism. Science 186: 531-533

Otterman J (1989) Enhancement of surface-atmosphere fluxes by desert.fringe vegetation through reduction of surface albedo and of soil heat flux. Theoretical and Applied Climatology 40: 67-79

Owe M, Van de Griend AA (1990) A daily surface moisture model for large area semi-arid land application with limited climate data. Journal of Hydrology 121: 119-132

Owe ML, Van de Griend AA, Chang ATC (1992) Surface soil moisture and satellite passive microwave observations in Semi-Arid Southern Africa. Water Resources Res.. 28(30): 829-839

Owe M, Van de Griend AA (1998) Comparison of soil moisture penetration depths for several bare soils at two frequencies and implications for remote sensing. Water Resources Research 34(9): 2319-2328

Owe M, Van de Griend AA (2001) On the relationship between thermodynamic surface temperature and high frequency (37 GHz) vertically polarized brightness temperature under semi-arid conditions. Int. J. Of Remote Sensing, 22(17): 3521-3532

Parlow E, Vogt R (1998) In: Remote Sensing of Mediterranean Desertification and Environmental Stability (RESMEDES), Final Report Project No. ENV4-CT95-0094, Firenze

Parlow E, Vogt R, Bing G (1999) In: Synthesis of Change Detection Parameters into a Land-surface Change Indicator for Long Term Desertification Studies in the Mediterranean Area (RESYSMED). Final Report Contract No. ENV4-CT97-0683, Firenze and München

Pelgrum H, Bastiaanssen WGM (1996) An intercomparison of techniques to determine the area-averaged latent heat flux from individual in situ observations: A remote sensing approach using the European Field Experiment in a Desertification-threatened Area data. Water Resources Research 32: 2775-2786

Pellarin T, Wigneron J-P, Calvet J-C, Waldteufel P (2003) Global soil moisture retrieval from a synthetic L-band brightness temperature data set. J. of Geophys. Res. 108 (D12): 4364

Penner SS (1959) Quantitative Molecular Spectroscopy and Gas Emissivities. Addison-Wesley Publishing Company, Inc., Reading, Massachusetts, U.S.A. and Pergamon Press, London, Paris

Perrini L, Brunetti A, Vento D (1995) Il Modello ad Area Limitata del Ministero delle Risorse Agricole, Alimentari e Forestali. AER 11: 21-25

Peters SWM (1995) Energy- and water-balance modeling of a semi-arid area using remote sensing. PhD. Thesis, Vrije Universiteit Amsterdam, pp 256

Peters SWM, Gouweleeuw BT, Van de Griend AA (1996) Regional evapotranspiration monitoring of the Iberian Peninsula during three months in 1991 using the DEMI-SEC model and METEOSAT observations; a multi-site calibration and validation study. Study carried out in support of EFEDA-II and RESMEDES. Institute of Environmental Studies, EFEDA-II/RESMEDES-Report, EC-contract EV5V-CT93-0284, IVM, Vrije Universiteit Amsterdam, The Netherlands, 43 pp

Petitcolin F, Nerry F, Bruckert D, Stoll MP (1997) Atmospheric Corrections of Infrared NOAA AVHRR Images Using Meteorological Data Extracted from Global Circulation Database. Proceedings of the Meteorological Satellite Data Users Conference, pp 497-501, Brussels, Belgium

Petitcolin F (1998) Mesure, analyse et cartographie de la reflectivité bidirectionnelle des surfaces terrestres à 3.7 μm à partir des données NOAA/AVHRR. Implication sur la mesure de l'émissivité et l'accès à la température de surface. Ph. D. Thesis, University of Strasbourg, France.

Pfister C (1999) Wetternachhersage. 500 Jahre Klimavariationen und Naturkatastrophen. Bern, Haupt, 304pp

Phulpin T, Derrien M, Brard A (1983) A Two-Dimensional Histogram Procedure to Analyse Cloud Cover from NOAA Satellites High Resolution Imagery. J. Clim. Appl. Meteor. 22: 1332-1345

Planet WG, ed. (1988) Data extraction and calibration of TIROS-N/NOAA radiometer, revised NOAA Technical Memorandum, NES 107

Pinto JG, Klawa M, Ulbrich U, Rudari R, and Speth P (2001) Extreme precipitation events over northwest Italy and their relationship with tropical-extratropical interactions over the Atlantic. Proceedings of the 3rd EGS Plinius Conference held in Baja Sardinia, Italy, October 2001

Pinty B, Verstraete MM (1992a) On the design and validation of surface bidirectional reflectance and albedo models. Remote Sens. Environ. 41: 155-167.

Pinty B, Verstraete MM (1992b) GEMI: a non-linear index to monitor global vegetation from satellites. Vegetatio 101: 15-20.

Portela A, Castro M (1996) Summer thermal lows in the Iberian peninsula: A three-dimensional simulation. Quart. J. Roy. Met. Society 122: 1-22

Potdar MB, Narayana A (1993) Determining short-wave planetary albedo from spectral signatures of land-ocean features and albedo mapping using NOAA-AVHRR data. Acta Astronautica 29: 687-690

Prata AJ (1993) Land surface temperatures derived from the advanced very high resolution radiometer and the Along-Track Scanning Radiometer - 1. Theory. Journal of Geophysical Research 98: 16689-16702.

Price JC (1992) Estimating vegetation amount from visible and near infrared reflectances. Remote Sens. Environ. 41: 29-34

Price JC (1983) Estimation of surface temperatures from satellite thermal infrared data- a simple formulation for the atmospheric effect. Remote Sensing of Environment, 13: 353-361.

Price JC (1984) Land surface temperature measurements from the split window channels of the NOAA-7 AVHRR. Journal of Geophysical Research, 89: 7231-7237.

Price JC (1989) Quantitative aspects of remote sensing in the thermal infrared. In: Asar G (ed.) Theory and Applications of Optical Remote Sensing. Wiley, New York, pp 578-603

Price JC (1977) Thermal inertia mapping: A new view of the earth. J. Geophas. Res. 82: 2582-2590

Priestley CHB, Taylor RJ (1972) On the assessment of surface heat flux and evaporation using large-scale parameters. Mon. Weather Rev. 100: 81-92

Prigent C, Rossow WB, Mathewa E (1997) Microwave land surface emissivities estimated from SSM/I observations. J. Geophys. Res. 102: 21867-21890

Prince SD, Tucker CJ (1986) Satellite remote sensing of rangelands in Botswana II: NOAA AVHRR and herbaceous vegetation. International Journal of Remote Sensing, 7: 1555-1570.

Prince S D (1990) High Temporal Frequency Remote Sensing of Primary Production using NOAA AVHRR. Applications of Remote Sensing in Agriculture. Ed. M. D. Steven and J. A. Clark, Butterworths, U.K.. 4: 169-183.

Prince SD (1991) Satellite remote sensing of primary production: comparison of results for Sahelian grasslands 1981- 1988. International Journal of Remote Sensing, 12: 1,301-1,312.

Prince SD, Kerr YH, Goutorbe J-P, Lebei T, Tinga A, Bessemoulin P, Brouwer J, Dolman AJ, Engman ET, Gash JHC, Hoepffner M, Kabat P, Monteny B, Said F, Sellers P, Wallace J (1995) Geographical, biological and remote sensing aspects of the Hydrologic Atmospheric Pilot Experiment in the Sahel (HAPEX-Sahel). Remote Sens. Environ., 51: 215 - 234

Puyou-Lascassies P, Podaire A, Gay M (1994) Extracting Crop Radiometric Responses from Simulated Low and High Spatial Resolution Satellite Data Using a Linear Mixing Model. Int. J. of Remote Sensing 15(18): 3767-3784

Qi J, Chehbouni A, Huete AR, Kerr YH, Sorooshian S (1994) A modified soil adjusted vegetation index. Remote Sens. Environ. 48:119-126.

Raffy M, Dugdale G (1990) In: Dugdale et al. (Eds.) Remote Sensing of Energy and Mass Exchanges at the Surface of the Earth, Twinning project no. ST 2 J 0222, Final Report

Raffy M (1992) Change of Scale in Models of Remote Sensing: A General Method for Spatialization of Models. Remote Sens. Environment 40: 101-112

Raffy M, Becker F (1986) A stable iterative procedure to obtain soil surface parameters and fluxes from satellite data. IEEE Transactions on Geoscience and Remote Sensing 24: 327-333

Rahman H (1996) Atmospheric optical depth and water vapour effects on the angular characteristics of surface reflectance in NOAA-AVHRR. Int. J. Of Remote Sensing 17(15): 2981-2999

Rahman H, Dedieu G (1994) SMAC: A simplified software for the atmospheric correction of satellite measurements in the solar spectrum. Intern. J. of Remote Sensing 15: 123-143

RahmanH, Pinty B, Verstraete M (1993) Coupled Surface Atmospheric Reflectance (CSAR) model, 2, Semiempirical surface model usable with NOAA Advanced Very High Resolution Radiometer data. J. Geophys. Res. 98: 20791-20801

Randel DL, Vonder Haar TH, Ringerud MA, Stephens GL, Greenwald TJ, Combs CL (1996) A new global water vapor dataset. Bull. Amer. Meteor. Soc., 7, no 6 (June).

Rao, C.R.N., 1993a. Nonlinearity corrections for the thermal infrared channels of the Advanced Very High Resolution Radiometer: assessment and recommendations. NOAA Technical Report NESDIS-69, NOAA/NESDIS. Washington, DC.

Rao, C.R.N., 1993b. Degradation of the visible and near infrared channels of the Advanced Very High Resolution Radiometer on the NOAAP9 spacecraft: assessment and recommendations for corrections. NOAA Technical Report NESDIS-70. NOAA/NESDIS. Washington, DC.

Rao CRN (1999) Post-launch calibration of the visible and near infrared channels of the AVHRR and GOES imager, CGMS XXVII

Rao CRN, Cao C, Zhang N (2003) Intercalibration of the Moderate-Resolution Imaging Spectroradiometer and the Along Track Scanning Radiometer-2. Int J. Remote Sensing, 24(9): 1913-1924

Rao CRN, Chen J (1993) Calibration of the visible and near-infrared channels of the Advanced Very High Resolution Radiometer (AVHRR) after launch. Proceedings of the SPIE Conference on Recent advances in sensors, radiometric calibration, and processing of remotely sensed data, 14-16 April 1993,SPIE,1938: 56-66.

Rao CRN, Chen J (1994) Post-launch calibration of the visible and near infrared channels of the Advanced Very High Resolution Radiometer on NOAA-7,-9, and -11 spacecraft. TR 78, NOAA/NESDIS, Washington,D.C.

Rao CRN, Chen JH (1995) Intersatellite calibration linkages for the visible and near-infrared channels of the advanced very high-resolution radiometer on the NOAA-7, NOAA-9, and NOAA-11 spacecraft. Int. J. Remote Sensing 16: 1931-1942

Rao N, Chen J (1996) Post-Iaunch calibration ofthe visible and near-infrared channels of the AVHRR on the NOAA-14 spacecraft. Int. J. Remote Sensing, V 01: 17:2743-2747.

Rao CRN, Chen J (1999) Revised post-launch calibration of the visible and near-infrared channels of the Advanced Very High Resolution Radiometer (AVHRR) on the NOAA-14 spacecraft. Int. J. Remote Sensing 20: 3485-3491.

Rapetti F, Vittorini S (1995) Carta climatica della Toscana Centro-Meridionale e Insulare. Pacini Editore, Pisa, Italy

Rasmusen MS (1998a) Developing simple, operational, consistent NDVI-Vegetation models by applying environmental and climatic information: Part I. Assessment of net primary production. International Journal of Remote Sensing, 19: 97-118

Rasmusen MS, 1998b. Developing simple, operational, consistent NDVI-Vegetation models by applying environmental and climatic information: Part II. Crop yield assessment. International Journal of Remote Sensing, 19: 119-140.

Rasool I (1984) On dynamics of deserts and climate. In: Houghton JT, ed., The Global Climate. Cambridge University Press, Cambridge

Rasool SI, Bolle H-J (1984) ISLSCP: International Satellite Land-Surface Climatology Project. Bull. Am. Met. Soc. 65: 143-144

Rayleigh s.Lord Rayleigh (JW Strutt)

RESRAPS (1994) Remote Sensing and Radiometric Properties of the Surface: Assessment of Desertification from Space. Final Report EFEDA Phase II, Group V Contract No.: EV5V-CT93-0284

RESMEDES (1998) Final Report of EC Contract Contract No.: ENV4-CT95-0094, European Commission, Directorate-General, Science, Research and Development

RESYSMED (1999) Synthesis of Change Detection Parameters into a Land-surface Change Indicator for Long Term Desertification Studies in the Mediterranean Area. Final Report Contract No.: ENV4-CT97-0683, Firenze and München

Richards LA (1931) Capillary conduction of liquids in porous mediums. Physics 1: 318-333

Richards JA (1993) Remote sensing digital image analysis: An introduction. 2nd Ed., Springer-Verlag Berlin

Richardson AJ, Wiegand CL (1977) Distinguishing vegetation from soil background information. Photogramm. Eng. Remote Sens. 43: 1541-1552.

Robinson N (1966) Solar Radiation. Elsevier Publishing Company, Amsterdam London New York

Rollin EM, Milton EJ (1994) Processing of high spectral resolution reflectance data for the retrieval of biophysical information. Proceedings of 6th International Symposium "Physical measurements and signatures in remote sensing", 17-21 January 1994, Val d'Isère, France, CNES, pp 357-364

Rosborough G, Baldwin D, Emery WJ (1994) Precise AVHRR Image Navigation. IEEE Geosci. Rem. Sens., 32: 644-657.

Ross RJ, Elliott WP (1996) Tropospheric water vapor climatology and trends over North America 1973-1993. J. Climate 9: 3561-3574

Ross RJ, Elliott WP (2001) Radiosonde-based Northern Hemisphere tropospheric water vapor trend. J. Climate 14: 1602-1612

Rossini P, Benedetti R (1993) On the Use of NDVI Profiles as a Tool for Agricultural Statistics: the Case Study of Wheat Yield Estimate and Forecast in Emilia Romagna. Remote Sensing of Environment, 45: 311-326.

Roujean JL, Leroy M, Podaire A, Deschamps PY (1992) Evidence of surface reflectance bidirectional effects from a NOAA/AVHRR multi-temporal data set. Int. J. Remote Sensing 13(4): 685-698.

Rouse JW Jr., Haas RH, Deering DW, Schell JA, Harlan JC (1974) Monitoring the Vernal Advancement and Retrogradation (Green Wave Effect) of Natural Vegetation. NASA/GSFC Type III Final report,

Greenbelt, MD., 371 pp

Royer A, Vincent P, Bonn F (1985) Evaluation and correction of viewing angle effects on satellite measurements of bi-directional reflectance. Photogrammetry Engineering and Remote Sensing 51: 1899-1914

Running SW, Coughlan JC (1988) A general model of forest ecosystem processes for regional applications. Ecological Modelling 42: 125-154

Running SW, Nemani RR, Peterson DL, Band LE, Potts DF, Pierce LL, Spanner MA (1989) Mapping regional forest evapotranspiration and photosynthesis by coupling satellite data with ecosystem simulation. Ecology 70(4): 1090-1101

Running SW, Gower ST (1991) Forest-BGC, a general model of forest ecosystem processes for regional applications. II. Dynamic carbon allocation and nitrogen budgets. Tree Physiol. 9: 147-160

Running SW, Hunt EWR (1993) Generalization of a forest ecosyswtem process model for other biomes, Biome-BGC, and an application for globval scale models. In: Ehleringer JR, Firld CB (Eds.) Scaling physiological processes: leaf to globe. Academic Press, San Diego, pp. 141-158

Russell MJ, Nunez M, Chladil MA, Valiente JA, Lopez-Baeza E (1997) Conversion of Nadir, Narrow-band Reflectance in Red and Near Infrared Channels to Hemispherical Surface Albedo. Remote Sensing of Environment 61: 16-23

Ryan EH, Mariano AJ, Olson DB, Evans RH (1996) Fall AGU Meeting Eos, Transactions, AGU, Vol 77, No. 46, November 12 (OP22A-16)

Saatchi SS, Lin D-S, Van Zyl JJ, Engman ET (1992) Soil moisture mapping using SAR imagery. Proc. IGARSS'92, p. 1750

Saatchi SS, Evans D, Dubois P, Van Zyl J (1993) SAR Experiment in Desertification-Threatened Areas: EFEDA-Spain. Jet Propulsion Laboratory, JPL D-10283, Pasadena

Sánchez Diaz J, Boluda Hernandez R, Artigao Ramirez A, Morel Cot C, Colomer Marco JC (1992) Study Soil Degradation. In: Boll H-J, Streckenbach B (eds) The ECHIVAL Field Experiment in a Desertification Threatened Area (EFEDA). First Annual Report. Berlin

Sánchez Diaz J, Boluda Hernandez R, Artigao Ramirez A, Morel Cot C, Colomer Marco JC, Guardado R (1994) Suelos. In: de Santa Olalla Mañas FM (coord.) Desertificación en Castilla-La Mancha . El Proyecto EFEDA. Ediciones de la Universidad de Castilla-La Mancha (Colleeión Estudios 23), Cuenca, cap. 5: 97-141

Sánchez Diaz J, Boluda Hernandez R, Artigao Ramirez A, Morel Cot C, Colomer Marco JC (1996) Assessment of Soil Degradation in Desertification Threatened Areas: A Case Study in Castilla-La Mancha (Spain). ECHIVAL Field Experiment in Desertification Threatened Areas. Departamento de Biologia Vegetal, Universitat de Valencia, UVALFF.DBV Group

Saunders RW, Pescod RW (1988) AVHRR Processing Over Land cLoud and Ocean, A.P.O.L.L.O. Manual, Meteorological Office Unit, Robert Hooke Institute für Cooperative Atmospheric Research, Clarendon Laboratory, Parks Road, Oxford OXI 3PU, England.

Saunders RW (1990) The determination of broad band surface albedo from AVHRR visible and near-infrared radiances. Int. J. Remote Sensing 11(1): 49-67.

Saunders RW, Kriebel KT (1988) An improved method for detecting clear sky and cloudy radiances from AVHRR data. Int. J. Remote Sensing 9(1): 123-150.

Schädlich S, Göttsche F-M, Olesen F-S (2001) Influence of land parameters and atmosphere on Meteosat brightness temperatures and generation of land surface temperature maps by temporally and spatially interpolating atmospheric correction. Remote Sensing of Environment, 75: 39-46.

Schaefer VJ (1968) The inadvertent modification of the atmosphere by air pollution. In: Singer SF (ed.) Global Effects of Environmental Pollution. Reidel Publ. Co. Dordrecht, pp.158-174

Schlittgen R, Streitberg BHJ (1994) Zeitreihenanalyse. Oldenbourg Verlag

Schmidt H, Karnieli A (2002) Analysis of the temporal and spatial vegetation patterns in a semi-arid environment observed by NOAA AVHRR imagery and spectral ground measurements. Int. J. Remote Sensing 23(19): 3971-3990

Schmidt H, Gitelson A (2000). Temporal and Spatial Vegetation Cover Changes in Israeli Transition Zone: AVHRR-based Assessment of Rainfall impact. International Journal of Remote Sensing, 21:997-1010.

Schmugge TJ, Jackson TJ, Kustar WP, Wang JR (1992) Passive microwave remote sensing of soil moisture: Results from HAPEX, FIFE, and MONSOON90. ISPRS J. of Photogrammetry and Remote Sensing 47: 127-143

Schmugge T (1985) Remote sensing of soil moisture. In: Anderson, Burt (eds) Hydrological Forecasting, John Wiley & Sons, pp 101-124

Schneider K (1999) Gekoppelte, flächenverteilte Modellierung von Pflanzenwachstum und Verdunstung im Ammer Einzugsgebiet mit dem prozeßorientierten Evapotranspirations- und Vegetationsmodell PROMET-V. Habilitationsschrift Geowissenschaftliche Fakultät der Ludwig-Maximilians-Universität München, pp. 263

Schnell RC (1975) The biogenic component of Sahelian eoloan dust: A possible drought factor. EOS 56: 994-995

Schnürch W (1968) Das richtungsabhängige Reflexionsvermögen natürlicher Materialien im kurzwelligen und mittleren Infrarot. Diplomarbeit für Physik, Ludwig-Maximilians-Universität München

Schroedter M, Olesen F, Fischer H (2003) Determination of LST distributions from single channel IR measurements: An effective spatial interpolation method for the use of TOVS, ECMWF and radiosonde profiles in the atmospheric correction scheme. International Journal of Remote Sensing 24: 1189-1196

Schwalb A (1978) The TIROS-N/NOAA A-G Satellite Series. NOAA/NESS Technical Memorandum NESS 95, Washington, D.C.

Schwiebus A (1998) Die Wärmebilanz einer landwirtschaftlichen Fläche am Beispiel eines Weizenfeldes. Diplomarbeit, Technische Universität Dresden, Fakultät für Forst-, Geo- und Hydrowissenschaften, Institut für Hydrologie und Meteorologie

Sellers PJ (1985) Canopy reflectance, photosynthesis and transpiration. Int. J. Remote Sensing.6: 1335-1372.

Sellers PJ, Mintz Y, Sud YC, and Dalcher A (1986) A simple Biosphere Model (SiB) for Use Within General Circulation Models. J. Atm. Sciences 43(6) 505-531

Sellers PJ (1989) Vegetation-canopy spectral reflectance and biophysical processes. In: Asrar G (Ed.) Theory and Applications of Optical Remote Sensing, Chapter 4, John Wiley and Sons, New York, pp. 733

Silva LF (1978) Radiation and instrumentation in remote sensing. In: Swain PH, Davis SM (eds) Remote Sensing: The Quantitative Approach, McGraw-Hill, New York, pp 21-135.

Sene KJ (1994) Parameterisations for energy transfers from a sparse vine crop. Agricultural and Forest Meteorology 71: 1-18

Shepard D (1968) A two-dimensional interpolation function for irregularly spaced data. Proc. 23rd National Conference of Association for Computing Machinery. Brandon/Systems Press Inc., Princeton, N.J., ACM Publication P-68, pp 617-524

Shettle EP (1984) Optical and radiative properties of a desert aerosol model. In: Fiocco G. (ed.) IRS '84: Current Problems in Atmospheric Radiation, Proceedings of the International Radiation Symposium Perugia, Italy, A. Deepak Publishing, Hampton, Virginia, U.S.A., pp. 74-77

Shukla J, Mintz Y (1982) Influence of land surfaceevapotranspiration on the Earth's climate. Science 215: 1498-1501.

Simonds JL (1963) Application of characteristic vector analysis to photographic and 0ptical response data. J. Opt. Soc. Am. 53: 968-974

Singh A (1989) Digital change detection techniques using remotely sensed data. Int. J. of Remote Sensing 10: 989-1003

Slater PN, Biggar SF, Holm RG, Jackson RD, Mao Y, Moran MS, Palmer JM, Yuan B (1987) Reflectance- and radiance-based methods for the in-flight absolute calibration of multispectral sensors. Remote Sensing of Environment 22: 11-37

Sobrino JA, Coll C, Caselles V (1991) Atmospheric correction for land surface temperature using NOAA-11 AVHRR channels 4 and 5. Remote Sensing of Environment, 38: 19-34.

Sobrino JA, Li Z, Stoll MP, Becker F (1996) Multi-channel and multi-angle algorithms for estimating sea and land surface temperature with ATSR data. International Journal of Remote Sensing, 17: 2089-2114

Sobrino JA, Caselles V, Coll C (1993) Theoretical split window algorithm for determining the actual surface temperature. Il Nuovo Cimento 16: 219-236

Sobrino JA, Li ZL, Stoll Mph, Becker F (1994) Improvement in the Split-Window technique for land surface temperature determination. IEEE Trans. Geosci. Remote Sensing 32: 243-253

Song J, Gao W (1999) An improved method to derive surface albedo from narrowband AVHRR satellite data (narrowband to broadband conversion). J. Appl. Meteor. 38: 239-249

Sotes V (1996) El sector vitivinícola regional y futuro dentro del marco europeo. La vid y el vino en Castilla-La Mancha. Salinas (ed) Junta de Comunidades de Castilla La-Mancha.

Spanner MA, Pierce LL, Petgersons DL, Running SW (1990) Remote sensing of temperate coniferus forest leaf area index. The influence of canopy closure, understorey vegetation and background reflectance. International J. Remote Sensing 11(1): 95-111

Spitters CJT, van Keulen H, Kraailingen V (1989) A simple and universal crop growth simulator: SUCROS87. In: Rabbinge R, Ward SA, van Laar HH (eds.) Simulation and systems management in crop protection. PUDOC, Wageningen, The netherlands, pp. 147-181

Squires P (1958) The microstructure and colloidal stability of warm clouds. 1. The relation between structure and stability. Tellus 10: 256-261

Staylor WF (1990) Degradation rates of the AVHRR visible channel for the NOAA 6, 7, and 9 Spacecraft. J. Atmos. Ocean. Tech. 7: 411-423

Steffen WL, Walker BH, Ingram JS, Koch GW (1992) Global change and terrestrial ecosystems. The Operational Plan. IGBP Report No. 21, Stockholm

Steinwand DR, Hutchinson JA, Snyder JP (1995) Map Projections for Global and Continental Data Sets and an Analysis of Pixel Distortion Caused by Reprojection. Photogrammetric Engineering & Remote Sensing 61: 1,487-1,497.

Strahler A N (1975) Physical Geography, Fourth Edition. John Wiley & Sons, Inc., New York, London, Sydney, Toronto, 643 pp.

Sud YC, Shukla aJ, Mintz Y (1988) Influence of Land Surface Roughness on Atmospheric Circulation and Precipiotation: A Sensitivity Study with a General Circulation Model. Journal of Applied Meteorology 27: 1036-1054

Sud YC, Smith WE (1985) The influence of surface roughness of deserts on the July circulation. Boundary-Layer Meteorology 33: 15-49

Susskind J, Rosenfeld J, Reuter D, Chahine M T (1984) Remote sensing of weather and climate parameters from HIRS2/Msu on TIROS-N. Journal of Geophysical Research 89: 4677-4697.

Swearingen, W., 1998. Agriculture and Environmental Constraints in North Africa, Conference Agricultural Development in Central Asia, Between Russia and the Middle East, November 20-22, 1998, University of Washington, Seattle. Proceedings

Swinbank WC (1963) Longwave radiation from clear skies. Quart. J. Roy. Meteor. Soc. 89: 339-348

Tamayo, J., 1994. Relación entre las precipitaciones y el Indice de Vegetacion Diferencia Normalizado (NDVI) en areas de riesgo de desertificacion. Aplicacion a la zona de estudio del proyecto EFEDA. Ph.D. Thesis. University of Valencia. Spain.

Tanré D, Holben RN, Kaufman YJ (1992) Atmospherie correetion algorithm for NOAA-AVHRR produets: theory and applieation. IEEE Transactions on Geosience and Remote Sensing, Vol.30, No. 2.

Tanré D, Kaufman YJ, Herman M,Mattoo S (1997) Remote sensing of aerosol properties over oceans using the MODIS/EOS spectral radiance. J. Geophys. Res. - Atmospheres 102(D14): 16971-16988

Tanrè D, Herman M, Deschamps PY, Leffe (1979) Atmospheric Modelling for Space Measurements of Ground Reflectances, Including Bidirectional Properties. Appl. Optics 18: 3587-3594

Tarpley JD, Schneider SR, Money RL (1984) Global Vegetation Indices from the NOAA-7 meteorological satellite. J. Clim. Appl. Meteor. 23: 491-494.

Tatarskii VI (1961) Wave propagation in a turbulnt Medium. McGraw Hill, N.Y.

Teillet PM (1986) Image correction for radiometric effects in remote sensing. International Journal of Remote Sensing 12: 1637-1651.

Thekaekara MP (1974) Extraterrestrial solar spectrum, 3000-6100 Å at 1-Å intervals. Appl. Opt. 13(3): 518-522.

Thiermann V, Grassl H (1992) The measurement of turbulent surface-layer fluxes by use of bicromatic scintillation. Boundary Layer Meteorology 58: 367-389

Thornton PE, Running SW (1999) An improved algerithm for estimating incident daily solar radiation from vmeasurements of temperature, humidity, and precipitation. Agricultural and Forest Meteorology 93: 211-228

Thornwaite CW (1944) Report of the Committee on Transpiration andEvaporation. Transactions of the American Geophysical Union 25: 683-693

Townshend JRG, Justice CO (1986) Analysis of dynamies of Afriean vegetation using the NDVI. Int. J. Remote Sensing 7: 1224-1242

Trishchenko AP, Cihlar J, Li Z (2002) Effects of spectral response function on surface reflectance and NDVI measured with moderate resolution satellite sensors. Remote sensing of Environment 81: 1-18

Tristan J (1997) Transformacion espectral del albedo de banda estrecha a banda ancha. Aplicacion a la obtencion del albedo a patir del canal visible de METEOSAT. Proyecto Fin de Carrera. Escuela Tecnica Superior de Ingenieros de Telecomunicacion, Universidad Politecnica de Valencia (Enero

1997). Dirigido por Dr E. Lopezz-Baeza. See also: Tristan J, Lopez-Baeza E, Fuster JM, Valiente JA (1997) Spectral transformation of METEOSAT-visible-band surface albedo to broad-band surface albedo and its dependency on solar zenith angle. VII Congreso Ncional de Teloedeteccion. Asociacion Espanola de Teledeteccion. Santiago de Compostela (Spain), 26-28 June

Tucker CJ, Vanpraet CL, Charman MJ, Van Ittersum G (1985) Satellite remote sensing of total herbaceous biomass productions in the Senegalese Sahel:1980-1984. Remote Sensing of Environment 17: 239-249.

Tucker CJ (1983) Satellite Remote Sensing of Total dry Matter Accumulation in Senegalese Sahel. Rem. Sens. Environment 13: 461-470

Tucker CJ (1979) Red and photographic infrared linear combinations for monitoring vegetation. Remote Sensing of Environment 8: 127-150

Tucker CJ, Van Praet CL, Boerwinkel E, Gaston A (1985a) Satellite remote sensing of total herbaceous biomass production in the Senegalese Sahel : 1980-1984. Remote Sensing of Environment, 13:461-474.

Tucker CJ, Townshend JRG, Goff TE (1985b) African Land-cover classification using satellite data. Science 227: 369-375

Twomey S (1977) Introduction to the Mathematics of Inversion in Remote Sensing and Indirect Measurements. Elsevier Scientific Publishing Company Amsterdam Oxford New York

Ulaby FT, Moore RK, Fung AK (1986) Microwave Remote Sensing - Active and Passive. Vol. I: Fundamentals and Radiometry, pp 231, Vol. III: Scattering and Emission Theory, Advanced System and Applications (From Theory to Applications. Chapter 19, p. 1522-1646). Artech House Publ., London

Ulivieri C, Castronuovo MM, Franconi R, Cardillo A (1994) A split-window algorithm for estimating land surface temperature from satellites. Adv. Space Res. 16: 59-65

United Nations (1992) Convention on Desertification, United Nations Conference on Environment and Development, Rio de Janeiro, Brazil, 3-14 June

Unsöld A (1938, 1955 2nd edition) Physik der Sternatmosphären. Berlin, Springer

Valiente JA, Nunez M, Lopez-Baeza E, Moreno JF (1995) Narrow-band to broad-band conversion for Meteosat-visible channel and broad-band albedo using bosh AVHRR-1 and -2 channels. International Journal Remote Sensing, 16(6): 1147-1166.

Vall-llossera M, Camps A, Corbella I, Torres F, Duffo N, Monerris A, Sabia R, Selva D, Antolín C, López-Baeza E, Ferrer JF, Saleh K (2004) SMOS REFLEX 2003: L-Band Emssivity Characterization of Vineyards. Transactions Geoscience and Remote Sensing (accepted)

Valor E, Caselles V (1996) Mapping land surface emissivity from NDVI: Application to European, African, and South American areas. Remote Sens. Environm. 57: 167-184

Valenzuela GR (1967) Depolarization of EM waves by slightly rough surfaces. IEEE Trans. Antennas Propagation AP-15(4): 552-557

Van de Griend AA (2001) The Effective Thermodynamic Temperature of the Emitting Surface at 6.6 GHz and Consequences for Soil Moisture Monitoring form Space. IEEE Geoscience and Remote Sensing, Vol. 39(8):1673-1679

Van de Griend AA, Gouweleeuw BT, Owe M, Seyhan E, de Vries JJ (1977) Contribution to the First Annual Report of RESMEDES, Faculty of Earth Sciences, Vrye Universitat Amsterdam, The netgherlands.

Van de Griend AA, Owe M, Vugts HF, Prince SD (1989) Water and Surface Energy Balance Modelling in Botswana. Bulletin of American Meteorological Society 70(11): 1404-1411.

Van de Griend AA, Van Boxel JH (1989) Water and Surface Energy Balance Model With a Multilayer Canopy Representation for Remote Sensing Purposes. Water resources Research 25, No. 5, 949-971

Van de Griend AA, Owe M (1993a) Determination of microwave vegetation optical depth and single scattering albedo from large-scale soil moisture and Nimbus/SMMR Satellite Observations. Int. Journal of Remote Sensing 14(10): 1875-1886

Van de Griend AA, Owe M (1993b) On the relationship between thermal emissivity and the normalized difference vegetation index for natural surfaces. Int. J. Remote Sensing 14: 1119-1131

Van de Griend AA, Owe M (1994a) Microwave Vegetation Optical depth and Inverse Modelling of Soil Emissivity using Nimbus/SMMR Satellite Observations. Journal of Meteorology and Atmospheric Physics 54: 225-239.

Van de Griend AA, Owe M (1994b) The Influence of Polarization on Canopy Transmission Properties at 6.6 GHz and Implications for Large Scale Soil Moisture Monitoring in Semi-Arid Environments. IEEE

Transactions on Geoscience and Remote Sensing 32(2): 409-415.
Van de Griend AA, Owe M (1994c) Bare soil surface resistance to evaporation by vapor diffusion under semiarid conditions. Water Resources research 30: 181-188
Van de Hulst HC (1980) Multiple Light Scattering. Academic Press Vol. 1 and 2, New York., London
Van den Hurk BJJM (1996) Sparce canopy parameterizations formeteorological models. PhD-thesis, Dept. of Meteorology, WageningenAgricultural University, 272 pp. [Available from Dept. of Meteorology, Duivendaal 2, 6701 AP Wageningen, The Netherlands.]
Van Oevelen PJ, Hoekman DH, Vissers MAM (1993) Soil moisture and surface roughness measurements during HAPEX-Sahel 1992. Ground data collection report. Technical Report, Dept. of Water Resources, Wageningen, Agricultural University, The Netherlands.
Van Oevcelen PJ, Hoekman DH (1996) Radar backscattering inversion techniques for estimation of surface soil moisture content. EFEDA-Spain and Hapex-Sahel case studies. IEEE Transactions on Geoscience and Remote Sensing 37(1): 113-123
Van Oevelen P (1998) Soil and vegetation assessment using ERS1/2 SAR and WSC. Final report RESMEDES ENV4
Van Oevelen PJ (2000) Estimation of Aeral Soil Water Content through Microwave Remote Sensing. PhD Thesis Wageningen University, The Netherlands
Vane G et al. (1993) The Airborne Visible/Infrared Imaging Spectrometer (AVIRIS). Rem. Sens. Environ., 44, 127-143
Van Zyl J (1989) Unsupervised classification of scattering behavior using polarimetry data. IEEE Trand. GRS, GE-27: 36-45
Vardavas I (2000) An Integrated Monitoring and Modelling Study of Desertification and Climatic Change Impacts in the Messara Valley of Crete. Final Scientific Report Contract EV5V-CT94-0466, Foundation for Research and Technology - Hellas (FORTH)
Velazquez A, Lopez-Baeza E, Smith GL,Szewczyk ZP (2004) CERES Operations for the Valencia Anchor Station in Support of GERB Validation Efforts. CERES SCALES Campaigns. 13th Conference on Satellite Meteorology and Oceanography, Norfolk, Virginia 20–23 September 2004
Verhoef A (1995) Surface energy balance of shrub vegetation in the Sahel. Thesis Landbouwuniversiteit Wageningen, pp 247
Verhoef W (1996) Application of Harmonic Analysis of NDVI Time Series (HANTS). In: Azzali S, Menenti M (eds.) Fourier analysis of temporal NDVI in the Southern African and American continents. Wageningen (The Netherlands), DLO Winand Staring Centre, Report 108: 19-24.
Vermonte E, Tanré D, Deuzé JL, Herman M, and Mocrette JJ (1994) Second simulation of the Satellite Signal in the Solar Spectrum (6S). User Guide Version 0. Laboratoire d'Optique Atmosphérique, U.S.T. de Lille, 54 p.
Vermote E, Kaufman YJ (1995) Absolute calibration of AVHRR visible and near-infrared channels using ocean and cloud views. Int. J. Remote Sensing 16(13): 2317-2340
Vermonte E, Tanré D, Deuze J, Herman M, Morcrette J (1995) Second Simulation of the Satellite Signal in the Solar Spectrum (6S) User Guiode, Greenbelt, MD, NASA Space Flight Center
Vermonte E, Tanré D, Deuzé JL,, Herman M, Mocrette J-J (1997) Seeond Simulation of the Satellite Signal in the Solar Speetrum (6S). IEEE Transactions on Geosience and Remote Sensing 35: 675-685 and also User guide version 2 Laboratoire d'Optique Atmosphérique, U.S.T. de Lille, 54 p.
Viedma O (1999) Caraterización espectral y espacial de los incendios forestales mediante técnicas de teledetección. Ph.D. Thesis. University of Alcalá de Henares. Spain.
Vigroux E (1953) Contribution à L'étude expérimentale de l'absorption de l'ozone. Annales de Phys. 8: 709
Veroustrate et al. 1998 4
Vicente GA, Scofield RA, Menzel WP (1996) The operational GOES infrared rainfall estimation technique. Bull. Amer. Meteor. Soc. 79: 1883-1898
Viterbo P (1996) The representation of surface processes in general circulation models. PhD-thesis, ECMWF, Reading, UK, 201 pp.
Viterbo P, Courtier P (1995) The importance of soil waterfor medium-range weather forecasting. Implications for dataassimilation. WMO workshop on imbalances of slowly varying componentsof predictable atmospheric motions, Beijing, China, March 1995; WMO/TD 652, pp 121-130.
Viterbo P, Beljaars ACM (1995) An improved land surfaceparametrization scheme in the ECMWF model and its validation. J. Climate 8: 2716-2748
Wagner H-G (2000) Mittelmeerraum - Geographie, Geschichte, Wirtschaft, Politik. Wiss.

Buchgesellschaft, Darmstadt, ISBN 3-534-12339-5, pp 381

Wigneron J-P, Waldteufel P, Chanzy A, Calvet J-C, Kerr Y (2000) Two-dimensional microwave interferometer retrieval capabilities over land surfaces (SMOS Mission). Remote Sensing of Environment 73: 270-282

Wigneron J-P, Chanzy A, Calvet J-C, Bruguier N (1995a) A simple algorithm ti retrieve soil moisture and vegetation biomass using passive microwave measurements over crop fields. Remote Sens. Environment 51: 331-341

Wigneron J-P, Calver J-C, Chancy A, Grosjean O, Laguerre L (1995b) A composite discrete-continuous approach to model the microwave emission of vegetation. IEEE Trans. Geosc. & Remote Sens. 33(1): 201-210

Walter-Shea EA, Biehl L (1990) Measuring vegetation spectral properties. In: Goel NS, Norman JM (eds) Instrumentation for studying vegetation canopies for remote sensing in optical and thermal infrared, Remote Sensing Reviews 5(1): 1-360.

Wan Z, Dozier J (1996) A generalized split-window algorithm for retrieving land-surface temperature from space. IEEE Trans. Geosc. Remote Sens. 35: 892-905

Wang JR, Schmugge TJ (1980) An empirical model for the complex dielectric permittivity of soils as a function of water content. IEEE Trans. Geosci. Remote Sensing GE-18: 288-295

Wang JR (1987) Microwave emission from smooth bare fields and soil moisture sampling depth. Remote Sens. Environm. 25: 616-622

Wang F (1990a) Fuzzy Supervised Classification of Remote Sensing Images. IEEE Transactions on Geoscience and Remoted Sensing 28: 194-201

Wang 1990 Wang, F. (1990b) Improving Remote Sensing Image Analysis through Fuzzy Information Representation. Photogrammetric Engineering and Remote Sensing 56: 1163-1169.

Waring RH, Running SW (1998) Forest Ecosystem. Analysis at multiple scale. Academic Press

Washington-Allen RA, Ramsey RD, Norton BE, West, NE (1998) Change detection of the effect severe drought on subsistence agropastoral communities on the Bolivian Altiplano. International Journal of Remote Sensing, 19: 1319-1334.

Weiss A, Hays CJ, Hu Q, Easterling WE (2001) Incorporating Bias Error in Calculating Solar Irradiance. Implication for Crop Yield Simulations. Agronomy Journal 93: 1321-1326

Whitlock CH, Staylor WF, Suttles JT, Smith et. al. (1990) AVHRR and VISSR satellite instrument calibration results for both cirrus and marine stratocumulus IFO periods. Fire Science Report 1988, NASA CP 3083

Wiegand CL, Mass SJ, Aase JK, Hatfield JL, Pinter PJ Jr., Jackson RD, Kanemasu ET, Lapitan RL (1990) Multisite analyses of spectral-biophysical data for wheat. Remote Sens. Environ. 42: 1-21

Wiegand CL, Richardson AJ, Escober DE, Gerbermann AH (1991) Vegetation indices in crop assessment. Remote Sensing of Environment, 35: 105-119

Windig W, Meuzelaar HLC (1984) Nonsupervised Numerical Extraction from Pyrolysis Mass Spectra of Complex Mixtures. Anal. Chem. 56: 2297-2303

Willmott CJ, JR Mather, Rowe CM (1981) Average monthly and annual surface air temperature and precipitation data for the world. Part 1: The eastern hemisphere. Part 2: The western hemisphere. Publ. in Climatology, 34: 395 and 378

Wittmeyer IL, Vonder Haar TH (1994) Analysis of the global ISCCP TOVS water vapor climatology. J. Climate, 7: 325-333

WMO World Meteorology Organisation (1975) Statistical analysis of the observations series (by R. Sneyers). No. 415. Geneva.

WMO (World Meteorological Organization) (1983) Guide to Meteorological Instruments and Methods of Observation. WMO - No 8, Geneva, Annex 9B.

WMO (World Meteorological Organization) – World Climate Programme (1986) A Preliminary Cloudless Standard Atmosphere for Radiation Computation. World Climate research Programme, International Association for Meteorology and Atmospheric Physics, Radiation Commission, WCP – 112, WMO/TD-No. 24, Boulder Colorado, U.S.A. (1984)

Woodcock CE, Strahler AH (1987) The factor of scale in remote sensing. Remote Sensing of Environment 21: 311-332

Wörle H, Rumpf H-J (1978) Taschenbuch der Mathematik, R. Oldenburg Verlag München Wien, p. 64

Wüthrich Treubig MP (1998) In Situ Measurements and Remote Sensing of Soil Moisture Using Time Domain Reflectometry, Thermal Infrared and Microwaves. In: Parlow E, Vogt R (eds) stratus 5, Technical Reports, Department Geowissenschaften, Universität Basel

Wydick JE, Davis PA, Gruber A (1987) Estimation of broadband planetary albedo from operational narrowband satellite measurements. NOAA Technical Report NESDIS 27

Wyngaard JC, Coté OR (1971) The budgets of turbulent kinetic energy and temperature variance in the atmospheric surface layer. J. Atmos. Sci. 28: 190-201

Yamamoto G (1952) On a radiation chart. Science Rept. Tohoku University, Series 5, Geophys. 4(1): 9

Young AT (1981) On the Rayleigh-Scattering Optical Depth of the atmosphere. J. Appl. Meteorology 20: 328-330

Younis MT, Gilabert AA, Meliá J, Bastida J (1997) Weathering process effects on spectral reflectance of rocks in a semi-arid environment. Int. J. Remote Sensing 18(16): 3361-3377

Zohary M. (1962) Plant life of Palestine. Ronald Press, New York.

Zink M, Schmullius C (1994) Activities and results from the MAC-Europe campaign at Oberpfaffenhofen. In: Wooding M, Lodge F, Attema E (eds.) MAC Europe ´91 Final Results Workshop. European Space Agency, Paris, 254 pp.

Glossary (Acronyms)

AATSR - Advanced Along Track Scanning Radiometer
ABL - atmospheric boundary layer
AERONET - Aerosol Robotic Network
AIRSAR - Aircraft Synthetic Aperture Radar
AIT - Advanced Intelligent Tapes
AMI - Active Microwave Instrument
AMSR - Advanced Microwave Scanning Radiometer
AMSR-E - Advanced Microwave Scanning Radiometer-Earth
AMSU - Advanced Microwave Sounding Unit
APAR - Absorbed Photosynthetic Active Radiation
APOLLO - AVHRR Processing scheme Over cLouds, Land and Ocean
ARVI - Atmospherically Resistant Vegetation Index
ASAR - Advanced Synthetic Aperture Radar
ASTER - Advanced Spaceborne Thermal Emission and Reflection Radiometer
ATOVS - Advanced TIROS Operational Vertical Sounder
ATSR - Along Track Scanning Radiometer
AVHRR - Advanced Very High Resolution Radiometer
AVIRIS - Airborne Visible-Infrared Imaging Spectrometer
BDRF - bidirectional reflectance factor
BGC - Bio-Geo-Chemical (cycles)
BDGP - Bi-Directional Gap Probability
BDRF - Bi-Directional Reflexion Factor
BRDF - Bidirectional Reflectance Distribution Function
BSRN - Baseline Surface Radiation Network
CAMELEO - Changes in Mediterranean Semi-arid Ecosystems on the Long Term through Earth Observation
CDA - Command and Data Acquisition
CEAM - Centro de Estudioa Ambientales del Mediterráneo
CEO - Centre of Earth Observation
CERES - Clouds and the Earth`s Radiant Energy System
CeSIA - Centro di Studio per l'Applicazione dell'Informatica in Agricoltura
CI - Climate Indicator
CLC - CORINE Landcover
CLEO - Cluster for Earth Observation
CONSURF - interpolation algorithm developed by D. Douglas at the University of Ottawa, Canada
CORINE - Coordinated Information on the Environment
CoSMOS - Campaign for the Operation of SMOS
CSAR - Coupled Surface-Atmosphere Reflectance Model
CTIV - Vegetation image processing center
DAIS - Digital Airborne Imaging Spectrometer
DAISEX - DAIS Experiment
DEM - Digital Elevation Model
DGVM - Dynamic Global Vegetation Models
DIAL - DIfferential Absorption Lidar
DLR - Deutsches Zentrum für Luft- und Raumfahrt

DLT - Digital Linear Tapes
DMSP - Defense Meteorological Satellite Program
DRS - Direct Receiving Station
DTC - Diurnal Surface Temperature Cycle
DTE - U.S. Defense Mapping Agency Digital Terrain Elevation Data
DTM - Digital Terrain Model
DVD - Digital Versatile Disc
EARLINET - European Aerosol Research Lidar Network to Establish an Aerosol Climatology
ECHIVAL - European International Project on Climatic and Hydrological Interactions between Vegetation, Atmosphere, and Land-surfaces
ECMWF - European Centre for Medium-Range Weather Forecast
EDC - EROS Data Center
EECF - Earthnet ERS-1 Central Facility
EFEDA - ECHIVAL Field Experiment in Desertification-threatened Areas
ENVISAT - European Environmental Satellite
EOS - Earth Observing System
EOSAT - Earth Observation Satellite Company
EROS - Earth Resources Observation and Science
EROS - Earth Resources Observation Satellites
EROS - Earth Resources Observation Systems
ERS - European Remote Sensing Satellite
ESM - Earth System Models
ESMR - Electrically/Electronically Scanned/ing Microwave Radiometer
ESRIN - European Space Research Institute
ETMP - Enhanced Thematic Mapper Plus
FAO - Food and Agriculture Organization
FOV - Field of View
FPAR - Fractional Absorbed Photosynthetic Active Radiation
G.R.T.R. - Groupment de Recherche en Télédétection et Radiometrie
GAC - Global Area Coverage
GACP - Global Aerosol Climatology Project
GCP - Ground Control Points
GEMI - Global Environment Monitoring Index
GERB - Geostationary Earth Radiation Budget
GEWEX - Global Energy and Water Cycle Experiment
GIMMS - Global Inventory Monitoring and Modelling System
GLOBE - Global Land One-km Base Elevation
GOES - Geostationary (Geosynchronous) Operational Environmental Satellite
GOM - Global Observation Mode
GOME - Global Ozone Monitoring Experiment
GOMOS - Global Ozone Monitoring by Occultation of Stars
GPCP - Global Precipitation Climatology Project
GPS - Global Positioning System
GSFC - Goddard Space Flight Center
GTS - Global Telecommunications Service
GVI - Green Vegetation Index
GVI - (NOAA) Global Vegetation Index
HANTS - Harmonic Analysis of Numerical Time Series
HIRLAM - High-Resolution Limited Area Model
HIRS - High Resolution Infrared Sounder
HIV - High Spatial-resolution Visible
HRPT - High Resolution Picture Transmission
HRPT - High Resolution Picture Transmission
HRV - High Resolution Visible (SPOT instrument)
HRV - (SPOT) High Resolution Visible
HRVIR - High Resolution Infrared (SPOT instrument)
IDRISI - GIS software named after Mohammed al-Idrisi, geographer and cartographer born 1099

IGBP - International Geosphere - Biosphere Programme
IHDP - International Human Dimensions Programme on Global Environmental Change
IKONOS - EOSAT satellite for one metre panchromatic and four metre multispectral imagery
INDOEX - Indian Ocean Experiment
IPCC - International Panel of Climate Change
JERS - Japanese Earth Resources Satellite
JPL - NASA/Jet Propulsion Laboratory
KWBC - National Weather Service Telecommunications Gateway
La.M.M.A - Laboratorio per la Meteorologia e la Modellistica Ambientale
LAC - Local Area Coverage
LAI - Leaf Area Index
LRR - Laser Retro Reflector
LSM - generally: Land Surface Model, specifically: NCAR Land Surface Model for Ecological, Hydrological, Atmospheric Studies developed by GB Bonan (http://daac.ornl.gov)
MAC - Multi-instrument Aircraft Campaign
MC-FUME - Median Composite of Fuzzy Multispectral Estimate
MEDOKADS - Mediterranean Extended Daily One Km AVHRR Data Set
MERIS - Medium Resolution Imaging Spectrometer
MGRS - Military Grid Reference System
MIPAS - Michelson Interferometer for Passive Atmospheric Sounding
MISR - Multiangle Imaging Spectro-Radiometer
MDIP - Morphology Dependent Interpolation Procedure
MODIS - Moderate Resolution Imaging Spectroradiometer
MODTRAN - Moderate Resolution Transmittance (code)
MOPITT - Measurements Of Pollution In The Troposphere)
MPDI - Microwave Polarization Difference Index
MPEF - Meteosat Products Extraction Facility
MSG - Meteosat Second Generation
MSS - Multispectral Scanner
MSU - Microwave Sounding Unit
MVC - Maximum Value Compositing
MVIRI - Meteosat Visible and InfraRed Imager
MWR - Microwave Radiometer
NCEP - U.S. National Climate and Environmental Prediction
NCF - Nadir Correction Function
NEE - Net Ecosystem Exchange
NVAP - NASA Water Vapor Project
NWP - Numerical Weather Prediction
PAPS - Programmable Azimuth Plane Scan
PAR - photosynthetic active radiation
PDUS - Primary Data User Station
POES - Polar Orbiting Satellite System
POLDER - Polarization and Directionality of Earth Reflectances
PVI - Perpendicular Vegetation Index
PWC - precipitable water content
QIV - Vegetation Image Quality Center
RA-2 - Radar Altimeter
RACMO - Regional Atmospheric Circulation Model
RISMOP - Radiometric Impact of Surface Moisture on Precipitation
SAR - Synthetic Aperture Radar
SAVI - Soil Adjusted Vegetation Index
SCIAMACHY -Scanning Imaging Absorption Spectrometer for Atmospheric CHartographY
SEBAL - Surface Energy Balance Algorithm
SEBASS - Spatially Enhanced Broadband Array Spectrograph System
SEVIRI - Spinning Enhanced Visible and Infrared Imager
SMMR - Scanning Multi-channel Microwave Radiometer
SMOS - Soil Moisture and Ocean Salinity

SPOT - Systeme Probatoire de l'Observation de la Terre
SPOT HRG - SPOT Haute Résolution Géométrique
SRF - spectral response function
SSC - Swedish Space Corporation
SSM/I - Special Sensor Microwave Imager
SST - sea surface temperature
SSU - Stratospheric Sounder Unit
SVAT - Soil-Vegetation-Atmosphere Transfer
SWIR - short-wave infrared
TBUS - NOAA APT Predict bulletins (identified by its message data identifier TBUS)
TDRS - Tracking Data and Relay Satellite
TIN - Triangulated Irregular Network
TIP - Tiros Information Processor
TIR - thermal infrared
TIROS - Television (&) Infrared Observation Satellite
TIROS-N - Advanced Television Infrared Observation Satellite
TISI - Temperature Independent Spectral Index
TM - (Landsat) Thematic Mapper
TMI - TRMM Microwave Imager
TMS - Thematic Mapper Simulator
TOA - top of atmosphere
TOVS - TIROS Operational Vertical Sounder
TRMM - Tropical Rainfall Measuring Mission
TSDIS - TRMM Science Data and Information System
TTHG - Henyey-Greenstein phase function
UNCCD - UNEP Convention to Combat Desertification
USGS - U. S. Geological Survey
UTC - Coordinated Universal Time
UTM - Universal Transverse Mercator Projection
VAS - Vertical Atmospheric Sounder
VCI - Vegetation Condition Index
VEGETATION (VGT) - instrument on board of SPOT
VGT-P products - VEGETATION products for scientific applications
VGT-S products - geocoded and corrected VEGETATION products
VIRS - Visible/Infrared Scanner
VNIR - visible and near infrared
WCRP - World Climate Research Programme
WDB - World Data Bank
WDVI - Weighted Difference Vegetation Index
WGS-84 - World Geodetic System 1984
WRS - Worldwide Reference System
WSC - Wind Scatterometer

Subject Index

Printing: Krips bv, Meppel
Binding: Stürtz, Würzburg